Woodhead Publishing in Materials

Polymeric Nanofibers and their Composites

Recent Advances and Applications

Edited by

Chandrabhan Verma

Yong X. Gan

Woodhead Publishing is an imprint of Elsevier
50 Hampshire Street, 5th Floor, Cambridge, MA 02139, United States
125 London Wall, London EC2Y 5AS, United Kingdom

Copyright © 2025 Elsevier Ltd. All rights are reserved, including those for text and data mining, AI training, and similar technologies.

Publisher's note: Elsevier takes a neutral position with respect to territorial disputes or jurisdictional claims in its published content, including in maps and institutional affiliations.

No part of this publication may be reproduced or transmitted in any form or by any means, electronic or mechanical, including photocopying, recording, or any information storage and retrieval system, without permission in writing from the publisher. Details on how to seek permission, further information about the Publisher's permissions policies and our arrangements with organizations such as the Copyright Clearance Center and the Copyright Licensing Agency, can be found at our website: www.elsevier.com/permissions.

This book and the individual contributions contained in it are protected under copyright by the Publisher (other than as may be noted herein).

Notices
Knowledge and best practice in this field are constantly changing. As new research and experience broaden our understanding, changes in research methods, professional practices, or medical treatment may become necessary.

Practitioners and researchers must always rely on their own experience and knowledge in evaluating and using any information, methods, compounds, or experiments described herein. In using such information or methods they should be mindful of their own safety and the safety of others, including parties for whom they have a professional responsibility.

To the fullest extent of the law, neither the Publisher nor the authors, contributors, or editors, assume any liability for any injury and/or damage to persons or property as a matter of products liability, negligence or otherwise, or from any use or operation of any methods, products, instructions, or ideas contained in the material herein.

ISBN: 978-0-443-14128-7 (print)
ISBN: 978-0-443-14129-4 (online)

For information on all Woodhead Publishing publications
visit our website at https://www.elsevier.com/books-and-journals

Publisher: Matthew Deans
Acquisitions Editor: Gwen Jones
Editorial Project Manager: Toni Louise Jackson
Production Project Manager: Fizza Fathima
Cover Designer: Christian Bilbow

Typeset by MPS Limited, Chennai, India

Contents

List of contributors — xiii

1 Nanofibers and their composites: fabrication, characterization, and structure with particular emphasis on their advantages and disadvantages — 1
Yong Xue Gan
- 1.1 Introduction to nanofibers and related composites — 1
- 1.2 Processing and fabrication technologies — 2
- 1.3 Structure assessment of nanofibers and their composites — 21
- 1.4 Property characterization — 23
- 1.5 Concluding remarks — 25
- References — 25

2 Synthetic nanofibers and natural nanofibers: A relatively study on physical, chemical, and mechanical properties — 31
Shveta Sharma, Richika Ganjoo, Alok Kumar and Ashish Kumar
- 2.1 Introduction — 31
- 2.2 Natural nanofibers — 34
- 2.3 Synthetic nanofibers — 40
- 2.4 Conclusion and future outlook — 45
- References — 46

3 Functionalization/modification of nanofibers and their impact on properties and applications — 51
Omar Dagdag, Rajesh Haldhar, Elyor Berdimurodov and Hansang Kim
- 3.1 Introduction — 51
- 3.2 Methods of surface functionalization of nanofibers — 52
- 3.3 Modification of polymer nanofibers — 56
- 3.4 Chemical functionalization, properties, and applications of nanocellulose — 56
- 3.5 Conclusion — 63
- Author contribution statement — 64
- Abbreviations — 64
- References — 64

4	**Processing of nanofibers for oil-spill cleaning application**	69
	Yong Xue Gan	
	4.1 Introduction	69
	4.2 Processing technologies	70
	4.3 Classical theories on contact angles of water and oil drops to solid surfaces	85
	4.4 Conclusions	86
	Acknowledgments	87
	References	87

5	**A detail account of natural nanofibers (as chitin/chitosan, cellulose, gelatin, alginate, hyaluronic acid, fibrin, collagen, etc.)**	91
	Heri Septya Kusuma, Ganing Irbah Al Lantip and Xenna Mutiara	
	5.1 Introduction	91
	5.2 Chitin/chitosan	93
	5.3 Cellulose	94
	5.4 Gelatin	96
	5.5 Alginate	98
	5.6 Hyaluronic acid	100
	5.7 Fibrin	101
	5.8 Collagen	102
	5.9 Conclusion	104
	References	104

6	**Green synthesis of nanofibers for energy and environmental applications**	113
	Nancy Elizabeth Davila-Guzman, J. Raziel Álvarez, M.A. Garza-Navarro and Alan A. Rico-Barragán	
	6.1 Introduction	113
	6.2 Biopolymers for the green synthesis of nanofibers	114
	6.3 Green solvents for nanofiber synthesis	117
	6.4 Environmental impact assessment	124
	6.5 Conclusions	126
	Reference	127

7	**Biomedical applications of nanofibers and their composites**	135
	Akbar Esmaeili	
	7.1 Introduction	135
	7.2 Synthesis of nanofibers	136
	7.3 Medical nanofibers	139
	7.4 Composite nanofibers	148
	7.5 Conclusions and future perspectives	152
	References	153

8	**Tissue engineering and drug delivery applications of nanofibers and their composites**	**157**
	Akbar Esmaeili	
	8.1 Introduction	157
	8.2 Delivery of new drugs	158
	8.3 Targeted delivery	160
	8.4 Codelivery	161
	8.5 Tissue engineering	165
	8.6 Nano scaffolds	165
	8.7 Controlled release	168
	8.8 Composites nanofibers	175
	8.9 Conclusions and future perspectives	179
	References	179
9	**Industrial wastewater treatment applications of nanofibers and their composites**	**185**
	Gianluca Viscusi	
	9.1 Introduction	185
	9.2 Modification of nanofibers for water remediation applications	186
	9.3 Fiber systems produced by electrospinning	188
	9.4 Ceramic fibers	197
	9.5 Use of cellulose nanofibers	205
	9.6 Conclusions and future perspectives	209
	References	209
10	**Wastewater treatment application of nanofibers and their composites**	**227**
	Akbar Esmaeili	
	10.1 Introduction	227
	10.2 The environmental threat of pollutants	229
	10.3 Sequential methods of removing contaminants from aqueous solutions by composites	231
	10.4 Pollutant absorption by plant composites	244
	10.5 Conclusions and future perspectives	248
	References	249
11	**Energy applications of nanofibers and their composites**	**255**
	Muhammad Tuoqeer Anwar, Raheela Naz, Arslan Ahmed, Saad Ahmed, Ghulam Abbas Ashraf and Tahir Rasheed	
	11.1 Introduction	255
	11.2 Brief history	256
	11.3 Fabrication methods	256
	11.4 Functional nanofibers	262
	11.5 Energy applications	262
	11.6 Outlook and conclusions	267
	References	268

12	**Nanofibers and their composites for battery, fuel cell and solar cell applications**	**273**
	Yong Xue Gan	
	12.1 Introduction	273
	12.2 Nanofiber processing technologies	273
	12.3 Nanofibers for batteries	280
	12.4 Nanofibers for solar cells	309
	12.5 Nanofibers for fuel cells	313
	12.6 Concluding remarks	316
	Acknowledgments	317
	References	317
13	**Tribological applications/lubricant additive applications of nanofibers and their composites**	**325**
	Muhammad Ullah, Sidra Subhan, Muhammad Shakir, Ata Ur Rahman and Muhammad Yaseen	
	13.1 Introduction	325
	13.2 Nanofibers and their composites	326
	13.3 Properties of nanofibers and their composites in tribology	330
	13.4 Lubricants	335
	13.5 Nanofibers and their composites as lubricant additives	344
	References	351
14	**Anticorrosive applications of nanofibers and their composites**	**357**
	Omar Dagdag, Rajesh Haldhar, Elyor Berdimurodov and Hansang Kim	
	14.1 Introduction	357
	14.2 Nanofiber composites and anticorrosive uses	358
	14.3 Conclusion	373
	Author contribution statement	374
	Abbreviations	374
	References	374
15	**Gas adsorption and storage of nanofibers and their composites**	**377**
	Nancy Elizabeth Davila-Guzman, Margarita Loredo-Cancino, Sandra Pioquinto-Garcia and Alan A. Rico-Barragán	
	15.1 Introduction	377
	15.2 Carbon dioxide capture	378
	15.3 Volatile organic compound adsorption for indoor air pollution control	386
	15.4 Gas adsorption for fuel purification	387
	15.5 Hydrogen storage	391
	15.6 Conclusions	395
	References	396

16 Catalysis and electrocatalysis application of nanofibers and their composites — 405
Elyor Berdimurodov, Khasan Berdimuradov, Ashish Kumar, Ilyos Eliboev, Nodira Eshmamatova, Bakhtiyor Borikhonov and Sardorbek Otajonov

- 16.1 Introduction — 405
- 16.2 Nanofiber synthesis methods — 407
- 16.3 Catalysis application of nanofiber composites — 411
- 16.4 Future suggestions — 416
- 16.5 Conclusion — 417
- Reference — 418

17 Sensors application of nanofibers and their composites — 423
Shveta Sharma, Richika Ganjoo, Elyor Berdimurodov, Alok Kumar and Ashish Kumar

- 17.1 Introduction — 423
- 17.2 Nanofibers in sensors — 425
- 17.3 Future outlook — 430
- 17.4 Conclusion — 431
- References — 432

18 Water splitting application of nanofibers and their composites — 437
Abhinay Thakur, Valentine Chikaodili Anadebe and Ashish Kumar

- 18.1 Introduction — 437
- 18.2 Fundamentals of nanofibers and composite — 439
- 18.3 Nanofibers and composites in water splitting — 444
- 18.4 Recent developments and advances of nanofibers in water splitting — 449
- 18.5 Challenges and future prospects — 455
- 18.6 Conclusion — 458
- References — 459

19 Water, air, and soil purification from the application of nanofibers and their composites — 471
Ainun Zulfikar, Marita Wulandari, Abdul Halim and Bimastyaji Surya Ramadan

- Abbreviations — 471
- 19.1 Introduction — 471
- 19.2 Water purification application of nanofibers and their composites — 473
- 19.3 Air purification application of nanofibers and their composites — 476
- 19.4 Soil purification applications of nanofibers and their composites — 482
- 19.5 Future direction — 487
- 19.6 Conclusion — 488
- Conflict of interest — 488
- References — 488

20	**Electronics application of nanofibers and their composites**	**497**
	Manoj Kumar Banjare, Kamalakanta Behera, Ramesh Kumar Banjare, Mamta Tandon, Siddharth Pandey and Kallol K. Ghosh	
	20.1 Introduction	497
	20.2 Terminology for nanotechnology	498
	20.3 The evolution of nanofiber technologies over time	499
	20.4 Nanofiber and nanofibrous material types	502
	20.5 Classification of nanofibers based on their chemical composition	503
	20.6 Synthesis methods of nanofibers	504
	20.7 Application of electronic-based nanofibers and their composites	504
	20.8 Applications in energy devices of electrospun NFs	509
	20.9 Conclusions	513
	Authors contribution	514
	Conflict of interest	514
	Acknowledgments	514
	References	514
21	**Additive application of nanofibers and their composites for enhanced performance**	**521**
	Sirsendu Sengupta*, Surya Sarkar* and Priyabrata Banerjee	
	21.1 Introduction	521
	21.2 Additive application of nanofibers and their composites for enhanced performance	522
	21.3 Energy conversion and storage	525
	21.4 Conclusions and future perspectives	533
	References	533
22	**Nanofibers and their composites for supercapacitor applications**	**539**
	Ishita Ishita, Shriram Radhakanth, Pradeep Kumar Sow and Richa Singhal	
	22.1 Introduction	539
	22.2 Performance metrics of supercapacitor	543
	22.3 Nanofibers in supercapacitors	546
	22.4 Nanofibers as electrodes	548
	22.5 Nanofibers as separators	559
	22.6 Conclusions and future outlook	560
	References	561
23	**Sources of natural fibers and their physicochemical properties for textile uses**	**569**
	Abhinay Thakur, Ashish Kumar and Valentine Chikaodili Anadebe	
	23.1 Introduction	569

23.2	Natural fibers: an overview	571
23.3	Sources of natural fibers	575
23.4	Physicochemical properties of natural fibers	580
23.5	Applications of natural fibers in textiles	597
23.6	Sustainability and eco-friendliness	600
23.7	Challenges and future directions	604
23.8	Conclusion	605
References		606

Index **617**

List of contributors

Arslan Ahmed Department of Mechanical Engineering, COMSATS University Islamabad, Wah Campus, Rawalpindi, Punjab, Pakistan

Saad Ahmed State Key Laboratory Breeding Base of Green Chemistry-Synthesis Technology, Zhejiang Province Key Laboratory of Biofuel, Biodiesel Laboratory of China Petroleum and Chemical Industry Federation, College of Chemical Engineering, Zhejiang University of Technology, Hangzhou, Zhejiang, P.R. China

J. Raziel Álvarez School of Chemistry, Autonomous University of Nuevo Leon, San Nicolas de los Garza, Nuevo Leon, Mexico

Valentine Chikaodili Anadebe Corrosion and Materials Protection Division, CSIR-Central Electrochemical Research Institute, Karaikudi, Tamil Nadu, India; Department of Chemical Engineering, Alex Ekwueme Federal University Ndufu Alike, Abakakili, Ebonyi State, Nigeria

Muhammad Tuoqeer Anwar Departemnt of Mechanical Engineering, COMSATS University Islamabad, Sahiwal Campus, Sahiwal, Punjab, Pakistan

Ghulam Abbas Ashraf Key Laboratory of Integrated Regulation and Resources Development on Shallow Lake of Ministry of Education, College of Environment, Hohai University, Nanjing, Jiangsu, P.R. China

Priyabrata Banerjee Electric Mobility and Tribology Research Group, CSIR-Central Mechanical Engineering Research Institute, Durgapur, West Bengal, India; Academy of Scientific and Innovative Research (AcSIR), Ghaziabad, Uttar Pradesh, India

Manoj Kumar Banjare MATS School of Sciences, MATS University, Raipur, Chhattisgarh, India

Ramesh Kumar Banjare Department of Chemistry(MSET), MATS University, Raipur, Chhattisgarh, India

Kamalakanta Behera Department of Chemistry, University of Allahabad, Prayagraj, Uttar Pradesh, India

Khasan Berdimuradov Faculty of Industrial Viticulture and Food Production Technology, Shahrisabz Branch of Tashkent Institute of Chemical Technology, Shahrisabz, Uzbekistan; Physics and Chemistry, Western Caspian University, Baku, Azerbaijan

Elyor Berdimurodov Faculty of Chemistry, National University of Uzbekistan, Tashkent, Uzbekistan; Akfa University, Tashkent, Uzbekistan; Chemical & Materials Engineering, New Uzbekistan University, Tashkent, Uzbekistan; University of Tashkent for Applied Sciences, Tashkent, Uzbekistan; Department of Physical Chemistry, National University of Uzbekistan, Tashkent, Uzbekistan

Bakhtiyor Borikhonov Faculty of Chemistry and Biology, Karshi State University, Karshi City, Uzbekistan

Omar Dagdag Department of Mechanical Engineering, Gachon University, Seongnam, Republic of Korea

Nancy Elizabeth Davila-Guzman School of Chemistry, Autonomous University of Nuevo Leon, San Nicolas de los Garza, Nuevo Leon, Mexico

Ilyos Eliboev Faculty of Chemistry, National University of Uzbekistan, Tashkent, Uzbekistan; Chemistry, Uzbek-Finnish Pedagogical Institute, Samarqand, Uzbekistan

Nodira Eshmamatova Faculty of Chemistry, National University of Uzbekistan, Tashkent, Uzbekistan

Akbar Esmaeili Department of Chemical Engineering, North Tehran Branch, Islamic Azad University, Tehran, Iran

Yong Xue Gan Department of Mechanical Engineering, California State Polytechnic University Pomona, Pomona, CA, United States

Richika Ganjoo Department of Chemistry, School of Chemical Engineering and Physical Sciences, Lovely Professional University, Phagwara, Punjab, India

M.A. Garza-Navarro School of Mechanical and Electrical Engineering, Autonomous University of Nuevo Leon, San Nicolas de los Garza, Nuevo Leon, Mexico; Center of Innovation, Research and Development in Engineering and Technology, Autonomous University of Nuevo Leon, Apodaca, Nuevo Leon, Mexico

Kallol K. Ghosh School of Studies in Chemistry, Pt. Ravishankar Shukla University, Raipur, Chhattisgarh, India

List of contributors

Rajesh Haldhar School of Chemical Engineering, Yeungnam University, Gyeongsan, Republic of Korea

Abdul Halim Department of Chemical Engineering, Universitas International Semen Indonesia, Gresik, Indonesia

Ishita Ishita Department of Chemical Engineering, BITS Pilani, Goa Campus, Zuarinagar, Goa, India

Hansang Kim Department of Mechanical Engineering, Gachon University, Seongnam, Republic of Korea

Alok Kumar Department of Mechanical Engineering, NCE, Bihar Engineering University, Department of Science, Technology and Technical Education, Nalanda, Bihar, India

Ashish Kumar Department of Science and Technology, NCE, Bihar Engineering University, Government of Bihar, Patna, Bihar, India; Department of Chemistry, NCE, Bihar Engineering University, Department of Science, Technology and Technical Education, Nalanda, Bihar, India

Heri Septya Kusuma Department of Chemical Engineering, Faculty of Industrial Technology, Universitas Pembangunan Nasional "Veteran" Yogyakarta, Indonesia

Ganing Irbah Al Lantip Department of Chemical Engineering, Faculty of Industrial Technology, Universitas Pembangunan Nasional "Veteran" Yogyakarta, Indonesia

Margarita Loredo-Cancino School of Chemistry, Autonomous University of Nuevo Leon, San Nicolas de los Garza, Nuevo Leon, Mexico

Xenna Mutiara Department of Chemical Engineering, Faculty of Industrial Technology, Universitas Pembangunan Nasional "Veteran" Yogyakarta, Indonesia

Raheela Naz School of Materials and Energy, Southwest University, Chongqing, P.R. China

Sardorbek Otajonov Faculty of Chemistry, National University of Uzbekistan, Tashkent, Uzbekistan

Siddharth Pandey Department of Chemistry, Indian Institute of Technology Delhi, New Delhi, India

Sandra Pioquinto-Garcia School of Chemistry, Autonomous University of Nuevo Leon, San Nicolas de los Garza, Nuevo Leon, Mexico

Shriram Radhakanth Department of Chemical Engineering, BITS Pilani, Goa Campus, Zuarinagar, Goa, India

Ata Ur Rahman Institute of Chemical Sciences, University of Peshawar, Peshawar, Khyber Pakhtunkhwa, Pakistan

Bimastyaji Surya Ramadan Department of Environmental Engineering, Faculty of Engineering, Environmental Sustainability Research Group (ENSI-RG), Universitas Diponegoro, Semarang, Indonesia

Tahir Rasheed Interdisciplinary Research Center for Advanced Materials, King Fahd University of Petroleum and Minerals (KFUPM), Dhahran, Eastern Province, Saudi Arabia

Alan A. Rico-Barragán Department of Environmental Engineering, National Technological of Mexico/Higher Technological Institute of Misantla, Misantla, Veracruz, Mexico

Surya Sarkar Electric Mobility and Tribology Research Group, CSIR-Central Mechanical Engineering Research Institute, Durgapur, West Bengal, India; Academy of Scientific and Innovative Research (AcSIR), Ghaziabad, Uttar Pradesh, India; Department of Physics, Durgapur Women's College, Durgapur, West Bengal, India

Sirsendu Sengupta Department of Chemistry, The university of Burdwan, Burdwan, West Bengal, India; Electric Mobility and Tribology Research Group, CSIR-Central Mechanical Engineering Research Institute, Durgapur, West Bengal, India

Muhammad Shakir Institute of Space Technology, Islamabad, Pakistan

Shveta Sharma Department of Chemistry, Government College Una, Affiliated to Himachal Pradesh University, Una, Himachal Pradesh, India

Richa Singhal Department of Chemical Engineering, BITS Pilani, Goa Campus, Zuarinagar, Goa, India

Pradeep Kumar Sow Department of Chemical Engineering, BITS Pilani, Goa Campus, Zuarinagar, Goa, India

Sidra Subhan Institute of Chemical Sciences, University of Peshawar, Peshawar, Khyber Pakhtunkhwa, Pakistan

Mamta Tandon MATS School of Sciences, MATS University, Raipur, Chhattisgarh, India

List of contributors

Abhinay Thakur Department of Chemistry, School of Chemical Engineering and Physical Sciences, Lovely Professional University, Phagwara, Punjab, India

Muhammad Ullah Institute of Chemical Sciences, University of Peshawar, Peshawar, Khyber Pakhtunkhwa, Pakistan

Gianluca Viscusi Department of Industrial Engineering, University of Salerno, Fisciano, Salerno, Italy

Marita Wulandari Graduate School of Science and Technology, University of Tsukuba, Tsukuba, Ibaraki, Japan; Department of Environmental Engineering, Institut Teknologi Kalimantan, Balikpapan, Indonesia

Muhammad Yaseen Institute of Chemical Sciences, University of Peshawar, Peshawar, Khyber Pakhtunkhwa, Pakistan

Ainun Zulfikar Graduate School of Science and Technology, University of Tsukuba, Tsukuba, Ibaraki, Japan; Department of Materials and Metalurgical Engineering, Institut Teknologi Kalimantan, Balikpapan, Indonesia

Nanofibers and their composites: fabrication, characterization, and structure with particular emphasis on their advantages and disadvantages

Yong Xue Gan
Department of Mechanical Engineering, California State Polytechnic University Pomona, Pomona, CA, United States

This chapter briefly introduces nanofibers and their composites. Technologies for the fabrication of nanofibers and composites will be presented first. Then, the structure characterization of nanofibers and their composites will be discussed. Chemical vapor deposition (CVD) as a very commonly used technology will be reviewed. Following that, the frequently used technology, electrospinning, followed by heat treatment, will be described. Hydrothermal synthesis, templating, liquid phase deposition (LPD), electrochemical oxidation (ECO), high-temperature annealing, and spraying pyrolysis will also be presented. The procedures and important processing parameters associated with each fabrication technology will be discussed. Examples of nanofibers processed with these technologies will be given. The advantages and disadvantages of various fabrication technologies will be delineated. Important instrumentation techniques for structure assessment and property characterization will be introduced as well.

1.1 Introduction to nanofibers and related composites

Nanofibers refer to the fibers with diameters ranging from 1 nm to 1000 nm. Nanofibers can be divided into two big categories: organic nanofibers and inorganic nanofibers. They may be classified in other ways such as synthetic and naturally formed nanofibers. Due to the diversity of nanofibers, they show different properties and have found many applications for energy storage, cell regeneration, catalysis, and so on. Nanofiber composites generally contain a continuous phase in the form of nanofiber and a discontinuous phase in the form of nanoparticle (NPs). For example, a copper sulfide NP-containing cellulose acetate nanofiber composite was made by Cota-Leal et al. (2024) for photocatalytic decomposition of methylene blue dye in wastewater. Another example is the silk fibroin and gelatin composite nanofiber added with silver and gold NPs for wound healing (Arumugam et al., 2023).

Nanofiber composites may be found with the core−shell structures as illustrated by Fang et al. (2023). The preparation of core−shell structured nanofiber composites can start with coaxial electrospinning of two polymers, which allows a hierarchical core and shell structure simultaneously formed in one processing step. The resulting composites contain two separate but continuous layers. The two layers have different functions. In addition to electrospinning, a coaxial airbrushing method to fabricate a wound-healing core−shell nanofiber composite was proposed by Singh et al. (2018). In their approach, relatively low-cost equipment was set up. The airbrushing facility was able to operate at the patient's bedside and run direct deposition of nanofiber composite on a wound. Coaxial electrospinning is the major technique for fabricating nanofiber composites because it offers various advantages. The disadvantages of electrospinning were also identified. They include high voltage requirements, relatively complex equipment, and a fairly low deposition rate. Besides the above-mentioned limitations, coaxial is not suitable for in situ and direct deposition of core−shell nanofibers on the wound at the patient bedside. The airbrushing method showed simplicity and could overcome some of the disadvantages associated with the electrospinning technique.

Nanofiber composites may be in heterogeneous structures consisting of different nanofiber segments or nanowires/nanorods. These nanofiber segments with different chemical compositions could be aligned alternatively just as we may found in the paper by Flynn et al. (2018). Heterogeneous structures are frequently formed by self-assembly. As shown in the literature (Park et al., 2008), the phase separation-induced self-assembly was used to generate arrays of two types of chemically distinct and regularly spaced nanofibers or nanorods with mutually coherent interfaces. One of the nanofibers or nanorods is the Ga-rich oxide; the other is the Mn-rich oxide.

Nanofibers and their composites are highly flexible. They also have a large surface area-to-volume ratio and high porosity. Depending on the materials, the prepared nanofibers and composites possess good mechanical strength. Therefore nanofibers and their composites are used for tissue repair, energy storage, wearable sensors, drug delivery, catalysis, and so on. The subsequent section will cover nanofibers and their composites processing technologies.

1.2 Processing and fabrication technologies

There are many ways to make nanofibers and their composites. Spinning, drawing, nanocasting, template synthesis, and phase separation are just some of the examples. This subsection will specifically focus on the commonly used CVD technology. Also, the electrospinning will be touched. Other techniques, such as hydrothermal synthesis, templating, LPD, ECO, high-temperature annealing, and spraying pyrolysis, will be presented. The procedures and important processing parameters associated with each processing or fabrication technology will be discussed. Examples of nanofibers processed with such technologies will be given. The advantages and disadvantages of various fabrication technologies will be delineated.

1.2.1 Chemical vapor deposition

CVD has been used for nanofiber processing for a very long time. During CVD, controlled heterogeneous reactions take place at the surface of a substrate. The procedures for CVD can be divided into several stages. First, reactants and inert dilute gases flow into a reaction chamber. The gaseous species move toward the substrate and reactants adsorb onto the substrate. The adsorbed substances migrate and participate in chemical reactions. Gaseous by-products generated by the reactions are desorbed from the substrate surface and taken away from the reaction chamber. The exhaust is subsequently trapped in a recycling container. To control the temperatures at different locations in the reactor, multizone resistance heaters may be installed.

The rate of CVD is temperature-dependent. In the high-temperature range, mass transport limits the rate of deposition. In the low-temperature range, the chemical reaction rate determines the whole rate of deposition. A transition temperature zone between mass transport and reaction limit can be found as well. In the mass transport−limited region, if CVD reactants arrive at the surface of substrates uniformly, the rate of deposition should not be so sensitive to the temperature variation. In the reaction rate−limited region, the rate of deposition is highly sensitive to temperature fluctuation. To achieve the demand for uniform deposition, the temperature distribution in the reaction zone has to be carefully controlled.

CVD can be divided into several subcategories. The first type of which is ambient pressure chemical vapor deposition (APCVD). APCVD is typically a mass transport−limited process. It is sensitive to gas phase reactions. The formation of particle by-products could cause nonuniformity of the deposited products in some of the APCVD processes. The second type of CVD is low-pressure CVD (LPCVD). In LPCVD, the pressure level is controlled to the level of less than 1.0 Torr. LPCVD is a reaction rate−limited process. The temperature for LPCVD is usually higher than that for APCVD. The third type of CVD is the plasma-enhanced CVD (PECVD). This is a surface reaction rate−limited process. During PECVD, the energy is supplied by plasma. There are several advantages of PECVD. These include low processing temperature, high deposition rate, and low porosity of the deposited products. Therefore PECVD leads to the high quality of the deposited nanofibers and their composites.

For example, both APCVD and LPCVD were used for the deposition of SiO_2. In a CVD chamber, silane and oxygen react with each other to generate SiO_2. PECVD for SiO_2 formation can also be achieved by the decomposition of tetraethoxysilane (Zhang et al., 2015). More options for the fabrication of SiO_2 via CVD were studied. One such option is through the reaction of dichlorosilane and nitrous oxide. The reaction products include nitrogen gas, hydrogen chloride gas, and silica solid. Another CVD example is the synthesis of carbon nanotubes (CNTs) in large quantities. Transition metals such as iron and nickel served as the catalysts. Various carbon-containing substances including carbon monoxide were used as the sources of carbon. The morphology of the produced CNTs is dependent on the catalyst type and the thermal conditions. In a typical process, metal catalysts such as Ni and Fe NPs were placed into a reaction chamber. The chamber was maintained in the temperature range from 750°C to 900°C. Carbon monoxide is introduced through a compressive pump. The pressure in the reaction chamber is kept at 10 atm. The chemical reaction from the decomposition of carbon monoxide generates CNTs at the surface of the catalysts. A product-collecting unit was used to harvest the CNTs. The by-product, carbon dioxide, was trapped in a storage tank (Zhang et al., 2015).

The development of delicate facilities increased the power of CVD for nanostructure fabrication. An integrated PECVD nanofabrication facility was used for assembling quantum structures and fabricating advanced functional nanomaterials (Xu et al., 2006). Multiple radio frequency (RF) magnetron sputtering plasma sources were integrated. Such an integrated facility can be used for various nanostructure fabrications, such as vertically aligned single-crystalline carbon nanofibers (CNFs) and ultrahigh-aspect-ratio semiconductor nanowires.

Exploration of new CVD technology leads to the emergence of the electron cyclotron resonance chemical vapor deposition. As an innovative approach, it can be used for the nanofabrication of GeSbTeSn phase-change alloy-ended CNTs (Wang et al., 2006). Cobalt-assisted CNTs can also be synthesized with H_2 and CH_4 as the gaseous reactants to grow CNTs at the surface of Co NPs. The as-grown CNTs can be treated in a hydrogen plasma atmosphere to remove the carbon layers on the Co catalyst. The catalyst can be subsequently removed from the tips of the CNTs in a diluted HNO_3 solution. The open-ended CNTs with bowl-like tips can be coated with a layer of GeSbTeSn phase-change alloy. The composition of the alloy at the CNT tips can be tuned from Te-rich to Ge-rich by heat treatment.

Microwave plasma chemical vapor deposition (MPCVD) is another unique CVD technique. The MPCVD was used for preparing well-oriented carbon nanocones (CNCs) (Wang et al., 2005). The structures of CNCs can be manipulated by adjusting the ratios of the gases (CH_4 and H_2) and the substrate bias. The formation of the cone-shaped nanostructures is the result of competition between the ion bombardment and the lateral growth of the nanostructures in the plasma. Because the ion bombardment is enhanced by the negative substrate bias, a higher substrate negative bias and a lower concentration of carbon species in the plasma are favorable conditions for growing highly oriented CNCs owing to a greater ion bombardment energy and a lower lateral growth rate of the nanostructures. The CNCs can be used as field emitters to provide high field emission current densities.

CVD offers the advantage of growing uniformly distributed nanofiber branches on nanofiber stems. An example was given by Xia et al. (2005). Through the carefully controlled CVD, the evenly distributed CNF branches on CNF trunks were formed. The whole process can be divided into several vapor reaction steps. In the first step, the stem CNFs as the starting materials or substrates were surface treated by oxygen plasma to attach oxygen-containing functional groups at the surface. In the second step, CVD of ferrocene was conducted in an oxygen oxygen-containing atmosphere. This allowed the formation of iron oxide particles at the nanofiber surface. Following that, the secondary CNFs (nanofiber branches) grew from cyclohexane catalyzed by the iron NPs from the reduction of iron oxide by hydrogen gas. The diameters of the CNF branches are in the range of 10–20 nm. The specific surface area of the nanocomposite was found to increase significantly due to the growth of the secondary nanofibers. The morphology of the nanofiber–nanofiber composites can be changed by varying the processing parameters. The CVD technique makes the synthesis of branched carbon fiber nanocomposites possible.

Reductive-catalytic chemical vapor deposition (RCCVD) was used to make CNFs capped by transition metal carbides (Huang et al., 2020). Such hierarchical assemblies are called carbon hybrids (CHs). The process can be briefly summarized by the following steps. The first step is to allow graphene oxide sheets to self-assemble into microfiber frames by wet spinning. Then, in situ growth of the carbon hybrid nanofibers occurred with the catalytic species generated from the reduction of transition metal ions including Fe^{3+}, Co^{2+}, and Ni^{2+}. By controlling the temperature, the products from the reductive catalyst-assisted CVD include nanocaps, nanofibers, and nanorods. The results show that RCCVD works very well for preparing carbon-based hierarchical structures.

Thermal CVD was used to catalytically grow sCNFs with diameters from 37 nm to 125 nm. The CNFs grew on the permalloy-coated substrates using ethanol containing a low concentration of carbon disulfide as the carbon source, as illustrated by Deno et al. (2007). The diameter and orientation of the CNFs varied depending on the size and distribution of the permalloy catalyst particles. By increasing the particle size and decreasing the space between adjacent particles, about 60% of the vertically aligned carbon nanofibers (VACNFs) contained Y-junctions. Composite structures of anatase-type TiO_2 NPs with 10–70 nm diameters supported by the VACNFs were made by the subsequent pulsed laser deposition.

Hybrid carbon nanofiber-graphene nanosheet (CNF–GN) materials with different morphologies were synthesized by CVD using ethyne (C_2H_2) as the carbon source and Ni as the catalyst [see the work described by Xu et al. (2015)]. The procedures and important parameters are shown in detail. By adjusting the C_2H_2 flow rate, hybrid materials with different CNF diameters and CNF/GN weight ratios were obtained. First, graphene oxide (GO) was synthesized by the modified Hummers method using natural graphite powder. In a typical experiment, 1 g of graphite powder, 92 mL of H_2SO_4, and 24 mL of HNO_3 were mixed in a 500 mL flask under vigorous stirring for 10 min. The temperature of the mixture in the flask was kept at 0°C–5°C by immersing the flask in an ice-water bath. Then 6 g of $KMnO_4$ was added slowly into the flask in 15 min. The mixture was stirred for another 15 min. Then the solution with suspensions was heated up in a water bath to 85°C and maintained at the temperature for 30 min. After that, 100 mL of distilled water was slowly added to the solution and kept at 85°C for half an hour until a yellow suspension was obtained. Immediately after the reaction, 10 mL of H_2O_2 was added into the flask and stirred for 10 min to reduce the residual permanganate. Finally, the suspension was filtered, washed with distilled water, and dried in a vacuum oven at 60°C for 12 h to obtain GO.

A typical process for synthesizing the GN-supported Ni material (Ni–GN) was also described by Xu et al. (2015). In brief, 1 g of sodium dodecyl sulfonate (SDS) was added to 40 mL of aqueous solution (5 mg/mL) and sonicated for 30 min to obtain a uniform suspension (solution I). Solution II was prepared by dissolving 0.8 g of $Ni(NO_3)_2 \cdot 6H_2O$ in 20 mL of distilled water. Solution II was added dropwise to solution I under stirring within 30 min. The resulting mixture was quickly heated to 80°C and maintained at that temperature for 12 h. The solvent was evaporated, and the product was dried in air at 60°C for 24 h to obtain the Ni–GN precursor. Such a precursor was transferred to a furnace and reduced with H_2 at 500°C for 30 min. It was then washed with distilled water several times to remove the surfactant SDS. This yielded Ni–GN with a Ni content of 78%. For comparison, GNs were synthesized using the same procedures except no nickel salt was added. Synthesis of hybrid CNF–GN materials was shown. The hybrid CNF–GN materials were synthesized by CVD using the prepared Ni–GN material as the catalyst.

CVD reactor consisting of a gas flow unit and a horizontally aligned quartz tube was used. The as-synthesized Ni–GN catalyst was uniformly dispersed using a sieve onto a quartz slide. The slide was placed into the quartz tube and heated up to 600°C under Ar gas protection. A mixture of C_2H_2 and Ar was induced into the quartz tube and maintained for 30 min. The rate of Ar flow was 100 mL/min. After the reactor was naturally cooled down to room temperature in Ar, the solid products were collected from the quartz plate. The process was repeated at varied C_2H_2 flow rates of 2, 5, 8, and 40 mL/min. The generated products were named Ni–CNF–GN-2, Ni–CNF–GN-5, Ni–CNF–GN-8, and Ni–CNF–GN-40, respectively. The purified CNF and graphene composite materials were obtained by removing nickel in nitric acid for 12 h at room temperature and then dried at 500°C in Ar for 1 h. These pure carbon hybrid composite specimens were called CNF–GN-2, CNF–GN-5, CNF–GN-8, and CNF–GN-40, respectively. Xu et al. (2015) considered that the diameter of the CNF increased with the increase in the C_2H_2 flow rate. Lin et al. (2007) found that the size of the nickel NP catalyst is one of the factors that determines the size of the grown CNF. The Ni NP also grew during the CVD process. The size of the nickel NP is obviously bigger than the diameter of the CNF. The reason for this increase in the particle size is the combining of two Ni particles at the high temperature for CVD. Earlier studies by Jian et al. (2010) indicated that the formation of CNFs started with the adsorption of C_2H_2 at the surface of the catalyst particles during the CVD process. Following that, the dissolution and diffusion of the released carbon atoms toward the catalyst allowed the precipitation of carbon and eventually the growth of CNF. The fast diffusion of carbon atoms away from the boundaries of different catalyst particles resulted in the fusion of catalyst particles. Therefore the diameter of the obtained CNFs became larger. This was also reported by Cao et al. (2007).

1.2.2 Electrospinning

As well described by Xue et al. (2019), electrospinning works due to electrostatic forces acting on electrified fluids (polymer solutions or sol-gel precursors). During electrospinning, the applied direct current (DC) voltage produces repulsions among molecules causing the fast evaporation of solvents. Fibers are then drawn out from the droplet at the end tip of a spinneret. For single-jet electrospinning, the spinneret could be a stainless-steel needle. The fibers deposit on an electrically grounded conductive collector. The conductive collector or target could be made of a metallic rotating disk, a roller, a piece of metallic foil, or simply a mesh or net. Electrospun fibers are much finer than those obtained from traditional methods such as solvent coagulation, melt spinning, dip casting, drawing, and extrusion. This is because the electrified solution droplet with the significantly reduced surface tension can be pulled out quickly and thrown into very thin fibers. The level of the DC voltage could be as high as several ten thousand volts. The intensity of the electric field is typically maintained at about 1 kV/cm. This is just a general guideline for the electrospinning voltage parameter setting. However, very low voltage or low electric field intensity may also be adopted. During the so-called electrohydrodynamic casting or extrusion, the electric filed intensity could be as low as several volts per centimeter. It should be mentioned that the diameter of the spun fiber is highly sensitive to the level of the applied voltage or the intensity of the electric field. The higher the applied voltage or stronger the electric field, the finer the fiber obtained.

The electrospinning jet may be aligned either horizontally or vertically (Rodoplu & Mutlu, 2012). For those electrospinning facilities with vertically aligned jets, the spun nanofibers are drawn toward collectors by both electric force and gravitational force. During traditional electrospinning, the spun nanofibers are randomly aligned on collectors and form meshes and networks. Recent development allows the nanofiber collectors to be metallic screens under controlled motions. For example, disks, rollers, or mandrels as collectors can be kept in rotation. During the stable electrospinning, the solvent used for making the solution evaporates rapidly. This allows the formation of charged solid fibers and the fibers accumulate on the collector with varying porosity and packing density.

In some cases, the concentration of the solution for electrospinning is relatively low. Then, nanobead formation could be observed (Fong et al., 1999). The bead-to-fiber formation conditions during electrospinning were given (Munir et al., 2009). A significantly low concentration or very low viscosity of the fluid could trigger the so-called electrospraying during which the fluid drops are separately drawn out and randomly deposited onto the collecting targets. It must be pointed out that there are other derived electrospinning techniques as compared to the traditional single-jet electrospinning. Coaxial spinning, triaxial spinning, needleless spinning, air bulb-promoted spinning, sol-gel spinning, and coelectrospinning are some of the examples. The electrospinning can be integrated into three-dimensional (3D) printing too. In sol-gel spinning, gelation chemical reactions happen. In cospinning process, NPs may be added into polymer solutions so that the spun products are composite fibers.

There are many factors influencing the electrospinning process besides the above-mentioned DC voltage level and viscosity of the polymer solution. For example, the electrical conductivity of the polymer solution is one of the significant factors (Angammana & Jayaram, 2011). Still, some other factors include the temperature, relative humidity of the environment, fluid flow rate, and the distance between the tip of the injection needle and the collecting target. The conductivity of a solution affects the Taylor cone formation. It also controls the diameter of the spun nanofibers. If a solution has a very low conductivity, the surface of the droplet will not have enough charge to form a Taylor cone. Consequently, electrospinning will not take place.

Increasing the conductivity of the solution over a critical value is necessary to trigger the electrospinning. A high solution conductivity will not only promote the charge accumulation at the surface of the droplet to form a Taylor cone but also reduce the nanofiber diameter. An ideal dielectric polymer solution or insulating polymer solution will not produce sufficient charges in the solution to move toward the surface of the fluid. As a result, the electrostatic force generated by the applied electric field will not be big enough to form a Taylor cone. Taylor cone formation is the prerequisite for starting the electrospinning process. On the contrary, a conductive polymer solution will have sufficient free charges to move onto the surface of the fluid and form a Taylor cone easily to initiate the electrospinning.

The conductivity of a polymer solution for electrospinning could be increased by adding an appropriate inorganic salt into the solution. A very small amount of addition of LiCl, KCl, or NaCl works effectively for this purpose (Matabola & Moutloali, 2013). The addition of such a salt influences the electrospinning process in two ways. One is adding the number of ions in the polymer solution, which helps to increase the surface charge density of the fluid. The electrostatic force generated along the direction of the applied DC electric field increases. The other is reducing the tendency of the transverse motion of nanofibers because the higher the conductivity of the polymer solution, the lower the tangential electric field along the transverse surface of the fluid. Therefore the nanofibers can be better stretched by the normal force and move directly toward the collecting target.

The distance between the tip of the spinneret and the collector is also a critical parameter that determines the morphology of the obtained nanofiber product (Hekmati et al., 2013). It controls the time of fiber flying and stretching, the extent of solvent evaporation, and the fiber whipping or instability interval. The nanofiber morphology could be readily tuned by varying the needle tip to the target distance. This distance should be long enough to generate the demanded morphology or structure and keep the quality of the spun fiber product. The least value for the distance between the metallic needle tip and collector varies with the nature of the solution used for electrospinning.

Electrospinning as a mature technology has found numerous applications in new product development. Constructing tissue scaffolds (Flores-Rojas et al., 2023), making nanofiber masks (Naragund & Panda, 2022), preparing wound dressings (Abrigo et al., 2014), and building wearable technology sensors (Xu et al., 2022) are some of the examples for practical applications. It is also used for making composite nanofibers for rechargeable batteries. In addition, electrospun nanofibers have been used for solar cells, hydrothermal solar evaporators, and photochemical fuel cells, as will be described in some of the later chapters.

1.2.3 Hydrothermal synthesis

Hydrothermal synthesis is an important technique for preparing discontinuous nanofibers. It is a solution reaction-based method. Hydrothermal synthesis of NPs, nanorods, nanotubes, hollow nanospheres, and graphene nanosheets has been extensively studied. In view of the major advantage of nanofiber preparation using the hydrothermal synthesis method, the formation of nanofibers during hydrothermal synthesis could occur in a wide temperature range from room temperature to elevated temperatures. To control the morphology of the nanofibers to be prepared, varying pressure conditions can be used depending on the vapor pressure of the main composition in the reaction. Many types of nanofibers have been successfully synthesized via this approach. There are other advantages of the hydrothermal synthesis method over others. Hydrothermal synthesis can generate nanofibers that are not stable at elevated temperatures. Nanofibers with high vapor pressures can be produced by the hydrothermal method with minimum loss of materials. The compositions of nanofibers to be synthesized can be well controlled in hydrothermal synthesis through liquid phase or multiphase chemical reactions. In this subsection, some of the recently researched examples for hydrothermal synthesis of typical oxide and CNFs will be presented as follows.

Hydrothermal synthesis was used by Wen et al. (2017) to make CNFs of 1–2 nm thick and micrometer length. Cellulose nanocrystals were taken as the raw materials. The hydrothermal synthesis temperature was set at 240°C. The self-assembly of cellulose nanocrystals before the carbonization produced the precursor for the CNFs. Zheng et al. (2013) showed the morphology control of hydrothermal synthesized WO_3 nanorod arrays. In another paper, the high aspect ratio WO_3 nanofiber bundles were illustrated (Zheng et al., 2018). The hexagonal WO_3 with nanofiber bundle morphology was successfully synthesized through Na_2SO_4 and Zn $(CH_3COO)_2 \cdot 2H_2O$-assisted hydrothermal method when the pH value was 0.8. The phase transformation of WO_3 from monoclinic to hexagonal type was observed when the pH value of the solution was increased. Adhikari and Sarkar (2014) revealed that hydrothermal synthesis facilitated the hierarchical growth of the hexagonal WO_3 nanofibers that consist of high interactive space and surface area. The effects of the major processing parameters including hydrothermal temperature, time, and the concentration of the directional growth reagent (sodium chloride) were examined. The hexagonal crystal structure and nanostructured 1D fibrous morphology of WO_3 were formed under different processing conditions. The hexagonal nanofiber WO_3 has a specific surface area of as high as 25.2 m^2/g. During the hydrothermal synthesis, sodium tungstate ($Na_2WO_4 \cdot 2H_2O$), structure directing agent sodium chloride (NaCl), and catalyst HCl were used to generate the WO_3 nanofibers. The sodium chloride was added to sodium tungstate aqueous solution and constantly stirred on a magnetic stirrer at 300 rpm for about 30 minutes to prepare a clear transparent solution. Then, concentrated hydrochloric acid was used to adjust the pH of the solution to about 2. After that, the solution was transferred to a 50 mL Teflon cylinder. The cylinder was placed inside an autoclave. The autoclave was tightly sealed and kept at the preset temperature and period in a hot air oven. The remaining NaCl and impurities were removed by centrifuging and subsequent hot water and isopropanol washing. The collected product was freeze-dried in vacuum. Varying NaCl molar concentration, hydrothermal duration, and reaction temperature were carried out to tune the morphology and crystal structure.

Li et al. (2015) obtained hydrothermally synthesized $Y(OH)_3$ nanofiber bundles. The hydrothermal synthesis for the nanofiber bundles includes dissolving Y_2O_3 in

6.0 M HNO$_3$ to form a transparent solution. Then 2.0 M NaOH was added dropwise under continuous stirring into the solution to generate a colloidal solution with a pH of 13. This colloidal solution was transferred to an autoclave and maintained at 140°C for 12 h. The Y(OH)$_3$ nanofiber bundles were generated as the white precipitates.

Hydrothermal synthesis can directly assemble CNF cryogels. Liang et al. (2018) used the glucose-derived monolithic carbon to construct porous nanofiber networks. The approach is claimed as simple, rapid, and sustainable. The obtained CNF cryogels have controllable structures and properties. To obtain monolithic nitrogen-doped carbon cryogels, polyaniline (PANI) was used as the crosslinking agent and the source of nitrogen. The crosslinking function of PANI allowed the formation of a 3D hydrogel network composed of interconnected coral-like nanofibers. The structures, properties, and surface chemistry can be tuned easily by changing the aniline concentration and carbonization temperature. Carbon cryogels showed a hierarchically porous structure with a nanofiber diameter of 60–200 nm and a high specific surface area of 900 m^2/g. The process for the monolithic nitrogen-doped carbon cryogels was demonstrated (Liang et al., 2018). After aniline was added to the solution containing glucose and ammonium persulfate (APS), the color of the solution was changed from clear to deep green, which is the color of PANI. Then, the mixture was treated hydrothermally to form dark brown monolithic hydrogels with a water content of about 93%. After the freeze-drying procedure, the light brown cryogels with densities less than 0.1 g/cm^3 were obtained. CNF cryogels with low densities in the order of 0.05 g/cm^3 were generated after carbonization at high temperatures. The carbonization led to a 40% reduction in the volume. In a typical experiment as shown (Liang et al., 2018), 10 g glucose was added to 25 mL 1 M HCl solution at the ambient temperature. The oxidant APS (0.306 g) and aniline (0.125 g) were added to the solution under continuous stirring. The molar ratio of APS to aniline was 1:1. After a short time (5 min) reaction, the dark and sticky solution was transferred to a 100 mL Teflon-lined autoclave and then kept at 160°C for 10 h. This resulted in the brown-colored and mechanically stable monolithic hydrogels. The hydrogels were washed using deionized (DI) water and ethanol. After that, changing the solvent back to water was done. The hydrogels were freeze-dried to obtain light brown cryogels. The light brown cryogels were calcined in a nitrogen stream for 2 h at different temperatures to generate carbon cryogels.

1.2.4 Template assembling

The demand for some special architectures or structures built upon nanofibers stimulated the research on scalable template-assisted chemical deposition method. This method has been used for preparing battery components (Qie et al., 2012; Wang et al., 2013) and for making photothermal solar cells as well (Shi et al., 2018). In the following part, the procedures for the template-assisted chemical deposition are briefly described. Two types of templates are introduced here. One is a polymer nanofiber template. The other is a quartz fiber template. Both templates can be used to generate carbon micro- and nanofibers with special structures. The polymer template can generate porous CNFs for the battery application (Qie et al., 2012), while the quartz template can produce hollow CNFs for the photothermal solar cell application (Shi et al., 2018).

To process porous CNFs for application in Li-S batteries, Qie et al. (2012) used polypyrrole (PPy) as the sacrificial template. PPy can be synthesized via chemical oxidation. The produced PPy nanofiber webs, with a high nitrogen (N) content of 16 wt.%, were collected. Then, carbonization and activation were performed to yield a porous carbon nanofiber web (CNFW) structure. The CNFWs were doped with nitrogen to achieve the high N content. The unique porous structure and high concentration of N in the CNFWs allowed the Li-S batteries to have the superhigh capacity and excellent rate capability.

Wang et al. (2013) showed the PPy-template method to prepare nitrogen-doped CNFWs in more detail. The PPy nanofibers were prepared by the modified oxidative template assembly approach, as described by Liu et al. (2010). In a typical experiment, 7.3 g cetrimonium bromide (CTAB; with a formula of $(C_{16}H_{33})N(CH_3)_3Br$) was dissolved in 120 mL 1.0 M HCl under the ice bath cooling condition. Then, 13.7 g APS was added to the solution. A white reactive template was instantly generated. Then, pyrrole monomer (8.3 mL) was added to the as-formed reactive template solution. After the reaction at 0°C—5°C for 24 h, the PPy nanofiber webs in the form of black precipitates were obtained. The nanofibers were washed with 1.0 M HCl and DI water several times until the filtrate became colorless and neutral. The product was oven-dried overnight at 80°C. To convert the PPy nanofiber webs into CNF networks, the as-prepared PPy was heated to 600°C at a heating rate of 5°C/min and held for 0.5—4 h to form nitrogen-doped CNFWs. The carbon yields were found to be close to 52%.

1.2.5 Liquid phase deposition

LPD is a wet solution method that can be used for making thin films, NPs, nanofibers, nanotubes, and nanowires. In sol-gel method, chemical precipitation, hydrothermal or solvothermal synthesis, and nanocasting, LPD occurs. However, this section is on the deposition relying on the metal-fluoro complex reactions. Specifically, discontinuous oxide nanofibers processed through the LPD technique will be introduced. The properties of typical liquid-phase deposited nanofiber materials will be briefly mentioned. The oxide nanofibers for selective ion transport, photocatalysis, sensing and self-powering, energy storage, and energy conversions will be shown. Combining LPD with other techniques such as hydrothermal processing, plasma synthesis, anodic oxidation, and in situ chemical reaction will also be addressed.

Deki et al. (2002) revealed the direct deposition mechanism for the LPD. The ligand exchange hydrolysis of a metal-fluoro complex generates metal oxide nanostructures. Specifically, the depletion of F^- through the reaction of a metal-fluoro complex with either boric acid or aluminum metal was demonstrated. Boric acid and aluminum were the fluorine ion scavengers in the reactions. Nagayama et al. (1988)

illustrated the deposition of silica (SiO_2) by this method. The silica (SiO_2) formed at the surface of soda lime silicate glass by immersing the glass into a hydrofluosilicic acid (H_2SiF_6) solution supersaturated with silica gel. Hishinuma et al. (1991) also made SiO_2 using a hexafluorosilicic acid (H_2SiF_6) mixture solution containing boric acid. More stable SiO_2 can be produced through the dissolution of aluminum in the H_2SiF_6 solution. The hydrolysis of hexafluorosilicic acid (H_2SiF_6) solution and the formation of silicon oxide occurred simultaneously through the fluorine ion consumption using the two F^- scavengers, that is, the boric acid and aluminum metal (Hishinuma et al., 1991).

Preparation of vanadium oxide nanostructures using the LPD method was shown by Deki et al. (1996). First, vanadium (V) oxide was dissolved in hydrofluoric acid (5%) to form a vanadium ion-saturated solution. The vanadium ion concentration reached 0.384 M. Then, the solution was diluted with distilled water to 0.15 M of vanadium ion concentration for the LPD of vanadium oxide. A cleaned glass slide was inserted vertically into the solution. A pure aluminum metal plate was used to consume the F^-. The aluminum plate was placed into the solution at a close distance to the glass slide. The reaction temperature was 30°C and the duration was from 20 to 40 h. After the reaction, the specimen was heat treated at various temperatures for 1 h in air to obtain the final product. The ligand exchange reaction mechanism for the deposition of vanadium (V) oxide was proposed.

Transition metal hydroxide, β-FeOOH, was deposited on borosilicate glass and Au wires, respectively, by the LPD method described by Deki et al. (2002). By adding NH_3 to adjust the pH value, FeOOH was precipitated by the hydrolysis of Fe $(NO_3)_2$ from an aqueous solution. After the separation, the precipitate was washed with distilled water and air-dried at room temperature. Then, the FeOOH precipitate was dissolved in $NH_4F \cdot HF$ to form the $FeOOH-NH_4F \cdot HF$ aqueous solution for β-FeOOH deposition. By adding H_3BO_3 into the solution, β-FeOOH films were uniformly deposited on the substrates. The deposition time was varied from 40 min to 20 h and the temperature was 30°C. Hydrolysis happened in the LPD process (Deki et al., 2002). First, the iron (III) ions, Fe^{3+}, were coordinated by fluorine ions producing $[FeF_6]^{3-}$ in the solution. The partial release of F^- ions from $[FeF_6]^{3-}$ in the liquid phase resulted in the formation of $[FeF_{6-n}(OH)_n]^{3-}$. Ligand exchange between $[FeF_6]^{3-}$ and $[FeF_{6-n}(OH)_n]^{3-}$ was accelerated by boric acid-promoted F^- consumption reaction in the solution. The final product at the substrate surface became the iron oxyhydroxide. Typically, the presence of Cl^- or F^- ions during hydrolysis of Fe^{3+} led to the formation of β-FeOOH instead of α-FeOOH, as indicated by Flynn Jr. (1984).

The LPD methodis not just limited to generating the above-mentioned metaloxides. It is also applicable for producing various ordered nanoparticels (NPs), for example, titanium dioxide and tin dioxide NPs (Deki, Nakata, et al., 2004). During the LPDprocess,reverse micelles (RMs) withinnerwaterpools served asbothnanoreactorsandtemplatesforthe NPs. As known, RMs are nanoscopic aggregates consisting of three components: a water pool, a surfactant, and a nonpolar solvent. To synthesize titanium dioxide and tin dioxide NPs, anionicsurfactant containing Triton X-100 (TX-100) and 1-hexanol was made in the work performed by Deki, Nakata, et al. (2004).

Cyclohexanewasused as theoil phaseforRMformation. The relative amount of TX-100 to 1-hexanol was set as 0.2. The ratio of water to TX-100, W_0, was kept at 10. Asa waterphaseinRM, the LPDreactionsolutioncontains $(NH_4)_2TiF_6$ and H_3BO_3 for titanium dioxide NP deposition. For tin dioxide NP deposition, the LPD reaction solution consists of tinfluorocomplex $(SnO_n \cdot HF)$ and H_3BO_3. TheLPD solutionswereinjected intotheinitially made T-100/1-hexanol/cyclohexane/water RMsolutionundervigorous stirringto make transparent RM solutions. ThetransparentRMsolutionswere placed into a warm bathat 30°Cfor 20 h. AftertheLPD reactions, theRMsamplesolutionswere removed by centrifugation. The generated precipitatesweredultrasonically cleanedin distilledwater. To remove the residual surfactant in NPs,theLPDsampleswerere-dispersed in cyclohexaneforpurification. Thepurifiedsampleswere dried,andthe final products of titanium dioxide and tin dioxide NPs were obtained. The average diameter andstandarddeviation for the titanium dioxide NPs were 59 nm and 0.84 nm, respectively. For SnO_2 NPs, the average size was found to be 3.0 nm. Its standard deviation was 0.71 nm (Deki, Nakata et al., 2004).

In general, the LPD behavior depends on the chemical equilibrium between a metal-fluoro complex and a metal oxide (Deki et al., 2002). No specific requirement on the substrate material is needed. Thus LPD products can be deposited on various substrates. The substrates include those with large surface areas and complex morphologies. In the subsequent section, the discussion will be on the deposition of ordered nanostructure.

LPD is applicable for ordered nanostructure synthesis. Ordered nanostructures have found various applications in surface engineering, for example, superhydrophobic surface design and metamaterial preparation. Jung et al. (2019) showed their work on making ultra-low light reflection surfaces consisting of aluminum oxyhydroxide nano lens and nanopillar arrays. Such nanostructures were generated by depositing aluminum oxyhydroxide on anodic oxidized aluminum via hydrothermal synthesis followed by platinum sputtering. The antireflective properties were evaluated on several types of nanostructures such as nano lenses and nanopillar arrays with different aspect ratios. It was found that the nanopillar arrays with an aspect ratio of 1:14 showed the reflectance as low as 0.18% to the light of 550 nm wavelength at all measured incident angles.

Nanorod and nanopore arrays can be prepared by the LPD, as shown by Deki, Iizuka, et al. (2004). First, electron lithography was used to make silicon templates with regularly aligned nanopores. Then, the liquid-phase deposition (LPD) method was used for preparing nanopillars and nanopores with a two-dimensional periodicity. The one-step approach resulted in the deposition of titanium dioxide into the pores of the silicon template. The etching of the template resulted in the nanopillars aligned on a solid oxide coating.

Because LPD reactions occur in a single phase of liquid, uniform mixing of various chemicals is relatively easy to achieve. This allows the LPD technique to be suitable for making multicomponent materials including nanofiber composite materials. In the work performed by Hsu et al. (2005), a composite of short nanofiber containing titanium dioxide nanotube sheath and cadmium sulfide nanofiber core (CdS@TiO$_2$) was made using the liquid phase deposited method. ECO of pure aluminum foil in an organic or inorganic acid solution resulted in the anodic aluminum oxide (AAO) template. The titanium dioxide nanotube was made by template-based deposition. Then cadmium sulfide nanofiber core was deposited onto the regularly aligned titanium dioxide nanotube array via chemical bath deposition, which can be considered as another name for the LPD. As known, both cadmium sulfide and titanium oxide are n-type semiconducting materials. The enhanced visible light absorption by such a hybrid composite nanofiber structure was observed.

Hsu et al. (2005) showed the procedures for making the CdS@TiO$_2$ hybrid coaxial nanofiber composite with the thickness of each layer controlled precisely using the LPD method with porous AAO as the template. During the LPD process, the thickness of the TiO$_2$ sheath was adjusted through the variation of the reaction conditions. The continuous and polycrystalline CdS nanotube or nanofiber core was deposited onto the titanium oxide nanotube sheath by chemical bath deposition. The structure and ultraviolet (UV)-visible light absorption property of the coaxial CdS@TiO$_2$ nanofibers were characterized. In a typical experiment on making the well-ordered CdS/TiO$_2$ composite nanofibers, the solution for LPD was made by mixing 0.05 M ammonium hexafluorotitanate, (NH$_4$)$_2$TiF$_6$, with 0.15 M boric acid, H$_3$BO$_3$, in a volume ratio of 1:3. Electrochemically processed anodic alumina porous membranes with a pore diameter of 200 nm were used as the template. Prior to the TiO$_2$ shell deposition, both side faces of the membrane were deposited with octadecyltetrachlorosilane-self-assembled monolayers (OTS-SAMs) by microcontact printing technique. The OTS-SAMs can prevent the deposition of TiO$_2$ on the top and bottom surfaces of the membrane. As a result, the TiO$_2$ thin coating can only be deposited on the inner wall of the pores but not on the side surfaces of the membrane. Without OTS-SAM modification, the side surfaces would be preferentially coated and the TiO$_2$ nanotubes cannot be obtained. These modified AAO templates were put vertically into a container with an appropriate amount of the solution containing hexafluorotitanate and boric acid. At the end of the reaction, the AAO template was taken out and washed with DI water several times and dried in nitrogen flow. Then the templates containing the precursor were subsequently heated at 500°C for 2 h. The subsequent thermal annealing allows the titanium dioxide to be crystallized. The CdS nanofiber core was deposited in the inner wall of the TiO$_2$ nanotube by soaking a solution of 0.2 M cadmium perchlorate first. After the TiO$_2$ nanotubes were impregnated with Cd^{2+} solution by capillary effect, the sample was washed with DI water and dried in N$_2$ flow. Then, the sample was inserted into a solution containing thioacetamide with the formula of C$_2$H$_5$NS. This procedure was repeated three to five times so that sufficient CdS nanocrystals grew inside the TiO$_2$ nanotube. The AAO template was eventually removed by selective etching in a 6.0 M NaOH solution. The final products of CdS@TiO$_2$ coaxial nanotubes or nanofibers were obtained.

It is a very common practice to integrate the LPD with other techniques. Sadeghzadeh-Attar (2020) reported the work on combining hydrothermal synthesis with LPD for making the SnO$_2$/V$_2$O$_5$ nanocomposite doped with Nb. The photocatalytic function of the nanocomposite was illustrated. As compared with the titanium dioxide NP (P25), pristine SnO$_2$ hollow nanofiber, and SnO$_2$/V$_2$O$_5$ composite nanostructure, the

Nb-doped SnO_2/V_2O_5 nanocomposite exhibited higher photocatalytic efficiency in decomposing Basic Red 46 (BR46) and producing H_2 production under visible light.

The synthesis is detailed in a study by Sadeghzadeh-Attar (2020). First, hydrous-SnO_2 was deposited in situ into the nanopores of alumina membranes using LPD method. Both a hydrolysis equilibrium reaction of $[SnF_6]^{2-}$ and an F^- consuming reaction were involved in the process. The nominal pore size, average thickness, and diameter of the alumina membranes were 100 nm, 60 μm, and 21 mm, respectively. The alumina membranes served as both the scavenger for F^- and the template for SnO_2 hollow nanofiber formation. Hydrous-SnO_2 nanofiber was subsequently converted to crystalline SnO_2 hollow nanofiber after annealing. Second, Nb-doped SnO_2 nanofibers were synthesized through the liquid phase codeposition from the mixed solutions consisting of 0.1 M ammonium hexafluorostannate, $(NH_4)_2SnF_6$, and several Nb-containing solutions with different concentrations. The Nb-containing solutions were made by dissolving niobium (V) oxide, Nb_2O_5, and ammonium fluoride, NH_4F, in DI water. The solutions containing Nb were added to the aqueous solution of $(NH_4)_2SnF_6$. The nominal concentration ratios of Nb dopant in the solutions were 0, 0.5, 1, 2, and 4 mol%. Porous alumina membranes were then inserted vertically into the treatment solutions. The LPD was sustained at room temperature for 4 h. After deposition, the specimens were taken out and rinsed with distilled water and ethanol before air-drying. Then, the samples were placed in 1 M NaOH solution for the removal of alumina membranes.

Next, the hydrothermal synthesis was used to add V_2O_5 onto the Nb-doped SnO_2 to obtain the Nb-doped SnO_2/V_2O_5 composite nanofiber. Briefly, a solution of 0.1 M ammonium metavanadate, NH_4VO_3, was prepared in distilled water as the source for V_2O_5. The synthesized Nb-doped SnO_2 nanofiber was added to the solution. The mixture was then transferred to a Teflon-lined stainless-steel autoclave. The autoclave was sealed and heated up to 120°C and for 10 h. Then the hydrothermal synthesis container was cooled down naturally to room temperature. The reaction products were centrifuged with ethanol and distilled water and naturally dried at ambient temperature. The amount of added vanadium was controlled at 6 wt.% approximately. The samples were placed into a muffle furnace and calcinated at 600°C for 2 h in air. Finally, the Nb-doped SnO_2/V_2O_5 heterogeneous nanostructured composites with 0, 0.5, 1, 2, and 4 mol% Nb were obtained.

Integrating LPD and carbonization was demonstrated by Gou et al. (2019). A porous ternary $MnTiO_3/TiO_2/C$ nanofiber composite from Mn_2O_3/polyaniline precursor through LPD and subsequent carbonization was obtained. The generated $MnTiO_3/TiO_2/C$ composite was used as the anode material for lithium-ion batteries (LIBs). It has been found that the composite has good electrochemical performance. The cycling stability is high (418 mAh/g after 200 cycles at 500 mA/g). The rate capability reached 270 mAh/g at 4 A/g. The coulombic efficiency is greater than 98% after 200 cycles. As compared with pristine Mn_2O_3 or Mn_2O_3/TiO_2 composite, the $MnTiO_3/TiO_2/C$ composite has better electrochemical performance. This is due to the unique porous structure and the synergistic effect of the three constituents. It is believed that this porous ternary $MnTiO_3/TiO_2/C$ composite may be used as a new generation of anode material for LIBs.

In the work performed by Yuan et al. (2018), integrating LPD and in situ chemical reaction was illustrated. Combining LPD and in situ chemical polymerization was carried out to make a CNT@SnO$_2$@PPy nanocomposite as the anode material for sodium-ion batteries. The CNT@SnO$_2$@PPy nanocomposite has the one-dimensional structure of the CNT. The diameter of the carbon nanotube is around 40 nm. The thickness of the PPy coating is about 7 nm. The sandwich-formed CNT@SnO$_2$@PPy electrode showed excellent rate capability. It maintained a high capacity of 226 mAh/g after 100 cycles at the current density of 100 mA/g. It was believed that the synergistic effects among CNT, SnO$_2$, and PPy enhanced the electrochemical performance of the anode. The good electrical conductivity could reduce the aggregation tendency of Sn during the charge/discharge cycles; thus the pulverization of the electrode was prevented.

The CNT@SnO$_2$@PPy was prepared using a similar route as for making the SnO$_2$/PPy (SnO$_2$/PPy) hollow nanospheres as reported in a study by Yuan et al. (2017). During the LPD, carbon was used as the template. Subsequent in situ chemical polymerization was conducted for PPy layer formation on carbon nanotube supported-SnO$_2$. The two steps of synthesis allowed the CNT@SnO$_2$@PPy nanofiber composite formation. This core/shell CNT@SnO$_2$ nanofiber composite was made first through the LPD approach. CNTs were ultrasonically dispersed in distilled water. Then, SnF$_2$ was added under stirring. The product from the reaction was centrifuged, cleaned, and air-dried. Calcination was conducted at 500°C for 2 h in an Ar atmosphere to crystalize the tin oxide (Yuan et al., 2018). The second step for preparing the CNT@SnO$_2$@PPy composite was the PPy coating deposition on the CNT@SnO$_2$ nanostructure through a chemical polymerization process. Such a chemical polymerization process was also demonstrated for making CuO/PPy composite (Yin et al., 2012). Briefly, the CNT@SnO$_2$ and surfactant SDS were added into the water by ultrasonication. Then, pyrrole monomer, HCl (1.0 M), and (NH$_4$)$_2$S$_2$O$_8$ (0.1 M) were added. The polymerization of pyrrole was maintained at 0°C for 3 h under magnetic stirring.

It is interesting to integrate the LPD with anodic oxidation for the preparation of oxide nanofiber composites. As shown by Mizuhata et al. (2015), the LPD method was combined with the anodic oxidation for making TiO$_2$/porous silicon (PSi) nanocomposites. TiO$_2$ was obtained by LPD. The anodization was used for the generation of the PSi. Scanning electron microscopy (SEM)-energy-dispersive X-ray spectroscopy showed that the TiO$_2$ was only deposited in the fine pores of anodized PSi (ca. 7.4 nm) when the PSi surface was anodized in the presence of Ti ions as F$^-$ scavengers. The TiO$_2$/PSi nanocomposites were fabricated by anodization of PSi in an H$_2$TiF$_6$ electrolyte at a constant potential. The amount of Ti deposited was maximum at 300 mV versus Ag/AgCl electrode, and the deposition process was controlled by varying the applied potential for the PSi anodization. The charge/discharge capacities of the fabricated TiO$_2$/PSi nanocomposites as Li-ion battery anodes were determined. Improvements in the charge/discharge capacity were achieved.

During the processing, the Si wafer or PSi working electrode, platinum mesh counter electrode, Ag/AgCl reference electrode, and a salt bridge were assembled into a three-electrode electrochemical cell. The H$_2$TiF$_6$/H$_2$O/EtOH solution was used as the electrolyte. The Si working electrode was anodized in the electrolyte at a constant potential from 0 to 550 mV versus the Ag/AgCl electrode. Anodization was performed for 30 min at room temperature. After anodization, the working electrode was removed from the electrolyte solution, rinsed with DI distilled water and methanol, and dried in an Ar atmosphere at room temperature. The TiO_2 on the Si wafer was characterized (Mizuhata et al., 2015).

It is possible to combine the LPD with plasma synthesis. Liquid-phase plasma synthesis integrates the plasma synthesis technique with the LPD. Wei et al. (2013) showed that silicon quantum dot/carbon (SiQD/C) can be successfully prepared by liquid-phase plasma synthesis. The synthesis procedures for SiQDs can also be found in more detail in earlier published papers by Kang et al. (Kang, Tsang, Wong, et al., 2007, 2009; Kang, Tsang, Zhang, et al., 2007). Absolute ethanol was used as the medium for preparing SiQD/C composites. The prepared SiQDs were mixed with the ethanol via ultrasonication. The mixed solution was then placed in the chamber with the RF generator setup, as shown by Yuan et al. (2008). Modulating the RF power allowed the plasma generation from the tungsten electrode tip of the RF microelectronic device. The SiQD/C nanocomposites were obtained after the sparking for a period of 20 min (Yuan et al., 2008). To remove the SiQDs from the nanocomposite, the SiQDs/C composite sample was added into a mixed solution of HF (50 mL, 10%) and H_2O_2 (5 mL, 30%) and stirred for 5 h. This allowed the Si to dissolve into the solution. The released carbon was separated from the solution by centrifugation. After that, the carbon material was washed with DI water and absolute ethanol and dried in a vacuum oven at 60°C.

The combination of LPD with other technologies, especially hydrothermal synthesis, has shown the capability of making various ordered nanofibers, nanowires, and/or nanorods (Zhang et al., 2018). LPDreactions generally happen in uniform solutions. No specific substrate material is needed to trigger the reactions. Homogeneous structures can also be deposited on various kinds of substrates. The substrates may include those with large surface areas and complex morphologies. Consequently, the LPD technique is especially suitable for ordered nanostructure deposition. Various nanocomposites can be made by LPD technique. The nanocomposites have been used for photocatalysis, Li- and Na-ion storage, photovoltaics, and thermoelectric energy conversion.

1.2.6 Electrochemical oxidation

ECO is also known as anodic oxidation or anodization. It is commonly used for making Cu-, Zn-, and Fe-based oxide nanostructures. In addition, almost all the so-called "valve metals," such as Ti, Al, Mg, Zr, Nb, and Ta (Cheng et al., 2020), show the unique ECO behavior so that oxide nanofibers or nanotubes from these metals can be generated via the ECO method. The most general layout of ECO consists of two electrodes, operating as anode and cathode, connected to a power source. Sometimes a third electrode is added as the reference electrode. When the energy input and sufficient supporting electrolytes are provided to the system, strong oxidizing species are formed. ECO has recently grown in popularity thanks to its ease of setup and effectiveness in generating in situ ECO to form the required reactive species at the anode surface. ECO has been applied to make oxide nanostructures based on the direct oxidation working principle. When voltage is applied crossing the anodic and cathodic electrodes, intermediates of oxygen evolution are formed at the anode. The surface of an "active" anode produces higher-state oxides or superoxides. The higher oxide then acts as an intermediate product for nanofiber or nanotube formation. In literature (Gan et al., 2012), a piece of titanium foil was used as the anode, and a platinum foil was used as the cathode. The electrolyte for anodization was made by mixing 7.5 wt.% water, 90 wt.% glycerol, and 1.5 wt.% ammonium fluoride (NH_4F). After the ECO, titanium dioxide hollow nanofibers were obtained.

The p-type copper oxide/hydroxide nanofibers were made by self-organized anodization of a high-purity copper in a 0.1 M Na_2CO_3 electrolyte (Stępniowski et al., 2020). The pure copper in the form of foil was cut into 25 × 10 (mm^2) coupons as the specimens for ECO study. The specimens were degreased using acetone and ethanol, respectively. Under the ambient temperature condition, the cleaned specimens were electrochemically polished in a 10.0 M H_3PO_4 at 7.5 V for 60 s. Then, the backside and edges of the specimens were coated with acid-resistant paint to keep an exposed working surface area of 1 cm^2. Self-organized anodization of the copper samples in the 0.1 M Na_2CO_3 was conducted at 20°C for 1 h in the voltage range from 3.0 to 31.0 V with a 4.0 V increment. Microscopic observation of the sample processed at 31.0 V reveals well-established and facetted nanofibers. At higher magnification, the selectively etched features of the oxide nanofibers were observed. Raman spectroscopic examination of the samples processed at various voltages reveals the presence of both CuO and Cu_2O oxides. Therefore the anodization of copper in sodium carbonate aqueous solutions can generate nanofiber composites consisting of CuO, Cu_2O, and $Cu(OH)_2$ (Stępniowski et al., 2020).

Experiments on ECO of copper were also performed in solutions containing strong bases such as NaOH (Caballero-Briones et al., 2010; Cheng et al., 2012; Jiang et al., 2015; Shooshtari et al., 2016; Stępniowski et al., 2017; Wu et al., 2005, 2013; Xiao et al., 2015). Jiang et al. (2015) used a de-aerated 1.0 M NaOH to oxidize copper at the current density of 0.06 mA/cm^2. The total anodization time was 5 min at a temperature of 25°C. Following that, the nanofibers were annealed in a tubular furnace under nitrogen protection. The obtained copper oxide nanofibers were surface modified by a 1.0 wt.% of 1H,1H,2H,2H-perfluorodecyltriethoxysilane (FAS-17) ethanol solution to achieve the superhydrophobic property. The electrochemical anodization of copper in potassium oxalate (Babu & Ramachandran, 2010) and potassium carbonate (Stępniowski et al., 2019) was also reported as typical examples of copper-containing oxide nanofiber formation.

1.2.7 High-temperature annealing

The high-temperature annealing technique works by heating glassy melt-quenched plates at high temperatures. A long-time slow annealing process allows the formation of tiny whiskers from the surface of the plate specimens. Typically, the sizes of the whiskers are in the nanometer and micrometer ranges. Using the glass annealing technique, Funahashi et al. (2000) prepared Ca-Co-O-based single-crystalline oxide whiskers with good thermoelectric properties at temperatures higher than 600 K in air. The composition of the whiskers is $Ca_2Co_2O_5$. $Ca_2Co_2O_5$ has a single-phase structure as determined by the X-ray diffraction (XRD) measurement and transmission electron microscopic (TEM) analysis. It is found that the whiskers have a layered structure in which Co-O layers of two different kinds alternate in the direction of the c-axis.

The measured Seebeck effect coefficient of the $Ca_2Co_2O_5$ whiskers is found higher than 100 μV/K at 100 K and increases close to 230 μV/K with the increase in temperature. Temperature dependence of electrical resistivity for the whiskers reveals the typical semiconducting-like behavior. The thermoelectric figure of merit (zT) of the $Ca_2Co_2O_5$ whiskers is estimated at a value between 1.2 and 2.7 at the temperature above 873 K, demonstrating high thermoelectric performance at high service temperatures (Funahashi et al., 2000).

In literature, Funahashi and Matsubara (2001) described the glass annealing method in synthesizing Ca- and Pb-doped oxide single-crystalline whiskers. In order to prepare $[(Bi, Pb)_2(Sr, Ca)_2O_4]_xCoO_2$, the starting materials, Bi_2O_3, PbO, $CaCO_3$, $SrCO_3$, and Co_3O_4 powders, were mixed with a cationic composition of Bi: Pb: Ca: Sr: Co = 1: 1: 1: 1: 2. The mixture was placed into an alumina crucible to melt at 1573 K for 30 min in air. The melt was quenched by rapid cooling between two copper plates to obtain glassy oxide plates. These glassy plates were heated in flowing O_2 gas at 1193 K for 600 h to grow Ca- and Pb-doped $(Bi_2Sr_2O_4)_xCoO_2$ whiskers at their surfaces. The scanning electron microscopic (SEM) image of the ribbon-like whiskers was shown (Funahashi et al., 2000). The whiskers show such measured dimensions: 1.0−3.0 μm in thickness, 20−100 μm in width, and about 1.0 mm in length. TEM analysis revealed the layered structure of the whiskers (Funahashi & Matsubara, 2001). The CoO_2 layers and rock salt $(Bi, Pb)_2(Sr, Ca)_2O_4$ layers alternate in the c-axis direction. Such a structure allows the whiskers to display superior thermoelectric properties at high temperatures in the air because the edge-sharing CoO_2 layers act as the conducting and thermoelectric units. Each rock salt layer consists of four ordered sublayers, that is, Sr(Ca)O-Bi(Pb)O-Bi(Pb)O-Sr(Ca)O. These sublayers could slow down the thermal transport in the whiskers. The average composition of the whiskers was found to be $(Bi, Pb)_{2.2}(Sr, Ca)_{2.8}Co_2O_y$ (named as BC-232).

The Seebeck coefficient, S, of the whiskers is around 100 μV/K at 100 K. The S value increases monotonically with temperature up to 773 K and reaches an asymptotic value of 190 μV/K at temperatures higher than 773 K. It is also revealed that the temperature dependence of electric resistivity follows a semiconducting-like behavior. However, its electrical conductivity value is much lower than that of ordinary semiconductors. The power factor of the BC-232 whiskers increases with the increase of test temperature. The estimated value is above 0.5 mW/(m · K^2) at temperatures higher than 650 K and reaches 0.9 mW/(m · K^2) at 973 K (Funahashi & Matsubara, 2001).

1.2.8 Spraying pyrolysis

As compared with other techniques, chemical spray pyrolysis deposition is a fast process, which is scalable in manufacturing. Although it is especially suitable for large-area coating preparation and thin film deposition, it is possible to make efficient thermoelectric nanofibers by carefully controlling the processing parameters. For example, n-type titanium dioxide nanofibers were made by spraying pyrolysis (Hussian et al., 2016). Briefly, the precursor solution for generating titanium dioxide was sprayed onto a glass substrate preheated at 350°C. The solution for the chemical spray pyrolysis consists of titanium chloride ($TiCl_3$), ethanol alcohol (C_2H_5OH), and distilled water. The water has the role of promoting the hydrolysis of $TiCl_3$ to get TiO_2 nanofibers. The thickness of the TiO_2 nanofiber layer was controlled at 350 nm. The nanofibers demonstrated good continuity and formed the 3D interlaced dense structure.

In addition to titanium dioxide nanofibers, zinc oxide nanofibers were deposited on glass substrates by the spray pyrolysis deposition technique (Ilican et al., 2006; Kumar et al., 2014; Maity et al., 2005; Sharmin & Bhuiyan, 2019). In a study by Kumar et al. (2014), the spraying pyrolysis solution was made by dissolving zinc acetate anhydrous $Zn(CH_3COO)_2$ into a mixture of solvent containing methanol and water. The ratio of methanol to water was 3:1. A small amount of acetic acid as a stabilizer was added to avoid the formation of $Zn(OH)_2$ precipitate. The concentration of the zinc acetate in the solution was about 0.05 M. This solution was forced to pass through a nozzle and sprayed onto the preheated glass substrate at 450°C under a constant air pressure of 0.2 Torr. The average diameter of the nanofibers in the as-grown state was about 300 nm. After 1 h of annealing at 450°C, the diameter of the nanofibers remained the same as that of the as-sprayed state. After 4 h annealing, the diameter of the fibers increased to about 800 nm. The further increase in the annealing time would not cause much change in the size of the nanofibers. Therefore the average diameter of the nanofibers after the 6 h annealing was still approximately 800 nm. The reason for the fiber coarsening with the increase of the annealing time was explored. It is probably due to the merging of those finer fibers during the annealing process, resulting in the formation of thicker fibers. It has been observed that the thickness of the fibers was dependent on the annealing temperature. The longer the annealing time, the better the arrangement of the ZnO nanofibers.

Sharmin and Bhuiyan (2019) prepared boron-doped zinc oxide nanofibers by spray pyrolysis of a zinc nitrate solution with the addition of boric acid. It was found that boron (B) increased the conductivity of the nanofibers. Decreasing in the nanofiber diameter due to B doping was also found. Without B doping, the average diameter of the ZnO nanofiber was about 500 nm. After doping B with the concentrations of 0.5, 0.75, 1.0, and 1.5 at%, the average diameter was 320, 240, 180, and 170 nm, respectively. It is evident that the higher the concentration of B, the smaller the diameter of the nanofibers. Maity et al. (2005) prepared ZnO nanofibers with an average diameter of 500 nm, which is consistent with the results reported in a study by Sharmin and Bhuiyan (2019). Indium doping can also reduce the fiber size as reported by Ilican et al. (2006). The indium-doped zinc oxide nanofibers have a uniform diameter of 200 nm.

The spray pyrolysis technique is not only suitable for processing TiO_2 and ZnO nanofibers but also can make other types of oxide nanofibers. For example, Zahan and Podder (2019) prepared Co_3O_4 via the spray pyrolysis technique. Cobalt oxide (Co_3O_4) nanofibers were deposited onto glass substrates by spraying the cobalt acetate $Co(CH_3COO)_2 \cdot 4H_2O$ precursor solution. Uniform and well-aligned Co_3O_4 nanofibers were obtained.

1.3 Structure assessment of nanofibers and their composites

The structure observation and assessment can be performed by using SEM, TEM, atomic force microscopy (AFM), XRD, and so on. In a study by Gan (2012), an overview of structural assessment of nanocomposite materials was presented. Brief descriptions of advanced structure characterization methods such as SEM, XRD, TEM, AFM, and scanning tunneling microscopy were introduced. Applications of these methods for the analysis of structures and compositions of typical nanofiber composites were mentioned. Assessment of the interface structures of nanocomposite materials using surface characterization techniques and mechanical damage models was discussed.

1.3.1 Scanning electron microscopy

For nanofiber morphology observation and composition analysis, the frequently used technique is SEM. A scanning electron microscope can produce images of a nanofiber by scanning its surface with a focused electron beam. The electrons interact with atoms in the sample, producing various signals that contain information about the surface topography and composition of the sample. The electron beam is scanned in a raster scan pattern, and the position of the beam is combined with the intensity of the detected signal to produce an image. In the most common SEM mode, secondary electrons emitted by atoms excited by the electron beam are detected using a secondary electron detector. The number of secondary electrons that can be detected, and thus the signal intensity, depends, among other things, on specimen topography. Some field emission scanning electron microscopes can achieve resolutions smaller than 1 nanometer. Typically, SEM operates under high vacuum conditions. However, low vacuum and/or wet conditions with variable pressures are adopted in the so-called environmental SEM. A wide range of cryogenic or elevated temperatures are created with specialized SEMs. The energy-dispersive XRD spectroscopy can be carried out with SEM. This allows the composition of the observed specimen to be determined as well.

1.3.2 Transmission electron microscopy

TEM is another powerful tool to analyze nanofibers and their composites. This technique uses a beam of high-energy electrons transmitted through specimens. It can offer even higher resolutions than the SEM. The crystal structure of the specimen can be identified. TEM has multiple operating modes including conventional imaging, scanning TEM (STEM) imaging, diffraction, spectroscopy, and combinations of these. The contrast of TEM images can be achieved by various mechanisms. Contrast can arise from position-to-position differences in the thickness or density (mass-thickness contrast), atomic number ("Z contrast", referring to the common abbreviation Z for atomic number), crystal structure or orientation ("crystallographic contrast" or "diffraction contrast"), the slight quantum-mechanical phase shifts that individual atoms produce in electrons that pass through them ("phase contrast"), the energy lost by electrons on passing through the sample ("spectrum imaging"), and more. Each mechanism tells the user a different kind of information, depending not only on the contrast mechanism but on how the microscope is used—the settings of lenses, apertures, and detectors. TEM is capable of returning an extraordinary variety of nanometer- and atomic-resolution information, in ideal cases revealing not only where all the atoms are but also what kinds of atoms they are and how they are bonded to each other. Therefore TEM is essential for structure identification and composition analysis of various nanofibers and their composites. As compared with SEM, TEM has the disadvantage of involving more expensive equipment and more complex procedures for sample preparations. The operation conditions for TEM are more rigorous in view of higher acceleration voltages, higher vacuum levels, and electron beam-generating unit cooling by liquid nitrogen.

1.3.3 X-ray diffraction

XRD technique has the capability of determining the compositions, measuring the crystallite sizes, and identifying the crystallographic structures of nanofibers and their composites. XRD works by irradiating the specimens with incident X-rays and measuring the intensities and scattering angles of the beams that are diffracted by the materials. From the diffraction patterns, the identifications of the nanofibers and composites can be determined. Based on the phase identification, XRD can further determine the grain sizes and the deviations of actual structures from the ideal ones caused by the residue stresses and defects. The major advantage of the XRD technique is that the measurement is nondestructive. A majority of studies can be found in determining atomic arrangements, crystalline phases, and orientations of nanocomposites, measuring structural properties including lattice parameters, strain, and grain sizes.

1.3.4 Atomic force microscopy

AFM is a very high-resolution type of scanning probe microscopy (SPM), with a demonstrated resolution of less than a nanometer. An atomic force microscope generates images by scanning a tiny cantilever over the surface of a nanofiber specimen in this case. The tip at the end of the cantilever interacts with the atoms at the surface. The associated atomic force causes the bending of the cantilever. A laser beam reflected by the top surface of the cantilever changes the intensity. A photodiode is used to acquire the signal of change in the laser beam intensity to provide information about the height of the cantilever. This information is converted into morphological features of the nanofiber composite surface. In addition to the topographic imaging and force measurement, the AFM has the ability to measure the mechanical and electrical properties of composites including modulus of elasticity, hardness, and electrical conductivity. AFM can also be used for nanoindentation tests. Lateral measurement of the atomic forces allows the AFM to measure the frictional coefficient of nanofiber composites at small length scales.

1.4 Property characterization

1.4.1 Thermoelectric behavior

Nanofibers and their composites demonstrate various physical and chemical properties. Thermoelectric behavior is one of these being characterized. As shown by Su et al. (2011), to measure the Seebeck coefficients of the TiO_2-CoO composite nanofiber samples, the two ends of each sample were bonded to strips of Al foils using a silver-based conductive adhesive. The aluminum foil strips can provide good conductive property at the composite/electrode interfaces. One end of the Al foil as the hot end was heated up to the required temperature. The hot end temperature ranged from 40°C to 130°C in the experiments. The other end (cold end) was kept at the ambient temperature of 25°C. The Seebeck coefficients at different temperatures were calculated for the four AAO-based composite nanofiber specimens (named TiO_2, CoO, TiO_2/Ag, and TiO_2/Ag/CoO) (Su et al., 2011). The calculated results show that the AAO-based TiO_2 with Ag NPs has the highest absolute value of the Seebeck coefficient, which may be due to the enhanced electrical property through the incorporation of the Ag NPs. These highly conductive metallic NPs possess a high density of electrons under thermal excitation. In addition, electron tunneling exists among the fine Ag NPs within the hollow nanofibers, which could improve the electrical conductivity of the nanocomposite significantly. The AAO-based CoO nanocomposite showed the lowest absolute value of the Seebeck coefficient. This could be due to the semiconductor type change of CoO. As known, CoO shows n- to p-type transition behavior, while TiO_2 is n-type. Another reason could be the reduced electrical conductivity of the $CoO@TiO_2$ nanofiber as compared with the TiO_2 nanotube. The positive influence of Ag NPs was found to be more intensive than the negative effect of the CoO for the Seebeck coefficient. This is why the AAO-based TiO_2/Ag/CoO nanocomposite has a higher absolute Seebeck coefficient value than the AAO-based TiO_2 nanocomposite. In view of the absolute values of the Seebeck coefficients, the nanocomposites are ranked from high to low as follows: TiO_2/Ag > TiO_2/Ag/CoO > TiO_2 > CoO. It is concluded that the oxide-based nanocomposites containing TiO_2 nanotubes and $CoO@TiO_2$ coaxial nanofibers possess a strong Seebeck effect. The absolute value of the Seebeck coefficient for the TiO_2 nanotube-filled AAO is 393 μV/K, while the $CoO@TiO_2$ coaxial nanofiber-filled AAO has a slightly lower absolute value of 300 μV/K. Both composites are n-type. The thermoelectric figure of merit of such nanocomposites could potentially be very high due to the low value of the thermal conductivity of the AAO matrix (Su et al., 2011).

1.4.2 UV-visible light sensitivity

The photosensitivity of the CdS/TiO_2 composite nanofiber was measured and compared with that of the pure TiO_2 nanotube (Hsu et al., 2005). Absorption of the two materials to both UV and visible light was measured. Their UV-visible absorption spectra are presented in the same paper (Hsu et al., 2005). The characteristic absorption band of TiO_2 falls only in the UV light spectrum range. The CdS/TiO_2 nanofiber composite showed absorption in both UV and visible light ranges. This indicates that after the CdS was deposited into the TiO_2 nanotubes, the obtained composite $CdS@TiO_2$ core−shell composite nanofiber had stronger light absorption capability. It is meaningful for the composite nanofiber to be used as a more effective solar energy harvesting material. Typical examples of the applications of $CdS@TiO_2$ on photocatalytic degradation of organics can be found in the papers published by Alizadeh, Fallah et al., and Al-Fandi et al. (see Synthesis and characterization of direct Z-scheme CdS/TiO_2 nanocatalyst and evaluate its photodegradation efficiency in wastewater treatment systems. *Chemical Papers* 74 (1), 133−143 (Alizadeh et al., 2020a), Visible light active $CdS@TiO_2$ core-shell nanostructures for the photodegradation of chlorophenols. *Journal of Photochemistry and Photobiology A: Chemistry*. 374, 75-83, and Photocatalytic degradation of dimethyl sulphoxide by CdS/TiO_2 core/shell catalyst: A novel measurement method. *Canadian Journal of Chemical Engineering* 98 (2), 491−502 in (Al-Fandi et al., 2019; Alizadeh et al., 2020b), respectively). For hydrogen generation through water splitting under sunlight, the $CdS@TiO_2$ core−shell composite nanofiber has taken an important role as shown by Wu et al. (2019).

1.4.3 Optoelectrical properties

The optoelectrical properties of the nanocomposites were studied and presented (Sadeghzadeh-Attar, 2020). A comparison of the light emission from various nanocomposites was made. The SnO_2 nanofibers exhibit an obvious PL emission signal around 568 nm. Introducing dopants to form nanocomposites resulted in a photon-quenching effect. This allows enhanced solar light absorption through the incorporation of the Nb- and V-containing oxides. This behavior can be further revealed by the photocurrent measurement experiments. The photo responses of selected nanocomposites were measured. The results clearly show that the light absorption of the liquid-phase deposited SnO_2 hollow nanofiber was enhanced by the hydrothermally synthesized V_2O_5 NPs. The doping of Nb can further increase the light absorption of the SnO_2/V_2O_5 nanocomposite. This is the major reason for the significant improvement in the photocatalysis performances of the Nb-doped SnO_2/V_2O_5 nanocomposite in hydrogen generation and organic dye decomposition, as shown in a study by Sadeghzadeh-Attar (2020).

1.5 Concluding remarks

Nanofibers and their composites represent advanced materials in various applications in energy conversions, sensing, tissue regeneration, drug delivery, and environment cleaning. Polymer nanofibers and their composites have the advantages of being flexible, easy to process, and relatively inexpensive. Among various processing techniques, electrospinning is particularly suitable for scalable manufacturing polymer nanofibers with various compositions and complex structures. The liquid deposition method has also shown the advantages of being fast, easy to control, and scalable. The structure analysis of nanofiber and their composites can be done by the use of high-resolution electron microscopic techniques such as SEM, TEM, and STEM. XRD becomes an essential method for lattice parameter measurement, phase composition identification, and crystal structure and orientation determination. In view of the property characterization, photonic, photoelectric, and thermoelectrical performances of some nanofibers and composites are lightly touched on in this chapter. More comprehensive work will be found in other chapters of this book.

References

Abrigo, M., McArthur, S. L., & Kingshott, P. (2014). Electrospun nanofibers as dressings for chronic wound care: Advances, challenges, and future prospects. *Macromolecular Bioscience, 14*(6), 772−792.

Adhikari, S., & Sarkar, D. (2014). High efficient electrochromic WO_3 nanofibers. *Electrochimica Acta, 138*, 115−123.

Al-Fandi, T., Al Marzouqi, F., Kuvarega, A. T., Mamba, B. B., Al Kindy, S. M. Z., Kim, Y., & Selvaraj, R. (2019). Visible light active CdS@TiO_2 core-shell nanostructures for the photodegradation of chlorophenols. *Journal of Photochemistry and Photobiology A: Chemistry, 374*, 75−83.

Alizadeh, S., Fallah, N., & Nikazar, M. (2020a). Synthesis and characterization of direct Z-scheme CdS/TiO_2 nanocatalyst and evaluate its photodegradation efficiency in wastewater treatment systems. *Chemical Papers, 74*(1), 133−143.

Alizadeh, S., Fallah, N., & Nikazar, M. (2020b). Photocatalytic degradation of dimethyl sulphoxide by CdS/TiO_2 core/shell catalyst: A novel measurement method. *Canadian Journal of Chemical Engineering, 98*(2), 491−502.

Angammana, C. J., & Jayaram, S. H. (2011). Analysis of the effects of solution conductivity on electrospinning process and fiber morphology. *IEEE Transactions on Industry Applications, 47*(3), 1109−1117.

Arumugam, M., Murugesan, B., Balasekar, P., Malliappan, S. P., Chinnalagu, D., Chinniah, K., Cai, Y., & Mahalingam, S. (2023). Silk fibroin and gelatin composite nanofiber combined with silver and gold nanoparticles for wound healing accelerated by reducing the inflammatory response. *Process Biochemistry*, *134*(2), 1−16.

Babu, T. G. S., & Ramachandran, T. (2010). Development of highly sensitive nonenzymatic sensor for the selective determination of glucose and fabrication of a working model. *Electrochimica Acta*, *55*, 1612−1618.

Caballero-Briones, F., Palacios-Padros, A., Calzadilla, O., & Sanz, F. (2010). Evidence and analysis of parallel growth mechanisms in Cu_2O films prepared by Cu anodization. *Electrochimica Acta*, *55*, 4353−4358.

Cao, Z. Y., Sun, Z., Guo, P. S., & Chen, Y. W. (2007). Effect of acetylene flow rate on morphology and structure of carbon nanotube thick films grown by thermal chemical vapor deposition. *Frontiers of Materials Science in China*, *1*, 92−96.

Cheng, Y. L., Zhu, Z. A., Zhang, Q. H., Zhuang, X. J., & Cheng, Y. L. (2020). Plasma electrolytic oxidation of brass. *Surface and Coatings Technology*, *385*, 125366.

Cheng, Z., Ming, D., Fu, K., Zhang, N., & Sun, K. (2012). pH-controllable water permeation through a nanostructured copper mesh film. *ACS Applied Materials & Interfaces*, *4*, 5826−5832.

Cota-Leal, M., García-Valenzuela, J. A., Borbon-Nunez, H. A., Cota, L., & Olivas, A. (2024). CuS/Cellulose acetate nanofiber composite: A study on adsorption and photocatalytic activity for water remediation. *Polymer*, *293*, 126627.

Deki, S., Aoi, Y., Miyake, Y., Gotoh, A., & Kajinami, A. (1996). Novel wet process for preparation of vanadium oxide thin film. *Materials Research Bulletin*, *31*(11), 1399−1406.

Deki, S., Iizuka, S., Horie, A., Mizuhata, M., & Kajinami, A. (2004). Liquid-phase infiltration (LPI) process for the fabrication of highly nano-ordered materials. *Chemistry of Materials*, *16*(9), 1747−1750.

Deki, S., Nakata, A., & Mizuhata, M. (2004). Fabrication of metal oxidenanoparticles by the liquidphase deposition methodintheheterogeneoussystem. *Electrochemistry*, *72*(6), 452−454.

Deki, S., Yoshida, N., Hiroe, Y., Akamatsu, K., Mizuhata, M., & Kajinami, A. (2002). Growth of metal oxide thin films from aqueous solution by liquid phase deposition method. *Solid State Ionics*, *151*(1−4), 1−9.

Deno, H., Sato, M., Tango, Y., Fukui, M., Kamiki, T., Koshio, A., & Kokai, F. (2007). Growth of vertically aligned carbon nanofibers and attachment of TiO2 nanoparticles. *New Diamond and Frontier Carbon Technology*, *17*(5), 243−252.

Fang, Z. P., Zhang, S. Y., Wang, H., Geng, X., Ye, L., Zhang, A. Y., & Feng, Z. G. (2023). Preparation and evaluation of core−shell nanofibers electrospun from PEU and PCL blends via a single-nozzle spinneret. *ACS Applied Polymer Materials*, *5*(4), 2382−2393.

Flores-Rojas, G. G., Gómez-Lazaro, B., López-Saucedo, F., Vera-Graziano, R., Bucio, E., & Mendizába, E. (2023). Electrospun scaffolds for tissue engineering: A review. *Macromol3*, *3*, 524−553.

Flynn, C. M., Jr. (1984). Hydrolysis of inorganic iron(III) salts. *Chemical Reviews*, *84*(1), 31−41.

Flynn, G., Stokes, K., & Ryan, K. M. (2018). Low temperature solution synthesis of silicon, germanium and Si−Ge axial heterostructures in nanorod and nanowire form. *Chemical Communications*, *54*, 5728−5731.

Fong, H., Chun, I., & Reneker, D. H. (1999). Beaded nanofibers formed during electrospinning. *Polymer*, *40*, 4585−4592.

Funahashi, R., & Matsubara, I. (2001). Thermoelectric properties of Pb- and Ca-doped $(Bi_2Sr_2O_4)_{(x)}CoO_2$ whiskers. *Applied Physics Letters*, *79*, 362−364.

Funahashi, R., Matsubara, I., Ikuta, H., Takeuchi, T., Mizutani, U., & Sodeoka, S. (2000). An oxide single crystal with high thermoelectric performance in air. *Japanese Journal of Applied Physics Part 2-Letters, 39*, L1127−1129.

Gan, Y. X., Gan, B. J., Clark, E., Su, L. S., & Zhang, L. H. (2012). Converting environmentally hazardous materials into clean energy using a novel nanostructured photoelectrochemical fuel cell. *Materials Research Bulletin, 47*, 2380−2388.

Gan, Y. X. (2012). Structural assessment of nanocomposites. *Micron, 43*(7), 782−817.

Gou, Q. Z., Li, C., Zhang, X. Q., Zhang, B., Huang, D. Y., & Lei, C. X. (2019). Facile synthesis of porous ternary $MnTiO_3/TiO_2/C$ composite with enhanced electrochemical performance as anode materials for lithium ion batteries. *Energy Technology, 7*(5), 1800761.

Hekmati, A. H., Rashidi, A., Ghazisaeidi, R., & Drean, J. Y. (2013). Effect of needle length, electrospinning distance, and solution concentration on morphological properties of polyamide-6 electrospun nanowebs. *Textile Research Journal, 83*(14), 1452−1466.

Hishinuma, A., Goda, T., Kitaoka, M., Hayashi, S., & Kawahara, H. (1991). Formation of silicon dioxide films in acidic solutions. *Applied Surface Science, 48−49*, 405−408.

Hsu, M. C., Leu, I. C., Sun, Y. M., & Hon, M. H. (2005). Fabrication of $CdS@TiO_2$ coaxial composite nanocables arrays by liquid-phase deposition. *Journal of Crystal Growth, 285*(4), 642−648.

Huang, T. Q., Chen, R. X., Hu, Y. F., Huang, A. M., Hu, K., Zhang, Y., Rui, K., Wang, N., Zhang, P., & Zhu, J. X. (2020). Rational design of hierarchical carbon hybrid microassemblies via reductive-catalytic chemical vapor deposition. *Carbon, 167*, 422−430.

Hussian, H. A. R. A., Hassan, M. A. M., & Agool, I. R. (2016). Synthesis of titanium dioxide (TiO_2) nanofiber and nanotube using different chemical method. *Optik, 127*, 2996−2999.

Ilican, S., Caglar, Y., Caglar, M., & Yakuphanoglu, F. (2006). Electrical conductivity, optical and structural properties of indium-doped ZnO nanofiber thin film deposited by spray pyrolysis method. *Physica E, 35*, 131−138.

Jian, X., Jiang, M., Zhou, Z. W., Yang, M. L., Lu, J., Hu, S. C., Wang, Y., & Hui, D. (2010). Preparation of high purity helical carbon nanofibers by the catalytic decomposition of acetylene and their growth mechanism. *Carbon., 48*(15), 4535−4541.

Jiang, W., He, J., Xiao, F., Yuan, S., Lu, H., & Liang, B. (2015). Preparation and antiscaling application of superhydrophobic anodized CuO nanowire surfaces. *Industrial & Engineering Chemistry Research, 54*, 6874−6883.

Jung, J. H., Han, E. D., Kim, B. H., Seo, Y. H., & Park, Y. M. (2019). Ultra-low light reflection surface using metal-coated high-aspect-ratio nanopillars. *Micro & Nano Letters, 14*(3), 313−316.

Kang, Z. H., Liu, Y., Tsang, C. H. A., Ma, D. D. D., Fan, X., Wong, N. B., & Lee, S. T. (2009). Water-soluble silicon quantum dots with wavelength-tunable photoluminescence. *Advanced Materials, 21*(6), 661−664.

Kang, Z. H., Tsang, C. H. A., Wong, N. B., Zhang, Z. D., & Lee, S. T. (2007). Silicon quantum dots: A general photocatalyst for reduction, decomposition, and selective oxidation reactions. *Journal of the American Chemical Society, 129*(40), 12090−12091.

Kang, Z. H., Tsang, C. H. A., Zhang, Z. D., Zhang, M. L., Wong, N. B., Zapien, J. A., Shan, Y. Y., & Lee, S. T. (2007). A polyoxometalate-assisted electrochemical method for silicon nanostructures preparation: From quantum dots to nanowires. *Journal of the American Chemical Society, 129*(17), 5326−5327.

Kumar, N. S., Bangera, K. V., & Shivakumar, G. K. (2014). Effect of annealing on the properties of zinc oxide nanofiber thin films grown by spray pyrolysis technique. *Applied Nanoscience, 4*, 209−216.

Li, L., Zou, L., Wang, H. R., & Wang, X. (2015). Converting $Y(OH)_3$ nanofiber bundles to YVO_4 polyhedrons for photodegradation of dye contaminants. *Materials Research Bulletin, 68*, 276−282.

Liang, L., Zhou, M. H., Li, K. R., & Jiang, L. L. (2018). Facile and fast polyaniline-directed synthesis of monolithic carbon cryogels from glucose. *Microporous and Mesoporous Materials, 265*, 26−34.

Lin, M., Tan, J. P. Y., Boothroyd, C., Loh, K. P., Tok, E. S., & Foo, Y. L. (2007). Dynamical observation of bamboo-like carbon nanotube growth. *Nano Letters, 7*, 2234−2238.

Liu, Z., Zhang, X., Poyraz, S., Surwade, S. P., & Manohar, S. K. (2010). Oxidative template for conducting polymer nanoclips. *Journal of the American Chemical Society, 132*, 13158.

Maity, R., Das, S., Mitra, M. K., & Chattopadhyaya, K. K. (2005). Synthesis and characterization of ZnO nano/microfibers thin films by catalyst free solution route. *Physica E, 25*, 605−612.

Matabola, K. P., & Moutloali, R. M. (2013). The influence of electrospinning parameters on the morphology and diameter of poly(vinyledene fluoride) nanofibers- effect of sodium chloride. *Journal of Materials Science, 48*, 5475−5482.

Mizuhata, M., Katayama, A., & Maki, H. (2015). On-site fabrication and charge-discharge property of TiO_2 coated porous silicon electrode by the liquid phase deposition with anodic oxidation. *Journal of Fluorine Chemistry, 174*(SI), 62−69.

Munir, M. M., Suryamas, A. B., Iskandar, F., & Okuyama, K. (2009). Scaling law on particle-to-fiber formation during electrospinning. *Polymer, 50*, 4935−4943.

Nagayama, H., Honda, H., & Kawahara, H. (1988). A new process for silica coating. *Journal of The Electrochemical Society, 135*(8), 2013−2016.

Naragund, V. S., & Panda, P. K. (2022). Electrospun nanofiber-based respiratory face masks- a review. *Emergent Materials, 5*, 261−278.

Park, S., Horibe, Y., Asada, T., Wielunski, L. S., Lee, N., Bonanno, P. L., O'Malley, S. M., Sirenko, A. A., Kazimirov, A., Tanimura, M., Gustafsson, T., & Cheong, S. W. (2008). Highly aligned epitaxial nanorods with a checkerboard pattern in oxide films. *Nano Letters, 8*(2), 720−724. (2008).

Qie, L., Chen, W. M., Wang, Z. H., Shao, Q. G., Li, X., Yuan, L. X., Hu, X. L., Zhang, W. X., & Huang, Y. H. (2012). Nitrogen-doped porous carbon nanofiber webs as anodes for lithium ion batteries with a superhigh capacity and rate capability. *Advanced Materials, 24*(15), 2047−2050.

Rodoplu, D., & Mutlu, M. (2012). Effects of electrospinning setup and process parameters on nanofiber morphology intended for the modification of quartz crystal microbalance surfaces. *Journal of Engineered Fibers and Fabrics, 7*(2), 118−123.

Sadeghzadeh-Attar, A. (2020). Enhanced photocatalytic hydrogen evolution by novel Nb-doped SnO_2/V_2O_5 heteronanostructures under visible light with simultaneous basic red 46 dye degradation. *Journal of Asian Ceramic Societies, 8*(3), 662−676.

Sharmin, M., & Bhuiyan, A. H. (2019). Modifications in structure, surface morphology, optical and electrical properties of ZnO thin films with low boron doping. *Journal of Materials Science-Materials in Electronics, 30*, 4867−4879.

Shi, Y. S., Zhang, C. L., Li, R. Y., Zhuo, S. F., Jin, Y., Shi, L., Hong, S. H., Chang, J., Ong, S. C., & Wang, P. (2018). Solar evaporator with controlled salt precipitation for zero liquid discharge desalination. *Environmental Science & Technology, 2018*(52), 11822−11830.

Shooshtari, L., Mohammadpour, R., & Zad, A. I. (2016). Enhanced photoelectrochemical processes by interface engineering, using Cu_2O nanorods. *Materials Letters, 163*, 81−84.

Singh, R., Ahmed, F., Polley, P., & Giri, J. (2018). Fabrication and characterization of core−shell nanofibers using a next-generation airbrush for biomedical applications. *ACS Applied Materials & Interfaces, 10*(49), 41924−41934.

Stępniowski, W. J., Paliwoda, D., Abrahami, S. T., Michalska-Domańska, M., Landskron, K., Buijnsters, J. G., Mol, J. M. C., Terryn, H., & Misiolek, W. Z. (2020). Nanorods grown by copper anodizing in sodium carbonate. *Journal of Electroanalytical Chemistry, 857*, 113628.

Stępniowski, W. J., Stojadinovic, S., Vasilic, R., Tadic, N., Karczewski, K., Abrahami, S. T., Buijnsters, J. G., & Mol, J. M. C. (2017). Morphology and photoluminescence of nanostructured oxides grown by copper passivation in aqueous potassium hydroxide solution. *Materials Letters, 198*, 89−92.

Stępniowski, W. J., Paliwoda, D., Chen, Z., Landskron, K., & Misiolek, W. Z. (2019). Hard anodization of copper in potassium carbonate aqueous solution. *Materials Letters, 252*, 182−185.

Su, L., Gan, Y. X., & Zhang, L. (2011). Thermoelectricity of nanocomposites containing TiO_2-CoO coaxial nanocables. *Scripta Materialia, 64*(8), 745−748.

Wang, W. H., Chao, K. M., Teng, I. J., & Kuo, C. T. (2006). Nanofabrication and the structure-property analyses of phase-change alloy-ended CNTs. *Surface Coating Technology, 200*(10), 3206−3210.

Wang, W. H., Lin, Y. T., & Kuo, C. T. (2005). Nanofabrication and properties of the highly oriented carbon nanocones. *Diamond Related Materials, 14*(3−7), 907−912.

Wang, Z. H., Xiong, X. Q., Qie, L., & Huang, Y. H. (2013). High-performance lithium storage in nitrogen-enriched carbon nanofiber webs derived from polypyrrole. *Electrochimica Acta, 106*, 320−326.

Wei, Y., Yu, H., Li, H. T., Ming, H., Pan, K. M., Huang, H., Liu, Y., & Kang, Z. H. (2013). Liquid-phase plasma synthesis of silicon quantum dots embedded in carbon matrix for lithium battery anodes. *Materials Research Bulletin, 48*(10), 4072−4077.

Wen, Y. M., Jiang, M. Z., Kitchens, C. L., & Chumanov, G. (2017). Synthesis of carbon nanofibers via hydrothermal conversion of cellulose nanocrystals. *Cellulose, 24*(11), 4599−4604. Available from https://doi.org/10.1007/s10570-017-1464-x.

Wu, J., Li, X., Yadian, B., Liu, H., Chun, S., Zhang, B., Zhou, K., Gan, C. L., & Huang, Y. (2013). Nano-scale oxidation of copper in aqueous solution. *Electrochemistry Communications, 26*, 21−24.

Wu, K. L., Wu, P. C., Zhu, J. F., Liu, C., Dong, X. J., Wu, J. N., Meng, G. H., Xu, K. B., Hou, J., Liu, Z. Y., & Guo, X. H. (2019). Synthesis of hollow core-shell CdS@TiO_2/Ni_2P photocatalyst for enhancing hydrogen evolution and degradation of MB. *Chemical Engineering Journal, 360*, 221−230.

Wu, X., Bai, H., Zhang, J., Chen, F., & Shi, G. (2005). Copper hydroxide nanoneedle and nanotube arrays fabricated by anodization of copper. *The Journal of Physical Chemistry B, 109*, 22836−22842.

Xia, W., Su, D. S., Birkner, A., Ruppel, L., Wang, Y. M., Woll, C., Qian, J., Liang, C. H., Marginean, G., Brandl, W., & Muhler, M. (2005). Chemical vapor deposition and synthesis on carbon nanofibers: Sintering of ferrocene-derived supported iron nanoparticles and the catalytic growth of secondary carbon nanofibers. *Chemistry of Materials, 17*(23), 5737−5742.

Xiao, F., Yuan, S., Liang, B., Li, G., Pehkonen, S. O., & Zhang, T. J. (2015). Superhydrophobic CuO nanoneedle-covered copper surfaces for anticorrosion. *Journal of Materials Chemistry, 3*, 4374−4388.

Xu, S., Ostrikov, K., Long, J. D., & Huang, S. Y. (2006). Integrated plasma-aided nanofabrication facility: Operation, parameters, and assembly of quantum structures and functional nanomaterials. *Vacuum, 80*(6), 621−630.

Xu, X., Wang, G., & Wang, H. (2015). Synthesis of hybrid carbon nanofiber-graphene nanosheet materials with different morphologies and comparison of their performance as anodes for lithium-ion batteries. *Chemical Engineering Journal, 266*, 222−232.

Xu, T. Z., Ji, G. J., Li, H., Li, J. D., Chen, Z., Awuye, D. E., & Huang, J. (2022). Preparation and applications of electrospun nanofibers for wearable biosensors. *Biosensors, 12*(3), 177.

Xue, J. J., Wu, T., Dai, Y. Q., & Xia, Y. N. (2019). Electrospinning and electrospun nanofibers: Methods, materials, and applications. *Chemical Reviews, 119*(8), 5298−5415.

Yin, Z. G., Ding, Y. H., Zheng, Q. D., & Guan, L. H. (2012). CuO/polypyrrole core−shell nanocomposites as anode materials for lithium-ion batteries. *Electrochemistry Communications, 20*, 40−43.

Yuan, J. J., Chen, C. H., Hao, Y., Zhang, X. K., Zou, B., Agrawal, R., Wang, C. L., Yu, H. J., Zhu, X. R., Yu, Y., Xiong, Z. Z., Luo, Y., & Xie, Y. M. (2017). SnO_2/polypyrrole hollow spheres with improved cycle stability as lithium-ion battery anodes. *Journal of Alloys and Compounds, 691*, 34−39.

Yuan, J. J., Hao, Y. C., Zhang, X. K., & Li, X. F. (2018). Sandwiched CNT@SnO_2@PPy nanocomposites enhancing sodium storage. *Colloids and Surfaces A: Physicochemical and Engineering Aspects, 555*, 795−801.

Yuan, Q. H., Xin, Y., Yin, G. Q., Huang, X. J., Sun, K., & Ning, Z. Y. (2008). Effect of low-frequency power on dual-frequency capacitively coupled plasmas. *Journal of Physics D: Applied Physics, 41*(20), 205209.

Zahan, M., & Podder, J. (2019). Surface morphology, optical properties and Urbach tail of spray deposited Co_3O_4 thin films. *Journal of Materials Science: Materials in Electronics, 30*, 4259−4269.

Zhang, T., Wu, S., Xu, J., Zheng, R., & Cheng, G. (2015). High thermoelectric figure-of-merits from large-area porous silicon nanowire arrays. *Nano Energy, 13*(4), 433−441.

Zhang, Y., Zhang, D. F., Xu, X. Y., & Zhang, B. (2018). Morphology control and photocatalytic characterization of WO_3 nanofiber bundles. *Chinese Chemical Letters, 29*(9), 1350−1354.

Zheng, F., Xi, C. P., Xu, J. H., Yu, Y., Yang, W. G., Hu, P. F., Li, Y., Zhen, Q., Bashir, S., & Liu, J. L. (2018). Facile preparation of WO3 nano-fibers with super large aspect ratio for high performance supercapacitor. *Journal of Alloys and Compounds, 772*, 933−942. Available from https://doi.org/10.1016/j.jallcom.2018.09.085.

Zheng, F., Zhang, M., & Guo, M. (2013). Controllable preparation of WO3 nanorod arrays by hydrothermal method. *Thin Solid Films, 534*, 45−53. Available from https://doi.org/10.1016/j.tsf.2013.01.102.

Synthetic nanofibers and natural nanofibers: A relatively study on physical, chemical, and mechanical properties

2

Shveta Sharma[1], Richika Ganjoo[2], Alok Kumar[3] and Ashish Kumar[4]
[1]Department of Chemistry, Government College Una, Affiliated to Himachal Pradesh University, Una, Himachal Pradesh, India, [2]Department of Chemistry, School of Chemical Engineering and Physical Sciences, Lovely Professional University, Phagwara, Punjab, India, [3]Department of Mechanical Engineering, NCE, Bihar Engineering University, Department of Science, Technology and Technical Education, Nalanda, Bihar, India, [4]Department of Chemistry, NCE, Bihar Engineering University, Department of Science, Technology and Technical Education, Nalanda, Bihar, India

2.1 Introduction

Nanofibers are defined as fibers having a diameter of 100 nm or less. Nanofibers have two similar nanoscale dimensions (diameter) and a third, much bigger dimension (length) (Eichhorn et al., 2010; He et al., 2008; Ahmed et al., 2020). Electrospinning a cellulose acetate solution is where it all began by Formhals in the year 1934, who created the first nanofibers. Prior to Formhals, the electrospinning method had been employed, but no one had been able to successfully create lengthy strands because of the inclusion of inflexible Newtonian fluids (Teo & Ramakrishna, 2006). Because the applied electric field induced a significant decrease in the fiber diameter owing to bending instability, the employment of viscoelasticity in the solutions led to the development of nanofibers. This phenomenon is subsequently reported by Reneker. At those times, the formation of nanofibers was taken as a failure in the research, and this invention got recognition by the research community with the advent of nanotechnology in the late 1990s. After about 60–70 years had passed, the work of Formhals was finally recognized, comprehended, and expanded (Ibrahim & Klingner, 2020). Because of their scales, nanofibers have a significant aspect ratio (length/diameter value) of over 200 and a large surface area. These two characteristics confer a number of benefits on nanofibers. Nanofibers may be selected and used in a wide variety of applications due to the fact that practically all of their characteristics can be altered. The fact that a broad variety of materials, including natural and synthetic polymers, composites, metals, metal oxides, carbon-based materials, and so on, are all suitable for use in

Polymeric Nanofibers and their Composites. DOI: https://doi.org/10.1016/B978-0-443-14128-7.00002-X
© 2025 Elsevier Ltd. All rights are reserved, including those for text and data mining, AI training, and similar technologies.

the fiber manufacturing process is one of the most significant advantages offered by nanofiber technology (Subbiah et al., 2005). Because of their nature, structure, and composition, several types of nanofibers may exist. Because of their structure, it is possible to generate nanofibers that are nonporous, porous, hollow, or core-shell structures. They can be either natural or designed nanofibers according to their nature. It is also possible to combine several kinds of fiber materials in order to produce a composition, which may either be organic or inorganic, based on carbon or a mixture of these things. Based upon the various beneficial properties, such as very good width and height ratio, well-adjustable properties, and capacity to form three-dimensional frameworks, nanofibers are considered a very suitable material in different biomedical applications, such as tissue engineering, regenerative medicine, drug delivery, nanoparticle delivery, etc. (Endo et al., 2002; Kajdič et al., 2019; Tan & Lim, 2006). Along with biomedical applications, nanofibers may also be applied in textiles (El-Aswar et al., 2022; Haider et al., 2018; Kenry & Lim, 2017). In oral drug delivery systems, nanofibers can be utilized in many ways: nanofiber frameworks, oral fast delivery systems in the form of electro sponge nanofibers, multilayered nanofiber loaded three-dimensional structures, and electrospun nanofibers having modified surfaces. Nanofibers range in size from 50 to 1000 nm and possess characteristics such as an extensive surface area, a high degree of porosity, minuscule pore size, and low density. There are many techniques mentioned in the literature for manufacturing nanofibers, including molecular assembly, thermally induced phase separation, and electrospinning. The electrospinning method is mostly used to form polymer nanofibers. when combined with pharmaceutical polymers, this method produced very usable tactics for creating drug delivery systems. All types of polymers can be employed, such as environmentally friendly hydrophilic, hydrophobic, and even amphiphilic polymers, for generating varied nanofibers. Because of the enormous surface area per unit of mass, electrospun nanofibers are frequently employed for the loading of insoluble pharmaceuticals in an effort to improve the drug dissolving capabilities (Fang et al., 2008; Xue et al., 2019). For instance, in addition to drugs that are insoluble in water, salt that is easily soluble in water may also be incorporated into the fibers. The examined delivery methods depend on changes in intraluminal pressure, gastrointestinal tract pH levels, enzymatic activity to initiate drug release, and temporal control. Both the reduction of azo-bonds in order to release an active agent and the breakdown of intestinal polymer films as a result of an alteration in the pH of the surrounding environment are used in commercially available products (Xue et al., 1976). Both in vitro and in vivo conducted studies have shown that the release rates of pharmaceuticals from compositions including nanofibers are significantly improved in comparison with the ones of the original drug components. This chapter is focused on the different physical, chemical, and mechanical properties of synthetic and natural nanofibers. On the basis of their origin, fibers are classified as natural fibers and synthetic fibers. Synthetic fibers can be produced in the laboratory and can be cheaper compared to natural fibers but natural fibers are much more comfortable.

2.1.1 Natural fibers

The term "natural fibers" refers to fibers that are derived from sources such as plants, animals, or minerals. Cotton, silk, and wool are some examples. Another way to classify natural fibers is according to their origin; plant fibers and animal fibers may be distinguished from one another (Badgar et al., 2022).

2.1.1.1 Animal fibers

These are the fibers that are harvested from animals, such as wool and silk.

Wool: Wool is a natural fiber that may be found in textiles and is derived from sheep, goats, and even camels. It is quite effective in retaining air. The transfer of heat via air is quite inefficient. Because of this, it is beneficial to wear woolen garments throughout the winter. Hard α-keratins are the primary components of wool fiber, which is mostly made up of proteins with sulfur and amino acid cysteine (Johnson & Russell, 2008; Lewis & Rippon, 2013).

2.1.1.2 Silk

Silk is another kind of natural fiber that is used in the textile industry and comes from the cocoons of silkworms. The cultivation of silkworms for the purpose of extracting silk is referred to as sericulture. The primary use of silk is in the making of clothing. Silk fibers that have been woven into different patterns may be found in parachutes and bicycle tires. It is made up of two distinct proteins, sericin and fibroin, and fibroin is one of the few materials that the FDA has given its approval for use in certain medical equipment (Altman et al., 2003; Gosline et al., 1986).

2.1.1.3 Plant fibers

These are the kinds that can be harvested from various plants. The plants are harvested for their fibers, which are then used to manufacture textiles.

2.1.1.4 Cotton

It is a kind of plant fiber that is used in the manufacturing of textiles. Cotton is a plant that produces a soft staple fiber that may be found coiled around the seeds of the cotton plant. Cotton is used to manufacture textiles that are not only comfortable but also breathable and long-lasting. Cellulose makes up anywhere from 88%–97% of a cotton fiber's composition, with waxes, proteins, and pectin making up the rest of the cotton fiber's components (Basra & Malik, 1984; Wilkins et al., 2000).

2.1.1.5 Jute

It is a vegetable fiber that has a smooth and lustrous appearance and can be braided into coarse threads that are quite strong. Jute fiber is used for the purpose of wrapping a broad variety of agricultural as well as industrial goods that call for usage in the form of bags, sacks, etc. Jute fiber is derived from the bark of the jute plant. Jute fiber has three primary categories of chemical compounds, namely cellulose (58%−63%), hemicellulose (20%−24%), and lignin (12%−15%), in addition to certain additional tiny amounts of elements such as lipids, pectin, and aqueous extract (Boopalan et al., 2013; Saleem et al., 2020).

2.1.2 Synthetic fibers

These are synthetic polymers that were meant to be woven into textiles. The chemical joining of a large number of smaller components results in the production of polymers.

Some of the examples of synthetic fibers are:

2.1.2.1 Rayon

It is created from wood pulp. Because it is so similar to silk in appearance and feel, it is often referred to as "artificial silk." The most common applications for rayon include carpeting, clothes, medicinal dressings, and insulating materials. Rayon fiber is a produced fiber made of 100% regenerated cellulose, or sometimes regenerated cellulose in which other reagents replaced not more than 15% of the hydrogens of the hydroxyl groups (Ingersoll, 1946; Knox & Hollander, 1966).

2.1.2.2 Nylon

This was the first synthetic fiber ever created. Ropes, sleeping bags, many kinds of clothing, and even parachutes may be made using this material. It is one of the most resilient materials that we are aware of (Kojima et al., 1993; Shakiba et al., 2021).

2.2 Natural nanofibers

Nanofibers made from natural materials have more natural biological and chemical characteristics than nanofibers made from synthetic materials, such as high capacity to behave as an exact host in many applications, especially medical, high affinity for water, nonpoisonous, *ability to easily decompose*, enhanced cell adhesion, easily producible, and reduced immune reaction. One more of its many appealing qualities is that they are made from naturally occurring and hence renewable resources. Natural raw materials are better for the environment than man-made ones because natural materials may be broken down by microbes. Proteins, such as collagen and gelatin, as well as polysaccharides make up natural polymers. For example, cellulose and chitosan will be discussed individually in the following paragraphs. Electrospinning is a method that may be used to transform these proteins and polysaccharides into natural nanofibers.

2.2.1 Natural polymer-based nanofibers

2.2.1.1 Structural, physicochemical, and mechanical properties hyaluronic acid

Because of its remarkable physicochemical and biological properties, the one-of-a-kind biopolymer known as "hyaluronan" (hyaluronic acid [HA]) has garnered a lot of interest in the fields of biomaterials, bioengineering processes, and medicine in recent years (Abatangelo et al., 2020; Pereira et al., 2018). Because of its biological compatibility, it has substantial moisture absorption ability, having both elasticity and viscosity in its nature, and great capacity to absorb moisture. Additionally, HA has an excellent hygroscopic nature, which enables it to attract and retain water. HA is essential for the control of key activities in the body, including tissue hydration, cell adhesion, and cell multiplication, and additional reactions such as wound repair, decreasing irritation, creating new blood vessels, and impaired tissue regeneration, which makes HA a great choice for bioengineering and building innovative devices for a wide variety of uses in the healthcare sector (Necas et al., 2008). The presence of the hydroxyl group (anionic in nature) in the structure of HA is the primary factor that contributes to the molecule's capacity to retain significant amounts of water. This polymeric arrangement, with its gel-like properties and negative charge on it, adds to HA's shock-absorber capabilities around surrounding tissues. In addition to allowing HA to function as a lubricant for joints. The existence of these polar as well as nonpolar groups in the HA polymeric matrix allows it to form bonds with other molecules, and the resultant compounds formed extend its uses in the area of biomedicine. Long-chain HA polymer occurs in solution in a polydispersed form and exhibits a three-dimensional structure that is thick fluid, nonuniform kind of structure and has a large hydration volume. This structure allows for the unhindered movement of micro-molecules through it and at the same time it stops the free flow of the macromolecules into the HA framework, therefore, it is a factor that adds to the impact of HA's omitted volume. HA solution shows specific flow behavior and HA play a special role in the field of the medical industry, cosmetics industry, and bioengineering applications. Additionally, it may also be used in all those processes that are dependent upon HA structure and its polyelectrolyte behavior. The time-dependent, non-Newtonian behavior of HA solution may be attributed to a number of different reasons, including an increase in hydrophobic effects in conjunction with an increased applied shearing deformation rate and the breaking of intramolecular hydrogen bonds. The water-repellent tendency is primarily caused by the involvement of distorted HA chain structure in the flow direction. This response lessens solution viscosity, and lowering the molecular weight of hyaluronan has been shown to shorten the amount of time required to untangle the three-dimensional HA network. In general, the electrospinning of HA solution is hampered by the rheological properties of HA, which include high viscoelasticity, high surface tension, and high hydrophilicity. In addition, the limited vaporability of the HA polymeric solution combined with its high conductivity might lead to a breakdown in the circuit that connects the needle and the collector, which in turn reduces the efficiency of the electrospinning process (Fallacara et al., 2018; Neuman et al., 2015). Molecular mass, as well as the circumstances under which HA is synthesized and degraded, are the primary parameters that influence the biological activity of HA. In general, the biological tasks that are performed by HMW (high molecular weight) and LMW (low molecular weight) HA are diametrically opposed to one another. Because HMW hyaluronan has elasticity as well as viscosity, it is most often found in synovial joints. As a result of its viscoelastic properties, it serves as a lubricant, protects the articulate cartilage, and is also useful in healing wounds, in the repair of tissues, etc. While low molecular weight hyaluronan may promote the growth of tumors by increasing extracellular matrix (ECM) remodeling. HMW HA was shown to inhibit apoptosis, increase cell quiescence, contribute to the inflammatory phase of the healing process, preserve tissue integrity, and stimulate the in vitro proliferation of fibroblasts. All of these effects were discovered.

2.2.1.2 Structural, physicochemical, and mechanical properties of chitosan

Chitosan is a biopolymer with a positive charge, constituted by 2-acetamide-2-deoxy-β-d-glucopyranose and 2-amino-2-deoxy-β-d-glucopyranose residues. The concentration of the second constituent should be at least 50% to be considered chitosan. Chitosan is formed by the deacetylation of chitin in an alkaline medium or through fermentation. It has been reported that chitosan has a wide variety of features, including its polycationic nature (Kou et al., 2022; Wang & Zhuang, 2022). Chitosan has a positive charge in acid solutions, so at low pH its ammonium group gets protonated. This leads to the creation of (NH3) + , which makes it possible for chitosan to engage electrostatically with negative-charged compounds. The mucoadhesive and gelling characteristics of the polysaccharide may be attributed to the cationic nature of chitosan. Chitosan can easily be attached to mucosal tissues, by electrostatic forces of attraction by using its positively charged site as mucus contains mucin glycoproteins (negatively charged) as the main constituent (Gao & Wu, 2022; Maleki et al., 2022). Hydrogen bonding and hydrophobic effects play a major role in this property. The hydrophilic groups of chitosan, which are responsible for chitosan's gelling capabilities, make it possible for hydrogel to form. Additionally, the film-forming capability is regarded to be very essential aspect of a biopolymer, and this ability is gained by both the mucoadhesive and gelling capabilities of the biopolymer. For the production of films, membranes, and fibers, the property of film formation is essential. Then, when new physical forms are produced, it is possible that the spectrum of uses for chitosan may expand.

2.2.1.3 Structural, physicochemical, and mechanical properties of alginate

Alginate is a naturally existing anionic polymer that is primarily derived from brown seaweed. Because of its less toxic nature and ease to get disintegrate, relatively cheap cost, and good gel-forming capacity upon the addition of divalent cations such as Ca^{2+}, alginate is considered as subject of research. Brown seaweed is the most common source of alginate. Alginate is now known to be a whole family of linear copolymers containing blocks of (1,4)-linked β-D-mannuronate (M) and α-L-guluronate (G) residues. The G-block content of the stems of Laminaria hyperborea is 60%, while the G-block concentration of other alginates that are commercially accessible ranges from 14.0%−31.0% (Hurtado et al., 2022; Varaprasad et al., 2022). Hydrogels are supposed to be made by G-blocks alginate only, as only these alginates react with divalent cations and form intermolecular cross-linking. Therefore, the ratio of M to G, the sequence, the length of the G-block, and the molecular weight are crucial elements that influence the physical characteristics of alginate and the hydrogels that are produced from it. Increasing the length of the G-block and the molecular weight of alginate gels often results in an improvement of the gel's mechanical characteristics. Alginates form polymers with different chemical properties, for example, bacterial alginates have a high G- blocks gel structure with a high level of stiffness and were produced by Azotobacter. The stability of the gels and the pace at which drugs are released from the gels, as well as the phenotypic and function of cells that are encapsulated in alginate gels, are all considerably controlled by the physical qualities. Further, the increased molecular weight may affect the physical properties of gels formed from alginates. On the other hand, an alginate solution that is generated from a polymer with a high molecular weight becomes very viscous, which is often not desired during the processing stage. For instance, proteins or cells that have been mixed with a high-viscosity alginate solution run the danger of being damaged as a result of the high shear pressures that are created during the mixing process and then injected into the body. Only the changes in the molecular weight and how the polymer has been distributed can somehow manage the viscosity issues in pregel solution and levels of stiffness in postgelling. Utilizing a mix of high and low molecular weight alginate polymers allows for the elastic modulus of gels to be greatly raised, while only a little rise in the solution's viscosity is seen as a result of this change.

2.2.1.4 Structural, physicochemical, and mechanical properties of collagen

The majority of the fibrous proteins that make up the ECM are collagen. It is responsible for 20%−30% of the total protein in the body and plays a significant part in the control of cellular function as well as the provision of structural support to organs and tissues. There has been a total of 29 different varieties of collagen discovered, each of which is made up of at least 46 different polypeptide chains. Collagen consists of three polypeptide α-chains, with each chain having a polyproline II-type helix (with 300 nm length and 1.5 nm in diameter) along with left-hand confirmation. The electrospinning of collagen has been the subject of a significant number of investigations. At present, electrospinning has only been successfully performed on types I, II, III, and IV of collagen (Matinong et al., 2022; Tang et al., 2022). Electrospinning has the capability of producing ordered as well as irregular collagen nanofibers for a variety of applications such as wound healing. When compared to other types of nanofibers made from polymers, the electrospun collagen nanofibers provide a number of advantages. To begin, collagen is the primary ECM protein found in various tissues throughout the body. Therefore, collagen nanofibers are imitating the ultrastructure of real tissues in the most accurate manner possible. Secondly, collagen has a poor tendency to provoke an immune response. The third advantage of collagen is its high level of biocompatibility. Because almost every cell enjoys collagen, collagen nanofibrous scaffold is an excellent choice for the creation of virtually all of the body's tissues. The collagen nanofibers' weak mechanical characteristics are the most glaringly visible deficiency in their design. Modified collagen nanofibers with natural or synthetic polymers or inorganic compounds, will improve the mechanical characteristics of collagen nanofibers. It has been discovered that the source of the collagen may change the qualities of the fibers that are produced. According to Matthews et al. (2002), electrospinning human placental collagen type I using conditions that had been optimized for calf skin collagen type I resulted in the creation of fibers that were less homogenous and had a wider range of diameters. Within the same body of research, the researchers discovered that the kind of collagen also has an effect on the structural qualities of the fiber. Many compounds such as 1,1,1,3,3,3-hexafluoro-2-propanol (HFP), 2,2,2-trifluoroethanol (TFE), acetic acid, and phosphate buffered saline/ethanol have selected for collagen electrospinning, but HFP has been considered as a preference. Preference has been given to HFP because of its high volatility and low boiling point. This makes HFP a suitable solvent for electrospinning. When compared with acetic acid, HFP is a much more effective solvent for collagen. Along the same lines as HFP, TFE is a very volatile fluoroalcohol that has the ability to promote the synthesis of collagen fibers. Despite the fact that HFP encourages the synthesis of collagen fibers, it is harmful to cells and causes collagen to become more denatured. Both HFP and TFE are fantastic solvents for collagen and are particularly useful in the process of fabricating electrospun collagen nanofibers. However, fluoroalcohols come with a high price tag and are harmful to the surrounding ecosystem. In addition, fluoroalcohols are responsible for the breakdown of collagen. Collagen nanofibers produced by electrospinning have low mechanical strength and are readily dissolving in water through cross-linking. Every possible way of establishing cross-links has certain drawbacks. The chemical treatment results in the introduction of chemicals that are hazardous to the substance. Because of this, its use in biological applications may not be optimal. Because the reaction only takes place on the surface of the material, the cross-linking produced by a physical treatment is of a modest degree. Enzymatic therapy can only target certain amino acids, and it is more difficult to exert control over the cross-linking process.

2.2.1.5 *Structural, physicochemical, and mechanical properties of cellulose nanofibers*

The term "cellulose nanofiber" refers to a material that is made up of nanosized cellulose fibrils that have a high aspect ratio (the ratio of length to width). The thicknesses of normal fibrils vary from 5 to 20 nm, and their lengths may be anywhere from a few hundred nanometers to several micrometers. These are pseudoplastic and exhibit thixotropy, the phenomenon in which some gels or fluids, which are normally thick (viscous), become less viscous when they are shaken or disturbed but regain their original state after the removal of shearing forces (Surendran & Sherje, 2022). The fibrils may be separated from any source containing cellulose, including wood-based fibers (pulp fibers). The dispersion property confirmed the values of storage (high value even as compared to CNCs) and loss modulus by varying the concentration of nanocellulose. The application of shear pressures causes a decrease in the viscosity of nanocellulose gels. The shear-thinning tendency is very helpful in a variety of various coating applications due to its wide-ranging applicability. Crystalline cellulose has a rigidity in the range of 140−220 GPa, and nanocellulose films have high strength (over 200 MPa) and high stiffness (around 20 GPa).

2.2.1.6 Structural, physicochemical, and mechanical properties of silk nanofibers

Silk is generated by a wide range of insects, one of which is the silkworm. Silk is a characteristic fibrous protein. Silk from silkworms, most often those of the domesticated *Bombyx mori* variety, has been utilized for a significant amount of time as a superior textile fiber and suture material. Two types of proteins are generally presented in *B. mori* silk, and that are fibroin and sericin. The protein known as fibroin is responsible for the formation of the strands in silkworm silk, which are responsible for giving silk its distinctive appearance along with specific chemical characteristics. The group of gelatinous proteins known as sericins is responsible for binding the fibroin strands. Depending on the specific application, silk fibroin (SF) can be utilized in a number of different forms, including gels, powders, fibers, and membranes. In the past few years, SF has been investigated for its usage in the medical field such as cutaneous wound healing, bone tissue regeneration, and vascular implants, among others. It was only possible because of its capacity to behave as an exact host in its medical application: it easily decomposes, has adequate mechanical performance, promotes cell growth and interaction, is not harmful to cells, and is compatible with blood. SF is often derived from the cocoons of superior silkworms; however, there are alternative sources, such as waste products made of silk fibers. The silk fiber is classified as a semicrystalline polymer with a distinct structure. This structure consists of highly orientated β-sheet crystallites organized in nanofibrils or fibrillar entities and connected with an amorphous framework along with weakly oriented β-sheets regions has also been claimed. Various spectral investigations revealed that the D_2O-inaccessible β sheets were found to be connected with crystallites, whilst the D_2O-accessible β sheets were found to be made of amorphous and β-sheets. In the case of spider dragline silk, crystalline components are bigger in size and less oriented. Both spider silk and *B. mori* silk have their own distinct physical characteristics, such as better mechanical capabilities in terms of toughness. The greatest strength of spider dragline silk has been measured up to

1.7 GPa so far; this is higher than the strength of steel (1.5 GPa) and places it into the realm of high-tech materials. When compared to man-made synthetic fibers such as Kevlar 49, the tensile strength of spider dragline silk is much higher. This is made possible by the silk's high extensibility than spider dragline silk. However, the mechanical characteristics of the particular *B. mori* silk have substantially enhanced to a level that is similar to the hardest spider silk when it is forced silking from immobilized silkworms artificially at a set spinning speed. The stress-strain curve profiles of silk fibers may be used to provide a description of their mechanical characteristics. These profiles are created by stretching the fibers at a certain strain rate. Both *B. mori* silkworm silk and spider dragline silk exhibit elastic behavior, which is thereafter followed by plastic deformation in typical stress-strain curves. Because of these capabilities, researchers have been motivated to investigate the structural basis of high-performance silk fiber, with the end objective of acquiring blueprints for the development of innovative materials with properties that are equivalent to those of silk (Liu & Zhang, 2014).

2.3 Synthetic nanofibers

2.3.1 *Synthetic polymer-based nanofibers*

2.3.1.1 *Polylactic acid fibers*

Polylactic acid, often known as PLA, has been identified as a possible polymer candidate that has desirable qualities for use in a variety of applications in the engineering and medical fields. PLA has a higher tensile modulus than many other synthetic polymers and a greater flexural strength, which indicates the possibilities of usage of PLA in the engineering field. According to research published over the course of the last two decades, PLA has a significant untapped potential for enhancement of both its physical and mechanical qualities. PLA is a polymer that may be processed by the use of a straightforward conventional technique that does not need a significant number of inputs of either energy or time, making it both an affordable and easily accessible material. The observable properties of PLA are significant in terms of the potential uses of the material in industry. It is bright and clear, maintains its stability at low temperatures, has only a moderate permeability to oxygen and water, and has a strong resistance to both grease and oil. Because of these characteristics, it is appropriate for use in the production of various materials used in daily life. However, properties such as surface energy, specific gravity, and glass transition temperature alter with D-content and alteration in molecular weight. For example, when the molecular weight of PLA is increased, the crystallinity of the material falls because this results in the formation of a longer polymer chain. On the other hand, as the chain becomes longer and more entangled, both its tensile strength and its shear viscosity will rise. PLA has distinct stereochemical structures and chemical structures, which may either be semicrystalline or amorphous, depending on the proportion of D- to L-containing molecules in the material. The chemical structure of PLA is what determines its physical and mechanical qualities, as well as the pace at which it biodegrades. An increase in the D-content of PLA may lead to a reduction in the rate of crystallization, and that in response may end up in the melting point being lowered. The D-content is an essential characteristic that enables the alteration of the properties of PLA in a number of different ways. Plasticization of PLA may also be done by combining it with other polymers, such as starch, and oligomeric lactic acid to increase its crystallinity (Pantani et al., 2014). Polylactic acid, a type of linear aliphatic thermoplastic polyester (Ranakoti et al., 2022), and poly(lactic acid) (PLA)/graphene oxide (GO) nanofiber membranes with different structures (single-axial and coaxial structure) were prepared by electrospinning by Mao et al. (2018). The characterization, comparison, and drug-releasing ability of the as-prepared nanofiber membranes with various shapes were examined by utilizing an organic dye (Rhodamine B (RhB)) as a drug model. According to the findings, the incorporation of GO not only increased the overall release of RhB from nanofiber membranes but also greatly improved the thermal endurance and mechanical properties of the PLA nanofiber membranes. The nanofiber membrane with the coaxial structure had a greater tensile strength and Young's modulus at the same concentration of GO, but it had a lower cumulative release and release rate than the nanofiber membrane without the coaxial structure. When the nanofiber membranes are formed into a coaxial structure, this helps to control the first burst release of RhB that occurs during the development of the membranes. For the enhancement of the properties of PLA, polybutylene adipate terephthalate was added and it resulted in two times increased tensile strength than pure PLA (Sis et al., 2013). PLA is modified with various stabilizers and compatibilizers, as shown in Fig. 2.1, and the mechanical properties may get effected (Ranakoti et al., 2022).

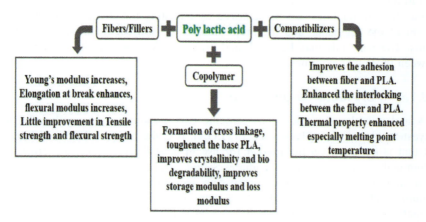

Figure 2.1 Mechanical properties of poly(lactic acid) after the addition of various reinforcements.
Source: From Ranakoti (2022). Properties, structure, processing, biocomposites, and nanocomposites (Vol. 15). (Original work published 2022).

2.3.1.2 Polycaprolactone fibers

Polycaprolactone (PCL) is a biodegradable polyester with a low melting point of around 60°C and a glass transition temperature of about −60°C. Ring-opening polymerization is used for the preparation of PCL, and for this reaction, ε-caprolactone has been taken as a repeating unit and the reaction is catalyzed with stannous octanoate. PCL got the attention of researchers to be used as implantable biomaterial, as it can be easily broken down by the process of hydrolysis (ester linkages are hydrolyzed) in the human body. Specifically, PCL has been used in manufacturing implantable devices, controlled release, and targeted drug delivery. PCL and starch are often combined in order to produce a material that breaks down quickly and that is of high quality at an affordable cost. Polycaprolactones with a low molecular weight may be mixed in with other types of plastic to make them more resistant to outdoor conditions.

2.3.1.3 Polyurethane fibers

Polyurethane (PU) fiber, often known as Spandex or Elastane, is a kind of synthetic fiber that is well-known for its remarkable resilience. In comparison to rubber, Spandex is more resilient, lightweight, adaptable, and long-lasting. Spandex fibers can really be stretched to approximately 500% of their original length without breaking (Otaigbe & Madbouly, 2009). PU copolymer, consisting of alternating soft polyester or polyether and hard PU−urea segments, is directly responsible for the unique springy feature of Spandex fibers and their extensive use in industrial settings. Below and above their glass transition temperatures, these two segments go through a process called microphase separation, in which they split into hard and soft phases, respectively. The remarkable elastomeric qualities of PU may be attributed to the microphase separation that occurs in the material. A wide variety of different physical qualities can be achieved by adjusting the structure, the molecular weight of the segments, and the proportion of the soft segments to the hard segments. The materials may be tough and brittle, sticky and squishy, or anything in between those two extremes.

2.3.1.4 Poly lacto-co-glycolic acid

PLA and polyglycolic acid (PGA) are the two monomers that combine to form polyester poly lacto-co-glycolic acid (PLGA). It is the most well-defined biomaterial that is currently on the market for use in drug delivery, both in terms of its design and its performance (Makadia & Siegel, 2011). PLGA may be formed practically into any shape or size by processing, and it can easily entrap molecules. Its solubility rate in different solvents is very high. In water, PLGA biodegrades through hydrolysis of its ester bonds as given in Fig. 2.2 (Makadia & Siegel, 2011). Because of the hydrophobic behavior of PLA, lactide-rich PLGA copolymers absorb less water and disintegrate at a slower rate. Because of the hydrolysis of PLGA, metrics that are normally thought of as unchanging characteristics of a solid formulation might vary throughout the course of its lifetime. These parameters

Figure 2.2 Poly lactic-co-glycolic acid undergoing hydrolysis.
Source: From Makadia, H.K., & Siegel, S.J. (2011). Poly lactic-co-glycolic acid (PLGA) as biodegradable controlled drug delivery carrier. *Polymers*, *3*(3), 1377−1397. https://doi.org/10.3390/polym3031377.

include the glass transition temperature (Tg), the amount of moisture present, and the molecular weight. Numerous studies have been conducted to determine the influence that various polymer characteristics have on the pace at which drugs are released from the biodegradable polymeric matrix. The rates at which integrated drug molecules are released into the environment and degraded are affected by how the characteristics of PLGA alter as the polymer degrades. It has been demonstrated that the physical characteristics of PLGA are dependent on a number of different conditions. These factors include the starting molecular weight, the exact proportion of both constituents, the dimension of the equipment, accessibility to water (surface shape), and the temperature at which it is stored. The molecular weight and polydispersity index, two examples of PLGA's physical qualities, both play a role in determining the material's mechanical strength. These features also influence the capability of the material to be constructed as a drug delivery device, and they may govern the rate of hydrolysis and degradation of the device. However, recent research has shown that the kind of medicine also has a role in determining how quickly the drug is released into the body. The degree of crystallinity of the PLGA, which is further dependent on the type and molar proportions of each of the monomers in the copolymer chain, has a direct impact on the material's mechanical endurance, swelling behavior, ability to go through hydrolysis, and afterward pace of decomposition of the polymer. These properties are directly affected by the crystalline nature of the PLGA. As a general rule, greater levels of PGA are associated with more rapid rates of deterioration except in the case of a ratio of 50:50 PLA/PGA, which shows the most rapid deterioration, with more PGA content resulting in increased degradation intervals below 50%. The molecular weight of a polymer has a direct bearing on the crystallinity degree of the polymer and the melting temperature of the polymer. According to reports, the glass transition temperature (Tg) of PLGA copolymers is higher than the physiological temperature of 37°C, making them glassy in nature and showing a rather stiff chain structure. It has also been observed that the glass transition temperature (Tg) of PLGAs drops when the amount of lactide present in the copolymer composition is reduced, as well as when the amount of molecular weight drops. PLGA polymers that are available for purchase often have their properties described in terms of their intrinsic viscosity, which is directly proportional to their respective molecular weights.

2.3.1.5 Poly(3-hydroxybutyrate-co-3-hydroxy valerate) fibers

Poly(3-hydroxybutyrate-co-3-hydroxy valerate (PHBV) is not toxic to living beings and is environmentally friendly. These qualities make it a possible contender for replacing polymers generated from petroleum. However, it does not possess desirable features such as mechanically not that strong, absorption of water and dispersion, electrical and/or thermal properties, antibacterial activity, and ability to be in contact with any liquid, which restricts its applicability in a variety of contexts. Because of this, a great number of academics from all over the globe are now focusing their attention on finding ways to address the limitations of this potentially useful substance (Rivera-Briso & Serrano-Aroca, 2018). PHBV is an aliphatic polyester with the chemical structure shown in Fig. 2.3.

In addition to this, it is safe to use, can easily be disintegrated, is safe to use biologically typified by its highly crystalline nature, and is resistant to UV light as well as appropriate levels of alcohols, fats, and oils. PHBV can strongly oppose oxygen, is chemically inactive, is viscous in nature, and can improve mechanical properties (large surface area and flexible in nature) in comparison to PHB. In addition to this, it has been manufactured on a large scale and has, in recent years, garnered the interest of both businesses and academic researchers as a potentially useful material. This is owing to its potential for use in biotechnology as well as its usefulness in the domains of medicine, agriculture, and product packaging.

2.3.1.6 Poly (ethylene-co-vinylacetate)

Poly (ethylene-co-vinylacetate) (PEVA) is a random copolymer made up of crystalline and amorphous parts, polyethylene (PE) segments, and poly (vinyl acetate), respectively. Fig. 2.4 shows the chemical structure of monomers and PEVA copolymer. PEVA is a thermoplastic polymer that finds widespread use across a variety of industries, including flexible packaging, footwear, hot melt adhesives, and cable sheathing, to name a few. The simplicity of handling and processing of PEVA, as well as its biocompatibility and potential for use in drug administration, contributes

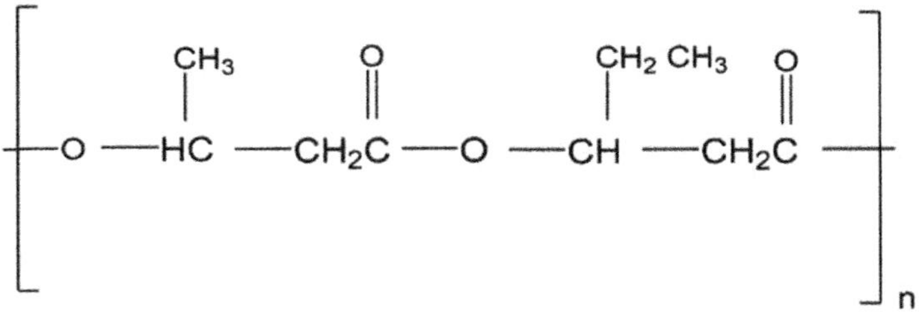

Figure 2.3 Chemical structure of the poly(3-hydroxybutyrate-co-3-hydroxyvalerate) copolymer.
Source: From Rivera-Briso, A.L., & Serrano-Aroca, Á. (2018). Poly(3-hydroxybutyrate-co-3-hydroxyvalerate): Enhancement strategies for advanced applications. *Polymers*, *10*(7). https://doi.org/10.3390/polym10070732.

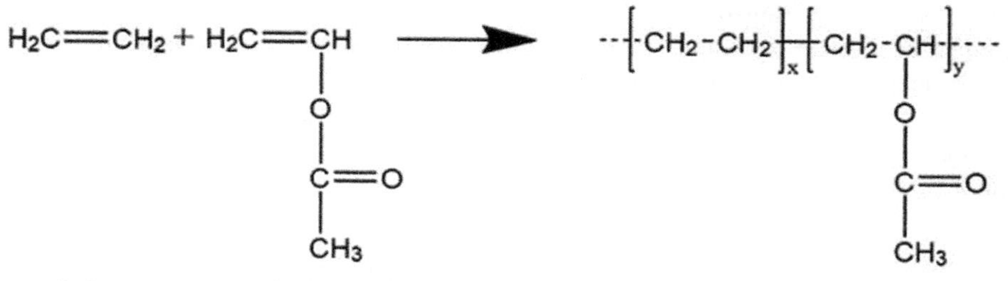

Figure 2.4 Chemical structure of monomers and poly(ethylene-co-vinyl acetate) random copolymers.
Source: From Wang, K., & Deng, Q. (2019). The thermal and mechanical properties of poly(ethylene-co-vinyl acetate) random copolymers (PEVA) and its covalently crosslinked analogues (cPEVA). *Polymers*, *11*(6), 1055. https://doi.org/10.3390/polym11061055.

to the material's standing as a strong contender for usage in the biomedical industry.

2.4 Conclusion and future outlook

Various nanofibers made up of natural materials, such as HA, chitosan, Alginate, Collagen, cellulose, and silk, and synthetic nanofibers made up of PLA, PLGA, PEVA, PHBV, etc. for their structural, physicochemical, and mechanical properties have been studied in detail in this chapter. HA plays a special role in the field of the medical industry, cosmetics industry, and bioengineering applications. Additionally, it may also be used in all those processes that are dependent upon HA structure and its polyelectrolyte behavior. Alginate has been the subject of a great deal of research and has been utilized for a variety of biomedical applications. Electrospinning has the capability of producing collagen nanofibers and it has more advantages over other polymers. Upgrading in the mechanical properties of collagen nanofibers can be achieved by incorporating natural or synthetic polymers or inorganic substances into the structure of the nanofibers. A past study also revealed that crystalline cellulose has a rigidity superior to that of glass fiber. In the same manner, synthetic nanofibers from polymeric material are of much importance in medicine industries, etc. The incorporation of GO greatly improved the thermal endurance and mechanical properties of the PLA nanofiber membranes. PCL is biodegradable polyester used as implantable biomaterial, but to enhance its properties, starch, etc. can be added. Polymeric nanofibers whether natural or synthetic find widespread use across a variety of industries, including flexible packaging, footwear, hot melt adhesives, and cable sheathing, to name a few, because of their excellent physical and mechanical properties. Researchers have been motivated to investigate the structural basis of high-performance silk fiber, with the end objective of acquiring blueprints for the development of innovative materials with properties that are equivalent to those of silk. Many polymeric materials still need a more detailed investigation, such as PHBV, a microbial biopolymer that does not possess desirable features such as mechanical strength, absorption of water and dispersion, and electrical and/or thermal properties. These limitations restrict its applicability in a variety of contexts.

References

Abatangelo, G., Vindigni, V., Avruscio, G., Pandis, L., & Brun, P. (2020). Hyaluronic acid: Redefining its role. *Cells*, *9*(7). Available from https://doi.org/10.3390/cells9071743.

Ahmed, J. Gultekinoglu, M. Edirisinghe, M. Bacterial cellulose micro-nano fibres for wound healing applications. Biotechnology Advances **41** (2020), Available from https://doi.org/10.1016/j.biotechadv.2020.107549.

Altman, G. H., Diaz, F., Jakuba, C., Calabro, T., Horan, R. L., Chen, J., Lu, H., Richmond, J., & Kaplan, D. L. (2003). Silk-based biomaterials. *Biomaterials*, *24*(3), 401–416. Available from https://doi.org/10.1016/S0142-9612(02)00353-8.

Badgar, K., Abdalla, N., El-Ramady, H., & Prokisch, J. (2022). Sustainable applications of nanofibers in agriculture and water treatment: A review. *Sustainability*, *14*(1). Available from https://doi.org/10.3390/su14010464.

Basra, A. S., & Malik, C. P. (1984). *Development of the cotton fiber* (pp. 65–113). Elsevier BV. Available from 10.1016/s0074-7696(08)61300-5.

Boopalan, M., Niranjanaa, M., & Umapathy, M. J. (2013). Study on the mechanical properties and thermal properties of jute and banana fiber reinforced epoxy hybrid composites. *Composites Part B: Engineering*, *51*, 54–57. Available from https://doi.org/10.1016/j.compositesb.2013.02.033.

Eichhorn, S. J., Dufresne, A., Aranguren, M., Marcovich, N. E., Capadona, J. R., Rowan, S. J., Weder, C., Thielemans, W., Roman, M., Renneckar, S., Gindl, W., Veigel, S., Keckes, J., Yano, H., Abe, K., Nogi, M., Nakagaito, A. N., Mangalam, A., Simonsen, J., Benight, A. S., Bismarck, A., Berglund, L. A., & Peijs, T. (2010). Review: Current international research into cellulose nanofibres and nanocomposites. *Journal of Materials Science*, *45*(1), 1–33. Available from https://doi.org/10.1007/s10853-009-3874-0.

El-Aswar, E. I., Ramadan, H., Elkik, H., & Taha, A. G. (2022). A comprehensive review on preparation, functionalization and recent applications of nanofiber membranes in wastewater treatment. *Journal of Environmental Management*, *301*. Available from https://doi.org/10.1016/j.jenvman.2021.113908, http://www.elsevier.com/inca/publications/store/6/2/2/8/7/1/index.htt.

Endo, M., Kim, Y. A., Hayashi, T., Fukai, Y., Oshida, K., Terrones, M., Yanagisawa, T., Higaki, S., & Dresselhaus, M. S. (2002). Structural characterization of cup-stacked-type nanofibers with an entirely hollow core. *Applied Physics Letters*, *80*(7), 1267–1269. Available from https://doi.org/10.1063/1.1450264.

Fallacara, A., Baldini, E., Manfredini, S., & Vertuani, S. (2018). Hyaluronic acid in the third millennium. *Polymers*, *10*(7). Available from https://doi.org/10.3390/polym10070701, http://www.mdpi.com/2073-4360/10/7/701/pdf.

Fang, J., Niu, H. T., Lin, T., & Wang, X. G. (2008). Applications of electrospun nanofibers. *Chinese Science Bulletin*, *53*(15), 2265−2286. Available from https://doi.org/10.1007/s11434-008-0319-0.

Gao, Y., & Wu, Y. (2022). Recent advances of chitosan-based nanoparticles for biomedical and biotechnological applications. *International Journal of Biological Macromolecules*, *203*, 379−388. Available from https://doi.org/10.1016/j.ijbiomac.2022.01.162, http://www.elsevier.com/locate/ijbiomac.

Gosline, J. M., DeMont, M. E., & Denny, M. W. (1986). The structure and properties of spider silk. *Endeavour*, *10*(1), 37−43. Available from https://doi.org/10.1016/0160-9327(86)90049-9.

Haider, A., Haider, S., & Kang, I. K. (2018). A comprehensive review summarizing the effect of electrospinning parameters and potential applications of nanofibers in biomedical and biotechnology. *Journal of Chemistry*, *11*(8), 1165−1188. Available from https://doi.org/10.1016/j.arabjc.2015.11.015, http://colleges.ksu.edu.sa/Arabic%20Colleges/CollegeOfScience/ChemicalDept/AJC/default.aspx, http://www.sciencedirect.com/science/journal/18785352.

He, J. H., Liu, Y., Mo, L. F., Wan, Y. Q., & Xu, L. (2008). *Electrospun Nanofibres and Their Applications*. Shawbury, UK: ISmithers.

Hurtado, A., Aljabali, A. A. A., Mishra, V., Tambuwala, M. M., & Serrano-Aroca, Á. (2022). Alginate: Enhancement strategies for advanced applications. *International Journal of Molecular Sciences*, *23*(9). Available from https://doi.org/10.3390/ijms23094486, https://www.mdpi.com/1422-0067/23/9/4486/pdf.

Ibrahim, H. M., & Klingner, A. (2020). A review on electrospun polymeric nanofibers: Production parameters and potential applications. *Polymer Testing*, *90*. Available from https://doi.org/10.1016/j.polymertesting.2020.106647, https://www.journals.elsevier.com/polymer-testing.

Ingersoll, H. G. (1946). Fine structure of viscose rayon. *Journal of Applied Physics*, *17*(11), 924−939. Available from https://doi.org/10.1063/1.1707665.

Johnson, N. A. G., & Russell, I. M. (2008). *Advances in wool technology advances in wool technology* (pp. 1−342). Australia: Elsevier Ltd. Available from http://www.sciencedirect.com/science/book/9781845693329, https://doi.org/10.1533/9781845695460.

Kajdič, S., Planinšek, O., Gašperlin, M., & Kocbek, P. (2019). Electrospun nanofibers for customized drug-delivery systems. *Journal of Drug Delivery Science and Technology*, *51*, 672−681. Available from https://doi.org/10.1016/j.jddst.2019.03.038, http://www.editionsdesante.fr/category.php?id_category = 48.

Kenry, C. T., & Lim. (2017). Nanofiber technology: Current status and emerging developments. *Progress in Polymer Science*, *70*, 1−17. Available from https://doi.org/10.1016/j.progpolymsci.2017.03.002, http://www.sciencedirect.com/science/journal/00796700.

Knox, R. L., & Hollander, S. (1966). The sources of increased efficiency: A study of dupont rayon plants. *Southern Economic Journal*, *32*(3). Available from https://doi.org/10.2307/1054889.

Kojima, Y., Usuki, A., Kawasumi, M., Okada, A., Fukushima, Y., Kurauchi, T., & Kamigaito, O. (1993). Mechanical properties of nylon 6-clay hybrid. *Journal of Materials Research*, *8*(5), 1185−1189. Available from https://doi.org/10.1557/JMR.1993.1185.

Kou, S. (G.), Peters, L., & Mucalo, M. (2022). Chitosan: A review of molecular structure, bioactivities and interactions with the human body and micro-organisms. *Carbohydrate Polymers*, *282*. Available from https://doi.org/10.1016/j.carbpol.2022.119132.

Lewis, D. M., & Rippon, J. A. (2013). *The coloration of wool and other keratin fibres*. Wiley. Available from 10.1002/9781118625118.

Liu, X., & Zhang, K.-Q. (2014). Silk fiber—Molecular formation mechanism, structure-property relationship and advanced applications. *Oligomerization of Chemical and Biological Compounds, 3*, 69–102.

Makadia, H. K., & Siegel, S. J. (2011). Poly lactic-co-glycolic acid (PLGA) as biodegradable controlled drug delivery carrier. *Polymers, 3*(3), 1377–1397. Available from https://doi.org/10.3390/polym3031377, http://www.mdpi.com/2073-4360/3/3/1377/pdf.

Maleki, G., Woltering, E. J., & Mozafari, M. R. (2022). Applications of chitosan-based carrier as an encapsulating agent in food industry. *Trends in Food Science and Technology, 120*, 88–99. Available from https://doi.org/10.1016/j.tifs.2022.01.001, http://www.elsevier.com/wps/find/journaldescription.cws_home/601278/description#description.

Mao, Z., Li, J., Huang, W., Jiang, H., Zimba, B. L., Chen, L., Wan, J., & Wu, Q. (2018). Preparation of poly(lactic acid)/graphene oxide nanofiber membranes with different structures by electrospinning for drug delivery. *RSC Advances, 8*(30), 16619–16625. Available from https://doi.org/10.1039/c8ra01565a, http://pubs.rsc.org/en/journals/journal/ra.

Matinong, A. M. E., Chisti, Y., Pickering, K. L., & Haverkamp, R. G. (2022). Collagen extraction from animal skin. *Biology, 11*(6). Available from https://doi.org/10.3390/biology11060905, https://www.mdpi.com/2079-7737/11/6/905/pdf?version = 1655103417.

Matthews, J. A., Wnek, G. E., Simpson, D. G., & Bowlin, G. L. (2002). Electrospinning of collagen nanofibers. *Biomacromolecules, 3*(2), 232–238. Available from https://doi.org/10.1021/bm015533u.

Necas, J., Bartosikova, L., Brauner, P., & Kolar, J. (2008). Hyaluronic acid (hyaluronan): A review. *Veterinarni Medicina, 53*(8), 397–411. Available from https://doi.org/10.17221/1930-VETMED, https://www.agriculturejournals.cz/publicFiles/02029.pdf.

Neuman, M. G., Nanau, R. M., Oruña-Sanchez, L., & Coto, G. (2015). Hyaluronic acid and wound healing. *Journal of Pharmacy and Pharmaceutical Sciences, 18*(1), 53–60. Available from https://doi.org/10.18433/j3k89d, http://ejournals.library.ualberta.ca/index.php/JPPS/article/download/23862/17888.

Otaigbe, J. U., & Madbouly, S. A. (2009). The processing, structure and properties of elastomeric fibers. *Handbook of Textile Fibre Structure, 1*. Available from https://doi.org/10.1533/9781845696504.2.325, http://www.sciencedirect.com/science/book/9781845693800.

Pantani, R., Volpe, V., & Titomanlio, G. (2014). Foam injection molding of poly(lactic acid) with environmentally friendly physical blowing agents. *Journal of Materials Processing Technology, 214*(12), 3098–3107. Available from https://doi.org/10.1016/j.jmatprotec.2014.07.002.

Pereira, H., Sousa, D. A., Cunha, A., Andrade, R., Espregueira-Mendes, J., Oliveira, J. M., & Reis, R. L. (2018). Hyaluronic acid. *Advances in Experimental Medicine and Biology, 1059*. Available from https://doi.org/10.1007/978-3-319-76735-2_6, http://www.springer.com/series/5584.

Ranakoti, L., Gangil, B., Mishra, S. K., Singh, T., Sharma, S., Ilyas, R. A., & El-Khatib, S. (2022). Critical Review on Polylactic Acid: Properties, Structure, Processing, Biocomposites, and Nanocomposites. *Materials (Basel), 15*(12), 4312. Available from https://doi.org/10.3390/ma15124312.

Rivera-Briso, A. L., & Serrano-Aroca, Á. (2018). Poly(3-hydroxybutyrate-co-3-hydroxyvalerate): Enhancement strategies for advanced applications. *Polymers, 10*(7). Available from https://doi.org/10.3390/polym10070732, http://www.mdpi.com/2073-4360/10/7/732/pdf.

Saleem, M. H., Ali, S., Rehman, M., Hasanuzzaman, M., Rizwan, M., Irshad, S., Shafiq, F., Iqbal, M., Alharbi, B. M., Alnusaire, T. S., & Qari, S. H. (2020). Jute: A potential candidate for phytoremediation of metals—A review. *Plants, 9*(2). Available from https://doi.org/10.3390/plants9020258, https://www.mdpi.com/2223-7747/9/2/258/pdf.

Shakiba, M., Rezvani Ghomi, E., Khosravi, F., Jouybar, S., Bigham, A., Zare, M., Abdouss, M., Moaref, R., & Ramakrishna, S. (2021). Nylon—A material introduction and overview for biomedical applications. *Polymers for Advanced Technologies, 32*(9), 3368−3383. Available from https://doi.org/10.1002/pat.5372, http://onlinelibrary.wiley.com/journal/10.1002/(ISSN)1099-1581.

Sis, A. L. M., Ibrahim, N. A., & Yunus, W. M. Z. W. (2013). Effect of (3-aminopropyl)trimethoxysilane on mechanical properties of PLA/PBAT blend reinforced kenaf fiber. *Polymer Journal (English Edition), 22*(2), 101−108. Available from https://doi.org/10.1007/s13726-012-0108-0, http://www.springerlink.com/content/1026-1265.

Subbiah, T., Bhat, G. S., Tock, R. W., Parameswaran, S., & Ramkumar, S. S. (2005). Electrospinning of nanofibers. *Journal of Applied Polymer Science, 96*(2), 557−569. Available from https://doi.org/10.1002/app.21481.

Surendran, G., & Sherje, A. P. (2022). Cellulose nanofibers and composites: An insight into basics and biomedical applications. *Journal of Drug Delivery Science and Technology, 75*. Available from https://doi.org/10.1016/j.jddst.2022.103601.

Tan, E. P. S., & Lim, C. T. (2006). Mechanical characterization of nanofibers − A review. *Composites Science and Technology, 66*(9), 1102−1111. Available from https://doi.org/10.1016/j.compscitech.2005.10.003.

Tang, C., Zhou, K., Zhu, Y., Zhang, W., Xie, Y., Wang, Z., Zhou, H., Yang, T., Zhang, Q., & Xu, B. (2022). Collagen and its derivatives: From structure and properties to their applications in food industry. *Food Hydrocolloids, 131*. Available from https://doi.org/10.1016/j.foodhyd.2022.107748.

Teo, W. E., & Ramakrishna, S. (2006). A review on electrospinning design and nanofibre assemblies. *Nanotechnology, 17*(14), R89−R106. Available from https://doi.org/10.1088/0957-4484/17/14/r01.

Varaprasad, K., Karthikeyan, C., Yallapu, M. M., & Sadiku, R. (2022). The significance of biomacromolecule alginate for the 3D printing of hydrogels for biomedical applications. *International Journal of Biological Macromolecules, 212*, 561−578. Available from https://doi.org/10.1016/j.ijbiomac.2022.05.157, http://www.elsevier.com/locate/ijbiomac.

Wang, J., & Zhuang, S. (2022). Chitosan-based materials: Preparation, modification and application. *Journal of Cleaner Production, 355*. Available from https://doi.org/10.1016/j.jclepro.2022.131825.

Wilkins, T. A., Rajasekaran, K., & Anderson, D. M. (2000). Cotton biotechnology. *Critical Reviews in Plant Sciences, 19*(6), 511−550. Available from https://doi.org/10.1080/07352680091139286.

Xue, J., Xie, J., Liu, W., & Xia, Y. (1976). Electrospun nanofibers: New concepts, materials, and applications. *Accounts of Chemical Research, 50*.

Xue, J., Wu, T., Dai, Y., & Xia, Y. (2019). Electrospinning and electrospun nanofibers: Methods, materials, and applications. *Chemical Reviews, 119*(8), 5298−5415. Available from https://doi.org/10.1021/acs.chemrev.8b00593, http://pubs.acs.org/journal/chreay.

Functionalization/modification of nanofibers and their impact on properties and applications

Omar Dagdag[1], Rajesh Haldhar[2], Elyor Berdimurodov[3] and Hansang Kim[1]
[1]Department of Mechanical Engineering, Gachon University, Seongnam, Republic of Korea, [2]School of Chemical Engineering, Yeungnam University, Gyeongsan, Republic of Korea, [3]Faculty of Chemistry, National University of Uzbekistan, Tashkent, Uzbekistan

3.1 Introduction

Because of their potential physicochemical characteristics and lack of a surfactant, as well as economic potential, nanofibers have received a great deal of interest and have grown into novel nanomaterials (Garkal et al., 2021; Suárez et al., 2022). Its ability to create an extremely porous mesh network having noticeable links between the pores allows for more sophisticated applications (Xue et al., 2019). The good qualities of nanofibers, which have fiber diameters of less than 1000 nm, include a greater surface area in terms of volume, nanoporous surface shape, enhanced drug load, and superior physicochemical and biological capabilities (Xie et al., 2020). Nanofibers may be used for a variety of things, including tissue engineering scaffolding, wound dressing, drug delivery, and biosensing (Ibrahim et al., 2020).

There are other ways to make nanofibers, but the electrospinning process is the most popular because it is regarded as the most palatable, affordable, and scalable method for producing nanofibers (Kumbar et al., 2008).

Because of the numerous potential uses for pharmaceutical treatments, medicinal apparatuses, and genetic engineering, the electrospun nanofiber is a versatile method that has drawn more interest (Elsadek et al., 2022). The electrospinning method uses an electric field to create polymer nanofibers by fusing or dissolving polymer solutions. For a number of applications, precise surface characteristics of nanofibers are still desired. Thus, surface-functionalized nanofibers having particular surface characteristics are intriguing and potentially broaden their uses (Thakkar & Misra, 2017). Furthermore, biocompatibility, optical qualities, conductivity, electrical properties, and wettability can all be affected by surface characteristics. Surface-functionalized nanofibers have been developed using a variety of techniques, including those from physics, chemistry, and biology. The surface graft polymerization, wet chemical methods, and plasma treatment, including coelectrospinning of surfactant agents as well as polymers, were all used to modify the nanofibers' surfaces (Sharma et al., 2014). Nondegradable as well as degradable synthetic nanofiber surfaces have been altered with a variety of bioactive compounds for advancements in biological and therapeutic applications (Wsoo et al., 2020). Additionally, electrospun nanofibers made of poly(lactic-co-glycolic acid), L-lactide-co-3-caprolactone, and caprolactone have demonstrated promising applications in human tissues, including bones, nerves, and ligaments. They also serve as an efficient scaffolding for tissue engineering and medication administration. Electrospun nanofibers have various applications: energy storage, energy conversion, energy production, water purification, environmental protection, intelligent textiles, surface coating, encapsulating bioactive species, and heterogeneous catalysis (Kajdič et al., 2019).

Electrospinning is a process frequently used to create nanofibers. It is straightforward, economical, and adaptable. The efficiency of nanofiber membranes is better than that of biological membranes in the separation of biological objects or environmental objects. It is also indicated that they have been exploited in tissue engineering and wound healing in particular. The nanofibers and nanofiber-based materials have good properties: high efficiency for cell development, good promoters for raped hemostasis, high resistance in the antibacterial activity, and electrospun nanofibers can be utilized to treat wounds. The surface of nanofibers is functionalized using a number of methods (Fig. 3.1).

This chapter aims to provide comprehensive information on electrospinning technologies, surface functionalization procedures, and reagents for functionalizing and assessing surfaces. Also, applications for nanofibers and the patent issue were examined.

3.2 Methods of surface functionalization of nanofibers

For the surface functionalization of nanofibers, a number of techniques have been documented in the literature, including the physical approach. These techniques contained plasma processing, ion beam implantation, and physical vapor deposition. Surface grafting, cross-linking, and chelation are examples of chemical surface functionalization techniques. Some techniques are not ideal and may have drawbacks.

3.2.1 Updates in surface modifications

3.2.1.1 Plasma treatment

Using the plasma treatment approach, it is simple to change the surface chemical composition of nanofibers. The literature reports changes in surface wettability, including biological characteristics after plasma treatment. The choice of plasma source is crucial for plasma processing because of the introduction of functional groups and their interactions with nanofibers. The plasma source ought to have a good force to immobilize the bioactive compounds via covalent bonding on the target surface. By plasma treating air or argon, nanofibers' surface hydrophilicity rises (Sofi et al., 2019). Plasma treatment of Poly(-caprolactone) (PCL) with discharge

Functionalization/Modification of Nanofibers and Their Impact

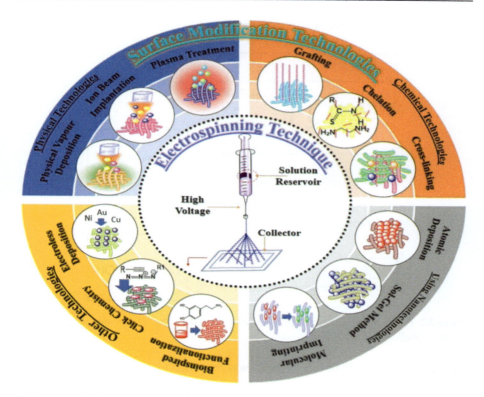

Figure 3.1 Nanofiber surface modification techniques (Kulkarni et al., 2022).

gases argon, ammonia, and helium conserved morphologies of nanofibers as well as drastically changed the surface's wettability compared to untreated plasma nanofibers (Yao et al., 2020). Altering the surface of nanofibers via plasma processing is clean, safe, and ecologically responsible without changing the bulk qualities. The findings of cold gas plasma processing on the surface of electrospun polyamide nanofibers revealed a total alteration in the surface's physical and chemical properties. Surface modification by plasma processing has enormous promise for biosensors, biomaterials, and medical devices (Wei et al., 2005). Poly (vinylidene fluoride) (PVDF) nanofibers were treated using argon plasma and the environment, as seen in Fig. 3.2, to create an active functional group that was grafted onto the surface using acrylic acid (Huang et al., 2012).

3.2.2 *Surface grafting, cross-linking, and chelation*
3.2.2.1 *Grafting*
Diallyl dimethyl ammonium chloride pretreated uniformly morphed poly (D, L-dairy product-co-glycolic acid) (PLGA) electro-extruded nanofibers with positive

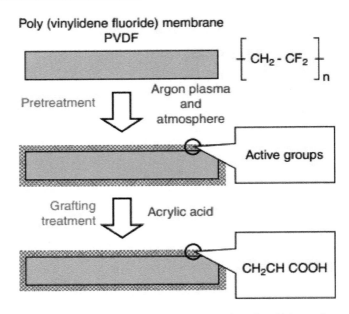

Figure 3.2 Pretreatment and grafting onto a poly (vinylidene fluoride) membrane made of nanofiber (Kulkarni et al., 2022).

and negative loading. Poly (acrylic acid) was then used to immobilize poly(amidoamine) (PAMAM) dendrimers (G5·NH$_2$) on the PLGA surface. Initially, as shown in Fig. 3.3, the PAMAM amino group (G5·NH$_2$) dendrimers, which were grafted onto the nanofibers PLGA scaffolding enabling gene transmission, are coupled to the poly carboxy group (acrylic acid) (Xiao et al., 2019).

3.2.2.2 Crosslinking

In the cross-linking tactic, different grades of chitosan nanofibers treated with crosslinked glutaraldehyde vapor, as well as polymeric solutions of various molecular weights, were transformed into nanofibers with basic Schiff functionality. As compared to nanofibers without cross-linking, this cross-linked chitosan demonstrated higher solubility. The nanofibres still included residual cross-linking, which had an impact on their biocompatibility and toxicity. Epichlorohydrin, glutaraldehyde, resimene, and gepinine, among others, are available as crosslinking agents. Fig. 3.4 illustrates how glutaraldehyde is used to cross-link zein-based nanofibres electroextruded containing poly (vinyl alcohol) in the literature. The adsorption capacity of nanofibres is greatly increased by the cross-linking of nanozein fibers (Yu et al., 2020).

3.2.2.3 Chelation

Electrospun polyacrylonitrile (PAN) nanofibers were cross-linked with diamine, ethylene glycol, and thioamide to undergo chemical modification. These nanofibres

Functionalization/Modification of Nanofibers and Their Impact 55

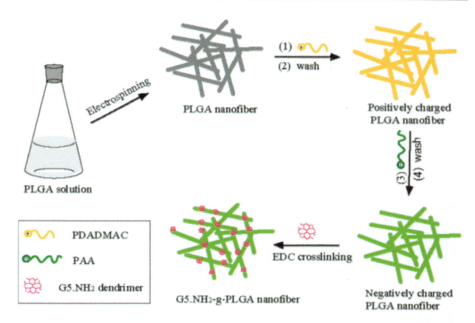

Figure 3.3 G5 NH$_2$-g-PLGA nanofiber preparation (Kulkarni et al., 2022).

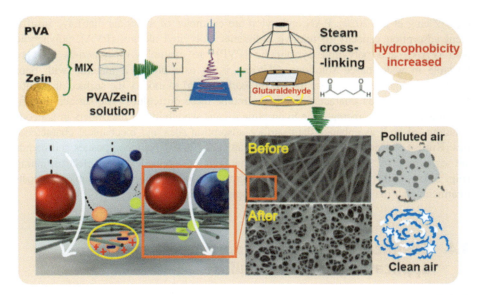

Figure 3.4 Hydrophobic covalently bonded zein nanofibers are used in the preparation and construction of the multifunctional nanofilter's schematic (Kulkarni et al., 2022).

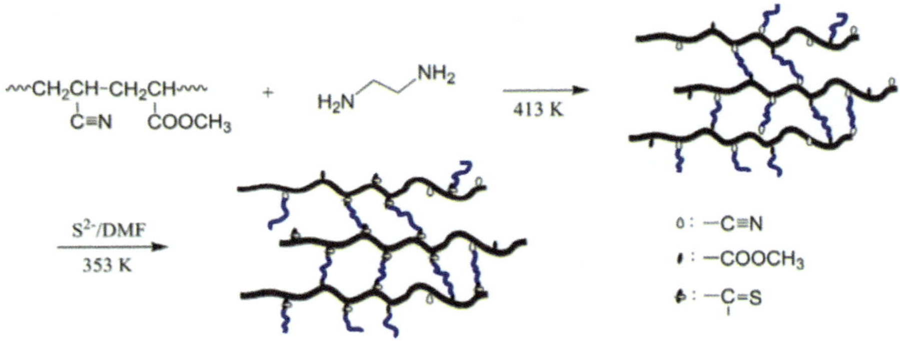

Figure 3.5 The synthetic diagram of the polyacrylonitrile nanofiber-based thioamide group's chelating nanofibres (Kulkarni et al., 2022).

can be used in biomedicine. The nanochelation fiber shown in Fig. 3.5 is presented schematically (Li et al., 2013).

3.3 Modification of polymer nanofibers

High surface-to-volume ratios, as well as high porosity of polymer nanofibers, are advantages, although certain virgin polymers still have constraints in the adsorption process (Zhang et al., 2020). PAN and nylon are examples of materials that have insufficient adsorption capacity to remove pollutants. Polyvinyl alcohol (PVA), polyacrylic acid (PAA), and polyvinyl pyrrolidone (PVP) are examples of materials that are unstable in water-based solutions. Chitosan is an example of a material with poor mechanical qualities. Scientists have been working hard to modify the surface of nanofibers to increase their characteristics to overcome these difficulties. The surface modification is intended to increase the hydrophilicity and wettability of nanofibers in aqueous solutions, as well as enhance their surface adsorption sites and mechanical characteristics (Kurusu & Demarquette, 2019; Nagarajan et al., 2019). Nanocomposites and blends are examples of single-step treatments that can be applied during electrospinning (Fig. 3.6a), while plasma, wet chemistry, grafting, and coating are examples of post-treatments that may be applied after electrospinning, as shown in Fig. 3.6b.

3.4 Chemical functionalization, properties, and applications of nanocellulose

3.4.1 Chemical treatment of nanocellulose

One of the key phases in creating a nanocomposite that can enhance the mechanical performance of cellulose nanocomposites is the superficial modification of

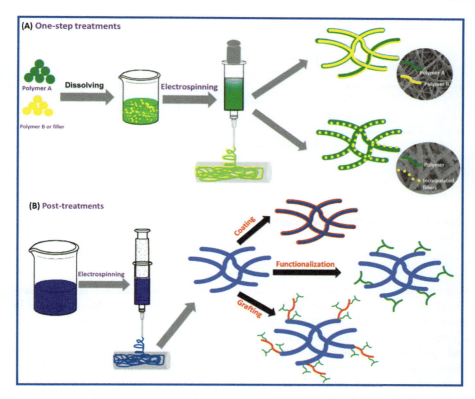

Figure 3.6 Surface modification of electroextruded polymer nanofibers (Thamer et al., 2020). (A) Nanocomposites and blends are examples of single-step treatments that can be applied during electrospinning; (B) plasma, wet chemistry, grafting, and coating are examples of post-treatments that may be applied after electrospinning

nanocellulose. Thus, it is required to pretreat with certain chemicals to strengthen the interfacial interaction seen between the matrix and the cellulosic material. The process also increases the moistening characteristics of the reinforcement to the matrix and supports the resistance for the aquatic treatments by forming a strong interfacial connection between the interface and materials (Crépy, et al., 2009; Long et al., 2021). Fig. 3.7 depicts the numerous chemical transformation processes. Attaching the chemical interaction of −OH to the interface of nanocellulose or directly chemically altering it are two common methods for surface modification of cellulose. Moreover, grafting polymers onto a biopolymer is a common way to modify grafted polymers and nanocomposites (Jiang et al., 2021). One of the primary applications for surface-modified nanocellulose is the creation of amphiphilic surfaces. Amphiphobic surfaces are often used for applications such as self-cleaning and antireflective materials. A surface that is amphiphilic may protect both polar and nonpolar liquids (Hayase et al., 2013). Adding −OH to the interface of nano cellulose changes the surface's wettability and makes it more hydrophobic

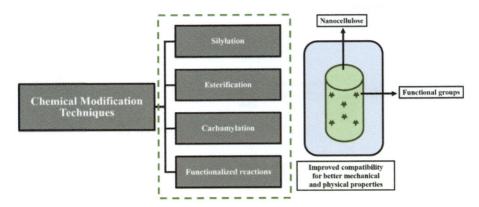

Figure 3.7 Diagram illustrating several nanocellulose modification techniques (Tahir et al., 2022).

when it comes into contact with chemicals. Other chemical reactions, such as etherification, carbonylation, and silylation, have an impact on the hydroxyl groups in nanocellulose (Habibi, 2014). Poly(glutamic acid) is often used to chemically modify cellulose nanocrystals (CNC). The cellulose is slightly oxidized to get the optimal qualities, and then the poly(glutamic acid) amino groups, as well as the aldehyde groups, react to provide those features. The hydrophobicity of cellulose is increased after being treated with poly(glutamic acid), which leads to the creation of a strong interfacial connection with a matrix (Averianov et al., 2019).

Silane coupling agents, including silane alkoxy, are the most effective chemical compounds to strengthen the interfacial interaction between cellulose and matrix. Fibers become more wettable due to silane hydrocarbon chains, which also improve their chemical affinity to the matrix. Compared to untreated fibers, silane modification is extremely effective for alkaline fibers. The great effectiveness results from the silane treatment making the areas more receptive. When processed without water and at high temperatures, cellulose fibers can be silylated. High temperatures prevent water from being present, which limits the interaction between both the Si-OR group and the cellulose OH group. This indicates that silane and cellulose OH are in touch (Abdelmouleh et al., 2005; Lu et al., 2013). Increasing the moisture content facilitates reaction here between cellulose Si-OR and OH groups. Chlorodimethyl isopropyl silane can be used to investigate the silylation impact on nanofibrillated cellulose (NFC) that has been isolated from bleached softwood mash. Silylation results in substituted silyl groups being added to the surface of cellulose nanofibers, which is advantageous for increasing hydrophobia.

Isocyanic acid attaches to functional groups within cellulose through the process of carboxylation to change the surface of the material. Butyl 4-(Boc-aminomethyl) phenyl isothiocyanate was employed by Navarro and Bergström (2014) to alter NFCs, and DMSO was utilized as the solvent. Moreover, NFCs were given a rhodamine B ester treatment, which was followed by the addition of N-hydroxysuccinimide to produce brilliant NFCs. These NFCs are frequently used in sensor applications.

Another intriguing method to chemically alter NFC surfaces is 2,2,6,6-tetramehylpiperidine-1-oxyl (TEMPO) oxidation technology, which has grown in prominence in recent publications. The NFC surface may initially be altered in this technique by adding carboxylic groups. The linked groups then provide assistance with the further functionalizing procedures required for the creation of customized NFCs. The hydrophobic films may be used as fluorescence detectors to identify the presence of nitroaromatics; recent research demonstrates that TEMPO-oxidized NFCs have found several uses (Niu et al., 2014; Orelma et al., 2012).

Important techniques for NFC surface modification include acetylation and esterification (Goudarzi et al., 2021). In organic settings, this calls for the utilization of aromatic and carboxylic reagents. A variety of studies have been conducted using the surface modification method acetylation (Bledzki et al., 2008; Goudarzi et al., 2021). Acetylation is the process by which cellulose's OH groups react with acetyl groups, causing lignocellulosic strands to become plastic (Bledzki et al., 2008). A combination of ethanol/solvent toluene, along with acid anhydride, is prepared, and then the NFC suspension is added. An acid anhydride is frequently employed as an acetylation medium. Bulota et al. examined how acetylation affects the characteristics of polylactic acid (PLA) and acetylated NFC compounds (Bulota et al., 2012). It was discovered that the acetylation operation, which was carried out at 105°C in toluene, is a useful way to improve cellulose dispersion in a nonpolar PLA solution. The degree of substitution (DS) for the acetyl content on the treated cellulose might be 0.43 and is significantly dependent on the reaction time. Also utilized for acetylation verification is fourier-transform infrared spectroscopy (FTIR). The results demonstrate a significant influence of higher DS nanofibers on the properties of PLA-acetylated NFC composites. To enhance the cellulose effectiveness of cellulose-reinforced composites, several anhydrides can be used (Jamaluddin et al., 2021). Propionic anhydride has recently been utilized in place of acetic anhydride because it offers more dimensional stability than cellulose. Also, it is reported that pyridine and sulfuric acid catalysts may be used to improve SD, and grafting acetyl halves to cellulose can increase the hydrophobicity of the material. Because of the cumulative effects of the aforementioned procedures, NFCs with a very hydrophobic surface have been produced, considerably reducing the passage of water to the interface. Missoum et al. introduced a novel technique for heterogeneous surface modification that results in the grafting of nanoscaled cellulose substrates (Missoum et al., 2012) using several anhydrides in an ionic fluid without changing the morphological properties of the substrates. With adequate consideration for environmental considerations, liquid−liquid extraction was performed to recycle the ionic liquid. Ionic liquid usage has the benefit of not generating volatile organic compounds.

Nanocomposites are reinforced using a type of synthetically modified cellulose called methylcellulose (MC). It is a cellulose ester with a maximum methoxy content of 32%. The DS for MC at 29.1% methoxy is 1.75 (Miki et al., 2020). The product's ductility may be improved by adding MC to nanocellulose or using it as a filler in the biopolymer. Nanocomposite hydrogels were wet spun into MC/CNC nanocomposite fibers in an ethanol coagulation bath by Hynninen et al. (2019). The treated fibers were ductile and had a high modulus. The best mechanical qualities were produced by fibers with an 80% MC and 20% CNC content. Increased CNC in fibers renders them brittle, but MC and CNC must be combined to get desired qualities. There is a synergistic impact on fibers when MC and CNC are combined, and no fibers can be made solely from CNC or MC.

3.4.2 Applications of nanocellulose

Nanocellulose is appealing for a variety of industrial uses, including as an oil recovery agent, a filler in tissues, and a thickener in cosmetics. The creation of translucent paper and functional nanocomposites is also made possible by its exceptional qualities and biodegradability (Abitbol et al., 2016). Composite nanomaterials with good lightness, transparency, high mechanical strength, and outstanding thermal conductivity are made using nanocellulose (Abitbol et al., 2016). Applications for cellulose-based nanocomposites are many, as demonstrated in Fig. 3.8. Nanocellulose has been utilized to make flexible batteries, lightweight armor, and extremely durable windmill blades, among other goods (Abitbol et al., 2016). The mechanical property of nanocomposites reinforced with soy-derived nanocellulose has been studied by Wang and Sain (2007), and the results show that the tensile strength of polymers is greatly increased compared to virgin polymers. Tensile strength in PVA reinforced with 5% nanocellulose fibers increased from 21 to

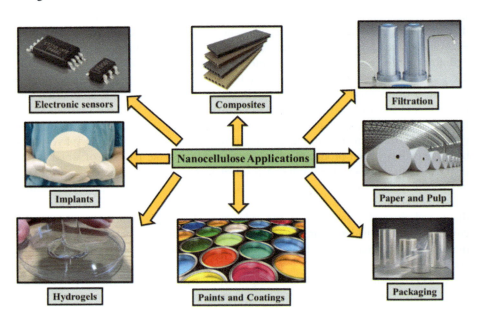

Figure 3.8 Nanocellulose and related derivatives have a variety of uses (Tahir et al., 2022).

103 MPa. The insertion of nanofibers alters the composite's stress-deformation behavior, enhancing its overall characteristics.

3.4.2.1 Biomedical applications

Because of its ability to degrade, low toxicity, and superior physical properties, nanocellulose has a wide range of uses in the medical sector (Nehra & Chauhan, 2021; Salas et al., 2014). In this work (Hakkarainen et al., 2016), it was discovered that NFC functions as a highly green composed using wound donor sites when used as a wound dressing and that the dressing may be readily removed after skin healing (Hu et al., 2013; Vismara et al., 2019). In the present times, nanocellulose is vastly used in medical engineering and technology, for example, in soft tissue implants, drug delivery, and cancer treatments (Vismara et al., 2019). A new type of hairy cellulose nanocrystalloid, made of CNC with functionalized chains at both ends, is also in high demand for its exceptional properties. Hosseinidoust et al. (2015) extract nanocellulose from softwood pulp sheets through a chemical process. This process results in nanocrystals with a very high carboxyl content (6.6 mmol/g) that continuously regulates surface load without altering reaction conditions. These nanocrystals are additionally used as a scaffold for tissue engineering and in nanocellulose composites (Si et al., 2016).

Hydrogen composites reinforced with nanocellulose, which can enhance the mechanical properties of polymer gel formulations having appropriate strengthening and matrix properties, represent the most effective biomedical use of nanocellulose (De France et al., 2017). Recently, a novel hydrogel with good mechanical properties and a network comprising semiinterpenetrators that can boost pH sensitivity was created (Sampath et al., 2017). New pharmaceutical and gene administration uses for this hydrogel are possible. The CNC, which implies carboxylated chains, is a carrier for nanomedicine because it can resist agglomeration and also be absorbed by numerous cells. By expanding the usage of hydrogels in the biomedical industry, cellulose boosts the gel shear modulus, decreases the freezing time, and improves cellular adhesion. In a 3D cell culture, nanocellulose hydrogels may replicate the extracellular matrix with very little cytotoxicity. These gels demonstrated superior cellular regeneration while offering the mechanical qualities required for tissue engineering scaffolding, making them advantageous for the treatment of wounds and cartilage repair. Moreover, nanocellulose can be used to encapsulate medications and deliver them to specific body areas.

3.4.2.2 Water treatment

Nanocelluloses are sustainable biomaterials with biodegradable properties that also have a large surface area, plenty of reactive sites, and scaffolding strength to support inorganic nanoparticles. These nanoparticles have several uses in the treatment of water. When employed as reinforcement, nanocellulose has special qualities, such as a nanometric dimension that offers composite strength and barriers to gas and water (Silva et al., 2020). These characteristics could help in developing nanocomposite films with a sturdy frame that can penetrate molecules (Nair et al., 2014). Wastewater treatment is very interesting in nanocellulose-based composites because of their renewable adsorption capacity. Numerous hydroxide groups on the nanocellulose's surface enable a variety of chemical transformations. These significant nanocellulose surface characteristics are used to build an ecological cellulose membrane that filters pollutants from water. A schematic of the nanocellulose membranes being used in water treatment may be found in Fig. 3.9. They are the finest option as an adsorbent for eliminating pollutants from damp environments, including heavy metals, organic pigments, oils, and medicines. The adsorption process is largely dependent on the interaction of adsorbent and adsorbent (Voisin et al., 2017). Various mechanisms, including ion exchange, dipole-dipole contact, surface and pore complexation, van der Waals forces, and hydrophobicity, can cause the absorbent to adsorbate interactions. Natural organic matter through biomass degradation can harm the environment and people's health. Some of the typical acids created from biomass that must be stopped from harming water include humic acid (HA) and fulvic acid. Using nanocellulose modified by the amine group, Jebali et al. (2015) examined the removal of HA from wastewater. Also, they discovered the role that electrostatic forces played in the adsorption of HA. The behavior of the amine-modified nanocellulose was altered by interactions between its amine group and the carboxyl as well as hydroxy functional groups of the acid (Hakami et al., 2019).

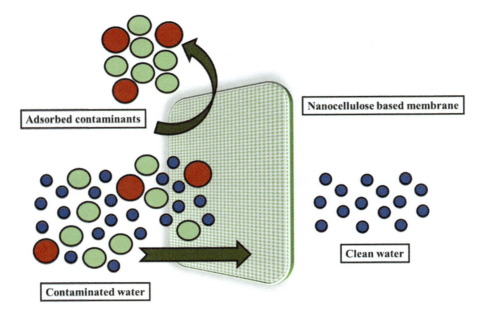

Figure 3.9 Illustration demonstrating the usage of nanocellulose membranes to filter out waterborne pollutants (Tahir et al., 2022).

3.4.2.3 Coatings

Several researchers have expressed interest in the usage of nanocellulose in the creation of nanocomposite films and coatings (Ge et al., 2021). One of the primary uses for nanocellulose, among its many other uses, is coating surfaces with diverse compositions and structural elements. To alter the permeability and filtering abilities of coatings, other variables may be changed, such as the porosity and thickness of the cellulose applied. The mechanical durability of nanocellulosic coatings, whether wet or dry, is crucial for many applications. Dufresne provided the first study on the enhancement of the mechanical properties of nanocomposite coatings utilizing cellulose nanofibres for reinforcement and potato starch as the polymer matrix (Dufresne, 2012). He discussed how these nanoparticles' outstanding mechanical qualities, capacity for reinforcement, low density, and biodegradability make them excellent choices for usage in nanocomposites. Even at modest filler loads, they have dramatically enhanced the mechanical characteristics of polymers, thanks to their Young module, which ranges from 100 to 130 GPa. Hydroxypropylated cellulose was employed by Dufresne (2012), who noted improved tensile characteristics, while Nakagaito and Yano (2008) investigated similar enhancement in tensile characteristics using nanofiber cellulose and phenol-formaldehyde resin (HPC).

It is economical and ecologically effective and its wide availability highlights the significance of nanocellulosic coatings. The researchers attempted to mimic the superhydrophobic properties of coatings made of nanocellulose to produce surfaces that would clean themselves. By adding micro or nanostructures, one may produce superhydrophobic materials that are rougher and have lower surface energies. The wetting characteristics of the coating are impacted by reduced surface energy, which leads to the realization of superhydrophobicity (Cherian et al., 2022). Moreover, wood is shielded by nanocellulosic coatings against various environmental hazards such as germs or microorganisms in a moist atmosphere. Natural extracts and organic and polymer compounds were once employed to preserve wood, but their usage creates certain volatile molecules that can have a negative impact on the environment. Thus, it is advised to apply a nanocellulose coating with little influence on the environment (Makarona et al., 2017).

3.5 Conclusion

Because of their exceptional engineering properties, nanofibers have become important in practically every industry, particularly in the packaging, purification, textile, pharmaceutical, and biomedical sectors. Nanofibers are made from a variety of polymers based on their intended function. The nanofibers were modified with various functional groups, such as amino, carboxyl, sulfur, imino, and hydroxyl. As a result, their various chemical and physical properties were developed. These modifications impacted the nanofibers' surface features, leading to changes in wettability, electrical conductivity, optical properties, and biocompatibility. Most of these nanofibers have been discovered to be biodegradable and biocompatible, with the ability to build a highly porous structure with superior properties. This makes them suitable for a wide range of applications, such as packaging, drug delivery, medical implants such as organ and tissue grafts, wound repair, and dressing materials in the pharmaceutical industry, and as a filter or adsorbent in water treatment. Consequently, the functionalized nanofiber materials were interestingly used in the following applications: medicine, pharmacology, genetic engineering, cancer treatments, tissue engineering, drug delivery, and others. Because of its ability to degrade, low toxicity, and superior physical properties, nanocellulose has a wide range of uses in the medical sector.

Author contribution statement

Equal contribution by all the authors.

Abbreviations

CNC	Cellulose nanocrystals
FTIR	Fourier-transform infrared spectroscopy
NFC	Nano-fibrillated cellulose
PCL	Poly(caprolactone)
PVDF	Poly (Vinylidene fluoride)
PAMAM	Poly(amidoamine)
PLGA	Poly (D, L-dairy product-co-glycolic acid)
PAN	Polyacrylonitrile
PVA	Polyvinyl alcohol
PAA	Polyacrylic acid
PVP	Polyvinyl pyrrolidone
TEMPO	2,2,6,6-tetramehylpiperidine-1-oxyl

References

Abdelmouleh, M., Boufi, S., Belgacem, M. N., Dufresne, A., & Gandini, A. (2005). Modification of cellulose fibers with functionalized silanes: Effect of the fiber treatment on the mechanical performances of cellulose−thermoset composites. *Journal of Applied Polymer Science, 98*, 974−984.

Abitbol, T., Rivkin, A., Cao, Y., Nevo, Y., Abraham, E., Ben-Shalom, T., Lapidot, S., & Shoseyov, O. (2016). Nanocellulose, a tiny fiber with huge applications. *Current Opinion in Biotechnology, 39*, 76−88.

Averianov, I. V., Stepanova, M. A., Gofman, I. V., Nikolaeva, A. L., Korzhikov-Vlakh, V. A., Karttunen, M., et al. (2019). Chemical modification of nanocrystalline cellulose for improved interfacial compatibility with poly (lactic acid). *Mendeleev Communications, 29*, 220−222.

Bledzki, A., Mamun, A., Lucka-Gabor, M., & Gutowski, V. (2008). The effects of acetylation on properties of flax fibre and its polypropylene composites. *Express Polymer Letters*, *2*, 413–422.

Bulota, M., Kreitsmann, K., Hughes, M., & Paltakari, J. (2012). Acetylated microfibrillated cellulose as a toughening agent in poly (lactic acid). *Journal of Applied Polymer Science*, *126*, E449–E458.

Cherian, R. M., Tharayil, A., Varghese, R. T., Antony, T., Kargarzadeh, H., Chirayil, C. J., et al. (2022). A review on the emerging applications of nano-cellulose as advanced coatings. *Carbohydrate Polymers*119123.

Crépy, L., Chaveriat, L., Banoub, J., Martin, P., & Joly, N. (2009). Synthesis of cellulose fatty esters as plastics—influence of the degree of substitution and the fatty chain length on mechanical properties. *ChemSusChem: Chemistry & Sustainability Energy & Materials*, *2*, 165–170.

Dufresne, A. (2012). Potential of nanocellulose as a reinforcing phase for polymers,". *J-For-Journal of Science & Technology for Forest Products and Processes*, *2*, 6–16.

Elsadek, N. E., Nagah, A., Ibrahim, T. M., Chopra, H., Ghonaim, G. A., Emam, S. E., et al. (2022). Electrospun nanofibers revisited: an update on the emerging applications in nanomedicine. *Materials*, *15*, 1934.

De France, K. J., Hoare, T., & Cranston, E. D. (2017). Review of hydrogels and aerogels containing nanocellulose,". *Chemistry of Materials*, *29*, 4609–4631.

Garkal, A., Kulkarni, D., Musale, S., Mehta, T., & Giram, P. (2021). Electrospinning nanofiber technology: A multifaceted paradigm in biomedical applications. *New Journal of Chemistry*, *45*, 21508–21533.

Ge, L., Yin, J., Yan, D., Hong, W., & Jiao, T. (2021). Construction of nanocrystalline cellulose-based composite fiber films with excellent porosity performances via an electrospinning strategy. *ACS Omega*, *6*, 4958–4967.

Goudarzi, Z. M., Behzad, T., Ghasemi-Mobarakeh, L., & Kharaziha, M. (2021). An investigation into influence of acetylated cellulose nanofibers on properties of PCL/Gelatin electrospun nanofibrous scaffold for soft tissue engineering. *Polymer*, *213*123313.

Habibi, Y. (2014). Key advances in the chemical modification of nanocelluloses. *Chemical Society Reviews*, *43*, 1519–1542.

Hakami, M. W., Alkhudhiri, A., Zacharof, M.-P., & Hilal, N. (2019). Towards a sustainable water supply: Humic acid removal employing coagulation and tangential cross flow microfiltration. *Water*, *11*, 2093.

Hakkarainen, T., Koivuniemi, R., Kosonen, M., Escobedo-Lucea, C., Sanz-Garcia, A., Vuola, J., et al. (2016). Nanofibrillar cellulose wound dressing in skin graft donor site treatment. *Journal of Controlled Release*, *244*, 292–301.

Hayase, G., Kanamori, K., Hasegawa, G., Maeno, A., Kaji, H., & Nakanishi, K. (2013). A superamphiphobic macroporous silicone monolith with marshmallow-like flexibility. *Angewandte Chemie*, *125*, 10988–10991.

Hosseinidoust, Z., Alam, M. N., Sim, G., Tufenkji, N., & van de Ven, T. G. (2015). Cellulose nanocrystals with tunable surface charge for nanomedicine. *Nanoscale*, *7*, 16647–16657.

Hu, L., Zheng, G., Yao, J., Liu, N., Weil, B., Eskilsson, M., et al. (2013). Transparent and conductive paper from nanocellulose fibers. *Energy & Environmental Science*, *6*, 513–518.

Huang, F., Wei, Q., & Cai, Y. (2012). Surface functionalization of polymer nanofibers,". *Functional Nanofibers and their Applications*, 92–118.

Hynninen, V., Mohammadi, P., Wagermaier, W., Hietala, S., Linder, M. B., & Ikkala, O. (2019). Methyl cellulose/cellulose nanocrystal nanocomposite fibers with high ductility. *European Polymer Journal, 112*, 334−345.

Ibrahim, N. A., Fouda, M. M., & Eid, B. M. (2020). *Functional nanofibers: Fabrication, functionalization, and potential applications* ed: *Handbook of functionalized nanomaterials fo*r industrial *applications* (pp. 581−609). Elsevier.

Jamaluddin, N., Hsu, Y.-I., Asoh, T.-A., & Uyama, H. (2021). Effects of acid-anhydride-modified cellulose nanofiber on poly (lactic acid) composite films. *Nanomaterials, 11*, 753.

Jebali, A., Behzadi, A., Rezapor, I., Jasemizad, T., Hekmatimoghaddam, S., Halvani, G. H., et al. (2015). Adsorption of humic acid by amine-modified nanocellulose: an experimental and simulation study. *International Journal of Environmental Science and Technology, 12*, 45−52.

Jiang, Y., Zhang, Y., Cao, M., Li, J., Wu, M., Zhang, H., et al. (2021). Combining 'grafting to' and 'grafting from' to synthesize comb-like NCC-g-PLA as a macromolecular modifying agent of PLA. *Nanotechnology, 32*385601.

Kajdič, S., Planinšek, O., Gašperlin, M., & Kocbek, P. (2019). Electrospun nanofibers for customized drug-delivery systems. *Journal of Drug Delivery Science and Technology, 51*, 672−681.

Kulkarni, D., Musale, S., Panzade, P., Paiva-Santos, A. C., Sonwane, P., Madibone, M., et al. (2022). Surface functionalization of nanofibers: The multifaceted approach for advanced biomedical applications,". *Nanomaterials, 12*, 3899.

Kumbar, S., James, R., Nukavarapu, S., & Laurencin, C. (2008). Electrospun nanofiber scaffolds: Engineering soft tissues. *Biomedical Materials, 3*034002.

Kurusu, R. S., & Demarquette, N. R. (2019). Surface modification to control the water wettability of electrospun mats. *International Materials Reviews, 64*, 249−287.

Li, X., Zhang, C., Zhao, R., Lu, X., Xu, X., Jia, X., et al. (2013). Efficient adsorption of gold ions from aqueous systems with thioamide-group chelating nanofiber membranes. *Chemical Engineering Journal, 229*, 420−428.

Long, S., Zhong, L., Lin, X., Chang, X., Wu, F., Wu, R., et al. (2021). Preparation of formyl cellulose and its enhancement effect on the mechanical and barrier properties of polylactic acid films. *International Journal of Biological Macromolecules, 172*, 82−92.

Lu, T., Jiang, M., Jiang, Z., Hui, D., Wang, Z., & Zhou, Z. (2013). Effect of surface modification of bamboo cellulose fibers on mechanical properties of cellulose/epoxy composites. *Composites Part B: Engineering, 51*, 28−34.

Makarona, E., Koutzagioti, C., Salmas, C., Ntalos, G., Skoulikidou, M.-C., & Tsamis, C. (2017). Enhancing wood resistance to humidity with nanostructured ZnO coatings. *Nano-Structures & Nano-Objects, 10*, 57−68.

Miki, K., Kamitakahara, H., Yoshinaga, A., Tobimatsu, Y., & Takano, T. (2020). Methylation-triggered fractionation of lignocellulosic biomass to afford cellulose-, hemicellulose-, and lignin-based functional polymers via click chemistry. *Green Chemistry, 22*, 2909−2928.

Missoum, K., Belgacem, M. N., Barnes, J.-P., Brochier-Salon, M.-C., & Bras, J. (2012). Nanofibrillated cellulose surface grafting in ionic liquid. *Soft Matter, 8*, 8338−8349.

Nagarajan, S., Balme, S., Kalkura, S. N., Miele, P., Bohatier, C., & Bechelany, M. (2019). *Various techniques to functionalize nanofibers*. *Handbook of Nanofibers* (pp. 347−372).

Nair, S. S., Zhu, J., Deng, Y., & Ragauskas, A. J. (2014). High performance green barriers based on nanocellulose. *Sustainable Chemical Processes, 2*, 1−7.

Nakagaito, A. N., & Yano, H. (2008). The effect of fiber content on the mechanical and thermal expansion properties of biocomposites based on microfibrillated cellulose. *Cellulose, 15*, 555–559.

Navarro, J. R., & Bergström, L. (2014). Labelling of N-hydroxysuccinimide-modified rhodamine B on cellulose nanofibrils by the amidation reaction. *RSC Advances, 4*, 60757–60761.

Nehra, P., & Chauhan, R. (2021). Eco-friendly nanocellulose and its biomedical applications: Current status and future prospect,". *Journal of Biomaterials Science, Polymer Edition, 32*, 112–149.

Niu, Q., Gao, K., & Wu, W. (2014). Cellulose nanofibril based graft conjugated polymer films act as a chemosensor for nitroaromatic. *Carbohydrate Polymers, 110*, 47–52.

Orelma, H., Filpponen, I., Johansson, L.-S., Österberg, M., Rojas, O. J., & Laine, J. (2012). Surface functionalized nanofibrillar cellulose (NFC) film as a platform for immunoassays and diagnostics. *Biointerphases, 7*, 61.

Salas, C., Nypelö, T., Rodriguez-Abreu, C., Carrillo, C., & Rojas, O. J. (2014). Nanocellulose properties and applications in colloids and interfaces. *Current Opinion in Colloid & Interface Science, 19*, 383–396.

Sampath, U. T. M., Ching, Y. C., Chuah, C. H., Singh, R., & Lin, P.-C. (2017). Preparation and characterization of nanocellulose reinforced semi-interpenetrating polymer network of chitosan hydrogel. *Cellulose, 24*, 2215–2228.

Sharma, R., Singh, H., Joshi, M., Sharma, A., Garg, T., Goyal, A., et al. (2014). Recent advances in polymeric electrospun nanofibers for drug delivery,". *Critical Reviews™ in Therapeutic Drug Carrier Systems, 31*.

Si, J., Cui, Z., Wang, Q., Liu, Q., & Liu, C. (2016). Biomimetic composite scaffolds based on mineralization of hydroxyapatite on electrospun poly (ε-caprolactone)/nanocellulose fibers. *Carbohydrate Polymers, 143*, 270–278.

Silva, F. A., Dourado, F., Gama, M., & Poças, F. (2020). Nanocellulose bio-based composites for food packaging. *Nanomaterials, 10*, 2041.

Sofi, H. S., Ashraf, R., Khan, A. H., Beigh, M. A., Majeed, S., & Sheikh, F. A. (2019). Reconstructing nanofibers from natural polymers using surface functionalization approaches for applications in tissue engineering, drug delivery and biosensing devices. *Materials Science and Engineering: C, 94*, 1102–1124.

Suárez, D. F., Pinzón-García, A. D., Sinisterra, R. D., Dussan, A., Mesa, F., & Ramírez-Clavijo, S. (2022). Uniaxial and coaxial nanofibers PCL/alginate or PCL/gelatine transport and release tamoxifen and curcumin affecting the viability of MCF7 cell line. *Nanomaterials, 12*, 3348.

Tahir, D., Karim, M. R. A., Hu, H., Naseem, S., Rehan, M., Ahmad, M., et al. (2022). Sources, Chemical functionalization, and commercial applications of nanocellulose and nanocellulose-based composites: A review. *Polymers, 14*, 4468.

Thakkar, S., & Misra, M. (2017). Electrospun polymeric nanofibers: New horizons in drug delivery. *European Journal of Pharmaceutical Sciences, 107*, 148–167.

Thamer, B. M., Aldalbahi, A., Moydeen A, M., Rahaman, M., & El-Newehy, M. H. (2020). Modified electrospun polymeric nanofibers and their nanocomposites as nanoadsorbents for toxic dye removal from contaminated waters: A review. *Polymers, 13*, 20.

Vismara, E., Bernardi, A., Bongio, C., Farè, S., Pappalardo, S., Serafini, A., et al. (2019). Bacterial nanocellulose and its surface modification by glycidyl methacrylate and ethylene glycol dimethacrylate. incorporation of vancomycin and ciprofloxacin. *Nanomaterials, 9*, 1668.

Voisin, H., Bergström, L., Liu, P., & Mathew, A. P. (2017). Nanocellulose-based materials for water purification. *Nanomaterials, 7*, 57.

Wang, B., & Sain, M. (2007). Isolation of nanofibers from soybean source and their reinforcing capability on synthetic polymers. *Composites Science and Technology, 67*, 2521−2527.

Wei, Q., Gao, W., Hou, D., & Wang, X. (2005). Surface modification of polymer nanofibres by plasma treatment. *Applied Surface Science, 245*, 16−20.

Wsoo, M. A., Shahir, S., Bohari, S. P. M., Nayan, N. H. M., & Abd Razak, S. I. (2020). A review on the properties of electrospun cellulose acetate and its application in drug delivery systems: A new perspective. *Carbohydrate Research, 491*107978.

Xiao, S., Peng, Q., Yang, Y., Tao, Y., Zhou, Y., Xu, W., et al. (2019). Preparation of [amine-terminated generation 5 poly (amidoamine)]-graft-poly (lactic-co-glycolic acid) electrospun nanofibrous mats for scaffold-mediated gene transfection. *ACS Applied Bio Materials, 3*, 346−357.

Xie, X., Chen, Y., Wang, X., Xu, X., Shen, Y., Aldalbahi, A., et al. (2020). Electrospinning nanofiber scaffolds for soft and hard tissue regeneration,". *Journal of Materials Science & Technology, 59*, 243−261.

Xue, J., Wu, T., Dai, Y., & Xia, Y. (2019). Electrospinning and electrospun nanofibers: Methods, materials, and applications. *Chemical Reviews, 119*, 5298−5415.

Yao, T., Baker, M. B., & Moroni, L. (2020). Strategies to improve nanofibrous scaffolds for vascular tissue engineering. *Nanomaterials, 10*, 887.

Yu, X., Li, C., Tian, H., Yuan, L., Xiang, A., Li, J., et al. (2020). Hydrophobic cross-linked zein-based nanofibers with efficient air filtration and improved moisture stability. *Chemical Engineering Journal, 396*125373.

Zhang, W., He, Z., Han, Y., Jiang, Q., Zhan, C., Zhang, K., et al. (2020). Structural design and environmental applications of electrospun nanofibers. *Composites Part A: Applied Science and Manufacturing, 137*106009.

Processing of nanofibers for oil-spill cleaning application

4

Yong Xue Gan
Department of Mechanical Engineering, California State Polytechnic University Pomona, Pomona, CA, United States

The increasing pollution by spilled oil on land and open seawater has been the motivation for developing high-efficiency oil absorption and oil/water separation materials. Nanofibers are promising materials for oil absorption and decomposition. In this chapter, several major nanofibers, including synthetic polymeric and natural nanofibers, for oil-spill cleaning will be presented. Considering the increased production of nanofibers, it is necessary to introduce their processing technologies. The emphasis of this chapter will be put on the processing technologies for making nanofiber-based sorbents. The performances of these kinds of nanofibers on oil-spill cleaning will be presented. The nature of contact between oil and solid surface in an oil/water environment will be discussed based on the classical theories on wettability. Methods for changing the wettability of solid surfaces to fluids through varying the interfacial properties and increasing the roughness will be explained. How to modify the surface properties of nanofibers to improve the affinity to oil by applying hydrophobic coatings will be discussed as well.

4.1 Introduction

During the production, transportation, and use of oils, the spilling of oil sometimes happens. Oil spills are pollutants that cause serious threats to health and the ecosystem. Therefore oil-spill cleaning is very important for environmental sustainability. Various technologies have been developed for oil-spill cleaning. Just taking the example of oil-spill cleaning in open seawater, besides burning in situ, oil booms, skimmers, dispersants, and porous sorbents are often used. Bioremediation is also considered, during which the oil removal is considerably slow. If oil-spills are located in small areas, hot water and high-pressure washing may work, but the flashed-away oil residual could result in the pollution of other areas. As compared with the above-mentioned methods, the application of composite nanofiber sorbents for oil spill cleanup can achieve a very higher percentage of oil recovery. Nanofiber oil sorbents are reusable due to their high mechanical strength and peculiar surface properties.

Composite nanofibers are multicomponent and highly porous materials. Some of the composite nanofibers are hydrophobic and show a strong affinity to oil. They are suitable for oil absorption. Oil absorption nanofibers have many uses. One of the

Polymeric Nanofibers and their Composites. DOI: https://doi.org/10.1016/B978-0-443-14128-7.00004-3
© 2025 Elsevier Ltd. All rights are reserved, including those for text and data mining, AI training, and similar technologies.

major applications is for oil-spill cleaning or water/oil separation. There are two different ways of cleaning spilled oil. One is by direct absorption and reclaiming the oil. The other is by decomposition of oil at the surface of nanofibers. In both approaches, the large surface areas of nanofibers enhance the oil spill cleaning, which makes nanofibers attractive. There are many kinds of composite aerogels for oil-spill cleaning including synthetic and natural composite nanofibers. Depending on the nature of the nanofibers, the processing methods could be very different. In the following section, the design and processing of various composite nanofibers for oil cleaning will be described. The currently used technologies, including electrospinning, water-mediated spinning, sol−gel electrospinning, chemical oxidation, deacetylation, delignification, and aerogel formation, will be introduced. The structural control through directional freezing will be discussed. Following that the wettability of oil to solid surfaces in air and water environments will be illustrated. Three classical models for analyzing the wetting of various interfaces will be briefly discussed.

4.2 Processing technologies

4.2.1 Electrospinning

In this subsection, electrospinning nanofiber membranes will be discussed. Nanofiber membranes with a high surface area to volume ratio and a complex pore structure are ideal for separation applications. Some oil-absorbing and separation membranes made by electrospinning are superhydrophobic. They possess stable properties and can be regenerated after being used. Electrospinning is triggered by the electric field force acting on a polymer solution or melt. During the spinning process, the applied voltage causes fibers to be pulled from the droplet at the end of a syringe tip onto a grounded electrically conductive collector. Fibers can be made much thinner through electrospinning than other methods because the electrified polymer solution droplets with significantly reduced surface tension can be drawn into very thin fibers. The voltage should be relatively high. Typical values are in the range from 5.0 to 30.0 kV. The diameter of the fiber is inversely proportional to the applied voltage. For some electrospinning facilities with vertically aligned jets, the nanofibers are drawn toward collectors by both gravitational and electrostatic forces. Generally, the spun nanofibers are randomly arranged on collectors and form meshed networks. However, the fiber collectors could be metallic screens, rotating disks, or mandrels. During stable electrospinning, the solvent evaporates rapidly. This allows the formation of charged polymer fibers and the fibers accumulate on the collector. But if the concentration of the polymer solution is relatively low, bead formation could be observed. A very low polymer concentration or low viscosity of the fluid results in electrospraying during which polymer liquid drops fly out and sporadically deposit onto the collectors. It must be pointed out that nanoparticles may be added into polymer solutions so that the spun products are composite nanofibers.

There are many factors influencing the electrospinning process besides the above-mentioned DC voltage level and viscosity of the polymer solution. For example, the electrical conductivity of the polymer solution is one of the significant factors. Still, some other factors include the temperature, relative humidity of the environment, fluid flow rate, and the distance between the tip of the injection needle and the collecting target. Solution conductivity not only affects the Taylor cone formation but also controls the diameter of nanofibers. For a solution with very low conductivity, the surface of the droplet will not have enough charge to form a Taylor cone. Consequently, no electrospinning will take place. Increasing the conductivity of the solution over a critical value is necessary to trigger the electrospinning. A high solution conductivity will not only promote the charge accumulation on the surface of the droplet to form Taylor cone but also cause a decrease in the nanofiber diameter. An ideal dielectric polymer solution or insulating polymer solution will not produce sufficient charges in the solution to move toward the surface of the fluid. As a result, the electrostatic force generated by the applied electric field will not be big enough to form a Taylor cone. Taylor cone formation is the prerequisite for initiating the electrospinning process. On the contrary, a conductive polymer solution will have sufficient free charges to move onto the surface of the fluid and form a Taylor cone easily to initiate the electrospinning. The conductivity of a polymer solution could be increased by adding an appropriate inorganic salt into the solution. A tiny amount of addition of LiCl, KCl, or NaCl works well for this purpose. The addition of such a salt affects the electrospinning process in two ways. One is the increase in the number of ions in the polymer solution, which helps increase the surface charge density of the fluid. The electrostatic force generated along the direction of the applied DC electric field increases. The other is to reduce the tendency of the transverse motion of nanofibers because the higher the conductivity of the polymer solution, the lower the tangential electric field along the surface of the fluid. Therefore the nanofibers can be better stretched and move directly toward the collecting target.

The distance is also a fundamental parameter that determines the morphology of the spun nanofiber. It controls the fiber deposition time, solvent evaporation rate, and the fiber whipping or instability interval. The nanofiber morphology could be easily affected by this distance. The distance between the spinneret tip and collector should be long enough to generate the required morphology and maintain the quality of spun fibers. It must be indicated that the critical value for the distance between the metallic needle tip and collector varies with the kind of polymer system.

Electrospinning as a mature technology has found various applications in making tissue scaffolds, masks, wound dressings, wearable technology sensors, and so on. It is also used for making composite nanofibers for oil spill cleaning. For example, a polystyrene (PS) composite nanofiber with embedded carbon nanotubes (CNTs) for oil-spill clean-ups was processed (Parangusan et al., 2019). Both carbon and PS are hydrophobic. They possess high surface areas in the nanostructure forms. The composite nanofiber with 0.5 wt.% of CNT was tested in treating oil-contaminated water. The oil absorption capacity was found to be 11.80 g/g. For PS nanofiber without the addition of CNT, the oil absorption capacity is 5.14 wt.%. CNT can increase the contact angle between the composite and water. Without CNT, the pure PS nanofiber showed a contact angle of 140 degrees with water. After the addition of 0.2 wt.% CNT, the contact angle was increased to 150 degrees. Increasing the content of CNT to 0.5 wt.% led to the contact angle reaching approximately 157 degrees. The maximum contact angle was found close to 160 degrees with 1.0 wt.% addition of CNT. The reason for the contact angle increase is due to the roughness change caused by the CNT. The procedures are shown in Fig. 4.1. In brief, PS pellets were dissolved in dimethylformamide (DMF) by magnetic stirring for 3 h at 60°C to form an 8% polymer solution (Parangusan et al., 2019). CNT was ultrasonically dispersed in DMF and then added to the PS solutions. Magnetic stirring was done overnight to obtain homogeneous dispersions of PS/CNT in DMF. The solution was electrospun at 12 kV with a solution-delivering rate of 1 mL/h. The spun nanofibers were collected on a cylindrical rotor moving at 500 rpm. The morphology of spun nanofiber is dependent on the distance between the needle tip to the collector. At a smaller needle tip-to-collector distance of 5 and 10 cm, defect-free fibers were obtained. However, at longer distances of 12 and 15 cm, beads formed on the fibers. The morphology of the fibers also depends on the solvent type and the applied voltage level. The composite nanofiber is suitable for oil cleaning because it is chemically inert. Its density is relatively low. Recyclability and selectivity are other advantages for oil absorption.

Carbon black particle-containing polymer nanofibers were made via electrospinning as can be found from the work performed (Elmaghraby et al., 2022). A mixture solution comprising cellulose acetate (CA), cellulose nitrate, and carbon black (CB) particles was mixed and spun into composite nanofibers with diameters ranging from 300 nm to 700 nm. The process is shown in Fig. 4.2. Cellulose nitrate is a cellulose derivative made by treating cellulose with strong nitric acid and replacing the −OH groups in the cellulose molecule with −ONO$_2$ groups (Chai et al., 2019). CA is made by esterifying cellulose. CA is a biodegradable polymer that has a negatively charged surface (Puls et al., 2011). The formation of CA-CN/CB solution

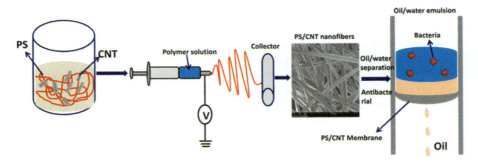

Figure 4.1 Schematic drawings and a scanning electron microscopic (SEM) image showing the carbon nanotube containing polystyrene matrix composite nanofiber mat made via electrospinning. The last drawing shows oil cleaning and bacteria filtering by the nanofiber membrane.
Source: Reproduced from Parangusan, H., Ponnamma, D., Hassan, M. K., Adham, S., & Al-Maadeed, M. A. (2019). Designing carbon nanotube-based oil absorbing membranes from gamma irradiated and electrospun polystyrene nanocomposites. *Materials*, *12*(5), 709. https://doi.org/10.3390/ma12050709

Processing of nanofibers for oil-spill cleaning application 73

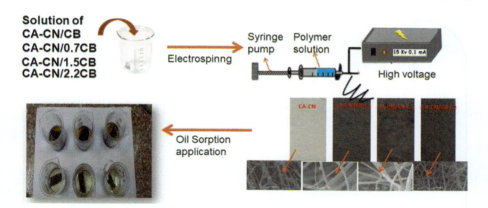

Figure 4.2 Schematic drawings, photos, and scanning electron microscopic (SEM) images showing electrospinning carbon black particles containing cellulose acetate and cellulose nitrate mixture solution to make nanofibers for oil sorption application. *CB*, Carbon black; *CA*, cellulose acetate; *CN*, cellulose nitrate.
Source: Reproduced from Elmaghraby, N. A., Omer, A. M., Kenawy, E. R., Gaber, M., & El Nemr, A. (2022). Fabrication of cellulose acetate/cellulose nitrate/carbon black nanofiber composite for oil spill treatment. *Biomass Conversion and Biorefinery*. https://doi.org/10.1007/s13399-022-03506-w

followed several steps. First, CA-CN polymer with a concentration of wt.10% was prepared by dissolving CA and CN at a mass ratio of 3:2 in a mixture solvent of tetrahydrofuran (THF) and DMF. The THF to DMF volume ratio is 4:1. Polyethylene glycol (PEG) was added as well to reduce the surface tension of the polymer solution. PEG was added to the solution with a PEG to CA mass ratio of 1:1. The CA-CN/CB nanofiber precursors were made by adding different amounts of CB into the polymer solution with the CB weight concentrations ranging from 0.7% to 2.2%. Electrospinning of the prepared CA-CN/CB solutions was done by using a 20 mL syringe with a blunt-end needle. The distance between the needle tip and the collector was 10 cm. The potential at the needle tip was + 26 kV and a negative potential of − 10 kV was kept at the collector. The obtained nanofibers were rinsed in water and ethanol to remove the PEG. Air drying the nanofibers at 50°C in an oven for 24 h was conducted to get the final product for oil absorption and cleaning tests. The oil adsorption maximum capacity for the composite nanofiber with 2.2 wt.% CB reached 13.67 g/g (Elmaghraby et al., 2022).

Modified electroplating was also used for oil-cleaning nanofiber processing. A water-mediated electrospinning approach can be considered as an example (Prasad et al., 2023). A highly intraporous fibrous poly(vinylidene fluoride) (PVDF) membrane was obtained. As shown in Fig. 4.3, the PVDF powder was dissolved into a mixture solvent of DMF and acetone. The ratio of DMF to acetone is 4:6. In a typical experiment, the PVDF concentration is 16 wt.% and the obtained nanofiber membrane was named M-16. The prepared PVDF solution was transferred into a 20 mL plastic syringe connected with a stainless-steel needle as the spinneret. A

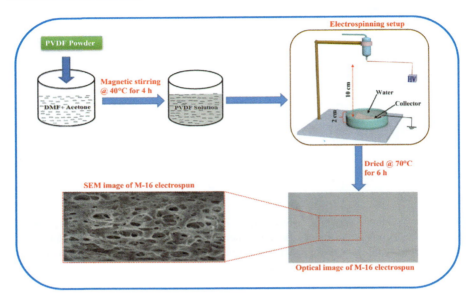

Figure 4.3 Schematic drawings and a scanning electron microscopic (SEM) image showing water-mediated electrospinning. M-16 refers to the nanofiber membrane made from 16 wt.% PVDF polymer solution. *PVDF*, Poly(vinylidene fluoride).
Source: Reproduced from Prasad, G., Lin, X., Liang, J., Yao, Y., Tao, T., Liang, B., & Lu, S. G. (2023). Fabrication of intra porous PVDF fibers and their applications for heavy metal removal, oil absorption and piezoelectric sensors. *Journal of Materiomics, 9*(1), 174–182. https://doi.org/10.1016/j.jmat.2022.08.003

digitally controlled fluid pump was used to deliver the PVDF polymer solution at a constant feed rate of 1 mL/h. A glass petri dish (200 mm in diameter and 30 mm in height) with the bottom covered with aluminum foil was filled with 20 mm deep deionized (DI) water. The petri dish along with the aluminum and DI water was grounded. The stainless-steel needle was positive, and the voltage added was +18 kV. During the electrospinning, the PVDF solution was drawn into nanofibers and the fibers were dispersed into DI water (Prasad et al., 2023).

With the accumulation of the PVDF nanofibers, a porous membrane was obtained. The membrane was dried at 60°C for 12 h to remove residues of solvent mixture and water before oil absorption characterization (Prasad et al., 2023). The water contact angle (WCA) measurement indicates that the PVDF nanofiber membrane is highly hydrophobic with the WCA ranging from 130 degrees to 138 degrees. The water-mediated electrospinning offers some advantages. For example, the solvent mixture can be washed away quickly by DI water. This prevents the fusion of fibers at the cross-points.

The above water-mediated electrospinning work (Prasad et al., 2023) just illustrated the modified electrospinning for pure polymer nanofiber membrane processing. However, it is possible to spin particle-containing nanofibers using sol–gel electrospinning, another modified electrospinning method. A silica nanoparticle/cellulose diacetate

(CDA) hybrid nanofiber was processed by the sol−gel electrospinning approach (Pirzada et al., 2020). As well known, during an ordinary electrospinning process, the conductivity and viscosity of the solution, chamber humidity and temperature of the environment, and applied voltage are important factors that affect the size and morphology of the spun nanofibers. In addition to those factors, one must consider the effects of gelation kinetics and the reaction between the sol and the polymer selected for the sol−gel electrospinning (Pirzada et al., 2012). To control the hydrolysis and condensation of tetraethylorthosilicate (TEOS) to generate the silica, hydrochloric acid (HCl) was used. The acid promotes the hydrolysis reaction while suppressing the condensation to allow the formation of long chains of siloxane network (Pirzada & Shah, 2014). The decrease in the condensation reaction rate resulted in a slower gelation process. This increased the spinnability of the silica-containing mixture within a period. To improve the compatibility of the silica sol with nonaqueous solvents such as dimethylacetamide (DMAC) and acetone for CDA, N,N-dimethylformamide (DMF)/water mixture instead of ethanol/water mixture was used as the solvents for TEOS hydrolysis and condensation. This provided the required sol−gel reaction environment, resulting in bead-free and flexible hybrid nanofibers consisting of uniformly distributed silica within the networks of CDA. The nanofibers contain a relatively low content of silica to maintain flexibility, mechanical strength, and thermal stability (Pirzada et al., 2020).

Gel electrospinning can produce nanofibers with special structures. For example, the titanium dioxide nanoparticle beads-on-PVDF nanofiber string structure was illustrated (Wang et al., 2016). Such a special structure shows the controllable oil/water separation functions. The as-made fiber mat with hierarchical roughness is highly hydrophobic. Oil can be absorbed quickly into the mat as shown by the drawing in the upper left corner of Fig. 4.4. If the beads-on-string nanofiber was illuminated by UV or sunlight, the filter made from the nanofiber absorbed water and blocked the oil. This is because of the photocatalytic property of the titanium dioxide nanoparticles inside and at the surface of the nanofiber, which allowed the water drops to pass through. After heat treatment, the hydrophobicity of the filter was recovered. Such a peculiar transition makes the fiber mat a smart product that is suitable for underwater oil/water separation applications.

The processing of the smart beads-on-string structure is divided into several steps, as illustrated in Fig. 4.5 (Wang et al., 2016). The TiO_2 gel was synthesized by the condensation of TiO_2 sol, which was prepared by the hydrolysis of a titanium tetra-isopropoxide (TTIP) hydrosol. The TTIP (4 g) was first added dropwise into a 10 mL solution containing 1 wt.% HCl and 0.5 wt.% HAc under continuous stirring to form a hydrosol. Then, the hydrosol was slowly dropped into a beaker with the addition of 90°C warm water. It took 4 h to complete this step. Then the obtained milk-like sol was cooled down to 60°C. After that, a vigorous stirring was maintained for 15 h until a transparent bluish sol was obtained. The mixture was then cooled down to room temperature and stored for 2 weeks. Eventually, the sol was obtained with a 10 wt.% solid content of TiO_2. The TiO_2 sol was condensed by evaporating the water to produce a 32 wt.% solid content of TiO_2.

According to Wang et al. (2016), the solution for the electrospinning was made by adding the TiO_2 nanogel (0.25 g) into 10 mL of DMAC/acetone first. The ratio

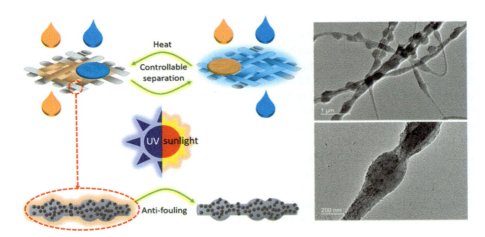

Figure 4.4 Schematic drawings and scanning electron microscopic (SEM) images showing the sol−gel spun beads-on-chain titanium oxide/PVDF composite nanofiber. The yellow drops are oil. The blue drops are water. In the right-hand side images, the beads contain titanium dioxide nanoparticles and particle clusters. The string refers to the PVDF nanofiber. *PVDF*, Poly(vinylidene fluoride).
Source: Reproduced with permission from Wang, Y., Lai, C., Wang, X., Liu, Y., Hu, H., Guo, Y., Ma, K., Fei, B., & Xin, J. H. (2016). Beads-on-string structured nanofibers for smart and reversible oil/water separation with outstanding antifouling property. *ACS Applied Materials and Interfaces*, 8(38), 25612−25620. https://doi.org/10.1021/acsami.6b08747

of DMAC to acetone is 3 to 1 in weight. The titanium dioxide gel-containing solution was ultrasonically treated for 2 h. Then, 1 g of PVDF powder was added to the gel solution at the ambient temperature of 23°C under continuous magnetic stirring for 12 h to produce the electrospinning solution. The solution was electrospun into nanofibers at the room temperature of 23°C. During the electrospinning, a positive voltage of 15 kV was applied to the needle and a negative voltage of 2 kV was imposed on the copper mesh collector. The distance between the spinneret and the collector was 20 cm. The flow rate of the solution was kept at 1 mL/h using a syringe pump. The relative humidity was about 60%. The as-spun nanofiber contains 8 wt.% of TiO_2 in the PVDF matrix. Following the electrospinning, the nanofiber was dried at 80°C in air to remove the solvent residues.

4.2.2 Vacuum filtration

Vacuum filtration is the one-step and simple method. It has the advantage of being cost-effective. A free-standing carbon nanofiber (CNF)-polydimethylsiloxane (PDMS) nanocomposite block was obtained using this approach (Abdulhussein et al., 2018). The composite nanofiber block showed excellent superhydrophobicity and superoleophilicity. Its WCA is as high as 163 degrees and its oil contact angle is close to 0 degrees. Also demonstrated in the paper are the high mechanical

Processing of nanofibers for oil-spill cleaning application 77

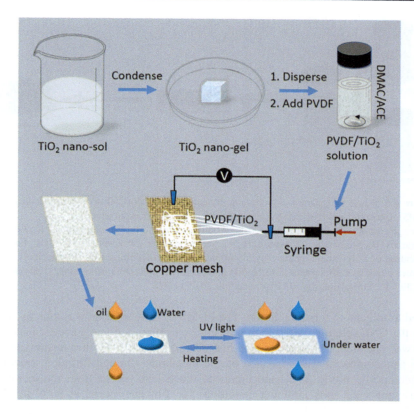

Figure 4.5 Schematic drawing showing the procedures for the sol—gel electrospinning of titanium dioxide nanoparticle beads-on-PVDF nanofiber. The sol was generated by the hydrolysis of titanium tetra-isopropoxide. *PVDF*, Poly(vinylidene fluoride).
Source: Reproduced with permission from Wang, Y., Lai, C., Wang, X., Liu, Y., Hu, H., Guo, Y., Ma, K., Fei, B., & Xin, J. H. (2016). Beads-on-string structured nanofibers for smart and reversible oil/water separation with outstanding antifouling property. *ACS Applied Materials and Interfaces*, *8*(38), 25612—25620. https://doi.org/10.1021/acsami.6b08747

stability, good absorbance capacity, and recyclability. Rapid and selective absorption of several types of oils and organic liquids floating on the water surface was demonstrated. The nanofiber block also shows the strong absorption of heavy oils sunken under water. The CNF/PDMS block is considered a promising material for practical applications of oil-spill cleanup. The processing can be briefly summarized as follows. The PDMS and curing agent (with a volume ratio of 10:1) were mixed to make the elastomer. The nanofibers were hollow with a carbon layer on the surface of the graphitic tubular core fiber. The diameter of the CNF was about 150 nm, and the estimated length was approximately 100 μm. PVDF microfiltration membrane with 0.45 μm was used for the vacuum infiltration experiment. The CNF was dispersed in solvents under ultrasonication. PDMS and curing agent were

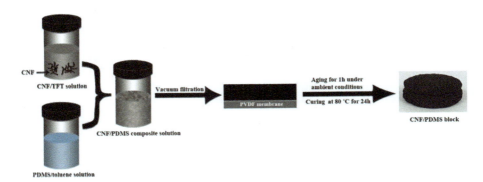

Figure 4.6 Schematic diagram of processing the carbon nanofiber (CNF)-polydimethylsiloxane (PDMS) nanocomposite block via vacuum infiltration. *TFT*, Trifluorotoluene.
Source: Reproduced with permission from Abdulhussein, A. T., Kannarpady, G. K., Ghosh, A., Barnes, B., Steiner, R. C., Mulon, P. Y., Anderson, D. E., & Biris, A. S. (2018). Facile fabrication of a free-standing superhydrophobic and superoleophilic carbon nanofiber-polymer block that effectively absorbs oils and chemical pollutants from water. *Vacuum*, *149*, 39−47. https://doi.org/10.1016/j.vacuum.2017.11.028

added to toluene to form a solution. The PDMS-toluene solution was directly mixed with the CNF solution. After 30 min of ultrasonication, a homogeneous black solution was obtained. The solution was filtered using the PVDF filter membrane. The CNF nanocomposite was deposited on the membrane through filtration, creating a nanocomposite block. The nanocomposite block was peeled from the membrane gently and aged under ambient conditions for 1 h, then cured at 80°C for 24 h. Fig. 4.6 schematically shows the process for making the CNF nanocomposite block (Abdulhussein et al., 2018).

4.2.3 Chemical oxidation

Micro-cellular polymer foam-supported polyaniline (PANI)-nanofiber was made by chemical oxidation as introduced (Ghorbankhani & Reza Zahedi, 2022). The fabricated samples consist of polyurethane (PU) foams and embedded PANI nanofibers. The samples were made for cleaning oil stains and separating oil from water. A field test on oil absorption and sea cleaning by the nanofiber foams was performed. The morphology of the PANI nanofibers was optimized to improve the oil absorption efficiency. The optimized sample of PU foam containing PANI nanofibers was formed with a molar ratio of aniline/ammonium persulfate (APS) of 1.43, aniline/acid ratio of 0.105, and nanofiber concentration of 0.433 in the foam. To make PANI, aniline was used as the raw material for the nanofibers, and APS was the oxidant. The chemical oxidation reaction of aniline and APS in an acidic environment produced the PANI nanofibers. Hydroxyl ammonium was used as the doping agent for the PANI nanofibers. Two components, polyol and isocyanate, at a mixing ratio of 2:1 in weight, were used to prepare the PU foam. In a typical experiment on pure PU foaming (Ghorbankhani & Reza Zahedi, 2022), 5 mg of polyol and 2.5 mg of isocyanate were mixed in a mold. The polymerization reaction began after 15–20 s, and the mixture began to foam until all the raw materials were consumed in the mold. After 10 min, the foam was taken out of the mold. The PU foam with 0.2 wt.% PANI nanofibers was made as follows. The polyol was diluted with acetone to reduce its viscosity. Then, the PANI nanofibers were added to the dilution of polyol. Following that step, isocyanate was added to the polyol and PANI complex to form the foam (Ghorbankhani & Reza Zahedi, 2022). The oil absorption coefficient achieved 5.04 g/g. The definition of the oil absorption coefficient was defined earlier (Stazi et al., 2018).

4.2.4 Delignification

Partial or full removal of lignin from the scaffold of wood without significantly changing the hierarchically aligned cellulosic structure is a scalable and economical way for obtaining oil-cleaning nanofibers. A hydrophobic oil–water separation composite consisting of cellulose nanofiber from the delignification of porous wood (PW) was illustrated by Chu et al. (2022). The nanofiber was decorated with a magnetic nickel (Ni) layer. This Ni layer was made by polymer-assisted metal deposition (PAMD). Finally, a hydrophobic PDMS coating was applied to the nanofiber through a dip coating procedure. Since the natural cell wall structure made of the nanofiber was maintained, the three-dimensional porous hydrophobic and oleophilic materials were obtained. Fig. 4.7 schematically illustrates the processing steps, intermediate PW, and the final product of PDMS-Ni-PW (Chu et al., 2022).

The electroless deposited nickel layer allowed the convenient magnetic recovery of the nanofiber filter made from the PDMS-Ni-PW during oil cleaning application. The PDMS-Ni-PW possessed excellent oil adsorption capacity, which was due to its high porosity, hydrophobicity, and lipophilicity. It also exhibited high cycle compressibility, maintaining its adsorption capacity after 200 cycles. An oil-collecting apparatus was designed to continuously separate various oils from water. During oil/water separation,

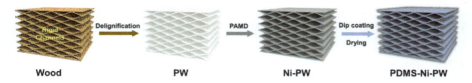

Figure 4.7 Schematic showing the three-dimensional cellulose nanofiber structure made from delignification followed by surface modification. *PW*, Porous wood; *PAMD*, polymer-assisted metal deposition.
Source: Reproduced with permission from Chu, Z., Li, Y., Zhou, A., Zhang, L., Zhang, X., Yang, Y., & Yang, Z. (2022). Polydimethylsiloxane-decorated magnetic cellulose nanofiber composite for highly efficient oil-water separation. *Carbohydrate Polymers*, 277, 118787. https://doi.org/10.1016/j.carbpol.2021.118787

unidirectional liquid transport occurred in the adsorbent, which allows rapid recovery of oil. All these properties make the PDMS and nickel-modified cellulose nanofiber ideal for oil-water separation (Chu et al., 2022).

The details about the delignification process are demonstrated by Chu et al. (2020). The generated product was a nano wood (NW). The NW was prepared from basswood as the starting material. The wood was cut into blocks. These wooden blocks were immersed in 100 mL of 20% (in volume) methanol solution containing 5 g NaOH and 15 g Na_2SO_3 for 4 h at 170°C. Then the wooden blocks were soaked in 60 mL DI water containing 0.5 g acetic acid and 0.6 g $NaClO_2$ until the color of the blocks completely turned white. After being cleaned with DI water, the wet blocks were freeze-dried. To obtain oxidized specimens, the dry wooden blocks with a weight of 1 g were placed in 100 mL DI water with a pH value of 10 adjusted with 0.5 M NaOH. To initiate the oxidation reaction, 5 mmol NaClO was added to the solution. During the reaction, the pH value of the solution was maintained at 10 with the addition of 0.5 M NaOH. After 2.5 h, the oxidation reaction stopped. The pH value of the solution was adjusted back to 7 by adding 0.5 M HCl into it. Continued washing of the specimens with 0.5 M HCl was performed. The wet blocks were rinsed with DI water until the pH value reached 5 or above. Next, the obtained NW containing cellulose nanofiber was freeze-dried for polyethyleneimine (PEI) grafting, as schematically shown in Fig. 4.8A. The cell-wall microstructure of the NW made of cellulose nanofibers can be seen from the scanning electron microscopic (SEM) images presented in Fig. 4.8B and C (Chu et al., 2020).

As also mentioned by Chu et al. (2020), PEI contains many primary, secondary, and tertiary amino groups; it can adsorb heavy metal ions. PEI, a strengthening agent, in cellulose is widely used in the papermaking industry. It is also used as an additive in many other products including detergents, adhesives, and cosmetics. The PEI can be grafted to cellulose nanofibers by immersing the dry NW into the PEI methanol solutions with varying concentrations at 35°C for 24 h during which the NW was gently pressed periodically. After that, the obtained PEI-NW was taken out and washed several times with DI water to remove the PEI residue. Then, the NW was immersed into the glutaraldehyde solutions with different concentrations and gently squeezed for 1 h at room temperature to cross-link the NW and PEI. The generated PEI-NW was washed thoroughly with DI water to remove unreacted glutaraldehyde and then freeze-dried. The morphology of the PEI-NW is revealed by SEM images in Fig. 4.8D and E (Chu et al., 2020).

4.2.5 Deacetylation

Deacetylation is the removal of an acetyl group from an organic chemical compound. CA can be converted to cellulose II or deacetylated cellulose acetate (d-CA) through a deacetylation process using NaOH solution. In the work performed by Wang et al. (2020), the d-CA membrane was generated by immersing CA nanofiber membranes in a 0.5 M NaOH aqueous solution. The deacetylation time was 3 h. Fig. 4.9A shows the initial CA nanofiber made by electrospinning. Fig. 4.9B illustrates the deacetylation process. The obtained d-AC was rinsed multiple times in DI

Processing of nanofibers for oil-spill cleaning application 81

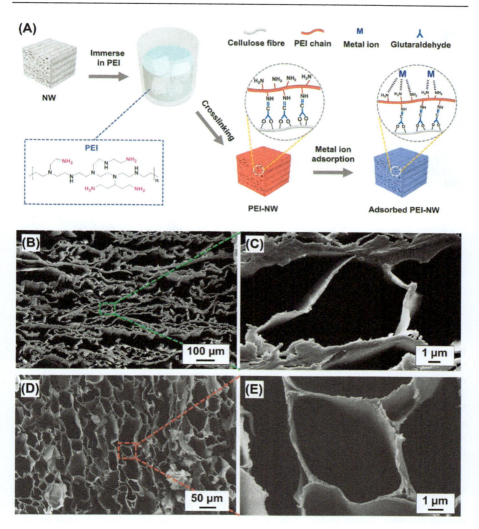

Figure 4.8 Schematic showing the fabrication process (A), SEM images at different magnifications showing the top views of the nano wood (B and C), and the polyethyleneimine-treated nano wood (D and E). *NW*, nano wood; *PEI*, polyethyleneimine; *SEM*, Scanning electron microscopic.
Source: Reproduced with permission from Chu, Z., Zheng, P., Yang, Y., Wang, C., & Yang, Z. (2020). Compressible nanowood/polymer composite adsorbents for wastewater purification applications. *Composites Science and Technology*, *198*, 108320. https://doi.org/10.1016/j.compscitech.2020.108320

water and dried in a vacuum oven. The oil/water separation performance of the membrane made from the d-AC was evaluated to show the multifunctionalities of the nanofiber. The d-CA nanofiber membrane was used for water-removal or oil removal from oil/water mixtures, as shown in Fig. 4.9C. The membrane can also

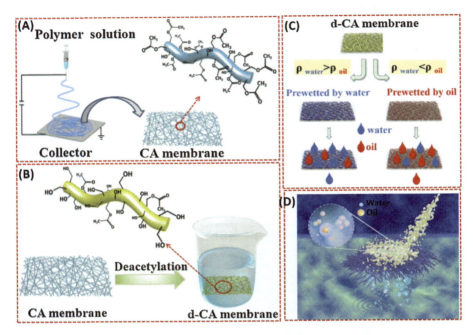

Figure 4.9 Illustration showing (A) cellulose acetate nanofiber preparation, (B) deacetylation, (C) oil/water separation from mixture, and (D) oil/water separation from emulsion. *CA*, cellulose acetate; *d-CA*, Deacetylated cellulose acetate.
Source: Reproduced with permission from Wang, W., Lin, J., Cheng, J., Cui, Z., Si, J., Wang, Q., Peng, X., & Turng, L. S. (2020). Dual super-amphiphilic modified cellulose acetate nanofiber membranes with highly efficient oil/water separation and excellent antifouling properties. *Journal of Hazardous Materials*, *385*. https://doi.org/10.1016/j.jhazmat.2019.121582

separate oil and water in emulsified oil/water, as shown in Fig. 4.9D. It must be indicated that only gravity served as the driving force for the oil/water separation during the tests shown in Fig. 4.9C and D. The d-CA nanofiber membranes possess a very high separation flux of 38,000 L/(m$^2 \cdot$h). The separation efficiency reached 99.97% for the separation of the chloroform/water mixture. It was reported that the separation flux was several times higher than that of commercial CA membranes (Wang et al., 2020). The new d-CA nanofiber membrane has excellent antipollution and self-cleaning capabilities. The cyclic stability and reusability of the membrane were also confirmed. The d-CA nanofiber membrane may find potential applications in chemical plants, textile mills, and the food industry. It was also recommended for cleaning offshore oil spills and separating oil from water.

During deacetylating, the acetyl group was replaced by the hydroxyl group (He, 2017). This reaction is illustrated in Fig. 4.10A. The deacetylation was dependent on the solvent type, NaOH concentration, and reaction time. In the initially carried out deacetylation experiments (He, 2017), a 0.2 M NaOH solution in 96 vol.

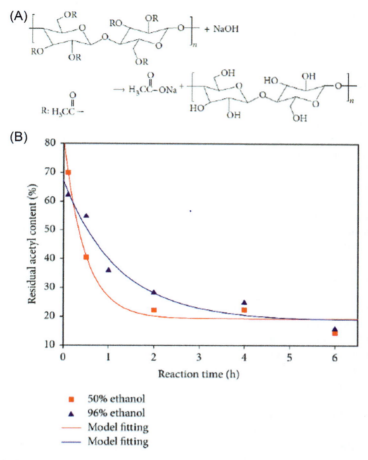

Figure 4.10 Deacetylation reaction of cellulose acetate in NaOH solution: (A) reaction formula in NaOH solution, (B) time-dependent residual acetyl content in the d-CA nanofiber (0.5 M NaOH in ethanol solutions with different concentrations). *d-CA*, Deacetylated cellulose acetate.
Source: Modified from He, X. (2017). Optimization of deacetylation process for regenerated cellulose hollow fiber membranes. *International Journal of Polymer Science*. https://doi.org/10.1155/2017/3125413

% ethanol at different reaction times was used to determine the complete deacetylation time. Then, 0.5 M NaOH solutions with different ethanol contents (50 vol.% and 96 vol.%) were used to investigate the effect of the solution on the deacetylation reaction kinetics. The residual acetyl content after deacetylation was determined by Fourier-transform infrared spectroscopy to assess the reaction rate. The reaction is very fast in the 0.5 M NaOH solution. Fig. 4.10B shows the relationship between the residual acetyl content in the CA nanofiber membranes and reaction time at different ethanol concentrations. It can be found that the reaction rate in a

50 vol.% ethanol solution was faster than that in a 96 vol.% ethanol solution (He, 2017). The completion of the deacetylation reaction takes about 2−4 hours in the 0.5 M NaOH solutions with different volume contents.

4.2.6 Aerogel formation

Aerogels are porous ultralight materials derived from gels. The liquid components for the gels are replaced by air without significant change in the gels' structures. Aerogels are prepared by extracting the liquid components of the gels through supercritical drying or freeze-drying. Because the liquid components are extracted out slowly, the structures of the solid matrices in the gels are kept. It must be mentioned that conventional evaporation causes the solid structures to collapse from capillary actions. Aerogels have found wide applications for chemical sorption, energy storage, catalysis, sensing, and thermal insulation. There are two important classes of nanofiber aerogels for oil−water separation including synthetic polymer-based and sustainable source-based aerogels. In the first category, polymer nanofibers made from polyacrylonitrile and polyvinyl alcohol (Ma et al., 2022); PU and PVDF (Cui et al., 2021); and carbon nitride and oxide graphene added PVDF (Shi et al., 2019) are some of the examples.

A typical aerogel formation method for producing a cellulose nanofiber aerogel is shown in more detail (Laitinen et al., 2017). The objective of the work is to get a low-cost, ultralight, highly porous, hydrophobic, and reusable superabsorbent cellulose nanofiber aerogel from recycled waste papers using a simple, environmentally friendly nanofiber treatment involving deep eutectic solvent and freeze-drying. Nanofiber reclamation from recycled box board was performed. The silylation or hydrophobic modification on the surface of the claimed waste cellulose nanofibers was carried out. After freeze-drying, nanofiber sponges with a very low density of 0.0029 g/cm^3 and a relatively high porosity of up to 99.81% were generated. These sponges were reusable, and they exhibited strong sorption to various oils and organic solvents. Specifically, the absorption tests on the nanofiber aerogels showed that they possess high selectivity to the absorption of marine diesel oil from an oil/water mixture. The ultrahigh absorption capacity of 142.9 g/g was achieved. This value is much higher than those of the commercial absorbents made from polypropylene-based materials (with an absorption capacity of 8.1−24.6 g/g). The absorbed oil was ready to be recovered by simple mechanical squeezing. The reusability of the nanofiber sponges was also tested. After being used for absorbing oil and squeezing for 30 cycles, the sponges kept 71.4%− 81.0% capacity of a fresh absorbent. All these advantages make the aerogel sponges from recycled cellulose nanofiber super absorbents for cleaning the spills from oil and certain chemicals (Laitinen et al., 2017). Fig. 4.11 consists of several photos and an SEM image. The SEM image in the middle of this figure reveals the highly porous cell wall structure of the recycled and silylated nanofiber aerogel. The photos illustrate the nanofiber aerogel sponges. The left-hand side photo demonstrates the hydrophobic nature of the aerogel sponge. The two photos on the right side of the figure show the oil absorption performance of the sponge.

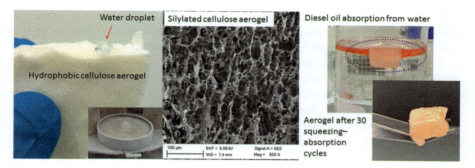

Figure 4.11 Stitched photos and SEM image showing the recycled and silylated nanofiber aerogel and its oil absorption. The absorbed oil was recovered by simple mechanical squeezing. *SEM*, Scanning electron microscopic.
Source: Reproduced with permission from Laitinen, O., Suopajärvi, T., Österberg, M., & Liimatainen, H. (2017). Hydrophobic, superabsorbing aerogels from choline chloride-based deep eutectic solvent pretreated and silylated cellulose nanofibrils for selective oil removal. *ACS Applied Materials and Interfaces*, 9(29), 25029−25037. https://doi.org/10.1021/acsami.7b06304

4.3 Classical theories on contact angles of water and oil drops to solid surfaces

Oil−water separation behavior is determined by the wettability at the interfaces between various phases including solid substrate (the composite aerogels in this case), vapor or gas (air), and liquid (oil or water). As the surface property of materials, the wettability of a liquid drop to the surface of a solid can be evaluated by the contact angle with the solid surface. A low contact angle of smaller than 90 degrees corresponds to the wetting state, while a high contact angle of larger than 90 degrees represents the unwetted state between the liquid and the solid interface. The value of the contact angle is determined by the well-known Young's Equation (Young, 1805). Considering a solid with an oleophobic surface underwater, the water−oil−solid interface exists. Thus the contact angle still can be found following the Young's equation for air−oil−solid one. The only change is the surface energy terms. Typically, the surface tension of oil is much lower than that of water. Therefore most hydrophilic surfaces can be made oleophobic at the solid−water−oil interface (Jung & Bhushan, 2009). For oil-water separation and self-cleaning of oil in a water environment, the oleophobic property is needed.

If the solid surface is rough, there are other models to describe the wetting states between a liquid drop and the solid surface. Among which Wenzel model (Wenzel, 1936) and the Cassie−Baxter model (Cassie & Baxter, 1944) have caught early attention. Without an air pocket as described by the Wentzel model, the contact angle of the liquid drop to the rough solid surface can be calculated using the surface roughness factor (Wenzel, 1936).

Wentzel model only considers the wetting to surface with uniform roughness details, which are smaller than the size of the liquid drop. Cassie and Baxter (1944) modified the Wentzel model to a solid surface with nonuniformity of roughness. In the Cassie—Baxter model, air bulbs are allowed to be trapped in the liquid-solid interfacial region. Under such conditions, the contact angle can be calculated from the solid—liquid contact area fraction in the region covered by the liquid drop and the air—liquid contact area fraction in the region covered by the liquid drop.

The wetting state of oil and water drops to solid surfaces is tunable by the environment of interfaces, materials' physical chemistry properties, and the geometrical roughness at the surface of the solids. Especially interesting is that an oil drop may show a low wetting angle at a nanofiber aerogel or membrane surface, but the nanofiber aerogel or membrane may be highly oleophobic in water. This is the foundation for achieving the antifouling and self-cleaning surface underwater for nanofiber materials. The super oleophobic property is also needed for efficient oil—water separation in many cases.

4.4 Conclusions

Several important polymer nanofibers for oil absorption and oil/water separation are presented. Their processing technologies are discussed. For the preparation of synthetic polymer nanofibers, electrospinning, in situ chemical oxidation reactions, and sol—gel spinning are the reliable technologies. Natural cellulose nanofiber can be processed by the delignification and deacetylation technologies, two mature and cost-effective methods. Applying coatings to nanofibers is very effective in changing the surface properties of the composites. Among various coating processes, silylation is especially useful for increasing the hydrophobicity of nanofibers. For controlling the microstructure with high porosity, aerogel formation is the most reliable method. Oriented porous cell-wall structures offer advantages including fast transport, good recovery capability, and high mechanical strength during the application in oil—water separation. Therefore unidirectional freeze-drying technology has been emphasized. There are many research activities focusing on the preparation of sustainable source-derived or bio-based composite nanofiber aerogels. Such natural composite nanofiber aerogels are highly absorptive, nontoxic, and recyclable. Wood, cotton, plant stems, and leaves are some of the most studied biomass sources for composite nanofiber aerogel manufacturing. Any cellulose and lignin-generating sources are considered to have value for oil-absorption nanofiber aerogel processing. Waste materials including used tire fabrics and recycled papers are studied for making oil sorption composite nanofiber aerogels. In view of the future research directions in the composite nanofiber field, carbon composite nanofibers with functional additives are promising because of the high-temperature stability and recyclability. Further studies on the durability of composite nanofibers under various spilled oil cleaning service conditions should be addressed. Cellulose nanofiber recovery from waste and/or recycled materials is the most economical way to generate high-performance oil sorbents.

Acknowledgments

"This research was performed under an appointment to the US Department of Homeland Security (DHS) Science & Technology (S&T) Directorate Office of University Programs Summer Research Team Program for Minority Serving Institutions, administered by the Oak Ridge Institute for Science and Education (ORISE) through an interagency agreement between the US Department of Energy (DOE) and DHS. ORISE is managed by ORAU under DOE contract number DE-SC0014664. All opinions expressed in this work are the author's and do not necessarily reflect the policies and views of DHS, DOE, or ORAU/ORISE." The support from the DHS Center of Excellence: Arctic Domain Awareness Center (ADAC) at the University of Alaska Anchorage is also gratefully acknowledged.

References

Abdulhussein, A. T., Kannarpady, G. K., Ghosh, A., Barnes, B., Steiner, R. C., Mulon, P. Y., Anderson, D. E., & Biris, A. S. (2018). Facile fabrication of a free-standing superhydrophobic and superoleophilic carbon nanofiber-polymer block that effectively absorbs oils and chemical pollutants from water. *Vacuum.*, *149*, 39−47. Available from https://doi.org/10.1016/j.vacuum.2017.11.028.

Cassie, A. B. D., & Baxter, S. (1944). Wettability of porous surfaces. *Transactions of the Faraday Society.*, *40*, 546−551. Available from https://doi.org/10.1039/tf9444000546.

Chai, H., Duan, Q., Jiang, L., Gong, L., Chen, H., & Sun, J. (2019). Theoretical and experimental study on the effect of nitrogen content on the thermal characteristics of nitrocellulose under low heating rates. *Cellulose.*, *26*(2), 763−776. Available from https://doi.org/10.1007/s10570-018-2100-0.

Chu, Z., Li, Y., Zhou, A., Zhang, L., Zhang, X., Yang, Y., & Yang, Z. (2022). Polydimethylsiloxane-decorated magnetic cellulose nanofiber composite for highly efficient oil-water separation. *Carbohydrate Polymers.*, *277*, 118787. Available from https://doi.org/10.1016/j.carbpol.2021.118787.

Chu, Z., Zheng, P., Yang, Y., Wang, C., & Yang, Z. (2020). Compressible nanowood/polymer composite adsorbents for wastewater purification applications. *Composites Science and Technology*, *198*, 108320. Available from https://doi.org/10.1016/j.compscitech.2020.108320.

Cui, Z., Wu, J., Chen, J., Wang, X., Si, J., & Wang, Q. (2021). Preparation of 3-D porous PVDF/TPU composite foam with superoleophilic/hydrophobicity for the efficient separation of oils and organics from water. *Journal of Materials Science.*, *56*(21), 12506−12523. Available from https://doi.org/10.1007/s10853-021-05995-y, http://www.springer.com/journal/10853.

Elmaghraby, N. A., Omer, A. M., Kenawy, E. R., Gaber, M., & El Nemr, A. (2022). Fabrication of cellulose acetate/cellulose nitrate/carbon black nanofiber composite for oil spill treatment. *Biomass Conversion and Biorefinery*. Available from https://doi.org/10.1007/s13399-022-03506-w.

Ghorbankhani, A., & Reza Zahedi, A. (2022). Micro-cellular polymer foam supported polyaniline-nanofiber: Eco-friendly tool for petroleum oil spill cleanup. *Journal of Cleaner Production.*, *368*, 133240. Available from https://doi.org/10.1016/j.jclepro.2022.133240.

He, X. (2017). Optimization of deacetylation process for regenerated cellulose hollow fiber membranes. *International Journal of Polymer Science*. Available from https://doi.org/10.1155/2017/3125413, http://www.hindawi.com/journals/ijps/.

Jung, Y. C., & Bhushan, B. (2009). Wetting behavior of water and oil droplets in three-phase interfaces for hydrophobicity/philicity and oleophobicity/philicity. *Langmuir, 25*(24), 14165−14173. Available from https://doi.org/10.1021/la901906h, http://pubs.acs.org/doi/pdfplus/10.1021/la901906h.

Laitinen, O., Suopajärvi, T., Österberg, M., & Liimatainen, H. (2017). Hydrophobic, superabsorbing aerogels from choline chloride-based deep eutectic solvent pretreated and silylated cellulose nanofibrils for selective oil removal. *ACS Applied Materials and Interfaces, 9* (29), 25029−25037. Available from https://doi.org/10.1021/acsami.7b06304, http://pubs.acs.org/journal/aamick.

Ma, W., Jiang, Z., Lu, T., Xiong, R., & Huang, C. (2022). Lightweight, elastic and superhydrophobic multifunctional nanofibrous aerogel for self-cleaning, oil/water separation and pressure sensing. *Chemical Engineering Journal., 430*, 132989. Available from https://doi.org/10.1016/j.cej.2021.132989.

Parangusan, H., Ponnamma, D., Hassan, M. K., Adham, S., & Al-Maadeed, M. A. (2019). Designing carbon nanotube-based oil absorbing membranes from gamma irradiated and electrospun polystyrene nanocomposites. *Materials., 12*(5), 709. Available from https://doi.org/10.3390/ma12050709, https://doi.org/10.3390/ma12050709.

Pirzada, T., Arvidson, S. A., Saquing, C. D., Shah, S. S., & Khan, S. A. (2012). Hybrid silica-PVA nanofibers via sol-gel electrospinning. *Langmuir: The ACS Journal of Surfaces and Colloids, 28*(13), 5834−5844. Available from https://doi.org/10.1021/la300049j.

Pirzada, T., Ashrafi, Z., Xie, W., & Khan, S. A. (2020). Cellulose silica hybrid nanofiber aerogels: From sol−gel electrospun nanofibers to multifunctional aerogels. *Advanced Functional Materials., 30*(5), 1907359. Available from https://doi.org/10.1002/adfm.201907359.

Pirzada, T., & Shah, S. S. (2014). Water-resistant poly(vinyl alcohol)-silica hybrids through sol-gel processing. *Chemical Engineering and Technology., 37*(4), 620−626. Available from https://doi.org/10.1002/ceat.201300295, http://www3.interscience.wiley.com/journal/10008333/home.

Prasad, G., Lin, X., Liang, J., Yao, Y., Tao, T., Liang, B., & Lu, S. G. (2023). Fabrication of intra porous PVDF fibers and their applications for heavy metal removal, oil absorption and piezoelectric sensors. *Journal of Materiomics., 9*(1), 174−182. Available from https://doi.org/10.1016/j.jmat.2022.08.003, https://www.journals.elsevier.com/journal-of-materiomics/.

Puls, J., Wilson, S. A., & Hölter, D. (2011). Degradation of cellulose acetate-based materials: A review. *Journal of Polymers and the Environment., 19*(1), 152−165. Available from https://doi.org/10.1007/s10924-010-0258-0.

Shi, Y., Huang, J., Zeng, G., Cheng, W., Hu, J., Shi, L., & Yi, K. (2019). Evaluation of self-cleaning performance of the modified g-C3N4 and GO based PVDF membrane toward oil-in-water separation under visible-light. *Chemosphere, 230*, 40−50. Available from https://doi.org/10.1016/j.chemosphere.2019.05.061, http://www.elsevier.com/locate/chemosphere.

Stazi, F., Tittarelli, F., Saltarelli, F., Chiappini, G., Morini, A., Cerri, G., & Lenci, S. (2018). Carbon nanofibers in polyurethane foams: Experimental evaluation of thermo-hygrometric and mechanical performance. *Polymer Testing., 67*, 234−245. Available from https://doi.org/10.1016/j.polymertesting.2018.01.028.

Wang, W., Lin, J., Cheng, J., Cui, Z., Si, J., Wang, Q., Peng, X., & Turng, L. S. (2020). Dual super-amphiphilic modified cellulose acetate nanofiber membranes with highly efficient oil/water separation and excellent antifouling properties. *Journal of Hazardous Materials, 385*, 121582. Available from https://doi.org/10.1016/j.jhazmat.2019.121582, http://www.elsevier.com/locate/jhazmat.2019.121582.

Wang, Y., Lai, C., Wang, X., Liu, Y., Hu, H., Guo, Y., Ma, K., Fei, B., & Xin, J. H. (2016). Beads-on-string structured nanofibers for smart and reversible oil/water separation with outstanding antifouling property. *ACS Applied Materials and Interfaces*, *8*(38), 25612−25620. Available from https://doi.org/10.1021/acsami.6b08747, http://pubs.acs.org/journal/aamick.

Wenzel, R. N. (1936). Resistance of solid surfaces to wetting by water. *Industrial and Engineering Chemistry*, *28*(8), 988−994. Available from https://doi.org/10.1021/ie50320a024.

Young, T. (1805). An essay on the cohesion of fluids. *Philosophical Transactions of the Royal Society of London.*, *95*, 65−87. Available from https://doi.org/10.1098/rstl.1805.0005.

A detail account of natural nanofibers (as chitin/chitosan, cellulose, gelatin, alginate, hyaluronic acid, fibrin, collagen, etc.)

Heri Septya Kusuma, Ganing Irbah Al Lantip and Xenna Mutiara
Department of Chemical Engineering, Faculty of Industrial Technology, Universitas Pembangunan Nasional "Veteran" Yogyakarta, Indonesia

5.1 Introduction

Nature provides a wide variety of materials, each with very broad benefits (Samyn, 2021). Examples of materials of natural origin are biopolymers, metal nanoparticles, proteins, and metal oxides (Wang et al., 2023). The implementation of these materials is very diverse and sustainable. Therefore in recent years, the development of natural materials has been very rapid, such as the development of fabrication methods and intelligent materials. In addition, several studies have developed interdisciplinary approaches between fields.

Biopolymers are materials that are often found in nature. Fig. 5.1 shows an example of a biopolymer from a natural origin. A widely used biopolymer product is nanofiber. Nanofibers are ultra-fine fibers that are usually cylindrical in shape and can be obtained from electrostatic forces (Estanqueiro et al., 2018). In addition, nanofibers have high porosity properties with very small interfiber pore diameters that prevent the entry of microorganisms, bacteria, viruses, and other particles by maintaining excellent vapor permeability (Vodseďálková et al., 2017). To obtain nanofibers used electrospinning method, this method is the most commonly used method. The electrospinning method forms nanofibers from polymer materials in the form of melts or solutions with high electric fields.

The polymers that can be used as materials for making nanofibers are usually derived from natural and synthetic polymers, as seen in Fig. 5.2 (Mohiti-Asli & Loboa, 2016). Natural polymers are usually produced from the extraction of plant or animal parts, as shown in Table 5.1. Natural polymers can be extracted by condensation polymerization to produce water as a by-product (Shrivastava, 2018; Thakur et al., 2022).

Natural polymers that are widely used as nanofibers are cellulose, collagen, and gelatin. Synthetic polymers are polymers produced artificially in the laboratory or man-made, usually derived from petroleum (Zainudin et al., 2020). In synthetic polymers, materials that can be used such as polylactic acid, polyglycolic acid, and

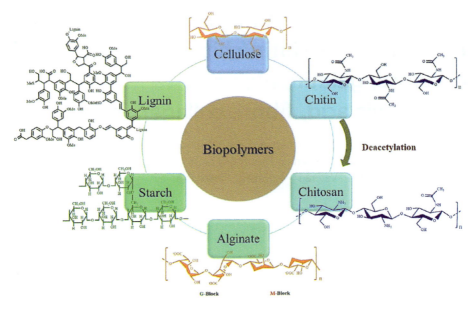

Figure 5.1 Scope of biopolymer (Wang et al., 2023).
Source: Data from Wang, X., Tarahomi, M., Sheibani, R., Xia, C., & Wang, W. (2023). Progresses in lignin, cellulose, starch, chitosan, chitin, alginate, and gum/carbon nanotube (nano)composites for environmental applications: A review. *International Journal of Biological Macromolecules*, 241, 124472. https://doi.org/10.1016/j.ijbiomac.2023.124472.

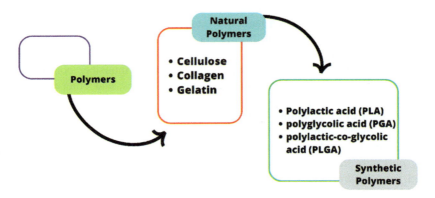

Figure 5.2 Natural and synthetic polymers.

polylactic-co-glycolic acid. The materials used from either natural polymers or synthetic polymers in the manufacture of nanofibers can have an effect on the diameter of the nanofibers produced (Rakhi et al., 2023).

This discussion will focus on nanofibers derived from natural polymers. Nanofibers themselves have benefits such as the prevention of bacterial invasion,

Table 5.1 Natural polymers and sources of getting it.

No	Natural polymers	Source
1	Cellulose	Plant cell wall
2	Collagen	Connective tissue found in human skin
3	Starch	Grains, cereals, and tubers
4	Chitin	Fishing industry waste
5	Alginate	Brown seaweed

very good use as a wound covering, tissue engineering, drug delivery, and others (Hayes & Su, 2011). However, these benefits depend on the materials used to make the nanofibers.

5.2 Chitin/chitosan

Chitin is a polysaccharide that makes up arthropods' exoskeleton, such as crab shells, shrimp, and insects (Jena et al., 2023). Chitin is the second most abundant biomass in nature (Zhao et al., 2023). The shell generally contains 15%–40% chitin (Lucas & Rode, 2023). Chitin monomers are n-acetyl-D-glucosamine (NAG) arranged repeatedly and bound by $\beta(1,4)$glycosidic bonds (Oyatogun et al., 2020). Chitin consists of three different allomorphs namely, α-chitin, β-chitin, and γ-chitin. The differences between these three allomorphs are the number of chains, the degree of hydration, and the size of the unit cell (Younes & Rinaudo, 2015). α-chitin is rigid and crystalline due to the presence of intramolecular and intermolecular hydrogen bonds. Intramolecular hydrogen bonds hold two dimer chains (N,N'diacetylchitobiose) in parallel directions, whereas intermolecular hydrogen bonds interact with N-H bonds and cause these chains to take the form of sheets. β-chitin has less rigid properties than α-chitin. This is due to the difference in the arrangement of polymer chains; there are intermolecular hydrogen bonds but not intramolecular hydrogen bonds, even though they have the same unit cells (Lucas & Rode, 2023). β-Chitin can be found in certain insects, cuttlefish, and certain fungi (Sulthan et al., 2023). While γ-chitin is a mixture of parallel and antiparallel chains in α-chitin and β-chitin. γ-chitin can be found in certain insects, fungi, and yeasts (Sulthan et al., 2023).

Chitin has several superior properties such as antitoxic, biodegradable, biocompatible, sustainable, and having low preparation costs. Therefore it is widely applied in the biomedical field for nanoparticles (Guo et al., 2019), nanofiber (Tao et al., 2020), hydrogel (Liao et al., 2022), and composites (Sun et al., 2022). In addition, chitin contains a fairly high nitrogen of 7.21% (dos Reis et al., 2023) so it can be applied as a metal absorbent and dye (Druzian et al., 2021). Chitin and its derivatives are antiinflammatory and good at being drug delivery agents in chemotherapy (Kumar et al., 2012) and good as an absorbent for heavy metals (Ahmed et al., 2020), uranium (Schleuter et al., 2013), and dyes (Dotto & Pinto, 2011).

Chitosan is a chitin derivative obtained from chitin extraction by the deacetylation process of the acetyl group in alkaline media. Chitosan consists of β-1,4-linked 2-amino-2-deoxy-β-d-glucose (deacetylated d-glucosamine) and NAG units (Kazemi Shariat Panahi et al., 2023). The difference between chitosan and chitin is the presence of a free amino group on the second carbon atom in the D-glucose unit, as seen in Fig. 5.3 (Zhou et al., 2021).

Chitosan gets a lot of attention because it has a significant share in the field of chemistry and nanomaterials (Eivazzadeh-Keihan et al., 2021). Its antitoxic and biodegradable properties make chitosan of high value (Zhou et al., 2021). In addition, its strong biological properties, such as antimicrobial, antifungal, and antioxidant, make chitosan widely used in the pharmaceutical and biomedical fields (Kazemi Shariat Panahi et al., 2023). Chitosan can also be applied as nanofibers, nanoparticles, films, nanocapsules, and hydrogels because it has biocompatibility properties and appropriate reactive functional groups (Yadav et al., 2023). Nanomaterials are very beneficial in drug delivery because they can penetrate the blood barrier in the brain (Dahaghin et al., 2021).

5.3 Cellulose

Cellulose is a linear polymer derived from glucose residues β-1,4 (Igarashi et al., 2012). Cellulose is also the most important polysaccharide found in plant cells (Godoy et al., 2018). So that the most cellulose can be taken from plants. Usually, cellulose taken from plants is carried out through the process of hydrolysis or the breakdown of molecules with water. Therefore during the hydrolysis process in cellulose, the β-1,4 bond will be released and will produce glucose, as seen in Fig. 5.4. In addition, cellulose can

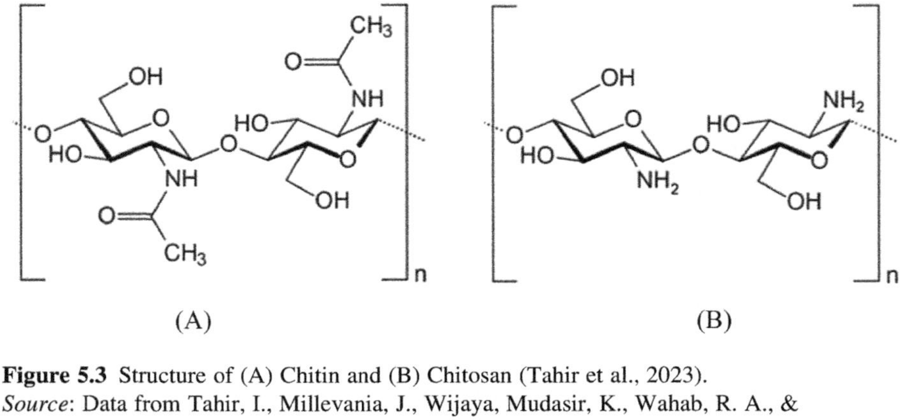

Figure 5.3 Structure of (A) Chitin and (B) Chitosan (Tahir et al., 2023).
Source: Data from Tahir, I., Millevania, J., Wijaya, Mudasir, K., Wahab, R. A., & Kurniawati, W. (2023). Optimization of thiamine chitosan nanoemulsion production using sonication treatment. *Results in Engineering, 17*, 100919. https://doi.org/10.1016/j.rineng.2023.100919.

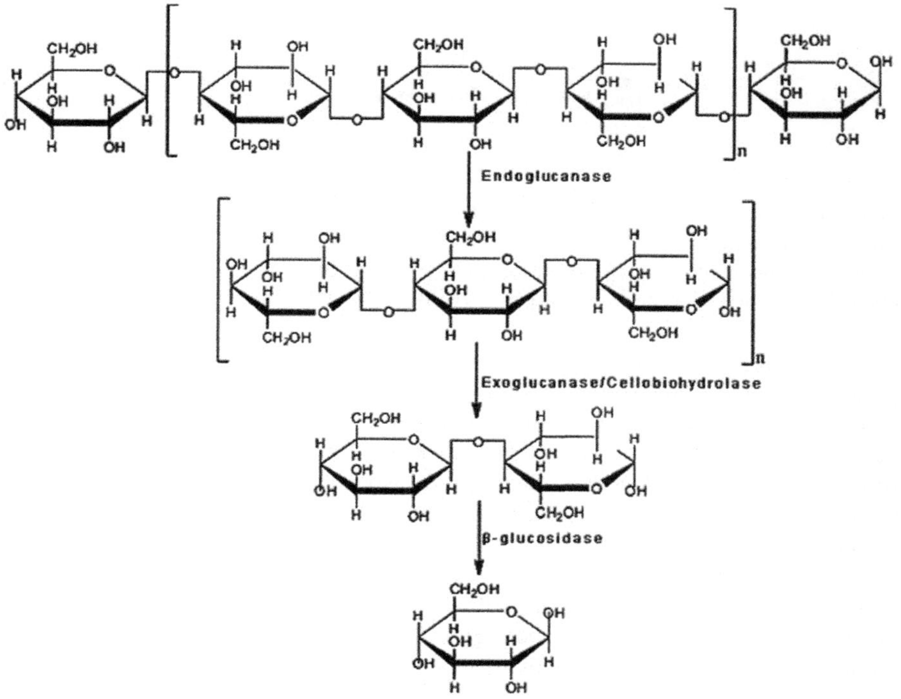

Figure 5.4 The structure of glucose production (Godoy et al., 2018).
Source: Data from Godoy, M. G., Amorim, G. M., Barreto, M. S., & Freire, D. M. G. (2018). Agricultural residues as animal feed: Protein enrichment and detoxification using solid-state fermentation. In A. Pandey, C. Larroche, & C. R. Soccol (Eds.), *Current developments in biotechnology and bioengineering* (pp. 235–256). Elsevier. https://doi.org/10.1016/B978-0-444-63990-5.00012-8.

also be produced using fermentation of certain bacteria. The metabolic process of microorganisms proceeds anaerobically or in the absence of oxygen (Rudin & Choi, 2013). The microorganisms used in fermentation are *Gluconacetobacter*, *Rhizobium*, *Agrobacterium*, and *Sarcina*. The resulting cellulose is known as bacterial cellulose by having no mixed compounds such as hemicellulose, lignin, or other extractives.

Microbial cellulose is the preferred type of cellulose over cellulose from plants due to its purity, crystallinity index, and high degree of polymerization. Usually, cellulose is observed in the form of microfibrils with a composition of amorphous and crystalline domains (Abhilash & Thomas, 2017). Bacterial cellulose belongs to hydrophilic compounds with a wider surface area with the molecular structure shown in Fig. 5.5. In addition, the most important thing is that bacterial cellulose is relatively cheaper to produce than other cellulose (Manoukian et al., 2019; Ulery et al., 2011).

Bacterial cellulose derived from *Gluconacetobacter xylinus* bacteria when modified will form a material similar to cartilage and is easy to form (Kowalska-Ludwicka et al., 2013).

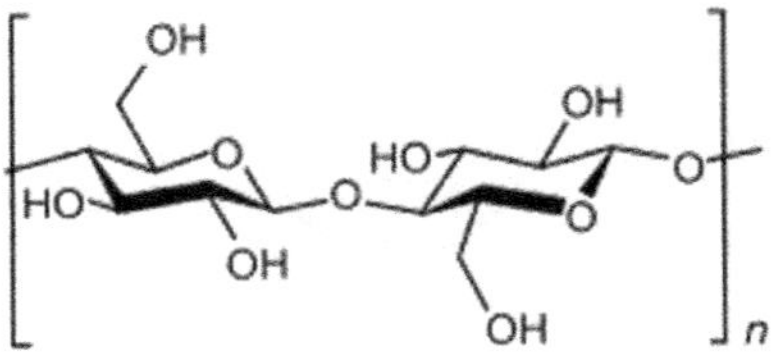

Figure 5.5 Structure of bacterial cellulose (Manoukian et al., 2019).
Source: Data from Manoukian, O. S., et al. (2019). Biomaterials for tissue engineering and regenerative medicine. In R. Narayan (Ed.), *Encyclopedia of biomedical engineering* (pp. 462–482). Elsevier. https://doi.org/10.1016/B978-0-1-801238-3.64098-9.

The bacterial cellulose produced from such bacteria aims to make a biocompatible version of natural cellulose. The resulting bacterial cellulose is nontoxic and has low cytotoxicity (Fu et al., 2013). The application of cellulose bacteria is often found on the skin as a wound dressing (Nandgaonkar et al., 2016). Wound dressings produced from bacterial cellulose can drain fluid into the wound, making it easier for the recovery process and easy to remove (Manoukian et al., 2019).

Cellulose can also be prepared as cellulose nanofibrils by extensive hydrolysis of the amorphous form of cellulose. Cellulose nanofibrils have several advantages such as nanoscale diameter (20 nm, 100–600 nm), a high surface area, low density, unique morphology, and are supported by good mechanical strength. The most important thing is the low cost of manufacture, easy chemical modification, and abundant availability. Usually, cellulose nanofibrils are used in the pharmaceutical industry such as stabilizers and drug excipients (Singh et al., 2016). In addition, cellulose nanofiber can be applied in the medical world, waste treatment such as heavy metal removal in water (Usmani et al., 2018), and oil-in-water emulsion stabilizer (Kosseva, 2020).

5.4 Gelatin

Gelatin is a protein derivative (Li et al., 2023) from animal body parts such as skin and bone tissue (Tan et al., 2023). Chemically, gelatin is a product of the hydrolysis of collagen in the skin, delicate connective tissue, bones of certain mammalian species (Sultana et al., 2018), and other animal sources such as fish and poultry (Rigueto et al., 2023). Generally, the highest percentage of gelatin comes from pigskin, accounting for 46%, followed by cow skin and bones as much as 29.4% and 23.1%, respectively, and 1.5% in fish (Sultana et al., 2018).

Gelatin is divided into two types, namely, type A and type B gelatin, based on its hydrolysis conditions. Type A gelatin is produced from acid hydrolysis, while type B gelatin is produced from base hydrolysis (Rigueto et al., 2023). Type A gelatin has a positive net charge with a high isoelectric point in the pH range of 7–9, while type B gelatin has a net negative charge with a low isoelectric point of pH 4.7–5.2 (Tan et al., 2023).

The main composition of gelatin is amino acids, specifically Gly-Pro-Hyp, which gives functional properties such as binding capacity, gel-forming ability, and foaming properties. Fig. 5.6 shows the structure of gelatin. In addition, gelatin is a soluble amorphous mixture and consists of different free chains. The free chains are α chains, β chains, and γ chains, which are then stabilized with hydrogen bonds (Noor et al., 2021). Although the amino acid composition in gelatin is similar to collagen, gelatin has different properties from collagen such as rheological, gel, and emulsifying properties (Duconseille et al., 2015).

The nature of gelatin is influenced by several factors such as the source of gelatin (pig, beef, fish, or poultry), type of gelatin (α, β, and γ), molecular weight, amino acid composition, and the extraction method used. Gelatin is one of the biomaterials that is widely applied in various fields such as food, pharmaceutical, cosmetics, and medical because of its superior properties, namely, biocompatibility, low antigenicity, water retaining, foam maker, gel and film former, emulsifier, (Lu et al., 2022; Zhang et al., 2020) and biodegradable (Minhas et al., 2023).

Gelatin can be extracted using a variety of different extraction parameters, such as extraction temperature, extraction time, and acid, base, and enzyme treatment (Lu et al., 2022). Gelatin extracted from animal body parts needs to be purified first to remove ingredients that can give color, taste, and smell that can interfere with gelatin quality (Bretanha et al., 2021). But before that, pretreatment is important to facilitate the breakdown of gelatin, which makes the gelatin extraction process lighter. The properties of gelatin will change due to inter- and intramolecular bond cleavage. This bond cleavage results in the breakdown of the triple helix structure of collagen and makes it swollen. This swelling will reduce the denaturation temperature allowing gelatin extraction under mild conditions (Chua et al., 2023).

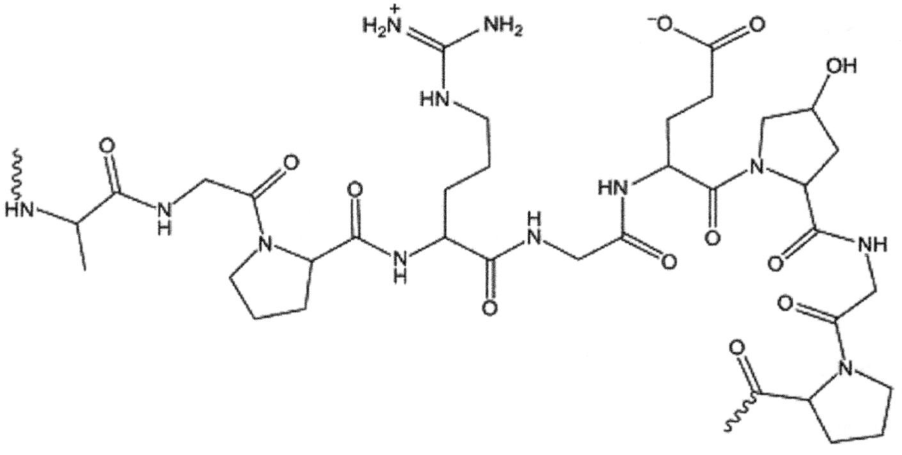

Figure 5.6 Structure of gelatin (Noor et al., 2021).
Source: Data from Noor, N. Q. I. M., et al. (2021). Application of green technology in gelatin extraction: A review. *Processes*, 9(12). https://doi.org/10.3390/pr9122227.

5.5 Alginate

Alginate is a biopolymer consisting of a linear copolymer of d-mannuronic acid and l-guluronic acid units bonded β-(1−4) (Alihosseini, 2016). The two copolymers are mannuronic acid (block M) and anionic guluronic acid (block G), which are arranged irregularly with the proportions of blocks GG, MG, and MM, as shown in Fig. 5.7. The proportion of blocks in alginate can affect its solubility for G blocks (GG) more rigid, while those in alternating blocks (GM) are more soluble in low pH. Alginate usually contains 40%−70% G content, which can affect the quality of the polymer. The molecular weight of alginate varies between 50 and 100,000 kDa. Usually, alginate containing the highest block G will be suitable for use in biomedical applications because it is easy to process and has low immunogenicity in the body. Therefore the composition of G and M blocks is very influential on the properties and application of alginate (Aravamudhan et al., 2014).

Alginate is usually produced from various kinds of brown seaweed such as *Laminaria* sp., *Macrocystis* sp., *Lessonia* sp., and others (Abhilash & Thomas, 2017).

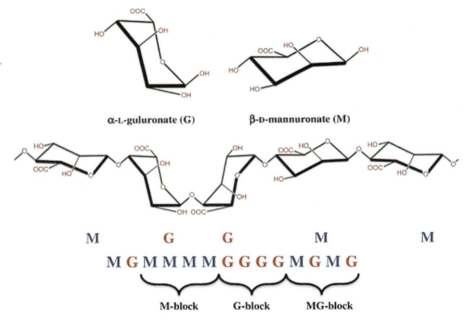

Figure 5.7 Structure of alginate (Angra et al., 2021).
Source: Data from Angra, V., Sehgal, R., Kaur, M., & Gupta, R. (2021). Commercialization of bionanocomposites. In S. Ahmed and Annu (Eds.), *Bionanocomposites in tissue engineering and regenerative medicine* (pp. 587−610). Woodhead Publishing. https://doi.org/10.1016/B978-0-12-821280-6.00017-9.

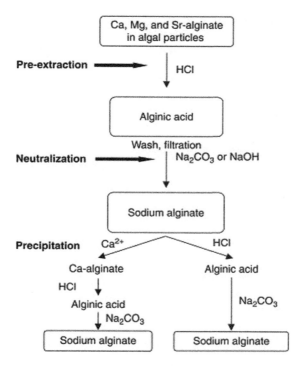

Figure 5.8 Diagram of alginate extraction using algae (Skjåk-Bræk & Draget, 2012).
Source: Data from Skjåk-Bræk, G., & Draget, K. I. (2012). Alginates: Properties and applications. In K. Matyjaszewski & M. Möller (Eds.), *Polymer science: A comprehensive reference* (pp. 213−220). Elsevier. https://doi.org/10.1016/B978-0-444-53349-4.00261-2.

Alginate can also be synthesized using alginate production tools with bacteria in biosynthetic polymers. Alginate synthesized using bacteria can have a composition of 100% mannuronate. Alginates with high guluronic acid content can be produced from specialized algal tissues such as the outer cortex of the *Laminaria hyperborea* layer (Skjåk-Bræk & Draget, 2012). Alginate extraction using algae can be seen in Fig. 5.8. Among the resulting alginate derivatives is propylene glycol alginate, which has a higher market value.

Alginate has many applications in the industrial world as stabilizers, viscosifiers, gellers, film formers, and water-binding agents (Abhilash & Thomas, 2017). In addition, in the medical field, alginate is used as a regulatable drug delivery capsule. The regulation of the rate of release of the drug depends on the molecular weight of the alginate used. Furthermore, in the empirical manufacture of teeth, alginate is used because it is easy to form, quickly hardens at room temperature, and is cost-effective (Abhilash & Thomas, 2017). Alginate is also widely used as a wound dressing and formulation to stop gastric reflux. This is because the gel formed from the exchange of calcium alginate ions and sodium ions can absorb moisture so that wound healing can run optimally (Qin et al., 2007).

5.6 Hyaluronic acid

Hyaluronic acid is an unsulfated polysaccharide and glycosaminoglycan and is a component of the extracellular matrix (Kuo & Prestwich, 2011). Hyaluronic acid consists of d-glucuronic acid and d-N-acetylglucosamine with repeatability of disaccharide molecules of about 5000 to 20,000,000 Da molecular weight (Zeng et al., 2011). Hyaluronic acid was first isolated in 1934 from the vitreous humor of the eye and in 1964 synthesized in vitro. In addition, hyaluronic acid on a large scale can be obtained from microbial fermentation (Zeng et al., 2011).

The physical properties of hyaluronic acid are viscosity, elasticity, and a high capacity to hold water (ASARI, 2004). The physiochemical and biological properties of hyaluronic acid depend on its molecular size, as shown in Fig. 5.9. Hyaluronic acid can be chemically modified to modulate the material and its biological properties (Wu & Elisseeff, 2014). A commonly used modification of hyaluronic acid is the esterification of carboxyl groups (Burd, 2004).

Modification of hyaluronic acid can be done chemically. Modifications are made by utilizing available functional groups (Lee et al., 2008). Each disaccharide unit can produce intermolecular and intramolecular hydrogen bonds because it contains four hydroxyl groups, one amide group, and a carboxyl group. The resulting hydrogen

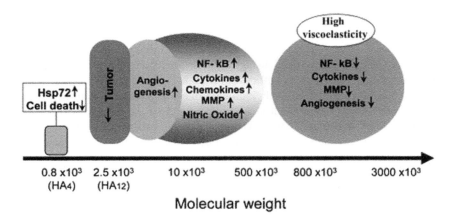

Figure 5.9 The physiochemical and biological properties of hyaluronic acid depend on its molecular size (ASARI, 2004).
Source: Data from ASARI, A. (2004). Medical application of hyaluronan. In H. G. Garg & C. A. Hales (Eds.), *Chemistry and biology of hyaluronan* (pp. 457–473). Elsevier Science Ltd. https://doi.org/10.1016/B978-008044382-9/50052-2.

bond is so strong that it can block hydrophobic modifications in organic solvents (Collins & Birkinshaw, 2013). This has resulted in restrictions on the wider implementation of hyaluronic acid, especially in the medical world (implants) (Zhang & James, 2005).

However, increased solubility in organic solvents has advantages for formulation development (Hill et al., 2015). There is research on hydroxyl groups (C-3) in GlcNAc to study alternative modifications (Tian et al., 2010). This modification takes place when the attachment of hydrophobic molecules cannot be carried out on a liquid medium. Other strategies that can be used are partial protection of functional groups through hydrophobic blocking agents and shielding of anionic groups with cationic groups. In this case, the carboxyl group is used for amidase and esterification, while the hydroxyl group provides ester and ether bonds. In addition, aldehyde groups can also be produced periodically and used as reductive aminations as in Fig. 5.10. After the derivatization process is complete, there will be functional unit activities that can change the intrinsic mechanical characteristics without changing the integrity of the polymer chain.

Hyaluronic acid has a variety of applications ranging from the field of tissue culture (medical) to nonmedical, such as hyaluronic acid modified into biometals that can be used for medical purposes such as eye surgery, injections for osteoarthritis, wound healing, and as cell therapy and tissue engineering (Kuo & Prestwich, 2011). The content of sodium hyaluronate and carmellose can be used for surgical adhesion because hyaluronic acid has low inflation and antigenic potential, making it suitable for the medical world.

5.7 Fibrin

Fibrin is a fibrous network that plays a key role in blood clotting. Fibrin tissue can form as a result of cutting two pairs of fibrinopeptides from blood protein fibrinogen with thrombin enzymes. After that, the resulting monomer binds to nearby monomers and forms protofibrils with double chains in the form of a three-dimensional network (Belcher et al., 2023). Fibrin tissue cannot dissolve if it is made of fibrinogen, but it can dissolve in plasma protein precursors. Fibrinogen is a 340 kDa protein consisting of three pairs of nonidentical chains (Aα2, Bβ2, and γ2) (Damiana et al., 2020), as seen in Fig. 5.11.

In the human body itself, the concentration of fibrinogen ranges from 2 to 4 g/L (Garyfallogiannis et al., 2023). However, if the concentration of fibrinogen is low, commonly called hypofibrinogenemia can occur due to traumatic injury (Hayakawa, 2017). Hypofibrinogenemia is a soft and porous blood clot, leading to the ineffectiveness of hemostasis (Hayakawa Satoshi et al., 2015). Conversely, high concentrations of fibrinogen have a risk for thrombotic conditions such as coronary artery disease, ischemic stroke, and venous thromboembolism (Sui et al., 2021).

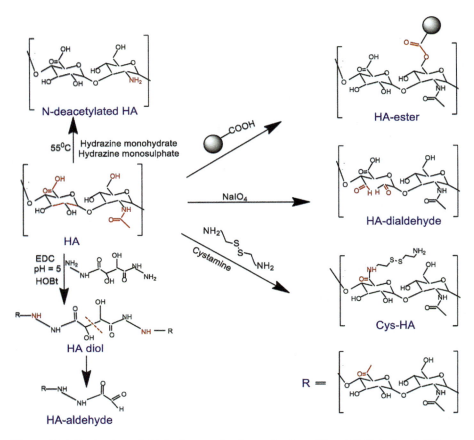

Figure 5.10 Production of aldehydes as reductive aminations (Tiwari & Bahadur, 2019). *Source*: Data from Tiwari, S., & Bahadur, P. (2019). Modified hyaluronic acid based materials for biomedical applications. International Journal of Biological Macromolecules, *121*, 556−571. https://doi.org/10.1016/j.ijbiomac.2018.10.049.

In addition, fibrin can form temporary structures that play a role in wound healing. Fibrinogen is a soluble blood plasma protein precursor that follows the activation scheme of the enzymatic cascade of blood clotting. Fibrin networks are formed from polymerized fibrinogen and increase hemostasis, providing a critical environment for initiating matrix remodeling (Jimenez et al., 2023). Therefore fibrin is a tissue that has a close relationship with human blood.

5.8 Collagen

Collagen is the main component of the dermis (Fu et al., 2023) and acts as a structural building block (Ding et al., 2023). Collagen is found in the skin, tendons, ligaments, and tissues of humans and animals (Bai et al., 2023). A good source of

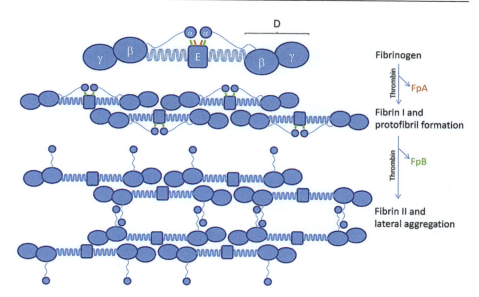

Figure 5.11 Three pairs of fibrinogen chains (Undas & Ariëns, 2011).
Source: Data from Undas, A., & Ariëns, R. A. S. (2011). Fibrin clot structure and function. *Arteriosclerosis, Thrombosis, and Vascular Biology*, *31*(12), e88–e99. https://doi.org/10.1161/ATVBAHA.111.230631.

collagen is found in animal skin such as cowhide, pigskin, fish skin, and chicken skin (Bai et al., 2023). The skin and tendons of cows are the main source of collagen because they produce more than 85% of collagen (Gao et al., 2022).

Collagen consists of a special hierarchical structure namely, triple helix (length ± 300 nm and diameter ± 1.5 nm), microfibrils (diameter ± 40 nm), fibrils (diameter ± 100–200 nm), and collagen fibers, as listed in Fig. 5.12. This hierarchy is beneficial for improving the mechanical properties of collagen. In addition, this structural hierarchy makes collagen have good biodegradability and biocompatibility properties (Zhang et al., 2023). The presence of a lot of hydrogen bonds is also influential because it makes collagen a tough biomaterial (Nuerjiang et al., 2023).

Collagen is divided into three types, namely, types I, II, and III. Type I collagen is found in many bones, tendons, and skin. This type of collagen consists of three helical α chains, two α1 subunits, and one α2 subunit. Collagen types II and III are commonly found in the skin, cartilage tissue, and very young organs (Ahmad et al., 2017).

The surface properties of collagen have high water absorption (Darbazi et al., 2023). Due to these special properties, collagen is widely applied in biomedical and health fields such as skin implants, cosmetics, and medicine, as well as the food and beverage industry. However, the amount of collagen that exists is underutilized even though collagen can be of high value if it is used as a biopolymer that can be degraded (Kristoffersen et al., 2023). Therefore the application of collagen depends on the quality of the extract because the source and processing of collagen can affect its physiochemical properties (Liu et al., 2015).

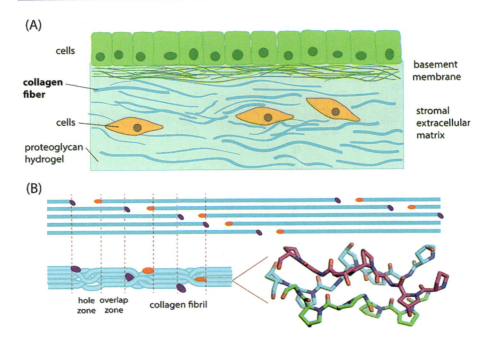

Figure 5.12 Scheme of (A) The extracellular matrix and (B) Arrangement of collagen molecules in fibrils (Goldberga et al., 2018).
Source: Data from Goldberga, I., Li, R., & Duer, M. J. (2018). Collagen structure–function relationships from solid-state NMR spectroscopy. *Accounts of Chemical Research, 51*(7), 1621–1629. https://doi.org/10.1021/acs.accounts.8b00092.

5.9 Conclusion

Polymers available from nature provide enormous benefits to human life. This polymer from nature is obtained from the extraction of plants or animals and is often called a biopolymer. Examples of such biopolymers are chitin/chitosan, cellulose, gelatin, alginate, hyaluronic acid, fibrin, and collagen. Recently, these materials and their derivatives have received a lot of attention because they have high values, notably, for being biodegradable and sustainable. In addition, its widespread functions and uses can be applied in various fields such as industry, food, medicine, and so on making humans increasingly switch to biopolymers.

References

ASARI, A. (2004). Medical application of hyaluronan. In H. G. Garg, & C. A. Hales (Eds.), *Chemistry and biology of hyaluronan* (pp. 457–473). Elsevier Science Ltd. Available from https://doi.org/10.1016/B978-008044382-9/50052-2.

Alihosseini, F. (2016). Plant-based compounds for antimicrobial textiles. In G. Sun (Ed.), *Antimicrobial textiles* (pp. 155−195). Woodhead Publishing. Available from https://doi.org/10.1016/B978-0-08-100576-7.00010-9.

Abhilash, M., & Thomas, D. (2017). Biopolymers for biocomposites and chemical sensor applications. In K. K. Sadasivuni, D. Ponnamma, J. Kim, J.-J. Cabibihan, & M. A. AlMaadeed (Eds.), *Biopolymer composites in electronics* (pp. 405−435). Elsevier. Available from https://doi.org/10.1016/B978-0-12-809261-3.00015-2.

Ahmad, T., et al. (2017). Recent advances on the role of process variables affecting gelatin yield and characteristics with special reference to enzymatic extraction: A review. *Food Hydrocolloids, 63*, 85−96. Available from https://doi.org/10.1016/j.foodhyd.2016.08.007.

Ahmed, M. J., Hameed, B. H., & Hummadi, E. H. (2020). Review on recent progress in chitosan/chitin-carbonaceous material composites for the adsorption of water pollutants. *Carbohydrate Polymers, 247*, 116690. Available from https://doi.org/10.1016/j.carbpol.2020.116690.

Angra, V., Sehgal, R., Kaur, M., & Gupta, R. (2021). Commercialization of bionanocomposites. In S. Ahmed, & Annu (Eds.), *Bionanocomposites in tissue engineering and regenerative medicine* (pp. 587−610). Woodhead Publishing. Available from https://doi.org/10.1016/B978-0-12-821280-6.00017-9.

Aravamudhan, A., Ramos, D. M., Nada, A. A., & Kumbar, S. G. (2014). Natural polymers: Polysaccharides and their derivatives for biomedical applications. In S. G. Kumbar, C. T. Laurencin, & M. Deng (Eds.), *Natural and synthetic biomedical polymers* (pp. 67−89). Elsevier. Available from https://doi.org/10.1016/B978-0-12-396983-5.00004-1.

Bai, L., et al. (2023). Antioxidant activity during in vitro gastrointestinal digestion and the mode of action with tannins of cowhide-derived collagen hydrolysates: The effects of molecular weight. *Food Bioscience, 53*, 102773. Available from https://doi.org/10.1016/j.fbio.2023.102773.

Belcher, H. A., Guthold, M., & Hudson, N. E. (2023). What is the diameter of a fibrin fiber? *Research and Practice in Thrombosis and Haemostasis*, 100285. Available from https://doi.org/10.1016/j.rpth.2023.100285.

Bretanha, C. C., Zin, G., Oliveira, J. V., & Di Luccio, M. (2021). Improvement of tangential microfiltration of gelatin solution using a permanent magnetic field. *Journal of Food Science and Technology, 58*(3), 1093−1100. Available from https://doi.org/10.1007/s13197-020-04623-y.

Burd, A. (2004). Hyaluronan and scarring. In H. G. Garg, & C. A. Hales (Eds.), *Chemistry and biology of hyaluronan* (pp. 367−394). Elsevier Science Ltd. Available from https://doi.org/10.1016/B978-008044382-9/50049-2.

Chua, L.-K., Lim, P.-K., Thoo, Y.-Y., Neo, Y.-P., & Tan, T.-C. (2023). Extraction and characterization of gelatin derived from acetic acid-treated black soldier fly larvae. *Food Chemistry Advances, 2*, 100282. Available from https://doi.org/10.1016/j.focha.2023.100282.

Collins, M. N., & Birkinshaw, C. (2013). Hyaluronic acid solutions—A processing method for efficient chemical modification. *Journal of Applied Polymer Science, 130*(1), 145−152. Available from https://doi.org/10.1002/app.39145.

Dahaghin, A., et al. (2021). A comparative study on the effects of increase in injection sites on the magnetic nanoparticles hyperthermia. *Journal of Drug Delivery Science and Technology, 63*, 102542. Available from https://doi.org/10.1016/j.jddst.2021.102542.

Damiana, T., et al. (2020). Citrullination of fibrinogen by peptidylarginine deiminase 2 impairs fibrin clot structure. *Clinica Chimica Acta, 501*, 6−11. Available from https://doi.org/10.1016/j.cca.2019.10.033.

Darbazi, M., Elmi, F., Elmi, M. M., Giglia, A., & Hoda, A. (2023). Synchrotron X-ray absorption spectroscopy and fluorescence spectroscopy studies of collagen type I-talc intercalated nanocomposites. *Journal of Molecular Structure*, *1285*, 135558. Available from https://doi.org/10.1016/j.molstruc.2023.135558.

Ding, C., Tian, M., Wang, Y., Cheng, K., Yi, Y., & Zhang, M. (2023). Governing the aggregation of type I collagen mediated through β-cyclodextrin. *International Journal of Biological Macromolecules*, *240*, 124469. Available from https://doi.org/10.1016/j.ijbiomac.2023.124469.

Dotto, G. L., & Pinto, L. A. A. (2011). Adsorption of food dyes onto chitosan: Optimization process and kinetic. *Carbohydrate Polymers*, *84*(1), 231−238. Available from https://doi.org/10.1016/j.carbpol.2010.11.028.

Druzian, S. P., et al. (2021). Chitin-psyllium based aerogel for the efficient removal of crystal violet from aqueous solutions. *International Journal of Biological Macromolecules*, *179*, 366−376. Available from https://doi.org/10.1016/j.ijbiomac.2021.02.179.

Duconseille, A., Astruc, T., Quintana, N., Meersman, F., & Sante-Lhoutellier, V. (2015). Gelatin structure and composition linked to hard capsule dissolution: A review. *Food Hydrocolloids*, *43*, 360−376. Available from https://doi.org/10.1016/j.foodhyd.2014.06.006.

Eivazzadeh-Keihan, R., et al. (2021). Chitosan hydrogel/silk fibroin/Mg(OH)$_2$ nanobiocomposite as a novel scaffold with antimicrobial activity and improved mechanical properties. *Scientific Reports*, *11*(1), 650. Available from https://doi.org/10.1038/s41598-020-80133-3.

Estanqueiro, M., Vasconcelos, H., Sousa Lobo, J. M., & Amaral, H. (2018). Delivering miRNA modulators for cancer treatment. In A. M. Grumezescu (Ed.), *Drug targeting and stimuli sensitive drug delivery systems* (pp. 517−565). William Andrew Publishing. Available from https://doi.org/10.1016/B978-0-12-813689-8.00014-8.

Fu, C., Shi, S., Tian, J., Gu, H., Yao, L., & Xiao, J. (2023). Non-denatured yak type I collagen accelerates sunburned skin healing by stimulating and replenishing dermal collagen. *Biotechnology Reports*, *37*, e00778. Available from https://doi.org/10.1016/j.btre.2022.e00778.

Fu, L., Zhang, J., & Yang, G. (2013). Present status and applications of bacterial cellulose-based materials for skin tissue repair. *Carbohydrate Polymers*, *92*(2), 1432−1442. Available from https://doi.org/10.1016/j.carbpol.2012.10.071.

Gao, Y., Wang, L., Qiu, Y., Fan, X., Zhang, L., & Yu, Q. (2022). Valorization of cattle slaughtering industry by-products: Modification of the functional properties and structural characteristics of cowhide gelatin induced by high hydrostatic pressure. *Gels*, *8*(4). Available from https://doi.org/10.3390/gels8040243.

Garyfallogiannis, K., et al. (2023). Fracture toughness of fibrin gels as a function of protein volume fraction: Mechanical origins. *Acta Biomaterialia*, *159*, 49−62. Available from https://doi.org/10.1016/j.actbio.2022.12.028.

Godoy, M. G., Amorim, G. M., Barreto, M. S., & Freire, D. M. G. (2018). Agricultural residues as animal feed: Protein enrichment and detoxification using solid-state fermentation. In A. Pandey, C. Larroche, & C. R. Soccol (Eds.), *Current developments in biotechnology and bioengineering* (pp. 235−256). Elsevier. Available from https://doi.org/10.1016/B978-0-444-63990-5.00012-8.

Goldberga, I., Li, R., & Duer, M. J. (2018). Collagen structure−function relationships from solid-state nmr spectroscopy. *Accounts of Chemical Research*, *51*(7), 1621−1629. Available from https://doi.org/10.1021/acs.accounts.8b00092.

Guo, X., et al. (2019). Electroassembly of chitin nanoparticles to construct freestanding hydrogels and high porous aerogels for wound healing. *ACS Applied Materials & Interfaces*, *11*(38), 34766−34776. Available from https://doi.org/10.1021/acsami.9b13063.

Hayakawa, M. (2017). Dynamics of fibrinogen in acute phases of trauma. *Journal of Intensive Care*, *5*, 3. Available from https://doi.org/10.1186/s40560-016-0199-3.

Hayakawa Satoshi, M. G., Yuichi, O., Takeshi, W., Yuichiro, Y., & Atsushi, S. (2015). Fibrinogen level deteriorates before other routine coagulation parameters and massive transfusion in the early phase of severe trauma: A retrospective observational study. *Seminars in Thrombosis and Hemostasis*, *41*(01), 35−42. Available from https://doi.org/10.1055/s-0034-1398379.

Hayes, T. R., & Su, B. (2011). Wound dressings. In L. A. Bosworth, & S. Downes (Eds.), *Electrospinning for tissue regeneration* (pp. 317−339). Woodhead Publishing. Available from https://doi.org/10.1533/9780857092915.20.317.

Hill, T. K., et al. (2015). Indocyanine green-loaded nanoparticles for image-guided tumor surgery. *Bioconjugate Chemistry*, *26*(2), 294−303. Available from https://doi.org/10.1021/bc5005679.

Igarashi, K., et al. (2012). Visualization of cellobiohydrolase i from trichoderma reesei moving on crystalline cellulose using high-speed atomic force microscopy. In H. J. Gilbert (Ed.), *Methods in enzymology* (pp. 169−182). Academic Press. Available from https://doi.org/10.1016/B978-0-12-415931-0.00009-4.

Jena, K., et al. (2023). Physical, biochemical and antimicrobial characterization of chitosan prepared from tasar silkworm pupae waste. *Environmental Technology & Innovation*, *31*, 103200. Available from https://doi.org/10.1016/j.eti.2023.103200.

Jimenez, J. M., Tuttle, T., Guo, Y., Miles, D., Buganza-Tepole, A., & Calve, S. (2023). Multiscale mechanical characterization and computational modeling of fibrin gels. *Acta Biomaterialia*, *162*, 292−303. Available from https://doi.org/10.1016/j.actbio.2023.03.026.

Kazemi Shariat Panahi, H., et al. (2023). Current and emerging applications of saccharide-modified chitosan: a critical review. *Biotechnology Advances*, *66*, 108172. Available from https://doi.org/10.1016/j.biotechadv.2023.108172.

Kosseva, M. R. (2020). Sources, characteristics and treatment of plant-based food waste. In M. R. Kosseva, & C. Webb (Eds.), *Food industry wastes* (2nd ed., pp. 37−66). Academic Press. Available from https://doi.org/10.1016/B978-0-12-817121-9.00003-6.

Kowalska-Ludwicka, K., et al. (2013). Special paper − New methods modified bacterial cellulose tubes for regeneration of damaged peripheral nerves. *Archives of Medical Science*, *9*(3), 527−534. Available from https://doi.org/10.5114/aoms.2013.33433.

Kristoffersen, K. A., et al. (2023). FTIR-based prediction of collagen content in hydrolyzed protein samples. *Spectrochimica Acta. Part A, Molecular and Biomolecular Spectroscopy*, *301*, 122919. Available from https://doi.org/10.1016/j.saa.2023.122919.

Kumar, P., Lakshmanan, V.-K., Biswas, R., Nair, S., & Jayakumar, R. (2012). Synthesis and biological evaluation of chitin hydrogel/nano ZnO composite bandage as antibacterial wound dressing. *Journal of Biomedical Nanotechnology*, *8*, 891−900. Available from https://doi.org/10.1166/jbn.2012.1461.

Kuo, J.-W., & Prestwich, G. D. (2011). Hyaluronic acid. In P. Ducheyne (Ed.), *Comprehensive biomaterials* (pp. 239−259). Elsevier. Available from https://doi.org/10.1016/B978-0-08-055294-1.00073-8.

Lee, H., Lee, K., & Park, T. G. (2008). Hyaluronic acid − paclitaxel conjugate micelles: Synthesis, characterization, and antitumor activity. *Bioconjugate Chemistry*, *19*(6), 1319−1325. Available from https://doi.org/10.1021/bc8000485.

Li, X., Sha, X.-M., Yang, H.-S., Ren, Z.-Y., & Tu, Z.-C. (2023). Ultrasonic treatment regulates the properties of gelatin emulsion to obtain high-quality gelatin film. *Food Chemistry: X*, *18*, 100673. Available from https://doi.org/10.1016/j.fochx.2023.100673.

Liao, J., Hou, B., & Huang, H. (2022). Preparation, properties and drug controlled release of chitin-based hydrogels: An updated review. *Carbohydrate Polymers*, *283*, 119177. Available from https://doi.org/10.1016/j.carbpol.2022.119177.

Liu, D., Nikoo, M., Boran, G., Zhou, P., & Regenstein, J. M. (2015). Collagen and gelatin. *Annual Review of Food Science and Technology*, *6*(1), 527−557. Available from https://doi.org/10.1146/annurev-food-031414-111800.

Lu, Y., et al. (2022). Application of gelatin in food packaging: A review. *Polymers*, *14*(3). Available from https://doi.org/10.3390/polym14030436.

Lucas, N., & Rode, C. V. (2023). Marine waste derived chitin biopolymer for N-containing supports, catalysts and chemicals. *Tetrahedron Green Chem*, *2*, 100013. Available from https://doi.org/10.1016/j.tgchem.2023.100013.

Manoukian, O. S., et al. (2019). Biomaterials for tissue engineering and regenerative medicine. In R. Narayan (Ed.), *Encyclopedia of biomedical engineering* (pp. 462−482). Elsevier. Available from https://doi.org/10.1016/B978-0-12-801238-3.64098-9.

Minhas, B., et al. (2023). The electrochemical and in-vitro study on electrophoretic deposition of chitosan/gelatin/hydroxyapatite coating on 316L stainless steel. *Carbohydrate Polymer Technologies and Applications*, *5*, 100322. Available from https://doi.org/10.1016/j.carpta.2023.100322.

Mohiti-Asli, M., & Loboa, E. G. (2016). 23 - Nanofibrous smart bandages for wound care. In M. S. Ågren (Ed.), *Wound healing biomaterials* (pp. 483−499). Woodhead Publishing. Available from https://doi.org/10.1016/B978-1-78242-456-7.00023-4.

Nandgaonkar, A. G., Krause, W. E., & Lucia, L. A. (2016). Fabrication of cellulosic composite scaffolds for cartilage tissue engineering. In H. Liu (Ed.), *Nanocomposites for musculoskeletal tissue regeneration* (pp. 187−212). Woodhead Publishing. Available from https://doi.org/10.1016/B978-1-78242-452-9.00009-1.

Noor, N. Q. I. M., et al. (2021). Application of green technology in gelatin extraction: A review. *Processes*, *9*(12). Available from https://doi.org/10.3390/pr9122227.

Nuerjiang, M., Bai, X., Sun, L., Wang, Q., Xia, X., & Li, F. (2023). Size effect of fish gelatin nanoparticles on the mechanical properties of collagen film based on its hierarchical structure. *Food Hydrocolloids*, *144*, 108931. Available from https://doi.org/10.1016/j.foodhyd.2023.108931.

Oyatogun, G. M., et al. (2020). Chitin, chitosan, marine to market. In S. Gopi, S. Thomas, & A. Pius (Eds.), *Handbook of chitin and chitosan* (pp. 335−376). Elsevier. Available from https://doi.org/10.1016/B978-0-12-817970-3.00011-0.

Qin, Y., et al. (2007). Combined use of chitosan and alginate in the treatment of wastewater. *Journal of Applied Polymer Science*, *104*(6), 3581−3587. Available from https://doi.org/10.1002/app.26006.

Rakhi, T., Pragati, M., Prerna, B., & Jyoti, S. (2023). Multifaceted approach for nanofiber fabrication. In N. Sharma, & B. S. Butola (Eds.), *Fiber and textile engineering in drug delivery systems* (pp. 253−283). Woodhead Publishing. Available from https://doi.org/10.1016/B978-0-323-96117-2.00012-1.

dos Reis, G. S., et al. (2023). Adsorption of yttrium (Y^{3+}) and concentration of rare earth elements from phosphogypsum using chitin and chitin aerogel. *Journal of Rare Earths*. Available from https://doi.org/10.1016/j.jre.2023.04.008.

Rigueto, C. V. T., et al. (2023). Steam explosion pretreatment for bovine limed hide waste gelatin extraction. *Food Hydrocolloids*, *142*, 108854. Available from https://doi.org/10.1016/j.foodhyd.2023.108854.

Rudin, A., & Choi, P. (2013). Biopolymers. In A. Rudin, & P. Choi (Eds.), *The elements of polymer science & engineering* (3rd ed., pp. 521−535). Academic Press. Available from https://doi.org/10.1016/B978-0-12-382178-2.00013-4.

Samyn, P. (2021). A platform for functionalization of cellulose, chitin/chitosan, alginate with polydopamine: A review on fundamentals and technical applications. *International Journal of Biological Macromolecules, 178*, 71–93. Available from https://doi.org/10.1016/j.ijbiomac.2021.02.091.

Schleuter, D., et al. (2013). Chitin-based renewable materials from marine sponges for uranium adsorption. *Carbohydrate Polymers, 92*(1), 712–718. Available from https://doi.org/10.1016/j.carbpol.2012.08.090.

Shrivastava, A. (2018). Introduction to plastics engineering. In A. Shrivastava (Ed.), *Introduction to plastics engineering* (pp. 1–16). William Andrew Publishing. Available from https://doi.org/10.1016/B978-0-323-39500-7.00001-0.

Singh, I., Sharma, A., & Park, B.-D. (2016). Drug-delivery applications of cellulose nanofibrils. In A. M. Holban, & A. M. Grumezescu (Eds.), *Nanoarchitectonics for smart delivery and drug targeting* (pp. 95–117). William Andrew Publishing. Available from https://doi.org/10.1016/B978-0-323-47347-7.00004-5.

Skjåk-Bræk, G., & Draget, K. I. (2012). Alginates: Properties and applications. In K. Matyjaszewski, & M. Möller (Eds.), *Polymer science: A comprehensive reference* (pp. 213–220). Elsevier. Available from https://doi.org/10.1016/B978-0-444-53349-4.00261-2.

Sui, J., Noubouossie, D. F., Gandotra, S., & Cao, L. (2021). Elevated plasma fibrinogen is associated with excessive inflammation and disease severity in COVID-19 patients. *Frontiers in Cellular and Infection Microbiology, 11*. Available from https://doi.org/10.3389/fcimb.2021.734005.

Sultana, S., Hossain, M. A. M., Zaidul, I. S. M., & Ali, M. E. (2018). Multiplex PCR to discriminate bovine, porcine, and fish DNA in gelatin and confectionery products. *LWT, 92*, 169–176. Available from https://doi.org/10.1016/j.lwt.2018.02.019.

Sulthan, R., et al. (2023). Extraction of β-chitin using deep eutectic solvents for biomedical applications. *Materials Today: Proceedings*. Available from https://doi.org/10.1016/j.matpr.2023.05.521.

Sun, C., et al. (2022). Chitin-glucan composite sponge hemostat with rapid shape-memory from Pleurotus eryngii for puncture wound. *Carbohydrate Polymers, 291*, 119553. Available from https://doi.org/10.1016/j.carbpol.2022.119553.

Tahir, I., Millevania, J., Wijaya, K., Mudasir., Wahab, R. A., & Kurniawati, W. (2023). Optimization of thiamine chitosan nanoemulsion production using sonication treatment. *Results in Engineering, 17*, 100919. Available from https://doi.org/10.1016/j.rineng.2023.100919.

Tan, Y., Zi, Y., Peng, J., Shi, C., Zheng, Y., & Zhong, J. (2023). Gelatin as a bioactive nanodelivery system for functional food applications. *Food Chemistry, 423*, 136265. Available from https://doi.org/10.1016/j.foodchem.2023.136265.

Tao, F., et al. (2020). Applications of chitin and chitosan nanofibers in bone regenerative engineering. *Carbohydrate Polymers, 230*, 115658. Available from https://doi.org/10.1016/j.carbpol.2019.115658.

Thakur, M., Sharma, A., Chandel, M., & Pathania, D. (2022). Modern applications and current status of green nanotechnology in environmental industry. In U. Shanker, C. M. Hussain, & M. Rani (Eds.), *Green functionalized nanomaterials for environmental applications* (pp. 259–281). Elsevier. Available from https://doi.org/10.1016/B978-0-12-823137-1.00010-5.

Tian, Q., Wang, X., Wang, W., Zhang, C., Liu, Y., & Yuan, Z. (2010). Insight into glycyrrhetinic acid: The role of the hydroxyl group on liver targeting. *International Journal of Pharmaceutics, 400*(1), 153–157. Available from https://doi.org/10.1016/j.ijpharm.2010.08.032.

Tiwari, S., & Bahadur, P. (2019). Modified hyaluronic acid based materials for biomedical applications. *International Journal of Biological Macromolecules*, *121*, 556−571. Available from https://doi.org/10.1016/j.ijbiomac.2018.10.049.

Ulery, B. D., Nair, L. S., & Laurencin, C. T. (2011). Biomedical applications of biodegradable polymers. *Journal of Polymer Science: Polymer Physics*, *49*(12), 832−864. Available from https://doi.org/10.1002/polb.22259.

Undas, A., & Ariëns, R. A. S. (2011). Fibrin clot structure and function. *Arteriosclerosis, Thrombosis, and Vascular Biology*, *31*(12), e88−e99. Available from https://doi.org/10.1161/ATVBAHA.111.230631.

Usmani, M. A., Khan, I., Gazal, U., Mohamad Haafiz, M. K., & Bhat, A. H. (2018). Interplay of polymer bionanocomposites and significance of ionic liquids for heavy metal removal. In M. Jawaid, & M. M. Khan (Eds.), *Polymer-based nanocomposites for energy and environmental applications* (pp. 441−463). Woodhead Publishing. Available from https://doi.org/10.1016/B978-0-08-102262-7.00016-7.

Vodseďálková, K., Vysloužilová, L., & Berezkinová, L. (2017). Coaxial electrospun nanofibers for nanomedicine. In T. Uyar, & E. Kny (Eds.), *Electrospun materials for tissue engineering and biomedical applications* (pp. 321−336). Woodhead Publishing. Available from https://doi.org/10.1016/B978-0-08-101022-8.00014-4.

Wang, X., Tarahomi, M., Sheibani, R., Xia, C., & Wang, W. (2023). Progresses in lignin, cellulose, starch, chitosan, chitin, alginate, and gum/carbon nanotube (nano)composites for environmental applications: A review. *International Journal of Biological Macromolecules*, *241*, 124472. Available from https://doi.org/10.1016/j.ijbiomac.2023.124472.

Wu, I., & Elisseeff, J. (2014). Biomaterials and tissue engineering for soft tissue reconstruction. In S. G. Kumbar, C. T. Laurencin, & M. Deng (Eds.), *Natural and synthetic biomedical polymers* (pp. 235−241). Elsevier. Available from https://doi.org/10.1016/B978-0-12-396983-5.00015-6.

Yadav, M., Kaushik, B., Rao, G. K., Srivastava, C. M., & Vaya, D. (2023). Advances and challenges in the use of chitosan and its derivatives in biomedical fields: A review. *Carbohydrate Polymer Technologies and Applications*, *5*, 100323. Available from https://doi.org/10.1016/j.carpta.2023.100323.

Younes, I., & Rinaudo, M. (2015). Chitin and chitosan preparation from marine sources. Structure, properties and applications. *Marine Drugs*, *13*(3), 1133−1174. Available from https://doi.org/10.3390/md13031133.

Zainudin, B. H., Wong, T. W., & Hamdan, H. (2020). Pectin as oral colon-specific nano- and microparticulate drug carriers. In M. A. A. AlMaadeed, D. Ponnamma, & M. A. Carignano (Eds.), *Polymer science and innovative applications* (pp. 257−286). Elsevier. Available from https://doi.org/10.1016/B978-0-12-816808-0.00008-1.

Zeng, Q., et al. (2011). Skin tissue engineering. In P. Ducheyne (Ed.), *Comprehensive biomaterials* (pp. 467−499). Elsevier. Available from https://doi.org/10.1016/B978-0-08-055294-1.00186-0.

Zhang, M., & James, S. P. (2005). Synthesis and properties of melt-processable hyaluronan esters. *Journal of Materials Science. Materials in Medicine*, *16*(6), 587−593. Available from https://doi.org/10.1007/s10856-005-0536-x.

Zhang, Q., Sun, Q., Duan, X., Chi, Y., & Shi, B. (2023). Effectively recovering catechin compounds in the removal of caffeine from tea polyphenol extract by using hydrophobically modified collagen fiber. *Separation and Purification Technology*, *322*, 124325. Available from https://doi.org/10.1016/j.seppur.2023.124325.

Zhang, T., Xu, J., Zhang, Y., Wang, X., Lorenzo, J. M., & Zhong, J. (2020). Gelatins as emulsifiers for oil-in-water emulsions: Extraction, chemical composition, molecular structure, and molecular modification. *Trends in Food Science & Technology*, *106*, 113−131. Available from https://doi.org/10.1016/j.tifs.2020.10.005.

Zhao, Q., Fan, L., Deng, C., Ma, C., Zhang, C., & Zhao, L. (2023). Bioconversion of chitin into chitin oligosaccharides using a novel chitinase with high chitin-binding capacity. *International Journal of Biological Macromolecules*, *244*, 125241. Available from https://doi.org/10.1016/j.ijbiomac.2023.125241.

Zhou, D.-Y., et al. (2021). Chitosan and derivatives: Bioactivities and application in foods. *Annual Review of Food Science and Technology*, *12*(1), 407−432. Available from https://doi.org/10.1146/annurev-food-070720-112725.

Green synthesis of nanofibers for energy and environmental applications

Nancy Elizabeth Davila-Guzman[1], J. Raziel Álvarez[1], M.A. Garza-Navarro[2,3] and Alan A. Rico-Barragán[4]

[1]School of Chemistry, Autonomous University of Nuevo Leon, San Nicolas de los Garza, Nuevo Leon, Mexico, [2]School of Mechanical and Electrical Engineering, Autonomous University of Nuevo Leon, San Nicolas de los Garza, Nuevo Leon, Mexico, [3]Center of Innovation, Research and Development in Engineering and Technology, Autonomous University of Nuevo Leon, Apodaca, Nuevo Leon, Mexico, [4]Department of Environmental Engineering, National Technological of Mexico/Higher Technological Institute of Misantla, Misantla, Veracruz, Mexico

6.1 Introduction

In recent years, the quest for sustainable and environmentally friendly materials has become paramount in the fields of energy and environmental applications (Zhang et al., 2023). Nanofibers, with their unique properties and versatile functionalities, have emerged as promising candidates for addressing the pressing challenges in these domains. The synthesis of nanofibers through green and sustainable methods has gained significant attention, aiming to minimize the environmental impact while maximizing performance and functionality.

Nanofibers have shown tremendous potential in energy applications, including energy conversion, storage, and harvesting (Sun et al., 2016). By utilizing green synthesis methods, the properties of nanofibers can be tailored to enhance their performance as catalysts, electrodes, and membranes, enabling efficient energy conversion processes and improved energy storage capabilities. Integrating nanofibers into sustainable energy systems can also lead to advancements in solar cells, fuel cells, and energy-efficient devices.

In the context of environmental applications, nanofibers synthesized through green routes offer exciting possibilities for resolving urgent environmental issues (Homaeigohar & Elbahri, 2014; Yoon et al., 2008). These nanofibers can be tailored to exhibit high efficiency in pollutant removal, water purification, and air filtration (Lv et al., 2018). By leveraging sustainable synthesis methods, we can minimize the use of hazardous chemicals and reduce energy consumption, leading to the development of eco-friendly nanofiber-based materials for environmental remediation.

This book chapter focuses on the green synthesis of nanofibers for energy and environmental applications. We explore the innovative strategies and approaches employed to fabricate nanofibers using environmentally benign materials, eco-friendly solvents, and energy-efficient processes. This chapter explores various green synthesis techniques, including electrospinning and bio-inspired approaches. We discuss the effect of neoteric solvents and mixed solvents on the environmental impacts of the nanofiber synthesis. Furthermore, the application of the life cycle assessment environmental tool to nanofiber synthesis is also discussed.

6.2 Biopolymers for the green synthesis of nanofibers

Green chemistry commonly refers to the practice of designing, developing, and implementing products and processes that minimize or eliminate the use and generation of harmful substances, aiming to protect human health and the environment (Moulton et al., 2010). Indeed, significant efforts have been made to replace petroleum-based products with environmentally friendly and sustainable alternatives. Specifically, petroleum-based polymers have limitations due to their dependence on finite fossil fuel resources, and their use provokes significant harm to the environment. Additionally, these polymers are often restricted to industrial applications because they lack biocompatibility and biodegradability (Baranwal et al., 2022; Syed et al., 2023).

As a result, biopolymers have emerged as a promising alternative. Biopolymers are organic species produced from natural and renewable resources such as agricultural waste, plant sources, seaweed, crustaceans, and insects. Examples of these natural biopolymers include cellulose (CL), chitin (CTN), chitosan (CHN), starch, gelatin (GE), and alginate (Baranwal et al., 2022). Natural biopolymers can be subjected to modification processes to enhance their chemical and thermal stability, as well as mechanical properties, resulting in the formation of semisynthetic biopolymers such as cellulose acetate (CA), carboxymethyl cellulose (CMC), and carboxymethyl chitosan. These polymers are commonly biocompatible and biodegradable, which makes them suitable for preparing edible films, drug delivery systems, medical implants, wound healing, and scaffolds for tissue engineering (Baranwal et al., 2022). Furthermore, there has been an exploration of the use of synthetic biopolymers, including poly(lactic acid) (PLA) and poly (ε-caprolactone) (PCL), which exhibit the required biocompatibility and stability under physiological conditions in medical applications (Luraghi et al., 2021).

Specifically, the nanofiber-based materials derived from biopolymers have been identified as suitable platforms for developing wound dressing membranes and scaffolds for tissue engineering (Hassiba et al., 2017; Keirouz et al., 2020; Mianehro, 2022; Stratton et al., 2016). These nanofiber-based materials are typically prepared using self-assembly, 3D printing, and electrospinning techniques. In particular, the electrospinning technique allows the production of nanofibrous materials that exhibit a morphology comparable to the extracellular matrix (ECM) (Keirouz et al.,

2020). This characteristic facilitates cell adhesion and proliferation and ultimately promotes tissue regeneration. The electrospinning technique produces nanofibers by applying a high electric potential to a polymeric solution held at the tip of a capillary tube (Vasita & Katti, 2006). This electric potential induces negative polarization in the polymeric solution, resulting in repulsive interactions among the charge within the solution. These repulsive interactions generate a force that counters the surface tension of the polymeric solution. As the electric potential increases, the solution's hemispherical surface at the tip of the capillary tube undergoes elongation, forming a conical shape known as the Taylor cone. Further increasing the electric potential leads to the elongation of this cone, giving rise to the polymeric jet, which is expelled toward a grounded conductive collector. The polymeric jet describes a chaotic movement during its trajectory toward the collector, increasing the distance traveled and the time taken to reach the collector. This process contributes to solvent evaporation and subsequent deposition of nanofibers onto the collector.

The electrospinning technique has been utilized to prepare two-layer nanofibrous membranes for wound dressing applications (Hassiba et al., 2017). The process involved the deposition of an upper layer comprising a nanofibrous mat composed of polyvinyl alcohol (PVA)/CHN/silver nanoparticles (AgNPs), followed by a lower layer consisting of nanofibers made from either polyvinylpyrrolidone (PVP) or polyethylene oxide (PEO) loaded with chlorhexidine. According to the authors, the PVA/CHN/AgNPs layer acts as a protective barrier against microbial invasion, thus reducing the risk of wound infection, whereas the lower layer permits controlled release of chlorhexidine at the wound site.

Moreover, significant efforts have been made to fabricate nanofibrous scaffolds using CHN-grafted-PLA (CHN-g-PLA) loaded with AgNPs and chondroitin sulfate (C_4S) for biomedical applications (Júnior et al., 2022). The CHN-g-PLA graft copolymer was synthesized by reacting oligomeric methyl lactate with CHN. Subsequently, an electrochemical approach was employed to synthesize AgNPs by using an aqueous solution of $AgNO_3$ containing CHN-g-PLA, resulting in the formation of CHN-g-PLA:AgNPs. The CHN-g-PLA:AgNPs material was then lyophilized, purified, and dispersed in an aqueous solution containing C_4S, yielding CHN-g-PLA:AgNPs:C_4S. Finally, the CHN-g-PLA:AgNPs:C_4S solution was electrospun to obtain the nanofibrous scaffolds. The experimental evidence reveal that these scaffolds are composed of nanofibers with a diameter of 318 nm. Furthermore, they exhibit antibacterial activity related to the release of Ag^+ from the CHN-g-PLA:AgNPs:C_4S nanofibers when exposed to *Escherichia coli* and *Staphylococcus aureus* bacterial media. Additionally, the CHN-g-PLA:AgNPs nanofibers demonstrate nontoxicity toward L929 fibroblast cells and provide a proper scaffold for their adhesion and colonization, which is attributed to the presence of C_4S.

The literature also discusses the preparation of nanofiber-based scaffolds for tissue engineering through electrospinning natural-synthetic copolymers, such as lignin-PCL (Wang et al., 2018). The lignin-PCL copolymer was synthesized via a solvent-free polymerization method by combining alkali lignin and ε-caprolactone. Afterward, the lignin-PCL copolymer was blended with PCL in 1,1,1,3,3,3-

hexafluoro-2-propanol, and the resulting polymeric blend was electrospun. According to the results, these scaffolds show a nanofiber diameter ranging from 120 to 500 nm, and they promote a higher proliferation of Schwann cells compared to the PCL-based scaffolds.

Additionally, incorporating CL and CA into electrospun PCL/GE nanofibers has significantly improved the mechanical properties and hydrophilicity of the resulting nanofibrous scaffolds (Moazzami Goudarzi et al., 2021). Accordingly, these scaffolds have demonstrated the ability to enhance the proliferation of L929 fibroblast cells, which can be attributed to the presence of amino groups and hydroxyl groups from GE and PCL, respectively. In fact, CL derivatives like CMC have been utilized in the fabrication of nanofiber-based scaffolds. Specifically, these scaffolds were prepared by electrospinning aqueous polymeric blends of CMC and PEO, resulting in a 176–398 nm diameter nanofiber. These nanofibrous scaffolds exhibited a significant improvement in the viability of NIH3T3 fibroblast cells (Basu et al., 2017).

Alternatively, the electrospinning technique has been employed for fabricating air filtration membranes based on nanofibers. The morphology of these membranes often facilitates the efficient capture of fine particles and gaseous waste, due to the high porosity and the distinctive physicochemical properties of the nanofibers' surface (Zhang et al., 2022). For instance, the fabrication of low-pressure drop PVA/CL-nanocrystals (CNCs) composite nanofibrous filters using a green electrospinning process has been documented (Zhang et al., 2019). The nanofibrous filters were obtained from the electrospinning of aqueous suspensions of PVA with different CNCs weight contents onto a metallic window screen for 2, 4, 6, or 8 min. The removal efficiency of these filters was evaluated against smoke generated by burning incense, which includes gases, such as CO, CO_2, SO_2, and NO_2, and volatile organic molecules, such as toluene, benzene, aldehydes, xylenes, and polycyclic aromatic hydrocarbons. According to the authors, the nanofiber diameter decreases from 209.4 to 127.6 nm as the weight content of CNCs in the nanofibrous filter increases. This feature enhances the efficiency of filters in removing both particulate matter 2.5 ($PM_{2.5}$) and 10 (PM_{10}) from the air. Moreover, they reported that this efficiency can reach 99% after 8 min of electrospinning for the nanofibrous filter with the highest CNC weight content, attributed to the lower nanofiber diameter and thickness of the filter.

Suitable nanofibrous membranes for air filtration have been also prepared using green electrospinning processes that involve biocompatible polymers such as CHN and PVA (Zhu et al., 2019). The preparation of these membranes includes SiO_2 nanoparticles to improve filtration efficiency and use a photo-sensitive crosslinker reagent to enhance the stability of the membranes in humid conditions. Furthermore, the authors also incorporated AgNPs into the membranes to provide them with antibacterial characteristics. The experimental evidence demonstrates that these composite air filtration membranes achieve efficiencies as high as 98.73% and 97.30% for filtering NaCl and diisooctyl sebacate aerosol particles, respectively. Additionally, they display a measurable in vitro antibacterial activity against *Bacillus subtilis* and *E. coli*, inducing a low cytotoxicity against L929 fibroblast cells.

In summary, biopolymers have emerged as a sustainable alternative to petroleum-based polymers. These organic materials can be derived from natural renewable resources and can be further modified or synthesized. Biopolymers are recognized for their biocompatibility and biodegradability, positioning them as promising candidates for the development of healthcare platforms. Notably, the electrospinning technique has demonstrated its efficacy in producing nanofibrous materials with a morphology similar to the ECM, thereby aiding tissue regeneration. Additionally, the morphology of these materials enables the efficient capture of fine particles and gaseous waste, making them effective in filtering hazardous particulate matter for human health. Moreover, the specialized literature highlights that combining natural/semisynthetic and synthetic biopolymers, inorganic compounds, and antibacterial species yields promising nanofiber-based platforms for wound dressing, tissue engineering, and air filtration.

6.3 Green solvents for nanofiber synthesis

The term nanofiber refers to nanostructures (<100 nm) with extremely high axial ratios and mechanical flexibility. Due to the properties of nanofibers, their use has been spread to a wide range of modern applications (e.g., clothing, telecommunications, tissue engineering, or water treatment) (Pisignano, 2013). Although researchers have developed different methods of fabrication, the most common still are self-assembly, phase separation, and electrospinning (Eatemadi et al., 2016). Recently, electrospinning has attracted more attention because of the straightforward experimental procedure and the promising approach to mass-scalable production.

Regardless of the method of nanofiber fabrication, two main components are usually employed: (1) the bulk material (e.g., polymers, carbon, ceramics, metal oxides, or even metals) and (2) solvent. From the Latin verb *solvō (which means loosen or untie)*, a solvent is usually defined as the major component of a solution that has dissolved another, known as a solute (Chaniago & Lee, 2022). Although not free of controversy, this definition is still practical because it focuses on the inherent properties that a solvent must own. For example, any polymer can be electrospun with the only condition that a melt or dissolution should be obtained (Xue et al., 2019). In this sense, it is common to use conventional organic solvents (e.g., N, N-dimethylformamide (DMF), ethanol, or dichloromethane) because they are composed of small and highly polar molecules (i.e., with relative permittivities elevated).

Although a solvent should not be characterized only by its macroscopic physical properties, these can be used to screen different solvent options to select the ideal one (Reichardt & Welton, 2010). Undoubtedly, solute−solvent interactions should not be forgotten, but from a macroscopic continuum point of view, properties such as boiling point, vapor pressure, relative permittivity, and viscosity are the most common to obtain on a quick search in chemical databases. With that information and setting aside considerations, such as price or commercial availability, a first attempt to compare different options of solvents can be made. A further understanding of the solubility of a solvent has been addressed by tools such as Kamlet−Taft (K−T) parameters, Reichard's parameter E_T, or a conductor-like screening model for real solvents (COSMO-RS) (Gao et al., 2020).

Nevertheless, an additional concern has strongly risen nowadays and is about the environmental impacts of chemicals used. Since 1988, the 12 green chemistry principles (GCP) have been used as guidelines to design sustainable chemical synthesis and processes (Anastas & Eghbali, 2010). Of particular interest is the GCP 5 because it is related to the auxiliary materials used (e.g., solvents, additives, or modulators). Although the starting materials are difficult to replace, the solvent used seems to be a perfect starting point for addressing the fundamental task of investigation (Fig. 6.1). However, finding ways to substantially reduce, change, or completely remove the use of substances with adverse effects on human health can be challenging.

The most common solvents are organic substances with well-known toxic effects. For example, DMF is a common aprotic polar solvent that is hepatotoxic and reprotoxic and causes gastric irritation (Gescher, 1993). DMF is classified as probably carcinogenic to humans in the 2A group of the International Agency for Research on Cancer (2018). Also, the problems with its recovery and ensuing waste

Figure 6.1 The use of eco-friendly alternatives to replace traditional organic solvents, in line with the first of the 12 Principles of Green Chemistry, is the first step to decreasing environmental contamination.

disposal should be mentioned. Despite all this information, DMF is one of the most popular solvents in chemical laboratories, with a market capitalization of $2.3 billion in 2022 (Dimethylformamide DMF Market Size, 2023). In this scenario, the development of alternatives or new kinds of solvents has attracted the attention of the scientific community.

Recently, the term *neoteric solvent*, coined by Prof. Kenneth R. Seddon in 1996, has been used to designate *novel* solvents that are developed as an alternative to traditional ones (i.e., water and common organic compounds) (Koel et al., 2001; Wilkes, 2002). This group of solvents includes supercritical fluids (SCFs), ionic liquids (ILs), deep eutectic solvents (DES), and fluorous solvents (e.g., perfluorohexane or PFMC) (Sheldon, 2019). These solvents are considered safer reaction media and more environmentally friendly than the usual options. However, it is worth mentioning that to be able to claim the greenness of any solvent used, a comprehensive environmental assessment should be done. Replacing an undesirable solvent is no simple task and warrants attention.

As mentioned above, the choice of a particular solvent based on its role in the synthesis or method of fabrication could limit the available options. Also, a first screening between the huge number of options can make this task slightly intimidating. Diverse efforts have been developed, such as Hansen solubility parameters theory, which is based on the total cohesive energy and the molar volume. This theory offers a more detailed approach to the *"like dissolves like"* rule of thumb (Sánchez-Camargo et al., 2019). In this way, it highlights the convenience of developing guidelines to make solvent selection a little easier. Consequently, creating a set of particular criteria to rank a large set of solvents sounds right. It is interesting to mention that mobile apps have made green solvent selection possible (Ekins et al., 2013).

The idea behind solvent ranking is to establish a method to compare their greenness and create a convenient solvent selection guide (Byrne et al., 2016). Due to the large set of available parameters, user-dependent variability, or different directions within the chemical industries, a universal ranking has not been constructed. Nevertheless, research on this topic is still in progress. Based on this, the European collaborative research project CHEM21 adopted the concept of scores based on waste disposal, environmental impact, and health and safety statements (Sheldon, 2019). With this, an overall ranking of commonly used solvents can be constructed (Table 6.1).

The use of solvent ranking gives quick alternatives in the laboratory. For example, polyacrylonitrile nanofibers can be fabricated by electrospinning using DMF as a solvent where the proportion of PAN/DMF influences the process (Pioquinto-García et al., 2023). The DMF is a traditional solvent election for PAN because of its extremely strong solubility and high relative permittivity. A prominent and common greener alternative is the dimethyl sulfoxide (DMSO) (Chen & Yu, 2010). This is because PAN has high resistance to most solvents (e.g., diphenyl ether or acrylonitrile) but can be solubilized in polar solvents (Nataraj et al., 2012). An additional choice is the use of concentrated acidic or inorganic salt solutions. Excellent results have been obtained with LiCl (Qin & Wang, 2008) or $ZnCl_2$ (Kamran et al.,

Table 6.1 Overall ranking of common solvents.

Family	Solvent	b.p. (°C)	Overall ranking[a]	Family	Solvent	b.p. (°C)	Overall ranking[a]
Water	Water	100	R	Hydrocarbons	Pentane	36	H
Alcohols	Methanol	65	R		n-Hexane	69	H
	Ethanol	78	R		n-Heptane	98	P
	i-Propanol	82	R		Cyclohexane	81	P
	t-Butanol	82	R		Benzene	80	HH
	n-Butanol	118	R		Toluene	111	P
	Ethylene glycol	198	R		Xylenes	140	P
Ketones	Acetone	56	R	Chlorinated hydrocarbons	Dichloromethane	40	H
	Methyl ethyl ketone	80	R		Chloroform	61	HH
	Methyl isobutyl ketone	117	R		1,2-Dichloroethane	84	HH
Esters	Methyl acetate	56	P	Polar aprotic	Carbon tetrachloride	77	HH
	Ethyl acetate	77	R		Acetonitrile	82	P
	i-Propyl acetate	89	R		Dimethylformamide (DMF)	153	H
	n-Butyl acetate	126	R		Dimethylacetamide (DMAc)	166	H
Ethers	Diethyl ether	34	HH		N-methylpyrrolidone	202	H
	Di-i-propyl ehter	69	H		Dimethyl sulfoxide (DMSO)	189	P
	Methyl-*tert*-butyl ether	55	P		Sulfolane	287	H
	Tetrahydrofuran (THF)	66	P		Hexamethylphosphoramide	>200	HH
	Methyl tetrahydrofuran	80	P		Nitromethane	101	HH
	1,4-Dioxane	101	H	Acids	Acetic acid	118	P

[a] R, recommend; P, problematic; H, hazardous; HH, highly hazardous.

2020). All the examples indicated the need to work with more elaborate options than the traditional ones. To overcome these kinds of problems, solvent mixtures have been an attractive option.

The effect of mixed solvents has been pointed at PVP nanofibers fabrication (Yang et al., 2004). The reason for the use of multicomponent solvent mixtures is to enhance the solubility of the interest solute due to the synergistic effects of each constituent. Of course, the physical properties of a mixture solvent are different from each of its components, and different methods have been developed to obtain it. On the other hand, an additional advantage is gained from using mixtures: reducing the volume of toxic solvents used. Back to the work of PVP nanofibers, it was demonstrated that using different binary mixed solvent systems, the diameter of nanofibers obtained was controlled.

Using solvent mixtures to reduce the impact on the GCP 5 (Safer Solvents and Auxiliaries) is a first attempt on the way to greener chemical synthesis. However, related complexities may arise due to the specific requirements for the mixture (e.g., selection and determination of the composition), miscibility (where the Hildebrand solubility parameter for nonaqueous solvents can be helpful to understand potential issues), or the liquid range at the desired temperature and composition (Marcus, 2002). Going beyond this alternative, the elimination of the solvent used is a greener option. This is not always an available option, but it is a promising research area.

On nanofiber fabrication, a set of methods known as solvent-free electrospinning (SFE) has been trending. Despite what its name might suggest, the total avoidance of solvent is not achieved. Instead, the SFE technique comprehends the high use ($>90\%$) of precursor solutions or the employment of neoteric solvents. Under this consideration, SFE comprises a wide group of techniques like melt electrospinning, supercritical CO_2-assisted electrospinning (CO_2-AE), anion-curing electrospinning (ACE), thermo-curing electrospinning (TCE), and UV-curing electrospinning (UV-CE) (Lv et al., 2018).

The quantification of residual solvent concentration into nanofibers could be a useful practice, not only in solvent-free techniques. Leidy & Maria Ximena (2019) have pointed to the relevance of ensuring lower concentrations as possible in nanofibers for food industries. Although this can imply additional effort, the benefits of implementing it far exceed the disadvantages. Common techniques, such as gas chromatography, have demonstrated an incredible potential to perform solvent quantification into nanofibers (Ricaurte et al., 2020). Optimization of the fabrication process and employing solvents wisely to avoid unnecessary waste are the best guidelines to follow in any chemical process.

Recently, Du et al. (2023) prepared nanofibers through electrospinning by using natural polymers and water as the solvent. The formation of inclusion complexes with hydroxypropyl-beta-cyclodextrin (a cyclic oligosaccharide) was a suitable option to overcome the poor solubility. This green approach focuses on a sustainable global solution due to consideration of both the bulk materials and the role of the solvent. Interestingly, Sharma et al. (2021) used the two-step annealing method to prepare metal-oxide perovskite nanofibers (PNFs). This solvent-free solid-state synthesis produced PNFs with stable optical and structural properties. Nonetheless, the use of solvent-free methods is not possible in every single process. In this way, the neoteric solvents are remarked are attractive options to overcome all this inconvenience. Within this group of solvents, ionic liquids are the most well-known due to the variety of available options, chemical tunability, and their increasing use in many areas of chemistry.

6.3.1 Ionic liquids

ILs, salts with melting points below 100°C (Kar et al., 2019), have been recognized as promising solvents in a wide range of reactions. The first IL known, ethylammonium nitrate, was synthesized in 1914, but it was not until several decades later that ILs began to be used as new reaction media (Wasserscheid & Keim, 2000). In the nanofiber area, their advantages took a little more time to be exploited. One of the first attempts to use an IL was on the synthesis of polyaniline nanofibrous networks (Miao et al., 2006). A long-chain imidazolium IL, the 1-hexadecyl-3-methylimidazolium chloride, was used as a template to direct the growth of the nanofibrous networks on the oxidative polymerization of aniline.

Depending on several factors, the usage of ILs in nanofibers requires a cosolvent. A cosolvent is a second solvent added in small amounts into a mixture to increase the solvent power of the primary solvent. This kind of modifier has been used to modulate the solubility of a poorly soluble substance through hydrophobic interactions and changing the stability of solute-solvent aggregates (Van Der Vegt & Nayar, 2017). For instance, DMSO has been used as a cosolvent along with 1-butyl-3-methylimidazolium chloride (BMIMCl) to prepare cellulose nanofibers (CNFs) by electrospinning (Márquez-Ríos et al., 2022). The BMIMCl enhanced the solubility, which was reflected in a more straightforward elaboration of electrospun nanofibers.

More than one thousand different ILs have been reported in the literature to date. ILs have a great capacity to dissolve several inorganic and organic compounds but still have disadvantages that limit their widespread use in nanofiber fabrication. Some ILs exhibit several drawbacks, such as preparation or high cost for use routinely. Regarding their ecological and environmental effects, ILs have been recently reviewed to remark on the relationship between structure and toxicity (Flieger & Flieger, 2020). This is a crucial topic because ILs have been considered environmentally friendly solvents for a long time. DES were developed to try to maintain all the benefits of ILs without negative impacts.

6.3.2 Deep eutectic solvents

DES, closely related in many aspects to ILs, can be defined as a eutectic solution formed by complexation between an ionic hydrogen-bond acceptor (HBA) and a molecular hydrogen-bond donor (HBD) (Binnemans & Jones, 2023; Hansen et al., 2021; Smith et al., 2014). DES are fluids at room temperature with a significant decrease in melting points compared to those of the individual components and are classified into five types (Hansen et al., 2021):

- Type I: quaternary ammonium salt + metal chloride
- Type II: quaternary ammonium salt + metal chloride hydrate
- Type III: quaternary ammonium salt + HBD
- Type IV: metal chloride hydrate + HBD
- Type V: molecular HBA + HBD

The use of DES in nanofiber fabrications has been evaluated. For example, lignin-containing CNFs were produced using a DES based on choline chloride (a quaternary ammonium salt used as HBA) and lactic acid. In addition, a microwave treatment was used on the DES. An additional step was used, which enhanced the DES's ionic properties and molecular polarity (Liu et al., 2020). A biodegradable polyester containing a DES was recently electrospun into a nanofiber (Basar et al., 2023). However, DES was not used as a typical solvent. The incorporation of the system choline chloride/urea/water (a DES model) caused an increase in the poly (3-hydroxybutyrate-co-3-hydroxyvalerate) (PHBV) solution conductivity of approximately 650-fold. The solution was prepared with 2,2,2-trifluoroethanol, a solvent that can cause severe eye damage but with a very low global warming potential (GWP) value compared to other halocarbons (Sellevåg et al., 2004). The PHBV/DES nanofibers are very interesting for air filtration membranes.

Using a one-step strategy with a natural DES as a media, Liu et al. (2021) fabricated a flexible CNF/reduced graphene oxide composite film. Due to its excellent electrical performance, the films have been suggested as substrates for supercapacitors or sensor applications. By the fibrillation method, lysozyme nanofibers were prepared, varying the DES (based on choline chloride and acetic acid) concentration in dissolution [with an optimal of 5 (v/v)%] (Silva et al., 2016). Only a few examples were mentioned, highlighting the popularity of choline chloride-based DES, but it is clear that DES are considered promising materials to use on different nanofiber techniques. However, exploring new alternatives is necessary to cover the different materials, conditions, and requirements of a particular nanofiber preparation.

6.3.3 Supercritical fluids

A popular neoteric solvent is the SCF, an interesting option to replace common toxic organic solvents. The supercritical state is defined using two parameters: critical pressure (p_c) and critical temperature (T_c). Above this critical point on the thermodynamic p-T state plane, which is characteristic of each substance, the behavior of fluids is between gas- and liquid-like. Changes in p or T can modify the properties of SCFs as density or viscosity (Noyori, 1999). From the list of possible SCFs (e.g., water, propane, n-butane, methylpropane, freon, nitrous oxide, or dimethyl ether), supercritical carbon dioxide (scCO$_2$) is one of the most used due to its low cost, availability, and environmental suitability. Furthermore, the critical point of CO$_2$ is at 304.25 K and 7.38 MPa, conditions that are readily attainable in the laboratory.

CTN, a natural linear polymer of N-acetylglucosamine, has a wide range of biomedical and pharmaceutical applications. However, its use as nanofibers has the inconvenience that common solvents do not solubilize it. In this case, a combination of 1,1,1,3,3,3-hexafluoroisopropanol (HFIP) and scCO$_2$ (as antisolvent, a solvent in which the compound is insoluble) has been used satisfactorily in the fabrication of CHN nanofibers (Louvier-Hernández et al., 2005). Although the use of a toxic organic solvent is not entirely avoided, PVP has been electrospun into nanofibers using scCO$_2$ to improve solvent devolatilization (Wahyudiono et al., 2012). Given that data, comparing with the aforementioned scCO$_2$ is also possible.

As mentioned before, it is necessary to be overly cautious with the control of temperature and pressure because they could have a large effect on fiber formation. This was observed in alumina nanofibers obtained by the one-pot sol-gel route in scCO$_2$ (Chowdhury et al., 2010). The nanofiber formation was facilitated at high temperatures (i.e., with the decrease of scCO$_2$ density) and pressure above 4000 psi (i.e., with the increase of CO$_2$ penetration in the drying state). Also, the scCO$_2$ was ideal in this kind of method in which water is not added at first. The hydrolysis and sol-gel reaction rate were controlled by the low solubility of scCO$_2$ in water (concentrations $\sim 0.3\% - 0.5\%$ w/w, in dependence on temperature and pressure). On the optimal conditions, a well-defined nanostructure formation of nanofibers was observed.

Finally, contrasting different solvents could be useful for understanding the difficulties associated with the selection process. The CTN nanofibers are suitable because they have also been fabricated using ILs and DES. Sambhudevan and collaborators (Sulthan et al., 2023) did an extensive literature review to understand the pros and cons of ILs and DES. Further investigation is needed to understand the interaction between CTN−DES, but it offers many advantages that cannot be underestimated (e.g., cost, feasibility, and environmental impact). Different mechanical and textural properties can be obtained with the variation of solvent employed. Subsequently, a wider perspective is essential to find the adequate solvent for each specific situation and requirement.

At this point, the need to comprehend the exact role of the solvent in the chosen method of nanofiber fabrication becomes clearer. The precise knowledge of the solvent (e.g., physicochemical properties, dissolving power, or toxicity) could facilitate solvent selection or search for an optimal alternative. However, focusing on only one part of the nanofiber manufacturing process is unsuitable. Transforming the predominant elaboration methods requires quantitative bases to be able to evaluate and compare different available options. The task can be addressed using methods such as life cycle assessment (LCA) or tools such as the free web-based DOZN 2.0 evaluator (Capello et al., 2007; Nowrouzi et al., 2021; Rico-Barragán et al., 2023; Sharma et al., 2022). The tools developed by green chemistry have effectively devised the most efficient and eco-friendly route possible. In this way, a comprehensive presentation about those tools and their application in research on nanofibers is mandatory.

6.4 Environmental impact assessment

Recently, nanofibers have received significant attention due to their unique properties and potential applications in various processes. However, like any new technology, emerging materials production must be evaluated in function of the environmental impacts of synthesis. The Environmental Impact Assessment (EIA) could be analyzed by LCA (Life cycle assessment); the methodology is considered the most effective method for determining the impacts of a process (Louvier-Hernández et al., 2005; Noyori, 1999), such as nanofibers production. To determine the environmental impact, a system boundary of the production of nanofibers must be presented. For instance, the CNF production was divided into three stages (Khanna et al., 2008): heating and ramp-up, catalytic pyrolysis/CNF production, and separation and purification. The raw materials and energy of every stage were added into the system boundaries; the environmental burdens and CNF product also were included (Fig. 6.2).

The amount of energy and chemical raw materials for material production is relevant to decreasing the environmental impact. The LCA of three different pretreatments (mechanical, enzymatic, and TEMPO oxidation) to produce cellulose micro and nanofibers (CNFs) were analyzed (Arfelis et al., 2023). Although enzymatic pretreatments seem to have less impact, LCA revealed higher environmental impacts in enzymatic procedure than mechanical and TEMPO oxidation; the increase in raw material use enhances the negative effect on acidification, climate change, eutrophication, land use, and photochemical ozone formation indicators.

Global Warming Potential (GWP) indicator represents one of the significant environmental adverse effects of CNF production (Khanna et al., 2008). Methane (700 kg CO_2 eq) and ethylene (500 kg CO_2 eq) are the main contributors to adverse impact in the previous indicator; both are determined as preferred material feedstocks in carbon production. Other environmental impacts related to CNF production are human toxicity potential, ozone layer depletion potential, and photochemical oxidation potential. Similar results were obtained by Berglund et al. (2020) in the LCA analysis of the manufacture of natural nanofibers from industrial

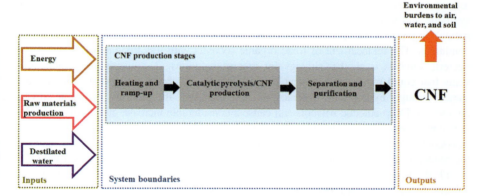

Figure 6.2 System boundaries to obtain cellulose nanofiber (CNF) materials.
Source: From Khanna, V., Bakshi, B. R., & James Lee, L. (2008). Carbon nanofiber production. *Journal of Industrial Ecology*, *12*(3), 394−410. https://doi.org/10.1111/j.1530-9290.2008.00052.x.

waste. The energy and chemicals' raw materials were observed as principal adverse effects in environmental indicators (Berglund et al., 2020).

New technologies, such as the use of food waste to obtain CNFs, also have been evaluated by LCA (Piccinno et al., 2015). The wet-spinning and electrospinning methods were studied; the energy wasted in the first method was lower compared with the other technology. Near 97.8% of the impact observed in the electrospinning process was electricity consumption.

Even though LCA is the robust method for environmental assessment, other alternatives have been applied to determine the EIA of nanofiber production; for example, Green Chemistry Principles (GCP) have been used to evaluate the fibers impregnated with the metal-organic framework DUT-4 using DOZN software tool (Pioquinto-García et al., 2023). The significant environmental impacts were related to the production of the DUT-4 rather than the nanofibers' production; nearly 90% of impacts observed in GCP were related to the toxic waste solvent (DMF) used in DUT-4 synthesis.

Some key environmental considerations are associated with nanofibers: chemical raw materials, resource consumption, energy consumption, and waste production. LCA and GCP have been applied to evaluate the EIA of nanofibers; climate change, photochemical, and ozone layer impacts are the indicators with more negative impacts. Additionally, it is necessary to continue determining the EIA of nanofibers due to their future applications.

6.5 Conclusions

GCP have been successfully applied to nanofiber synthesis to reduce environmental impacts by different approaches. Replacing petroleum-based products with biopolymers as a sustainable and ecologically friendly polymer alternative. Combining natural/semisynthetic and synthetic biopolymers with inorganic compounds can reduce the environmental impacts while at the same time improving the chemical stability in humid conditions. The replacement of toxic organic solvents with less toxic substances such as neoteric solvents is another strategy for greener solutions. IL solvents are neoteric solvents that could require a cosolvent to modulate the solubility of a poorly soluble substance through hydrophobic interactions and changing the stability of solute-solvent aggregates. Deep eutectic solvents can cause a considerable increase in the polymer solution conductivity, improving fiber morphology and enhancing spinnability. In addition, multicomponent solvent mixtures have been demonstrated to increase solubility, causing the reduction of the volume of toxic solvents and the environmental impact on the GCP 5 with the purpose of obtaining a greener chemical synthesis. Methods or tools such as solvent ranking, DOZN green chemistry evaluator, and LCA can be used to devise the most efficient and eco-friendly synthesis route. LCA can be implemented to analyze the nanofiber production process.

Despite significant advancements in the development of green nanofiber synthesis, further research is needed to advance its progress and enable large-scale industrial applications. One key area for future exploration is the use of ILs as solvents in nanofiber production. In this regard, there is an urgent need for a more comprehensive study on the environmental effect of nanofiber production with ILs. Even though these solvents have unique characteristics, such as the capacity to dissolve a wide range of substances, their sustainability and environmental friendliness must be properly assessed. This research is critical to ensuring that the use of ILs in nanofiber synthesis follows GCP. Furthermore, efforts should be directed toward maximizing the use of green solvents to reduce costs. Therefore future research should take a more holistic approach, for example, by using multicriteria decision analysis, which offers a systematic framework for evaluating several criteria, including cost, performance, and environmental impacts. Through these efforts, the field of nanofiber synthesis for energy and environmental applications can realize its full potential and contribute to a more sustainable future.

Reference

Anastas, P., & Eghbali, N. (2010). Green chemistry: Principles and practice. *Chemical Society Reviews*, *39*(1), 301–312. Available from https://doi.org/10.1039/b918763b.

Arfelis, S., Aguado, R. J., Civancik, D., Fullana-i-Palmer, P., Pèlach, M. À., Tarrés, Q., & Delgado-Aguilar, M. (2023). Sustainability of cellulose micro-/nanofibers: A comparative life cycle assessment of pathway technologies. *Science of the Total Environment*, *874*. Available from https://doi.org/10.1016/j.scitotenv.2023.162482, http://www.elsevier.com/locate/scitotenv.

Baranwal, J., Barse, B., Fais, A., Delogu, G. L., & Kumar, A. (2022). Biopolymer: A sustainable material for food and medical applications. *MDPI, India Polymers*, *14*(5). Available from https://doi.org/10.3390/polym14050983, https://www.mdpi.com/2073-4360/14/5/983/pdf.

Basu, P., Repanas, A., Chatterjee, A., Glasmacher, B., NarendraKumar, U., & Manjubala, I. (2017). PEO–CMC blend nanofibers fabrication by electrospinning for soft tissue engineering applications. *Materials Letters*, *195*, 10–13. Available from https://doi.org/10.1016/j.matlet.2017.02.065, http://www.journals.elsevier.com/materials-letters/.

Berglund, L., Breedveld, L., & Oksman, K. (2020). Toward eco-efficient production of natural nanofibers from industrial residue: Eco-design and quality assessment. *Journal of Cleaner Production*, *255*. Available from https://doi.org/10.1016/j.jclepro.2020.120274.

Binnemans, K., & Jones, P. T. (2023). Ionic liquids and deep-eutectic solvents in extractive metallurgy: Mismatch between academic research and industrial applicability. *Springer Science and Business Media Deutschland GmbH, Belgium Journal of Sustainable Metallurgy*, *9*(2), 423–438. Available from https://doi.org/10.1007/s40831-023-00681-6, https://www.springer.com/journal/40831.

Byrne, F. P., Jin, S., Paggiola, G., Petchey, T. H. M., Clark, J. H., Farmer, T. J., Hunt, A. J., R.McElroy, C., & Sherwood, J. (2016). Tools and techniques for solvent selection: Green solvent selection guides. *Sustainable Chemical Processes*, *4*(1). Available from https://doi.org/10.1186/s40508-016-0051-z.

Capello, C., Fischer, U., & Hungerbühler, K. (2007). What is a green solvent? A comprehensive framework for the environmental assessment of solvents. *Green Chemistry*, *9*(9), 927–993. Available from https://doi.org/10.1039/b617536h.

Chaniago, Y. D., & Lee, M. (2022). *Recovery of solvents and fine chemicals sustainable separation engineering: Materials. Techniques and process development* (pp. 483−518). South Korea: Wiley. Available from https://onlinelibrary.wiley.com/doi/book/10.1002/9781119740117, https://doi.org/10.1002/9781119740117.ch13.

Chen, H. M., & Yu, D. G. (2010). An elevated temperature electrospinning process for preparing acyclovir-loaded PAN ultrafine fibers. *Journal of Materials Processing Technology, 210*(12), 1551−1555. Available from https://doi.org/10.1016/j.jmatprotec.2010.05.001.

Chowdhury, M. B. I., Sui, R., Lucky, R. A., & Charpentier, P. A. (2010). One-pot procedure to synthesize high surface area alumina nanofibers using supercritical carbon dioxide. *Langmuir: The ACS Journal of Surfaces and Colloids, 26*(4), 2707−2713. Available from https://doi.org/10.1021/la902738y.

Dimethylformamide (DMF) Market Size, Share. 2023. *Markets and Markets,* 2022−2027. https://www.marketsandmarkets.com/Market-Reports/dimethylformamide-market-129340374.html. Accessed: 2023-07-11.

Du, Z., Lv, H., Wang, C., He, D., Xu, E., Jin, Z., Yuan, C., Guo, L., Wu, Z., Liu, P., & Cui, B. (2023). Organic solvent-free starch-based green electrospun nanofiber mats for curcumin encapsulation and delivery. *International Journal of Biological Macromolecules, 232.* Available from https://doi.org/10.1016/j.ijbiomac.2023.123497.

Eatemadi, A., Daraee, H., Zarghami, N., Yar, H. M., & Akbarzadeh, A. (2016). Nanofiber: Synthesis and biomedical applications. *Artificial Cells, Nanomedicine and Biotechnology, 44* (1), 111−121. Available from https://doi.org/10.3109/21691401.2014.922568, http://www.tandfonline.com/loi/ianb20#.VmugQbfovcs.

Ekins, S., Clark, A. M., & Williams, A. J. (2013). Incorporating green chemistry concepts into mobile chemistry applications and their potential uses. *ACS Sustainable Chemistry and Engineering, 1*(1), 8−13. Available from https://doi.org/10.1021/sc3000509.

Flieger, J., & Flieger, M. (2020). Ionic liquids toxicity—Benefits and threats. *International Journal of Molecular Sciences,* 2020. Available from https://doi.org/10.3390/IJMS21176267.

Gao, F., Bai, R., Ferlin, F., Vaccaro, L., Li, M., & Gu, Y. (2020). Replacement strategies for non-green dipolar aprotic solvents. *Green Chemistry, 22*(19), 6240−6257. Available from https://doi.org/10.1039/d0gc02149k, http://pubs.rsc.org/en/journals/journal/gc.

Gescher, A. (1993). Metabolism of N,N-dimethylformamide: Key to the understanding of its toxicity. *Chemical Research in Toxicology, 6*(3), 245−251. Available from https://doi.org/10.1021/tx00033a001.

Hansen, B. B., Spittle, S., Chen, B., Poe, D., Zhang, Y., Klein, J. M., Horton, A., Adhikari, L., Zelovich, T., Doherty, B. W., Gurkan, B., Maginn, E. J., Ragauskas, A., Dadmun, M., Zawodzinski, T. A., Baker, G. A., Tuckerman, M. E., Savinell, R. F., & Sangoro, J. R. (2021). Deep eutectic solvents: A review of fundamentals and applications. *Chemical Reviews, 121*(3), 1232−1285. Available from https://doi.org/10.1021/acs.chemrev.0c00385, http://pubs.acs.org/journal/chreay.

Hassiba, A. J., El Zowalaty, M. E., Webster, T. J., Abdullah, A. M., Nasrallah, G. K., Khalil, K. A., Luyt, A. S., & Elzatahry, A. A. (2017). Synthesis, characterization, and antimicrobial properties of novel double layer nanocomposite electrospun fibers for wound dressing applications. *International Journal of Nanomedicine, 12,* 2205−2213. Available from https://doi.org/10.2147/IJN.S123417, https://www.dovepress.com/getfile.php?fileID = 35586.

Homaeigohar, S., & Elbahri, M. (2014). Nanocomposite electrospun nanofiber membranes for environmental remediation. *Materials, 7*(2), 1017−1045. Available from https://doi.org/10.3390/ma7021017Germany, http://www.mdpi.com/1996-1944/7/2/1017/pdf.

International Agency for Research on Cancer. (2018). *Some Industrial Chemicals*, World Health Organization.

Júnior, A. F., Ribeiro, C. A., Leyva, M. E., Marques, P. S., Soares, C. R. J., & Alencar de Queiroz, A. A. (2022). Biophysical properties of electrospun chitosan-grafted poly(lactic acid) nanofibrous scaffolds loaded with chondroitin sulfate and silver nanoparticles. *SAGE Publications Ltd, Brazil Journal of Biomaterials Applications, 36*(6), 1098−1110. Available from https://doi.org/10.1177/08853282211046418, http://jba.sagepub.com/.

Kamran, U., Choi, J. R., & Park, S. J. (2020). A role of activators for efficient CO_2 affinity on polyacrylonitrile-based porous carbon materials. *Frontiers in Chemistry, 8*. Available from https://doi.org/10.3389/fchem.2020.00710, http://journal.frontiersin.org/journal/chemistry.

Kar, M., Plechkova, N. V., Seddon, K. R., Pringle, J. M., & MacFarlane, D. R. (2019). Ionic liquids-further progress on the fundamental issues. *Australian Journal of Chemistry, 72* (2), 3−10. Available from https://doi.org/10.1071/CH18541, http://www.publish.csiro.au/nid/52/aid/49.htm.

Keirouz, A., Chung, M., Kwon, J., Fortunato, G., & Radacsi, N. (2020). 2D and 3D electrospinning technologies for the fabrication of nanofibrous scaffolds for skin tissue engineering: A review. *WIREs Nanomedicine and Nanobiotechnology, 12*(4). Available from https://doi.org/10.1002/wnan.1626.

Khanna, V., Bakshi, B. R., & James Lee, L. (2008). Carbon nanofiber production. *Journal of Industrial Ecology, 12*(3), 394−410. Available from https://doi.org/10.1111/j.1530-9290.2008.00052.x.

Koel, M., Ljovin, S., Hollis, K., & Rubin, J. (2001). Using neoteric solvents in oil shale studies. *Pure and Applied Chemistry, 73*(1), 153−159. Available from https://doi.org/10.1351/pac200173010153.

Leidy, R., & Maria Ximena, Q. C. (2019). Use of electrospinning technique to produce nanofibres for food industries: A perspective from regulations to characterisations. *Trends in Food Science and Technology, 85*, 92−106. Available from https://doi.org/10.1016/j.tifs.2019.01.006, http://www.elsevier.com/wps/find/journaldescription.cws_home/601278/description#description.

Liu, C., Li, M. C., Chen, W., Huang, R., Hong, S., Wu, Q., & Mei, C. (2020). Production of lignin-containing cellulose nanofibers using deep eutectic solvents for UV-absorbing polymer reinforcement. *Carbohydrate Polymers, 246*. Available from https://doi.org/10.1016/j.carbpol.2020.116548, http://www.elsevier.com/wps/find/journaldescription.cws_home/405871/description#description.

Liu, Q., Sun, W., Yuan, T., Liang, Sb, Peng, F., & Yao, Cl (2021). Green and cost-effective synthesis of flexible, highly conductive cellulose nanofiber/reduced graphene oxide composite film with deep eutectic solvent. *Carbohydrate Polymers, 272*. Available from https://doi.org/10.1016/j.carbpol.2021.118514, http://www.elsevier.com/wps/find/journaldescription.cws_home/405871/description#description.

Louvier-Hernández, J. F., Luna-Bárcenas, G., Thakur, R., & Gupta, R. B. (2005). Formation of chitin nanofibers by supercritical antisolvent. *Journal of Biomedical Nanotechnology, 1*(1), 109−114. Available from https://doi.org/10.1166/jbn.2005.002.

Luraghi, A., Peri, F., & Moroni, L. (2021). Electrospinning for drug delivery applications: A review. *Journal of Controlled Release, 334*, 463−484. Available from:. Available from http://www.elsevier.com/locate/jconrel, 10.1016/j.jconrel.2021.03.033.

Lv, D., Zhu, M., Jiang, Z., Jiang, S., Zhang, Q., Xiong, R., & Huang, C. (2018). Green electrospun nanofibers and their application in air filtration. *Macromolecular Materials and*

Engineering, *303*(12). Available from https://doi.org/10.1002/mame.201800336, http://onlinelibrary.wiley.com/journal/10.1002/(ISSN)1439-2054.

Marcus, Y. Solvent mixtures: Properties and selective solvation, solvent mix. (2002), Available from https://doi.org/10.1201/9781482275834.

Mianehro, A. (2022). Electrospun bioscaffold based on cellulose acetate and dendrimer-modified cellulose nanocrystals for controlled drug release. *Carbohydrate Polymer Technologies and Applications*, *3*. Available from https://doi.org/10.1016/j.carpta.2022.100187.

Miao, Z., Wang, Y., Liu, Z., Huang, J., Han, B., Sun, Z., & Du, J. (2006). Synthesis of polyaniline nanofibrous networks with the aid of an amphiphilic ionic liquid. *Journal of Nanoscience and Nanotechnology*, *6*(1), 227−230. Available from https://doi.org/10.1166/jnn.2006.17935.

Moazzami Goudarzi, Z., Behzad, T., Ghasemi-Mobarakeh, L., & Kharaziha, M. (2021). An investigation into influence of acetylated cellulose nanofibers on properties of PCL/Gelatin electrospun nanofibrous scaffold for soft tissue engineering. *Polymer*, *213*. Available from https://doi.org/10.1016/j.polymer.2020.123313.

Moulton, M. C., Braydich-Stolle, L. K., Nadagouda, M. N., Kunzelman, S., Hussain, S. M., & Varma, R. S. (2010). Synthesis, characterization and biocompatibility of "green" synthesized silver nanoparticles using tea polyphenols. *Nanoscale*, *2*(5), 763−770. Available from https://doi.org/10.1039/c0nr00046a.

Márquez-Ríos, E., Robles-García, M. Á., Rodríguez-Félix, F., Aguilar-López, J. A., Reynoso-Marín, F. J., Tapia-Hernández, J. A., Cinco-Moroyoqui, F. J., Ceja-Andrade, I., González-Vega, R. I., Barrera-Rodríguez, A., Aguilar-Martínez, J., Omar-Rueda-Puente, E., & Del-Toro-Sánchez, C. L. (2022). Effect of ionic liquids in the elaboration of nanofibers of cellulose bagasse from *Agave tequilana* Weber var. azul by electrospinning technique. *Nanomaterials*, *12*(16). Available from https://doi.org/10.3390/nano12162819, http://www.mdpi.com/journal/nanomaterials.

Nataraj, S. K., Yang, K. S., & Aminabhavi, T. M. (2012). Polyacrylonitrile-based nanofibers—A state-of-the-art review. *Progress in Polymer Science*, *37*(3), 487−513. Available from https://doi.org/10.1016/j.progpolymsci.2011.07.001.

Nowrouzi, M., Abyar, H., Younesi, H., & Khaki, E. (2021). Life cycle environmental and economic assessment of highly efficient carbon-based CO_2 adsorbents: A comparative study. *Journal of CO2 Utilization*, *47*, 101491. Available from https://doi.org/10.1016/J.JCOU0.2021.101491.

Noyori, R. (1999). Supercritical fluids: Introduction. *Chemical Reviews*, *99*(2), 353−354. Available from https://doi.org/10.1021/cr980085a.

Ozan Basar, A., Prieto, C., Pardo-Figuerez, M., & Lagaron, J. M. (2023). Poly(3-hydroxybutyrate- co -3-hydroxyvalerate) electrospun nanofibers containing natural deep eutectic solvents exhibiting a 3D rugose morphology and charge retention properties. *ACS Omega*, *8*(4), 3798−3811. Available from https://doi.org/10.1021/acsomega.2c05838.

Piccinno, F., Hischier, R., Seeger, S., & Som, C. (2015). Life cycle assessment of a new technology to extract, functionalize and orient cellulose nanofibers from food waste. *ACS Sustainable Chemistry & Engineering*, *3*(6), 1047−1055. Available from https://doi.org/10.1021/acssuschemeng.5b00209.

del Pilar Sánchez-Camargo, A., Bueno, M., Parada-Alfonso, F., Cifuentes, A., & Ibáñez, E. (2019). Hansen solubility parameters for selection of green extraction solvents. *TrAC Trends in Analytical Chemistry*, *118*, 227−237. Available from https://doi.org/10.1016/j.trac.2019.05.046.

Pioquinto-García, S., Álvarez, J. R., Rico-Barragán, A. A., Giraudet, S., Rosas-Martínez, J. M., Loredo-Cancino, M., Soto-Regalado, E., Ovando-Medina, V. M., Cordero, T., Rodríguez-Mirasol, J., & Dávila-Guzmán, N. E. (2023). Electrospun Al-MOF fibers as D4 Siloxane adsorbent: Synthesis, environmental impacts, and adsorption behavior. *Microporous and Mesoporous Materials, 348*. Available from https://doi.org/10.1016/j.micromeso.2022.112327, http://www.elsevier.com/inca/publications/store/6/0/0/7/6/0.

Pisignano, D. (2013). *Soft Matter Nanotechnologies. Polymer nanofibers: Building blocks for nanotechnology* (pp. 1−49). Royal Society of Chemistry.

Qin, X. H., & Wang, S. Y. (2008). Interior structure of polyacrylonitrile(PAN) nanofibers with LiCl. *Materials Letters, 62*(8-9), 1325−1327. Available from https://doi.org/10.1016/j.matlet.2007.08.047.

Reichardt, C., & Welton, T. (2010). *Solvents and Solvent Effects in Organic Chemistry. In Solute-Solvent Interactions* (pp. 7−64). Wiley-VCH Verlag GmbH & Co. KGaA. Available from https://doi.org/10.1002/9783527632220.ch2.

Ricaurte, L., Tello-Camacho, E., & Quintanilla-Carvajal, M. X. (2020). Hydrolysed gelatin-derived, solvent-free, electrospun nanofibres for edible applications: Physical, chemical and thermal behaviour. *Food Biophysics, 15*(1), 133−142. Available from https://doi.org/10.1007/s11483-019-09608-9, http://www.springer.com/sgw/cda/frontpage/0,11855,1-40517-70-72976008-0,00.html?changeHeader = true.

Rico-Barragán, A. A., Álvarez, J. R., Pioquinto-García, S., Rodríguez-Hernández, J., Rivas-García, P., & Dávila-Guzmán, N. E. (2023). Cleaner production of metal-organic framework MIL-101(Cr) for toluene adsorption. *Sustainable Production and Consumption, 40*, 159−168. Available from https://doi.org/10.1016/J.SPC0.2023.06.011.

Sellevåg, S. R., Nielsen, C. J., Søvde, O. A., Myhre, G., Sundet, J. K., Stordal, F., & Isaksen, I. S. A. (2004). Atmospheric gas-phase degradation and global warming potentials of 2-fluoroethanol, 2,2-difluoroethanol, and 2,2,2-trifluoroethanol. *Atmospheric Environment, 38*(39), 6725−6735. Available from https://doi.org/10.1016/j.atmosenv.2004.09.023.

Sharma, A. K., Huang, W. S., Pandey, S., & Wu, H. F. (2021). Solvent-free low-temperature annealing method for aqueous stable perovskite nanofibers and its role in peroxide catalysis. *ACS Sustainable Chemistry and Engineering, 9*(22), 7678−7686. Available from https://doi.org/10.1021/acssuschemeng.1c02676, http://pubs.acs.org/journal/ascecg.

Sharma, P., Ponnusamy, E., Ghorai, S., & Colacot, T. J. (2022). DOZN™ 2.0: A quantitative green chemistry evaluator for a sustainable future. *Journal of Organometallic Chemistry, 970−971*. Available from https://doi.org/10.1016/j.jorganchem.2022.122367.

Sheldon, R. A. (2019). The greening of solvents: Towards sustainable organic synthesis. *Current Opinion in Green and Sustainable Chemistry, 18*, 13−19. Available from https://doi.org/10.1016/j.cogsc.2018.11.006, http://www.journals.elsevier.com/current-opinion-in-green-and-sustainable-chemistry.

Silva, N. H. C. S., Pinto, R. J. B., Freire, C. S. R., & Marrucho, I. M. (2016). Production of lysozyme nanofibers using deep eutectic solvent aqueous solutions. *Colloids and Surfaces B: Biointerfaces, 147*, 36−44. Available from https://doi.org/10.1016/j.colsurfb.2016.07.005. Available from, http://www.elsevier.com/locate/colsurfb.

Smith, E. L., Abbott, A. P., & Ryder, K. S. (2014). Deep eutectic solvents (DESs) and their applications. *Chemical Reviews, 114*(21), 11060−11082. Available from https://doi.org/10.1021/cr300162p, http://pubs.acs.org/journal/chreay.

Stratton, S., Shelke, N. B., Hoshino, K., Rudraiah, S., & Kumbar, S. G. (2016). Bioactive polymeric scaffolds for tissue engineering. *Bioactive Materials, 1*(2), 93−108. Available from https://doi.org/10.1016/j.bioactmat.2016.11.001, http://www.keaipublishing.com/en/journals/bioactive-materials/.

Sulthan, R., Reghunadhan, A., & Sambhudevan, S. (2023). A new era of chitin synthesis and dissolution using deep eutectic solvents- comparison with ionic liquids. *Journal of Molecular Liquids*, *380*. Available from https://doi.org/10.1016/j.molliq.2023.121794.

Sun, G., Sun, L., Xie, H., & Liu, J. (2016). Electrospinning of nanofibers for energy applications. *Nanomaterials*, *6*(7). Available from https://doi.org/10.3390/nano6070129, http://www.mdpi.com/2079-4991/6/7/129/pdf.

Syed, M. H., Zahari, M. A. K. M., Khan, M. M. R., Beg, M. D. H., & Abdullah, N. (2023). An overview on recent biomedical applications of biopolymers: Their role in drug delivery systems and comparison of major systems. *Journal of Drug Delivery Science and Technology*, *80*. Available from https://doi.org/10.1016/j.jddst.2022.104121, http://www.editionsdesante.fr/category.php?id_category = 48.

Vasita, R., & Katti, D. S. (2006). Nanofibers and their applications in tissue engineering. *International Journal of Nanomedicine*, *1*(1), 15−30. Available from https://doi.org/10.2147/nano.2006.1.1.15India, http://www.dovepress.com/getfile.php?fileID = 178.

Van Der Vegt, N. F. A., & Nayar, D. (2017). The hydrophobic effect and the role of cosolvents. *Journal of Physical Chemistry B*, *121*(43), 9986−9998. Available from https://doi.org/10.1021/acs.jpcb.7b06453, http://pubs.acs.org/journal/jpcbfk.

Wahyudiono., Murakami, K., Machmudah, S., Sasaki, M., & Goto, M. (2012). Production of nanofibers by electrospinning under pressurized CO_2. *High Pressure Research*, *32*(1), 54−59. Available from https://doi.org/10.1080/08957959.2011.645474.

Wang, J., Tian, L., Luo, B., Ramakrishna, S., Kai, D., Loh, X. J., Yang, I. H., Deen, G. R., & Mo, X. (2018). Engineering PCL/lignin nanofibers as an antioxidant scaffold for the growth of neuron and Schwann cell. *Colloids and Surfaces B: Biointerfaces*, *169*, 356−365. Available from https://doi.org/10.1016/j.colsurfb.2018.05.021, http://www.elsevier.com/locate/colsurfb.

Wasserscheid, P., & Keim, W. (2000). Ionic liquids − New 'solutions' for transition metal catalysis. *Angewandte Chemie − International Edition*, *39*(21), 3772−3789. Available from https://doi.org/10.1002/1521-3773(20001103)39:21 < 3772::aid-anie3772 > 3.0.co;2-5.

Wilkes, J. S. (2002). A short history of ionic liquids − From molten salts to neoteric solvents. *Green Chemistry*, *4*(2), 73−80. Available from https://doi.org/10.1039/b110838g, http://pubs.rsc.org/en/journals/journal/gc.

Xue, J., Wu, T., Dai, Y., & Xia, Y. (2019). Electrospinning and electrospun nanofibers: Methods, materials, and applications. *Chemical Reviews*, *119*(8), 5298−5415. Available from https://doi.org/10.1021/acs.chemrev.8b00593, http://pubs.acs.org/journal/chreay.

Yang, Q., Li, Z., Hong, Y., Zhao, Y., Qiu, S., Wang, C., & Wei, Y. (2004). Influence of solvents on the formation of ultrathin uniform poly(vinyl pyrrolidone) nanofibers with electrospinning. *Journal of Polymer Science Part B: Polymer Physics*, *42*(20), 3721−3726. Available from https://doi.org/10.1002/polb.20222.

Yoon, K., Hsiao, B. S., & Chu, B. (2008). Functional nanofibers for environmental applications. *Journal of Materials Chemistry*, *18*(44), 5326−5334. Available from https://doi.org/10.1039/b804128h.

Zhang, F., Si, Y., Yu, J., & Ding, B. (2023). Electrospun porous engineered nanofiber materials: A versatile medium for energy and environmental applications. *Chemical Engineering Journal*, *456*. Available from https://doi.org/10.1016/j.cej.2022.140989.

Zhang, Q., Li, Q., Young, T. M., Harper, D. P., & Wang, S. (2019). A novel method for fabricating an electrospun poly(vinyl alcohol)/cellulose nanocrystals composite nanofibrous filter with low air resistance for high-efficiency filtration of particulate matter. *ACS*

Sustainable Chemistry and Engineering., *7*(9), 8706−8714. Available from https://doi.org/10.1021/acssuschemeng.9b00605, http://pubs.acs.org/journal/ascecg.

Zhang, X., Ru, Z., Sun, Y., Zhang, M., Wang, J., Ge, M., Liu, H., Wu, S., Cao, C., Ren, X., Mi, J., & Feng, Y. (2022). Recent advances in applications for air pollutants purification and perspectives of electrospun nanofibers. *Journal of Cleaner Production, 378*. Available from https://doi.org/10.1016/j.jclepro.2022.134567.

Zhu, M., Xiong, R., & Huang, C. (2019). Bio-based and photocrosslinked electrospun antibacterial nanofibrous membranes for air filtration. *Carbohydrate Polymers, 205*, 55−62. Available from https://doi.org/10.1016/j.carbpol.2018.09.075.

Biomedical applications of nanofibers and their composites

Akbar Esmaeili
Department of Chemical Engineering, North Tehran Branch, Islamic Azad University, Tehran, Iran

7.1 Introduction

Nanofibers are 1D-nanomaterials in many industrial applications due to their unique physicochemical properties. The cross-sectional diameter of nanofibers can vary from tens to hundreds of nanometers. For this reason, it has created a high specific surface area and surface-to-volume ratio. Nanofibers are a suitable option for making highly porous networks. This structure can be used as a host for various applications. Nanofibers can use materials such as natural and synthetic polymers, carbon materials, semiconductor materials, and composite materials to synthesize.

Along with the rapid progress in synthesizing nanofibers during the last few decades, many efforts have been made to find the potential of nanofibers in various applications. This nanomaterial can be used in many fields and improve the device's performance. Sources of energy production and storage, water and environment purification, healthcare, and medicine are diverse areas where nanofibers can be very effective.

One of the essential types of nanostructures is nanofibers. Generally, fiber with a diameter of less than one micron is considered nanofiber. The American National Science Foundation calls a thread a nanofiber with at least one dimension of one hundred nanometers or less. By reducing the diameter from microns to several hundreds of nanometers, nanofibers can provide excellent properties such as a very high surface-to-volume ratio, flexibility, and good mechanical efficiency compared to conventionally known materials. These outstanding properties introduce nanofiber as a suitable choice for many essential tasks, including tissue engineering.

Nanofibers can be made from different materials, the most important of which are polymeric nanofibers and ceramic nanofibers. Polymeric nanofibers can be spun using various polymers such as organic polymers, high-efficiency polymers, mixed polymers, and biopolymers. Ceramic nanofiber is also prepared using electrospinning and precursors that give it the ability to spin. Characterized by low density, high porosity, excellent mechanical and structural properties, increased flexibility, an extremely high surface area to mass ratio, large volume, potential to interact with chemical agents, filtering capabilities, thinness, high permeability, lightweight, and versatility. The reduction and increase of disorder have made nanofibers suitable for various applications including medical applications, advanced aerospace technology, capacitors, transistors, drug releasers, batteries, fuel cells, catalysts, reinforcing composites, protective clothing, sensors, insulation, and storage (Aliahmadi & Esmaeili, 2022). These composites are suitable for energy, power, electronics, security, defense, and, more recently, clothing.

Nanocomposite fibers are suitable for playing the role of natural extracellular substrate in laboratory conditions. In recent years, these electrospun fibers have received a lot of attention due to the closeness of their structure to the fibrous structure of body tissues and extracellular matrix (ECM), as well as the highly effective surface for cell adhesion and growth, and studies on these scaffolds have significantly expanded (Esmaeili & Kalantari, 2012).

In describing the medical applications of nanofibers, we can mention drug release, artificial blood vessels, medical masks, tissue engineering, wound dressing, etc. For example, carbon nanotubes can carry drugs to blood cells.

Composite nanofibers can be used for many purposes, such as filling soft tissues such as breasts and damaged organs or building veins and ducts. Also, polymer fiber compatible with the body makes temporary body parts.

The human body's cells can grow and connect around fibers smaller than themselves. This way, nanoscale fiber scaffolds can create a suitable environment for cell growth, reproduction, and propagation. In tissue engineering, this biological structure is called a scaffold.

Composite nanofibers can be directly electrospun from biocompatible polymers with drugs. Thus a large surface area is created for drug particles to be placed. This way, the efficiency of the effective transfer of the drug to the target tissue will be increased.

Nanofiber prepared from body-compatible polymers with unique properties and structures can be used to treat wounds or burns. The numerous pores of the nanofiber layer, while facilitating oxygen penetration into the damage, prevent the accumulation of fluids in the injury. On the other hand, the small size of the holes prevents the penetration of bacteria and makes it suitable for use as a wound cover. In addition, the flexibility of the electrospinning process allows polymers to be spun together with drugs or proteins. This way, it is possible to prepare fiber layers from a mixture of polymer and medicine, which have a medical structure that can pass air and wet vapors caused by the wound. At the same time, they prevent the entry of bacteria, and the release of the drug improves the healing process of the damage. It increases the intensity (Tadayon et al., 2015).

7.2 Synthesis of nanofibers

With the help of this method, it is possible to produce a fiber with a long length and a diameter of several nanometers. After exiting the micro-pipette, each nanofiber is stretched during solvent evaporation or at the moment of freezing (the freezing step is done in the melt spinning process by cooling and in dry spinning by solvent evaporation). In the next step, the micropipette is removed from the solution, and nanofibers are produced by adjusting the optimal conditions. The diameter of the fiber produced will depend on the drawing rate, the cooling or evaporation rate, the exact composition of the raw material, etc. Standard drawing processes have not yet been able to produce fibers with a diameter of less than 200 nm because no material has been found that can reach the peak of the curve. Short molecules should be used instead of long polymer chains to obtain a fiber with a diameter of less than 100 nm. Also, stretching thread during the evaporation of the solvent at room temperature leads to the improvement of the raw material. It makes it possible to reach the optimal viscosity for stretching the fibers (peak curve). The stretching technique requires the use of materials with appropriate viscoelastic behavior to withstand high deformation, and there is a need for sufficient adhesion to withstand the pressure applied during the stretching operation regarding the material used (Esmaeili & Haseli, 2017a, 2017b).

In the drawing method for producing nanofiber, it is possible to have a thread with a long length and a diameter of several nanometers. After exiting the micro-pipette, each nanofiber is stretched during solvent evaporation or at the moment of freezing (the freezing step is done in the melt spinning process by cooling and in dry spinning by solvent evaporation).

A micropipette (with a diameter of several microns) is inserted into a droplet of the desired polymer solution. Nanofibers are produced using this technique by inclining the arm and positioning it near the contact line.

In the next step, the micropipette is removed from the solution, and nanofibers are produced by adjusting the optimal conditions.

The diameter of the produced fibers will depend on the drawing rate, the cooling or evaporation rate, the exact composition of the raw material, etc.

As seen in Fig. 7.1, standard stretching processes have not yet been able to produce fibers with a diameter of less than 200 nm because no material has been found that can reach the peak of the curve.

It is explained by the method of stretching and the diameter of the obtained fiber. It mentions the production of (1) polycaprolactone (PCL), (2) polyvinyl alcohol, (3) polyethylene oxide, and (4) polymethyl methacrylate (PMMA) nanofiber (Table 7.1).

Short molecules should be used instead of long polymer chains to obtain a fiber with a diameter of less than 100 nm. Also, stretching the thread during the evaporation of the solvent at room temperature leads to the raw material. It makes it possible to reach the optimal viscosity for testing the fiber (peak curve).

The stretching technique requires using materials with appropriate viscoelastic behavior to withstand high deformation. There is a need for sufficient adhesion to withstand the pressure applied during the stretching operation regarding the material used (Bahramimehr & Esmaeili, 2019).

Nanofibers can be produced from two or more component fibers of different polymers using various methods such as Island-in-the-sea, Side-by-Side, and Sheath-core. In the second step, the fiber structure must remove one of the two polymers. Although the production of two-component fiber with different shapes and cross-sections in micro dimensions is possible with the existing methods, the production of nanofiber with the help of this method entails many problems.

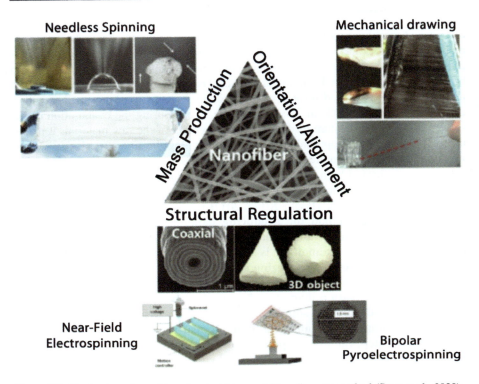

Figure 7.1 Production of nanofibers by stretching and fiber diameter method (Song et al., 2020). *Source*: Permission of MDPI publishing.

Table 7.1 Type materials and diameters of individual nanofibers obtained from the drawing process.

Type of materials	Diameter of nanofibers ± variation (nm)
Polycaprolactone (PCL)	270 ± 30
Polyvinyl alcohol (PVA)	200 ± 40
Blend of hyaluronic acid (HA) and fish gelatin (FG)	470 ± 70
Polyethylene oxide (PEO)	530 ± 60
Polyvinyl butyrate (PVB)	600 ± 90
Polymethyl methacrylate (PMMA)	560 ± 40

For example, a nonwoven layer of two-component fibers is produced with the shape of an island in the sea, in which nylon 6 is the island component. Polylactic acid is the sea component, which has been reported by the spun-bond method; the final nanofiber is obtained after removing the sea component by dissolving it in a suitable solvent (Fig. 7.2).

Extensive research has been conducted on two-component nanofibers derived from the cerebral cortex due to their high potential for use in thermal sensors and composites (Senthamizhan et al., 2016).

7.3 Medical nanofibers

7.3.1 Nanomedicine

One of the challenges medical science has faced in the past is the treatment of damaged body tissues. Tissue engineering uses biological science and engineering to produce physical replacements for damaged tissues. In 1993, a scientist named Bob Langer defined tissue engineering: "Tissue engineering is an interdisciplinary science that uses engineering and biological sciences to improve biological components so that the tissue can be regenerated and maintained and the function improve it." To form a new tissue, cells (stem cells or differentiated cells of the desired tissue) must be cultured on a surface that can simulate the ECM. The ECM is a network of glucosamine glycan chains containing several protein fibrils and fibers. The ECM has specific binding sites (ligands) that affect cell adhesion. The basement membrane also creates a connection between connective tissue and other tissues. The basement membrane includes ECM components such as hyaluronic acid, collagen fibers, laminin, fibronectin, and proteoglycans. Therefore, simulating the structure of the matrix and the composition of its natural materials is one of the primary needs in the construction of tissue engineering scaffolds.

One of the critical approaches in tissue engineering is the construction of a stand that can provide a platform for the growth and differentiation of cells. In three dimensions, the essential things for tissue engineering scaffolds should be considered: (1) cell adhesion, (2) filtration, (3) proliferation, (4) differentiation, and (5) new tissue production scaffolds. It must also degrade over time, matching the rate of degradation speed of tissue regeneration. Today, electrospun nanofibers have become one of the main options for making scaffolds in tissue engineering. It is due to their (1) high surface-to-volume ratio, (2) biocompatibility, (3) biodegradability (according to electrospun polymer), (4) mechanical strength, and (5) porosity. These nanofibrous scaffolds provide the environment for cell

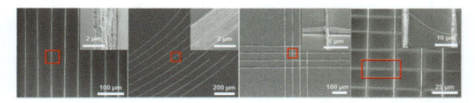

Figure 7.2 Production of nanofibers (Song et al., 2020).
Source: Permission of MDPI publishing.

growth and attachment by simulating the structure of the ECM. Also, nanofiber composites containing materials such as collagen and glucosamine glycans can improve cell-cell and cell-ECM interactions (Eatemadi et al., 2016; Huang et al., 2001; Kelleher & Vacanti, 2010; Sridhar et al., 2011).

Various cells have been cultured on nanofiber scaffolds, and successful results have been obtained. Changes in the diameter, orientation, and composition of nanofibers have essential effects on the adhesion and growth of cells. Achieving better outcomes in cell culture can be achieved through well-controlled fiber diameter, porosity, and direction (random or aligned) of nanofibers. So far, nanofiber scaffolds have been used to reconstruct various tissues such as nerves, bones, veins, cartilage, tendon/ligament, and skin (Sridhar et al., 2011). In a study to make a random and aligned nanofiber scaffold on which muscle cells were cultured, the combination of chitosan and PCL was used. For comparison, the film of this composition was also used for cell culture (Fig. 7.3). The results showed that cell adhesion and morphology on nanofiber were better than on film. Also, it was clear that the results of culture on aligned scaffolds were better than random ones because, in the natural muscle matrix, the orientation of cells and filaments are aligned (Cooper et al., 2010).

Studies have shown that the size of nanofibers affects the adhesion and proliferation of cells. For example, carbon fibers with a length of 60−200 nm lead to increased osteoblast proliferation and attachment, increased alkaline phosphatase activity, and ECM secretion. Core-shell nanofibrous structures can also encapsulate biological substances such as growth factors or influential factors on cell differentiation. Substances are removed from the fiber and placed within reach of cultured cells on the scaffold (Webster et al., 1999; Zhang et al., 2007).

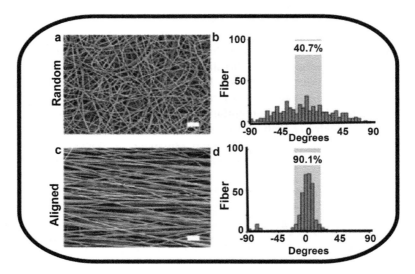

Figure 7.3 Chitosan−PCL fiber morphology and alignment: SEM analysis and quantitative assessment of random and aligned fibers (Cooper et al., 2011).
Source: Permission of Elsevier.

Stem cells are undifferentiated cells that can make cells similar to themselves for a long time and differentiate into a specific cell line (when the conditions for differentiation into a particular type of cell are met). Due to the unique conditions that they have in terms of reproduction, stem cells have been highly considered in tissue engineering. Below are the stages of stem cell culture on nanofibers for skin tissue engineering. Deep wounds/wounds caused by diabetes can be treated using nanofibrous scaffolds containing stem cells and injecting stem cells into the wound Fig. 7.4) (Zhang et al., 2007).

7.3.2 Tissue engineering

The idea behind tissue engineering is the creation of two types of engineered autografts, one by growing one's cells in a laboratory environment on a scaffold and the other by implanting a noncellular scaffold into the cells of the patient's body—the damaged tissue. Repair the sight by guiding the scaffolding. In both cases, at the same time, as the tissue grows, the scaffold must be destroyed so that after the development and growth of the tissue, the platform will no longer exist, and the newly produced tissue can act like the lost tissue (Patrick et al., 1998).

In tissue engineering, a porous material is prepared as an ECM or scaffold for the growth of cells, and then growth factors are placed on it. After the proper development of cells in the space of the pores, the scaffold is transferred from the laboratory environment to the living organism's body. Gradually, the vessels penetrate the platform to feed the cells. Soft tissues of the body must destroy the scaffold, and new tissue replaces it, but in complex tissues, materials that are not necessarily degradable can be used.

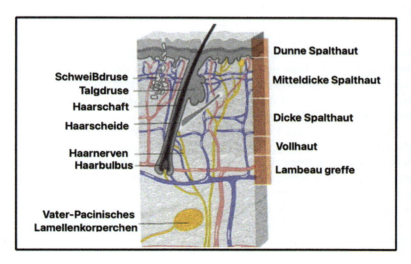

Figure 7.4 Types of skin grafts based on thickness Hierner et al., 2005.
Source: Permission of Elsevier.

Cultured cells can be special cells of that tissue or stem cells. Today, stem cells are one of the most attractive fields of research in biology. A stem cell has a unique feature that can self-regenerate and differentiate into other types of cells. This property of stem cells allows the use of these cells in regenerative medicine or cell therapy. For this reason, these cells are used in Tissue engineering and have received a lot of attention (Misra et al., 2006).

The most critical concern for any tissue engineering application is patient safety. Bulk materials and materials resulting from the destruction of a scaffold must be biocompatible and able to be cleaned and removed by the body. The main task of a stand is to guide the growth and migration of cells from adjacent tissues to the defective site or to facilitate the development of cells implanted on the platform before transplantation. The surface should be chemically favorable for cell adhesion and proliferation. The high diameter of the pores and the high communication of the pores are necessary for forming tissue and transferring nutrients and metabolic wastes. As the porosity and diameter of the holes increase, it leads to an increase (Vacanti & Vacanti, 2014).

In tissue engineering, many engineering sciences are used to achieve this goal. Cell and molecular biologists, medical material engineers, computer simulation designers, microscopic imaging specialists, robotics engineers, and much-advanced equipment, such as bioreactors where tissues are grown and fed, contribute to tissue engineering research. Artificial human tissues such as skin, liver, bone, muscle, cartilage, tendon, and blood vessels are among the things that have been studied so far. The primary purpose of tissue engineering implants is to identify, repair, and regenerate tissue defects and failures, for which engineering principles and biological principles are combined to produce complete replacements of human tissues.

The most critical concern for any tissue engineering application is patient safety. Bulk materials and materials resulting from the destruction of a scaffold must be biocompatible and able to be cleaned and removed by the body. The main task of a stand is to guide the growth and migration of cells from adjacent tissues to the defective site or to facilitate the development of cells implanted on the platform before transplantation. The surface should be chemically favorable for cell adhesion and proliferation. The high diameter of the pores and the high communication of the pores are necessary for forming tissue and transferring nutrients and metabolic wastes. As the porosity and diameter of the holes increase, it leads to an increase (Vacanti & Vacanti, 2014).

As stated by Langer and Vacanti, a typical definition of tissue engineering is "an interdisciplinary field that applies the principles of engineering and life sciences to the development of biological alternatives that regenerate, preserve, or improve tissue engineering." It is defined as "understanding the principles of tissue growth and using it to produce functional replacement tissue for clinical use." The description further states that "the basic hypothesis of tissue engineering is that applying the natural biology of this system will lead to greater success in developing therapeutic strategies. The goal is replacing, repairing, maintaining, or improving tissue function. Key developments in the multidisciplinary field of tissue engineering have yielded a new set of tissue replacement components and implementation strategies. It has created unique opportunities to construct tissues in the laboratory from combinations of extracellular matrix ("scaffolds"), cells, and biologically active molecules. The continued success of tissue engineering and the eventual development of actual human units has stemmed from the convergence of engineering advances and basic research in tissue, matrix, growth factor, stem cell, developmental biology, materials, and life sciences.

A researcher in the field of tissue engineering replaces damaged organs or tissues with applied engineering in the body. Stem cells have opened a new way in tissue engineering. They can regenerate and enter the relevant cells to stimulate and restore conditions with a very high power, which often leads to the use of engineered tissue. Biology and pathology show that many diseases are caused by the wrong Cells. The science of biology and pathology shows presently that the improper functioning of cells causes many diseases (Jiang et al., 2017). The difference between stem cells with different types of tissues and organs is still an essential limiting factor in the field of tissue engineering, which mainly results in multicellular and complex structures of tissues and organs (Esmaeili & Mondal, 2024; Jiang et al., 2017).

7.3.3 Relationship between nanomedicine and tissue engineering

Tissue engineering and nanomedicine are new branches of technology, and the combination of both has a good effect on health. There is a strong need for drug delivery systems that carry growth factors, biological signals, biomaterials, and tissue engineering scaffolds. Regenerative scaffolds are essential in inflammation, angiogenesis, and physical construction (Stayton et al., 2005). Nanoscale materials are cellular assemblies' main structures and aggregates, including subcellular organelles and extracellular matrices (Tuan, 2007). The application of nanomedicine is not limited to drug delivery and tissue engineering. Instead, it has many applications in the health system (Sudhakar et al., 2015; Tuan, 2007).

Nanomedicines are engineering tools for monitoring, repairing, constructing, and controlling biological systems at the molecular scale (Venugopal et al., 2008). Nanotechnology in drug delivery can be unique in the natural system. Drug delivery systems (liposomes and dendrimers) that are used with materials (polymeric scaffolds and hydrogels) in the tissue engineering method are used as drug carriers in the three-dimensional system for tissue engineering (Fig. 7.5).

A model used in tissue engineering is the use of scaffolds in combination with cells and other extracellular agents in simulating the extracellular environment at the vulnerable site.

There are two methods for tissue engineering to produce tissue or organ regeneration. The first method is tissue or organ regeneration using biomolecules with biomaterial scaffolds. The second method is tissue or organ regeneration using donor cells or the cell itself with biomaterial scaffolds (Fig. 7.6). In the method used in tissue engineering, tissues are vital to optimize any tissue engineering technique to produce an equivalent tissue. The source of the cells is the underlying biomaterial to reach the cells in a specific anatomical location, necessitating a regenerative process (Venugopal et al., 2008). Both methods require 3D scaffolds or biomaterials for suturing and regenerating tissue.

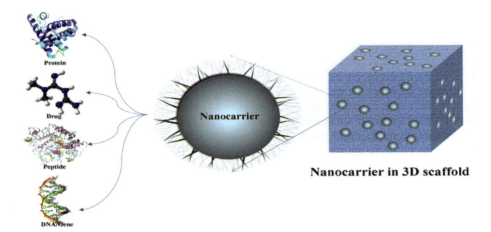

Figure 7.5 The use of 3D carriers in tissue engineering (Thomas et al., 2015).
Source: Permission of William Andrew.

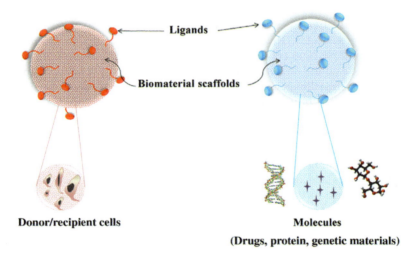

Figure 7.6 Patterns of tissue engineering (Thomas et al., 2015).
Source: Permission of William Andrew.

7.3.4 Nanotechnology methods in bone tissue engineering

Bones include regular collagen fibers, hydroxyapatite, and proteoglycans. These bone structure compounds exist in the macroscopic state (in the centimeter range) to all smaller molecular states (in the nanometer range). Polymers (macromolecules) are the raw materials for scaffolds in various tissue engineering applications. It includes bone and other mineralized tissues. As a scaffold, these polymers have a weak integration with bones and tissue structure. As we know, nanoscale materials show different properties when compared to bulk materials.

Nanostructured materials are designed to simulate natural bones. Nanomaterial compounds in the range of 1–100 nm show better and higher electrical and mechanical properties and cell compatibility on a scale with conventional materials on the corresponding micro scale.

A type of new material, spiral nanotubes, is pristine organic nanotubes with a natural nanostructure of collagen and other compounds found in bone. Helical nanotubes are new materials that grow in spontaneous DNA assembly based on base pair structure in the body solution. They are primarily coil-shaped nanotubes that act similarly to natural collagen fields (Chen et al., 2011). Helical nanotubes of a guanine-cytosine structure halt the process by which key elements perform spontaneous sequential assembly of themselves to produce stable nanotubes.

Helical nanotubes have unique physical and chemical properties that make them significant for drug delivery and tissue engineering applications.

Helical nanotube hydrogels are used as new tissue engineering injectable scaffold materials for orthopedic applications (Zhang et al., 2006). They are used to deliver drugs due to their compatibility with the body, low cytotoxicity, and ability to produce a biologically favorable environment. It is for cell attachment and growth (Chen et al., 2011). They are very suitable as a model of biological simulation for inserting helical nanotubes (Zhang et al., 2008). They can be modified chemically with various peptides for tissue engineering applications and physically by entrapping drugs without their cores and structurally by synthetically changing their dimensions (length, thickness) for a wide range of therapeutic qualities (Chen et al., 2011). More about this source text is required for additional translation information to send feedback on side panels.

Hydroxyapatite nanophase has excellent cell compatibility properties with osteoblasts (Chang et al., 2001). The use of implants with nanoporous titanium oxide-modified surface in dental treatment study showed that a thin film of titanium oxide increased bonding in faster healing of human oral mucosa and reduced marginal bone resorption (Wennerberg et al., 2011; Zhang et al., 2008). Structured nanosurface implants are known to enhance osteoblast activity.

Scaffolds based on nanocomposites are very common in complex tissue engineering, especially for bone tissue regeneration (Mouriño & Boccaccini, 2010). Improving the biological activity and implementation of bone substrate materials and scaffolds is one of the main concerns in bone regeneration (Mouriño & Boccaccini, 2010). Bone tissue requires the action of growth factors at the wound sites so that both stimulated and previously stimulated cells can move and contribute to the healing process (Furth et al., 2007). Growth factors or other signaling molecules must be released from the scaffolds to be released in a controlled manner in the amount required for tissue regeneration.

Drug delivery plays a vital role in maintaining the release rate in a controlled manner. Control of tissue concentration and concentration in drug delivery is essential for safety and effectiveness. Drug delivery can complement the demand for drug engineering; for example, controlling the release of growth factor for optimal concentration creates positive feedback. Drug release control carries out drug delivery guidance and supervision of development processes in the tissue and detection of particular tissue and morphology. Nanoporous drug delivery systems act to control the release of growth factors or molecules for tissue engineering (Thomas et al., 2015).

7.3.5 Heart tissue engineering

Cardiac tissue engineering is used to produce contractile heart muscle tissues to replace lost parts (due to congenital heart defects) or features that do not function properly, causing heart attacks, and therefore used for heart reconstruction (Amin et al., 2020). Objectives in heart tissue engineering are blood vessels, heart tissue, and heart valves. Often, the common biomaterials used for heart tissue engineering are hydrogels, biodegradable polymer scaffolds, and noncellular tissues. Scaffold materials capable of forming a matrix structure, mimicking the extracellular environment, are essential factors. The rapid development of platforms in the past ten years has led to a new perspective and progress in biomedical research and conventional treatments. The microscopic and nanoscale structures of the scaffold surface play a crucial role in the adhesion and growth of heart cells (Zhang et al., 2011).

Electrospun nanofibrous scaffolds have been extensively used to control cardiac and vascular tissue engineering architecture. It has resulted in specific physicochemical properties (Oh & Lee, 2013).

The need for heart tissue engineering production is that it should be able to withstand the dynamic disturbance of high blood pressure for blood tissue engineering and also be able to operate in a very dynamic environment for heart valve tissue engineering. Heart attack treatment can be achieved using tissue-engineered heart sections.

Integrating gold nanosystems with alginate scaffolds or polylactic acid scaffolds can bridge the porous walls of alginate electrical resistance and improve the electrical communication between adjacent heart cells (Jaconi, 2011). Combining nanofiber carriers and stem cell therapy for tissue regeneration seems to have great potential for treating heart diseases, including arteriosclerosis and myocardial infarction (Oh & Lee, 2013). 3D inkjet methods are also used to develop artificial valves produced by nanomaterials for cardiovascular tissue engineering.

7.3.6 Dendrimer

Nanodendrimers are a group of nanoparticles being considered in many biomedical fields today. Dendrimers are a unique branch of polymer macromolecules with multiple stems growing from a center and creating an almost complete three-dimensional pattern (Bharali et al., 2009). These particles have a diameter between 10 and 200 Å. Dendrimers are composed of three parts of the core and side branches (Bharali et al., 2009).

On the other hand, they can connect and transport all kinds of molecules to their character. It used nano dendrimers for a specific tissue.

Dendrimers also have holes inside. Drug molecules can be attached to these particles' surfaces and carried inside (Klajnert & Bryszewska, 2001). Another unique feature of dendrimers is the multifunctional nature of these particles so that they simultaneously contain drug molecules, targeting parts, and dissolvable groups. The unique structure of these particles allows for conjugating molecules on the surface and encapsulating them. Pegylated dendrimers (dendrimers coated with polyethylene glycol) are one of the classes of dendrimers that have attracted the attention of many researchers due to their long circulation time in the blood, less toxic level, and relatively less accumulation in different organs.

Dendrimers also have holes inside. Drug molecules can be attached to these particles' surfaces and carried inside (Klajnert & Bryszewska, 2001). Another unique feature of dendrimers is the multifunctional nature of these particles so that they simultaneously contain drug molecules, targeting parts, and dissolvable groups. The unique structure of these particles allows for conjugating molecules on the surface and encapsulating them. Pegylated dendrimers (dendrimers coated with polyethylene glycol) are one of the classes of dendrimers that have attracted the attention of many researchers due to their long circulation time in the blood, less toxic level, and relatively less accumulation in different organs. However, their effectiveness cannot be used as anticancer drugs due to their high toxicity. The therapeutic effects of any drug are related to its solubility in water. Most drug molecules cannot be used due to a lack of dissolution in suitable solvents. Dendrimers can form bonds with these molecules and dissolve hydrophobic molecules (Klajnert & Bryszewska, 2001).

Cancer tissues have unique properties that are permeable to macromolecules up to 400 nm in diameter (compared to healthy tissues with a permeability of 2−6 nm). The process of entering molecules through the leaky structure of cancerous tissue is called the enhanced permeability and retention (EPR) phenomenon. It is well proven that pegylated nanoparticles accumulate in tumor tissues due to the EPR phenomenon. This natural pattern of nanoparticle distribution is called passive targeting, but the EPR phenomenon is not always effective and efficient. The phenomenon may not exist in all parts of a large tumor. Also, due to the nonspecificity of the carriers, it raises the necessity of targeted drug delivery.

Nanosystems will cause targeted drug delivery, increasing efficiency, and reducing side effects. (Bharali et al., 2009). Folic acid or folate is an essential vitamin required for the functioning of cancer cells. For cancer cells to receive this substance as much as possible, many receptors are expressed on the cell surface. As mentioned, dendrimers can attach many meanings to their character and carry them due to having multiple functional groups on the surface. Folic acid is conjugated on the surface of dendrimer nanoparticles. Studies have shown that the folic acid receptor binds to folic acid with a very high affinity and guides it into the cell through endocytosis (Bharali et al., 2009).

7.3.7 Nanofibers as three-dimensional scaffolds for tissue regeneration

Researchers presented a new strategy that converts nanofiber membranes into 3D scaffolds. These scaffolds have application potential in fields such as engineering.

In the movie Transformers, cars turn into robots and jets can turn into machines. A group of researchers used a similar concept to produce complex 3D structures using bubble-forming chemicals that could operate in 3D modeling technologies in biomedicine.

In an article published by this research group in Applied Physics Reviews, a new strategy shows a significant improvement in the speed and quality of producing complex three-dimensional structures compared to standard methods.

This project is the first successful demonstration of the production of 3D neural tissue structures using stable structures in which differentiated stem cells are placed on the surface of 3D nanofiber scaffolds.

The physical basis of this technology is based on the application of electric current and overcoming the surface tension of a solution, which causes the solution to become fragile fibers, done after the solvent's evaporation (Esmaeili & Mousavi, 2017).

Due to the inherent properties of electrospinning, nanofibers are primarily created as two-dimensional membranes on the surface, which have a dense structure and tiny dimensions, usually smaller than cells. Therefore, the cells are not able to penetrate these fibers.

This group used gas foaming and 3D molding to produce 3D nanofibrous membranes, where the nanofibers can form a predetermined shape due to being confined in a specific space.

One of the advantages of this project is that the production process can be done in 1 hour, while in other methods, it takes 12 hours.

Woven nanofibers can have many applications in tissue engineering due to their similarity with the intercellular matrix. One of the most exciting features of the findings of this group is that the three-dimensional structure produced with nanofibers, when covered with gelatin, has high superelastic properties and quickly returns to its original shape when deformed (Tahmasebi et al., 2022).

7.4 Composite nanofibers

7.4.1 Polymer nanotechnology: nanocomposites

A composite material, also known simply as a composite, is made from two or more components. These constituents have distinctly different chemical or physical properties and combine to create a substance with different properties, unlike the individual elements. In the final structure, the individual components remain separate and distinct, distinguishing composites from mixtures and solid solutions.

Nanofiber composites are a relatively new, unique, and versatile class of nanomaterials. The nanofiber composite approach significantly increased cell attachment and cellular functions compared to the single-phase nanofiber approach.

Research shows that nanofibrous composites produced using electrospinning techniques have superior mechanical properties, bioactivity, and biochemical properties compared to their constituent phases. Hence, the fabrication of nanofiber composite using electrospinning techniques has attracted wide attention in recent years.

With the complexity of life in the environment, the demand for air filter materials has gradually increased. There are many methods of making air filter materials by electrospinning. Multifluid electrospinning processes and research on preparing composite materials with multiple polymers have never stopped. The environment in which a person lives is often not monistic and usually includes various needs. For example, air filters in coal power plants need high-temperature resistance and flame-retardant performance, and chemical plants need moisture resistance and chemical corrosion resistance, etc. Filter membranes made from a single polymer are often monofunctional and cannot meet the demands of complex environments. Therefore, it is concluded that a single polymer can improve its performance after preparation by another polymer. TPP and nylon core-shell fibers are applied through a coaxial electrospinning process, combining the flame-retardant properties of TPP and the filtration properties of nylon. Research on multiple polymers is continuously developing (Correa et al., 2017).

The world has been facing severe problems related to air, land, and water pollution by several hazardous chemicals released into the environment, including those caused by industrial activities, car wear, food production and consumption, and population growth. Such pollutants, including heavy metals, dyes, and emerging contaminants, can be hazardous to human health and terrestrial and aquatic environments. In this context, new technologies aimed at reducing water pollution have been sought, such as those based on filter and absorption techniques, in which the selection or combination of distinctive materials plays a vital role. It has been shown that polymer nanofibers are very suitable for such purposes due to their high absorption, including small diameters (on the micro and nanoscale). High porosity and large surface area are available for chemical functionalization. Among the fiber forming techniques, electrospinning is capable of producing long, porous fibers with small diameters and a large surface area suitable for water pollutant removal. It offers several advantages, such as a much higher output rate and no need for a high-voltage source after applying very high pressure (Fig. 7.7) (Vega-Cázarez et al., 2018).

As wound healing continues to be a challenge for the medical field, wound management has become an essential factor for healthcare systems. Nanotechnology is a field that can offer different new approaches to regenerative medicine. In the past years, nano research has taken steps to develop molecular engineering strategies for self-assembling various biocompatible nanoparticles. It is well known that nanomaterials can heal and delay wound healing, as current wound healing therapies usually do not provide an excellent clinical outcome either structurally or functionally. Therefore, using nanomaterials in wound management represents a unique tool that can be specifically designed to reflect the basic physiological processes in tissue repair (Stoica et al., 2020).

Chitosan has medical applications due to its natural origin and properties, biocompatibility, nontoxicity, and antimicrobial capacity. Electrospinning produces

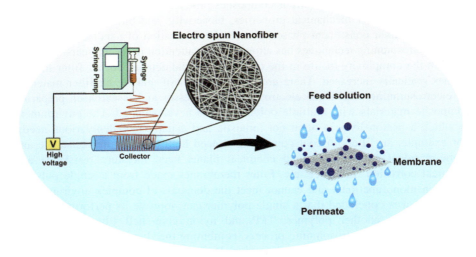

Figure 7.7 Composite nanofibers to remove pollutants (Radoor et al., 2024).
Source: Permission of Elsevier.

nonwoven nanofibers for wound dressings with high specific surface area and tiny pores. These properties help absorb secretions, prevent the penetration of bacteria, and, as a result, heal the wound. For this reason, chitosan composition is used to produce nanofibrous dressings, and its structural, mechanical, and biological properties are promising for further studies. Researchers are looking for biomaterials that provide modern dressings with many qualities designed for wound healing (Stoica et al., 2020).

7.4.2 Comments and ideas for the development of polymer matrix-based nanocomposites

The essential characteristic of composite pipes has made them superior in the transmission lines of various products. Corrosion resistance is caused by fluids (liquids and gases) in internal and external walls. Due to their polymer structure, composite pipes are entirely safe from this phenomenon and can work without repair in active chemical and electrochemical environments for 25–50 years. Eliminating the heavy maintenance and repair costs of corroded oil or gas pipes and damages caused by service interruptions to industrial centers are the most critical factors that have made composite pipes surpass their traditional competitors.

The diagram above shows an overview of different countries and their position in this industry. 8.95% of the composites used in Iran are polymer-based composites, and the most significant volume of their consumption is specific to the construction and transportation industries. In the opposite diagram, the distribution of the frequency of composites in various industrial sectors in our country is visible.

Given the widespread use of conventional composites domestically and the projected increase in polymer production by the National Petrochemical Industries Company in the coming years, there is a growing need to expand the utilization of these polymers. Producing polymer nanocomposites (http://www.jozveha.ir) is one of the most suitable approaches to meet this demand of polymer nanocomposites is one of the most appropriate ways to met this demand. The market aims to improve the properties and expand the scope of application for domestic polymers. Presently, the total consumption of polymer mixtures inside the country is about 150 thousand tons per year. These are supplied through imports from countries such as the Netherlands, Italy, Taiwan, Sweden, and Germany, and domestic producers providing the other part. These mixtures are mainly used in automotive, home and office appliances, and rubber industries. They are considering the superior properties of polymer nanocomposites compared to ordinary polymer mixtures and the downward trend of the global price of nanoparticles. Due to competitive pricing possibilities, it is expected that producing nanocomposites within the country can replace a significant portion of conventional polymer blends.

Polymers do not have much strength despite having lightness and processability at low temperatures. Therefore, they cannot be a suitable substitute for metals.

Reinforcement of polymers by expected reinforcements, such as glass fibers, damages the two main characteristics of polymers, that is, lightness and ease of processability (Beni & Esmaeili, 2020).

Polymer nanocomposites are a new generation of materials that contain a polymer base (matrix) and a small percentage (less than 10% by weight) of a nanometer reinforcement mixed with a suitable method. Due to small dimensions and a very high contact surface, nanoparticles improve the desired properties at a lower loading rate. Problems related to expected reinforcements, such as weight gain, surface defects, and processability problems, are less common in them.

Existing polymer gels are unsuitable due to their low strength and poor stability in harsh conditions of oil tanks. For this reason, in the first phase of this project, using nanotechnology, polymer hydrogels' strength and efficiency were increased for use in harsh conditions of oil tanks.

The project's third phase investigated the production of polymer nanocomposites with unique properties such as resistance to gas penetration, bacterial growth, and utilizing nanotechnology.

The nanocomposites prepared at the Oil Industry Research Institute can develop and increase scale and industrial production (Rajabi & Esmaeili, 2020).

Nanocomposites find applications in various areas of the oil industry, including water management of oil and gas wells, increased extraction from oil reservoirs, purification and separation of gases, lightening and increasing the flame resistance of oil structures, facilitating the transfer of crude oil, and reducing the environmental pollution (Mahmoudi et al., 2022).

7.5 Conclusions and future perspectives

The scope of nanomedicine is infinite and shared with several traditional sciences. Our understanding of nanomedicine and nanomaterials still leaves many obstacles in achieving the goal. The advancement of the power of new analytical tools and research at the nanoscale makes it possible to detect the mechanical and chemical properties of cells and to measure the properties of single molecules. There is still much research in psychiatry and tissue engineering, and the field of nanomedicine is providing new solutions and products that will solve the treatment challenges of the next decade.

In the last decade, much research and experiments have been done to investigate nanofibers for medical applications. Nanofibers are precisely similar to the natural ECM. They protect cell proliferation and adhesion and maintaining their physical shape and growing directly according to the direction of the nanofibers. This review has discussed recent research utilizing nanofibers in tissue engineering applications for skin, bone, heart, cartilage, vascular, and nerve tissues, as well as in drug delivery systems and wound dressings. In each application, some basic principles about the role of nanofibers in performing the desired function are discussed. This review has noted new insights for medical applications, focusing on the promising benefits of using electrospun nanofibers. In general, with the development of science and technology, we believe that shortly we will see the use of electrospun nanofiber scaffolds in extensive clinical applications.

Predicting the future of superior technology is challenging due to the possibility of underestimating the effect of the technology. Currently, the development of nanotechnology has surpassed the predictions made in 2000. Every power graph eventually reaches a point where the growth rate becomes infinite. What will happen after 2020? According to the current conditions, it is predicted that the acceleration of technology growth will lead to products that look like science fiction today. The primary driving force for technological advancement is the constant demand for new materials and the competitive pressure. The progress of artificial intelligence is such that computers help technological developments and scientific discoveries. In other words, intelligent machines will make discoveries that will be very complicated for humans. Finally, there is a hypothesis that the solution to many of today's problems, including material scarcity, human health, and environmental degradation, will be solved by technology. Technological progress has been much faster than the best predictions in the last few years. Although the development of science is speedy, the pace of change in technology and everyday life is much slower for several reasons. The first reason is that it takes time to transform scientific discoveries into new products, especially when the target product market is unclear. The second reason is the resistance of individuals and institutions to change because any new technology needs organizational and cost changes to be effective. For example, computer technology was not economically profitable until its use every day in offices and various activities.

Different fields of nanotechnology are dedicated to cancer diagnosis and treatment. One of the main goals of nanotechnology in medicine is to make devices that act as drug-release systems inside the body. Radiotherapy and chemotherapy impose many side effects on the patient because they destroy healthy and cancerous cells. Nanotechnology can cure all types of cancer by targeting cancer cells.

Super-sticky complex material: The lizard's ability to stick to surfaces and walk on walls led researchers to simulate the elastic fibers in the soles of the creature's feet. The subject led to the creation of a material with an adhesion ten times stronger than the adhesion of lizard legs using carbon nanotubes. When these nanotubes are placed parallel to the surface, they stick much more to the surface than when they are vertical. Attaching a heavy weight to a vertical surface while being able to detach it from the surface quickly used the result of this design. The existence of nanotechnology has made it possible to implement the idea of making clothing that allows humans to walk on the vertical surface of the wall.

Energy production and its use: the new generation of nanosensors, catalysts, and materials has reduced energy consumption. The use of engineering nanomaterials increases efficiency in energy transfer. Nanotechnology in renewable energy helps to improve efficiency and reduce cost. It is possible to take advantage of the nanoscale surface's properties and new nanomaterial manufacturing methods.

References

Aliahmadi, M., & Esmaeili, A. (2022). Preparation of nanocapsules chitosan modified with selenium extracted from the *Lactobacillus acidophilus* and their anticancer properties. *Archives of Biochemistry and Biophysics* 109327.

Bahramimehr, F., & Esmaeili, A. (2019). Producing hybrid nanofiber-based on/PAN/$F_{e3}O_4$/zeolite/nettle plant extract/urease and a deformed coaxial natural polymer to reduce toxicity materials in the blood of dialysis patients. *Journal of Biomedical Materials Research Part A, 107*, 1736–1743.

Beni, A. A., & Esmaeili, A. (2020). Biosorption, an efficient method for removing heavy metals from industrial effluents: A review. *Environmental Technology & Innovation, 17* 100503.

Bharali, D. J., Khalil, M., Gurbuz, M., Simone, T. M., & Mousa, S. A. (2009). Nanoparticles and cancer therapy: A concise review with emphasis on dendrimers. *International Journal of Nanomedicine, 4*, 1.

Chang, C., Wu, J., Mao, D., & Ding, C. (2001). Mechanical and histological evaluations of hydroxyapatite-coated and noncoated Ti6Al4V implants in tibia bone. *Journal of Biomedical Materials Research: An Official Journal of The Society for Biomaterials, The Japanese Society for Biomaterials, and The Australian Society for Biomaterials and the Korean Society for Biomaterials, 56*, 17–23.

Chen, Y., Song, S., Yan, Z., Fenniri, H., & Webster, T. J. (2011). Self-assembled rosette nanotubes encapsulate and slowly release dexamethasone. *International Journal of Nanomedicine, 6*, 1035.

Cooper, A., Jana, S., Bhattarai, N., & Zhang, M. (2010). Aligned chitosan-based nanofibers for enhanced myogenesis. *Journal of Materials Chemistry, 20*, 8904–8911.

Cooper, A., Bhattarai, N., & Zhang, M. (2011). Fabrication and cellular compatibility of aligned chitosan–PCL fibers for nerve tissue regeneration. *Carbohydrate Polymers, 85*, 149–156.

Correa, D. S., Mercante, L. A., Schneider, R., Facure, M. H., & Locilento, D. A. (2017). Composite nanofibers for removing water pollutants: Fabrication techniques. *Handbook of Ecomaterials* (pp. 1−29).

Eatemadi, A., Daraee, H., Zarghami, N., Melat Yar, H., & Akbarzadeh, A. (2016). Nanofiber: Synthesis and biomedical applications. *Artificial Cells, Nanomedicine, and Biotechnology, 44*, 111−121.

Esmaeili, A., & Haseli, M. (2017a). Electrospinning of thermoplastic carboxymethyl cellulose/poly (ethylene oxide) nanofibers for use in drug-release systems. *Materials Science and Engineering: C, 77*, 1117−1127.

Esmaeili A., Haseli M. (2017b). Optimization, synthesis, and characterization of coaxial electrospun sodium carboxymethyl cellulose-graft-methyl acrylate/poly (ethylene oxide) nanofibers for potential drug-delivery applications. *Carbohydrate Polymers, 173*, 645−653.

Esmaeili, A., & Kalantari, M. (2012). Bioremoval of an azo textile dye, Reactive Red 198, by *Aspergillus flavus*. *World Journal of Microbiology and Biotechnology, 28*, 1125−1131.

Esmaeili, A., & Mondal, M. I. (2024). *Smart textiles from natural resources* (1, pp. 3−30). Elsevier.

Esmaeili, A., & Mousavi, S. N. (2017). Synthesis of a novel structure for the oral delivery of insulin and the study of its effect on diabetic rats. *Life Sciences, 186*, 43−49.

Furth, M. E., Atala, A., & Van Dyke, M. E. (2007). Smart biomaterials design for tissue engineering and regenerative medicine. *Biomaterials, 28*, 5068−5073.

Hierner, R., Degreef, H., Vranckx, J. J., Garmyn, M., Massagé, P., & van Brussel, M. (2005). Skin grafting and wound healing—The "dermato-plastic team approach". *Clinics in Dermatology, 23*, 343−352.

Huang, L., Apkarian, R. P., & Chaikof, E. L. (2001). High-resolution analysis of engineered type I collagen nanofibers by electron microscopy. *Scanning, 23*, 372−375.

Jaconi, M. E. (2011). Gold nanowires to mend a heart. *Nature Nanotechnology, 6*, 692−693.

Jiang, W., Von Roemeling, C. A., Chen, Y., Qie, Y., Liu, X., Chen, J., & Kim, B. (2017). Designing nanomedicine for immuno-oncology. *Nature Biomedical Engineering, 1*, 1−11.

Kelleher, C. M., & Vacanti, J. P. (2010). Engineering extracellular matrix through nanotechnology. *Journal of the Royal Society Interface, 7*, S717−S729.

Klajnert, B., & Bryszewska, M. (2001). Dendrimers: Properties and applications. *Acta Biochimica Polonica, 48*, 199−208.

Mahmoudi, R., Esmaeili, A., & Nematollahzadeh, A. (2022). Preparation of $Fe_3O_4/Ag_3VO_4/$ Au nanocomposite coated with *Caerophyllum macropodum* extract modified with oleic acid for theranostics agent in medical imaging. *Journal of Photochemistry and Photobiology A: Chemistry, 425*113724.

Misra, S. K., Valappil, S. P., Roy, I., & Boccaccini, A. R. (2006). Polyhydroxyalkanoate (PHA)/ inorganic phase composites for tissue engineering applications. *Biomacromolecules, 7*, 2249−2258.

Mouriño, V., & Boccaccini, A. R. (2010). Bone tissue engineering therapeutics: Controlled drug delivery in three-dimensional scaffolds. *Journal of the Royal Society Interface, 7*, 209−227.

Oh, B., & Lee, C. H. (2013). Nanofiber for cardiovascular tissue engineering. *Expert Opinion on Drug Delivery, 10*, 1565−1582.

Patrick, C. W., Mikos, A. G., & McIntire, L. V. (1998). Section IV: Tissue engineering applied to organs. *Frontiers in tissue engineering*, (Volume 1, p.561). Elsevier.

Radoor, S., Karayil, J., Jayakumar, A., & Siengchin, S. (2024). Efficient removal of dyes, heavy metals and oil-water from wastewater using electrospun nanofiber membranes: A review. *Journal of Water Process Engineering, 59*, 104983.

Rajabi, A., & Esmaeili, A. (2020). Preparation of three-phase nanocomposite antimicrobial scaffold BCP/Gelatin/45S5 glass with drug vancomycin and BMP-2 loading for bone regeneration. *Colloids and Surfaces A: Physicochemical and Engineering Aspects*, *606*125508.

Amin, D. R., Sink, E., Narayan, S. P., Abdel-Hafiz, M., Mestroni, L., & Peña, B. (2020). Nanomaterials for cardiac tissue engineering. *Molecules*, *25*, 5189.

Senthamizhan, A., Balusamy, B., & Uyar, T. (2016). Glucose sensors based on electrospun nanofibers: A review. *Analytical and Bioanalytical Chemistry*, *408*, 1285−1306.

Song, J., Kim, M., & Lee, H. (2020). Recent Advances on nanofiber fabrications: Unconventional state-of-the-art spinning techniques. *Polymers*, *12*, 1386.

Sridhar, R., Venugopal, J., Sundarrajan, S., Ravichandran, R., Ramalingam, B., & Ramakrishna, S. (2011). Electrospun nanofibers for pharmaceutical and medical applications. *Journal of Drug Delivery Science and Technology*, *21*, 451−468.

Stayton, P., El-Sayed, M., Murthy, N., Bulmus, V., Lackey, C., Cheung, C., & Hoffman, A. (2005). 'Smart' delivery systems for biomolecular therapeutics. *Orthodontics & Craniofacial Research*, *8*, 219−225.

Stoica, A. E., Chircov, C., & Grumezescu, A. M. (2020). Nanomaterials for wound dressings: An up-to-date overview. *Molecules*, *25*, 2699.

Sudhakar, C., Upadhyay, N., Verma, A., Jain, A., Charyulu, R. N., & Jain, S. (2015). Nanomedicine and tissue engineering. *Nanotechnology applications for tissue engineering* (pp. 1−19). Elsevier.

Tadayon, A., Jamshidi, R., & Esmaeili, A. (2015). Delivery of tissue plasminogen activator and streptokinase magnetic nanoparticles to target vascular diseases. *International Journal of Pharmaceutics*, *495*, 428−438.

Tahmasebi, M., Esmaeili, A., & Bambai, B. (2022). New method of identifying morphine in urine samples using nanoparticle-dendrimer-enzyme hybrid system. *Arabian Journal of Chemistry*, *15*103630.

Thomas, S., Grohens, Y., & Ninan, N. (2015). Chapter 2. Biomaterials: design, development, and biomedical applications. *Nanotechnology applications for tissue engineering*, (Volume 1, p.21). William Andrew.

Tuan, R. (2007). Nanomaterials and stem cells in skeletal tissue engineering and regeneration. *Journal of Stem Cells & Regenerative Medicine*, *2*, 69.

Vacanti, J. P., & Vacanti, C. A. (2014). The history and scope of tissue engineering. *Principles of tissue engineering* (pp. 3−8). Elsevier.

Vega-Cázarez, C. A., Sánchez-Machado, D. I., & López-Cervantes, J. (2018). Overview of electrospinned chitosan nanofiber composites for wound dressings. *Chitin-Chitosan-Myriad Functionalities Science and Technology*, 157−181.

Venugopal, J., Prabhakaran, M. P., Low, S., Choon, A. T., Zhang, Y., Deepika, G., & Ramakrishna, S. (2008). Nanotechnology for nanomedicine and delivery of drugs. *Current Pharmaceutical Design.*, *14*, 2184−2200.

Webster, T. J., Siegel, R. W., & Bizios, R. (1999). Osteoblast adhesion on nanophase ceramics. *Biomaterials*, *20*, 1221−1227.

Wennerberg, A., Fröjd, V., Olsson, M., Nannmark, U., Emanuelsson, L., Johansson, P., Josefsson, Y., Kangasniemi, I., Peltola, T., & Tirri, T. (2011). Nanoporous TiO_2 thin film on titanium oral implants for enhanced human soft tissue adhesion: A light and electron microscopy study. *Clinical Implant Dentistry and Related Research*, *13*, 184−196.

Zhang, L., Chen, Y., Rodriguez, J., Fenniri, H., & Webster, T. J. (2008). Biomimetic helical rosette nanotubes and nanocrystalline hydroxyapatite coatings on titanium for improving orthopedic implants. *International Journal of Nanomedicine*, *3*, 323.

Zhang L., Ramsaywack S., Fenniri H., Webster T. (2006). Helical rosette nanotubes as a biomimetic tissue engineering scaffold material. *Proceedings of AIChE 2006 Annual Meeting.*

Zhang, Y., Su, B., Venugopal, J., Ramakrishna, S., & Lim, C. (2007). Biomimetic and bioactive nanofibrous scaffolds from electrospun composite nanofibers. *International Journal of Nanomedicine, 2*, 623–638.

Zhang, Y., Tang, Y., Wang, Y., & Zhang, L. (2011). Nanomaterials for cardiac tissue engineering application. *Nano-Micro Letters, 3*, 270–277.

Tissue engineering and drug delivery applications of nanofibers and their composites

8

Akbar Esmaeili
Department of Chemical Engineering, North Tehran Branch, Islamic Azad University, Tehran, Iran

8.1 Introduction

Many tissues of the human body cannot regenerate themselves. Therefore, damage to these tissues is irreparable (Badylak et al., 2011). In addition, the limited ability to heal tissues such as nerves, tendons, cartilage, and fat tissue, if they are damaged, leads to various disabilities. Even with multiple surgeries, the tissue may not regain its function. Despite the numerous successes seen in clinical studies, surgeries encounter challenges stemming from individual immune responses and significant limitations in finding suitable donors. Many patients hope to recover their abilities for an extended period on the waiting list. Tissue engineering scaffolds are a desirable physicochemical framework for growing and regenerating biological tissues. Their primary role is the mechanical support of the construction and expansion of cells to restore, protect, or change the use of a tissue or organ. Scaffold design and construction is one of the most critical factors of tissue engineering affecting the cell population's environmental conditions (Yang et al., 2001).

An ideal scaffold should provide a nutrient microenvironment to support attachment, migration, expansion, proliferation, and differentiation. During the last two decades, there have been various methods for making, washing, and connecting scaffold strands. These include the self-assembly method, dry ice emulsion, particle foam gas, solvent molding, and melt molding, which is a thermal method. Electrospinning is a multipurpose technique that allows fibers with different diameters suitable for making three-dimensional and porous scaffolds that have a similar structure to be made. The porous extracellular matrix (ECM) is one of the most critical applications of electrospun platforms for directing and controlling the differentiation of human mesenchymal stem cells (MSCs) in soft and hard tissue engineering (Chen et al., 2013; Cipitria et al., 2011; Huang et al., 2003; Jang et al., 2009; Teo & Ramakrishna, 2006).

The main components of the scaffold are biological materials that must have biocompatibility, appropriate porosity, desirable mechanical properties, and a suitable degradation pattern (Gunn & Zhang, 2010). Polycaprolactone (PCL) is well-known for its approval by the US Food and Drug Administration and is recognized as a biodegradable polymer suitable for both soft and hard tissues. PCL is highly favored for designing and fabricating scaffolds in tissue engineering. (Agrawal & Ray, 2001; Li et al., 2014; Woodruff & Hutmacher, 2010). On the other hand, due to its low hydrophilicity and delay in degradation, the weak mechanical properties of porous scaffolds, and the limitation of biological activities in complex tissue engineering, its effect is soft, inducing a cellular response (Lam et al., 2009; Liao et al., 2012; Pitt et al., 1984; Zhu et al., 2002). In addition, the dispersion of hydroxyapatite nanoparticles in the PCL matrix increased the bioactivity (Wutticharoenmongkol et al., 2006). Nano minerals belong to a large group of minerals that are widely used in the manufacture of polymer-clay nanocomposites with suitable properties for specific applications. They are also synthetic silicates with a two-sided surface morphology and high surface charge, leading to the interaction between clay and polymer. The combination of nano-clays with a polymer base can affect the physicochemical properties. Nanocomposites stimulate specific cellular responses for tissue engineering applications (Ambre et al., 2013; Gaharwar et al., 2010; Marras et al., 2008; Nitya et al., 2012).

In other studies, the scaffold was made using 16 PCLs reinforced with electrospinning. Observations showed that the presence of nano-clays causes the fiber distribution to be thinner, and the mechanical strength is improved without reducing the torque (Marras et al., 2008). Anionic nano-clays, called layered double hydroxide nanoparticles (LDH), have positively charged layers. Various unique properties, such as pH-dependent solubility and favorable biocompatibility, have attracted more attention. Numerous reports on the biomedical applications of these double-layered hydroxides include drug delivery, gene generation, molecular therapy, and tissue engineering (Choy et al., 2004; Rives et al., 2014; Shafiei et al., 2013; Xia et al., 2008).

Cation of PCL composites containing double-layer hydroxide prepared antiinflammatory molecules (diclofenac sodium). It has been shown that double-layer hydroxide in the PCL matrix makes the diameter of the fibers uniform. Also, PCL fibrous membrane can be used as an antiinflammatory scaffold in tissue engineering (Tammaro et al., 2009).

Although most research is focused on the construction of polycaprolactone nanocomposites reinforced with double-layer hydroxides for drug delivery applications, the use of different proportions of these double hydroxides in polymer composites can stimulate the modification of physical and chemical properties (Miao et al., 2012; Shafiei et al., 2021; Zhao et al., 2008).

8.2 Delivery of new drugs

Traditional drug delivery systems, which necessitate patients to take multiple doses at specific intervals, fail to meet global drug delivery needs adequately. In today's world, we are dealing with many recombinant peptide and protein drugs and hormone analogs in the body, most of which are made with genetic engineering techniques. Most drugs treat viral diseases such as cancer, diabetes, and autoimmune diseases. Considering the extensive range of sensitive protein and peptide drugs, the need to design new drug delivery systems seems necessary. With traditional drug delivery systems, there is practically no control over drug release time, place, and speed. The concentration of the drug in the blood constantly fluctuates in the form of valleys and peaks and may exceed the therapeutic range. They result in less efficacy and more side effects. With novel drug delivery systems, also called controlled drug delivery systems, we can control the three areas of drug release—speed, time, and place.

The work on drug delivery systems with controlled release started with the design of slow-release drug delivery systems. With these systems, it was possible to create a constant and uniform plasma concentration of the drug in the blood for a certain period and eliminate the peaks and troughs of the traditional drug administration, resulting in fewer side effects, greater efficiency, and patient comfort. After some time, we observed that this method of drug administration does not work for all drugs. For many medications, we need to release the drug when necessary and stop it when it is unnecessary. For example, insulin in the treatment of diabetes must be increased when blood sugar levels rise, and its release must be stopped when blood sugar returns to its normal level.

On the other hand, in many cases, our therapeutic objective is to minimize the drug's side effects while maximizing its efficacy. To achieve this, the drug should be delivered in its maximum amount to the target organ. Therefore, with these interpretations, the need to design a drug delivery system that can release the drug in response to the body's needs at the required time and place is necessary.

In general, drug delivery systems that control the time and place of drug release are called intelligent and automated drug delivery systems (Khodaei & Esmaeili, 2020).

The method by which the drug is delivered to the body affects the effectiveness of the treatment. Some medicines have an optimum concentration range in which the maximum efficiency of the drug is achieved, and concentrations higher and lower than this limit are either toxic or have no therapeutic effect. For example, for the treatment of cancer, there was a need for a targeted drug delivery system that could release the drug in the target organ. The studies have been conducted on pharmacokinetic control, pharmacodynamics, immunogenicity, and efficiency of drugs. These new strategies are often called (1) drug delivery systems, (2) based on combining polymer science, (3) pharmaceutics, (4) bioconjugate chemistry, (5) molecular biology, and (6) genetic engineering. To reduce the destruction and loss of the drug, prevent adverse side effects, increase the bioavailability of the drug, and increase the fraction of the drug that accumulates in a required part, different drug delivery systems, including stimuli-sensitive drug delivery systems and targeted drug delivery systems are under investigation. Among the drug carriers mentioned are soluble polymers, microcapsules, solutions, lipoproteins, liposomes, micelles, and liquid crystals. These carriers can be biodegradable, stimuli-sensitive (hydrogels and liquid crystals), and even targeted (by conjugating them with specific antibodies against specific components at the site of action). Targeting is the ability to deliver most of the drug to the desired location and target organ.

Controlled drug release and subsequent degradability are essential factors for a drug formulation with controlled release. The potential mechanisms of drug release are as follows: (1) The return and release of the bound drug to the surface, (2) Diffusion through the carrier matrix, (3) Diffusion from the carrier wall for microparticles and microcapsules, (4) Erosion and destruction of the carrier matrix, (5) A combination mechanism of the process of erosion/diffusion (Tahmasebi et al., 2022).

The type of drug, the choice of the method of administration, and the type of drug delivery system are very effective in the success of the treatment. Continuous drug release from a polymer system includes the slow diffusion of the drug out of the matrix or the slow erosion of the polymer over time. Oscillatory or stimulus-responsive clearance systems are often superior drug delivery systems over slow release drug delivery systems because the former mimics the precise pattern in which the body releases hormones such as insulin. Drug-carrying polymers, such as hydrogels, respond to specific stimuli (such as temperature, pH, and electricity) (Esmaeili & Khoshnevisan, 2016; Esmaeili & Loghmani, 2016).

For more than twenty years, researchers have received the potential benefits of nanotechnology in improving the quality of drug delivery systems and therapeutic purposes. Improving drug delivery techniques that reduce toxicity and increase the efficacy of drugs has many potential benefits for patients and has opened new markets for pharmaceutical companies. Other recent findings in the field of new drug delivery systems emphasize the passage of the drug through particular physical barriers, such as the blood−brain barrier, to deliver the drug to the target organ better and increase its efficiency. Another huge group of new studies in drug delivery systems deals with finding acceptable side routes for peptide and protein drugs other than injection routes and oral routes, which cause the destruction of protein peptides and pain for the patient.

8.3 Targeted delivery

The global market for advanced and new drug delivery systems was more than 37.9 billion euros in 2000, estimated to reach 75 billion euros in 2005 (e.g., 19.8 billion euros for drug delivery systems). The market for drug delivery with controlled release is as follows: (1) 0.8 billion euros for needle-free injection, (2) 5.4 billion euros for implantable and injectable polymeric systems, (3) 9.6 billion euros for skin drug delivery systems, (4) 0.12 billion euros for intranasal drug delivery systems, (5) 0.17 billion euros for pulmonary systems, (6) 4.9 billion euros for mucosal systems, (7) 0.9 billion euros for rectal systems, (8) 2.5 billion euros for liposomal drug delivery systems, (9) 3.8 billion euros for cell/gene therapies, etc.

The market for new drug delivery systems is progressing at a dizzying speed; this progress is awe-inspiring in the peptide and protein drug delivery systems, resulting from genetic engineering and other biological treatments (Tadayon et al., 2015).

8.4 Codelivery

Drug delivery systems containing colloidal carriers (such as mucilage, blister, and liquid crystal solutions) and nanoparticle dispersions, including small particles with a diameter of 10–400 nm, are promising futures in the field of drug delivery systems. In designing and developing such drug delivery systems, the goal is to reach a system with appropriate drug loading, desired release properties, high half-life, and low toxicity (Tadayon et al., 2016). The selected drug is deposited in the carrier structure (Fig. 8.1).

Micelles are spontaneous aggregations of amphiphile copolymers in aqueous solutions with particle diameters of usually 5–50 nm, which have received much attention for drug delivery purposes. The drug is physically trapped in the core of the copolymer micelles, and thus it can be formulated in soluble forms at concentrations higher than its intrinsic solubility. In addition, these hydrophilic copolymers can form hydrogen bonds with the aqueous environment around them, and create a solid shell around the mucilaginous core; as a result, the hydrophobic content of the body can be effectively protected from hydrolysis and enzymatic destruction (Rajabi & Esmaeili, 2020; Soltanabadi et al., 2022). In addition, the capillary crown prevents the reticuloendothelial system from recognizing and harvesting the capillaries. Another feature that makes amphiphilic copolymer micelles suitable for drug delivery is total molecular weight, and copolymer chain length ratios allow their size and morphology to be easily controlled. Combining these copolymers with factors and cross-linking groups increases the stability of the micelles and improves the control process of drug release from them. Substituting copolymers with particular ligands is a promising strategy for delivering the drug to a wide range of target sites and organs (Fig. 8.1).

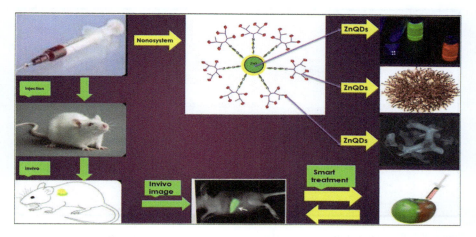

Figure 8.1 Pharmaceutical carriers with quantum dots (Khateri and Esmaeili, 2024).

Liposomes are forms of vesicles that are composed of one or several lipid bilayers similar to what is seen in the cell membrane. The polar character of the liposome core allows polar drugs to be well encapsulated in it. On the other hand, lipophilic and amphiphilic molecules are easily dissolved in two lipid layers. Because they tend to be phospholipids, they are replaced and encapsulated in this region. Channel proteins can quickly enter the hydrophobic part of vesicle membranes, act as a filter based on size, and allow only small molecules such as ions, nutrients, and antibiotics to enter. Therefore, drugs encapsulated in nano-sized carriers containing channel proteins are effectively protected from early degradation by proteolytic enzymes. In contrast, the drug molecule is quickly absorbed due to the concentration gradient inside and outside the system space, and it can spread outside the channel (Fig. 8.2).

Dendrimers are macromolecules with a narrow, branched, and nano-sized particle size spectrum with a symmetrical design. From a central core, branched units are formed as tree branches and several functional groups (Esmaeili & Ebrahimzadeh, 2015, 2016).

The main body and its internal teams determine the environment inside the cavity and subsequent solubility properties. In contrast, these external units determine these polymers' solubility properties and chemical behavior. The therapeutic goal is realized by attaching targeted functional groups to the surface of the dendrimers. In contrast, by connecting the functional groups containing polyethylene glycol (PEG) chains to the surface of the dendrimers, their stability can be increased, and the reticuloendothelial system can remove them. They are prevented to a great extent. Liquid crystals are interesting drug carriers. These materials are between solid and liquid in terms of molecular order; as a result, they have both liquid and

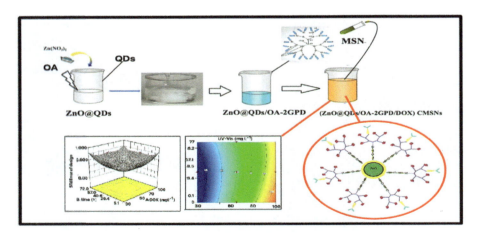

Figure 8.2 Procedures followed by different polymers with quantum dots (Khateri and Esmaeili, 2024).

stable properties. These materials have different molecular orders with varying geometries. These phases are easily interchangeable. The drug can be encapsulated between the molecules of liquid crystals and released from the system by changing the stage as a result of stimulating action.

Nanoparticles (nanospheres and nanocapsules 10−200 nm) are solid and amorphous or crystalline. These carriers can absorb and encapsulate the drug, thus protecting it against enzymatic and chemical degradation. At the same time, nanospheres are matrix systems in which the drug is physically and uniformly dispersed in the carrier. Nanoparticles as drug carriers are made from both biodegradable and nonbiodegradable polymers. In recent years, biodegradable polymeric nanoparticles have received considerable attention as potentially suitable systems for drug delivery. Nanoparticles targeting specific organs and tissues, as carriers of DNA in gene therapy and protein therapy, have received much attention from dietary routes (Hormozi & Esmaeili, 2019; Jadidi et al., 2020). Hydrogels are three-dimensional polymeric networks of hydrophilic and hydrophilic polymers that can absorb water and biological fluids, sometimes up to several times their volume and weight. These networks are made up of homopolymers or copolymers, which are insoluble in water because they are cross-linked. Hydrogels show thermodynamic compatibility with water, which allows them to swell in aqueous environments. Hydrogels are used as carriers to control the release of drugs in systems where the swelling phenomenon governs drug release. As one of the most interesting and advanced systems in modern drug delivery, some hydrogels are sensitive to environmental stimuli and can act under the influence of a specific stimulation like an on-and-off switch and release the drug under control. These ecological stimuli can be pH, temperature, ionic strength, electricity, magnetism, light, or the concentration difference of a specific chemical compound. In all these systems, drug release occurs in certain places of the body (such as the pH of the digestive tract or the more acidic pH of cancer cells). Combining hydrogels as drug delivery systems with molecular labeling techniques can make them unique materials, promising for future drug delivery systems (Fig. 8.3).

Conjugates, which include the conjugation of synthetic polymers with biological polymers (such as proteins and peptides), are efficient and effective in improving the drug release process. Conjugating appropriate biocompatible polymers with natural peptides and proteins reduces the toxicity risk. It increases blood flow time and improves solubility. Changing and modifying synthetic polymers with appropriate oligopeptide sequences, in other words, prevents the random distribution of drugs throughout the patient's body and helps the target site and organ to be treated. The ability of cationic peptide sequences to complex with DNA and subsequently compress DNA and nucleotides brings good hopes for developing nonviral DNA carriers in gene therapy (Esmaeili & Rafiee, 2015; Esmaeili et al., 2013).

The field of in situ forming implants has grown a lot in recent years. Liquid formulations that produce a semisolid drug reservoir in place after subcutaneous injection are interesting and preferred drug delivery systems for injectable use because

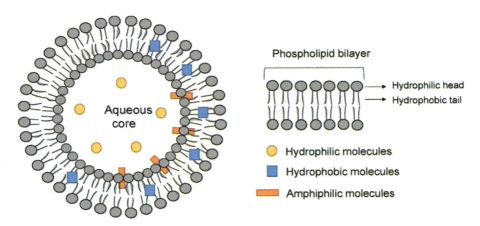

Figure 8.3 Representation of the general structure of liposomes (Guimarães et al., 2021).

they are less invasive compared to other implant systems, with a painless injection. The subcutaneous layer of this medicinal reservoir is created under the skin. These systems can realize systemic or local drug delivery for one to several months. Most of the implant systems formed in place by injection can reduce its side effects by creating a constant drug concentration in the plasma, similar to intravenous infusion, and are especially suitable for protein drugs with a narrow therapeutic index. The production of these systems is straightforward compared to other polymer systems. These systems are divided into four main categories: (1) thermoplastic pastes, (2) cross-curing polymer systems in place, (3) polymer deposition in place, and (4) thermal gelling systems. The final goal in developing drug delivery systems with controlled release has been the development of devices and tools that can store and release chemicals when needed. Recent advances in the development of mechanical-electrical systems in micro size have provided a unique opportunity to make tiny biomedical devices as drug delivery systems. Implantable microchips containing medicine have the following advantages: (1) Several chemical substances in different forms (solid and liquid) can be kept inside the microchip and released from it. (2) Chemical substances are released from such devices following the destruction of the barrier membrane as a result of electric current. (3) A wide range of solid drugs with a narrow therapeutic index can be reliably and precisely made available to the body in this way. (4) These systems can simultaneously release the drug with several release patterns (e.g., both fluctuating and constant release patterns and can have continuously). (5) These systems are made small enough to be placed at the site of effect and create high concentrations of the drug while keeping the drug's systemic attention low. (6) Due to the impossibility of water penetration into these systems, the stability of the protein drugs in them is still maintained (Zecheru, 2008).

8.5 Tissue engineering

There is ambiguity in the classification of nanomedicine. (1) Personal health care, (2) diagnosis and treatment of diseases through nanoscience, (3) drug design, and (4) targeted drug delivery are some categories based on nanomedicine for treatment. Based on the literature, the types of nanomedicine are the following: (1) drug delivery, (2) nanomedicine, (3) repair, and (4) disease treatment through nanoscience (Fig. 8.4).

Drug delivery systems are engineered for targeted drug delivery technology or controlled release of therapeutic agents. In this process, they control the speed at which a drug is released and absorbed into the body—the introduction of nanotechnology in drug delivery substantially affected tissue engineering applications. Drug delivery can have unique properties in connection, control, or planning of compounds and the nature of cooperation with molecules in tissue engineering. Drug delivery applications are critical because of their ability to trap hydrophobic or hydrophilic drugs (Gaharwar et al., 2014; Tabata, 2000). Drug delivery is often used in tissue engineering, which includes nanoparticles, liposomes, dendrimers, nanotubes, and hydrogels (Fig. 8.5).

8.6 Nano scaffolds

Contrary to popular belief, tissue engineering is not only limited to repairing or replacing damaged tissues in the body but also includes other things such as study and educational models, identification of drugs, and even food production. For example, the process of forming an engineered tissue can contain essential information regarding the many factors affecting tissue formation in the body and develop

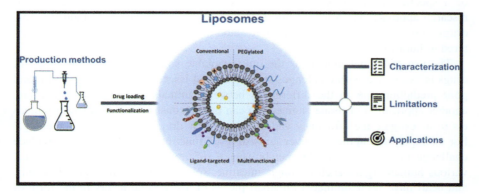

Figure 8.4 Design of liposomes for drug delivery systems in therapeutic applications (Guimarães et al., 2021).

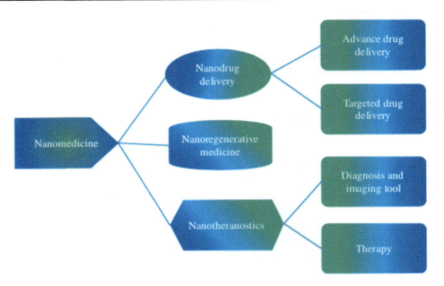

Figure 8.5 Types of drug delivery systems in tissue engineering (Thomas et al., 2015).

the knowledge of developmental biology. Also, engineered tissues can be suitable laboratory models for studying various diseases and identifying pathogenic agents. They can even be designed to be ideal for training medical students. For example, today, surgical training is performed on a human corpse before entering the operating room. Still, in cases such as separating a cancerous tumor from damaged tissue, the condition of the corpse is very different from reality, even though it can use engineered tissues. They are made in such a way that they include a tumor or a specific defect. Based on this, tissue engineering can have an educational research approach and a therapeutic approach. In the pharmaceutical industry, new drugs must be evaluated by appropriate laboratory models before being administered to humans.

The use of animal models as the only available solution to study the effect of drugs is associated with challenges such as ethical issues, long time, high cost, and poor prediction due to the differences between animals and humans. Therefore, another application of tissue engineering can be the construction of laboratory models to investigate the efficiency and effects of new drugs, called "drug screening." For example, an engineered liver consisting of functional hepatocytes can replace challenge models to study the metabolism of different drugs. Based on this idea, various human organs such as lungs, intestine, spleen, liver, kidney, blood vessels, and heart have been simulated on small pieces of a few centimeters in the form of electronic chips. These models, known as "organ on a chip", can perform several critical tasks of the desired organ and provide outputs that a computer can analyze by entering materials through micro-scale channels.

One of the limitations of these chips is the independent examination of an organ without considering its interactions with other body organs. Recently (2012), a project with the financial support of the US Department of Defense has been launched. The aim is to include ten organs, from the essential organs affected by substances entering the body. They are on a small chip, so they are connected and form a model of the "whole body on a chip" (Fig. 8.6). In short, such a chip can provide detailed information about the possible effects of unknown substances on different body parts. Therefore, it is considered a suitable tool for testing drugs in the pharmaceutical industry or identifying suspicious substances on battlefields.

Another application for tissue engineering is the production of healthy and fresh engineered meats, similar to common edible hearts. Based on this idea, in August 2013, in a TV show, an engineered meat container was unveiled, which was claimed to be very similar to beef and entirely safe for humans to consume. A strange claim that has never been proven and is highly questionable. Apart from the dangers that the consumption of this meat can have for humans, the cost of producing a small dish is so high that such ideas' success seems impossible. However, with the further development of cell culture technologies, engineered meats may one day replace slaughtered meats.

Because some cells were considered the origin of other body cells, they were called original cells or stem cells. Stem cells are known for their two essential characteristics: self-renewal and differentiation. These two features are called fundamental. In self-renewal, the stem cell produces its cell during cell division, and in differentiation, it creates a different type of cell. Differentiated cells have other proteins and, as a result, will have different functions.

Adult stem cells exist in different adult organs, and their task is to repair and regenerate the body. These cells have limited proliferation and self-renewal capabilities, and therefore are not tumorigenic. Also, they cannot differentiate into all types

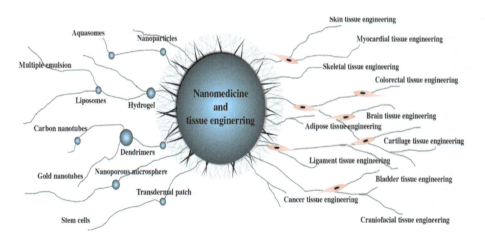

Figure 8.6 Classification of nanomedicine (Thomas et al., 2015).

of body cells and can only become cells related to their tissue, so they are multipotent. Among adult stem cells, we can refer to hematopoietic stem cells, found in the bone marrow or umbilical cord, and can turn them into all types of blood cells. MSCs are also mainly extracted from the bone marrow and can differentiate into tissue-forming cells such as bone, cartilage, muscle, and fat. ES are obtained from the inner cell mass of the embryos that are created in the blastocyst stage of embryo development (5th−7th day). ES cells are pluripotent 16; they can differentiate into cells of all three embryonic layers, ectoderm, mesoderm, and endoderm. This way, they can turn into all types of specialized cells in the body. Also, these cells can multiply indefinitely in the laboratory environment. This feature has caused the ability of tumor formation in these cells. iPS are mature cells (such as fibroblasts) reprogrammed into embryonic-like stem cells through genetic manipulation by certain factors. Mouse and human fibroblast cells produced iPS cells in 2006 and 2007, respectively. These cells are qualitatively equal to ES, so they can multiply indefinitely and have all three layers of the embryo and, as a result, all types of body cells.

8.7 Controlled release

8.7.1 Skin tissue engineering

The skin plays a vital role in human health, and defects in its natural physiology process can be a significant problem; it is the thinnest organ in the body. Moreover, the skin, being the largest organ of the body, is increasingly recognized as vulnerable to damage from various factors, including environmental exposure and anatomical vulnerabilities between its inner and outer layers. It is among the most frequently injured organs compared to others. Given its critical role, extensive efforts are dedicated to its restoration through various methods, including advanced skin tissue engineering. One significant advantage of engineered skin over traditional skin grafts is its potential for customization, a feature that underscores its superiority. However, challenges remain, such as resource limitations and the risk of immune reactions.

Today, biocompatible polymers like silicone are widely utilized in skin engineering due to their favorable biological properties. Tissue substitutes are highly sought-after and pivotal in medical applications. Modifying the interaction between tissue and these materials before their medical use is crucial. Surface grafting of monomers onto silicone is particularly critical as it can enhance its properties for applications in skin tissue engineering (Esmaeili et al., 2015).

Skin, as the most extensive tissue in the human body, plays a significant protective role in the body, and any damage to this protective barrier leads to the creation of all kinds of wounds in this tissue. Today, wound healing, especially chronic wounds resulting from various injuries or diseases, is one of the fundamental problems in the field of health, which strongly affects the patients' quality of life. Until today, many efforts have been made to find the most effective treatment for this purpose, but still, the long waiting list of patients shows the need for a new method to overcome the limitations of old remedies. Tissue engineering is an interdisciplinary science and one of the new branches of medical science, which uses the principles of engineering (materials engineering and biomedicine) and biological sciences (biochemistry, genetics, and cellular and molecular biology) to develop natural alternatives to repair, preserve, or improve tissue function. Broadly defined, tissue engineering involves using scaffolds, cells, and biologically active molecules to produce tissues with specific functions.

In the field of tissue engineering, one of the essential topics is the design and production of scaffolds for implantation. These scaffolds should facilitate migration, differentiation, and normal function of cells, as well as support tissue regeneration, maintenance, and improvement. These scaffolds can be made from natural materials such as collagen or synthetic polymers. In addition, the cellular components of their composition may be of human or animal origin (with or without genetic modification). Accordingly, one notable natural scaffold is the amniotic membrane, which has gained the attention of many researchers and clinicians for its ability to repair various types of wounds. The amniotic membrane is a biodegradable scaffold that can accelerate tissue repair and stimulate cell attachment, proliferation, and differentiation (Esmaeili et al., 2015).

Based on this, Aria Reconstructive Medicine Knowledge Base Company has started producing woven engineering products based on the principles and supervised concepts to ensure high-quality, safe, and effective outcomes. The selection of the correct cell source is crucial in this category of tissue engineering products. Because it is one of the most critical issues, human placenta tissue has been used as a valuable and rich source of MSCs. This tissue provides a rich and available source of MSCs, which enables its production on a large scale to be used for the rapid treatment of patients at a lower cost. In other words, using this cell source will be easier and faster to achieve our goal: to provide a "shelf-the-off" cell-based product with commercialization capabilities (Zahedi et al., 2019).

8.7.2 Dendrimer

Today, cancer affects many people in different age groups, both men and women. Tumors often require high doses of treatment, resulting in increased toxicity. In addition, these drugs lack specificity and cause severe damage to noncancerous tissues. Most available chemotherapeutic agents have low molecular weight and high distribution in the body, both of which will cause toxicity. The low molecular weight of the drug molecules has made them quickly excreted. As a result, a higher concentration of the drug is needed; on the other hand, the nonspecificity of these drugs causes severe damage to noncancerous tissues.

Therefore, many research groups are working worldwide to improve the distribution of drugs in the body, reduce side effects, and achieve the best pharmacological effect. Nanodendrimers are a group of nanoparticles that have been considered in many biomedical fields today. Dendrimers are a unique branch of polymer macromolecules with multiple stems growing from a center and forming an almost complete three-dimensional pattern (Bharali et al., 2009). These particles have a diameter between 10 and 200 Å (Tahmasebi et al., 2022). Dendrimers are composed of three parts: a core, lateral branches, and terminal functional groups (Bharali et al., 2009). Due to the presence of multiple functional groups on the surface, they can connect and transport all kinds of molecules to their character; on the other hand, they can use this feature for active targeting. They used nanodendrimers for a specific tissue.

Dendrimers also have holes inside. Drug molecules can be attached to these particles' surfaces and carried inside (Klajnert & Bryszewska, 2001). Another unique feature of dendrimers is the multifunctional nature of these particles so that they simultaneously contain drug molecules, targeting parts, and dissolvable groups. The unique structure of these particles allows for conjugating molecules on the surface and encapsulating them. Pegylated dendrimers (dendrimers coated with PEG) are one of the classes of dendrimers that have attracted the attention of many researchers due to their long circulation time in the blood, less toxic level, and relatively less accumulation in different organs.

Some anticancer drugs, despite their effectiveness, cannot be used due to high toxicity. The therapeutic effects of any drug are related to its solubility in water. Most drug molecules cannot be used due to a lack of dissolution in suitable solvents. Dendrimers can form bonds with these molecules and dissolve hydrophobic molecules (Klajnert & Bryszewska, 2001).

Cancer tissues have unique properties that are permeable to macromolecules up to 400 nm in diameter (compared to healthy tissues with a permeability of 2−6 nm). The process of entering molecules through the leaky structure of cancerous tissue is called the enhanced permeability and retention (EPR) phenomenon. It is well proven that pegylated nanoparticles accumulate in tumor tissues due to the EPR phenomenon. This natural pattern of nanoparticle distribution is called passive targeting. The EPR phenomenon is not always effective and efficient. The nonspecificity of the carriers raises the necessity of targeted drug delivery.

Nanosystems will cause targeted drug delivery, increasing efficiency and reducing side effects (Bharali et al., 2009). Folic acid or folate is an essential vitamin required for the functioning of cancer cells. For cancer cells to receive this substance as much as possible, many receptors are expressed on the cell surface. As mentioned, dendrimers can attach many meanings to their character and carry them due to having multiple functional groups on the surface. Studies have shown that the folic acid receptor binds to folic acid with a very high affinity and guides it into the cell through endocytosis (Bharali et al., 2009).

If these nanoparticles contain drug molecules, after entering the cell specifically, they can release the drug and destroy the cancer cell. The multifunctional nature of these nanoparticles has attracted the attention of many researchers.

Quintana and colleagues synthesized PAMAM-G5 dendrimers containing folic acid, fluorescein, and methotrexate. This complex created targeted drug delivery and imaging, and the ability to deliver drugs with 100 times less toxicity than methotrexate in free form.

Conventional methods of drug administration, such as oral ingestion and injection, often result in the distribution of drugs throughout the entire body, impacting the entire system and potentially causing side effects. As a result, large quantities of medicine may be required to achieve the desired therapeutic effect. However, alternative methods offer the potential for more targeted and effective drug delivery, providing hope for improved treatment outcomes. Nanotechnology makes it possible to achieve targeted drug delivery and control the time, place, and speed of drug release in the body. Consuming fewer drug doses and reducing side effects can benefit the patient. The field of oncology will soon come into action with new strategies for diagnosis and treatment based on dendrimers.

8.7.3 Nanofibers as three-dimensional scaffolds for tissue regeneration

3D bioprinting is a transformative technology that brings the development and progress of tissue engineering processes, regenerative medicine, and thus the future of medicine. Tissue engineering, controlled drug release, and production of replacement cells require scaffolds and cell models with precise geometry and predesigned internal structure, which can only be achieved with 3D bioprinting technology.

Recent developments have made it possible to 3D print biocompatible materials, cells, and supporting components with the help of a bio 3D printer. 3D printing addresses the need for suitable tissues and organs for organ transplantation in regenerative medicine.

3D bioprinting is a new technology used to design and build 3D cell structures to replace lost organs and test drugs and cosmetics. The most attractive advantage of this technology is its ability to create three-dimensional structures with living biological components such as cells and their food.

In 3D bioprinting, precise and regular placement of physical, biochemical, and living cells layer by layer with a spatial structure similar to the lost tissue is used to create a 3D design. Researchers worldwide are trying to produce a 3D biological system with appropriate physical and mechanical properties to replace lost tissues or organs. One of the most critical challenges in 3D printing is updating the technology to print polymers and metals previously used. However, the ultimate goal of the research groups is to produce structures that are very similar to the ECM in micro dimensions and contain an appropriate and diverse number of cells.

Inkjet: This technology is one of the first used for biological 3D printing. This technology is divided into two subcategories: (1) thermal and (2) piezoelectric. Both approaches follow the same methodology: a syringe or print cartridge is filled with bio-ink and then ejected from a nozzle via force. Bio-thermal inkjet printers use the technology found in conventional commercial inkjet printers, with the difference that in these printers, the ink has been replaced by bio-ink. In the thermal type, the built-in heating element creates a bubble in the bio-ink, which leads to an increase in pressure and thus ejects the bio-ink from the nozzle head. Among the problems of this technology, we can mention the fast closing of the nozzle head and low efficiency. Also, the temperature of the thermal element in this method increases to 300 degrees; as a result, the ink temperature rises to more than 10 degrees.

For this reason, some research groups consider the need for a recovery phase for cells after printing to prevent cell death. In contrast, the piezoelectric type uses acoustic waves or a physical component to change shape or size to create pressure instead of changing temperature, thereby increasing the pressure in the cartridge and expelling the bio-ink. Due to the high ability to control the duration, magnitude, and frequency of the acoustic waves and, as a result, prevent the size of the drops, the accuracy of printing in this type of printer is very high. Also, cell survival in this method is between 80% and 90%.

Although this type of printer has advantages over the thermal type, both approaches face limitations regarding the bio-ink used. One of the biggest problems is the viscosity of bio-based ink when printing. The bio-ink must be liquid in the cartridge and quickly change to a solid or semisolid state after exiting the nozzle (Fig. 8.7).

Extrusion: Printing based on extrusion is known as one of the most common and cheapest methods in bioprinting and is widely used in tissue engineering. Extrusion-based printing consists of syringe-like components that move in XYZ three-dimensional space, and the release of bio-ink from the moving syringe head results in 3D bioprinting. This movement in 3D space is controlled using robotic parts. In this method, pneumatic pressure or mechanical movement (piston or screw) causes the material to exit the nozzle. One of the essential advantages of this method is its ability to print materials with high viscosity and cell density at a relatively high speed. However, this method usually creates a relatively low resolution (around 200 μm).

In this method, the materials coming out of the nozzle must have proper printability, which means that the bio-ink can exit from the tip of the nozzle quickly.

It must be designed to mechanically maintain a three-dimensional structure after leaving the nozzle. In other words, the rheological properties of the bio-ink should be adjusted so that it has a high elastic modulus and can maintain its shape after exiting the nozzle. At the same time, this material must have a sufficient loss modulus so that the exit of the material from the nozzle head is possible (Fig. 8.8).

The process of the 3D printer is a layer-by-layer process, that is, the materials are stacked like a brick wall construction, different methods harden them, and they become strong enough that nozzles usually do the process of laying layers of materials and solidifying them. The raw materials this technology requires are filaments, which are used for printing by melting them using heat (Fig. 8.9).

The natural organs of the body are artificially made and designed by 3D printing techniques, and its primary purpose is in the field of organ transplantation. Bioprinting is a potential solution to the shortage of donated organs worldwide.

Tissue engineering and drug delivery applications of nanofibers and their composites 173

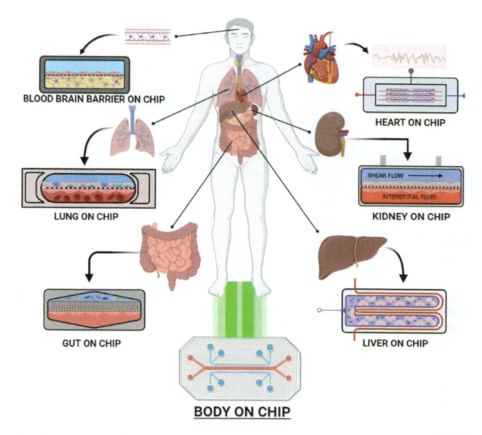

Figure 8.7 A glimpse of the whole body on a chip, A set of 10 interconnected organs (Park et al., 2020).
Source: Permission of Elsevier.

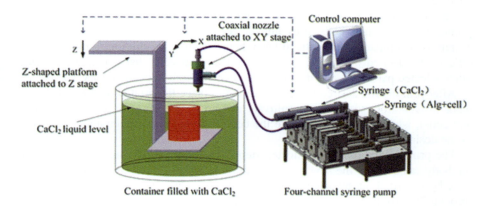

Figure 8.8 Schematic of bioprinter nozzle (Gao et al., 2015).
Source: Permission of Elsevier.

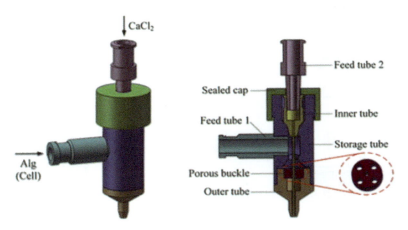

Figure 8.9 Schematic of the nozzle of bioprinters based on extrusion (Gao et al., 2015).
Source: Permission of Elsevier.

Organs prepared for transplantation are usually made from the host's cells. The organs produced by this device, which are used in the medical program of some countries, may be like hollow bladders and veins.

Inkjet printers are used to produce 3D biological tissue. A suspension of living cells and bright gel is placed inside the cartridge of this printer, and they are used to prepare the structure. Alternating patterns of living cells are placed on the desired substrate. They used a 3D printing nozzle. After the process, the gel is cooled and washed, and the living cell remains.

3D bioprinting was first introduced in 2003 when Thomas Boland invented droplet cell printers. Since then, the 3D printing of biological structures, known today as bioprinting, has made significant progress in producing tissues and organs.

In addition, new techniques, such as extrusion bioprinting, were created, which have been successful in this field.

Instead of using plastic or metal as the primary material of the print, the 3D printer of the human body uses human cells depending on the type of organ. Materials accompany these cells to keep them together. In addition to the cells specific to the desired organ, it is possible to use stem cells that can transform into other selected partitions, and the human body can also accept such cells.

After printing the desired organs, it is time to place the printed organ in the incubator (a laboratory tool used to cultivate and grow live samples such as microbes and cells), which is used to activate the cells and lead to the beginning—the work of the desired member.

The primary and most complicated step is to place it in the human body so that the body can adapt to the new organ and not reject it. Until now, countless activities have been carried out by scientists in the field of producing and creating body parts through 3D printing, and we can say that most of them have failed due to the lack of testing the printed samples in the natural environment. In the following, we introduce several human body parts made by these printers. In 2017, Chinese researchers created a miniature model of a kidney using human cells and a function similar to an actual kidney. These kidneys were also designed in small dimensions. Accurate measurements in 2018 made the original model. It was transplanted to a child in America for the first time. These kidneys can function for several months without any problems in the recipient's body.

In 2017, researchers in Weill Cornell Medical College in New York succeeded in 3D printing an artificial ear. This artificial ear has the same function and appearance as the human ear, which can replace the ear in different patients. To make this artificial ear, the researchers took a digital 3D image of the ear and gave it to a 3D printer to produce the mold of the ear shape. In the next step, the gel made of living cow ear cells and collagen was injected into the mold. This process lasted two days, including design, printing, gel injection, and drying for 15 minutes.

8.8 Composites nanofibers

8.8.1 Polymer nanotechnology: nanocomposites

One of the problems in the construction of composites is the need to add a large number of reinforcing components to the background, which causes the properties of the resulting composite piece to be significantly different from the properties of the environment, and this difference is considered a disadvantage in many cases. For example, by adding metal powder such as copper to the polymer matrix, we seek to create electrical conductivity in the resulting composite. Creating high electrical conductivity in this composite requires high amounts of copper powder. In this case, the obtained composite piece will inevitably have a much higher weight than the base polymer, which is very undesirable in places where polymers are used. Also, in this case, the mechanical properties of the resulting composite will decrease compared to the polymer base; for example, the elongation and impact ability in this composite piece is significantly less than its polymer base. Now, this problem does not occur if, instead of a typical amplifier with large dimensions, electrically conductive nanostructures such as carbon nanotubes are used. In this case, high electrical conductivity is achieved with much smaller amounts of the reinforcing component (e.g., about 2% by weight compared to 20% by weight of metal powder). As a result, due to the lower amount of the reinforcing component in the nanocomposite case, the background's desired intrinsic properties are maintained, and only the intended electrical conductivity property is added to it. Therefore, increasing electrical conductivity reduces tensile strength and impact resistance while the weight of the nanocomposite remains unchanged, which is crucial for industrial applications.

In this case, significantly lower values of the desired property can be achieved at the nanometer scale. Using smaller amounts of the reinforcing element reduces the finished price in most patients because the reinforcing member usually has a higher price than the background component. Reducing its quantity is very economical.

In general, nanocomposites have higher unique properties compared to composites. Special properties mean the amount of a property divided by the density of that material. The specific property states that considering the same weight between two or more materials, it has the desired property. For example, the specific tensile strength between two materials means which one has a higher tensile strength at the same weight. As mentioned, in nanocomposites, the weight is usually lighter due to the use of much smaller amounts of the reinforcing component (Ajayan et al., 2006; Astrom, 2018; Lubin, 2013).

However, what is the reason for using smaller amounts of the reinforcing component in the nano state? The reason for this is the high surface-to-volume ratio of nanomaterials. In this case, the interaction between the amplifier component and the background is very high due to the very high surface-to-volume ratio. With shallow values, sufficient interaction between the amplifier and the environment occurs (of course, on the uniform distribution condition)—unique properties in nanomaterials, such as the ballistic transmission of electrons or photons. Unique optical and magnetic properties result from increasing surface area to volume or converting energy bands into energy. The levels have also created special applications of nanomaterials in the nanocomposite.

Nanocomposites can be classified from several aspects. From one part, they can be divided into two categories: natural nanocomposites and artificial nanocomposites. Some of the most famous natural nanocomposites are wood and bone. Bone can be considered a collection of collagen fibrils. These collagen fibrils are composed of spirals of collagen molecules on which hydroxyapatite nanocrystals are placed as components that enhance mechanical properties. The bone's unique structure has achieved excellent mechanical properties such as high strength and impact resistance (Ajayan et al., 2006; Venkatesan & Kim, 2014).

Also, wood can be mentioned as another natural nanocomposite. In wood, there is a matrix of amorphous cellulose and natural glue reinforced by cellulose nanocrystals. Unlike amorphous cellulose, these nanocrystals have high mechanical strength and give high impact resistance, elastic modulus, and high mechanical strength.

Another aspect of the division of nanocomposites (and composites), which is very common, is their division based on the type of context. Adding a nano-reinforcement component to each of these areas is due to eliminating their primary defects. In polymers, these defects can be low mechanical strength, low thermal resistance, low electrical and thermal conductivity, and destruction against ultraviolet sunlight; this purpose uses different nano reinforcements.

In metals, these defects can be low elastic modulus or mechanical strength, low creep resistance, or low fracture strength, which usually applies to light metals such as aluminum, titanium, or magnesium. In ceramics, the primary defect is related to the low toughness or toughness of the ceramic base, which is tried to be improved by adding nanomaterials (Ajayan et al., 2006; Astrom, 2018).

In polymer nanocomposites, the purpose of using the nano component, as mentioned above, is to eliminate the inherent defects of polymers, such as low mechanical strength, low thermal resistance, low electrical and thermal conductivity, and destruction against ultraviolet waves. Nanomaterials are used in both types of thermoplastic (Ajayan et al., 2006) and thermosetting polymers (Astrom, 2018). The most common materials are polyethylene, polypropylene, polyamide, polyamide, polycarbonate, polystyrene, epoxy, polyester, and vinyl esters, which are also used as thermosetting materials. Carbon nanotubes, graphene, carbon black, titania, calcium carbonate, and nano-clay are among the nanomaterials primarily used in these polymer fields. For example, with carbon nanotubes and graphene, the polymer matrix can be given electrical conductivity, thermal conductivity, slow-burning properties, and, in some cases, improved mechanical properties. Also, nano titania is used as an ultraviolet absorber of sunlight and bleaching in polymer fields. Also, calcium carbonate nanoparticles can be used as a filler (to reduce the use of polymer), which in some cases leads to the improvement of mechanical properties such as mechanical strength. Also, nano-clay is used to improve the mechanical properties of polymeric materials and make them flame retardant due to their barrier properties (Lubin, 2013) (e.g., preventing oxygen from entering food packaging).

In many applications that require low weight and unique properties, these nanocomposites are the best option because they are very light due to the polymer base. On the other hand, a small amount of reinforcement is used in them. The properties are improved due to the nanostructure. Its presence is significant (Venkatesan & Kim, 2014).

The use of nanostructures in metal fields primarily involves light metals, such as aluminum, titanium, and magnesium. Due to their lightweight, these metals are a very suitable option in fields where there is movement, such as aerospace and automotive. Among them, the most crucial option is aluminum, which has a better price. However, the problem these metals have is that they do not have favorable mechanical properties for applications such as steel, and they have low yield strength, modulus, and tensile strength. Using nanomaterials such as carbon nanotube, silicon carbide, alumina, and graphene can improve the mechanical properties of these metals.

In ceramic fields, the most critical problem is its low toughness. For this reason, ceramics are very fragile and have low plastic deformation ability. One of the ways to improve the toughness of ceramics is to use nanomaterials in them. These nanomaterials help increase the ceramic base's toughness by preventing crack growth and postponing its failure. Among these cases, we can mention using alumina or silica nanoparticles in the zirconia ceramic field or the zirconia-alumina two-component ceramic (Astrom, 2018).

8.8.2 Comments and ideas for the development of polymer matrix-based nanocomposites

The ever-increasing use of flexible packaging materials costs $38 billion globally. With an average annual increase of 3.5% in demand in the industrial sector, flexible materials used in the packaging industry must be able to meet and exceed the high expectations of consumers and supply chain stressors. As the competition between suppliers increases within the framework of government regulations, these materials become innovations in producing packaging films that improve product performance and address the global community's concerns about packaging waste. One of these innovations is polymer nanocomposite technology, which is considered the key to future developments in the flexible packaging industry. According to Aaron Brody's findings in his December 2003 Food Technology article, nanocomposites appear to transform plastic into a superbarrier equivalent to glass or metal without regulators (Mahmoudi et al., 2022).

Nanocomposites, defined as polymers linked with nanoparticles to produce materials with improved properties, have existed for many years but have recently entered the commercial trend in the packaging industry. The United States leads technology research with more than 400 research centers and companies with more than $3.4 billion in funding. Europe has more than 175 companies and organizations active in nanoscience research with a budget of 1.7 billion dollars. In addition, Japan also operates with more than 100 companies in this field. The global nanocomposites market is expected to increase to 250 million dollars by 2008, and this growth rate is likely to be between 18% and 25% per year.

Polymer nanocomposites are made by adding fillers of nanoparticles that form flat platelets. These platelets are then dispersed in a polymer matrix to create multiple parallel layers. This allows gases to pass through an "indirect pathway" in the polymer, forming complex barriers to gases and water vapor. (Tadayon et al., 2016).

More dispersion of nanoparticles in a polymer significantly reduces its permeability. According to the US Army Natick Soldier Systems Center, the amount of nanoparticles dispersed in the polymer is related to the improvement of mechanical and barrier properties in the obtained nanocomposite films compared to pure polymer films. Nanoparticles have much lower loading levels than traditional fillers to achieve optimal performance. Usually, the additive amount of nanofillers is less than 5%, significantly affecting nanocomposite films' weight reduction. This dispersion process increases the dimensions' ratio to the surface, which increases the plastic performance compared to conventional fillers. Various filters are used, the most common of which is a nanoclay called Montmorillonite, used as layered clay. It will be in its natural state; clay is hydrophilic, unlike hydrophobic polymers. For the compatibility of these two materials, the clay's polarity must be modified to become more "organic" and successfully interact with polymers. One of the methods of clay modification is the exchange of organic ammonium cations with inorganic cations from the clay surface. Carbon nanotubes are more expensive than nanoclay nanofillers, which are readily available and have excellent electrical and thermal conductivity (Zahedi et al., 2019).

The two leading suppliers of nano-clays are the American companies, Nanocore and Kelly Southern. The melt composition or processing of nanofillers into the polymer is done simultaneously if the polymer is processed through an extruder, injection molding, or other processing devices. Polymer pellets and filler (clay) are compressed using shear forces in the peeling process. In addition, in in situ polymerization, the filler is added directly to the liquid monomer during the polymerization stage. Fillers are added to the polymer solution with the help of solvents such as toluene, chloroform, and acetonitrile (Aliahmadi & Esmaeili, 2022; Amand & Esmaeili, 2020; Bahramimehr & Esmaeili, 2019).

8.9 Conclusions and future perspectives

Based on the evaluations, by 2009, the consumption of nanocomposite materials in the food and beverage industry was 5 million. This figure reached 100 million pounds by 2011. The beer industry appeared to be the largest consumer of nanocomposites by 2006. With 3 million pounds, soda bottles used 50 million nanocomposites by 2011. Polymer nanocomposites guarantee the future of the global packaging industry. Is:Based on evaluations, the consumption of nanocomposite materials in the food and beverage industry reached 5 million pounds by 2009 and surged to 100 million pounds by 2011. The beer industry emerged as the largest consumer of nanocomposites by 2006, while soda bottles alone utilized 50 million nanocomposites by 2011. Polymer nanocomposites are poised to shape the future of the global packaging industry. Polymer nanocomposites guarantee the future of the global packaging industry. By reducing production costs and raw materials, companies will use this technology to maximize the lifespan and survival of their products through the supply chain; in addition to providing quality products, they will also save on consumption costs. The benefits of nanocomposites are usually more than the costs and concerns ahead, and with time, this technology has changed and developed more. Research continues in the area of other fillers (e.g., carbon nanotubes). That allows new nanocomposite structures to increase nanocomposites in many packaging applications (Esmaeili & Khoshnevisan, 2016).

References

Agrawal, C. M., & Ray, R. B. (2001). Biodegradable polymeric scaffolds for musculoskeletal tissue engineering. *Journal of Biomedical Materials Research: An Official Journal of The Society for Biomaterials, The Japanese Society for Biomaterials, and The Australian Society for Biomaterials and the Korean Society for Biomaterials*, 55, 141–150.

Ajayan, P. M., Schadler, L. S., & Braun, P. V. (2006). *Nanocomposite science and technology* (pp. 44–46). John Wiley & Sons.

Aliahmadi, M., & Esmaeili, A. (2022). Preparation nanocapsules chitosan modified with selenium extracted from the *Lactobacillus acidophilus* and their anticancer properties. *Archives of Biochemistry and Biophysics*, 727, 109327.

Amand, F. K., & Esmaeili, A. (2020). Investigating the properties of electrospun nanofibers made of hybride polymer containing anticoagulant drugs. *Carbohydrate polymers*, 228, 115397.

Ambre, A. H., Katti, D. R., & Katti, K. S. (2013). Nanoclays mediate stem cell differentiation and mineralized ECM formation on biopolymer scaffolds. *Journal of Biomedical Materials Research. Part A*, 101, 2644–2660.

Astrom, B. T. (2018). *Manufacturing of polymer composites: Imprint* (1st Edition, p. 44) Routledge. Available from https://doi.org/10.1201/9780203748169.

Badylak, S. F., Taylor, D., & Uygun, K. (2011). Whole-organ tissue engineering: Decellularization and recellularization of three-dimensional matrix scaffolds. *Annual Review of Biomedical Engineering, 13*, 27−53.

Bahramimehr, F., & Esmaeili, A. (2019). Producing hybrid nanofiber-based on/PAN/Fe$_3$O$_4$/zeolite/nettle plant extract/urease and a deformed coaxial natural polymer to reduce toxicity materials in the blood of dialysis patients. *Journal of Biomedical Materials Research. Part A, 107*, 1736−1743.

Bharali, D. J., Khalil, M., Gurbuz, M., Simone, T. M., & Mousa, S. A. (2009). Nanoparticles and cancer therapy: A concise review with emphasis on dendrimers. *International Journal of Nanomedicine, 4*, 1.

Chen, L., Bai, Y., Liao, G., Peng, E., Wu, B., Wang, Y., Zeng, X., & Xie, X. (2013). Electrospun poly (L-lactide)/poly (ε-caprolactone) blend nanofibrous scaffold: Characterization and biocompatibility with human adipose-derived stem cells. *PLoS One, 8*, e71265.

Choy, J.-H., Jung, J.-S., Oh, J.-M., Park, M., Jeong, J., Kang, Y.-K., & Han, O.-J. (2004). Layered double hydroxide as an efficient drug reservoir for folate derivatives. *Biomaterials, 25*, 3059−3064.

Cipitria, A., Skelton, A., Dargaville, T., Dalton, P., & Hutmacher, D. (2011). Design, fabrication and characterization of PCL electrospun scaffolds—A review. *Journal of Materials Chemistry, 21*, 9419−9453.

Esmaeili, A., Afshari, S., & Esmaeili, D. (2015). Formation of harmful compounds in biotransformation of lilial by microorganisms isolated from human skin. *Pharmaceutical Biology, 53*, 1768−1773.

Esmaeili, A., & Ebrahimzadeh, M. (2015). Polymer-based of extract-loaded nanocapsules *Aloe vera* L. delivery. *Synthesis and Reactivity in Inorganic, Metal-Organic, and Nano-Metal Chemistry, 45*, 40−47.

Esmaeili, A., & Ebrahimzadeh, M. (2016). Optimization and preparation of methylcellulose edible film combined with of ferulago angulata essential oil (FEO) nanocapsules for food packaging applications. *Flavour and Fragrance Journal, 31*, 341−349.

Esmaeili, A., & Khoshnevisan, N. (2016). Optimization of process parameters for removal of heavy metals by biomass of Cu and Co-doped alginate-coated chitosan nanoparticles. *Bioresource Technology, 218*, 650−658.

Esmaeili, A., & Loghmani, K. (2016). Removal of monoethylene glycol from gas field wastewater using *Aspergillus tubingensis* and a new bioreactor. *Waste and Biomass Valorization, 7*, 151−156.

Esmaeili, A., & Rafiee, R. (2015). Preparation and biological activity of nanocapsulated *Glycyrrhiza glabra* L. var. *glabra*. *Flavour and Fragrance Journal, 30*, 113−119.

Esmaeili, A., Rahnamoun, S., & Sharifnia, F. (2013). Effect of O/W process parameters on *Crataegus azarolus* L. nanocapsule properties. *Journal of Nanobiotechnology, 11*, 1−9.

Gaharwar, A. K., Peppas, N. A., & Khademhosseini, A. (2014). Nanocomposite hydrogels for biomedical applications. *Biotechnology and Bioengineering, 111*, 441−453.

Gaharwar, A. K., Schexnailder, P., Kaul, V., Akkus, O., Zakharov, D., Seifert, S., & Schmidt, G. (2010). Highly extensible bio-nanocomposite films with direction-dependent properties. *Advanced Functional Materials, 20*, 429−436.

Gao, Q., He, Y., Fu J-z., Liu, A., & Ma, L. (2015). Coaxial nozzle-assisted 3D bioprinting with built-in microchannels for nutrients delivery. *Biomaterials, 61*, 203−215.

Guimarães, D., Cavaco-Paulo, A. & Nogueira, E. (2021). Design of liposomes as drug delivery system for therapeutic applications. *International Journal of Pharmaceutics, 601*, 120571.

Gunn, J., & Zhang, M. (2010). Polyblend nanofibers for biomedical applications: Perspectives and challenges. *Trends in Biotechnology, 28*, 189−197.

Hormozi, N., & Esmaeili, A. (2019). Synthesis and correction of albumin magnetic nanoparticles with organic compounds for absorbing and releasing doxorubicin hydrochloride. *Colloids and Surfaces B: Biointerfaces, 182*, 110368.

Huang, Z.-M., Zhang, Y.-Z., Kotaki, M., & Ramakrishna, S. (2003). A review on polymer nanofibers by electrospinning and their applications in nanocomposites. *Composites Science and Technology, 63*, 2223−2253.

Jadidi, K., Esmaeili, M., Kalantari, M., Khalili, M., & Karakouzian, M. (2020). A review of different aspects of applying asphalt and bituminous mixes under a railway track. *Materials, 14*, 169.

Jang, J.-H., Castano, O., & Kim, H.-W. (2009). Electrospun materials as potential platforms for bone tissue engineering. *Advanced Drug Delivery Reviews, 61*, 1065−1083.

Khateri, M. & Esmaeili, A. (2024). Synthesis and characterization of ZnO quantum dot-functionalized mesoporous nanocarriers for controlled drug delivery. *Ceramics International, 7*.

Khodaei, M., & Esmaeili, A. (2020). Capsulation of methadone in polymeric layers based on magnetic nanoparticles. *Inorganic and Nano-Metal Chemistry, 50*, 278−285.

Klajnert, B., & Bryszewska, M. (2001). Dendrimers: Properties and applications. *Acta Biochimica Polonica, 48*, 199−208.

Lam, C. X., Hutmacher, D. W., Schantz, J. T., Woodruff, M. A., & Teoh, S. H. (2009). Evaluation of polycaprolactone scaffold degradation for 6 months in vitro and in vivo. *Journal of Biomedical Materials Research Part A: An Official Journal of The Society for Biomaterials, The Japanese Society for Biomaterials, and The Australian Society for Biomaterials and the Korean Society for Biomaterials, 90*, 906−919.

Li, D., Wu, T., He, N., Wang, J., Chen, W., He, L., Huang, C., Ei-Hamshary, H. A., Al-Deyab, S. S., & Ke, Q. (2014). Three-dimensional polycaprolactone scaffold via needleless electrospinning promotes cell proliferation and infiltration. *Colloids and Surfaces B: Biointerfaces, 121*, 432−443.

Liao, G., Jiang, S., Xu, X., & Ke, Y. (2012). Electrospun aligned PLLA/PCL/HA composite fibrous membranes and their in vitro degradation behaviors. *Materials Letters, 82*, 159−162.

Lubin, G. (2013). *Handbook of composites* (p. 12) Springer Science & Business Media.

Mahmoudi, R., Esmaeili, A., & Nematollahzadeh, A. (2022). Preparation of $Fe_3O_4/Ag_3VO_4/$Au nanocomposite coated with *Caerophyllum macropodum* extract modified with oleic acid for theranostics agent in medical imaging. *Journal of Photochemistry and Photobiology A: Chemistry, 425*, 113724.

Marras, S. I., Kladi, K. P., Tsivintzelis, I., Zuburtikudis, I., & Panayiotou, C. (2008). Biodegradable polymer nanocomposites: The role of nanoclays on the thermomechanical characteristics and the electrospun fibrous structure. *Acta Biomaterialia, 4*, 756−765.

Miao, Y.-E., Zhu, H., Chen, D., Wang, R., Tjiu, W. W., & Liu, T. (2012). Electrospun fibers of layered double hydroxide/biopolymer nanocomposites as effective drug delivery systems. *Materials Chemistry and Physics, 134*, 623−630.

Nitya, G., Nair, G. T., Mony, U., Chennazhi, K. P., & Nair, S. V. (2012). In vitro evaluation of electrospun PCL/nanoclay composite scaffold for bone tissue engineering. *Journal of Materials Science: Materials in Medicine, 23*, 1749−1761.

Park, D., Lee, J., Chung, J. J., Jung, Y., & Kim, S. H. (2020). Integrating organs-on-chips: Multiplexing, scaling, vascularization, and innervation. *Trends in Biotechnology, 38*, 99−112.

Pitt, C. G., Hendren, R. W., Schindler, A., & Woodward, S. C. (1984). The enzymatic surface erosion of aliphatic polyesters. *Journal of Controlled Release, 1*, 3−14.

Rajabi, A., & Esmaeili, A. (2020). Preparation of three-phase nanocomposite antimicrobial scaffold BCP/gelatin/45S5 glass with drug vancomycin and BMP-2 loading for bone regeneration. *Colloids and Surfaces A: Physicochemical and Engineering Aspects, 606*, 125508.

Rives, V., del Arco, M., & Martín, C. (2014). Intercalation of drugs in layered double hydroxides and their controlled release: A review. *Applied Clay Science, 88*, 239−269.

Shafiei, S., Shavandi, M., & Nickakhtar, Y. (2021). Effect of nanoclay addition on the properties of polycaprolactone nanocomposite scaffolds containing adipose derived mesenchymal stem cells used in soft tissue engineering. *Journal of Advanced Materials in Engineering (Esteghlal), 39*, 45−59.

Shafiei, S., Solati-Hashjin, M., Rahim-Zadeh, H., & Samadikuchaksaraei, A. (2013). Synthesis and characterisation of nanocrystalline Ca−Al layered double hydroxide {[Ca$_2$Al(OH)$_6$]NO$_3$.nH$_2$O}: In vitro study. *Advances in Applied Ceramics, 112*, 59−65.

Soltanabadi, Z., Esmaeili, A., & Bambai, B. (2022). Fabrication of morphine detector based on quartz@ Au-layer biosensor. *Microchemical Journal, 175*, 107127.

Tabata, Y. (2000). The importance of drug delivery systems in tissue engineering. *Pharmaceutical Science & Technology Today, 3*, 80−89.

Tadayon, A., Jamshidi, R., & Esmaeili, A. (2015). Delivery of tissue plasminogen activator and streptokinase magnetic nanoparticles to target vascular diseases. *International Journal of Pharmaceutics, 495*, 428−438.

Tadayon, A., Jamshidi, R., & Esmaeili, A. (2016). Targeted thrombolysis of tissue plasminogen activator and streptokinase with extracellular biosynthesis nanoparticles using optimized *Streptococcus equi* supernatant. *International Journal of Pharmaceutics, 501*, 300−310.

Tahmasebi, M., Esmaeili, A., & Bambai, B. (2022). New method of identifying morphine in urine samples using nanoparticle-dendrimer-enzyme hybrid system. *Arabian Journal of Chemistry, 15*, 103630.

Tammaro, L., Russo, G., & Vittoria, V. (2009). Encapsulation of diclofenac molecules into poly (ε-caprolactone) electrospun fibers for delivery protection. *Journal of Nanomaterials*.

Teo, W. E., & Ramakrishna, S. (2006). A review on electrospinning design and nanofibre assemblies. *Nanotechnology, 17*, R89.

Thomas, S., Grohens, Y., & Ninan, N. (2015). *Nanotechnology applications for tissue engineering: William Andrew* (p. 24) Elsevier.

Venkatesan, J., & Kim, S.-K. (2014). Nano-hydroxyapatite composite biomaterials for bone tissue engineering—A review. *Journal of biomedical nanotechnology, 10*, 3124−3140.

Woodruff, M. A., & Hutmacher, D. W. (2010). The return of a forgotten polymer— Polycaprolactone in the 21st century. *Progress in polymer science, 35*, 1217−1256.

Wutticharoenmongkol, P., Sanchavanakit, N., Pavasant, P., & Supaphol, P. (2006). Preparation and characterization of novel bone scaffolds based on electrospun polycaprolactone fibers filled with nanoparticles. *Macromolecular Bioscience, 6*, 70−77.

Xia, S.-J., Ni, Z.-M., Xu, Q., Hu, B.-X., & Hu, J. (2008). Layered double hydroxides as supports for intercalation and sustained release of antihypertensive drugs. *Journal of Solid State Chemistry, 181*, 2610−2619.

Yang, S., Leong, K.-F., Du, Z., & Chua, C.-K. (2001). The design of scaffolds for use in tissue engineering. Part I. Traditional factors. *Tissue Engineering, 7*, 679−689.

Zahedi, E., Esmaeili, A., Eslahi, N., Shokrgozar, M. A., & Simchi, A. (2019). Fabrication and characterization of core-shell electrospun fibrous mats containing medicinal herbs for wound healing and skin tissue engineering. *Marine Drugs, 17*, 27.

Zecheru, T. (2008). *New biopolymers with possible use in dentistry and orthopaedics.* Université d'Angers, p. 10.

Zhao, N., Shi, S., Lu, G., & Wei, M. (2008). Polylactide (PLA)/layered double hydroxides composite fibers by electrospinning method. *Journal of Physics and Chemistry of Solids, 69*, 1564−1568.

Zhu, Y., Gao, C., Liu, X., & Shen, J. (2002). Surface modification of polycaprolactone membrane via aminolysis and biomacromolecule immobilization for promoting cytocompatibility of human endothelial cells. *Biomacromolecules, 3*, 1312−1319.

Industrial wastewater treatment applications of nanofibers and their composites

9

Gianluca Viscusi
Department of Industrial Engineering, University of Salerno, Fisciano, Salerno, Italy

9.1 Introduction

The contamination of water sources has increased over the years, and it represents a serious environmental and human health concern. Therefore, there is a great need for developing novel technologies in wastewater purification (Marinho et al., 2021). Nanotechnology can play a significant role in water purification because nanomaterials show unique physical and chemical properties (Qu et al., 2013). These materials can be applied for in situ treatment of polluted water effluents because, due to their small size, they contribute to the economical aspect of treatment processes (Ambashta & Sillanpää, 2010) To date, different nanosized materials have been fabricated for water purification (Nassar, 2010).

Among all, nanofibers (NFs) are an interesting class of materials, which could be used for wastewater remediation. They are defined as fibers with diameters in the nanometric range and specific advantages such as high surface area, high porosity, ease of separation, and size uniformity, which allow for applying NFs in many industrial applications (Girijappa et al., 2019; Xu et al., 2012).

Because of the high specific surface area and the ability to incorporate different functional groups, molecules, and nanoparticles, NFs have been already applied for different applications such as air filtration (Naragund & Panda, 2022), water treatment (Sakib et al., 2021), antimicrobial treatment (Yavari Maroufi et al., 2021), environmental sensing (Sonwane & Kondawar, 2021), and agricultural/environmental remediation (Agrawal et al., 2021; Badgar et al., 2022). As far as water remediation is concerned, NFs have been used as adsorbents, photocatalytic materials, electrochemical electrodes, and membranes for removing different classes of pollutants such as heavy metals, organic pollutants, persistent organic pollutants, contaminants of emerging concern, and particulate matter, as well as for desalination (Fig. 9.1) (Kenry & Lim, 2017).

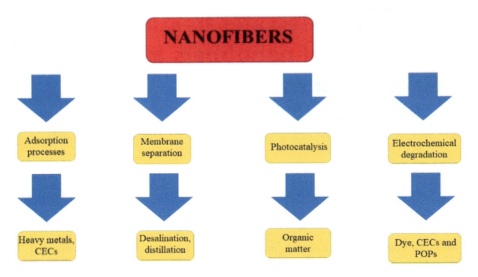

Figure 9.1 Water treatment applications of nanofiber.

9.2 Modification of nanofibers for water remediation applications

NFs can be specifically functionalized with different functional groups, catalysts and additives with high selectivity for adsorbing pollutants.

To increase the adsorption properties, it is possible to design functionalized materials by surface modification with functional groups (i.e., $-NH_2$, $-SH$, and $-SO_3H$ for removing heavy metals) (Wu et al., 2010). The functional groups with their specific properties should be chosen to control the adsorption performances. Previous studies have already shown that adsorbent materials can be modified by chemically bonding groups to confer functional ends acting as chelating agents (Stephen et al., 2011).

For example, fibers modified with tetrazine, carboxyl, amidoxime-hydroxam, imidazoline, amino, and phosphoric groups are used to improve the metal ions adsorption (Coşkun et al., 2000).

Zhang et al. prepared carbonyl groups containing hydrazine-modified polyacrylonitrile fiber for removing heavy metals. The maximum adsorption capacities of the modified adsorbent for Cu^{2+}, Cd^{2+}, Zn^{2+}, Co^{2+}, Pb^{2+}, Cr^{3+}, Ni^{2+}, and Hg^{2+} were 1.33, 1.30, 1.03, 1.02, 0.98, 0.96, 0.95, and 0.63 mmol/g (dry weight), respectively (Zhang et al., 1994).

Hydrolyzed polyacrylonitrile fibers were tested to remove copper ions from an aqueous solution. The amount of adsorbed copper is dependent on solution pH (from 2.8 mg/g at pH = 3 up to 12.4 mg/g at pH = 4). The adsorbed copper amount reached 27.95 mg/g with a copper ion concentration of 132 mg/L (Deng et al., 2003).

The poly(acrylaminophosphonic-carboxyl-hydrazide) chelating ion exchange fibers were fabricated through amination and phosphorization reaction of the hydrazine-modified polyacrylonitrile fibers. The produced adsorbent was tested for removing different heavy ions such as Cu^{2+}, Pb^{2+}, Zn^{2+}, Co^{2+}, Ni^{2+}, Hg^{2+}, Cd^{2+}, Mn^{2+}, Cr^{3+}, and Ag^+ (Liu et al., 1999).

Thiosemicarbazide-modified polyacrylonitrile fiber (PANMW-TSC) adsorbent was synthesized under microwave irradiation (Fig. 9.2). The which adsorbent was applied for the uptake of Cd^{2+} and Pb^{2+}.

The adsorption capacities for Cd^{2+} and Pb^{2+} were dependent on pH, and the maximum values were recorded at pH 6.4 (1.47 and 1.01 mmol/g for Cd^{2+} and Pb^{2+}, respectively). The obtained results can be ascribed to the introduction of nitrogen and sulfur atoms onto the surface of the adsorbent. Finally, the adsorbent showed good stability and reusability (Deng et al., 2016).

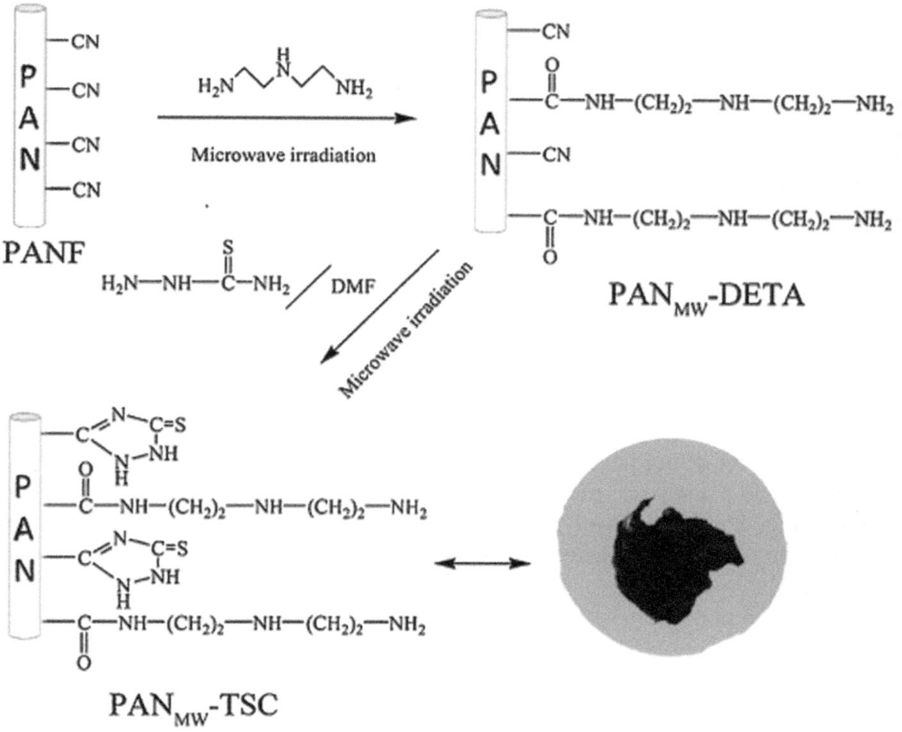

Figure 9.2 Scheme of synthesis routine of PANMW-TSC.
Source: From Deng, S., Wang, P., Zhang, G., & Dou, Y. (2016). Polyacrylonitrile-based fiber modified with thiosemicarbazide by microwave irradiation and its adsorption behavior for Cd(II) and Pb(II). *Journal of Hazardous Materials*, *307*, 64−72. https://doi.org/10.1016/j.jhazmat.2016.01.002.

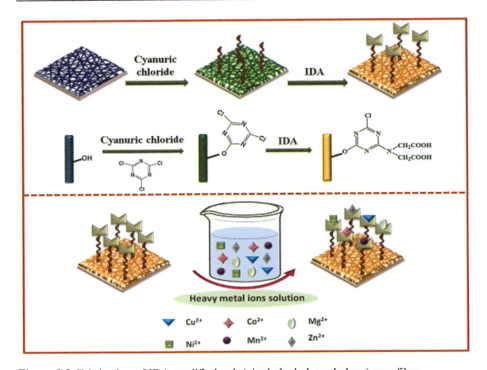

Figure 9.3 Fabrication of IDA-modified poly(vinyl alcohol-coethylene) nanofiber membranes for removing heavy metal ions.
Source: From Lu, Y., Wu, Z., Li, M., Liu, Q., & Wang, D. (2014). Hydrophilic PVA-co-PE nanofiber membrane functionalized with iminodiacetic acid by solid-phase synthesis for heavy metal ions removal. *Reactive and Functional Polymers*, 82, 98–102. https://doi.org/10.1016/j.reactfunctpolym.2014.06.004.

Lu et al. produced hydrophilic poly(vinyl alcohol-coethylene) (PVA-coPE) NF membrane for removing heavy metal ions through the solid-phase synthesis of iminodiacetic acid (IDA) on NF membrane surfaces after activation with cyanuric chloride (Fig. 9.3). The novel-produced adsorbent demonstrated excellent adsorption capability of different heavy metals such as Cu^{2+}, Co^{2+}, Zn^{2+}, and Ni^{2+} (Lu et al., 2014).

Table 9.1 reports some examples of modified fibers for different wastewater treatment applications.

9.3 Fiber systems produced by electrospinning

NFs are nanostructures that may be produced by using several methodologies such as the drawing method, solvothermal approach, thermal-induced phase separation method, vapor phase technique, template method, self-assembly, and electrospinning

Table 9.1 Use of modified fibers for wastewater treatment applications.

Adsorbent	Removed substance	Adsorption capacity	References
Epichlorohydrin-crosslinked chitosan	Cr^{6+}	1400 mg/g	Baroni et al. (2008)
Thiosemicarbazide modified polyacrylonitrile	Cd^{2+}	1.47 mmol/g	Deng et al. (2016)
	Pb^{2+}	1.01 mmol/g	Deng et al. (2016)
Polyamino-polycarboxylic acid ligands modified polyacrylonitrile	Cd^{2+}	1.34 mmol/g	Zhang et al. (2009)
Nitro-oxidized carboxy-cellulose	Hg^{2+}	257.07	Chen et al. (2021b)
TEMPO-oxidized cellulose	Cu^{2+}	56.5	Fiol et al. (2021)
CeO_2-Fe_3O_4 decorated polyaniline	F^-	117.64	Chigondo et al. (2018)
Polypyrrole self-assembled on EVOH	Cr^{6+}	90.74	Xu et al. (2015)
PVA-coPE functionalized with iminodiacetic acid	Cu^{2+}	101.87	Lu et al. (2014)
PVDF/e-TiO_2	Bisphenol A	Degradation = 85%	Nor et al. (2016)
Fe/N-doped carbon	Bisphenol A	Degradation = 100%	Chen et al. (2021a)
Fe/Co-CNFs	Tetracycline	Degradation = 97.55%	Xie et al. (2021)
Poly(p-phenylene terephthalamide) modified by salicylaldehyde.	Pb^{2+}	19.6 mg/g	Qu et al. (2012)
	Cu^{2+}	4.06 mg/g	Qu et al. (2012)

(Gundloori et al., 2019). Among all, electrospinning is a versatile technique able to generate ultrathin fibers (Marinho et al., 2020) at the nano- and micro-scale (Araújo et al., 2013) with specific characteristics such as porosity, fine diameters (up to several nanometers), large surface area per unit mass, high gas permeability, and small interfibrous pore size (Saeed et al., 2008; Saber-Samandari et al., 2013). This technique presents many advantages such as low cost, reliability, repeatability, high effectiveness, control of fiber dimensions, and scale production (Pasini et al., 2019). The fibers are produced by electrospinning due to the production of an electrostatically driven jet of the polymer solution from a needle to a collector.

During the transit toward the collector, solidification occurs, allowing for recovering polymeric thin fibers. Due to its versatility, electrospinning allows for producing different kinds of nanomaterials (both inorganic and organic materials) with characteristic features favourable for water remediation, energy conversion, and storage (LeCorre-Bordes et al., 2016).

For example, polyacrylonitrile fibers modified with phenolate groups favor the removal of cationic dyes (Xiao et al., 2020). NiO and ZnO incorporated into electrospun fibers are able to remove anionic species (Chen et al., 2019), while porous membranes of poly(vinyl alcohol) and chitosan show high adsorption capacity toward Direct red 80 (Hosseini et al., 2017). Then, the electrospun fiber systems are even used as filtrating membranes for removing oil traces at the water/oil interface (i.e., nanoporous fibers of polystyrene for oil spill clean-up (Lin et al., 2012) or dopamine-modified polyacrylonitrile fibers for removing motor oil and diesel fuel (Almasian et al., 2017). Other examples concern the removal of heavy metals by adding the required properties to the NFs. For example, the modified electrospun polystyrene fibers with the incorporation of TiO_2 nanoparticles were applied for the adsorption capacity for Cu^{2+} ions (Wanjale et al., 2016), or polyacrylonitrile/graphene oxide nanofibers for removing Cr^{6+} (Feng et al., 2020).

9.3.1 Electrospun synthetic fibers

Electrospinning is an interesting technique for the purification of water effluents. Synthetic polymers are commonly used for their low cost, mechanical performance, and ease of production. From now on, some works concerning the use of electrospun synthetic fibers for water remediation have been reported.

Zhao et al. produced bPEI-grafted PAN fiber produced by electrospinning as adsorbents of Cr^{6+} (q_m = 637.46 mg/g). Batch adsorption and dynamic filtration could also decrease the chromium ion concentration below the drinking water standard from a high initial concentration. Finally, the adsorption capacities for Cr^{2+}, Hg^{2+}, Cu^{2+}, and Cd^{2+} were also investigated (Zhao et al., 2017).

Pasini et al. produced submicrometric electrospun polyetherimide (PEI) fibers incorporating TiO_2 nanoparticles before submitting them to cold plasma in order to improve adhesion and photocatalytic performance. Methylene blue (MB) was completely decolorized after a time exposure of 40 minutes. Moreover, 75% of methyl orange (MO) and tetracycline were degraded after 120 minutes (Pasini et al., 2019).

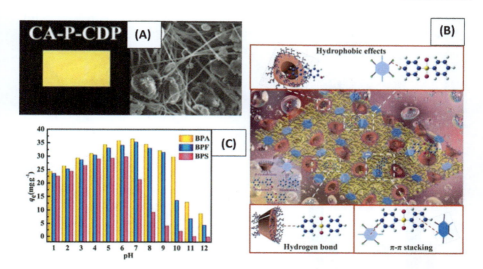

Figure 9.4 (A) Image and SEM micrograph of porous CA-P-CDP membrane; (B) adsorption mechanisms between adsorbent and bisphenol pollutants; and (C) effect of pHs on the adsorption performances.
Source: From Lv, Y., Ma, J., Liu, K., Jiang, Y., Yang, G., Liu, Y., Lin, C., Ye, X., Shi, Y., Liu, M., & Chen, L. (2021). Rapid elimination of trace bisphenol pollutants with porous β-cyclodextrin modified cellulose nanofibrous membrane in water: adsorption behavior and mechanism. *Journal of Hazardous Materials*, 403. https://doi.org/10.1016/j.jhazmat.2020.123666.

Lv et al. produced porous β-cyclodextrin-modified cellulose NFs (CA-P-CDP) to remove bisphenol A (BPA), bisphenol S (BPS), and bisphenol F (BPF) (Fig. 9.4). The maximum adsorption capacities were 50.37 mg/g for BPA, 48.52 mg/g for BPS, and 47.25 mg/g for BPF. The adsorption mechanisms between adsorbates and adsorbents can be ascribed to hydrophobic effects, hydrogen-bonding interactions, and π–π stacking interactions (π for BPF) (Lv et al., 2021).

Saeed et al. produced electrospun polyacrylonitrile (PAN) NFs chemically modified with amidoxime groups. The saturation adsorption capacities for Cu^{2+} and Pb^{2+} were 52.70 and 263.45 mg/g, respectively. Finally, after the adsorption, over 90% of metals were recovered in a nitric acid solution after 1 hour, proving the potential use of the PAN-oxime—based adsorbent to recycle metals from wastewater (Saeed et al., 2008).

Iron alkoxide/thermal plastic elastomer ester solution was prepared to produce a nanofibrous membrane by combining electrospinning and hydrothermal strategy. The produced adsorbent showed high efficiency for the removal of Cr^{6+}, attributable to two processes: electrostatic adsorption and redox reaction of Cr(VI)-Cr (III). Fig. 9.5 reports the Cr variation over time, the removal percentages of different membranes, and the adsorption mechanism (Xu et al., 2012).

Table 9.2 reports some examples of electrospun synthetic fibers for different wastewater treatment applications.

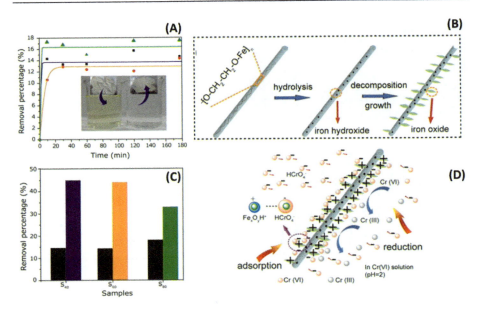

Figure 9.5 (A) Removal % of chromium ions with increasing time; (B) preparation of the obtained nanofibrous membrane; (C) the Cr removal efficiency in 3 days for the produced membrane; and (D) schematic illustration of the adsorption and reduction of Cr^{6+}.
Source: From Xu, G. R., Wang, J. N. & Li, C. J. (2012). Preparation of hierarchically nanofibrous membrane and its high adaptability in hexavalent chromium removal from water. *Chemical Engineering Journal, 198−199,* 310−317. https://doi.org/10.1016/j.cej.2012.05.104.

9.3.2 Electrospun natural fibers

Nowadays, because there is a need for novel eco-friendly methodologies in water treatment processes, the use of natural fibers could be a potential solution. In fact, fossil fuel-derived polymers can cause harsh health and environmental concerns because of their nonbiodegradability, persistence, and toxic nature. Therefore the minimization of waste generation or the use of recyclable and biodegradable materials appears to be a need (Benelli et al., 2017). For example, the use of natural NFs has been widely studied because of their abundance, availability, low cost, and potential industrial applications (Haider & Park, 2009), while the application of electrospun protein-based fibers is attracting the interest of the scientific community because they can be used for heavy metals removal due to a large number of polar groups of protein (Ki et al., 2007), which guarantees a high affinity toward metals, organic substances, and other materials (Kruppa et al., 2006). For example, keratin biofiber protein can be considered an efficient adsorbent for removing heavy metals (Kar & Misra, 2004).

Pan et al. designed and synthesized alginate fibers containing graphene oxide for the removal of heavy metal ions. High affinity to Pb^{2+} ion (q_m = 386.2 mg/g) was observed. The high number of adsorption sites and favorable transport channels allowed to obtain maximum performance in a short time (Pan et al., 2019).

Table 9.2 Use of synthetic polymer–based electrospun nanofibers for water treatment applications.

Adsorbent	Removed substances	q_{max}, mg/g	References
Boehmite NPs impregnated PCL	Cd^{2+}	0.20	Hota et al. (2008)
Polystyrene	Diesel oil	7.13	Wu et al. (2012)
	Silicon oil	81.40	Wu et al. (2012)
	Peanut oil	112.3	Wu et al. (2012)
	Motor oil	131.6	Wu et al. (2012)
Polypropylene	Motor oil	129	Li et al. (2014a)
	Peanut oil	80	Li et al. (2014a)
Polystyrene	Motor oil	84.41	Lin et al. (2011)
	Sunflower oil	79.62	Lin et al. (2011)
Poly(acrylo-amidino ethylene amine)	As(V)	76.92	Vu et al. (2013)
Thioamide-group chelating polyacrylonitrile	Au(III)	34.60 (mmol/g)	Li et al. (2013)
β-cyclodextrin-based poly(acrylic acid)	Methylene blue	826.45	Zhao et al. (2015a)
Vinyl-modified mesoporous poly(acrylic acid)/SiO_2	Malachite green	220.49	Xu et al. (2012)
Ethylenediamine-grafted-polyacrylonitrile	Methylene blue	94.07	Haider et al. (2014)
	Safranin T	110.62	Haider et al. (2014)
	Rhodamine B	138.69	Haider et al. (2014)
Polydopamine-coated poly(vinyl alcohol)/poly(acrylic acid)	Methylene blue	1147.6	Yan et al. (2015)
Poly(vinyl alcohol)/SiO_2	Cu^{2+}	504.89	Wu et al. (2010)
Amidoxime-modified polyacrylonitrile	Pb^{2+}	263.45	Saeed et al. (2008)
Polystyrene	Motor oil	113.87	Lin et al. (2012)
	Bean oil	111.80	Lin et al. (2012)
	Sunflower seed oil	96.89	Lin et al. (2012)
Poly(vinylidene fluoride)/reduced graphene oxide/TiO_2	Oil	98.46	Lou et al. (2020)
Magnetic poly(vinylidene fluoride)/Fe_3O_4	Oil	35–46	Jiang et al. (2015)
Amidoxime-modified polyindole	Cr^{6+}	404.86	Zhijiang et al. (2017)

(Continued)

Table 9.2 (Continued)

Adsorbent	Removed substances	q_{max}, mg/g	References
Diethylenetriamine modified polyacrylonitrile	Cu^{2+}	87.77	Aung et al. (2020)
Polydopamine-polyacrylonitrile	Heavy motor oil	148.58	Almasian et al. (2017)
	Diesel fuel	62.53	Almasian et al. (2017)
Polyacrylonitrile-polyamidoamine composite	Direct red 80	1666.66	Almasian et al. (2015)
	Direct red 23	2000	Almasian et al. (2015)
Polyacrylonitrile/carbon nanotube/titanium dioxide	Cr^{6+}	527	Mohamed et al. (2017a)
Polyester/polyacrylonitrile/GO/Fe$_3$O$_4$	Pb^{2+}	799.4	Koushkbaghi et al. (2016)
	Cr^{6+}	911.9	Koushkbaghi et al. (2016)
Amidoxime-modified polyacrylonitrile	Cu^{2+}	52.70	Saeed et al. (2008)
	Pb^{2+}	263.45	Saeed et al. (2008)
Polystyrene/TiO$_2$ composite	Cu^{2+}	522	Wanjale et al. (2016)
Amidoxime modified polyindole	Cr^{6+}	340.14	Zhijiang et al. (2017)
Polyacrylonitrile-CNT/TiO$_2$-NH$_2$	As(III)	251	Mohamed et al. (2017b)
	As(V)	249	Mohamed et al. (2017a)
Poly(acrylonitrile comaleic acid)	Ni^{2+}	243.2	Allafchian et al. (2017)
	Cr^{6+}	173.9	Allafchian et al. (2017)
Polystyrene-dithizone	Pb^{2+}	0.016	Deng et al. (2011)
Polyacrylonitrile/polypyrrole	Cr^{6+}	44.95	Wang et al. (2013)
Polyacrylonitrile/graphene oxide	Cr^{6+}	382.5	Feng et al. (2020)
Polyacrylonitrile/γ-AlOOH composite	Pb^{2+}	233.10	Sun et al. (2016)
	Cu^{2+}	43.90	Sun et al. (2016)
	Cd^{2+}	144.30	Sun et al. (2016)
MOFs/polyacrylonitrile and polyvinylidene fluoride	Hg^{2+}	53.09	Efome et al. (2018)
	Pb^{2+}	50.88	Efome et al. (2018)
Poly(methyl methacrylate)/zeolite	Methyl orange	95.33	Lee et al. (2017)

Material	Pollutant	Capacity	Reference
Aminated polyacrylonitrile (PAN)/γ-AlOOH	Pb^{2+}	180.83	Sun et al. (2016)
	Cu^{2+}	48.68	Sun et al. (2016)
	Cd^{2+}	114.94	Sun et al. (2016)
Thioether groups functionalized mesoporous polyvinylpyrrolidone/SiO_2	Hg^{2+}	4.26 (mmol/g)	Teng et al. (2011b)
Mercapto groups modified poly(vinyl alcohol)/SiO_2	Cu^{2+}	324.72	Wu et al. (2010)
Amidoxime-modified polyacrylonitrile	Cu^{2+}	52.70	Saeed et al. (2008)
	Pb^{2+}	and 263.45	Saeed et al. (2008)
PAN/PPy/MnO_2	Pb^{2+}	0.83 mmol/g	Luo et al. (2015)
Phosphorylated PAN	Cd^{2+}	0.17 mmol/g	Zhao et al. (2015b)
	Pb^{2+}	0.47 mmol/g	Zhao et al. (2015a)
Aminated polyacrylonitrile	Cu^{2+}	150.6	Kampalanonwat & Supaphol (2010)
	Ag^+	155.5	Kampalanonwat & Supaphol (2010)
	Fe^{2+}	116.5	Kampalanonwat & Supaphol (2010)
	Pb^{2+}	60.6	Kampalanonwat & Supaphol (2010)
PVA/TiO_2 nanofiber modified with mercapto groups	U^{6+}	196.1	Abbasizadeh et al. (2013)
	Th^{4+}	238.1	Abbasizadeh et al. (2013)
Polyacrylonitrile/polypyrrole core/shell nanofiber	Cr^{6+}	64.5 mg/g	Wang et al. (2014)
PVA/TiO_2/ZnO functionalized with mercapto groups	Th^{4+}	333.33	Alipour et al. (2016)
PVA/TEOS/APTES	Cd^{2+}	327.3	Irani et al. (2012)
SiO_2-APTES modified polyacrylonitrile	Th^{4+}	249.4	Dastbaz & Keshtkar (2014)
	U^{6+}	193.1	Dastbaz & Keshtkar (2014)
	Cd^{2+}	69.5	Dastbaz & Keshtkar (2014)
	Ni^{2+}	138.7	Dastbaz & Keshtkar (2014)
Magnesium silicate functionalized PAN	Diquat herbicide	197.53	Li et al. (2017c)
La_2O_3 doped PAN	Phosphate	77.76 mg/g La	He et al. (2016)
Polyacrylonitrile modified with β-cyclodextrin	Bromophenol blue	1.197	Chabalala et al. (2021)
	Atrazine	0.817	Chabalala et al. (2021)

(Continued)

Table 9.2 (Continued)

Adsorbent	Removed substances	q_{max}, mg/g	References
Aminated-polyacrylonitrile	Cu^{2+}	116.522	Neghlani et al. (2011)
Hydrazine-modified polyacrylonitrile nanofibers	Cu^{2+}	114	Saeed et al. (2011)
Polyacrylonitrile/polypyrrole/manganese dioxide	Pb^{2+}	251.90	Luo et al. (2015)
MnO_2/polydopamine/PAN	Pb^{2+}	185.19	Li et al. (2018)
Meldrum's acid (2,2-dimethyl-1,3-dioxane-4,6-dione)-modified cellulose NFs/PVDF	Crystal violet	3.984	Gopakumar et al. (2017)

Wang et al. produced poly(acrylic acid)-sodium alginate nanofibrous hydrogels by electrospinning and thermal cross-linking treatment. The resultant adsorbent exhibited good adsorption performances toward Cu^{2+} (q_m = 591.7 mg/g) (Wang et al., 2018).

Li et al. fabricated electrospun chitosan nanofiber membranes crosslinked by glutaraldehyde vapor for removing heavy metals. The effect of different parameters on Cr^{6+} adsorption was studied. The module was able to remove some metal ions in the following order: $Cr^{6+} > Cu^{2+} > Cd^{2+} > Pb^{2+}$, while the adsorption capacities at a breakthrough point were 7.56, 3.36, 3.35, and 2.43 mg/g, respectively (Li et al., 2017a).

Haider et al. produced chitosan NF mats as adsorbent for metal ions. The adsorption tests demonstrated that the adsorption capacities for Cu^{2+} and Pb^{2+} at equilibrium were 485.44 and 263.15 mg/g, respectively (Haider & Park, 2009).

Ghani et al. studied the adsorption performances of NFs based on alginate for anionic (AR14) and cationic (BB41) dyes. The effect of some parameters such as solution pH, membrane dosage, contact time, and dye concentration on the adsorption performances was investigated. The research proved that the adsorption is pH-dependent and the maximum adsorption capacity at the solution pH of 1 and 9 was 93% and 71% for AR14 and BB41, respectively (Ghani et al., 2016).

Ma et al. fabricated calcium alginate and gelatin-calcium alginate composite membranes for MB removal (Fig. 9.6). The maximum adsorption capacity decreased from 2046 to 1937 mg/g after the modification with gelatin. The regeneration and reusability were improved by gelatin because of the higher electrostatic repulsions due to the formation of NH_3^+ groups from amino groups, which were able to desorb MB molecules. The proposed methodology opens the route for the development of efficient adsorbents for cationic dyes (Ma et al., 2019).

Table 9.3 reports some examples of electrospun natural fibers for different wastewater treatment applications.

9.4 Ceramic fibers

Ceramic fibers appear to be an interesting class of materials, which can be used for water remediation applications. They are known to possess greater chemical and physical stability in extreme conditions, making them suitable to be used in wastewater treatment applications (Ashaghi et al., 2007). Different technologies have been applied to develop ceramic NFs such as template-mediated growth of NFs, chemical vapor deposition, laser-mediated NF synthesis, hydrothermal technique, and electrospinning. Among all techniques, electrospinning appears to be one of the most interesting and effective methodologies to produce low-cost ceramic NFs with unique physicochemical properties, such as high porosity and high surface area, for water remediation. Besides, NFs can be easily separated from the treated water (Malwal & Gopinath, 2015).

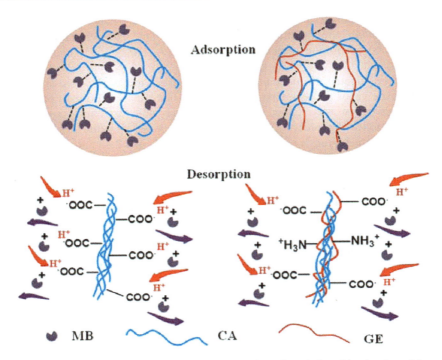

Figure 9.6 Schematization of adsorption and desorption of methylene blue by the calcium alginate and gelatin-calcium alginate composite membranes.
Source: From Ma, Y., Qi, P., Ju, J., Wang, Q., Hao, L., Wang, R., Sui, K., & Tan, Y. (2019). Gelatin/alginate composite nanofiber membranes for effective and even adsorption of cationic dyes. *Composites Part B: Engineering*, *162*, 671–677. https://doi.org/10.1016/j.compositesb.2019.01.048.

To produce ceramic NFs, a polymer as carrier must be blended with the ceramic melt to produce electrospun NFs. Moreover, it is crucial to prepare a precursor solution with adequate rheological properties. The typical procedure to produce ceramic NFs is reported in Fig. 9.7 and it involves the following steps:

- Dissolution of precursor salt and a polymer as matrix in a solvent or mixture of solvents;
- Electrospinning of the prepared solution;
- Calcination or sintering of the electrospun NFs to remove the organic components (Zhang et al., 2008).

In the last years, ceramic NFs have been used for disinfection, decomposition of organic pollutants, and removal of inorganic pollutants attracting a great interest for use in water treatment applications. Their action is based on adsorption and photocatalysis processes. As far as photocatalysis is concerned, complex ceramic-based structures have been synthesized in order to enhance the photocatalytic activity, for example, cadmium sulfide NPs encapsulated in cadmium titanate NFs to enhance the photocatalytic activity of MB (Pant et al., 2014), and

Table 9.3 Use of electrospun natural nanofibers for wastewater treatment applications.

Adsorbent	Removed substances	q_m, mg/g	References
Calcium crosslinked sodium alginate	Cu^{2+}	285.5	Wang et al. (2022)
Sodium alginate/graphene oxide	Pb^{2+}	386.2	Pan et al. (2019)
Poly(vinyl alcohol)/chitosan	Pb^{2+}	266.12	Karim et al. (2019)
	Cd^{2+}	148.78	Karim et al. (2019)
PEO/chitosan	Ni^{2+}	357.1	Aliabadi et al. (2013)
	Cu^{2+}	310.2	Aliabadi et al. (2013)
	Cd^{2+}	248.1	Aliabadi et al. (2013)
Wool keratose/silk fibroin	Cu^{2+}	2.88	Ki et al. (2007)
Chitosan/polyacrylonitrile/magnetic ZSM-5	Motor oil	99.40	Samadi et al. (2017)
	Lubricating oil	95.30	Samadi et al. (2017)
	Pump oil	88.10	Samadi et al. (2017)
PVA/chitosan	Direct red 80	790	Hosseini et al. (2017)
	Cu^{2+}	43.90	Sun et al. (2016)
	Cd^{2+}	144.30	Sun et al. (2016)
Polyacrylic acid	Hg^{2+}	Removal = 91%	Xiao et al. (2010)
Chitosan	Cu^{2+}	485.44	Haider & Park (2009)
	Pb^{2+}	263.15	Haider & Park (2009)
Cellulose acetate/titanium oxide	Pb^{2+}	31.9	Gebru & Das (2017)
	Cu^{2+}	31.4	Gebru & Das (2017)
Cellulose acetate/poly (dimethyldiallylammonium chloride—acrylamide)	Acid black 172	231	Xu et al. (2020)
Carboxymethylated and polydopamine-coated deacetylated cellulose acetate	Methylene blue	69.89	Chen et al. (2020)
Cellulose acetate/silica	Cr^{6+}	19.46	Taha et al. (2012)
Alkali lignin/PVA	Fluoxetine	Adsorption = 70%	Camiré et al. (2020)
Chitosan-grafted porous poly (L-lactic acid)	Cu^{2+}	270.27	Zia et al. (2021)
Cyclodextrin-functionalized mesoporous polyvinyl alcohol/SiO_2	Indigo carmine	495	Teng et al. (2011a)
PVA/Sodium alginate	Cd^{2+}	67.05	Ebrahimi et al. (2017)
Chitosan/(polyvinyl alcohol)/zeolite	Cr^{6+}	0.17 (mmol/g)	Habiba et al. (2017)
	Fe^{3+}	0.11 (mmol/g)	Habiba et al. (2017)
	Ni^{2+}	0.03 (mmol/g)	Habiba et al. (2017)
Polyacrylic acid sodium/chitosan	Cr^{6+}	78.96	Jiang et al. (2018)

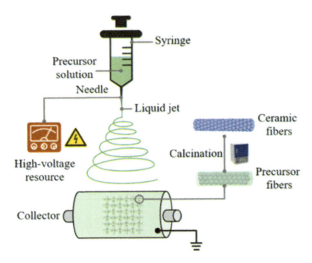

Figure 9.7 Preparation of ceramic nanofibers by electrospinning technique.
Source: From Xing, Y., Cheng, J., Li, H., Lin, D., Wang, Y., Wu, H., & Pan, W. (2021). Electrospun Ceramic Nanofibers for Photocatalysis. *Nanomaterials, 11*(12). https://doi.org/10.3390/nano11123221.

$Bi_4Ti_3O_{12}$ NFs assembled with Ag and Au for the photocatalytic degradation of rhodamine B (Zhao & Yang, 2019b).

Li et al. produced electrospun α-Fe_2O_3—γ-Al_2O_3 core-shell NFs combined with vapor deposition and heat treatment techniques. Electrostatic forces between different charged species are supposed to occur. Besides, the electron-hole pair provided by Fe_2O_3 induced the Cr^{6+} reduction to Cr^{3+} (Li et al., 2014b).

Singh et al. fabricated mats of mesoporous electrospun ZnO NFs by decomposing PAN polymer as a carrier. The mats were tested for the complete degradation of naphthalene and anthracene dyes (Singh et al., 2013).

Nickel oxide NFs were fabricated by electrospinning and calcination of poly(ethylene oxide)/nickel acetate tetrahydrate NFs. Photodegradation of Congo red (CR) under visible light irradiation was studied. The results proved that about 50% of the dye was degraded after 15 minutes and 98% within 6 hours. Moreover, the amount of NFs directly affected the degradation time (Malwal & Gopinath, 2015).

Table 9.4 focuses on some works concerning the use of ceramic NFs for water remediation applications.

9.4.1 Application of TiO₂ fibers for water remediation

The use of some traditional processes developed for wastewater treatments is restricted due to the mechanical filtration, use of chemicals, high costs, and energy-expensive processes. To overcome that, photocatalysis appears to be a potential candidate. It is an advanced oxidative process to degrade organic and recalcitrant

Table 9.4 Use of ceramic nanofibers for water treatment applications.

Adsorbent	Removed substances	q_{max}, mg/g	References
α-Fe$_2$O$_3$-Al$_2$O$_3$	Cu^{2+}	4.98	Mahapatra et al. (2013)
	Pb^{2+}	23.75	Mahapatra et al. (2013)
	Ni^{2+}	32.36	Mahapatra et al. (2013)
	Hg^{2+}	63.69	Mahapatra et al. (2013)
Nickel ferrite NPs anchored onto silica	Methylene blue	1.279	Hong et al. (2015)
ZrO$_2$	F$^-$	297.7	Yu et al. (2018)
Phosphonate-TiO$_2$/ZrO$_2$	Cd^{2+}	25.4 (μmol/g)	Choi et al. (2013)
	Pb^{2+}	69.4 (μmol/g)	Choi et al. (2013)
	Ni^{2+}	<1 (μmol/g)	Choi et al. (2013)
	Cu^{2+}	89.9 (μmol/g)	Choi et al. (2013)
	Zn^{2+}	11.6 (μmol/g)	Choi et al. (2013)
Cerium-doped TiO$_2$	Rhodamine B	Efficiency = 99.59%	Xiao et al. (2013)
Magnesium silicate	Methylene blue	609.75	Zhao et al. (2019b)
α-Fe$_2$O$_3$	Methyl orange	Decolorization efficiency >99%	Ghasemi et al. (2015)
WO$_3$/Fe(III)	Methyl orange	Decolorization efficiency = 94.6%	Ma et al. (2017)
Bi(VO$_4$)$_{1-m}$(PO$_4$)$_m$	Methylene blue	Decolorization efficiency = 91.3%	Liu et al. (2014)
BiVO$_4$	Rhodamine B	87.1	Liu et al. (2016)
ZnO	Acid fuchsine	Photocatalytic activity = 99%	Gupta et al. (2015)
Bi$_2$WO$_6$	Methylene blue	Efficiency = 94.8%	Zhao et al. (2012)
Nd^{3+}-doped TiO$_2$	Rhodamine 6G	Efficiency = 80%	Hassan et al. (2012)
Hydrogen-treated WO$_3$	Rhodamine B	312.5	Tahmasebi et al. (2020)
MgO	Pb^{2+}	983.4	Xu et al. (2019)
	Cd^{2+}	1824.0	Xu et al. (2019)
LaCoO$_3$	Rhodamine B	Decolorization efficiency = 99%	Dong et al. (2010)

(Continued)

Table 9.4 (Continued)

Adsorbent	Removed substances	q_{max}, mg/g	References
CdTiO$_3$	Rhodamine 6G	Degradation efficiency = 97%	Shamshi Hassan et al. (2014)
Bi$_2$Fe$_4$O$_9$	Methyl orange	Photocatalytic efficiency = 70%	Qi et al. (2013)
α-Fe$_2$O$_3$	Cr^{6+}	16.17	Ren et al. (2013)
ZnO/Bi$_2$O$_3$	Rhodamine B	Photocatalytic efficiency = 95%	Yang et al. (2014)
Ag/TiO$_2$	Methylene blue	Dye degradation = 80%	Liu et al. (2012)
Ag/TiO$_2$	Parathion	Degradation = 64%	Li et al. (2012)
α-Fe$_2$O$_3$	Methyl orange	Decolorization efficiency > 99%	Ghasemi et al. (2015)
C-doped 1D ZnO-C	Caffeine	Degradation = 80.4%	Gadisa et al. (2020)
Ce-doped MoO$_3$	Safranin T	Degradation = 98%	Li et al. (2009)

pollutants (Malwal & Gopinath, 2015; Sedghi & Heidari, 2016). The methodology is based on the oxidation of pollutants due to the photogenerated holes or oxygen-containing radicals (Pasini et al., 2021). Among the different metal oxide-based photocatalysts, TiO$_2$ is the most utilized in photocatalytic applications (Armstrong et al., 2020) because of its properties such as the following:

- Photocatalytic activity,
- Chemical stability (in both acid and basic environment),
- Abundance,
- No toxicity,
- Cheapness.

TiO$_2$-based NFs can be classified as follows (Marinho et al., 2020):

- Pristine TiO$_2$ NFs,
- TiO$_2$ coated NFs,
- Modified TiO$_2$ NFs

Generally, TiO$_2$ NFs are prepared by mixing a TiO$_2$ precursor with a polymer before carrying out an electrospinning technique to process it. Then, the calcination allows the modification of the TiO$_2$ amorphous phase into a crystalline phase (Mahltig et al., 2007). The use of TiO$_2$ NFs has attracted some interest because they possess interesting properties such as high surface area, flexibility, and improved optical and photocatalytic properties (Pascariu et al., 2019). Photocatalysis plays a key role in water purification because photocatalytic oxidation is a photon-assisted advanced oxidation process to obtain the complete abatement and mineralization of organic compounds.

It is already proposed that the mechanism of photodegradation of pollutants induced by TiO$_2$ is based on the generation of conduction-band electrons (e$^-$) and valence-band holes (h$^+$) when irradiated with UV light. The highly oxidative photo-induced holes are able to oxidize contaminants generating species with high reactivity, for example, O$_2^-$, OH•, and H$_2$O$_2$, or react with adsorbed water and hydroxyl anions to form hydroxyl radicals. Meanwhile, the e$^-$ can also react with proper electron acceptors, such as O$_2$, to yield oxidative radicals. Research has shown that the degradation mechanism occurs also through oxygen derivatives, as shown below (Turchi & Ollis, 1990):

$$TiO_2 + h\nu \rightarrow e^- + h^+(TiO_2)$$

$$h^+ + OH^-_{ads} \rightarrow HO^\bullet$$

$$h^+ + H_2O_{ads} \rightarrow HO^\bullet + H^+$$

$$e^- + O_2 \rightarrow O_2^{\bullet-}$$

$$O_2^{\bullet-} + H^+ \rightarrow HO_2^\bullet$$

Among the designed methods to produce TiO_2, electrospinning is an interesting and versatile technique to fabricate NF-based catalysts (Altaf et al., 2020; Marinho et al., 2021).

The electrospun TiO_2 NFs show several advantages as the potential capacity for degrading pollutants compared to filtration systems and appear to be an interesting class of materials because of their ease of separation and better photocatalytic activity compared to dispersed TiO_2 nanoparticles (Choi et al., 2010).

For example, polyaniline-modified titanium dioxide/polyacrylonitrile composite NF was prepared through electrospinning for the photodegradation of MO, which was decolorized up to 90% in less than 1 hour in the presence of visible light in comparison with the neat NFs (about 10%) (Sedghi & Heidari, 2016).

Li et al. fabricated a photocatalyst based on TiO_2 supported on poly(methyl methacrylate) NFs for the degradation of MO. Based on the reported results, 0.1 g of the adsorbent could completely degrade 100 mL of dye solution (10 mg/L) within 50 minutes under UV illumination (Li et al., 2017b).

Li et al. produced cadmium sulfide/TiO_2 fibers for the photodegradation of congo red (CR), methylene blue (MB), and eosin red (ER). Results proved that the degradation efficiencies were 96% for CR, 71% for ER, and 56% for MB (Li et al., 2015).

Cao et al. produced core/sheath TiO_2/SiO_2 NFs through electrospinning and subsequent calcination at 500°C. The effect of SiO_2 sheath and its thickness on the photodegradation of MB under UV light irradiation. The degradation efficiency was found to be at a maximum of 94.9% (Cao et al., 2013).

Norouzi et al. studied phenol degradation using Ag/TiO_2 electrospun NFs photocatalyst. The results proved that 5% of silver content on TiO_2 and a calcination temperature is 450°C are the optimal conditions. Maximum phenol degradation was 82.65% (pH = 7, catalyst dosage = 1.5 g/L, and phenol concentration = 5 ppm under a low power visible light [18 W]) (Norouzi et al., 2022) (Fig. 9.8).

Sedghi et al. produced polyaniline-modified TiO_2/polyacrylonitrile composites. The decolorization rate of MO was 97% for the first time (Sedghi et al., 2017).

Other works concern the use of C-doped TiO_2/HNT NFs (Jiang et al., 2015), CNT-embedded hollow TiO_2 NFs (Jung et al., 2015), C/TiO_2 NFs (Song et al., 2020), reduced graphene oxide/titanium dioxide composite NFs (Nasr et al., 2017), electrospun TiO_2 NFs with Pt NPs (Formo et al., 2008), TiO_2/CuO composite NFs, electrospun TiO_2/CuO, TiO_2 NPs-PA6 NFs (Blanco et al., 2019), TiO_2-PAN NFs (Im et al., 2008), carbon NFs@TiO_2 (Xu et al., 2016), Ag-AgI-TiO_2 NPs supported on carbon NFs (Yu et al., 2015), TiO_2 NPs loaded on graphene/carbon composite NFs (Kim et al., 2012), and hollow TiO_2 NFs by microemulsion electrospinning method (Choi et al., 2013).

Table 9.5 reports an overview of the use of titania NFs for water remediation applications.

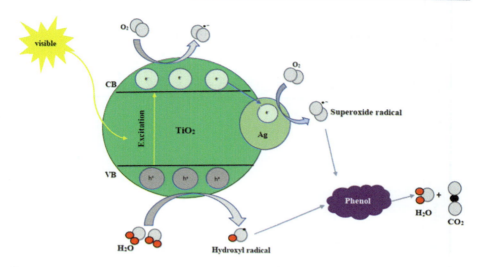

Figure 9.8 A suggested photocatalytic mechanism for Ag/TiO$_2$ nanofibers.
Source: From Norouzi, M., Fazeli, A., & Tavakoli, O. (2022). Photocatalytic degradation of phenol under visible light using electrospun Ag/TiO$_2$ as a 2D nano-powder: Optimizing calcination temperature and promoter content. *Advanced Powder Technology*, *33*(11). https://doi.org/10.1016/j.apt.2022.103792.

9.5 Use of cellulose nanofibers

To overcome the drawbacks of synthetic materials, the development of green, cheap, and eco-friendly solutions for water remediation is now attracting the interest of the research community (Carpenter et al., 2015). For example, the use of natural fibers can be a potential solution. It is known that cellulose NFs have been used for heavy metal remediation. Their nanostructure can guarantee high surface area, porosity, and good mechanical properties (Lee et al., 2018). Despite their interesting properties, cellulose nanofibers (CNFs) exhibit poor adsorption behavior. A potential solution could be the blending with other adsorbing materials, the use of modification processes (Phan et al., 2019), or the inclusion of specific groups such as carboxylate, phosphate, and amine groups (Liu et al., 2016).

For example, CNFs obtained from 2,2,6,6-tetramethylpiperidinyloxyl (TEMPO)-mediated oxidation showed to possess high adsorption capacity toward heavy metals due to the presence of carboxylic groups on the cellulose backbone (Melone et al., 2015). TEMPO/NaBr/NaClO system can oxidize the alcoholic hydroxyl group at the C6 position of cellulose to the carboxyl group with high selectivity (Lal & Mhaske, 2018).

Betaine-modified cationic cellulose has been produced through the reaction of cellulose with betaine hydrochloride for removing Reactive Red 24 (q_m = 95.2 mg/g) and Reactive Red 195 (q_m = 243.9 mg/g) (Ma et al., 2014).

Table 9.5 Use of TiO$_2$ nanofiber-based systems for water remediation applications.

System	Removed substances	Degradation	References
TiO$_2$-polyacrylonitrile	Rhodamine B	80%	Im et al. (2008)
Cds/TiO$_2$	Methylene blue	56%	Li et al. (2015)
	Congo red	96%	Li et al. (2015)
	Eosin red	71%	Li et al. (2015)
TiO$_2$/SiO$_2$	Methylene blue	94.9%	Cao et al. (2013)
Ag/TiO$_2$	Phenol	82.65%	Norouzi et al. (2022)
Ag/TiO$_2$	Dairy effluent	60%	Kanjwal et al. (2016)
Graphene/TiO$_2$	Methanol	81%	Roso et al. (2015)
TiO$_2$/PAN	Methylene blue	91.6%	Nguyen & Deng (2012)
Sb-doped SnO$_2$/RuO$_2$	Bisphenol A	Degradation = 100%	Kim et al. (2019)
Ag/TiO$_2$	Methylene blue	80%	Cui et al. (2019)
CNT-embedded hollow TiO$_2$	Methylene blue	62%	Jung et al. (2015)
C/TiO$_2$	Methylene blue	94.98%	Song et al. (2020)
TiO$_2$ NPs wrapped in carbon nanofibers	Rhodamine B	95.4%	Liang et al. (2018)
Reduced graphene oxide/TiO$_2$	Methyl orange	90%	Nasr et al. (2017)
TiO$_2$/CuO	Acid orange 7	35.3%	Lee et al. (2013)
TiO$_2$ NPs-PA$_6$	Reactive black 5	80%	Blanco et al. (2019)
SnO$_2$@TiO$_2$	Rhodamine B	100%	Hwang et al. (2011)
TiO$_2$/g-C$_3$N$_4$	Rhodamine B	99%	Tang et al. (2018)
ZnFe$_2$O$_4$@TiO$_2$	Methylene blue	98%	Nada et al. (2017)
Ag-doped TiO$_2$	Methylene blue	78%	Park et al. (2011)
Amorphous TiO$_2$	As(III)	0.0469 mmol/g	Vu et al. (2013)
rGO@TiO$_2$	Propranolol	100%	Gao et al. (2020)

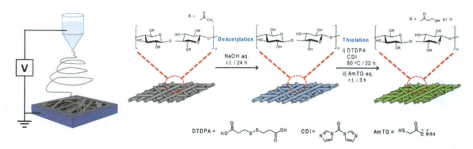

Figure 9.9 Schematic process for thiol-functionalized cellulose nanofiber membrane.
Source: From Choi, H. Y., Bae, J. H., Hasegawa, Y., An, S., Kim, I. S., Lee, H., & Kim, M. (2020). Thiol-functionalized cellulose nanofiber membranes for the effective adsorption of heavy metal ions in water. *Carbohydrate Polymers*, 234. https://doi.org/10.1016/j.carbpol.2020.115881.

Microfibrillated cellulose modified with aminopropyltriethoxysilane has been used for the treatment of aqueous solutions containing Ni^{2+}, Cu^{2+}, and Cd^{2+} ions (2.734, 3.150, and 4.195 mmol/g, respectively) (Hokkanen et al., 2014), while amphoteric cellulose with amine and carboxylic groups showed good adsorption performances for removing Cu^{2+} ($q_m = 73.53$ mg/g) and Cr^{6+} ($q_m = 227.3$ mg/g) ions from wastewaters (Zhong et al., 2014).

Choi et al. reported the production of biocompatible, nontoxic, and sustainable thiol-functionalized cellulose NF membrane for heavy metal ions adsorption (Fig. 9.9), such as Cu(II) (49 mg/g), Cd(II) (45.9 mg/g), and Pb(II) (22 mg/g). Chemisorption of metals is supposed to occur with two thiol groups on the surface.

Liu et al. reported the fabrication of CNF membranes via TEMPO-mediated oxidation of cellulose sludge-producing NFs in the size range of 18–40 nm for the remediation of copper ions. Copper ions were adsorbed due to the interactions with surface carboxylate groups. Moreover, a conversion in copper oxide nanoparticles with a size ranging from 200–300 nm was observed. The maximum copper ion adsorption capacity was 75 mg/g (Liu et al., 2016b).

Snyder et al. investigated the performances of eucalyptus pulp-derived CNFs doped with TiO_2, Au, and Ag particles for removing MB. The photocatalytic activity of TiO_2-CNF improved due to the surface functionalization of Au and Ag particles in the presence of simulated sunlight (Snyder et al., 2013).

Sehaqui et al. used cellulose NFs derived from waste pulp for removing PO_4^{3-}, SO_4^{2-}, F^-, and NO_3^- from polluted water sources. Cationic CNFs were produced through the surface functionalization of quaternary ammonium as high potential adsorbent of nitrate (Sehaqui et al., 2016).

Table 9.6 reports an overview of the use of CNF-based systems for water remediation applications.

Table 9.6 Use of cellulose nanofibers-based systems for water remediation applications.

System	Removed substances	q_m, mg/g	References
Thiol-functionalized CNFs	Cu^{2+}	49	Choi et al. (2020)
	Cd^{2+}	45.9	Choi et al. (2020)
	Pb^{2+}	22	Choi et al. (2020)
Chitosan/cellulose	As(V)	39.4	Phan et al. (2019)
	Pb(II)	57.3	Phan et al. (2019)
	Cu^{2+}	112.6	Phan et al. (2019)
Carboxylate groups modified CNFs	UO_2^{2+}	167	Ma et al. (2012)
Chitosan/alginate/CNFs	Eriochrome black-T	2297	Mokhtari et al. (2021)
TEMPO-oxidized CNFs modified with 3-aminopropyl sulfonic acid	Methylene blue	526	El-Sayed et al. (2022)
Pomelo-CNF	Malachite green	530	Tang et al. (2020)
	Cu^{2+}	74.2	Tang et al. (2020)
TEMPO-mediated oxidized cellulose nanofibrils modified with PEI	Cu^{2+}	52.32	Zhang et al. (2016)
(TEMPO)-mediated oxidized cellulose nanofibers and graphene oxide nanocolloid	Cu^{2+}	68.1	Zhu et al. (2017)
TEMPO-oxidized CNFs with branched polyethyleneimine	p-nitrophenol	1630	Melone et al. (2015)
	2,4,5-trichlorophenol	205	Melone et al. (2015)
	Amoxicillin	556	Melone et al. (2015)
Oxolane-2,5-dione-modified electrospun CNFs aerogels	Pb^{2+}	1 mmol/g	Stephen et al. (2011)
	Cd^{2+}	2.91 mmol/g	Stephen et al. (2011)
Quaternary ammonium-functionalized CNFs	Cr(VI)	Removal >99%	He et al. (2014)
Nitro-oxidized carboxy cellulose	Hg^{2+}	Removal = 88.9%	Chen et al. (2021b)
ZrO_2 modified ball-milled cellulose	Cr(VI)	Removal = 54%	Barbosa et al. (2022)
TEMPO-cellulose/silica/calcium carbonate	Methylene blue	520	Salama et al. (2023a)
MgS-doped CNFs	Cd^{2+}	333.33	Sankararamakrishnan et al. (2019)
Fe_3O_4 NPs on TEMPO-oxidized cellulose nanofiber	Methylene blue	303	Salama et al. (2023b)
Carboxycellulose nanofibers prepared by nitro-oxidation method	Pb^{2+}	2270	Salama et al. (2023a)

9.6 Conclusions and future perspectives

This chapter summarizes the use of NF-based materials for environmental applications. The focus was on the use of natural and synthetic fibers derived adsorbents for the removal of different contaminants from wastewater. Performances and properties of different fiber-based systems have been investigated and compared by gathering scientific works related to the investigated topic. The advantages of the electrospinning technique to produce high-performance nanofibrous adsorbents with tuneable adsorption properties have also been discussed. Finally, fiber-based systems as photocatalytic adsorbents for the removal of pollutants have been investigated. Besides, the use of cellulose NFs for the investigated topic has been discussed. Moreover, despite the performances of NFs and NF-based composites, they suffer from some drawbacks such as mechanical strength, water solubility, and cost-effectiveness. The lack of mechanical strength is crucial, for example, in some applications such as separation and filtration. Therefore the structure and morphology of NFs should be controlled and tailored by developing functional tailoring technologies or carrying out a modification with functional compounds. Besides, other disadvantages concern the recyclability, the reusability, and as a consequence, the global cost of the fabrication process. Finally, the use of NFs is limited because the production on a large scale is not completely developed. In fact, scaling up manufacturing is probably the major impediment to the wide use of NF-based systems in water treatment applications because of the limited mass production, the imperfect control of properties during the fabrication process, and the environmental considerations and safety concerns. Therefore it remains crucial to investigate novel solutions to overcome the previously reported problems to sustain and support the economic, technical, and sustainable development of water treatment processes.

References

Abbasizadeh, S., Keshtkar, A. R., & Mousavian, M. A. (2013). Preparation of a novel electrospun polyvinyl alcohol/titanium oxide nanofiber adsorbent modified with mercapto groups for uranium(VI) and thorium(IV) removal from aqueous solution. *Chemical Engineering Journal, 220*, 161−171. Available from https://doi.org/10.1016/J.CEJ.2013.01.029.

Agrawal, S., Ranjan, R., Lal, B., Rahman, A., Singh, S. P., Selvaratnam, T., & Nawaz, T. (2021). Synthesis and water treatment applications of nanofibers by electrospinning. *Processes, 9*(10). Available from https://doi.org/10.3390/pr9101779.

Aliabadi, M., Irani, M., Ismaeili, J., Piri, H., & Parnian, M. J. (2013). Electrospun nanofiber membrane of PEO/chitosan for the adsorption of nickel, cadmium, lead and copper ions from aqueous solution. *Chemical Engineering Journal, 220*, 237−243. Available from https://doi.org/10.1016/j.cej.2013.01.021.

Alipour, D., Keshtkar, A. R., & Moosavian, M. A. (2016). Adsorption of thorium(IV) from simulated radioactive solutions using a novel electrospun PVA/TiO2/ZnO nanofiber adsorbent functionalized with mercapto groups: Study in single and multi-component systems. *Applications of Surface Science, 366,* 19−29. Available from https://doi.org/10.1016/J.APSUSC.2016.01.049.

Allafchian, A. R., Shiasi, A., & Amiri, R. (2017). Preparing of poly(acrylonitrile co maleic acid) nanofiber mats for removal of Ni(II) and Cr(VI) ions from water. *Journal of the Taiwan Institute of Chemical Engineers, 80,* 563−569. Available from https://doi.org/10.1016/J.JTICE.2017.08.029.

Almasian, A., Jalali, M. L., Fard, G. C., & Maleknia, L. (2017). Surfactant grafted PDA-PAN nanofiber: Optimization of synthesis, characterization and oil absorption property. *Chemical Engineering Journal, 326,* 1232−1241. Available from https://doi.org/10.1016/j.cej.2017.06.040.

Almasian, A., Olya, M. E., & Mahmoodi, N. M. (2015). Synthesis of polyacrylonitrile/polyamidoamine composite nanofibers using electrospinning technique and their dye removal capacity. *Journal of the Taiwan Institute of Chemical Engineers, 49,* 119−128. Available from https://doi.org/10.1016/J.JTICE.2014.11.027.

Altaf, A. A., Ahmed, M., Hamayun, M., Kausar, S., Waqar, M., & Badshah, A. (2020). Titania nano-fibers: A review on synthesis and utilities. *Inorganica Chimica Acta, 501.* Available from https://doi.org/10.1016/j.ica.2019.119268.

Ambashta, R. D., & Sillanpää, M. (2010). Water purification using magnetic assistance: A review. *Journal of Hazardous Materials, 180,* 38−49. Available from https://doi.org/10.1016/J.JHAZMAT.2010.04.105.

Araújo, E. S., Luis, M., Nascimento, F., & De Oliveira, H. P. (2013). Influence of Triton X-100 on PVA Fibres Production by the Electrospinning Influence of Triton X-100 on PVA Fibres Production by the Electrospinning Technique. Technique FIBRES & TEXTILES in Eastern. *Europe, 21,* 39−43.

Armstrong, M., Nealy, S., Severino, C., Maniukiewicz, W., Modelska, M., Binczarski, M., Witonska, I., Chawla, K. K., & Stanishevsky, A. (2020). Composite materials made from glass microballoons and ceramic nanofibers for use as catalysts and catalyst supports. *Journal of Materials Science, 55*(27), 12940−12952. Available from https://doi.org/10.1007/s10853-020-04956-1.

Ashaghi, K. S., Ebrahimi, M., & Czermak, P. (2007). Ceramic ultra- and nanofiltration membranes for oilfield produced water treatment: A mini review. *The Open Environmental Journal, 1,* 1−8. Available from https://doi.org/10.2174/1874233507010111053.

Aung, K. T., Hong, S.-H., Park, S.-J., & Lee, C.-G. (2020). Removal of Cu(II) from aqueous solutions using amine-doped polyacrylonitrile fibers. *Applied Sciences, 10*(5). Available from https://doi.org/10.3390/app10051738.

Badgar, K., Abdalla, N., El-Ramady, H., & Prokisch, J. (2022). Sustainable applications of nanofibers in agriculture and water treatment: A review. *Sustainability, 14*(1). Available from https://doi.org/10.3390/su14010464.

Barbosa, R. Fd. S., Zanini, N. C., Mulinari, D. R., & Rosa, Dd. S. (2022). Hexavalent chromium sorption by modified cellulose macro and nanofibers obtained from eucalyptus residues. *Journal of Polymers and the Environment, 30*(9), 3852−3864. Available from https://doi.org/10.1007/s10924-022-02469-3, http://www.kluweronline.com/issn/1566-2543/.

Baroni, P., Vieira, R. S., Meneghetti, E., da Silva, M. G. C., & Beppu, M. M. (2008). Evaluation of batch adsorption of chromium ions on natural and crosslinked chitosan membranes. *Journal of Hazardous Materials, 152*(3), 1155−1163. Available from https://doi.org/10.1016/j.jhazmat.2007.07.099.

Benelli, G., Pavela, R., Maggi, F., Petrelli, R., & Nicoletti, M. (2017). Commentary: Making green pesticides greener? The potential of plant products for nanosynthesis and pest control. *Journal of Cluster Science*, *28*(1), 3−10. Available from https://doi.org/10.1007/s10876-016-1131-7.

Blanco, M., Monteserín, C., Angulo, A., Pérez-Márquez, A., Maudes, J., Murillo, N., Aranzabe, E.-b, Ruiz-Rubio, L., & Vilas, J. L. (2019). TiO$_2$-doped electrospun nanofibrous membrane for photocatalytic water treatment. *Polymers*, *11*(5). Available from https://doi.org/10.3390/polym11050747.

Camiré, A., Espinasse, J., Chabot, B., & Lajeunesse, A. (2020). Development of electrospun lignin nanofibers for the adsorption of pharmaceutical contaminants in wastewater. *Environmental Science and Pollution Research*, *27*(4), 3560−3573. Available from https://doi.org/10.1007/s11356-018-3333-z.

Cao, H., Du, P., Song, L., Xiong, J., Yang, J., Xing, T., Liu, X., Wu, R., Wang, M., & Shao, X. (2013). Co-electrospinning fabrication and photocatalytic performance of TiO$_2$/SiO$_2$ core/sheath nanofibers with tunable sheath thickness. *Materials Research Bulletin*, *48*(11), 4673−4678. Available from https://doi.org/10.1016/j.materresbull.2013.08.035.

Carpenter, A. W., De Lannoy, C. F., & Wiesner, M. R. (2015). Cellulose nanomaterials in water treatment technologies. *Environmental Science and Technology*, *49*, 5277−5287. Available from https://doi.org/10.1021/ES506351R.

Chabalala, M. B., Al-Abri, M. Z., Mamba, B. B., & Nxumalo, E. N. (2021). Mechanistic aspects for the enhanced adsorption of bromophenol blue and atrazine over cyclodextrin modified polyacrylonitrile nanofiber membranes. *Chemical Engineering Research and Design*, *169*, 19−32. Available from https://doi.org/10.1016/j.cherd.2021.02.010, http://www.elsevier.com/wps/find/journaldescription.cws_home/713871/description#description.

Choi, J., Ide, A., Truong, Y. B., et al. (2013). High surface area mesoporous titanium−zirconium oxide nanofibrous web: a heavy metal ion adsorbent. *Journal of Materials Chemistry A*, *1*, 5847−5853. Available from https://doi.org/10.1039/C3TA00030C.

Choi, K. Il., Ho, L. S., Park, J. Y., et al. (2013). Fabrication and characterization of hollow TiO2 fibers by microemulsion electrospinning for photocatalytic reactions. *Materials Letters*, *112*, 113−116. Available from https://doi.org/10.1016/J.MATLET.2013.08.101.

Chen, S., Li, M., & Zhang, M. (2021a). Metal organic framework derived one-dimensional porous Fe/N-doped carbon nanofibers with enhanced catalytic performance. *Journal of Hazardous Materials*, *416*, 126101. Available from https://doi.org/10.1016/J.JHAZMAT.2021.126101.

Chen, H., Sharma, S. K., & Sharma, P. R. (2021b). Nitro-oxidized carboxycellulose nanofibers from moringa plant: effective bioadsorbent for mercury removal. *Cellulose*, *28*, 8611−8628. Available from https://doi.org/10.1007/S10570-021-04057-5/TABLES/3.

Chen, W., Ma, H., & Xing, B. (2020). Electrospinning of multifunctional cellulose acetate membrane and its adsorption properties for ionic dyes. *International Journal of Biological Macromolecules*, *158*, 1342−1351. Available from https://doi.org/10.1016/J.IJBIOMAC.2020.04.249.

Chen, H., Wageh, S., Al-Ghamdi, A. A., Wang, H., Yu, J., & Jiang, C. (2019). Hierarchical C/NiO-ZnO nanocomposite fibers with enhanced adsorption capacity for Congo red. *Journal of Colloid and Interface Science*, *537*, 736−745. Available from https://doi.org/10.1016/j.jcis.2018.11.045, http://www.elsevier.com/inca/publications/store/6/2/2/8/6/1/index.htt.

Chigondo, M., Paumo, H. K., Bhaumik, M., Pillay, K., & Maity, A. (2018). Hydrous CeO$_2$-Fe$_3$O$_4$ decorated polyaniline fibers nanocomposite for effective defluoridation of drinking water. *Journal of Colloid and Interface Science*, *532*, 500−516. Available from https://doi.org/10.1016/j.jcis.2018.07.134.

Choi, H. Y., Bae, J. H., Hasegawa, Y., An, S., Kim, I. S., Lee, H., & Kim, M. (2020). Thiol-functionalized cellulose nanofiber membranes for the effective adsorption of heavy metal ions in water. *Carbohydrate Polymers, 234*. Available from https://doi.org/10.1016/j.carbpol.2020.115881.

Choi, S. K., Kim, S., Lim, S. K., & Park, H. (2010). Photocatalytic comparison of TiO_2 nanoparticles and electrospun TiO_2 nanofibers: Effects of mesoporosity and interparticle charge transfer. *Journal of Physical Chemistry C, 114*(39), 16475–16480. Available from https://doi.org/10.1021/jp104317x.

Coşkun, R., Yiğitoğlu, M., & Saçak, M. (2000). Adsorption behavior of copper(II) ion from aqueous solution on methacrylic acid-grafted poly(ethylene terephthalate) fibers. *Journal of Applied Polymer Science, 75*, 766–772. https://doi.org/10.1002/(SICI)1097-4628(20000207)75:6

Cui, L., Song, Y., Wang, F., Sheng, Y., & Zou, H. (2019). Electrospinning synthesis of SiO_2-TiO_2 hybrid nanofibers with large surface area and excellent photocatalytic activity. *Applied Surface Science, 488*, 284–292. Available from https://doi.org/10.1016/j.apsusc.2019.05.151.

Dastbaz, A., & Keshtkar, A. R. (2014). Adsorption of Th4 + , U6 + , Cd2 + , and Ni2 + from aqueous solution by a novel modified polyacrylonitrile composite nanofiber adsorbent prepared by electrospinning. *Applications of Surface Science, 293*, 336–344. Available from https://doi.org/10.1016/J.APSUSC.2013.12.164.

Deng, S., Bai, R., & Chen, J. P. (2003). Behaviors and mechanisms of copper adsorption on hydrolyzed polyacrylonitrile fibers. *Journal of Colloid and Interface Science, 260*, 265–272. Available from https://doi.org/10.1016/S0021-9797(02)00243-6.

Deng, J., Kang, X., Chen, L., Wang, Y., Gu, Z., & Lu, Z. (2011). A nanofiber functionalized with dithizone by co-electrospinning for lead (II) adsorption from aqueous media. *Journal of Hazardous Materials, 196*, 187–193. Available from https://doi.org/10.1016/j.jhazmat.2011.09.016.

Deng, S., Wang, P., Zhang, G., & Dou, Y. (2016). Polyacrylonitrile-based fiber modified with thiosemicarbazide by microwave irradiation and its adsorption behavior for Cd(II) and Pb(II). *Journal of Hazardous Materials, 307*, 64–72. Available from https://doi.org/10.1016/j.jhazmat.2016.01.002.

Dong, B., Li, Z., Li, Z., Xu, X., Song, M., Zheng, W., Wang, C., Al-Deyab, S. S., & El-Newehy, M. (2010). Highly efficient $LaCoO_3$ nanofibers catalysts for photocatalytic degradation of rhodamine B. *Journal of the American Ceramic Society, 93*(11), 3587–3590. Available from https://doi.org/10.1111/j.1551-2916.2010.04124.x.

Ebrahimi, F., Sadeghizadeh, A., Neysan, F., & Heydari, M. (2017). Fabrication of nanofibers using sodium alginate and Poly(Vinyl alcohol) for the removal of Cd2 + ions from aqueous solutions: adsorption mechanism, kinetics and thermodynamics. *Heliyon, 5*, 02941. Available from https://doi.org/10.1016/j.heliyon.2019.e02941.

Efome, J. E., Rana, D., Matsuura, T., & Lan, C. Q. (2018). Metal-organic frameworks supported on nanofibers to remove heavy metals. *Journal of Materials Chemistry A, 6*(10), 4550–4555. Available from https://doi.org/10.1039/c7ta10428f, http://pubs.rsc.org/en/journals/journal/ta.

El-Sayed, N. S., Salama, A., & Guarino, V. (2022). Coupling of 3-aminopropyl sulfonic acid to cellulose nanofibers for efficient removal of cationic dyes. *Materials, 15*, 6964. Available from https://doi.org/10.3390/MA15196964.

Feng, Z. Q., Yuan, X., & Wang, T. (2020). Porous polyacrylonitrile/graphene oxide nanofibers designed for high efficient adsorption of chromium ions (VI) in aqueous solution. *Chemical Engineering Journal, 392*, 123730. Available from https://doi.org/10.1016/J.CEJ.2019.123730.

Fiol, N., Tarrés, Q., Vásquez, M. G., Pereira, M. A., Mendonça, R. T., Mutjé, P., & Delgado-Aguilar, M. (2021). Comparative assessment of cellulose nanofibers and calcium alginate beads for continuous Cu(II) adsorption in packed columns: the influence of water and surface hydrophobicity. *Cellulose*, *28*(7), 4327−4344. Available from https://doi.org/10.1007/s10570-021-03809-7.

Formo, E., Lee, E., Campbell, D., & Xia, Y. (2008). Functionalization of electrospun TiO_2 nanofibers with Pt nanoparticles and nanowires for catalytic applications. *Nano Letters*, *8*(2), 668−672. Available from https://doi.org/10.1021/nl073163v.

Gadisa, B. T., Kassahun, S. K., Appiah-Ntiamoah, R., & Kim, H. (2020). Tuning the charge carrier density and exciton pair separation in electrospun 1D ZnO-C composite nanofibers and its effect on photodegradation of emerging contaminants. *Journal of Colloid and Interface Science*, *570*, 251−263. Available from https://doi.org/10.1016/j.jcis.2020.03.002, http://www.elsevier.com/inca/publications/store/6/2/2/8/6/1/index.htt.

Gao, Y., Yan, N., Jiang, C., Xu, C., Yu, S., Liang, P., Zhang, X., Liang, S., & Huang, X. (2020). Filtration-enhanced highly efficient photocatalytic degradation with a novel electrospun rGO@TiO_2 nanofibrous membrane: Implication for improving photocatalytic efficiency. *Applied Catalysis B: Environmental*, *268*. Available from https://doi.org/10.1016/j.apcatb.2020.118737.

Gebru, K. A., & Das, C. (2017). Removal of Pb (II) and Cu (II) ions from wastewater using composite electrospun cellulose acetate/titanium oxide (TiO2) adsorbent. *Journal of Water Process Engineering*, *16*, 1−13. Available from https://doi.org/10.1016/J.JWPE.2016.11.008.

Ghani, M., Rezaei, B., Ghare Aghaji, A., & Arami, M. (2016). Novel cross-linked superfine alginate-based nanofibers: Fabrication, characterization, and their use in the adsorption of cationic and anionic dyes. *Advances in Polymer Technology*, *35*(4), 428−438. Available from https://doi.org/10.1002/adv.21569.

Ghasemi, E., Ziyadi, H., Afshar, A. M., & Sillanpää, M. (2015). Iron oxide nanofibers: A new magnetic catalyst for azo dyes degradation in aqueous solution. *Chemical Engineering Journal*, *264*, 146−151. Available from https://doi.org/10.1016/j.cej.2014.11.021.

Girijappa, T., Y., Rangappa, M., Parameswaranpillai., & Siengchin, J. (2019). Natural fibers as sustainable and renewable resource for development of eco-friendly composites: A comprehensive review. *Frontiers in Materials*, *6*. Available from https://doi.org/10.3389/FMATS.2019.00226/BIBTEX.

Gopakumar, D. A., Pasquini, D., Henrique, M. A., de Morais, L. C., Grohens, Y., & Thomas, S. (2017). Meldrum's acid modified cellulose nanofiber-based polyvinylidene fluoride microfiltration membrane for dye water treatment and nanoparticle removal. *ACS Sustainable Chemistry & Engineering*, *5*(2), 2026−2033. Available from https://doi.org/10.1021/acssuschemeng.6b02952.

Gundloori, R. V. N., Singam, A., & Killi, N. (2019). Nanobased intravenous and transdermal drug delivery systems. Applications of targeted nano drugs and delivery systems. *Nanoscience and Nanotechnology in Drug Delivery*, 551−594. Available from https://doi.org/10.1016/B978-0-12-814029-1.00019-3.

Gupta, A., Nandanwar, D. V., & Dhakate, S. R. (2015). Electrospun self-assembled ZnO nanofibers structures for photocatalytic activity in natural solar radiations to degrade acid fuchsin dye. *Advanced Materials Letters*, *6*, 706−710. Available from https://doi.org/10.5185/AMLETT.2015.5834.

Habiba, U., Afifi, A. M., Salleh, A., & Ang, B. C. (2017). Chitosan/(polyvinyl alcohol)/zeolite electrospun composite nanofibrous membrane for adsorption of Cr^{6+}, Fe^{3+} and

Ni^{2+}. *Journal of Hazardous Materials, 322,* 182−194. Available from https://doi.org/10.1016/j.jhazmat.2016.06.028.

Haider, S., Binagag, F. F., Haider, A., Mahmood, A., Shah, N., Al-Masry, W. A., Khan, S. U.-D., & Ramay, S. M. (2014). Adsorption kinetic and isotherm of methylene blue, safranin T and rhodamine B onto electrospun ethylenediamine-grafted-polyacrylonitrile nanofibers membrane. *Desalination and Water Treatment, 55*(6), 1609−1619. Available from https://doi.org/10.1080/19443994.2014.926840.

Haider, S., & Park, S. Y. (2009). Preparation of the electrospun chitosan nanofibers and their applications to the adsorption of Cu(II) and Pb(II) ions from an aqueous solution. *Journal of Membrane Science, 328,* 90−96. Available from https://doi.org/10.1016/J.MEMSCI.2008.11.046.

Hassan, M. S., Amna, T., Yang, O. B., Kim, H. C., & Khil, M. S. (2012). TiO$_2$ nanofibers doped with rare earth elements and their photocatalytic activity. *Ceramics International, 38*(7), 5925−5930. Available from https://doi.org/10.1016/j.ceramint.2012.04.043.

He, J., Wang, W., Shi, W., & Cui, F. (2016). La$_2$O$_3$ nanoparticle/polyacrylonitrile nanofibers for bacterial inactivation based on phosphate control. *RSC Advances, 6*(101), 99353−99360. Available from https://doi.org/10.1039/C6RA22374E.

He, X., Cheng, L., Wang, Y., Zhao, J., Zhang, W., & Lu, C. (2014). Aerogels from quaternary ammonium-functionalized cellulose nanofibers for rapid removal of Cr(VI) from water. *Carbohydrate Polymers, 111,* 683−687. Available from https://doi.org/10.1016/j.carbpol.2014.05.020.

Hosseini, S. A., Vossoughi, M., & Mahmoodi, N. M. (2017). Preparation of electrospun affinity membrane and cross flow system for dynamic removal of anionic dye from colored wastewater. *Fibers and Polymers, 18,* 2387−2399. Available from https://doi.org/10.1007/S12221-017-7530-Z.

Hokkanen, S., Repo, E., Suopajärvi, T., Liimatainen, H., Niinimaa, J., & Sillanpää, M. (2014). Adsorption of Ni(II), Cu(II) and Cd(II) from aqueous solutions by amino modified nanostructured microfibrillated cellulose. *Cellulose, 21*(3), 1471−1487. Available from https://doi.org/10.1007/s10570-014-0240-4.

Hong, F., Yan, C., Si, Y., He, J., Yu, J., & Ding, B. (2015). Nickel ferrite nanoparticles anchored onto silica nanofibers for designing magnetic and flexible nanofibrous membranes. *ACS Applied Materials and Interfaces, 7*(36), 20200−20207. Available from https://doi.org/10.1021/acsami.5b05754, http://pubs.acs.org/journal/aamick.

Hota, G., Kumar, B. R., Ng, W. J., & Ramakrishna, S. (2008). Fabrication and characterization of a boehmite nanoparticle impregnated electrospun fiber membrane for removal of metal ions. *Journal of Materials Science, 43*(1), 212−217. Available from https://doi.org/10.1007/s10853-007-2142-4.

Hwang, S. H., Kim, C., & Jang, J. (2011). SnO2 nanoparticle embedded TiO2 nanofibers — Highly efficient photocatalyst for the degradation of rhodamine B. *Catalysis Communications, 12,* 1037−1041. Available from https://doi.org/10.1016/J.CATCOM.2011.02.024.

Im, J. S., Kim, M., Il, & Lee, Y. S. (2008). Preparation of PAN-based electrospun nanofiber webs containing TiO2 for photocatalytic degradation. *Materials Letters, 62,* 3652−3655. Available from https://doi.org/10.1016/J.MATLET.2008.04.019.

Irani, M., Keshtkar, A. R., & Moosavian, M. A. (2012). Removal of cadmium from aqueous solution using mesoporous PVA/TEOS/APTES composite nanofiber prepared by sol−gel/electrospinning. *Chemical Engineering Journal, 200−202,* 192−201. Available from https://doi.org/10.1016/J.CEJ.2012.06.054.

Jiang, M., Han, T., Wang, J., Shao, L., Qi, C., Zhang, X. M., Liu, C., & Liu, X. (2018). Removal of heavy metal chromium using cross-linked chitosan composite nanofiber mats. *International Journal of Biological Macromolecules, 120*, 213−221. Available from https://doi.org/10.1016/j.ijbiomac.2018.08.071.

Jiang, Z., Tijing, L. D., Amarjargal, A., Park, C. H., An, K. J., Shon, H. K., & Kim, C. S. (2015). Removal of oil from water using magnetic bicomponent composite nanofibers fabricated by electrospinning. *Composites Part B: Engineering, 77*, 311−318. Available from https://doi.org/10.1016/j.compositesb.2015.03.067.

Jung, J. Y., Lee, D., & Lee, Y. S. (2015). CNT-embedded hollow TiO2 nanofibers with high adsorption and photocatalytic activity under UV irradiation. *Journal of Alloys and Compounds, 622*, 651−656. Available from https://doi.org/10.1016/J.JALLCOM.2014.09.068.

Kampalanonwat, P., & Supaphol, P. (2010). Preparation and adsorption behavior of aminated electrospun polyacrylonitrile nanofiber mats for heavy metal ion removal. *ACS Applied Materials & Interfaces, 2*, 3619−3627. Available from https://doi.org/10.1021/AM1008024.

Kanjwal, M. A., Alm, M., Thomsen, P., Barakat, N. A. M., & Chronakis, I. S. (2016). Hybrid matrices of TiO_2 and TiO_2-Ag nanofibers with silicone for high water flux photocatalytic degradation of dairy effluent. *Journal of Industrial and Engineering Chemistry, 33*, 142−149. Available from https://doi.org/10.1016/j.jiec.2015.09.026, http:http://www.sciencedirect.com/science/journal/1226086X.

Kar, P., & Misra, M. (2004). Use of keratin fiber for separation of heavy metals from water. *Journal of Chemical Technology & Biotechnology, 79*, 1313−1319. Available from https://doi.org/10.1002/jctb.1132.

Karim, M. R., Aijaz, M. O., Alharth, N. H., Alharbi, H. F., Al-Mubaddel, F. S., & Awual, M. R. (2019). Composite nanofibers membranes of poly(vinyl alcohol)/chitosan for selective lead(II) and cadmium(II) ions removal from wastewater. *Ecotoxicology and Environmental Safety, 169*, 479−486. Available from https://doi.org/10.1016/j.ecoenv.2018.11.049, http://www.elsevier.com/inca/publications/store/6/2/2/8/1/9/index.htt.

Kenry., & Lim, C. T. (2017). Nanofiber technology: current status and emerging developments. *Progress in Polymer Science, 70*, 1−17. Available from https://doi.org/10.1016/J.PROGPOLYMSCI.2017.03.002.

Ki, C. S., Gang, E. H., Um, I. C., & Park, Y. H. (2007). Nanofibrous membrane of wool keratose/silk fibroin blend for heavy metal ion adsorption. *Journal of Membrane Science, 302*(1−2), 20−26. Available from https://doi.org/10.1016/j.memsci.2007.06.003.

Kim, J.-C., Oh, S.-I., Kang, W., Yoo, H.-Y., Lee, J., & Kim, D.-W. (2019). Superior anodic oxidation in tailored Sb-doped SnO_2/RuO_2 composite nanofibers for electrochemical water treatment. *Journal of Catalysis, 374*, 118−126. Available from https://doi.org/10.1016/j.jcat.2019.04.025.

Koushkbaghi, S., Jafari, P., Rabiei, J., Irani, M., & Aliabadi, M. (2016). Fabrication of PET/PAN/GO/Fe_3O_4 nanofibrous membrane for the removal of Pb(II) and Cr(VI) ions. *Chemical Engineering Journal., 301*, 42−50. Available from https://doi.org/10.1016/j.cej.2016.04.076.

Kruppa, M., Frank, D., Leffler-Schuster, H., & König, B. (2006). Screening of metal complex−amino acid side chain interactions by potentiometric titration. *Inorganica Chimica Acta, 359*(4), 1159−1168. Available from https://doi.org/10.1016/j.ica.2005.12.008.

Lal, S. S., & Mhaske, S. T. (2018). AgBr and AgCl nanoparticle doped TEMPO-oxidized microfiber cellulose as a starting material for antimicrobial filter. *Carbohydrate Polymers, 191*, 266−279. Available from https://doi.org/10.1016/J.CARBPOL.2018.03.011.

LeCorre-Bordes, D., Tucker, N., Huber, T., Buunk, N., & Staiger, M. P. (2016). Shear-electrospinning: Extending the electrospinnability range of polymer solutions. *Journal of Materials Science, 51*(14), 6686−6696. Available from https://doi.org/10.1007/s10853-016-9955-y.

Lee, H., Nishino, M., Sohn, D., Lee, J. S., & Kim, I. S. (2018). Control of the morphology of cellulose acetate nanofibers via electrospinning. *Cellulose, 25*(5), 2829−2837. Available from https://doi.org/10.1007/s10570-018-1744-0.

Lee, J. J. L., Ang, B. C., Andriyana, A., Shariful, M. I., & Amalina, M. A. (2017). Fabrication of PMMA/zeolite nanofibrous membrane through electrospinning and its adsorption behavior. *Journal of Applied Polymer Science, 134*(6). Available from https://doi.org/10.1002/app.44450.

Lee, S. S., Bai, H., Liu, Z., & Sun, D. D. (2013). Novel-structured electrospun TiO$_2$/CuO composite nanofibers for high efficient photocatalytic cogeneration of clean water and energy from dye wastewater. *Water Research, 47*(12), 4059−4073. Available from https://doi.org/10.1016/j.watres.2012.12.044.

Li, W., Zhao, S., Qi, B., Du, Y., Wang, X., & Huo, M. (2009). Fast catalytic degradation of organic dye with air and MoO$_3$:Ce nanofibers under room condition. *Applied Catalysis B: Environmental, 92*(3−4), 333−340. Available from https://doi.org/10.1016/j.apcatb.2009.08.012.

Li, X., Chen, X., Niu, H., Han, X., Zhang, T., Liu, J., Lin, H., & Qu, F. (2015). The synthesis of CdS/TiO$_2$ hetero-nanofibers with enhanced visible photocatalytic activity. *Journal of Colloid and Interface Science, 452*, 89−97. Available from https://doi.org/10.1016/j.jcis.2015.04.034.

Li, X., Wang, F., Qian, Q., Liu, X., Xiao, L., & Chen, Q. (2012). Ag/TiO$_2$ nanofibers heterostructure with enhanced photocatalytic activity for parathion. *Materials Letters, 66*(1), 370−373. Available from https://doi.org/10.1016/j.matlet.2011.08.090.

Li, L., Zhang, J., Li, Y., & Yang, C. (2017a). Removal of Cr (VI) with a spiral wound chitosan nanofiber membrane module via dead-end filtration. *Journal of Membrane Science, 544*, 333−341. Available from https://doi.org/10.1016/J.MEMSCI.2017.09.045.

Li, X., Zhang, C., Zhao, R., Lu, X., Xu, X., Jia, X., Wang, C., & Li, L. (2013). Efficient adsorption of gold ions from aqueous systems with thioamide-group chelating nanofiber membranes. *Chemical Engineering Journal, 229*, 420−428. Available from https://doi.org/10.1016/j.cej.2013.06.022.

Li, Y., Zhao, R., & Chao, S. (2017c). A flexible magnesium silicate coated electrospun fiber adsorbent for high-efficiency removal of a toxic cationic herbicide. *New Journal of Chemistry, 41*, 15601−15611. Available from https://doi.org/10.1039/C7NJ03168H.

Li, Y., Zhao, R., Chao, S., Sun, B., Wang, C., & Li, X. (2018). Polydopamine coating assisted synthesis of MnO$_2$ loaded inorganic/organic composite electrospun fiber adsorbent for efficient removal of Pb^{2+} from water. *Chemical Engineering Journal, 344*, 277−289. Available from https://doi.org/10.1016/j.cej.2018.03.044.

Li, H., Wu, W., Bubakir, M. M. et al. (2014a). Polypropylene fibers fabricated via a needleless melt-electrospinning device for marine oil-spill cleanup. *Journal of Applied Polymer Science 131*. Available from https://doi.org/10.1002/APP.40080.

Li, X., Zhao, R., Sun, B., et al. (2014b). Fabrication of α-Fe$_2$O$_3$−γ-Al$_2$O$_3$ core−shell nanofibers and their Cr(vi) adsorptive properties. *RSC Advances, 4*, 42376−42382. Available from https://doi.org/10.1039/C4RA03692A.

Li, Y., Zhao, H., & Yang, M. (2017b). TiO2 nanoparticles supported on PMMA nanofibers for photocatalytic degradation of methyl orange. *Journal of Colloid and Interface Science, 508*, 500−507. Available from https://doi.org/10.1016/J.JCIS.2017.08.076.

Liang, Y., Zhou, B., Li, N., Liu, L., Xu, Z., Li, F., Li, J., Mai, W., Qian, X., & Wu, N. (2018). Enhanced dye photocatalysis and recycling abilities of semi-wrapped TiO$_2$@carbon nanofibers formed via foaming agent driving. *Ceramics International, 44* (2), 1711−1718. Available from https://doi.org/10.1016/j.ceramint.2017.10.101.

Lin, J., Shang, Y., Ding, B., et al. (2012). Nanoporous polystyrene fibers for oil spill cleanup. *Marine Pollution Bulletin, 64,* 347−352. Available from https://doi.org/10.1016/J.MARPOLBUL.2011.11.002.

Lin, J., Ding, B, Yang, J., et al. (2011). Subtle regulation of the micro- and nanostructures of electrospun polystyrene fibers and their application in oil absorption. *Nanoscale, 4,* 176−182. Available from https://doi.org/10.1039/C1NR10895F.

Liu, H., Hou, H., Gao, F., Yao, X., & Yang, W. (2016). Tailored fabrication of thoroughly mesoporous BiVO$_4$ nanofibers and their visible-light photocatalytic activities. *ACS Applied Materials & Interfaces, 8*(3), 1929−1936. Available from https://doi.org/10.1021/acsami.5b10086.

Liu, L., Liu, Z., Bai, H., & Sun, D. D. (2012). Concurrent filtration and solar photocatalytic disinfection/degradation using high-performance Ag/TiO$_2$ nanofiber membrane. *Water Research, 46*(4), 1101−1112. Available from https://doi.org/10.1016/j.watres.2011.12.009.

Liu, G., Liu, S., Lu, Q., et al. (2014). Synthesis and characterization of Bi(VO4)1−m(PO4)m m nanofibers by electrospinning process with enhanced photocatalytic activity under visible light. *RSC Advances, 4,* 33695−33701. Available from https://doi.org/10.1039/C4RA04107K.

Liu, R. X., Zhang, B. W., & Tang, H. X. (1999). Synthesis and characterization of poly(acrylaminophosphonic-carboxyl-hydrazide) chelating fibre. *Reactive and Functional Polymers, 39,* 71−81. Available from https://doi.org/10.1016/S1381-5148(97)00174-0.

Lou, L., Kendall, R. J., Smith, E., & Ramkumar, S. S. (2020). Functional PVDF/rGO/TiO$_2$ nanofiber webs for the removal of oil from water. *Polymer, 186.* Available from https://doi.org/10.1016/j.polymer.2019.122028.

Lu, Y., Wu, Z., Li, M., Liu, Q., & Wang, D. (2014). Hydrophilic PVA-co-PE nanofiber membrane functionalized with iminodiacetic acid by solid-phase synthesis for heavy metal ions removal. *Reactive and Functional Polymers, 82,* 98−102. Available from https://doi.org/10.1016/j.reactfunctpolym.2014.06.004.

Luo, C., Wang, J., Jia, P., Liu, Y., An, J., Cao, B., & Pan, K. (2015). Hierarchically structured polyacrylonitrile nanofiber mat as highly efficient lead adsorbent for water treatment. *Chemical Engineering Journal., 262,* 775−784. Available from https://doi.org/10.1016/j.cej.2014.09.116.

Lv, Y., Ma, J., Liu, K., Jiang, Y., Yang, G., Liu, Y., Lin, C., Ye, X., Shi, Y., Liu, M., & Chen, L. (2021). Rapid elimination of trace bisphenol pollutants with porous β-cyclodextrin modified cellulose nanofibrous membrane in water: Adsorption behavior and mechanism. *Journal of Hazardous Materials, 403.* Available from https://doi.org/10.1016/j.jhazmat.2020.123666.

Ma, G., Chen, Z., Chen, Z., Jin, M., Meng, Q., Yuan, M., Wang, X., Liu, J.-M., & Zhou, G. (2017). Constructing novel WO$_3$/Fe(III) nanofibers photocatalysts with enhanced visible-light-driven photocatalytic activity via interfacial charge transfer effect. *Materials Today Energy, 3,* 45−52. Available from https://doi.org/10.1016/j.mtener.2017.02.003.

Ma, W., Yan, S., Meng, M., & Zhang, S. (2014). Preparation of betaine-modified cationic cellulose and its application in the treatment of reactive dye wastewater. *Journal of Applied Polymer Science, 131*(15), 40522. Available from https://doi.org/10.1002/app.40522.

Ma, H., Hsiao, B. S., & Chu, B. (2012). Ultrafine cellulose nanofibers as efficient adsorbents for removal of UO22 + in water. *American Chemical Society*, *1*, 213−216. Available from https://doi.org/10.1021/MZ200047Q/ASSET/IMAGES/LARGE/MZ-2011-00047Q_0004.JPEG.

Ma, Y., Qi, P., Ju, J., Wang, Q., Hao, L., Wang, R., Sui, K., & Tan, Y. (2019). Gelatin/alginate composite nanofiber membranes for effective and even adsorption of cationic dyes. *Composites Part B: Engineering*, *162*, 671−677. Available from https://doi.org/10.1016/j.compositesb.2019.01.048.

Mahapatra, A., Mishra, B. G., & Hota, G. (2013). Electrospun Fe2O3−Al2O3 nanocomposite fibers as efficient adsorbent for removal of heavy metal ions from aqueous solution. *Journal of Hazardous Materials*, *258−259*, 116−123. Available from https://doi.org/10.1016/J.JHAZMAT.2013.04.045.

Mahltig, B., Gutmann, E., Meyer, D. C., Reibold, M., Dresler, B., Günther, K., Faßler, D., & Böttcher, H. (2007). Solvothermal preparation of metallized titania sols for photocatalytic and antimicrobial coatings. *Journal of Materials Chemistry*, *17*(22), 2367−2374. Available from https://doi.org/10.1039/b702519j.

Malwal, D., & Gopinath, P. (2015). Fabrication and applications of ceramic nanofibers in water remediation: A review. *46*, 500−534. Available from https://doi.org/10.1080/10643389201511099137 13.

Malwal, D., & Gopinath, P. (2015). Fabrication and characterization of poly(ethylene oxide) templated nickel oxide nanofibers for dye degradation. *Environmental Science: Nano*, *2*, 78−85. Available from https://doi.org/10.1039/C4EN00107A.

Marinho, B. A., Souza, S. G., de Souza, A. A. U., & Hotza, D. (2020). Electrospun TiO2 nanofibers for water and wastewater treatment: a review. *Journal of Materials Science*, *56*. Available from https://doi.org/10.1007/S10853-020-05610-6.

Marinho, B. A., de Souza, S. M. A. G. U., de Souza, A. A. U., & Hotza, D. (2021). Electrospun TiO$_2$ nanofibers for water and wastewater treatment: A review. *Journal of Materials Science*, *56*(9), 5428−5448. Available from https://doi.org/10.1007/s10853-020-05610-6.

Melone, L., Rossi, B., Pastori, N., Panzeri, W., Mele, A., & Punta, C. (2015). TEMPO-oxidized cellulose cross-linked with branched polyethyleneimine: Nanostructured adsorbent sponges for water remediation. *ChemPlusChem*, *80*(9), 1408−1415. Available from https://doi.org/10.1002/cplu.201500145.

Mohamed A., Nasser W. S, Osman T. A., et al. (2017a). Removal of chromium (VI) from aqueous solutions using surface modified composite nanofibers. *Journal of Colloid and Interface Science 505*, 505:682−691. Available from https://doi.org/10.1016/J.JCIS.2017.06.066.

Mohamed, A., Osman, T. A., Toprak, M. S., et al. (2017b). Surface functionalized composite nanofibers for efficient removal of arsenic from aqueous solutions. *Chemosphere*, *180*, 108−116. Available from https://doi.org/10.1016/J.CHEMOSPHERE.2017.04.011.

Mokhtari, A., Sabzi, M., & Azimi, H. (2021). 3D porous bioadsorbents based on chitosan/alginate/cellulose nanofibers as efficient and recyclable adsorbents of anionic dye. *Carbohydrate Polymers*, *265*, 118075. Available from https://doi.org/10.1016/J.CARBPOL.2021.118075.

Nada, A. A., Nasr, M., Viter, R., Miele, P., Roualdes, S., & Bechelany, M. (2017). Mesoporous ZnFe$_2$O$_4$@TiO$_2$ nanofibers prepared by electrospinning coupled to PECVD as highly performing photocatalytic materials. *Journal of Physical Chemistry C*, *121* (39). Available from https://doi.org/10.1021/acs.jpcc.7b08567.

Naragund, V. S., & Panda, P. K. (2022). Electrospun polyacrylonitrile nanofiber membranes for air filtration application. *International Journal of Environmental Science and Technology*, *19*, 10233−10244. Available from https://doi.org/10.1007/S13762-021-03705-4.

Nasr, M., Balme, S., Eid, C., Habchi, R., Miele, P., & Bechelany, M. (2017). Enhanced visible-light photocatalytic performance of electrospun rGO/TiO$_2$ composite nanofibers. *Journal of Physical Chemistry C*, *121*(1), 261−269. Available from https://doi.org/10.1021/acs.jpcc.6b08840.

Nassar, N. N. (2010). Rapid removal and recovery of Pb(II) from wastewater by magnetic nanoadsorbents. *Journal of Hazardous Materials*, *184*, 538−546. Available from https://doi.org/10.1016/J.JHAZMAT.2010.08.069.

Neghlani, P. K., Rafizadeh, M., & Taromi, F. A. (2011). Preparation of aminated-polyacrylonitrile nanofiber membranes for the adsorption of metal ions: Comparison with microfibers. *Journal of Hazardous Materials*, *186*, 182−189. Available from https://doi.org/10.1016/J.JHAZMAT.2010.10.121.

Nguyen, H. Q., & Deng, B. (2012). Electrospinning and in situ nitrogen doping of TiO2/PAN nanofibers with photocatalytic activation in visible lights. *Materials Letters*, *82*, 102−104. Available from https://doi.org/10.1016/J.MATLET.2012.04.100.

Nor, N. A. M., Jaafar, J., Ismail, A. F., Mohamed, M. A., Rahman, M. A., Othman, M. H. D., Lau, W. J., & Yusof, N. (2016). Preparation and performance of PVDF-based nanocomposite membrane consisting of TiO$_2$ nanofibers for organic pollutant decomposition in wastewater under UV irradiation. *Desalination*, *391*, 89−97. Available from https://doi.org/10.1016/j.desal.2016.01.015.

Norouzi, M., Fazeli, A., & Tavakoli, O. (2022). Photocatalytic degradation of phenol under visible light using electrospun Ag/TiO$_2$ as a 2D nano-powder: Optimizing calcination temperature and promoter content. *Advanced Powder Technology*, *33*(11). Available from https://doi.org/10.1016/j.apt.2022.103792.

Pan, L., Wang, Z., Zhao, X., & He, H. (2019). Efficient removal of lead and copper ions from water by enhanced strength-toughness alginate composite fibers. *International Journal of Biological Macromolecules*, *134*, 223−229. Available from https://doi.org/10.1016/j.ijbiomac.2019.05.022.

Pant, B., Pant, H. R., Barakat, N. A. M., Park, M., Han, T. H., Lim, B. H., & Kim, H. Y. (2014). Incorporation of cadmium sulfide nanoparticles on the cadmium titanate nanofibers for enhanced organic dye degradation and hydrogen release. *Ceramics International*, *40*(1), 1553−1559. Available from https://doi.org/10.1016/j.ceramint.2013.07.042.

Park, J. Y., Hwang, K. J., Lee, J. W., & Lee, I. H. (2011). Fabrication and characterization of electrospun Ag doped TiO$_2$ nanofibers for photocatalytic reaction. *Journal of Materials Science*, *46*(22), 7240−7246. Available from https://doi.org/10.1007/s10853-011-5683-5.

Pascariu, P., Airinei, A., Iacomi, F., Bucur, S., & Suchea, M. P. (2019). *Electrospun TiO$_2$-based nanofiber composites and their bio-related and environmental applications. Functional Nanostructured Interfaces for Environmental and Biomedical Applications* (pp. 307−321). Romania: Elsevier. Available from https://doi.org/10.1016/B978-0-12-814401-5.00012-8.

Pasini, S. M., Valério, A., Guelli Ulson de Souza, S. M. A., Hotza, D., Yin, G., Wang, J., & Ulson de Souza, A. A. (2019). Plasma-modified TiO$_2$/polyetherimide nanocomposite fibers for photocatalytic degradation of organic compounds. *Journal of Environmental Chemical Engineering*, *7*(4). Available from https://doi.org/10.1016/j.jece.2019.103213.

Pasini, S. M., Valério, A., Yin, G., Wang, J., de Souza, S. M. A. G. U., Hotza, D., & de Souza, A. A. U. (2021). An overview on nanostructured TiO$_2$−containing fibers for photocatalytic degradation of organic pollutants in wastewater treatment. *Journal of Water Process Engineering*, *40*. Available from https://doi.org/10.1016/j.jwpe.2020.101827.

Phan, D.-N., Lee, H., Huang, B., Mukai, Y., & Kim, I.-S. (2019). Fabrication of electrospun chitosan/cellulose nanofibers having adsorption property with enhanced mechanical property. *Cellulose, 26*(3), 1781−1793. Available from https://doi.org/10.1007/s10570-018-2169-5.

Qi, S., Zuo, R., Wang, Y., & Chan, H. W. L. W. (2013). Synthesis and photocatalytic performance of the electrospun $Bi_2Fe_4O_9$ nanofibers. *Journal of Materials Science, 48*(11), 4143−4150. Available from https://doi.org/10.1007/s10853-013-7227-7.

Qu, R., Sun, X., Sun, C., Zhang, Y., Wang, C., Ji, C., Chen, H., & Yin, P. (2012). Chemical modification of waste poly(p-phenylene terephthalamide) fibers and its binding behaviors to metal ions. *Chemical Engineering Journal, 181-182*, 458−466. Available from https://doi.org/10.1016/j.cej.2011.12.001.

Qu, X., Brame, J., Li, Q., & Alvarez, P. J. J. (2013). Nanotechnology for a safe and sustainable water supply: Enabling integrated water treatment and reuse. *Accounts of Chemical Research, 46*(3), 834−843. Available from https://doi.org/10.1021/ar300029v.

Ren, T., He, P., Niu, W., Wu, Y., Ai, L., & Gou, X. (2013). Synthesis of α-Fe_2O_3 nanofibers for applications in removal and recovery of Cr(VI) from wastewater. *Environmental Science and Pollution Research, 20*(1), 155−162. Available from https://doi.org/10.1007/s11356-012-0842-z.

Roso, M., Lorenzetti, A., Boaretti, C., Hrelja, D., & Modesti, M. (2015). Graphene/TiO_2 based photo-catalysts on nanostructured membranes as a potential active filter media for methanol gas-phase degradation. *Applied Catalysis B: Environmental, 176-177*, 225−232. Available from https://doi.org/10.1016/j.apcatb.2015.04.006.

Saber-Samandari, S., Saber-Samandari, S., & Gazi, M. (2013). Cellulose-graft-polyacrylamide/hydroxyapatite composite hydrogel with possible application in removal of Cu (II) ions. *Reactive and Functional Polymers, 73*, 1523−1530. Available from https://doi.org/10.1016/j.reactfunctpolym.2013.07.007.

Saeed, K., Haider, S., Oh, T. J., & Park, S. Y. (2008). Preparation of amidoxime-modified polyacrylonitrile (PAN-oxime) nanofibers and their applications to metal ions adsorption. *Journal of Membrane Science, 322*(2), 400−405. Available from https://doi.org/10.1016/j.memsci.2008.05.062.

Saeed, K., Park, S. Y., & Oh, T. J. (2011). Preparation of hydrazine-modified polyacrylonitrile nanofibers for the extraction of metal ions from aqueous media. *Journal of Applied Polymer Science, 121*, 869−873. Available from https://doi.org/10.1002/APP.33614.

Sakib, M. N., Mallik, A. K., & Rahman, M. M. (2021). Update on chitosan-based electrospun nanofibers for wastewater treatment: A review. *Carbohydrate Polymer Technologies and Applications, 2*. Available from https://doi.org/10.1016/J.CARPTA.2021.100064.

Salama, A., Abouzeid, R., Prelot, B., Diab, M., Assaf, M., & Hesemann, P. (2023a). Oxidized cellulose nanofibers decorated with magnetite as efficient bioadosrbent for organic dyes. *Chemistry Africa*. Available from https://doi.org/10.1007/s42250-023-00669-5.

Salama, A., Abouzeid, R., Prelot, B., Diab, M., Assaf, M., & Hesemann, P. (2023b). Preparation and adsorption performance of cellulose nanofibers/silica/calcium carbonate composites for water purification. *Chemistry Africa*. Available from https://doi.org/10.1007/s42250-023-00741-0.

Sankararamakrishnan, N., Singh, R., & Srivastava, I. (2019). Performance of novel MgS doped cellulose nanofibres for Cd(II) removal from industrial effluent − mechanism and optimization. *Scientific Reports, 9*, 1−8. Available from https://doi.org/10.1038/s41598-019-49076-2.

Samadi, S., Yazd, S. S., Abdoli, H., Jafari, P., & Aliabadi, M. (2017). Fabrication of novel chitosan/PAN/magnetic ZSM-5 zeolite coated sponges for absorption of oil from water surfaces. *International Journal of Biological Macromolecules*, *105*, 370−376. Available from https://doi.org/10.1016/j.ijbiomac.2017.07.050.

Sedghi, R., & Heidari, F. (2016). A novel & effective visible light-driven TiO2/magnetic porous graphene oxide nanocomposite for the degradation of dye pollutants. *RSC Advances*, *6*, 49459−49468. Available from https://doi.org/10.1039/C6RA02827F.

Sedghi, R., Moazzami, H. R., Hosseiny Davarani, S. S., Nabid, M. R., & Keshtkar, A. R. (2017). A one step electrospinning process for the preparation of polyaniline modified TiO$_2$/polyacrylonitile nanocomposite with enhanced photocatalytic activity. *Journal of Alloys and Compounds*, *695*, 1073−1079. Available from https://doi.org/10.1016/j.jallcom.2016.10.232.

Sehaqui, H., Mautner, A., Perez de Larraya, U., Pfenninger, N., Tingaut, P., & Zimmermann, T. (2016). Cationic cellulose nanofibers from waste pulp residues and their nitrate, fluoride, sulphate and phosphate adsorption properties. *Carbohydrate Polymers*, *135*, 334−340. Available from https://doi.org/10.1016/j.carbpol.2015.08.091.

Shamshi Hassan, M., Amna, T., & Khil, M. S. (2014). Synthesis of High aspect ratio CdTiO3 nanofibers via electrospinning: characterization and photocatalytic activity. *Ceramics International*, *40*, 423−427. Available from https://doi.org/10.1016/J.CERAMINT.2013.06.018.

Singh, P., Mondal, K., & Sharma, A. (2013). Reusable electrospun mesoporous ZnO nanofiber mats for photocatalytic degradation of polycyclic aromatic hydrocarbon dyes in wastewater. *Journal of Colloid and Interface Science*, *394*, 208−215. Available from https://doi.org/10.1016/J.JCIS.2012.12.006.

Snyder, A., Bo, Z., Moon, R., Rochet, J.-C., & Stanciu, L. (2013). Reusable photocatalytic titanium dioxide−cellulose nanofiber films. *Journal of Colloid and Interface Science*, *399*, 92−98. Available from https://doi.org/10.1016/j.jcis.2013.02.035.

Song, M., Cao, H., Zhu, Y., Wang, Y., Zhao, S., Huang, C., Zhang, C., & He, X. (2020). Electrochemical and photocatalytic properties of electrospun C/TiO$_2$ nanofibers. *Chemical Physics Letters*, *747*. Available from https://doi.org/10.1016/j.cplett.2020.137355.

Sonwane, N. D., & Kondawar, S. B. (2021). Enhanced room temperature ammonia sensing of electrospun nickel cobaltite/polyaniline composite nanofibers. *Materials Letters*, *303*. Available from https://doi.org/10.1016/J.MATLET.2021.130566.

Stephen, M., Catherine, N., Brenda, M., Andrew, K., Leslie, P., & Corrine, G. (2011). Oxolane-2,5-dione modified electrospun cellulose nanofibers for heavy metals adsorption. *Journal of Hazardous Materials*, *192*(2), 922−927. Available from https://doi.org/10.1016/j.jhazmat.2011.06.001.

Sun, B., Li, X., Zhao, R., Yin, M., Wang, Z., Jiang, Z., & Wang, C. (2016). Hierarchical aminated PAN/γ−AlOOH electrospun composite nanofibers and their heavy metal ion adsorption performance. *Journal of the Taiwan Institute of Chemical Engineers*, *62*, 219−227. Available from https://doi.org/10.1016/j.jtice.2016.02.008.

Taha, A. A., Wu, Yn, Wang, H., & Li, F. (2012). Preparation and application of functionalized cellulose acetate/silica composite nanofibrous membrane via electrospinning for Cr(VI) ion removal from aqueous solution. *Journal of Environmental Management*, *112*, 10−16. Available from https://doi.org/10.1016/j.jenvman.2012.05.031.

Tang, F., Yu, H., Yassin Hussain Abdalkarim, S., Sun, J., Fan, X., Li, Y., Zhou, Y., & Chiu Tam, K. (2020). Green acid-free hydrolysis of wasted pomelo peel to produce carboxylated cellulose nanofibers with super absorption/flocculation ability for environmental remediation materials. *Chemical Engineering Journal, 395*. Available from https://doi.org/10.1016/j.cej.2020.125070.

Tahmasebi, N., Sezari, S., & Zaman, P. (2020). Fabrication and characterization of hydrogen-treated tungsten oxide nanofibers for cationic dyes removal from water. *Solid State Sciences, 100*, 106073. Available from https://doi.org/10.1016/J.SOLIDSTATESCIENCES.2019.106073.

Tang, Q., Meng, X., Wang, Z., Zhou, J., & Tang, H. (2018). One-step electrospinning synthesis of TiO_2/g-C_3N_4 nanofibers with enhanced photocatalytic properties. *Applied Surface Science, 430*, 253−262. Available from https://doi.org/10.1016/j.apsusc.2017.07.288.

Teng, M., Li, F., Zhang, B., & Taha, A. A. (2011a). Electrospun cyclodextrin-functionalized mesoporous polyvinyl alcohol/SiO2 nanofiber membranes as a highly efficient adsorbent for indigo carmine dye. *Colloids and Surfaces A: Physicochemical and Engineering Aspects, 385*, 229−234. Available from https://doi.org/10.1016/J.COLSURFA.2011.06.020.

Teng, M., Wang, H., Li, F., & Zhang, B. (2011b). Thioether-functionalized mesoporous fiber membranes: Sol−gel combined electrospun fabrication and their applications for Hg2 + removal. *Journal of Colloid and Interface Science, 355*, 23−28. Available from https://doi.org/10.1016/J.JCIS.2010.11.008.

Turchi, C. S., & Ollis, D. F. (1990). Photocatalytic degradation of organic water contaminants: Mechanisms involving hydroxyl radical attack. *Journal of Catalysis, 122*, 178−192. Available from https://doi.org/10.1016/0021-9517(90)90269-P.

Vu, D., Li, X., Li, Z., & Wang, C. (2013). Phase-structure effects of electrospun TiO_2 nanofiber membranes on As(III) adsorption. *Journal of Chemical & Engineering Data, 58*(1), 71−77. Available from https://doi.org/10.1021/je301017q.

Wang, H., Wang, Y., Li, C., & Jia, L. (2022). Fabrication of eco-friendly calcium crosslinked alginate electrospun nanofibres for rapid and efficient removal of Cu(II). *International Journal of Biological Macromolecules, 219*, 1−10. Available from https://doi.org/10.1016/j.ijbiomac.2022.07.221.

Wang, J., Luo, C., Qi, G., Pan, K., & Cao, B. (2014). Mechanism study of selective heavy metal ion removal with polypyrrole-functionalized polyacrylonitrile nanofiber mats. *Applied Surface Science, 316*(1), 245−250. Available from https://doi.org/10.1016/j.apsusc.2014.07.198.

Wang, J., Pan, K., He, Q., & Cao, B. (2013). Polyacrylonitrile/polypyrrole core/shell nanofiber mat for the removal of hexavalent chromium from aqueous solution. *Journal of Hazardous Materials, 244-245*, 121−129. Available from https://doi.org/10.1016/j.jhazmat.2012.11.020.

Wang, M., Li, X., Zhang, T., Deng, L., Li, P., Wang, X., & Hsiao, B. S. (2018). Eco-friendly poly(acrylic acid)-sodium alginate nanofibrous hydrogel: A multifunctional platform for superior removal of Cu(II) and sustainable catalytic applications. *Colloids and Surfaces A: Physicochemical and Engineering Aspects, 558*, 228−241. Available from https://doi.org/10.1016/j.colsurfa.2018.08.074.

Wanjale, S., Birajdar, M., Jog, J., Neppalli, R., Causin, V., Karger-Kocsis, J., Lee, J., & Panzade, P. (2016). Surface tailored PS/TiO_2 composite nanofiber membrane for copper removal from water. *Journal of Colloid and Interface Science, 469*, 31−37. Available from https://doi.org/10.1016/j.jcis.2016.01.054.

Wu, J., Wang, N., Wang, L., Dong, H., Zhao, Y., & Jiang, L. (2012). Electrospun porous structure fibrous film with high oil adsorption capacity. *ACS Applied Materials & Interfaces, 4*(6), 3207−3212. Available from https://doi.org/10.1021/am300544d.

Wu, S., Li, F., Wu, Y., Xu, R., & Li, G. (2010). Preparation of novel poly(vinyl alcohol)/ SiO$_2$ composite nanofiber membranes with mesostructure and their application for removal of Cu^{2+} from waste water. *Chemical Communications*, *46*(10), 1694–1696. Available from https://doi.org/10.1039/b925296g.

Xiao, G., Huang, X., Liao, X., & Shi, B. (2013). One-pot facile synthesis of cerium-doped TiO$_2$ mesoporous nanofibers using collagen fiber as the biotemplate and its application in visible light photocatalysis. *Journal of Physical Chemistry C*, *117*(19), 9739–9746. Available from https://doi.org/10.1021/jp312013m.

Xiao, J., Wang, L., Ran, J., Zhao, J., Tao, M., & Zhang, W. (2020). Highly selective removal of cationic dyes from water by acid-base regulated anionic functionalized polyacrylonitrile fiber: Fast adsorption, low detection limit, reusability. *Reactive and Functional Polymers*, *146*. Available from https://doi.org/10.1016/j.reactfunctpolym.2019.104394.

Xiao, S., Shen, M., Ma, H., Guo, R., Zhu, M., Wang, S., & Shi, X. (2010). Fabrication of water-stable electrospun polyacrylic acid-based nanofibrous mats for removal of copper (II) ions in aqueous solution. *Journal of Applied Polymer Science*, *116*(4), 2409–2417. Available from https://doi.org/10.1002/app.31816, T.

Xie, W., Shi, Y., Wang, Y., Zheng, Y., Liu, H., Hu, Q., Wei, S., Gu, H., & Guo, Z. (2021). Electrospun iron/cobalt alloy nanoparticles on carbon nanofibers towards exhaustive electrocatalytic degradation of tetracycline in wastewater. *Chemical Engineering Journal*, *405*. Available from https://doi.org/10.1016/j.cej.2020.126585.

Xu, C., Yu, Z., Yuan, K., Jin, X., Shi, S., Wang, X., Zhu, L., Zhang, G., Xu, D., & Jiang, H. (2019). Improved preparation of electrospun MgO ceramic fibers with mesoporous structure and the adsorption properties for lead and cadmium. *Ceramics International*, *45*(3), 3743–3753. Available from https://doi.org/10.1016/j.ceramint.2018.11.041.

Xu, D., Zhu, K., Zheng, X., & Xiao, R. (2015). Poly(ethylene- co -vinyl alcohol) functional nanofiber membranes for the removal of Cr(VI) from water. *Industrial & Engineering Chemistry Research*, *54*(27), 6836–6844. Available from https://doi.org/10.1021/acs.iecr.5b00995.

Xu, Q., Peng, J., Zhang, W., Wang, X., & Lou, T. (2020). Electrospun cellulose acetate/P (DMDAAC-AM) nanofibrous membranes for dye adsorption. *Journal of Applied Polymer Science*, *137*(15). Available from https://doi.org/10.1002/app.48565.

Xu, R., Jia, M., Zhang, Y., & Li, F. (2012). Sorption of malachite green on vinyl-modified mesoporous poly(acrylic acid)/SiO$_2$ composite nanofiber membranes. *Microporous and Mesoporous Materials*, *149*(1), 111–118. Available from https://doi.org/10.1016/j.micromeso.2011.08.024.

Xu, Z., Li, X., Wang, W., Shi, J., Teng, K., Qian, X., Shan, M., Li, C., Yang, C., & Liu, L. (2016). Microstructure and photocatalytic activity of electrospun carbon nanofibers decorated by TiO$_2$ nanoparticles from hydrothermal reaction/blended spinning. *Ceramics International*, *42*(13), 15012–15022. Available from https://doi.org/10.1016/j.ceramint.2016.06.150.

Yan, J., Huang, Y., Miao, Y. E., Tjiu, W. W., & Liu, T. (2015). Polydopamine-coated electrospun poly(vinyl alcohol)/poly(acrylic acid) membranes as efficient dye adsorbent with good recyclability. *Journal of Hazardous Materials*, *283*, 730–739. Available from https://doi.org/10.1016/j.jhazmat.2014.10.040.

Yang, Y., Xu, L., Su, C., Che, J., Sun, W., & Gao, H. (2014). Electrospun ZnO/Bi$_2$O$_3$ nanofibers with enhanced photocatalytic activity. *Journal of Nanomaterials.*, *2014*, 1–7. Available from https://doi.org/10.1155/2014/130539.

Yavari Maroufi, L., Ghorbani, M., Mohammadi, M., & Pezeshki, A. (2021). Improvement of the physico-mechanical properties of antibacterial electrospun poly lactic acid nanofibers by incorporation of guar gum and thyme essential oil. *Colloids and Surfaces A:*

Physicochemical and Engineering Aspects, 622. Available from https://doi.org/10.1016/j.colsurfa.2021.126659.

Yu, D., Bai, J., Liang, H., Wang, J., & Li, C. (2015). Fabrication of a novel visible-light-driven photocatalyst Ag-AgI-TiO$_2$ nanoparticles supported on carbon nanofibers. *Applied Surface Science*, *349*, 241−250. Available from https://doi.org/10.1016/j.apsusc.2015.05.019.

Yu, Z., Xu, C., Yuan, K., Gan, X., Feng, C., Wang, X., Zhu, L., Zhang, G., & Xu, D. (2018). Characterization and adsorption mechanism of ZrO$_2$ mesoporous fibers for health-hazardous fluoride removal. *Journal of Hazardous Materials*, *346*, 82−92. Available from https://doi.org/10.1016/j.jhazmat.2017.12.024.

Zhang, B. W., Fischer, K., Bieniek, D., & Kettrup, A. (1994). Synthesis of carboxyl group containing hydrazine-modified polyacrylonitrile fibres and application for the removal of heavy metals. *Reactive Polymers*, *24*(1), 49−58. Available from https://doi.org/10.1016/0923-1137(94)90136-8.

Zhang, L., Zhang, X., Li, P., & Zhang, W. (2009). Effective Cd^{2+} chelating fiber based on polyacrylonitrile. *Reactive and Functional Polymers*, *69*(1), 48−54. Available from https://doi.org/10.1016/j.reactfunctpolym.2008.10.008.

Zhang, N., Zang, G. L., Shi, C., Yu, H. Q., & Sheng, G. P. (2016). A novel adsorbent TEMPO-mediated oxidized cellulose nanofibrils modified with PEI: Preparation, characterization, and application for Cu(II) removal. *Journal of Hazardous Materials*, *316*, 11−18. Available from https://doi.org/10.1016/j.jhazmat.2016.05.018.

Zhang, Y., He, X., Li, J., Miao, Z., & Huang, F. (2008). Fabrication and ethanol-sensing properties of micro gas sensor based on electrospun SnO$_2$ nanofibers. *Sensors and Actuators, B: Chemical*, *132*(1), 67−73. Available from https://doi.org/10.1016/j.snb.2008.01.006.

Zhao, G., Liu, S., Lu, Q., & Song, L. (2012). Controllable synthesis of Bi$_2$WO$_6$ nanofibrous mat by electrospinning and enhanced visible photocatalytic degradation performances. *Industrial & Engineering Chemistry Research*, *51*(31), 10307−10312. Available from https://doi.org/10.1021/ie300988z.

Zhao, R., Li, X., Sun, B., Li, Y., Li, Y., Yang, R., & Wang, C. (2017). Branched polyethylenimine grafted electrospun polyacrylonitrile fiber membrane: A novel and effective adsorbent for Cr(vi) remediation in wastewater. *Journal of Materials Chemistry A*, *5*(3), 1133−1144. Available from https://doi.org/10.1039/C6TA09784G.

Zhao, R., Li, X., Sun, B., et al. (2015a). Preparation of phosphorylated polyacrylonitrile-based nanofiber mat and its application for heavy metal ion removal. *Chemical Engineering Journal*, *268*, 290−299. Available from https://doi.org/10.1016/J.CEJ.2015.01.061.

Zhao, R., Wang, Y., & Li, X. (2015b). Synthesis of β-cyclodextrin-based electrospun nanofiber membranes for highly efficient adsorption and separation of methylene blue. *ACS Applied Materials & Interfaces*, *7*, 26649−26657. Available from https://doi.org/10.1021/ACSAMI.5B08403.

Zhao, R., Li, Y., Sun, B., Chao, S., Li, X., Wang, C., & Zhu, G. (2019a). Highly flexible magnesium silicate nanofibrous membranes for effective removal of methylene blue from aqueous solution. *Chemical Engineering Journal*, *359*, 1603−1616. Available from https://doi.org/10.1016/j.cej.2018.11.011.

Zhao, X., & Yang, H. (2019b). Synergistically enhanced photocatalytic performance of Bi4Ti3O12 nanosheets by Au and Ag nanoparticles. *Journal of Materials Science: Materials in Electronics*, *30*, 13785−13796. Available from https://doi.org/10.1007/S10854-019-01762-7.

Zhijiang, C., Xianyou, S., Qing, Z., & Yuanpei, L. (2017). Amidoxime surface modification of polyindole nanofiber membrane for effective removal of Cr(VI) from aqueous solution. *Journal of Materials Science*, *52*(9), 5417–5434. Available from https://doi.org/10.1007/s10853-017-0786-2.

Zhong, Q. Q., Yue, Q. Y., Li, Q., Gao, B. Y., & Xu, X. (2014). Removal of Cu(II) and Cr (VI) from wastewater by an amphoteric sorbent based on cellulose-rich biomass. *Carbohydrate Polymers*, *111*, 788–796. Available from https://doi.org/10.1016/j.carbpol.2014.05.043, http://www.elsevier.com/wps/find/journaldescription.cws_home/405/871description#description.

Zhu, C., Liu, P., & Mathew, A. P. (2017). Self-assembled TEMPO cellulose nanofibers: graphene oxide-based biohybrids for water purification. *ACS Applied Materials & Interfaces*, *9*, 21048–21058. Available from https://doi.org/10.1021/ACSAMI.7B06358/SUPPL_FILE/AM7B06358_SI_001.PDF.

Zia, Q., Tabassum, M., Meng, J., Xin, Z., Gong, H., & Li, J. (2021). Polydopamine-assisted grafting of chitosan on porous poly (L-lactic acid) electrospun membranes for adsorption of heavy metal ions. *International Journal of Biological Macromolecules*, *167*, 1479–1490. Available from https://doi.org/10.1016/j.ijbiomac.2020.11.101.

Wastewater treatment application of nanofibers and their composites

Akbar Esmaeili
Department of Chemical Engineering, North Tehran Branch, Islamic Azad University, Tehran, Iran

10.1 Introduction

The increasing world population will inevitably increase the pressure on water resources. As a result, there is an urgent need worldwide for more efficient and cost-effective water purification technologies. The use of nanotechnology in the industry is a significant development, and zero-valent iron nanoparticles (INPs) have been comprehensively studied for various purification applications. However, the application of soluble INP suspensions has been limited, while INP reaction mechanisms, transport properties, and environmental toxicity are still under study. From a theoretical point of view, they are developing INP-containing nanocomposites is a logical step toward developing nanomaterials that have broad applications in the water industry. This chapter examines a wide range of nanocomposites in volumetric and static water treatment containing INPs while emphasizing the limitations of each method. It also examines future studies to optimize nanocomposite water purifying systems to achieve commercial maturity.

Water pollution is one of the critical international issues caused by industrial processes, domestic use, and also environmental factors. The United Nations has estimated that 300–500 million tons of heavy metals, solvents, and other wastes enter the world's water resources yearly (Beni & Esmaeili, 2019). Water pollution can have a natural origin. For example, arsenic contamination is a severe problem in countries such as Bangladesh, West Bengal, and Nepal due to the weathering of sinks that naturally contain arsenic (Beni & Esmaeili, 2020a). Moreover, as the world's population continues to grow, the pressure on water resources has intensified.

During the last decade, nanotechnology has increasingly become a potential alternative to traditional treatment methods and reactive agents to provide safe water at a low cost while meeting global water quality standards.

In 2010, the Joint Research Center of the European Commission (JCR) published a report that provided international definitions. In Britannia, two explanations for the term nanoscale have been reported. The Department for Environment, Rural Affairs, and Food defines it as 200 nm, while other organizations consider 100 nm.

Based on the recommendations provided by the JCR, in October 2011, the European Commission adopted the following definition of nanomaterials for regulatory purposes (Beni & Esmaeili, 2020b).

A natural, random, or produced substance is considered to contain particles in a limited state or a combined state, and 50% of the particles in the size distribution have dimensions corresponding to 1—100 nm. Due to their tiny size, nanomaterials have shown different physical, chemical, and biological characteristics compared to more petite, micro, and large-scale types (Beni & Esmaeili, 2020a). Nanomaterials have a surface-to-volume ratio, and as a result, there is a high density of surface reaction zones per unit weight. Surface-free energy for nanomaterials is more significant than that of micro- and macro-scale materials. Therefore, nanomaterials show high reactivity for surface processes. However, as the particle size approaches the electron path and is on the wavelength scale (below 30 nm), quantum size effects become more pronounced, and fundamental physical properties change significantly.

Due to their tiny size, nanomaterials have shown different physical, chemical, and biological characteristics compared to more petite, micro, and large-scale types (Esmaeili & Khoshnevisan, 2016). Nanomaterials have a surface-to-volume ratio, and as a result, there is a high density of surface reaction zones per unit weight. Therefore, nanomaterials show high reactivity for surface processes. However, as the particle size approaches the electron path and is on the wavelength scale (below 30 nm), quantum size effects become more pronounced, and fundamental physical properties change significantly.

When nanomaterials are placed in the framework of the optimal size range, they are a suitable and more efficient alternative to the current materials used for water purification (Sajadi et al., 2010). A newly emerging and commercial technology in America is the injection of nanoparticles (NPs). NPs, usually zero-valent INPs, are injected into the ground as a dry powder or two pods for natural water treatment. NPs can be immobilized directly and converted into a PRB subsurface permeable reactive barrier or dispersed so that the NPs can migrate with the contaminated water body.

However, this method has disadvantages in using free NPs for purification, and the behavior of NPs is not well understood. It is well known that different processes limit the release of NPs in the groundwater system: adsorption of minerals, microbiological activity, aggregation, and formation of corrosion products. INPs are particularly susceptible to deposition due to their strong magnetic properties. NP—NP electrostatic attractions act effectively in concentrated particle solutions (slurries).

Surfactants or polymers can be added to the surfaces of NPs to increase the steric hindrance and change the surface charge to prevent electrostatic attraction. NPs can be placed in other mobile structures, such as carbon, silica, and colloidal clays. However, each treatment scenario's underground transmission and delay mechanisms are unique. They are associated with several variable factors, including soil composition, flow rates, acidity balance, and EH and bacterial communities. It is difficult to predict these variables and requires the use of NP for each condition. Changes in underground water can cause contaminants to be absorbed into the surface of NPs. In this way, considering the difficulty of removing NPs from the ground should also be taken into account. In addition, there is relatively little information about the long-term toxic effects of NPs in the environment. These qualitative characteristics make them harmful to living organisms.

Because the purification methods must have nontoxic reactive agents to provide long-term and stable removal mechanisms, their disadvantages have made using this technology difficult. Therefore, although there is no argument for toxicity, the UK is using approaches to introduce engineered NPs into the environment. This practice follows reports from the Royal Society, the Royal Academy of Engineering, and CL:AIRE for the Jahandust and Esmaeili (2024). Both reports indicate the need for fundamental research on NP behavior and nanotoxicology in the subsurface environmental system.

As mentioned limitations, it can be very beneficial to develop a purification method to use the activity of NPs while avoiding the release of NPs in the environment. A possible route and method is the development of a nanocomposite, which is defined as follows:

- A multiphase material where at least one of the constituent phases has a dimension less than 100 nm.

Recent research has presented changes in different nanocomposites, in which NPs are generally combined with micro- and macro-materials. The reactivity of the nanoparticle is demonstrated and complemented by the properties of the accompanying materials.

This article examines iron oxide and nanocomposites used in static water treatment systems, including permeable reactive barriers, batch reactor systems, and point filters. These systems should avoid issues related to uncontrolled NP release; therefore, the adsorbents should be in the framework of a stable structure. INPs and iron oxide are of great importance because iron has been used in purification methods for a long time and NPs have been used entirely for purification purposes. Most importantly, they purify a wide range of pollutants, from heavy metals through absorption to the decomposition of chlorine solvents through chemical reduction.

Although this type of technology appears to be very useful, this paper discusses areas of research and development that require further development, especially if nanocomposites are considered a proper water purification technology. One of the critical issues in this paper is that there is no consistency in performance testing for nanocomposites developed by different groups. In this way, comparing products and making decisions for funding and development is necessary (Bahramimehr & Esmaeili, 2019; Beni & Esmaeili, 2019, 2020a, 2020b).

10.2 The environmental threat of pollutants

As we read earlier, environmental threats are divided into different types. Therefore, the factors of their emergence are also varied and scattered. One of the ecological threats is the destruction of forests. As you know, forests and forest trees are vital for the survival of some animal generations. With the collapse of the woods, some species of animals have also become extinct. In the meantime, it found that two critical economic and cultural factors have significantly impacted the occurrence of such environmental threats.

If we want to examine it from an economic point of view, we must say that when people are in poverty and unemployed, they destroy forests and the environment by cutting down trees and other things. At this time, they only care about their profit and meeting the needs of themselves and their families, and nothing else. From a cultural point of view, our investigation concluded that people do not have the necessary awareness when doing bad things. When people cut or burn trees, they do not know how much they damage the environment and ecosystem, and this lack of awareness has made them act clumsily.

If we can raise the level of information and awareness of the people, we can essentially prevent the severe damage of such threats. Air heat and pollution are considered among the most dangerous environmental threats. This ecological threat has caused the death of many people

Cutting down a tree and turning it into timber, beams, or logs for making wooden products and structures, only about 50% of this tree can be used, and the rest will be wasted. By incorporating these wastes into plastic materials, a composite that can withstand outdoor conditions for over 20 years without requiring special protection or maintenance can be produced. Furthermore, the composite's recyclability minimizes environmental harm, ensuring it does not enter the biome (Beni & Esmaeili, 2019, 2020a).

In the production of wooden products, a large amount of wood waste is generated annually (about 63 million tons in the United States in 2002), and the reuse of these wastes in the production lowers the finished price of the product. Therefore, wood waste is used more compared to other natural fibers. In addition, wood-plastic composites have a significant feature of recyclability due to the use of thermoplastics as the base material. This allows these products to be reprocessed like unfilled plastic materials if they are not used, enhancing their sustainability.

Biodegradable plastics degrade when buried or exposed to light, oxygen, and moisture. Without these conditions, they can persist for years. Of course, wood plastics produced from plant polymers are not 100% degradable, biodegradable, or compostable; such wood plastics are obtained by adding some additives to traditional plastics. The suitable additives cause the obtained plastic to turn into smaller pieces after a while, although these small pieces, like conventional plastics, may remain in nature for a long time.

The mechanical properties of these composites depend on their constituent components, fibers, and matrix and how they are connected. A review of the existing models for composites with short fibers shows that in these models, the failure occurs in the thread and the matrix accompanies the stretching of the fiber. Although natural fibers such as wood have lower strength and modulus than some existing synthetic fibers, these fibers have a favorable strength-to-weight ratio. For this reason, emphasizing this point, artificial fibers are replaced with natural fibers.

However, there is a challenge in combining natural fibers with plastics due to the inherent hydrophilicity of natural fibers and the hydrophobic nature of polymers.

The size of the particles, the adhesion of the interface between the matrix and the particles, the particles' volume percentage, and the particles' shape are among the factors that affect the mechanical properties of fiber-reinforced composites. Many factors affect the performance of composites reinforced with natural fibers, but better contact between fibers and the matrix will increase the strength of the reinforced composite. Adhesion is improved by adding a compatible agent because it is a chemical substance that creates compatibility with its polar and nonpolar groups in the structure of natural fibers and polymers, respectively (Esmaeili & Beni, 2014).

10.3 Sequential methods of removing contaminants from aqueous solutions by composites

NPs can produce healthy drinking water, eliminate pathogenic microorganisms, and improve the quality of produced water and treated wastewater—a problem of conventional wastewater treatment methods. The most crucial feature of NPs that makes their use possible in this matter is the effect of these particles in preventing microbial growth and destruction of microbial tissue. On the other hand, using common disinfectants increases microorganisms' resistance to disinfectants and their side effects. Silver has received more attention among the NPs due to its many applications and low production cost. It is currently used in pharmaceutical and medical sciences, water purification, disinfection, paint and coating industries, textiles, and food packaging.

NPs with antimicrobial properties can reduce biological clogging in combination with membrane bioreactors. NPs have two properties that make them attractive, like "absorbents." They generally have a higher specific surface area than most particles, and they can also increase their use by various chemical groups to increase their effectiveness against the target compounds. One of the unique properties of NPs is the development of high capacity and selective absorption for metal ions and anions. Monovalent silver and silver compounds have been used as antimicrobial compounds for coliform in water and wastewater (Beni & Esmaeili, 2020a, 2020b; Dadashi & Esmaeili, 2021).

The nanocomposite performance of silver nanocrystals includes antimicrobial, antibiotic, and antifungal properties in combination with coatings, nanofibers, first-aid bandages, plastics, soap, and textiles. Additionally, in the treatment of certain viruses, self-cleaning fabrics incorporating nanocomposites.

Conductive fillers and other specific catalysts are also used in the nanowire. Nanosilver particles are antimicrobial against Gram-positive and Gram-negative bacteria. These bacteria include *Klebsiella pneumoniae*, *Staphylococcus*, *Escherichia coli*, and *Pseudomonas aeruginosa* (Kim & Van der Bruggen, 2010). Although zinc oxide mass cannot absorb arsenic, zinc oxide NPs have been used to remove arsenic from water. Studies based on the specific surface area have shown that the absorption capacity of media with NPs is more than twice that of iron hydroxide media, which is generally used in water purification (Nieto-Suárez et al., 2009).

Magnetic NPs can trap bacteria completely and sufficiently because their high surface area simply provides more contact area. Iron hydroxide is a general term for the group of magnetic compounds of iron oxide.

Iron hydroxides have self-magnetized properties and are crystalline solvent materials that only dissolve in strong acids. Many other metal ions can replace iron atoms without altering the spinel structure of iron hydroxide. This property allows iron hydroxides and natural iron to remove actinides and heavy metals from wastewater using nanodots. In most wastewater treatment operations, metals are removed as hydroxide metals because it has low solubility. Iron oxide and titanium dioxide are suitable absorbents for metal pollutants. Reducing the transfer of metal salts is the oldest, easiest, and most widespread method for preparing metal NPs (Tiwari et al., 2008).

With the ability to bond with the cell wall and the plasma of the bacterial membrane and penetrate the bacterial cell, nanosilver leads to changes and destruction in the structure of the cell and, finally, its death. Metal oxide NPs show different antibacterial properties based on the ratio of surface area to volume.

Compared to Gram-negative bacteria, Gram-positive bacteria show more resistance to metal NPs. It can be related to the structure of the cell wall. Numerous types of research have been conducted based on possible reactions between NPs and macromolecules of living organisms.

The difference between the microorganism's negative charge and the NP's positive charge acts as an absorbent electromagnet between the microbe and the NP. It causes the NP to connect to the cell surface and, as a result, can cause cell death. Finally, many of these contacts lead to the oxidation of surface molecules of microbes and their rapid death (Hajipour et al., 2012).

Ions released from nanomaterials may react with thiol (SH-) groups of surface proteins of bacterial cells. A number of these bacterial cell membrane proteins are responsible for transferring mineral materials from the wall's surface.

Deactivation of membrane permeability eventually causes cell death. Also, nanomaterials delay bacterial cell adhesion and biofilm formation, which prevents a group of bacteria from being able to stabilize and multiply.

The effect of NPs on bacteria has been studied, and it has been shown that combining nanotechnology and biology can produce new bacteria. Bacterial cells in contact with silver become functionally impaired and damaged by consuming it (Li et al., 2013).

Bacteria must not develop resistance to these particles. Therefore, it will be possible to affect a wide range of bacteria. In addition, these particles affect other microorganisms after the effect at the target point (Franci et al., 2015).

In the research on silver polymer nanocomposite, these particles' advantages, problems, applications, and functional improvement through changing their properties have been investigated (Dallas et al., 2011).

Metal oxide NPs show different antibacterial properties based on the ratio of surface area to volume. Several types of research based on possible reactions between NPs and macromolecules of living organisms have been carried out.

NPs have been widely studied to improve antibacterial properties and reduce biological fouling, and favorable results have been obtained (Kim & Van der Bruggen, 2010). In the meantime, silver has been widely used due to its good antibacterial performance and low risk to human health (Li et al., 2013). The results of the investigations show that in the presence of silver NPs on the agar plate, the growth of *E. coli* bacteria is wholly prevented. Of course, this inhibitory effect depends on the concentration of silver NPs.

By using CFU close to reality in the previous research, these particles have an antimicrobial effect and high efficiency in preventing microbial growth. Sandi et al. believe that the mechanism of silver's effect on bacteria is due to the inability of DNA replication and cellular protein inactivation in the presence of silver NPs.

In other research results, the germicidal properties of NPs have been declared due to the electrostatic attraction between the bacteria's negative charge and the NP's positive charge. Changes and severe damage to the microbial membrane in the presence of metal NPs, under the name of cavity formation on the surface of the membrane, lead to structural changes and destruction and, ultimately, cell death (Franci et al., 2015).

10.3.1 Chemical precipitation and filtration

Water use in factories and industrial and residential sectors gradually increases the number of heavy metals. It exists in the form of a solution in the effluents and endangers the health of the environment. For this reason, different methods are used in the treatment plants to separate these metals from the wastewater, turn them into solids, and finally settle, collect, and drain them before creating sediment in the treatment plant. In removing deposits from wastewater, various heavy metals are converted into hydroxides and sulfides, each of which has advantages and disadvantages and has different applications.

The most common methods to remove soluble heavy metals in industrial and sanitary wastes are separating chemicals and sedimentation. With the help of different processes such as alum, Fe-sulfate and lime, ferric chloride reaction, and lime reaction, 80%–90% of the suspended materials in the wastewater can be converted into a solid state so that they settle. Only 30%–60% of suspended heavy metals can be fixed in normal conditions and without chemical methods. According to the type of chemicals used in removing sediments from wastewater with the help of sedimentation, the deposited metals will be different. For example, using alum in wastewater sedimentation, the aluminum in the wastewater is converted into aluminum hydroxide and gradually settles on the bottom of the treatment pool with a gelatinous structure. Of course, adjusting the pH after precipitation and controlling the alkalinity is necessary. The type of salt used for the chemical sedimentation of wastewater is directly related to the metals present in the wastewater, and efforts are being made to remove them. For example, in sewage that contains a high amount of phosphorus, precipitation can be done with lime and metal salts such as ferric chloride and aluminum sulfate. For example, with the help of rainfall with calcium, you can separate the phosphate in the wastewater. These methods are widely used in recycling water from sewage. Coagulation or flocculation is another method of removing sediments from wastewater that act as a bridge to connect uncharged particles. By clicking these particles to each other and carrying out the process of solidifying suspended particles in sewage, it is necessary to allocate time for separation and sedimentation. Finally, with the help of this method, a large amount of existing heavy metals can be precipitated and settled. This method checks the wastewater for pH and adjusts it to a neutral or alkaline state.

The coagulant materials are transferred to all parts of the treatment pond with the help of large mechanical stirrers. Separating heavy metal particles and organic substances is carried out over time. Other methods of removing sediments from sewage include sewage flotation. This process, which is in the group of physical forms and the pretreatment stage, uses gravity flotation and aeration and flotation with the help of air bubbles to move many substances suspended in the wastewater to the highest level of the treatment plant. One of the most common methods for wastewater descaling is air bubble aeration. It is done in different ways, such as injecting air at atmospheric pressure, injecting air at atmospheric pressure and entering it into vacuum conditions, and injecting air at a pressure higher than atmospheric. Finally, it enters the ponds with atmospheric pressure. In fact, by injecting air into the wastewater, suspended particles stick to the air bubbles and eventually float on the surface.

With the complexity of life in the environment, the demand for air filter materials has gradually increased; there are many methods of making air filter materials by electrospinning. Research on preparing composite materials, including those with multiple polymers, continues unabated. The environment in which a person lives is often not monistic and includes various needs. For example, coal power plants' air filters require high-temperature and flame-retardant performance, and chemical plants require moisture resistance and chemical corrosion resistance, etc. Filter membranes made of a single polymer are often monofunctional and cannot meet the demands of complex environments. Therefore, it is concluded that a single polymer can improve its performance after preparation by another polymer. TPP and nylon core-shell fibers are applied through a coaxial electrospinning process, combining the flame-retardant properties of TPP and the filtration properties of nylon. Research on multiple polymers is continuously developing (Esmaeili & Beni, 2014, 2015).

10.3.2 Reverse osmosis

Among the membrane processes, reverse osmosis can remove more than 99% of insoluble salts, colloids, organic substances, and bacteria based on their size and load from the feed flow. It significantly reduces water hardness (Jafarinejad, 2017). Reverse osmosis is favored for its low energy consumption, minimal process steps, high separation efficiency, excellent product quality, and absence of solvents. It is also preferred because of the ease and speed with which systems can be developed and defective membranes replaced without interrupting operations. The main reasons for the popularity of reverse osmosis are compared to other membrane separation methods (Ravanchi et al., 2009). On the one hand, the lack of water and, on the other hand, the increase in population increase the ever-increasing need for water and wastewater treatment. It is predicted that the value of the global reverse osmosis membrane market will reach 5 billion dollars by 2026 from 1.3 billion dollars in 2021.

Osmosis is a natural process in which the solvent (water) moves from the dilute side to the thick side through a semipermeable boundary material such as a membrane. This movement continues until the equilibrium (Ghernaout & El-Wakil, 2017). In the normal osmosis process of water and salt solution, the difference in concentration leads to water flowing to the side with a higher concentration in the membrane, which causes the difference in osmotic pressure. In reverse osmosis systems, the presence of a driving force, such as pressure, reverses the flow; so that the water moves from the salt solution to the pure water in the membrane. The applied pressure must be greater than the osmotic pressure difference (Zirehpour & Rahimpour, 2016).

A large surface area is required for large-scale membrane processes, such as industrial or commercial applications. These levels are classified according to economic considerations, known as modules. The module is a completely engineered unit that arranges the membrane and increases the surface area. The module, as much as the membrane itself, is decisive in the efficiency of the membrane, and in general, the membrane process determines the module suitable for the operation (Obotey Ezugbe & Rathilal, 2020; Zirehpour & Rahimpour, 2016). Hollow, tubular, spiral, plate, and frame fibers are the four general categories of membrane modules. However, two types of open and spiral fiber modules are used in reverse osmosis membranes (Ismail et al., 2018; Obotey Ezugbe & Rathilal, 2020). Both modules are designed to increase the surface-to-volume ratio, but the hollow fiber module has a higher surface-to-volume ratio than the spiral module (Khulbe & Matsuura, 2017).

Hollow fiber membranes were developed in the 1960s for reverse osmosis applications and are used in water purification, desalination, cell culture, and in the medical and pharmaceutical fields. This module consists of hollow fiber arms in a closed and open container under pressure and includes a porous support layer and a selectable active layer. The active layer needs a support layer to withstand the hydrostatic pressure. Hollow fiber membranes are viral due to low energy requirement, simplicity of operation, and large surface area (Khulbe & Matsuura, 2018; Obotey Ezugbe & Rathilal, 2020).

10.3.3 Ion exchange

Ion exchange, one of the forms of surface adsorption, is a process in which various ions replace specific ions in solid and nonsoluble substances in a solution. In this process, the main focus of the reaction is ion exchange. In this process, unwanted ions from water are exchanged with desired ions by an ion exchange material (porous resins or nanomaterials). Ion exchangers are usually porous resins with a high specific surface area and the ability to absorb and desorb ions. They can also consist of some nanostructures such as zeolites or clay nanoplates. Ion exchange resins are water-insoluble polymers that can exchange their ions with water-soluble ions. Ion exchange resins are divided into two categories: cationic resins and anionic resins. Cationic resins exchange positive ions and anionic resins exchange negative ions.

Ion exchange is the exchange of ions between two electrolytes or between an electrolyte solution and a compound. In most cases, this term refers to the purification, separation, and disinfection processes of aqueous solutions and other solutions containing ions with a solid polymer or ion exchange mineral. One of the water purification methods is the use of ion exchange resins. Resin is produced in different types and grades and used in various industries to make water. The ion exchange process is one of the most appropriate ways to improve the quality of nitrate-contaminated water due to its ease of implementation, low cost, and high efficiency.

Ion-balancing resins are solid particles that can replace undesirable ions in solution with an equivalent amount of desirable ions of similar electrical charge. Ion exchange resins contain a cationic positive charge and anionic negative charge that are electrically neutral. Balancers differ from electrolyte solutions because only one of the two ions is mobile and replaceable.

These mobile cations can participate in an ion exchange reaction like an anion exchanger has immobile cation points, to which mobile anions such as —Cl or —OH are attached. As a result of ion exchange, the cations or anions in the solution are exchanged. The key and the resin remain electrically neutral with the cations and anions in the wax. Here we are dealing with solid—liquid equilibrium without the solid being dissolved in the solution.

Recently, the introduction of pollutants of human origin, such as heavy metals, has increased significantly, posing a severe health and environmental risk (Padervand & Gholami, 2013). Paper and cardboard manufacturing, oil refineries, and electrical, rubber, and chemical fertilizers introduce various toxic metals. Examples include arsenic, cadmium, mercury, chromium, nickel, zinc, cobalt, and copper. These toxic substances enter the food chain through drinking water and agricultural products (Suopajärvi et al., 2015).

By producing cellulose with cost-effective and environmentally friendly methods, making a product based on nanotechnology with unique and special efficiency is possible. This material, called nanocellulose, has a promising future and is prepared from lignocellulosic raw materials. Cellulose nanofibers are one of the thinnest and newest fibers in the natural world. They have exciting features, such as renewability, low price, high specific resistance, high length-to-diameter ratio, and very high resistance compared to nanoviscose cellulose giving them a promising future (Yousefi et al., 2013). Specifically, the development research related to this material started at the beginning of this century and will be global and updated shortly. In general, nano cellulose is categorized into two mechanisms: top-down (such as acid hydrolysis and super grinding) and bottom-up (such as bacterial synthesis). Nanocellulose produced through bacterial synthesis typically exhibits higher purity, strength, and crystallinity than nanocellulose obtained through top-down methods. Therefore, bacterial synthesis is more advantageous in specific applications than obtaining nano cellulose from lignocellulosic materials such as wood. It is noteworthy that, according to Yousefi et al. (2013), there have been no documented reports on the use of bacterial cellulose nanofiber gel in scientific literature.

10.3.4 Adsorption

10.3.4.1 Physical surface absorption

The term adsorption was coined in 1881 by Heinrich Kaiser, a German physicist. Surface adsorption means the physical absorption of molecules, atoms, and ions from the gas or liquid phase by the surface of a solid adsorbent. In this process, the absorbed particles form a film on the absorbent surface of the solid body. This process has distinguished the surface absorption process from the chemical absorption process in which a chemical solvent is used as an adsorbent. Among the advantages of surface absorption compared to chemical absorption, we can point out renewability, high selectivity of adsorbents, less environmental pollution, and absorption of particles by a particular surface in a substrate. The subject is that the adsorbent cannot be regenerated and selected in chemical absorption. As a result, the entire volume of the used solvent is involved in the absorption process. Also, the disposal process is only possible in surface absorption systems.

To prevent the adverse effects caused by the presence of moisture in distribution and transmission lines, including condensation, freezing, corrosion, rust, and contamination of catalysts, it is necessary to remove moisture and other gaseous fluids from airflow. In many gas streams, in addition to humidity, some pollutants and impurities cause many problems in the relevant processes, among the most important examples of the use of absorbent materials in industrial applications. We can mention moisture removal from the airflow in air conditioning systems. It is the dehumidification of natural gas and drying of compressed air required by industries.

Active alumina is undoubtedly considered the most widely used moisture-absorbent material in the compressed air industry. Active alumina is made of aluminum hydroxide and is a very porous material. Due to its high specific surface area (around 415–345 m^2/g), this absorbent material has a higher moisture absorption capacity than silica gel. For this reason, it plays a very prominent role in removing moisture from compressed air streams and other gases. They have high relative humidity. Among the reasons for the acceptance of active alumina in moisture removal processes, we can point out its cheapness, increased physical and chemical resistance, proper regeneration and porosity, specific surface, and high moisture absorption capacity.

In addition to the mentioned cases, active alumina is chemically an inert and neutral substance, which makes it easy to use in various processes. The very high resistance of activated alumina against precipitation by the polymerization of olefins has caused this material to be widely used in the petrochemical industry. Also, other uses of active alumina include its role as a filter in drinking water purification and removal of impurities such as fluoride, arsenic, and selenium.

This absorbent can dehumidify up to the dew point of $-70°C$.

Physical surface adsorption is created due to intermolecular forces between the adsorbent and the adsorbed. These physical electrostatic forces include the van der Waals force caused by dipole—dipole interactions and hydrogen bonding. Dipolar—dipolar interactions are achieved due to the orientation of polar compounds according to their charges, resulting in the lower free energy of the mixture. Hydrogen bonding is a particular case of dipole—dipole interactions in which the hydrogen atom, which has a partial positive charge in its molecule, attracts other atoms or molecules with a partial negative charge. In liquid phase systems, the van der Waals force is the primary physical force effective in surface adsorption. Physical absorption is an easily reversible reaction and includes monolayer and multilayer coating. Because physical absorption does not involve the electronic structure change, it will have lower absorption energy. The intermolecular forces between chemical molecules in the liquid flow are not explicitly determined. A solid (adsorbent) is greater than the forces between the molecules in the fluid flow. The chemicals are adsorbed onto the surface of the adsorbent. It is assumed that the particles physically adsorbed are accessible for movement on the adsorbent surface. Surface adsorption is multilayered, where each new molecular layer is formed on the previously adsorbed layers. If activated carbon is used as an adsorbent, it is assumed to be physical surface adsorption on polar surface carbon particles. These surfaces are inherently homogeneous and do not contain functional groups because the electrons of carbon atoms have covalent bonds. Therefore, a large amount of the surface area within the micropores of carbon particles is probably of the polar surface type. Many surface adsorption processes in wastewater treatment are not purely physical or chemical processes but are a combination of both. The distinction between these two processes is complex; fortunately, this distinction is not necessary for the field analysis and design of surface adsorption processes (Beni & Esmaeili, 2020b; Limousin et al., 2007).

10.3.4.2 Chemical surface absorption

Chemical surface adsorption is based on electrostatic forces. Activated adsorption occurs due to chemical interactions between the solid and the adsorbent. The chemical composition of the material cannot identify the irreversible process.

It is produced and has a broader temperature range than the physical state. The heat of absorption is significantly more than physical absorption, which causes the absorbed materials to change due to a chemical reaction. Chemical absorption only includes single-layer coating, and the particular site of the response occurs at the site of special functional groups (the reaction is carried out through the special functional groups that exist). Functional groups are different atoms in organic compounds with unique physical and chemical properties (Bailey et al., 1999; Cooney, 1999).

Agricultural residues, peels of fruits and vegetables, and materials such as straw can be used as cheap absorbents after fast processing (Singh et al., 2018). Agricultural wastes mainly comprise lignin, hemicellulose, lipids, proteins, simple sugars, water, hydrocarbons, and starch. Due to their unique structure and chemical properties, they are an attractive absorbent with a strong absorption capacity. They work for different pollutants. Special functional groups such as alcohol, phenol, aldehyde, carboxyl, and ketone are present in the polymer chains of these compounds and help remove various water pollutants (Singh et al., 2018). A large number of agricultural wastes include lemon peels (Kannan & Veemaraj, 2010), banana peels (Ponou et al., 2016), rice husks (Lakshmi et al., 2009), wheat bran (Singh et al., 2006), flax powder (Ahalya et al., 2005) coconut skin (Hameed et al., 2008), waste of branches and sesame leaves (Cheraghi et al., 2015), and garlic skin (Liang et al., 2013). Agricultural wastes are used in natural and beneficial ways. The product is washed and sieved until the particles' size reaches the desired size, and Sis is used in absorbent tests. While undergoing modifications, products use well-known modification techniques and are pre-filtered. The purpose of this pretreatment is to improve the physicochemical characteristics, such as the specific surface area, size, and volume distribution of pores; improve and strengthen the potential of functional groups; and, as a result, increase the number of active sites for water absorption.

Among these methods, modification with chemical compounds to increase the absorption capacity of the adsorbent is used more. These compounds often include organic and inorganic acids, essential solutions, and carbon dioxide. Many other mineral substances and organic chemical compounds can be found in water. Several gaps need more attention. Absorbents based on agricultural waste should be made more efficient, reusable, and suitable for removing real industrial effluents and leachate treatment with various pollutants (Hajjizadeh et al., 2020; Singh et al., 2018).

10.3.4.3 Exchange surface adsorption

It is a process in which ions accumulate on the surface and make the surface charged. The determining factor of exchange surface absorption is the charge of ions. For example, ions with three charges have a faster adsorption speed to the opposite surface than ions with one account.

If the charge of the ions is the same, the absorption intensity depends on the size of the molecule. The smaller the molecule size, the easier it is to penetrate more profoundly than the absorbent surface. This type of absorption, like chemical surface absorption, requires much energy to carry out the photo, which is irreversible.

It is a process in which ions accumulate on the surface and make the surface charged. The determining factor of exchange surface absorption is the charge of ions. For example, ions with three charges have a faster adsorption speed to the opposite surface than ions with one account.

If the charge of the ions is the same, the absorption intensity depends on the size of the molecule. The smaller the molecule size, the easier it is to penetrate more profoundly than the absorbent surface. This type of absorption, such as chemical surface absorption, requires much energy to carry out the photoreaction, making it irreversible. The cation exchange capacity of the soil is a measure of the ability of the earth to hold positively charged ions. This fundamental property of the soil reflects its physical stability, nutrient accessibility, soil pH, and the earth's reaction to fertilizers and modifiers.

The surface of a clay particle or organic colloidal particle consists of negative (−) charges, which results in the absorption of positive ions or cations. Water is added to the soil, and the cations can move into the solution. They absorb the surface of the clay or colloidal particles, and as a result, they crowd around these particles.

The mechanism of adsorption and desorption of cations on harmful sites is essential. However, less than 1% of cations do this at any time. The plant absorbs separated cations. On the other hand, when the cations are absorbed into the soil particles, they become stable, and the leaching rate of these nutrient cations decreases (Esmaeili & Kalantari, 2012).

10.3.5 Bio-absorption and factors affecting

The process of biological absorption is one of the methods of removing environmental pollution.

Bioaccumulation depends on the cell's energy, vital mechanism, and biological metabolism.

Basically, in biological absorption, microorganisms are used in the nonliving form. In the process of natural accumulation, the absorption mechanism can act as surface absorption. Bio-absorption is based on separating metal ions from aqueous solutions, so different microorganisms such as bacteria, fungi, algae, and yeasts play a significant role in such processes. Recently, biological adsorption has been developed as a solution for industrial wastewater treatment (Chojnacka, 2010; Volesky & Holan, 1995).

The exciting properties of biological adsorbents have made physical adsorption a practical, economical approach in wastewater treatment and recycling of precious metals to compete well with conventional methods.

Some bio-sorbents can remove a wide range of heavy metals, while others only absorb certain metals. The complex structure of microorganisms enables them to absorb metals in different ways. These processes can be classified in two aspects: based on dependence on cellular metabolism (metabolism-dependent and metabolism-independent) and based on the site of metal absorption from the solution (metal accumulation outside the cell, surface absorption, and intracellular metal assembly).

The transfer of metal through the cell wall causes its accumulation inside the microorganism. This absorption type is related to the active defense system of the organism that reacts in the presence of heavy metal. Potassium, magnesium, and sodium transporters transport heavy metals through the microbial cell membranes. The balance of this system is disturbed in the presence of metals with the same charge and ionic radius as other ions. In case of a reaction between the metal and the structural groups on the surface of the microbial cell, the absorption process is carried out independently of the metabolism. This absorption type is relatively fast and reversible (Masoudzadeh et al., 2011).

Factors affecting biosorption include temperature, pH (this factor affects the properties of the solution and can change the activity of structural groups in the biological mass or the competition of metal ions), and the concentration of the pack, with increasing the attention from a particular range. The interference of the binding sites causes a decrease in absorption and retention time (Esmaeili et al., 2015). In addition to these three critical factors, factors such as ionic strength, initial solute concentration, stirring (Das, 2010), and adsorbent size (Vijayaraghavan & Yun, 2008) are also involved in biosorption.

Using microorganisms is mainly related to the ligands in their polymer wall molecules. Biosorption includes a combination of several mechanisms, such as electrostatic attraction, complexation, ion exchange, and surface adsorption. The absorption process of metal ions by biological adsorbent occurs in two stages. The first step is moving from the mass of the solution containing ions to reaching the surface of the adsorbent, which is a relatively fast process. The speed of this step can be increased using unique methods (for example, by stirring the solution). The second step is transferring the dissolved component from the absorbent surface to the internal active sites and creating a link between the dissolved part and the active sites, which is a relatively slow process. Cell wall polymers include many chemical groups such as hydroxyl, phosphonate, and phosphodiester (Schiewer & Volesky, 1996). These chemical groups of biopolymers are placed in the binding sites and provide ligand atoms to form complexes with metal ions.

The amount of absorption obtained using absorbents depends on various factors. It contains several sites in the biosorbent: the chemical state of the sites, the accessibility of the sites, and the attraction between the site and the metal (Schiewer & Volesky, 1996). Fig. 10.1 can be defined as the bio-absorption mechanism as follows (Chojnacka, 2010): (1) penetration on the cell surface, (2) absorption, (3) connection with surface groups, and (4) transfer of metal ions to active or inactive sites inside the cell.

A large industrial scale to remove and recover heavy metals soluble in water used bio-absorption technology. Biomasses as natural absorbent materials are a new group of absorbents with good potential. As mentioned before, to use an adsorbent on an industrial scale, it must have criteria, such as fast and efficient absorption and removal; ability to be regenerated and reused; suitable shape, size, and mechanical properties for use in continuous systems; high selectivity, cost-effectiveness; and availability (Hawari & Mulligan, 2006).

This process is classified into three general groups according to the structure and performance: fluid bed columns, fixed bed columns, and stirred tanks.

The bioabsorbent particles in these columns become liquid due to the continuous upward flow of the liquid. In this system, mixing is done better than in other systems. Due to the system's fluidity, absorbents occupy a large volume, so they need more space than fixed bed systems. Also, a large part of absorbent particles is wasted due to high wear in fluidized bed systems. Among the other disadvantages

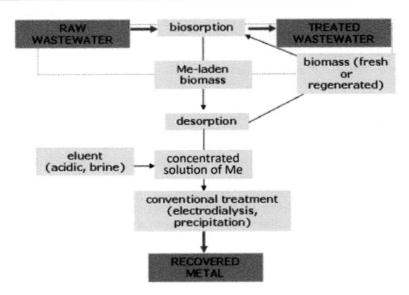

Figure 10.1 Schematic diagram of biosorption.
Source: With permission from Chojnacka, K. (2010). Biosorption and bioaccumulation — The prospects for practical applications. *Environment International, 36*, 299–307.

of these systems, we can point out that it is not easy to control their operating conditions. It is low due to complete mixing in this system; the group's driving force is also joint. As a result, the solution leaving the column is free of metal ions. This issue is perhaps the biggest flaw of these systems (Hawari & Mulligan, 2006). The advantages of these systems include the high contact surface between solid and fluid and the possibility of working with solutions containing foreign suspended substances. Therefore, in these systems, there is no need to separate the solid particles from the solution (Hawari & Mulligan, 2006).

In a fluidized bed column, the bed volume increases by more than 30%–40%. Fluidized bed columns usually have a height of about 3 meters, with more than half filled naturally with the absorbent. With these systems, it is possible to work at very high currents. The adsorbent inside can be restored by removing the column from the process. While the queue is working, the saturated adsorbent can be removed from the bottom of the column, and the new adsorbent can be inserted from the top—Kurdish column (Hawari & Mulligan, 2006).

In the fixed bed columns, the holder settled the solid particles in their position. The disadvantages can be mentioned in the problems of replacing after their inactivation. In addition to the pressure drop problem in these systems, sometimes the phenomenon of fluid channeling while passing through the column reduces the residence time required for the reaction, which is another limitation of this type of equipment.

In this type of adsorbent or solution contacting system, the amount of concentration difference, the mass transfer factor between the solution and the adsorbent, is minimal. In one step, the concentration of the adsorbed component in the tank's total volume equals its value in the outlet flow concentration (Hawari & Mulligan, 2006).

- Output concentration is not essential.
- Bioabsorbent in the form of powder or granules.
- There is a need to separate the solid from the liquid.
- In this case, several systems of this type can be put together as an opposite flow system.

Stirring in the tank is done mechanically or pneumatically. The amount of stirring is adjusted so that a homogeneous state is always maintained in the tank, and mass transfer is carried out well. The volume of the tank is large enough for the suspension solution. Several tanks are used in series for handling large volumes of the input solution and reaching the expected state at the output. The number of steps and the residence time of the solution for each stage are determined by considering the optimal system performance conditions (Hawari & Mulligan, 2006). Stirred tanks are called continuous stirred tank reactor (CSTR) reactors when used in continuous flow. These reactors can be used as counterflow and crossflow. In the case of opposite flow between the solution containing the absorbent and the absorbent itself, the efficiency is higher. In this case, the new adsorbent entering the system is in contact with the solution containing the lowest amount of adsorbent, and the adsorbent exiting the system, which has absorbed a large amount of adsorbent, is in connection with the input solution containing the most significant amount of adsorbent at the exit. This process causes the concentration gradient that determines the driving force of mass transfer to have the highest value during the process. To have a counterflow system, it is necessary to separate the adsorbent from the solution between steps (usually using a filter or sedimentation process). The adsorbent must be in granular form (Hawari & Mulligan, 2006).

In terms of performance, if the number of CSTR reactor stages approaches infinity, they behave like a fixed bed column. Also, fluidized bed columns are theoretically stable when mixing CSTR reactors and selected bed columns (Hawari & Mulligan, 2006).

The membrane bioreactor is an effective technology for wastewater treatment and water recycling, and due to its many advantages, it has increasing applications in the treatment of various wastewater types. This process has attracted the attention of researchers and related organizations in the last decade, leading to its everyday industrial-scale use in developing and advanced countries, including Japan, England, Ireland, France, Canada, and America (Severn, 2003).

Biological wastewater treatment is an effective method to remove pollutants in wastewater. The most common biological wastewater treatment process is the activated sludge process. The combination of membrane technology with biological wastewater treatment was first reported in 1969 by Smith and his colleagues (Brindle & Stephenson, 1996). In this system, a membrane was used to separate activated sludge from the output stream of the bioreactor by returning the microorganism to the aeration tank. The first submerged bioreactor on the laboratory scale was launched in 1989. Even though the construction and start-up of the first membrane bioreactor go back several decades, extensive research on the performance of this system has been started since 1996. Today, more than 900 membrane bioreactors have been used on a large industrial scale in developed countries for industrial and urban wastewater treatment (Brindle & Stephenson, 1996).

10.4 Pollutant absorption by plant composites

10.4.1 Influential functional groups in biosorption by plant composites

Today, in many industries, such as textile and dyeing, a significant amount of wastewater, including a wide range of pollutants such as microorganisms, heavy metals, and organic metal materials, is produced during various processes. The low efficiency of the usual wastewater treatment methods and the resulting high costs have led to the fact that part of the polluted water is not thoroughly treated and enters the consumption cycle with some pollutants.

It can use chemical oxidation and biological methods to remove pollutants, which do not have the necessary efficiency for some reasons. Chemical oxidation methods cause the decomposition and breaking of contaminants, but this issue requires very complex and expensive processes. Therefore, it is inevitable to use efficient and cost-effective ways to purify and recover industrial wastes to reuse them in the consumption cycle. A group of pollutants in the chemical and textile industries are pigmented. Most of the dyes have carcinogenic properties, indicating the importance of effectively treating colored wastes in dye factories. Among the types of pigments, methylene blue is one of the most important pollutants.

Today, using inexpensive absorbent materials with high absorption capacity to reduce the concentration of organic pollutants in wastewater and minimize waste-related problems has attracted researchers' attention. Among the porous materials used, we can mention activated carbon, chitin and chitosan biopolymers, ion exchange resins, clay, and low-risk mineral metal oxides. In recent decades, many researchers have focused their attention and research on using metal oxide NPs such as TiO_2, ZnO, and Fe_2O_3. Due to their multiple applications, zinc oxide NPs are one of the most widely used metal oxides. It has antibacterial, antifungal, anticancer, photocatalytic, and high absorption properties.

Today, environmental pollution has become a global issue and has attracted the attention of all politicians. Water is considered one of the essential components in the ecological cycle, and its quality, maintenance, and health are of great importance (Inobeme et al., 2022; Xu et al., 2022). Pesticides are substances used in agriculture as insecticides to kill insects and arthropods or as herbicides to fight weeds. These compounds are considered one of the primary environmental pollutants whose presence in ecological components, especially water, has caused great concern. Sources can contaminate water with these poisons in various ways, including direct washing or irrigation from the places of consumption. Because most pesticides are used during the spring season, due to the high rainfall in this season, they are washed away by the rain. In addition, pesticides can find their way to underground water tables through water penetration in soil layers and cause pollution of underground water resources. Most agricultural pesticides are phosphorous and chlorinated, each of which has its characteristics and can significantly affect food and living organisms, both human and animal. Chlorpyrifos is an organophosphorus poison used in many countries in different fields, including citrus fruits, summer vegetables, fruit trees, soybean leaves, and ornamental flowers. It has a good effect on them. However, chlorpyrifos is relatively toxic to humans. Poisoning with it may affect the central nervous, cardiovascular, and respiratory systems. The pesticide chlorpyrifos harms birds, freshwater fish, sea creatures, and rivers. For this reason, providing measures to remove this compound from water resources is in the programs of environmental organizations (Foong et al., 2020).

Due to the resistance of these compounds to decomposition in nature and their high toxicity, various processes are used today to purify organic pollutants in aqueous solutions, including absorption, chemical coagulation, and membrane processes. However, some mentioned methods do not altogether remove contaminants, especially in small amounts. Also, some water purification systems are expensive, and the membranes are easily clogged and destroyed. Therefore, the need to provide new methods to remove agricultural pollutants seems necessary.

As one of the ways to remove pollutants, nanotechnology has dramatically helped to separate toxins from water, air, and soil. Nanotechnology is the science of efficient production of materials and tools by controlling their dimensions on the nanometer scale and exploiting new features and phenomena that appear on this small scale. Various researchers have used nanotechnology to purify water and remove pollutants, and have achieved good results. Nanofibers are fragile strands with a long length compared to their diameter and are divided into natural and artificial categories. Nanofibers are defined as fibers with a diameter of less than 100 nm. Among their characteristics, high porosity and surface-to-volume ratio can be mentioned (Han et al., 2022).

Electrospinning is one of the possible methods in the production of continuous nanofibers. Electrospun nanofibers exhibit unique features that make them attractive for various applications, including medicine, drug release systems, wound dressings, and filtration. Filtration and purification of wastewater using nanofibers have many advantages over conventional filters. Nonwoven membranes produced from nanofibers have tiny holes with high porosity, which increases their filtration efficiency. This characteristic has caused researchers to use these materials for various filtration applications in various environmental or industrial fields (Chen et al., 2022).

Iron oxide NPs have attracted much attention due to their unique characteristics compared to other NPs. This material has wide applications (Jabbar et al., 2022). Due to these excellent features, iron oxide NPs are used by researchers in the field of purification of water and wastewater from pollutants such as heavy metals, dyes, and toxins (Babacan et al., 2022).

Several physical and chemical methods are used to prepare iron oxide NPs (Sachdeva et al., 2022). Using biological organisms, such as microorganisms, plant extracts, or plant biomass, can replace chemical and physical methods to produce NPs in an environmentally friendly way (Begum et al., 2020; Guan et al., 2022).

Among the standard methods, using different plant extracts has positive results in forming NPs. Plant extracts are composed of various metabolites such as terpenoids, phenols, or carbohydrates, which act as stabilizers and coating agents in the preparation of NPs, controlling the growth of crystals, for which these compounds are responsible. Direct oxidation-reduction reactions of NPs act as stabilizing and covering agents. Unlike time-consuming chemical and physical methods with complications, using the green form is much simpler and safer.

10.4.2 Removal of pollutants using biopolymers

With the global consumption of plastics and their natural resistance to decomposition, their accumulation in the environment is very worrying. This study aims to provide an overview of the current knowledge in the related fields, especially polyethylene terephthalate (PET). Also, this study includes an overview of the problems associated with plastic pollution in the marine environment; a description of the properties, commercial production, and degradability of PET; and an overview of the biodegradability potential of conventional and biodegradable polymers being produced.

Due to plastic flexibility against its decomposition and reproduction in industry, the issue of plastic pollution has become a threat to global ecology. Plastic pollution comes from both terrestrial and marine sources. The continuous flow of polluting plastics is maintained by two methods: intentionally and unintentionally. The intentional method is the illegal and improper discharge of domestic and industrial waste, and the accidental process is static waste and poor transportation. The elements carry plastic waste on land into waterways, combined with waste from ships and offshore oil platforms. Such pollution will undoubtedly have many harmful consequences.

There are many biodegradable polymers, synthetic and natural. However, there are still two main obstacles to their integration in current plastic-based applications, increasing production costs and low material properties such as low resistance. It can minimize production costs through the continuous development of production protocols and increased efficiency. However, extensive research is still needed to produce biodegradable polymers with physical properties comparable to conventional plastics. Despite these advances in biodegradable plastics, several strategies for their development have also been presented. One of the methods of producing biodegradable plastic is to make materials based on ordinary plastics with more excellent biodegradability without removing the material's properties. For example, polymers with different functional groups on polymer chains have been produced by postpolymerization treatment and polymerization with equivalent functionalized monomers. It is an opportunity for microbial enzymes to attack polymer chains. However, the biodegradation of these polymers is still relatively limited, and for the decomposition to take place, significant energy input is needed, especially for the treated materials after polymerization. A better decomposition rate is obtained in producing block copolymers of ordinary plastics with easily hydrolyzable polymer molecules. Unfortunately, these polymers are significantly less durable than most conventional plastics. In many cases, it is unclear whether these polymers are truly biodegradable or just break down into smaller pieces. The development of plastics based on biological molecules has been a significant research area. Polymers have been produced entirely or partially based on starch. In many cases, their mechanical resistance has been improved by adding lubricants and NPs or carefully controlling the production conditions. One of the most important groups of biopolymers is poly (hydroxyalkanoates) (PHAs). PHAs are polymeric substances naturally produced by many bacteria. Some archaea can be processed for packaging, coatings, and biomedical applications.

PHAs are commercially produced through bacterial fermentation; however, their quantities are somewhat limited due to high production costs compared to conventional plastics and the lack of practical applications. Producers have reduced costs by using cheaper food materials for bacterial metabolism. However, the main obstacle is the extraction method to recover the polymer (Naderi & Esmaeili, 2020). Extraction methods include solvent extraction, chemical digestion, enzymatic treatment, mechanical disruption, supercritical fluid disruption, flotation, gamma radiation, and two-phase systems. Therefore, no inexpensive extraction techniques have yet been developed to allow PHAs to compete with conventional plastics on the market. Despite the extensive studies conducted in developing biodegradable polymers, it is unlikely that any polymer can compete with traditional synthetic plastics in terms of properties and widespread use when exposed to significant environmental degradation. The popularity of ordinary plastics in many applications is their physical and chemical resistance, which can also be considered the main reason for their nonbiodegradability. Finally, any attempt to increase biodegradability will likely compromise the material's physical properties. However, according to the intended application, it is possible to compare degradability and resistance on a case-by-case basis (Tadayon et al., 2015).

The accumulation of plastic, especially in the oceans, is a growing environmental concern. One of the main components of plastic waste is PET, a polymer often used in many applications, including textiles and food packaging. PET is highly resistant to biodegradation and, therefore, creates many environmental concerns related to its accumulation, including the absorption and concentration of environmental pollutants, threats to the marine environment, and the release of potentially invasive species into new territories. To date, only three methods of plastic disposal —landfill, incineration, and recycling—are routinely used on a large scale. Each technique has advantages and disadvantages. Landfilling and incineration both lead to the release of hazardous secondary pollutants into the environment. Landfill also has a significant drawback: it requires more land space. Recycling addresses the environmental concerns of landfilling and incineration; however, this process is relatively inefficient, and reducing quality of the produced polymer is considered a limiting factor.

On the other hand, this process is economically viable and there is less incentive to invest in recycling facilities. Biodegradation is a suitable option to dispose of the biocompatibility of plastic waste. To date, no protocol has been developed for the biodegradation of PET on a commercial scale. However, extensive research is being done on the biodegradation of polymers due to the vast metabolic potential of microorganisms. It is expected that the development of sustainable biodegradation processes is only related to the time issue (Esmaeili & Asgari, 2015; Esmaeili & Behzadi, 2017; Esmaeili & Beni, 2014).

10.5 Conclusions and future perspectives

Industrial pollution, domestic sewage, and agricultural and urban runoff contaminate rivers and make it difficult to treat the water to remove all pollutants. This pollution can affect the quality of waterways, aquatic animals' quality of life, and human health through water consumption. Nanotechnology has been studied as an alternative for better removal of pollutants such as heavy metals, separation of oily water, and antimicrobial activity. In addition, with the increase of industrialization and the pollution of rivers, seawater can be an attractive alternative source of drinking water after adequate treatment. Nanomaterials are being studied as a possibility to remove salt from seawater and make it drinkable. However, there is a need to clarify the potential risks these nanomaterials can cause to the environment. Therefore, this review aims to describe the application of nanotechnology in wastewater treatment regarding metals, oil removal from water, antimicrobial activity, and desalination to improve water quality and discuss their potential risk to the environment.

As a result of the industrialization of today's world, rivers have been polluted with the discharge of hazardous chemicals, including heavy metals, into the environment. Along with industrial pollution, domestic sewage and agricultural and urban runoff are also a double concern for the quality of rivers. This pollution may lead to the bioaccumulation of metals in water and aquatic animals, threatening human and animal health. Water pollutants, such as heavy metal ions and dyes, harm living organisms and can affect the ecosystem. Conventional water purification methods are not very efficient in removing many contaminants. Based on this problem, nanotechnology can improve water treatment due to the size of nanomaterials, which have a larger surface area. High reactivity, fast kinetics, and nanomaterials' cost are essential advantages of using these materials.

It is estimated that approximately 663 million people do not have access to safe drinking water, mostly in developing countries. Therefore, providing raw water for these people is essential in situations where there is sometimes no water purification. Removing pollutants from polluted water is necessary to prevent damage to human health and the environment. Based on the abovementioned problems, this paper aims to use nanomaterials to improve water quality. It discusses the potential risks that these nanomaterials can cause to the environment. In this way, we have investigated the application of nanotechnology in water and wastewater treatment.

Nano adsorbents are NPs composed of organic or inorganic materials with a high affinity to absorb substances; in other words, they can remove many pollutants. These NPs are developed and used to remove various types of contaminants. These materials have essential features such as catalytic potential, small size, high reactivity, and significant surface energy. They can be classified based on their absorption process, which includes metal NPs, nanostructured mixed oxides, magnetic NPs, and metal oxide NPs.

Nanocatalysts are based on the interaction of light energy with metal NPs. This type of purification has been considered due to its extensive photocatalytic activities. Photocatalyst activities are based on destroying bacteria and organic matter by reacting with hydroxyl radicals. Materials used in nanocatalysts are usually inorganic materials such as semiconductors and metal oxides. However, to be considered nano photocatalytic, they must meet some requirements.

References

Ahalya, N., Kanamadi, R., & Ramachandra, T. (2005). Biosorption of chromium (VI) from aqueous solutions by the husk of Bengal gram (*Cicer arientinum*). *Electronic Journal of Biotechnology*, *8*(3), 258–264.

Babacan, T., Doğan, D., Erdem, Ü., & Metin, A. Ü. (2022). Magnetically responsive chitosan-based nanoparticles for remediation of anionic dyes: Adsorption and magnetically triggered desorption. *Materials Chemistry and Physics*, *284*, 126032.

Bahramimehr, F., & Esmaeili, A. (2019). Producing hybrid nanofiber-based on/PAN/Fe$_3$O$_4$/zeolite/nettle plant extract/urease and a deformed coaxial natural polymer to reduce toxicity materials in the blood of dialysis patients. *Journal of Biomedical Materials Research Part A*, *107*, 1736–1743.

Bailey, S. E., Olin, T. J., Bricka, R. M., & Adrian, D. D. (1999). A review of potentially low-cost sorbents for heavy metals. *Water Research*, *33*, 2469–2479.

Begum, Q., Kalam, M., Kamal, M., & Mahboob, T. (2020). Biosynthesis, characterization, and antibacterial activity of silver nanoparticles derived from *Aloe barbadensis* miller leaf extract. *Iranian Journal of Biotechnology*, *18*, e2383.

Beni, A. A., & Esmaeili, A. (2019). Design and optimization of a new reactor based on biofilm-ceramic for industrial wastewater treatment. *Environmental Pollution*, *255*, 113298.

Beni, A. A., & Esmaeili, A. (2020a). Biosorption, an efficient method for removing heavy metals from industrial effluents: A review. *Environmental Technology & Innovation*, *17*, 100503.

Beni, A. A., & Esmaeili, A. (2020b). Fabrication of 3D hydrogel to the treatment of moist air by solar/wind energy in a simulated battery recycle plant salon. *Chemosphere*, *246*, 125725.

Brindle, K., & Stephenson, T. (1996). The application of membrane biological reactors for the treatment of wastewaters. *Biotechnology and Bioengineering*, *49*, 601–610.

Chen, K., Hu, H., Zeng, Y., Pan, H., Wang, S., Zhang, Y., Shi, L., Tan, G., Pan, W., & Liu, H. (2022). Recent advances in electrospun nanofibers for wound dressing. *European Polymer Journal*, 111490.

Cheraghi, E., Ameri, E., & Moheb, A. (2015). Adsorption of cadmium ions from aqueous solutions using sesame as a low-cost biosorbent: Kinetics and equilibrium studies. *International Journal of Environmental Science and Technology, 12*, 2579–2592.

Chojnacka, K. (2010). Biosorption and bioaccumulation – The prospects for practical applications. *Environment International, 36*, 299–307.

Cooney, D. (1999). *Adsorption design for wastewater treatment* (p. 10). Boca Raton, Florida, USA: CRC Pres. INC.

Dadashi, F., & Esmaeili, A. (2021). Optimization, in-vitro release and in-vivo evaluation of bismuth-hyaluronic acid-melittin-chitosan modified with oleic acid nanoparticles computed imaging-guided radiotherapy of cancer tumor in eye cells. *Materials Science and Engineering: B, 270*, 115197.

Dallas, P., Sharma, V. K., & Zboril, R. (2011). Silver polymeric nanocomposites as advanced antimicrobial agents: Classification, synthetic paths, applications, and perspectives. *Advances in Colloid and Interface Science, 166*, 119–135.

Das, N. (2010). Recovery of precious metals through biosorption—A review. *Hydrometallurgy, 103*, 180–189.

Esmaeili, A., & Asgari, A. (2015). In vitro release and biological activities of *Carum copticum* essential oil (CEO) loaded chitosan nanoparticles. *International Journal of Biological Macromolecules, 81*, 283–290.

Esmaeili, A., & Behzadi, S. (2017). Performance comparison of two herbal and industrial medicines using nanoparticles with a starch/cellulose shell and alginate core for drug delivery: In vitro studies. *Colloids and Surfaces B: Biointerfaces, 158*, 556–561.

Esmaeili, A., & Beni, A. A. (2014). A novel fixed-bed reactor design incorporating an electrospun PVA/chitosan nanofiber membrane. *Journal of Hazardous Materials, 280*, 788–796.

Esmaeili, A., & Beni, A. A. (2015). Novel membrane reactor design for heavy-metal removal by alginate nanoparticles. *Journal of Industrial and Engineering Chemistry, 26*, 122–128.

Esmaeili, A., & Kalantari, M. (2012). Bioremoval of an azo textile dye, reactive red 198, by *Aspergillus flavus*. *World Journal of Microbiology and Biotechnology, 28*, 1125–1131.

Esmaeili, A., & Khoshnevisan, N. (2016). Optimization of process parameters for removal of heavy metals by biomass of Cu and Co-doped alginate-coated chitosan nanoparticles. *Bioresource Technology, 218*, 650–658.

Esmaeili, A., Saremnia, B., & Kalantari, M. (2015). Removal of mercury (II) from aqueous solutions by biosorption on the biomass of *Sargassum glaucescens* and *Gracilaria corticata*. *Arabian Journal of Chemistry, 8*, 506–511.

Foong, S. Y., Ma, N. L., Lam, S. S., Peng, W., Low, F., Lee, B. H., Alstrup, A. K., & Sonne, C. (2020). A recent global review of hazardous chlorpyrifos pesticide in fruit and vegetables: Prevalence, remediation and actions needed. *Journal of Hazardous Materials, 400*, 123006.

Franci, G., Falanga, A., Galdiero, S., Palomba, L., Rai, M., Morelli, G., & Galdiero, M. (2015). Silver nanoparticles as potential antibacterial agents. *Molecules, 20*, 8856–8874.

Ghernaout, D., & El-Wakil, A. (2017). Requiring reverse osmosis membranes modifications —An overview. *American Journal of Chemical Engineering, 5*, 81–88.

Guan, Z., Ying, S., Ofoegbu, P. C., Clubb, P., Rico, C., He, F., & Hong, J. (2022). Green synthesis of nanoparticles: Current developments and limitations. *Environmental Technology & Innovation*, 102336.

Hajipour, M. J., Fromm, K. M., Ashkarran, A. A., de Aberasturi, D. J., de Larramendi, I. R., Rojo, T., Serpooshan, V., Parak, W. J., & Mahmoudi, M. (2012). Antibacterial properties of nanoparticles. *Trends in Biotechnology, 30*, 499−511.

Hajjizadeh, M., Ganjidoust, H., & Farsad, F. (2020). Review the types of adsorbents in water and wastewater treatment. *Journal of Environmental Science Studies, 5*, 3173−3182.

Hameed, B., Tan, I., & Ahmad, A. (2008). Adsorption isotherm, kinetic modeling and mechanism of 2, 4, 6-trichlorophenol on coconut husk-based activated carbon. *Chemical Engineering Journal, 144*, 235−244.

Han, W.-H., Wang, M.-Q., Yuan, J.-X., Hao, C.-C., Li, C.-J., Long, Y.-Z., & Ramakrishna, S. (2022). Electrospun aligned nanofibers: A review. *Arabian Journal of Chemistry*, 104193.

Hawari, A. H., & Mulligan, C. N. (2006). Heavy metals uptake mechanisms in a fixed-bed column by calcium-treated anaerobic biomass. *Process Biochemistry, 41*, 187−198.

Inobeme, A., Nayak, V., Mathew, T. J., Okonkwo, S., Ekwoba, L., Ajai, A. I., Bernard, E., Inobeme, J., Agbugui, M. M., & Singh, K. R. (2022). Chemometric approach in environmental pollution analysis: A critical review. *Journal of Environmental Management, 309*, 114653.

Ismail, F., Khulbe, K. C., & Matsuura, T. (2018). *Reverse osmosis*. Elsevier.

Jabbar, K. Q., Barzinjy, A. A., & Hamad, S. M. (2022). Iron oxide nanoparticles: Preparation methods, functions, adsorption and coagulation/flocculation in wastewater treatment. *Environmental Nanotechnology, Monitoring & Management, 17*, 100661.

Jafarinejad, S. (2017). A comprehensive study on the application of reverse osmosis (RO) technology for the petroleum industry wastewater treatment. *Journal of Water and Environmental Nanotechnology, 2*, 243−264.

Jahandust, M., & Esmaeili, A. (2024). Construction of a new membrane bed biofilm reactor and yttria-stabilized zirconia for removing heavy metal pollutants. *RCS Advance, 14*, 8150−8160.

Kannan, N., & Veemaraj, T. (2010). Dynamics and equilibrium studies for the removal of Cd 2 + and Cd 2 + EDTA onto lemon peel carbon. *Indian Journal of Environmental Protection, 30*, 26−33.

Khulbe, K., & Matsuura, T. (2017). Recent progresses in preparation and characterization of RO membranes. *Journal of Membrane Science and Research, 3*, 174−186.

Khulbe, K. C., & Matsuura, T. (2018). Thin film composite and/or thin film nanocomposite hollow fiber membrane for water treatment, pervaporation, and gas/vapor separation. *Polymers, 10*, 1051.

Kim, J., & Van der Bruggen, B. (2010). The use of nanoparticles in polymeric and ceramic membrane structures: review of manufacturing procedures and performance improvement for water treatment. *Environmental Pollution, 158*, 2335−2349.

Lakshmi, U. R., Srivastava, V. C., Mall, I. D., & Lataye, D. H. (2009). Rice husk ash as an effective adsorbent: Evaluation of adsorptive characteristics for Indigo Carmine dye. *Journal of Environmental Management, 90*, 710−720.

Li, J.-H., Shao, X.-S., Zhou, Q., Li, M.-Z., & Zhang, Q.-Q. (2013). The double effects of silver nanoparticles on the PVDF membrane: Surface hydrophilicity and antifouling performance. *Applied Surface Science, 265*, 663−670.

Liang, S., Guo, X., & Tian, Q. (2013). Adsorption of Pb^{2+}, Cu^{2+} and Ni^{2+} from aqueous solutions by novel garlic peel adsorbent. *Desalination and Water Treatment, 51*, 7166−7171.

Limousin, G., Gaudet, J.-P., Charlet, L., Szenknect, S., Barthes, V., & Krimissa, M. (2007). Sorption isotherms: A review on physical bases, modeling and measurement. *Applied Geochemistry, 22*, 249−275.

Masoudzadeh, N., Zakeri, F., bagheri Lotfabad, T., Sharafi, H., Masoomi, F., Zahiri, H. S., Ahmadian, G., & Noghabi, K. A. (2011). Biosorption of cadmium by *Brevundimonas* sp. ZF12 strain, a novel biosorbent isolated from hot-spring waters in high background radiation areas. *Journal of Hazardous Materials*, *197*, 190−198.

Naderi, S., & Esmaeili, A. (2020). Preparation of 3D-printed (Cs/PLA/PU) scaffolds modified with plasma and hybridization by Fe@ PEG-CA for treatment of cardiovascular disease. *New Journal of Chemistry*, *44*, 12090−12098.

Nieto-Suárez, M., Palmisano, G., Ferrer, M. L., Gutiérrez, M. C., Yurdakal, S., Augugliaro, V., Pagliaro, M., & del Monte, F. (2009). Self-assembled titania−silica−sepiolite based nanocomposites for water decontamination. *Journal of Materials Chemistry*, *19*, 2070−2075.

Obotey Ezugbe, E., & Rathilal, S. (2020). Membrane technologies in wastewater treatment: A review. *Membranes*, *10*, 89.

Padervand, M., & Gholami, M. R. (2013). Removal of toxic heavy metal ions from waste water by functionalized magnetic core−zeolitic shell nanocomposites as adsorbents. *Environmental Science and Pollution Research*, *20*, 3900−3909.

Ponou, J., Wang, L. P., Dodbiba, G., Matuo, S., & Fujita, T. (2016). Effect of carbonization on banana peels for removal of cadmium ions from aqueous solution. *Environmental Engineering & Management Journal (EEMJ)*, *15*(4), 851.

Ravanchi, M. T., Kaghazchi, T., & Kargari, A. (2009). Application of membrane separation processes in petrochemical industry: A review. *Desalination*, *235*, 199−244.

Sachdeva, V., Monga, A., Vashisht, R., Singh, D., Singh, A., & Bedi, N. (2022). Iron oxide nanoparticles: The precise strategy for targeted delivery of genes, oligonucleotides and peptides in cancer therapy. *Journal of Drug Delivery Science and Technology*, 103585.

Sajadi, G., Shojaei, A., Fazeli, M., Amini, J., & Jamalifar, H. (2010). Extracellular synthesis of silver nanoparticles by *Fusarium exispurium* in laboratory scale. *Donyay-e-Microbeha Journal*, *1*, 44−47.

Schiewer, S., & Volesky, B. (1996). Modeling multi-metal ion exchange in biosorption. *Environmental Science & Technology*, *30*, 2921−2927.

Severn, R. (2003). Long term operating experience with submerged plate MBRs. *Filtration & Separation*, *40*, 28−31.

Singh, K., Singh, A., & Hasan, S. (2006). Low cost bio-sorbent 'wheat bran' for the removal of cadmium from wastewater: Kinetic and equilibrium studies. *Bioresource Technology*, *97*, 994−1001.

Singh, N., Nagpal, G., & Agrawal, S. (2018). Water purification by using adsorbents: A review. *Environmental Technology & Innovation*, *11*, 187−240.

Suopajärvi, T., Liimatainen, H., Karjalainen, M., Upola, H., & Niinimäki, J. (2015). Lead adsorption with sulfonated wheat pulp nanocelluloses. *Journal of Water Process Engineering*, *5*, 136−142.

Tadayon, A., Jamshidi, R., & Esmaeili, A. (2015). Delivery of tissue plasminogen activator and streptokinase magnetic nanoparticles to target vascular diseases. *International Journal of Pharmaceutics*, *495*, 428−438.

Tiwari, D. K., Behari, J., & Sen, P. (2008). Application of nanoparticles in waste water treatment. *World Applied Sciences Journal*, *3*(3), 417−433.

Vijayaraghavan, K., & Yun, Y.-S. (2008). Bacterial biosorbents and biosorption. *Biotechnology Advances*, *26*, 266−291.

Volesky, B., & Holan, Z. (1995). Biosorption of heavy metals. *Biotechnology Progress*, *11*, 235−250.

Xu, H., Jia, Y., Sun, Z., Su, J., Liu, Q. S., Zhou, Q., & Jiang, G. (2022). *Environmental pollution, a hidden culprit for health issues, Eco-Environment & Health* (1, pp. 31−45).

Yousefi, H., Faezipour, M., Hedjazi, S., Mousavi, M. M., Azusa, Y., & Heidari, A. H. (2013). Comparative study of paper and nano paper properties prepared from bacterial cellulose nanofibers and fibers/ground cellulose nanofibers of canola straw. *Industrial Crops and Products, 43*, 732−737.

Zirehpour, A., & Rahimpour, A. (2016). *Membranes for wastewater treatment, Nanostructured polymer membranes* (2, pp. 159−207). London, UK: John Wiley & Sons Ltd.

Energy applications of nanofibers and their composites

Muhammad Tuoqeer Anwar[1], Raheela Naz[2], Arslan Ahmed[3], Saad Ahmed[4], Ghulam Abbas Ashraf[5] and Tahir Rasheed[6]
[1]Departemnt of Mechanical Engineering, COMSATS University Islamabad, Sahiwal Campus, Sahiwal, Punjab, Pakistan, [2]School of Materials and Energy, Southwest University, Chongqing, P.R. China, [3]Department of Mechanical Engineering, COMSATS University Islamabad, Wah Campus, Rawalpindi, Punjab, Pakistan, [4]State Key Laboratory Breeding Base of Green Chemistry-Synthesis Technology, Zhejiang Province Key Laboratory of Biofuel, Biodiesel Laboratory of China Petroleum and Chemical Industry Federation, College of Chemical Engineering, Zhejiang University of Technology, Hangzhou, Zhejiang, P.R. China, [5]Key Laboratory of Integrated Regulation and Resources Development on Shallow Lake of Ministry of Education, College of Environment, Hohai University, Nanjing, Jiangsu, P.R. China, [6]Interdisciplinary Research Center for Advanced Materials, King Fahd University of Petroleum and Minerals (KFUPM), Dhahran, Eastern Province, Saudi Arabia

11.1 Introduction

It has been found that the size reduction of the materials can lead to enhanced properties when compared with their counterparts. Recently, nanomaterials have been the topic of research for energy, health, food, and environmental applications (Anwar et al., 2023; Rasheed & Anwar, 2023; Raza et al., 2021; Rim et al., 2013; Wang et al., 2021; Wang, Kaneti, et al., 2018). Nanomaterials can be defined as the materials that have external dimensions, surface structure, or internal structure in the range of 100 nm (Jeevanandam et al., 2018). The nanomaterials are classified on the basis of structure (hollow, solid, core-shell, and spongy), composition (composites, organic, inorganic, and carbon-based), rigidity (stiff and flexible), and nature (engineered or naturally occurring). The different types of nanomaterials include nanocrystals, nanofibers, nanofilms, nanocages, nanorods, nanotubes, and nanospheres with potential technological applications (Anwar et al., 2017; Anwar, Yan, Asghar, Husnain, Shen, Luo, Cheng, et al., 2019; Burda et al., 2005; Jayaraman et al., 2004; Richardson et al., 2015; Su et al., 2014; Wang et al., 2014; Wu, 2020; Zhang, Qiao, et al., 2015). Nanofibers possess diameters in the nanoscale, while the length of the nanofiber is relatively larger. They have been extensively researched among all the nanostructural materials and can be defined as one-dimensional materials with less than 1 micron diameter (1000 times less than a human hair) and corresponding length to diameter ratio (aspect ratio) of 50. They are preferable due to the unique properties such as smaller number of defects;

Polymeric Nanofibers and their Composites. DOI: https://doi.org/10.1016/B978-0-443-14128-7.00011-0
© 2025 Elsevier Ltd. All rights are reserved, including those for text and data mining, AI training, and similar technologies.

tailorable porosity; 3D topography; high specific surface area; excellent thermal, electrical, and mechanical (e.g., tensile strength and stiffness) attributes; and larger pore volumes owing to their unique nanostructure (Huang et al., 2019).

Nanofibers find a number of promising applications such as in electronics, including stretchable sensors (Wang, Zhang, et al., 2018), photodetectors, thermos and piezoelectric generators, and photovoltaic devices; lithium-ion batteries (Persano et al., 2015); biomedical applications, including drug delivery, protective clothing, tissue engineering, biosensors, wound dressing, and antibacterial research studies (Haider et al., 2018); and environmental applications, including filtration, absorption, distillation, catalysis, and water purification (Wang & Hsiao, 2016). In addition to the aforesaid applications, the incorporation of nanofibers into two-dimensional and three-dimensional structures further enhances their performances. For instance, the introduction of nanofibers in bulk materials may provide electronic pathways for electronic applications.

11.2 Brief history

The human civilization has always been inspired by nature for the discovery and invention of different things. Spiderweb (since millions of years) and silk fibers (since 2700 BCE) are examples of naturally occurring fibers, which have been there for a number of millennia. Fibers were fabricated using cotton and wool with the invention of the spindle and trace back to 1880s. The first synthetic fibers came into the limelight after the introduction of nylon fibers by DuPont. Later, the synthesis procedures were further modified to reduce the demand for naturally occurring fibers. Mainly, spinning methods, namely dry, wet, gel, and melt spinning, have been used for the synthesis of synthetic fibers (Xue et al., 2019). Among these techniques, electrospinning is the most widely used technique for fabrication of the nanofibers.

11.3 Fabrication methods

The fabrication techniques can be categorized into the following types: top-down and bottom-up techniques, as shown in Fig. 11.1. In top-down methods, the starting bulk material is converted into nanofibers. The specific techniques include milling, grinding, refining, and step-by-step cutting. For instance, cellulose nanofibers can be synthesized using this technique. In the latter technique, nanofibers are fabricated using the constituent materials. It includes methods such as drawing, phase separation, self-assembly, template-based fabrication, physical vapor deposition, chemical vapor deposition, and spinning (electrospinning and centrifugal spinning) (Soltani et al., 2020). Among all aforesaid techniques, electrospinning is preferred due to its facile synthesis, affordability, simplicity, and tunability of the properties (Barhoum et al., 2019). These fabrication techniques are briefly explained in the following section (Fig. 11.1).

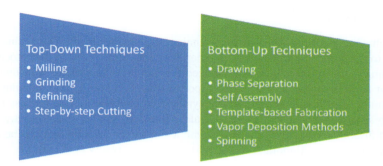

Figure 11.1 Fabrication techniques for nanofibers.

11.3.1 Milling

This process is mainly used for the synthesis of CNF and is one of the top-to-bottom approaches. For example, Zhang and coworkers produced cellulose nanofibers employing ball milling at room temperature and ambient pressure conditions. It was interesting to note that ball size, milling time, and mass ratio of cellulose and ball had a significant effect on the morphology. In a particular experiment, 100 nm nanofibers were synthesized using doped-zirconia-based balls through gentle mechanical milling (Zhang, Tsuzuki, et al., 2015).

11.3.2 Grinding

Grinding is another technique that is being used for the synthesis of nanofibers. It has been reported that several characteristics such as morphology and size of the nanofiber can be tailored by varying material quantity, grinding medium, speed, duration, grinding conditions, and energy exchanged between material and grinding media (Vasita & Katti, 2006).

11.3.3 Refining

In the present times, cellulose nanofibers have been extensively studied, which are usually synthesized using fibrillation after the chemical treatment. For instance, Nobuta et al. used wise treatments for the detachment of hemicellulose. Grinding process was carried out to fibrillate the bast fibers. Vacuum filtration was further employed to produce nanosheets of cellulose nanofibers. It was observed that the nanosheets with wise treatment exhibited better mechanical properties as compared to the untreated ones, demonstrating the importance of alkaline treatment (Nobuta et al., 2016).

There are some issues associated with aforesaid processes. For instance, these methods cause shredding of the fibers instead of their fibril delamination. In addition to that, CNF thus produced exhibit poor mechanical attributes realized through lower degree of polymerization, smaller aspect ratios, and lower level of crystallinity (Samyn et al., 2018).

11.3.4 Drawing

This technique can be defined as dry spinning at a smaller level. This technique is limited to the viscoelastic materials that can bear the stresses during the processing. For example, the surface of SiO_2 is utilized for this technique along with micromanipulator and micropipette. It has been reported that the viscosity and drawing speed significantly affect the dimension of the synthesized nanofibers. This process is used on a lab scale rather than an industrial scale. Moreover, smaller quantities are possible with discontinuous yield. The fibers with a size larger than 100 nm can be drawn using this technique (Alghoraibi & Alomari, 2018; Almetwally et al., 2017).

11.3.5 Phase separation

In this type, the gelation process is carried out once the mixture of solvent and polymer is prepared. Different phases thus prepared can be separated due to being physically inconsistent. Consequently, different phases can be collected separately. For instance, Zhao et al. synthesized chitosan-based nanofibers using this technique. The structure of the resulting nanofibers was found to be influenced by different parameters such as temperature and concentration. Increasing the concentration of the acetic acid resulted in the formation of micro/nanoarchitecture, whereas decreasing the temperature led to the formation of nanostructures. Interestingly, both structures showed different levels of crystallinity. Moreover, the presence of nitrogen resulted in the transformation of the crystals from one shape to another (Zhao et al., 2011).

11.3.6 Self-assembly

It has been reported that PANI can be prepared by electrochemical/chemical polymerization. Previously, chemical methods were used for the synthesis of polyaniline nanofibers. In recent times, it has been produced without using protonic acid and directly from the aqueous media. Chiou et al. prepared highly uniform, longer nanotubes/nanofibers of polyaniline through a self-assembly technique using an excess amount of the oxidant. This method paved the path for the creation of 3D nanostructures employing surfactants (Chiou et al., 2007).

11.3.7 Template-based fabrication

This technique is usually employed for the synthesis of inorganic nanofibers such as polypyrrole, carbon nanofibers, and polyaniline. More specific examples include the duplication of DNA and casting techniques. Different types of templates can be used for the synthesis of nanofibers, including SiO_2 and ZnO. For instance, Ma et al. synthesized carbon nanofibers employing carbon precursor and additive (zinc nitrate hexahydrate), as shown in Fig. 11.2. The microporous nanofibers thus prepared had sufficient enough specific surface area (1363 m^2/g), flexibility, and self-sustaining capability accompanied by excellent electrochemical performance when used in supercapacitors (Ma et al., 2021).

Energy applications of nanofibers and their composites 259

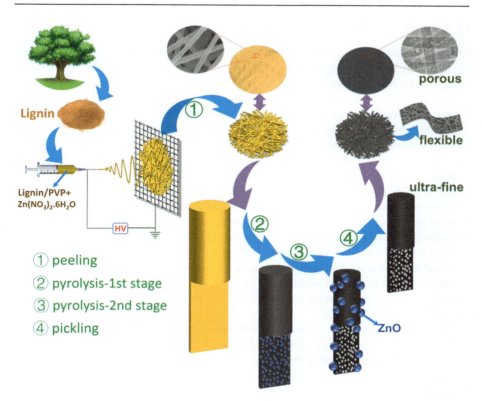

Figure 11.2 The fabrication route for microporous carbon nanofibers.
Source: From Ma, C., Wu, L., Dirican, M., Cheng, H., Li, J., Song, Y., Shi, J., & Zhang, X. (2021). ZnO-assisted synthesis of lignin-based ultra-fine microporous carbon nanofibers for supercapacitors. *Journal of Colloid and Interface Science*, *586*, 412–422, https://doi.org/10.1016/j.jcis.2020.10.105 (reused with the permission from Elsevier).

11.3.8 Vapor deposition methods

These methods are a type of bottom-up method and are employed for the synthesis of metal oxide and carbon-based nanofibers by using vapor phase. Physical vapor deposition is further subdivided into sputtering, electron beam evaporation, deposition through vacuum arc, and pulsed laser deposition. As far as chemical vapor deposition methods are concerned, transition metal elements along with hydrocarbon chains are used for the synthesis of carbon-based nanofibers. The final structure of the nanofiber depends upon the morphology and size of the catalyst (Zhi et al., 2017). Likewise, functionalized CNFs are also being synthesized through catalytic chemical vapor deposition for a wide variety of applications (Lu et al., 2017).

11.3.9 Spinning

Spinning is one of the oldest techniques that is used for the synthesis of yarn using fibers obtained from animals, plants, and synthetic sources. Basically, spinning is categorized into two types, namely electrospinning and non-electrospinning techniques. The main difference between the two types arises on the basis of applied force (i.e., electrostatic force *vs.* pneumatic, centrifugal, and gravitational forces).

The history of electrospinning dates back to 1887, when Charles and coworkers first demonstrated that fibers could be synthesized using a viscoelastic liquid under the action of an electric field. Later, in 1934 first patent was filed by Formhals (Anton, 1934; Boys, 1987). This is a preferable technique due to the facile synthesis of ultrathin fibers (diameter below 1 nm) and delivers excellent volume/area ratios.

Electrospinning can be defined as an electrohydrodynamic process in which droplets of the liquid are employed to form stretched/elongated fibers under the action of electricity. A typical setup consists of a syringe-based pump, a conductive collector, and a high-voltage electric supply (DC/AC). The liquid coming out of the spinneret forms a pendant droplet due to surface tension. Afterward, the application of electricity leads to the formation of Taylor cone. The jet thus produced exhibits bending instabilities followed by solidification and collection at the grounded collector (Xue et al., 2019).

There is a variety of electrospun nanofibers such as polyvinylidene fluoride (PVDF), polyvinylpyrrolidone (PVP), and polyvinylacetate (PVA). In addition to that, hybrid nanofibers comprising both nanofibers and nanoparticles have also been reported. For instance, carbon-based materials, metals, ceramics, and metal oxides can be incorporated into the nanofibers to form the composite. The synthesis routes usually consist of multichannel electrospinning or blended mixtures. The composites thus produced exhibit tremendous properties leading to diverse applications (Huang et al., 2019). Initially, electrospinning process was used for the fabrication of the fibers from the solution of the polymers. Later, the combination with sol-gel method led to its applicability to the synthesis of ceramic and composite nanofibers. Usually, there are three steps involved, which are briefly described as follows: (1) the formation of stable colloidal suspension, (2) the synthesis of hybrid nanofibers via electrospinning, and (3) formation of ceramic nanofibers using solvent extraction or calcination and by removing organic components. The nanofibers of CeO_2, TiO_2, ZrO_2, SnO_2, SiO_2, Fe_3O_4, $BaTiO_3$, and $NiFe_2O_4$ can be prepared by employing the above-mentioned approach (Xue et al., 2017).

Electrospun nanofibers demonstrate exceptional properties as compared to microfibers and bulk materials. These properties include molecular orientation, pore size, aspect ratio, specific surface area, optical transparency, and mechanical attributes. The nanofibers produced by electrospinning show a higher degree of molecular orientation to high strain rates and draw ratios. Generally, rigid rod-like nanofibers have higher degree of molecular orientation as compared to the soft molecular structure, and this is realized through chain entanglement. The pore size has been reported to decrease linearly with reduced fiber diameter. As far as aspect ratio is concerned, it is far better than the microfibers (hundreds of times better). This particular property is useful when fibers are employed as a reinforcement as mechanical performance is directly proportional to the aspect ratio. It is found to be better as compared to whiskers, nanoparticles, and microfibers. The specific surface area of the nanofibers is also significantly higher as compared to the counterparts due to the smaller diameters. The typical range for the size of the spun nanofibers is 10–1000 nm, suggesting a specific surface area of 3–300 m^2/g. The increased surface area will ultimately offer higher contact areas, leading to uniform distribution of the applied load. The optical transparency is more evident when nanofibers are used. When an incident light falls, it may refract, transmit, or reflect depending upon the situation. The reflection leads to the loss of light, which can be countered using nanofibers as they offer smaller interfacial area. Lastly, the mechanical properties demonstrated by the nanofibers are exceptional, which are realized through higher levels of molecular orientation and crystallinity. For instance, the Young's modulus can be achieved as high as 3200 MPa, accompanied by tensile strength of 220 MPa, when nanofibers are used in place of microfibers (Jiang et al., 2018).

There are certain factors that affect the process of electrospinning. These factors include environmental conditions (such as temperature, air flow in the solution, and chamber humidity), attributes of the solution (such as surface tension, viscosity, elasticity, and conductivity), and controlling variables (capillary tube hydrostatic pressure, spinneret voltage, and distance from tip to ground) (Doshi & Reneker, 1995).

Although electrospinning offers a plethora of advantages, there are some avenues that still need attention. For example, there is a need for high voltage electric supply that may be a cause of an accident. Additionally, when multiple needles are used, the electric field may influence the overall process. Also, the typical yield for this process is small compared to the other processes. In case of application to high-salt mixtures, the process is infeasible due to the requirement of dielectric constant. Against this backdrop, other techniques have been introduced to counter the aforesaid issues (Ramakrishna et al., 2005).

Centrifugal spinning is the analogy of the production process of cotton candy. The process consists of spinning the nanofibers from the solution of the polymer. It has been observed that the length and diameter of the nozzle dominantly affect the morphology and size of the nanofiber, whereas the distance between the nozzle and collector has no clear effect on the diameter of the fiber (Gholipour-Kanani & Daneshi, 2022). This type of spinning offers special incentives as compared to electrospinning such as being environmentally benign, flexibility, higher yields, lower voltage requirements, facile synthesis, and provision to accommodate larger quantities of the material. There are also some drawbacks associated with this technique such as the production of beaded fibers because of vibration of the device and perturbation of the air. Therefore, the problems of electrospinning and centrifugal spinning can be overcome through the introduction of electro-centrifugal spinning (Xu et al., 2023).

11.4 Functional nanofibers

Usually, nanofibers do not exhibit any function unless they are modified by providing surface structures, adding nanoparticles, or incorporating drugs. The enhanced functionality is realized through the larger surface area of the nanofibers. In case of drugs, nanofibers constitute a composite for the transportation of the drug at a specified location into the human body followed by controlled release of the therapeutic effect. Such systems can be further divided into single, dual, or multidrug delivery systems. The nanoparticle/nanofiber composites have proven to enhance the functionality of the nanoparticles. The reason behind such functionality is the availability of a larger surface and contact area for nanoparticles. Another advantage that is being offered by such a combination is that the toxic effects of the nanoparticles can be reduced as sole nanoparticles will adversely affect the environment and human beings. Such composites can be synthesized by adding nanoparticles into the solutions of different polymers, by creating core-shell structures, through adhesion, and through reduction reactions. Interestingly, the application areas are in the fields of lithium-ion batteries, solar cells, supercapacitors, and fuel cells (Mittal & Mittal, 2016). The structure also plays a pivotal role in the functionality of the materials. For instance, the lotus effect can lead to the superhydrophobic nature of the materials, which in turn enhances the self-cleaning abilities of the surface (Lou et al., 2020).

11.5 Energy applications

11.5.1 Fuel cells

Proton exchange membrane fuel cells are green energy conversion technology for transportation and other applications. In order to realize the widespread applications of fuel cells, the main obstacles are the poor durability of Pt/C catalyst and high cost associated with it. PEMFCs are prone to electrochemical degradation of the Pt/C catalysts due to carbon corrosion under specific conditions such as startups and shutdowns, thus leading to compromised durability. In this backdrop, the research has been focused on introducing more robust support materials for PEMFCs. One such effort is to incorporate graphitic carbon components. It has been reported that carbon nanofiber-supported catalysts have performed satisfactorily when employed in PEMFCs (Anwar, Yan, Asghar, Husnain, Shen, Luo, & Zhang, 2019; Peera et al., 2021). For example, Bae et al. utilized Pd-polyaniline hybrid for the synthesis of nitrogen-doped, carbon-caged Pd catalysts supported on carbon nanofibers. It was observed that the durability and catalytic activity strongly depended upon the synthesis route and heat treatment temperature. The optimized temperature for the heat treatment was found to be 500°C and the catalyst thus produced demonstrated higher ORR activity and stability as compared to the Pd counterpart, as demonstrated in Fig. 11.2 (Bae et al., 2019).

Nanofibers can also be utilized as a substitute for ionomer-based membranes. For example, Sharma et al. synthesized nitro-oxidized carboxycellulose nanofibers that were derived from jute and employed them as a sustainable replacement for conventionally used membranes. The membranes thus produced possessed two functionalities, namely carboxylate and carboxylic acid, and were tested in PEMFC environment. The membrane with carboxylic acid performed better in terms of power density (19.1 mW/cm^2) and proton conductivity (14.2 mS/cm) at 80°C as compared to the other membrane followed by outstanding durability tested for a day (Sharma et al., 2022).

In addition to the PEMFCs, nanofibers have found application in microbial fuel cells. For example, Li and coworkers reported a composite consisting of layered double hydroxide and CoNi. The hybrid showed excellent ORR activity when employed in microbial fuel cells. The maximum power density was found to be 1.5 times as compared to that of Pt/C. The durability was also found to be promising when tested in single-chamber MFCs (Li et al., 2022).

11.5.2 Lithium-ion batteries

Nanofibers have found applications in lithium-ion batteries as they exhibit high mechanical strength, excellent electrochemical activity, and adequate specific surface area. It has been reported that the performance of nanofibers produced through electrospinning can be improved using a number of techniques: (1) by synthesizing network of carbon fibers, (2) designing carbon nanostructures, and (3) enhancing the percentage of active materials, including Sn and Si. CNF-based electrode demonstrates high surface area realized through small diameter of the nanofiber, adequate electrical conductivity, high carbon content, thin-web-based morphology, structural integrity, excellent reversible capacity, and better voltage profile (Chinnappan et al., 2017).

Recent research has focused on the performance of the separator to make Li-ion batteries a viable option in energy storage systems and electric vehicles. In this regard, nanofibers are being investigated as separators. Interestingly, they have been found to assist in transportation of lithium ions through interlinked porous structures and excellent surface-to-volume ratios. The nanofiber-based separators have been categorized as modified separators, gel polymer electrolytes, composite separators, monolayer separators, and multilayer separators (Li et al., 2019). For instance, Liu and colleagues prepared modified electrolytes by the introduction of aramid nanofibers as nanoadditives. The modified PEO electrolyte exhibited less agglomeration of ANF, suppression of crystallization of PEO, prolonged paths of ion transport, and facilitation of LiTFSI breakdown, as depicted in Fig. 11.2. It was interesting to note that high room temperature conductivity (8.8×10^{-5} S/cm) was achieved. The cell based on aforesaid electrolyte showed improved cycling stability and rate performance (Liu et al., 2020) as demonstrated in Figs. 11.3 and 11.4.

Nanofibers have also been utilized as electrode materials in metal-ion batteries. For example, Xia and coworkers reported a composite comprising tin, selenium, and

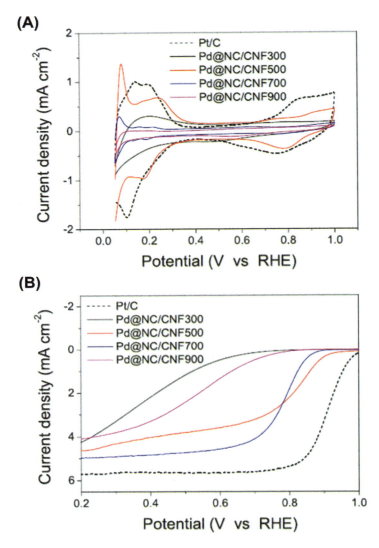

Figure 11.3 (A) Cyclic voltammograms and (B) ORR plots of the catalysts synthesized at different temperatures.
Source: From Bae, H. E., Park, Y. D., Kim, T. H., Lim, T., & Kwon, O. J. (2019). Carbon-caged palladium catalysts supported on carbon nanofibers for proton exchange membrane fuel cells. *Journal of Industrial and Engineering Chemistry*, 79, 431−436, http://www.sciencedirect.com/science/journal/1226086X. https://doi.org/10.1016/j.jiec.2019.07.018 (reused with permission from Elsevier).

Energy applications of nanofibers and their composites 265

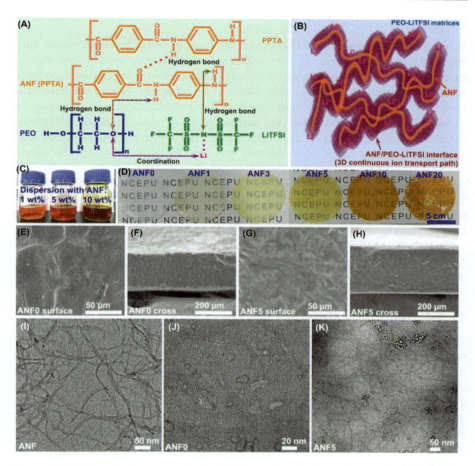

Figure 11.4 Schematic diagram depicting (A) interactions between the constituent materials, (B) paths of ion transport, (C) solution with different ANF %, (D) films produced, (E and G) surficial SEM micrographs, (F and H) cross-sectional view of CPEs, and (I, J, and K) TEM micrographs of CPEs with different ANF wt.%.
Source: From Liu, L., Lyu, J., Mo, J., Yan, H., Xu, L., Peng, P., Li, J., Jiang, B., Chu, L., & Li, M. (2020). Comprehensively-upgraded polymer electrolytes by multifunctional aramid nanofibers for stable all-solid-state Li-ion batteries. *Nano Energy, 69*, https://doi.org/10.1016/j.nanoen.2019.104398 (reused with the permission from Elsevier).

carbon as a binder-free anode for metal ion batteries. The nanoparticles of SnSe were found to be uniformly distributed over the network of carbon nanofibers. The presence of such a network not only worked as a conductive framework but also acted as a buffer to accommodate the volume expansion. Outstanding discharge capacity even after 500 cycles (405 mAh/g @ 1000 mA/g) was observed for Li-ion battery demonstrating the SnSe/C composite as flexible anode material (Xia et al., 2020).

11.5.3 Supercapacitors

In recent times, supercapacitors have been the topic of research due to unique attributes such as excellent cycling life, sufficient enough energy density, low maintenance cost, and high-power density. Basically, there are two types of supercapacitors, namely pseudocapacitors and electric double-layer capacitors. Nanofibers have been used as electrode materials in supercapacitors. For instance, Gedela et al. prepared a composite comprising polyaniline and reduced graphene oxide nanofibers employing chemical oxidative polymerization. The synthesized composite demonstrated adequate specific capacitance (approximately 655 F/g) and it was found to be almost constant at different current densities. This excellent performance was attributed to the interaction between the constituent nanofibers (Gedela et al., 2015). Likewise, Yan and coworkers prepared carbon nanofibrous membranes based on polyamide, and employed it as an electrode material in supercapacitor. The synthesis went through H_2O_2 activation (facile green method), which led to the porous architecture accompanied by the introduction of oxygen-rich species. The prepared electrode exhibited excellent specific mass capacitance (339.9 F/g) and durability (98.4% retention) even after 50,000 cycles (Yan, 2022).

11.5.4 Hydrogen storage

The environmental issues and the need for sustainable solutions have emphasized the need for hydrogen energy, as it is a zero-carbon emission paradigm. Additionally, it offers exceptional chemical energy density (142 MJ/kg) as compared to the other available options. Indeed, there is going to be a shift toward hydrogen energy replacing the traditional fossil fuel-based supremacy. Therefore, efforts have been placed for the production and storage of the hydrogen. One way to store the hydrogen is with the help of interstices/spaces in between the metal atoms, which is achieved through the porosity of the materials. The materials with small pores of 1 nm or below have been found to be more efficient for the storage of hydrogen. The common materials used for this purpose are fullerenes, activated carbon, and exfoliated graphite as they are easily available in large amounts and inexpensive. The quantity of hydrogen adsorbed depends on the pore volume and surface area. In addition to the aforesaid materials, carbon nanofibers, and carbon nanotubes have been found to be promising candidates for high hydrogen storage capacity. By varying different process parameters, nanofibers with required pore size, ordered structure, and high aspect ratio can be synthesized with attractive morphology useful for the storage of hydrogen (Thavasi et al., 2008).

11.5.5 Solar cells

Solar cells are the devices that are being used for the conversion of solar energy into electrical energy. Due to environmental concerns, the demand for solar cells has increased due to low maintenance, efficiency, portability, and durability. Nanofibers have demonstrated improved efficiencies along with miniaturization, leading to large-scale production. There are different types of solar cells that are being used these days, including dye-sensitized solar cells and perovskite solar cells. For instance, the modification of noble metals is of prime importance for the light absorption capacity of titania films. It has been reported that the composite of titania nanofibers/nanoparticles was synthesized with variable Ag content achieved through different reduction times. It was interesting to note that the prepared composite exhibited an increment of 18% in power conversion efficiency (Sun et al., 2020). In another study, Patil et al. reported a composite comprising titania and reduced graphene oxide nanofibers and employed it as an electron-transferring material in perovskite solar cells. Interestingly, the power conversion efficiency of the cell was found to be 17.66% accompanied by OCV of 1.070 V, FF of 0.754, and current density (short circuit) of 22.16 mA/cm^2. This efficiency was higher in comparison with mesoporous TiO_2 (Patil et al., 2019).

11.5.6 Field effect transistors

During the past decade, there has been an advancement in the field of metal oxide semiconductors with their application in field effect transistors. Features such as facile charge transport, robust carrier confinement, and excellent surface-to-volume ratios have paved the path for optoelectronics applications. For example, Qin et al. synthesized ternary-cation IAZO nanostructures through electrospinning with lesser surface defects, enhanced uniformity, and outstanding metal-oxide-metal lattice at relatively higher annealing temperatures. The prepared devices exhibited outstanding electron mobility (approximately 10 cm^2/V extendable up to 30 10 cm^2/V) and better on-off current ratio (10^7) (He et al., 2021). Likewise, Zhu and colleagues synthesized p-type copper oxide nanofibers via facile electrospinning and employed them in field effect transistors, along with alumina-based dielectric layer. Interestingly, FET demonstrated quick switching speed and higher mobility of hole, accompanied by modulation of light emission on external LED (Zhu et al., 2017).

11.6 Outlook and conclusions

In the last few decades, extensive efforts have been made in the development and engineering of nanofibers and their composites for different applications. These efforts focused on achieving better understanding of different manufacturing/fabrication techniques of the nanofibers, including top-down and bottom-up approaches. Interestingly, a wide range of applications exists for the nanofibers, especially in energy conversion and storage devices. There is a dire need to improve energy storage and conversion technologies, which ultimately depends on the fabrication methods and the nanostructures. In this regard, one-dimensional nanomaterials, i.e., nanofibers, have proved themselves as excellent candidates due to exceptional attributes such as high surface area, small diameter, small pore size, and greater aspect ratios. They also exhibit the provision for the incorporation of nanoparticles, which is useful for many applications and reduces the likelihood of pollution caused by pristine nanoparticles. Therefore, there is great scope for the composites of nanofibers in combination with nanoparticles. Electrospinning is the most commonly used fabrication technique that is facile, and limitless possibilities of nanostructures are there. However, there are some issues associated with electrospinning, namely safety hazards and scale-up production, so the rise of other fabrication techniques is inevitable. As far as energy applications are concerned, they are dominantly affected by factors such as fiber diameter and its distribution, fiber morphology, surface area, porosity, pore size and its distribution, surface characteristics such as hydrophilicity and hydrophobicity, and mechanical and chemical properties. Therefore, it is anticipated that nanofibers and their composites will be more lucrative and competitive owing to excellent performances and will have a substantive share in the global market.

References

I. Alghoraibi, S. Alomari, (2018). Different methods for nanofiber design and fabrication. Springer Science and Business Media LLC, 1−46, doi: 10.1007/978-3-319-42789-8_11-2.

Almetwally, A. A., El-Sakhawy, M., Elshakankery, M. H., & Kasem, M. H. (2017). Technology of nano-fibers: Production techniques and properties − Critical review. Textile Association (India). *Egypt Journal of the Textile Association*, 78(1), 5−14. Available from http://www.textileassociationindia.org/.

Anton, F. (1934). US Patent No. 1, 975, 504.

Anwar, M. T., Asghar, M. R., Ahmed, A., Fareed, S., Khan, H. I., & Rasheed, T. (2023). Metal organic frameworks-carbon based nanocomposites for environmental sensing and catalytic applications. *Advances in Chemical Pollution, Environmental Management and Protection*, 9. Available from https://doi.org/10.1016/bs.apmp.2022.12.002, http://www.sciencedirect.com/bookseries/advances-in-chemical-pollution-environmental-management-and-protection.

Anwar, M. T., Yan, X., Asghar, M. R., Husnain, N., Shen, S., Luo, L., Cheng, X., Wei, G., & Zhang, J. (2019). MoS$_2$-rGO hybrid architecture as durable support for cathode catalyst in proton exchange membrane fuel cells. *Chinese Journal of Catalysis*, 40(8), 1160−1167. Available from https://doi.org/10.1016/S1872-2067(19)63365-6, http://www.elsevier.com.

Anwar, M. T., Yan, X., Asghar, M. R., Husnain, N., Shen, S., Luo, L., & Zhang, J. (2019). Recent advances in hybrid support material for Pt-based electrocatalysts of proton exchange membrane fuel cells. *International Journal of Energy Research*, 43(7), 2694−2721. Available from https://doi.org/10.1002/er.4322, http://onlinelibrary.wiley.com/journal/10.1002/(ISSN)1099-114X.

Anwar, M. T., Yan, X., Shen, S., Husnain, N., Zhu, F., Luo, L., & Zhang, J. (2017). Enhanced durability of Pt electrocatalyst with tantalum doped titania as catalyst support. *International Journal of Hydrogen Energy*, 42(52), 30750−30759. Available from https://doi.org/10.1016/j.ijhydene.2017.10.152, http://www.journals.elsevier.com/international-journal-of-hydrogen-energy/.

Bae, H. E., Park, Y. D., Kim, T. H., Lim, T., & Kwon, O. J. (2019). Carbon-caged palladium catalysts supported on carbon nanofibers for proton exchange membrane fuel cells. *Journal of Industrial and Engineering Chemistry*, 79, 431−436. Available from https://doi.org/10.1016/j.jiec.2019.07.018, http://www.sciencedirect.com/science/journal/1226086X.

Barhoum, A., Pal, K., Rahier, H., Uludag, H., Kim, I. S., & Bechelany, M. (2019). Nanofibers as new-generation materials: From spinning and nano-spinning fabrication techniques to emerging applications. *Applied Materials Today*, *17*, 1−35. Available from https://doi.org/10.1016/j.apmt.2019.06.015.

Boys C.V. (1887). On the production, properties, and some suggested uses of the finest threads, *Proceedings of the Physical Society of London*.

Burda, C., Chen, X., Narayanan, R., & El-Sayed, M. A. (2005). Chemistry and properties of nanocrystals of different shapes. *Chemical Reviews*, *105*(4), 1025−1102. Available from https://doi.org/10.1021/cr030063a.

Chinnappan, A., Baskar, C., Baskar, S., Ratheesh, G., & Ramakrishna, S. (2017). An overview of electrospun nanofibers and their application in energy storage, sensors and wearable/flexible electronics. *Journal of Materials Chemistry C*, *5*(48), 12657−12673. Available from https://doi.org/10.1039/c7tc03058d.

Chiou, N. R., Lee, L. J., & Epstein, A. J. (2007). Self-assembled polyaniline nanofibers/nanotubes. *Chemistry of Materials*, *19*(15), 3589−3591. Available from https://doi.org/10.1021/cm070847v.

Doshi, J., & Reneker, D. H. (1995). Electrospinning process and applications of electrospun fibers. *Journal of Electrostatics*, *35*(2−3), 151−160. Available from https://doi.org/10.1016/0304-3886(95)00041-8.

Gedela, V., Puttapati, S. K., Nagavolu, C., & Venkata Satya Siva Srikanth, V. (2015). A unique solar radiation exfoliated reduced graphene oxide/polyaniline nanofibers composite electrode material for supercapacitors. *Materials Letters*, *152*, 177−180. Available from https://doi.org/10.1016/j.matlet.2015.03.113.

Gholipour-Kanani, A., & Daneshi. (2022). A review on centrifugal and electro-centrifugal spinning as new methods of nanofibers fabrication. *Journal of Textiles and Polymers*, *10*(1), 41−55.

Haider, A., Haider, S., & Kang, I. K. (2018). A comprehensive review summarizing the effect of electrospinning parameters and potential applications of nanofibers in biomedical and biotechnology. *Arabian Journal of Chemistry*, *11*(8), 1165−1188. Available from https://doi.org/10.1016/j.arabjc.2015.11.015, http://colleges.ksu.edu.sa/Arabic%20Colleges/CollegeOfScience/ChemicalDept/AJC/default.aspx, ScienceDirect, http://www.sciencedirect.com/science/journal/18785352.

He, J., Liu, X., Song, L., Li, H., Zu, H., Li, J., Zhang, H., Zhang, J., Qin, Y., & Wang, F. (2021). High annealing stability of InAlZnO nanofiber field-effect transistors with improved morphology by Al doping. *The Journal of Physical Chemistry Letters*, *12*(4), 1339−1345. Available from https://doi.org/10.1021/acs.jpclett.1c00030.

Huang, Y., Song, J., Yang, C., Long, Y., & Wu, H. (2019). Scalable manufacturing and applications of nanofibers. *Materials Today*, *28*, 98−113. Available from https://doi.org/10.1016/j.mattod.2019.04.018.

Jayaraman, K., Kotaki, M., Zhang, Y., Mo, X., & Ramakrishna, S. (2004). Recent advances in polymer nanofibers. *Journal of Nanoscience and Nanotechnology*, *4*(1−2), 52−65.

Jeevanandam, J., Barhoum, A., Chan, Y. S., Dufresne, A., & Danquah, M. K. (2018). Review on nanoparticles and nanostructured materials: History, sources, toxicity and regulations. *Beilstein Journal of Nanotechnology*, *9*(1), 1050−1074. Available from https://doi.org/10.3762/bjnano.9.98, https://www.beilstein-journals.org/bjnano/content/pdf/2190-4286-9-98.pdf.

Jiang, S., Chen, Y., Duan, G., Mei, C., Greiner, A., & Agarwal, S. (2018). Electrospun nanofiber reinforced composites: A review. *Polymer Chemistry*, *9*(20), 2685−2720. Available from https://doi.org/10.1039/c8py00378e.

Li, H., Sun, Y., Wang, J., Liu, Y., & Li, C. (2022). Nanoflower-branch LDHs and CoNi alloy derived from electrospun carbon nanofibers for efficient oxygen electrocatalysis in microbial fuel cells. *Applied Catalysis B: Environmental*, *307*. Available from https://doi.org/10.1016/j.apcatb.2022.121136.

Li, Y., Li, Q., & Tan, Z. (2019). A review of electrospun nanofiber-based separators for rechargeable lithium-ion batteries. *Journal of Power Sources*, *443*. Available from https://doi.org/10.1016/j.jpowsour.2019.227262.

Liu, L., Lyu, J., Mo, J., Yan, H., Xu, L., Peng, P., Li, J., Jiang, B., Chu, L., & Li, M. (2020). Comprehensively-upgraded polymer electrolytes by multifunctional aramid nanofibers for stable all-solid-state Li-ion batteries. *Nano Energy*, *69*. Available from https://doi.org/10.1016/j.nanoen.2019.104398.

Lou, L., Osemwegie, O., & Ramkumar, S. S. (2020). Functional nanofibers and their applications. *Industrial and Engineering Chemistry Research*, *59*(13), 5439−5455. Available from https://doi.org/10.1021/acs.iecr.9b07066, http://pubs.acs.org/journal/iecred.

Lu, W., He, T., Xu, B., He, X., Adidharma, H., Radosz, M., Gasem, K., & Fan, M. (2017). Progress in catalytic synthesis of advanced carbon nanofibers. *Journal of Materials Chemistry A*, *5*(27), 13863−13881. Available from https://doi.org/10.1039/C7TA02007D.

Ma, C., Wu, L., Dirican, M., Cheng, H., Li, J., Song, Y., Shi, J., & Zhang, X. (2021). ZnO-assisted synthesis of lignin-based ultra-fine microporous carbon nanofibers for supercapacitors. *Journal of Colloid and Interface Science*, *586*, 412−422. Available from https://doi.org/10.1016/j.jcis.2020.10.105.

V. Mittal, V. Mittal, (2016). Nanoparticle- and nanofiber-based polymer nanocomposites: An overview spherical and fibrous filler composites. Wiley, 1−38, doi: 10.1002/9783527670222.ch1.

Nobuta, K., Teramura, H., Ito, H., Hongo, C., Kawaguchi, H., Ogino, C., Kondo, A., & Nishino, T. (2016). Characterization of cellulose nanofiber sheets from different refining processes. *Cellulose*, *23*(1), 403−414. Available from https://doi.org/10.1007/s10570-015-0792-y.

Patil, J. V., Mali, S. S., Patil, A. P., Patil, P. S., & Hong, C. K. (2019). Highly efficient mixed-halide mixed-cation perovskite solar cells based on rGO-TiO$_2$ composite nanofibers. *Energy*, *189*. Available from https://doi.org/10.1016/j.energy.2019.116396.

Peera, S. G., Koutavarapu, R., Akula, S., Asokan, A., Moni, P., Selvaraj, M., Balamurugan, J., Kim, S. O., Liu, C., & Sahu, A. K. (2021). Carbon nanofibers as potential catalyst support for fuel cell cathodes: A review. *Energy and Fuels*, *35*(15), 11761−11799. Available from https://doi.org/10.1021/acs.energyfuels.1c01439, http://pubs.acs.org/journal/enfuem.

Persano, L., Camposeo, A., & Pisignano, D. (2015). Active polymer nanofibers for photonics, electronics, energy generation and micromechanics. *Progress in Polymer Science*, *43*, 48−95. Available from https://doi.org/10.1016/j.progpolymsci.2014.10.001, http://www.sciencedirect.com/science/journal/00796700.

Ramakrishna, S., Fujihara, K., Teo, W. E., Lim, T. C., & Ma, Z. (2005). *An introduction to electrospinning and nanofibers* (pp. 1−382). Singapore: World Scientific Publishing Co. Available from http://www.worldscientific.com/worldscibooks/10.1142/5894#t = toc, 10.1142/5894.

Rasheed, T., & Anwar, M. T. (2023). Metal organic frameworks as self-sacrificing modalities for potential environmental catalysis and energy applications: Challenges and perspectives. *Coordination Chemistry Reviews*, *480*. Available from https://doi.org/10.1016/j.ccr.2022.215011.

Raza, Z. A., Munim, S. A., & Ayub, A. (2021). Recent developments in polysaccharide-based electrospun nanofibers for environmental applications. *Carbohydrate Research*, *510*. Available from https://doi.org/10.1016/j.carres.2021.108443.

Richardson, J. J., Björnmalm, M., & Caruso, F. (2015). Technology-driven layer-by-layer assembly of nanofilms. *Science (New York, N.Y.), 348*(6233). Available from https://doi.org/10.1126/science.aaa2491, http://www.sciencemag.org/content/348/6233/aaa2491.full.pdf.

Rim, N. G., Shin, C. S., & Shin, H. (2013). Current approaches to electrospun nanofibers for tissue engineering. *Biomedical Materials, 8*(1). Available from https://doi.org/10.1088/1748-6041/8/1/014102.

Samyn, P., Barhoum, A., Öhlund, T., & Dufresne, A. (2018). Review: Nanoparticles and nanostructured materials in papermaking. *Journal of Materials Science, 53*(1), 146−184. Available from https://doi.org/10.1007/s10853-017-1525-4.

Sharma, S. K., Sharma, P. R., Wang, L., Pagel, M., Borges, W., Johnson, K. I., Raut, A., Gu, K., Bae, C., Rafailovich, M., & Hsiao, B. S. (2022). Nitro-oxidized carboxylated cellulose nanofiber based nanopapers and their PEM fuel cell performance. *Sustainable Energy & Fuels, 6*(15), 3669−3680. Available from https://doi.org/10.1039/d2se00442a.

Soltani, S., Khanian, N., Choong, T. S. Y., & Rashid, U. (2020). Recent progress in the design and synthesis of nanofibers with diverse synthetic methodologies: Characterization and potential applications. *New Journal of Chemistry, 44*(23), 9581−9606. Available from https://doi.org/10.1039/d0nj01071e.

Su, Z., Ding, J., & Wei, G. (2014). Electrospinning: A facile technique for fabricating polymeric nanofibers doped with carbon nanotubes and metallic nanoparticles for sensor applications. *RSC Advances, 4*(94), 52598−52610. Available from https://doi.org/10.1039/c4ra07848a, http://pubs.rsc.org/en/journals/journalissues.

Sun, J., Yang, X., Zhao, L., Dong, B., & Wang, S. (2020). Ag-decorated TiO_2 nanofibers for highly efficient dye sensitized solar cell. *Materials Letters, 260*. Available from https://doi.org/10.1016/j.matlet.2019.126882.

Thavasi, V., Singh, G., & Ramakrishna, S. (2008). Electrospun nanofibers in energy and environmental applications. *Energy and Environmental Science, 1*(2), 205−221. Available from https://doi.org/10.1039/b809074m.

Vasita, R., & Katti, D. S. (2006). Nanofibers and their applications in tissue engineering. *International Journal of Nanomedicine, 1*(1), 15−30. Available from https://doi.org/10.2147/nano.2006.1.1.15India, http://www.dovepress.com/getfile.php?fileID = 178.

Wang, C., Kaneti, Y. V., Bando, Y., Lin, J., Liu, C., Li, J., & Yamauchi, Y. (2018). Metal-organic framework-derived one-dimensional porous or hollow carbon-based nanofibers for energy storage and conversion. *Materials Horizons, 5*(3), 394−407. Available from https://doi.org/10.1039/c8mh00133b, http://pubs.rsc.org/en/journals/journal/mh.

Wang, J., Zheng, B., Liu, H., & Yu, L. (2021). A two-factor theoretical model of social media discontinuance: Role of regret, inertia, and their antecedents. *Information Technology and People, 34*(1), 1−24. Available from https://doi.org/10.1108/ITP-10-2018-0483, http://www.emeraldinsight.com/info/journals/itp/itp.jsp.

Wang, X., & Hsiao, B. S. (2016). Electrospun nanofiber membranes. *Current Opinion in Chemical Engineering, 12*, 62−81. Available from https://doi.org/10.1016/j.coche.2016.03.001, http://www.elsevier.com/wps/find/journaldescription.cws_home/725837/description#description.

Wang, X., Li, Z., Shi, J., & Yu, Y. (2014). One-dimensional titanium dioxide nanomaterials: Nanowires, nanorods, and nanobelts. *Chemical Reviews, 114*(19), 9346−9384. Available from https://doi.org/10.1021/cr400633s.

Wang, X., Zhang, Y., Zhang, X., Huo, Z., Li, X., Que, M., Peng, Z., Wang, H., & Pan, C. (2018). A highly stretchable transparent self-powered triboelectric tactile sensor with metallized nanofibers for wearable electronics. *Advanced Materials, 30*(12). Available from https://doi.org/10.1002/adma.201706738.

Wu, Q. (2020). Carbon-based nanocages: A new platform for advanced energy storage and conversion. *Advanced Materials*, *32*(27).

Xia, J., Yuan, Y., Yan, H., Liu, J., Zhang, Y., Liu, L., Zhang, S., Li, W., Yang, X., Shu, H., Wang, X., & Cao, G. (2020). Electrospun SnSe/C nanofibers as binder-free anode for lithium−ion and sodium-ion batteries. *Journal of Power Sources*, *449*. Available from https://doi.org/10.1016/j.jpowsour.2019.227559.

Xu, H., Yagi, S., Ashour, S., Du, L., Hoque, M. E., & Tan, L. (2023). A review on current nanofiber technologies: Electrospinning, centrifugal spinning, and electro-centrifugal spinning. *Macromolecular Materials and Engineering*, *308*(3). Available from https://doi.org/10.1002/mame.202200502.

Xue, J., Wu, T., Dai, Y., & Xia, Y. (2019). Electrospinning and electrospun nanofibers: Methods, materials, and applications. *Chemical Reviews*, *119*(8), 5298−5415. Available from https://doi.org/10.1021/acs.chemrev.8b00593.

Xue, J., Xie, J., Liu, W., & Xia, Y. (2017). Electrospun nanofibers: New concepts, materials, and applications. *Accounts of Chemical Research*, *50*(8), 1976−1987. Available from https://doi.org/10.1021/acs.accounts.7b00218.

Yan, B. (2022). Green H_2O_2 activation of electrospun polyimide-based carbon nanofibers towards high-performance free-standing electrodes for supercapacitors. *Diamond and Related Materials*.

Zhang, L., Tsuzuki, T., & Wang, X. (2015). Preparation of cellulose nanofiber from softwood pulp by ball milling. *Cellulose*, *22*(3), 1729−1741. Available from https://doi.org/10.1007/s10570-015-0582-6.

Zhang, P., Qiao, Z. A., & Dai, S. (2015). Recent advances in carbon nanospheres: Synthetic routes and applications. *Chemical Communications*, *51*(45), 9246−9256. Available from https://doi.org/10.1039/c5cc01759a, http://pubs.rsc.org/en/journals/journal/cc.

Zhao, J., Han, W., Chen, H., Tu, M., Zeng, R., Shi, Y., Cha, Z., & Zhou, C. (2011). Preparation, structure and crystallinity of chitosan nano-fibers by a solid−liquid phase separation technique. *Carbohydrate Polymers*, *83*(4), 1541−1546. Available from https://doi.org/10.1016/j.carbpol.2010.10.009.

Zhi, Y., Li, Z., Feng, X., Xia, H., Zhang, Y., Shi, Z., Mu, Y., & Liu, X. (2017). Covalent organic frameworks as metal-free heterogeneous photocatalysts for organic transformations. *Journal of Materials Chemistry A*, *5*(44), 22933−22938. Available from https://doi.org/10.1039/c7ta07691f.

Zhu, H., Liu, A., Liu, G., & Shan, F. (2017). Electrospun p-type CuO nanofibers for low-voltage field-effect transistors. *Applied Physics Letters*, *111*(14). Available from https://doi.org/10.1063/1.4998787.

Nanofibers and their composites for battery, fuel cell and solar cell applications

Yong Xue Gan
Department of Mechanical Engineering, California State Polytechnic University Pomona, Pomona, CA, United States

12.1 Introduction

This chapter deals with applications of nanofibers for energy conversion and storage devices or components including batteries, photovoltaic (PV) solar cells, photothermal solar evaporators, and fuel cells. The first part introduces three important nanofiber processing technologies: electrospinning, template-assisted chemical deposition, and spray pyrolysis. The second part deals with nanofibers for anodes, cathodes, and separators in batteries. The third part is on nanofiber solar cells. Nanofiber PV and photothermal solar cells are introduced. The photosensitive property of nanofibers is described and the concept of solar energy harvesting by composite nanofiber fuel cells is demonstrated. The fabrication and characterization of flexible biophotofuel cell anodes using oxide nanoparticle-loaded composite carbon nanofibers (CNFs) are shown.

12.2 Nanofiber processing technologies

Although many technologies can be used for nanofiber processing, the emphasis will be put on three important ones that are related to the focus of this chapter: the applications of nanofibers for energy storage and conversions. One key technology is the electrospinning (Zhang et al., 2011). Others include the template assisted chemical deposition (Qie et al., 2012) and spray pyrolysis (Dhamodharan et al., 2021).

12.2.1 Electrospinning

As well known, electrospinning is initiated mainly by an electrostatic force acting on an electrified fluid (a polymer solution or sol−gel precursor). During electrospinning, the applied voltage generates repulsions among molecules leading to the fast evaporation of the solvent. Fibers are then drawn out from the droplet at the end tip of a spinneret. For single-jet electrospinning, the spinneret could be a stainless-steel needle. The fibers deposit on an electrically grounded conductive collector. The conductive collector or target could be made of a metallic rotating disk, a roller, a piece of metallic foil, or a metallic mesh. Electrospun fibers are much finer than those obtained from traditional methods such as solvent coagulation, melt dipping, drawing, and extrusion. This is because the electrified solution droplet with the significantly reduced surface tension can be pulled out quickly and thrown into very thin fibers. The direct current (DC) voltage could be as high as several ten thousand volts. The intensity of the electric field is typically maintained at about 1 kV/cm. This is just a general guideline for the electrospinning voltage parameter setting. However, very low voltage or low electric field intensity may also be adopted. During the so-called electrohydrodynamic casting or extrusion, the electric field intensity could be as low as several volts per centimeter. It should be mentioned that the diameter of the spun fiber is highly sensitive to the level of the applied voltage or the intensity of the electric field. The higher the applied voltage or stronger the electric field, the finer the fiber obtained.

The electrospinning jet may be aligned either horizontally or vertically. For those electrospinning facilities with vertically aligned jets, the spun nanofibers are drawn toward collectors by both electric force and gravitational force. During traditional electrospinning, the spun nanofibers are randomly aligned on collectors and form meshed networks. Recent development allows the nanofiber collectors to be metallic screens under controlled motions. For example, disks, rollers, or mandrels as collectors can be kept in rotation. During the stable electrospinning, the solvent used for making the solution evaporates rapidly. This allows the formation of charged solid fibers, and the fibers accumulate on the collector with varying porosity and packing density.

In some cases, the concentration of the solution for electrospinning is relatively low. Then, beads formation could be observed. A significantly low concentration or very low viscosity of the fluid could trigger the so-called electrospraying during which the fluid drops are separately drawn out and randomly deposited onto the collecting targets. It must be pointed out that there are other derived electrospinning techniques as compared to the traditional single-jet electrospinning. Coaxial spinning, triaxial spinning, needleless spinning, air bulb-promoted spinning, sol−gel spinning, and coelectrospinning are some of the examples. The electrospinning can be integrated with three dimensional (3D) printing too. In sol−gel spinning, chemical reactions could happen. While in the cospinning process, nanoparticles can be added into polymer solutions to produce composite nanofibers.

There are many factors influencing the electrospinning process besides the above-mentioned DC voltage level and viscosity of the polymer solution. For example, the electrical conductivity of the polymer solution is one of the significant factors. Still, some other factors include the temperature, relative humidity (RH) of the environment, fluid flow rate, and the distance between the tip of the injection needle and the collecting target. Solution conductivity affects the Taylor cone formation. It also controls the diameter of nanofibers. For a solution with very low conductivity, the surface of the droplet will not have enough charge to form a Taylor cone. Consequently, no electrospinning will take place. Increasing the conductivity of the solution over a critical value is necessary to trigger the electrospinning. A high solution conductivity will not only promote the charge accumulation on the surface of the droplet to form a Taylor cone but also cause a decrease in the nanofiber diameter. An ideal dielectric polymer solution or insulating polymer solution will not produce sufficient charges in the solution to move toward the surface of the fluid. As a result, the electrostatic force generated by the applied electric field will not be strong enough to form a Taylor cone. Taylor cone formation is the prerequisite for initiating the electrospinning process. On the contrary, a conductive polymer solution will have sufficient free charges to move onto the surface of the fluid and form a Taylor cone easily to initiate the electrospinning.

The conductivity of a polymer solution for electrospinning could be increased by adding an appropriate inorganic salt into the solution. A tiny amount of addition of LiCl, KCl, or NaCl works well for this purpose. The addition of such a salt affects the electrospinning process in two ways. One is the increase in the number of ions in the polymer solution, which helps increase the surface charge density of the fluid. The electrostatic force generated along the direction of the applied DC electric field increases. The other is to reduce the tendency of the transverse motion of nanofibers because the higher the conductivity of the polymer solution, the lower the tangential electric field along the surface of the fluid. Therefore, the nanofibers can be better stretched by the normal force and move directly toward the collecting target.

The distance between the tip of the spinneret and the collector is also a critical parameter that determines the morphology of the obtained nanofiber product. It controls the time of fiber flying and stretching, the extent of solvent evaporation, and the fiber whipping or instability interval. The nanofiber morphology could be readily tuned by varying the needle tip to the target distance. This distance should be long enough to generate the demanded morphology or structure and keep the quality of the spun fiber product. The least value for the distance between the metallic needle tip and collector varies with the nature of the solution used for electrospinning.

Electrospinning as an old technology has found numerous new applications in product development. It has been noticed that constructing tissue scaffolds, making nanofiber masks, preparing wound dressings, and building wearable technology sensors are some of the useful application examples. It is also used for making composite nanofibers for rechargeable batteries. In the review presented by Shi et al. (2021), advances in electrospun nanofiber materials for lithium batteries were presented. In addition to rechargeable battery applications, nanofibers have been used for solar cells, hydrothermal solar evaporators, and photochemical fuel cells, as will be described in some later subsections of this chapter.

12.2.2 Template-assisted chemical deposition

The demand for some special architectures or structures built upon nanofibers stimulated the research on a scalable template-assisted chemical deposition method. This method has been used for preparing battery components (see Qie et al. (2012) and Wang et al. (2013)) and for making photothermal solar cells as well (see Shi et al. (2018)). In the following part, the procedures for the template-assisted chemical deposition are briefly described. Two types of templates are introduced here. One is a polymer nanofiber template. The other is a quartz fiber template. Both templates can be used to generate carbon micro- and nanofibers with special structures. The polymer template can generate porous CNFs for the battery application (Qie et al., 2012), while the quartz template can produce hollow CNFs for the photothermal solar cell application (Shi et al., 2018).

12.2.2.1 Direct carbonization from polymer nanofiber template

To process porous CNFs for application in lithium-sulfur (Li−S) batteries, Qie et al. (2012) used polypyrrole (PPy) as the sacrificial template. PPy can be synthesized via chemical oxidation (Qie et al., 2012). The produced PPy nanofiber webs with a high nitrogen (N) content of 16 wt.% were collected. Then, carbonization and activation were performed to yield a porous carbon nanofiber web (CNFW) structure. The CNFWs were doped with nitrogen to achieve the high N content. The unique porous structure and high concentration of N in the CNFWs allowed the Li−S batteries to have the superhigh capacity and excellent rate capability.

Wang et al. (2013) showed the PPy-template method to prepare nitrogen-doped CNFWs in more detail (Wang et al., 2013). The PPy nanofibers were prepared by the modified oxidative template assembly approach (Liu et al., 2010). In a typical experiment, 7.3 g cetrimonium bromide (with the formula of $(C_{16}H_{33})N(CH_3)_3Br$) was dissolved in 120 mL 1.0 M HCl under the ice bath cooling condition. Then, 13.7 g ammonium persulfate was added to the solution. A white reactive template was generated immediately. After being stirred for 0.5 h, 8.3 mL pyrrole monomer was added into the as-formed reactive template solution. After the reaction at 0−5°C for 24 h, the PPy nanofiber webs in the form of black precipitates were obtained. The precipitates were washed with 1.0 M HCl solution and deionized water several times until the filtrate became colorless and neutral. The product was dried overnight at 80°C in an oven. To convert the PPy nanofiber webs into CNF networks, the as-prepared PPy was heated to 600°C at a heating rate of 5°C/min and held for 0.5−4 h to form nitrogen-doped CNFWs. The carbon yields for multiple specimens were close to about 52%. The corresponding CNF products were named N-CNFWs-0.5 h, N-CNFWs-1 h, N-CNFWs-2 h, and N-CNFWs-4 h, depending on their different carbonization periods.

The scanning electron microscopic (SEM) and transmission electron microscopic (TEM) images of the prepared hollow CNFs are shown by Wang et al. (2013). In Fig. 12.1A, the SEM image of the PPy nanofiber template is presented. The cross-linked PPy nanofibers with a diameter of around 80 nm formed a porous membrane. The carbonized nanofibers are seen in Fig. 12.1B through E. Fig. 12.1B represents the morphology of the nanofiber web after high-temperature treatment for 0.5 h. The specimen was named N-CNFWs-0.5 h. Fig. 12.1C, the image of the specimen being treated for 1 h (N-CNFWs-1 h) is revealed. In Fig. 12.1D, the image of the specimen treated for 2 h (N-CNFWs-2 h) is given. Fig. 12.1E is the image for the specimen being treated for 4 h with the designation of N-CNFWs-4 h. These SEM images showed that the carbonization of PPy at different times did not change the initial morphology of PPy. Therefore, the templating procedure was successful for generating the N-doped CNFs. Fig. 12.1F is the TEM image for the specimen under 2 h heat

Nanofibers and their composites for battery, fuel cell and solar cell applications

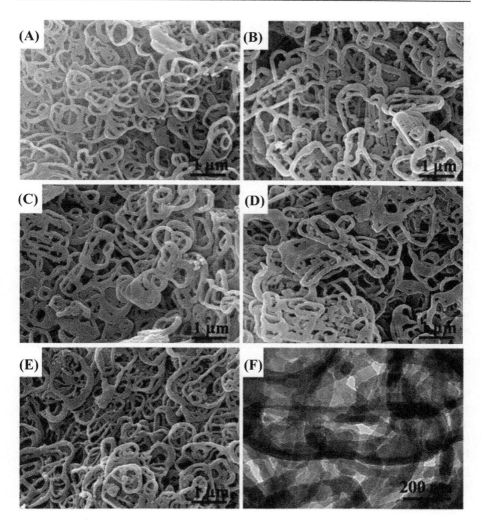

Figure 12.1 Scanning electron microscopic (SEM) images of (A) PPy nanofiber webs, (B) N-CNFWs-0.5 h, (C) N-CNFWs-1 h, (D) N-CNFWs-2 h, (E) N-CNFWs-4 h, and (F) transmission electron microscopic (TEM) image of N-CNFWs-2 h. PPy is a semiconducting polymer.
Source: Reproduced with permission from Wang, Z., Xiong, X., Qie, L., & Huang, Y. (2013). High-performance lithium storage in nitrogen-enriched carbon nanofiber webs derived from polypyrrole. *Electrochimica Acta*, *106*, 320–326. https://doi.org/10.1016/j.electacta.2013.05.088. © 2013 Elsevier Ltd.

treatment (N-CNFWs-2 h). Large quantities of pores were observed within the CNF networks. These pores came from the pyrolysis of the PPy precursor during the high-temperature heat treatment. As the anode material for lithium-ion batteries, the N-CNFWs show high capacity and good rate capability. The reversible capacity is up to

668 mAh/g at a current density of 0.1 A/g and 238 mAh/g at 5 A/g. The porous nanofiber network with high nitrogen content is the reason for the high performance (Wang et al., 2013).

12.2.2.2 Hollow carbon nanofiber from quartz fiber template

Shi et al. (2018) used quartz glass fiber wool as the template for making silica, carbon, and silica multilayered composite fibers. The silica carbon composite fiber has a coaxial three-layered structure. A quartz glass fibrous wool, as shown in Fig. 12.2A, served as the template or matrix material. Phenol formaldehyde (PF) was used as a carbon precursor. The carbonization of the phenolic resin allowed the formation of the carbonaceous middle layer. The outer silica layer was obtained by hydrolysis and condensation of tetraethyl orthosilicate (TEOS). The pristine quartz glass fiber wool with a nonwoven structure was binder-free. It was made of quartz glass fibers with diameters ranging from 200 nm to more than 2 μm. A very thin layer of phenolic resin and an additional silica layer were sequentially coated onto the quartz glass fibers by the dip coating method. The intermediate product, silica/phenolic resin/silica coaxial fiber, was then converted to the silica/carbon/silica coaxial fiber through high-temperature treatment in protective gas, as shown by the photo images in the second raw of Fig. 12.2A. It is possible to make 3D structures by using the obtained multilayered composite fiber mats. The photo images in Fig. 12.2B illustrate the procedures for making the 3D fiber architectures. This product is a photothermal solar cell for brine desalination (Shi et al., 2018).

According to Shi et al. (2018), the carbon layer is amorphous. The carbon content in the tri-layered composite fiber was only about 1.2 wt.%. A smooth tube-like structure of the carbon layer was observed. The SEM and TEM images of the quartz fiber template, the fiber with carbonized layer, and the hollow carbon fiber (CF) after all the silica was etched away by hydrofluoric acid (HF) are shown in Fig. 12.3. The images in Fig. 12.3E−H reveal that the phenolic coating is uniformly placed onto the glass fiber surface. The phenolic resin coating on the glass fiber was converted to carbon in the subsequent calcination procedure.

12.2.3 Spray pyrolysis

Dhamodharan et al. (2021) showed the spray pyrolysis approach for making solar cell films consisting of indium (In) doped ZnO nanofibers. In a typical experiment, 0.1 M of $ZnC_{10}H_{14}O_5$ was dissolved in ethanol. Indium nitrate as the In source with varying concentrations from 0 to 2.0 wt.% was added into the solution. The temperature of the indium tin oxide (ITO) substrate was kept at 350°C. The solution was atomized into a spray of fine droplets using a spray nozzle. Compressed air was used as the carrier gas. The spray rate was set as 3 mL/min, and the spray time was 0.5 h. The distance from the tip of the vertically aligned nozzle to the ITO substrate was 23 cm. The grown nanofiber films had good adhesion to the substrate. To make the uniform coating, the

Nanofibers and their composites for battery, fuel cell and solar cell applications 279

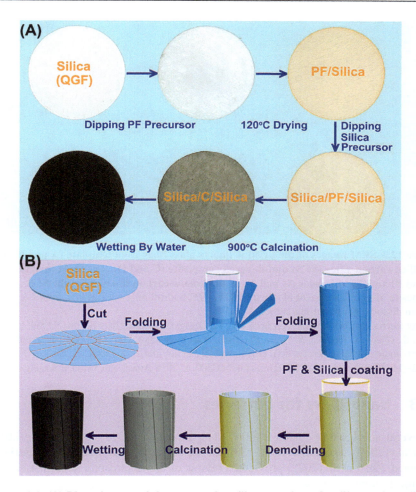

Figure 12.2 (A) Photo images of the quartz glass fiber template, the silica-carbon-silica (SCS) tri-layer fiber membrane, and the associated products. The gray SCS membrane turns to deep dark when wetted by water. (B) Schematic drawings of the fabrication process for the 3D SCS structure. By using a glass cup as the mold, a photothermal solar evaporator cell was made for the desalination of brine water.
Source: Reproduced with permission from Shi, Y., Zhang, C., Li, R., Zhuo, S., Jin, Y., Shi, L., Hong, S., Chang, J., Ong, C., & Wang, P. (2018). Solar evaporator with controlled salt precipitation for zero liquid discharge desalination. *Environmental Science and Technology*, 52(20), 11822−11830. https://doi.org/10.1021/acs.est.8b03300. © 2018 American Chemical Society.

spray nozzle was controlled in X−Y planar motion by step motors. The spray nozzle scanned a 200 × 200 mm^2 area with X and Y motions at 20 mm/s and 5 mm/s simultaneously. After spray pyrolysis deposition, the nanofiber film was used as a photoanode in a dye-sensitized solar cell (DSSC) (Dhamodharan et al., 2021).

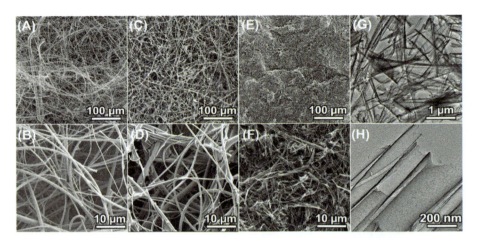

Figure 12.3 SEM images of the pristine quartz glass fiber wool template (A and B) and the silica-carbon-silica (SCS) multilayered composite (C and D). The SEM images (E and F) and TEM images (G and H) of the hollow carbon fibers from the SCS composite after the removal of silica by HF. *HF*, hydrofluoric acid. It is a reductive acid. HF can dissolve glass, but not carbon.
Source: Reproduced with permission from Shi, Y., Zhang, C., Li, R., Zhuo, S., Jin, Y., Shi, L., Hong, S., Chang, J., Ong, C., & Wang, P. (2018). Solar evaporator with controlled salt precipitation for zero liquid discharge desalination. *Environmental Science and Technology*, 52(20), 11822−11830. https://doi.org/10.1021/acs.est.8b03300. © 2018 American Chemical Society.

12.3 Nanofibers for batteries

In several articles (see Zhang et al. (2011), Liu et al. (2019), Gong et al. (2014), Li et al. (2019), and Zhao et al. (2022)), the status of nanofibers for batteries is reviewed. Zhao et al. (2022) provided a review of the electrospun nanofiber electrodes for lithium-ion batteries (LIBs). Both single-jet electrospinning and coaxial electrospinning are discussed (Zhao et al., 2022). The electrochemical performance of the nanofiber electrodes is dealt with. Zhang et al. (2011), introduced various electrospun nanofibers for anodes, cathodes, and separators in advanced LIBs. The electrospinning technology for preparing novel composite nanofibers was introduced. The emphasis was put on the processing and characterization of silicon/carbon (Si/C) nanofiber anodes, lithium iron phosphate/carbon nanofiber (LiFePO$_4$/C) cathodes, and lithium lanthanum titanate oxide/polyacrylonitrile (LLTO/PAN) nanofiber separators. The Si/C nanofiber anodes took the advantages of both carbon and silicon. Carbon has a long cycle life and Si possesses high lithium-storage capacity. The LiFePO$_4$/C nanofiber cathodes demonstrated required electrochemical performance including large capacity and high cycling stability. The LLTO/PAN nanofiber separators showed large electrolyte uptake, high ionic conductivity, and low interfacial resistance with lithium, which increased the capacity and improved the cycling stability of LIBs. The results revealed that electrospinning is a feasible approach to preparing high-performance composite nanofiber-based anodes, cathodes, and separators. Composite nanofibers could potentially replace some of the existing lithium-ion battery materials.

In the review performed by Liu et al. (2019), the electrospun nanofibers for Li−S batteries are highlighted. The review also covers processing, structure analysis, morphology control, and electrochemical performance characterization of the electrospun nanofibers. Designing new electrospun nanofibers for Li−S batteries is discussed. Proposed directions are given to stimulate thinking on the rational design of electrospun nanofibers for making better Li−S batteries.

Gong et al. (2014) reviewed electrospinning nanofibers for lithium battery applications with an emphasis on battery separators. From the requirements of a power source, lithium-ion battery stands out due to the high-capacity density, high-rate capacity, and good thermal stability, as well as excellent cyclic stability. The electrospinning technique has played an important role in preparing nanofibers that are used in LIBs. For example, nanofiber separators have a high surface area and porosity. Such properties are very important for the electrochemical performance of power LIBs. Electrospinning is especially capable of preparing both intrafiber and intrinsic porous nanofibers. The process allows the formation of both polymer blending membranes and inorganic-polymer composite membranes. As separators, these nanofiber membranes possess good mechanical properties and high thermal stabilities. In addition to separators, the electrospun nanofiber cathodes and anodes with enhanced electrochemical performance were also briefly introduced. The existing problems and the corresponding improvements were discussed as well.

Recently, extensive efforts have been made to improve battery performances for future energy applications such as electric vehicles and energy storage systems. Separators, as a crucial component in lithium-ion batteries, also gained rapid developments to achieve advanced properties. Li et al. (2019) reviewed electrospinning nanofibers for the separator application in LIBs (Li et al., 2019). Electrospun nanofiber-based membrane is a promising candidate for separators in LIB to enhance lithium ions transportation efficiency due to its ideal features such as interconnected porous structures, high porosity, and large surface-to-volume ratio. Different types of electrospun separators, including monolayer separators, multilayer separators, modified separators, and composite separators, were presented. The development of polymer gel electrolytes was discussed. In several of the subsections shown below, the nanofiber anodes, cathodes, and separators will be presented in more detail. A separate part on Li−S batteries will be given. Then, the recent development of flexible batteries will be mentioned. Finally, other miscellaneous topics such as oxide nanofibers in batteries, special nanofiber structure design, electrolytes, and extracting nanofibers from various sustainable sources for battery applications will be discussed briefly.

12.3.1 Nanofiber anodes

LIBs have generated a significant impact on our daily lives via powering smartphones, wearable devices, and electric vehicles. The demands for further improving battery performance have increased with the continuous development of new electronic devices.

Currently used anode materials suffer from serious volume changes and nanoparticle aggregations during lithium intercalation and extraction, which causes rapid pulverization and capacity loss. Nanofibers could hold anode materials during repeated cycling. The cycling stability can be maintained. Finding new nanofiber composite anodes becomes one of the ways to resolve this problem and mediate other issues present in existing graphite anodes, such as a low theoretical capacity and poor rate capabilities. As indicated in Lee's review (see Lee (2020)), significant improvements in existing batteries have been achieved through the electrospinning process. This technology can produce various nanofibers using spinnable materials. Recent work on nanofiber anode materials reveals positivity in implementing the technology for potential lithium-ion battery production. The nanofiber anode materials are highly possible to be used in the next-generation batteries. Lee (2020) summarized the previous research on various nanofiber anode materials (Lee, 2020). The electrochemical reactions on the nanofiber anodes were presented. How to improve conventional lithium-ion battery performances through novel anode design was dealt with. Related manufacturing routes for producing the next-generation batteries were proposed.

Li et al. (2013) illustrated a composite anode consisting of CNFs and silicon particles. A succinic anhydride (SA) was used as an electrolyte additive. The Si/C composite nanofibers were made by electrospinning and carbonization. Polyacrylonitrile (PAN) was the polymer for spinning and generating carbon. The effect of the electrolyte additive, SA, on the electrochemical performance of the Si/C composite nanofiber anode was examined. After 50 charging-discharging cycles, the discharge capacity of the Si/C composite nanofiber anode with the SA-added electrolyte was 34% higher than that of the anode without the SA additive in the electrolyte. At the 150th cycle, the Si/C composite nanofiber anode with SA-added electrolyte kept 82% discharge capacity. Therefore, adding additive SA in the electrolyte could be an economic and effective way to improve the cyclability of the high-capacity Si/C composite nanofiber anodes for next-generation high-energy LIBs.

Instead of incorporating additives into the electrolyte, doping the anode made from CNFs became another way to improve the performance of lithium batteries. Qie et al. (2012) doped CFs with nitrogen during the high-temperature heat treatment carbonization process. The nitrogen-doped porous CNFWs as anodes for LIBs with a superhigh capacity and rate capability were confirmed by Qie et al. (2012). The porous CNFWs were made by using an organic fiber template. PPy nanofiber was the template. Carbonization at 650°C in nitrogen gas was performed to generate porous CF and doping the carbon fiber with nitrogen.

Wang et al. (2010) also made lithium-ion battery anode using electrospun carbon−silicon composite nanofibers. The varying in the C/Si ratio can tune the morphology of the composite nanofibers so that the silicon particles can be dispersed uniformly within the carbon matrix. The fiber with a C/Si mass ratio of 77:23, -exhibited a high reversible capacity of 1240 mAh/g. This fiber also showed excellent capacity retention. During a typical discharge/charge cycle, the morphology of the fiber was unchanged, indicating no significant volume expansion during the insertion of lithium into silicon. The nanofiber prevents the failure of the electrode under mechanical stress very effectively. However, there are some limitations in using the nanofiber as well. The charge/discharge rate of the nanofiber electrode was found still low as revealed by the alternating current (AC) impedance spectroscopic analysis. Therefore, the challenge remains in making high-power LIBs using nanofibers. That is why a continued effort has been put into the new anode research.

Besides adding Si into CNFs, Ge and Sn were loaded into CNFs for anode fabrication. Li et al. (2014) conducted comparative studies on Si/C, Ge/C, and Sn/C composite nanofiber anodes for lithium-ion battery applications. It was found that with the increase in the amount of Si, Ge, and Sn, the cycling stability became worse due to the formation of large aggregates in the CNFs. Compared with Si/CNFs, Ge/carbon and Sn/carbon exhibited better cycling performance due to the more uniform particle distribution and smaller volume changes. The failure mechanism of the Si/carbon structure was discussed in this article as well. During the processing, PAN with a molecular weight of 150,000, M (M = Si, Ge, Sn) nanoparticles with sizes ranging from 70 to 120 nm, and N, N-dimethylformamide (DMF) solvent were mixed to form the electrospinning solution for making M/PAN composite nanofibers. Following the electrospinning, carbonization of the nanofibers was carried out. The electrospun M/PAN composite nanofibers were transformed into M/carbon composite nanofibers. The obtained carbon composite nanofiber mats showed high porosity and large specific surface areas. Li et al. (2014) indicated that the composite CNF mats could prevent the significant volume change of the Si, Ge, and Sn nanoparticles. The procedures for composite nanofiber processing were given and the associated parameters were provided. The solutions consisting of 8 wt.% PAN in DMF were added with different amounts of Si, Ge, and Sn nanoparticles (10, 30, and 50 wt.% relative to PAN) at 60°C. After being stirred for at least 24 h, the nanoparticles dispersed into the polymer solutions uniformly. The electrospinning was performed at 15 kV. The spun Si/PAN, Ge/PAN, and Sn/PAN nanofibers were oxidized and stabilized in air at 280°C for 5.3 h with a heating rate of 5°C/min. The carbonization was carried out at 700°C for 1 h in argon atmosphere. The heating rate for carbonization was 2°C/min. After that, the Si/C, Ge/C, and Sn/C nanofibers were obtained. A Perkin Elmer Elemental Analyzer was used to determine the compositions of these carbonized nanofibers. The carbon contents for the Si/C, Ge/C, and Sn/C nanofibers were 73.8, 74.2, and 75.7 wt.%, respectively, when they were made from the 10 wt.% Si/PAN, Ge/PAN, and Sn/PAN precursors. The carbon contents of the Si/C, Ge/C, and Sn/C nanofibers were found to be 51.8, 52.9, and 54.2 wt.%, respectively, when they were prepared from the 30 wt.% Si/PAN, Ge/PAN, and Sn/PAN precursors. The carbon contents were only 30.3, 33.8, and 34.5 wt.%, respectively, when these fibers were produced from the 50 wt.% Si/PAN, Ge/PAN, and Sn/PAN precursors. The Si/C, Ge/C, and Sn/C nanofibers were observed by scanning electron microscopy (SEM) at an acceleration voltage of 5 kV. Representative SEM images are given in Fig. 12.4A, C, and E for the nanofibers prepared from 10 wt.% Si/PAN, Ge/PAN, and Sn/PAN precursors. The size distributions of typical composite nanofibers are shown in Fig. 12.4B, D, and F.

Carbon-based composite nanofibers with more than one element addition were investigated. For example, Zhang et al. (2016) synthesized the Si/Ni$_3$Si-loaded CNF

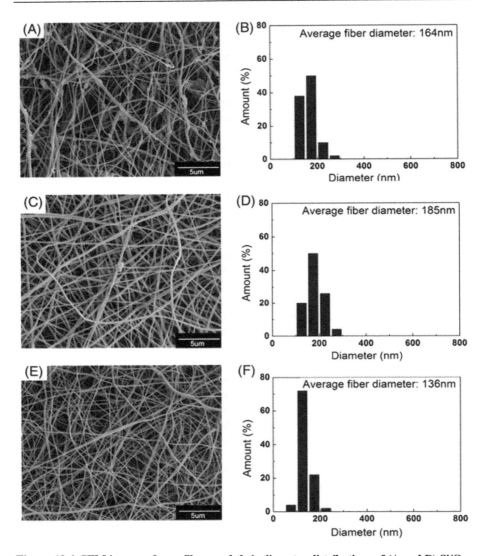

Figure 12.4 SEM images of nanofibers and their diameter distributions of (A and B) Si/C, (C and D) Ge/C, and (E and F) Sn/C. The nanofibers were prepared from 10 wt.% Si/PAN, Ge/PAN, and Sn/PAN precursors, respectively.
Source: Reproduced with permission from Li, S., Chen, C., Fu, K., Xue, L., Zhao, C., Zhang, S., Hu, Y., Zhou, L., & Zhang, X. (2014). Comparison of Si/C, Ge/C and Sn/C composite nanofiber anodes used in advanced lithium-ion batteries. *Solid State Ionics*, *254*, 17−26. https://doi.org/10.1016/j.ssi.2013.10.063. © 2013 Elsevier B.V.

composites using the one-pot electrospinning method. The starting materials for Si/Ni₃Si included Si nanoparticles and nickel acetate. The two materials (with mass ratios of 1:0, 1:1, 1:2, 1:3, 1:4, and 1:5) were dispersed in a 10 wt.% poly-vinylpyrrolidone (PVP) aqueous solution. After stirring for 7 h, the homogenous precursor solution for electrospinning was obtained. Electrospinning was conducted at a voltage of 11 kV. The solution feeding rate was 0.8 mL/h. The needle tip-to-aluminum nanofiber collector distance was 20 cm. The spun nanofibers collected on the aluminum foil were dried in a vacuum oven at 120°C for 12 h. The morphology can be seen from Fig. 12.5A. The obtained nanofibers were oxidized and stabilized at 230°C for 3 h at a heating rate of 5°C/min. The morphology is revealed in Fig. 12.5B. Carbonization was performed at 600°C for 3 h. The heating rate for carbonization was 2°C/min. During the carbonization, the Si/Ni₃Si nanoparticles were produced. The Si/Ni₃Si-encapsulated CNF

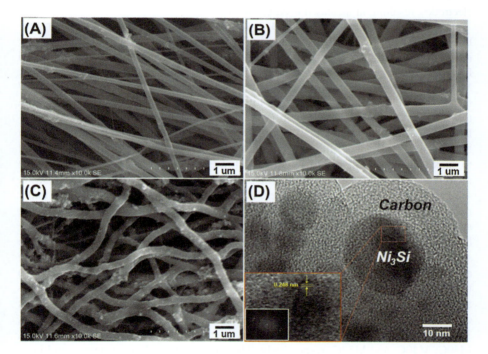

Figure 12.5 Scanning electron microscopic (SEM) images of the Si/Ni ₃Si-encapulated carbon nanofiber composites (A) as-spun, (B) after oxidation and stabilization at 230°C, (C) carbonized for 3h at 600°C, and (D) the high-resolution transmission electron microscopic (HRTEM) image of the composite. The inset of (D) is the selected area electron diffraction (SAED) pattern. The stabilization allowed the polymer to keep high-temperature stability during carbonization.
Source: Reproduced with permission from Zhang, P., Huang, L., Li, Y., Ren, X., Deng, L., & Yuan, Q. (2016). Si/Ni₃Si-encapulated carbon nanofiber composites as three-dimensional network structured anodes for lithium-ion batteries. *Electrochimica Acta, 192,* 385–391. https://doi.org/10.1016/j.electacta.2016.01.223. © 2016 Elsevier Ltd.

composites were examined by both SEM and TEM. In Fig. 12.5C, the SEM image shows the particle-loaded nanofibers. The distribution of the nanoparticles is uniform. The high-resolution transmission electron microscopic (HRTEM) image, Fig. 12.5D, reveals multiple nickel silicate nanoparticles embedded in the CNF. The prepared composite nanofibers served as three-dimensional network structured anode materials for LIBs. The composite nanofiber anodes increased the cycling performance of the batteries. A reversible capacity of 1132.4 mAh/g after 200 cycles was achieved. Such a value is 45% higher than that of the anode with Si as the only additive. The improved battery performance is believed from the alloying of Si with Ni. The Ni–Si alloy was found very effective in alleviating the volume expansion during lithium intercalation and extraction. Besides that, the incorporation of Si/Ni$_3$Si nanoparticles into the one-dimensional (1D) CNF with uniform distribution increased the electrical conductivity of the anode. The use of the nanofibers also enhanced the mechanical strength of the electrodes, resulting in increased cycling stability.

Many other polymers can be used as CNF-generating sources. In the paper published by Wang et al. (2015), polyimide (PI) was used for CNF production because it has a high carbon yield of 70%. It must be mentioned that the commonly used PAN for making CF has a much lower carbon yield in a range from 40% to 50%. MnO-carbon hybrid nanofiber composites were prepared by electrospinning PI/manganese acetylacetonate precursor followed by a carbonization procedure. In a typical experiment, 2.2 g pyromellitic dianhydride (PMDA) and 2 g 4,4-oxydianilline (ODA) were dissolved in 30.8 g of N,N-dimethylacetamide (DMAc) to form a homogeneous polymer solution of polyamic acid (PAA). This solution can be used to make the CNF via electrospinning followed by carbonization. For composite nanofiber preparation, different amounts of manganese acetylacetonate (15, 25, 30, 40, and 50 wt.% relative to PAA) were added into the PAA polymer solution under magnetic stirring for 12 h. The mixtures were used as precursors for electrospinning at a high voltage of 25 kV. A flow rate of 1 mL/h was maintained. The distance between the tip of the injecting needle and the collector was 25 cm. The stable electrospinning generated nanofiber mats on the aluminum foil collector. The as-spun nanofibers were treated by the imidization process (see Nan et al. (2013)) to form a PI matrix. The nanofibers were carbonized at 600°C in an argon atmosphere for 2 h. The heating rate was 5°C/min. After that, the final products in black color were obtained. The CNF composites made from the precursors with 15, 25, 30, 40, and 50 wt.% of manganese acetylacetonate were named M15C, M25C, M30C, M40C, and M50C, respectively. The composition, phase structure, and morphology of the composites were characterized by scanning and transmission electron microscopy, X-ray diffraction (XRD), and thermogravimetric analysis. The results obtained by Wang et al. (2015) indicated that the composites exhibited good nanofibrous morphology with MnO nanoparticles uniformly encapsulated by CNFs as can be seen from the images in Fig. 12.6. The TEM images of the composite nanofibers with the designations of M15C, M30C, and M50C were presented in Fig. 12.6A, B, and C, respectively. All the three images confirmed the nanofibrous feature of the composites. The inset of Fig. 12.6C is the corresponding selected area electron diffraction (SAED) pattern of the nanofiber. Fig. 12.6D, a HRTEM

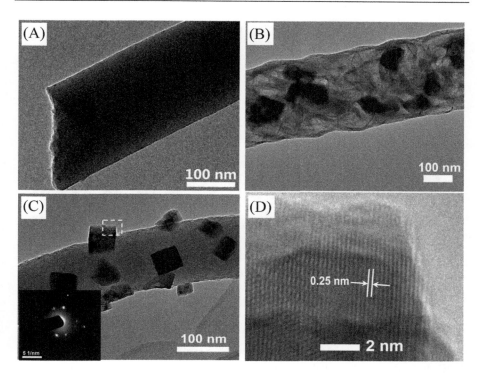

Figure 12.6 Transmission electron microscopic (TEM) images of the (A) M15C, (B) M30C, and (C) M50C composite nanofibers. (D) high-resolution transmission electron microscopic (HRTEM) image of the MnO nanocrystal. The inset in subfigure (C) is the corresponding selected area electron diffraction (SAED) pattern.
Source: Reproduced with permission from Wang, J. G., Yang, Y., Huang, Z. H., & Kang, F. (2015). MnO-carbon hybrid nanofiber composites as superior anode materials for lithium-ion batteries. *Electrochimica Acta*, *170*, 164–170. https://doi.org/10.1016/j.electacta.2015.04.157. © 2015 Elsevier Ltd.

image, reveals the MnO nanocrystal with a typical lattice parameter shown. The hybrid nanofiber composites were used directly as freestanding anodes for LIBs to evaluate their electrochemical properties. An optimized MnO-CNF composite can deliver a high reversible capacity of 663 mAh/g, along with excellent cycling stability and good rate capability. The superior performance indicated that the composites could be alternative anode materials for high-performance LIBs.

Wang et al. (2015) also evaluate the Li-ion storage capacity of the MnO/CNF composite nanofiber anodes. The as-prepared composite CNFs were directly used as anode materials without adding any polymer binders or conducting additives. The specific capacity was measured using galvanostatic charge/discharge tests at a current density of 50 mA/g. Fig. 12.7A shows the specific capacity and cycling performance of the freestanding electrodes of pure CNF, M15C, M25C, M30C, M40C, and M50C at a current density of 50 mA/g. As can be found from the results shown in Fig. 12.7A, the

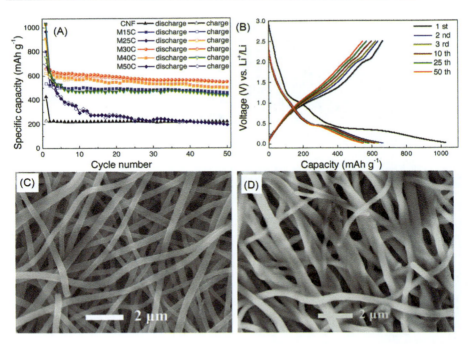

Figure 12.7 (A) Cycling performance of the CNF, M15C, M25C, M30C, M40C, and M50C electrodes (solid symbols: discharge capacity and open symbols: charge capacity). (B) Galvanostatic charge/discharge curves of the M30C composite electrode. (C and D) scanning electron microscopic (SEM) images of (C) M30C and (D) M50C nanocomposites after 50 cycling tests. *M*, MnO; *C*, carbon.
Source: Reproduced with permission from Wang, J. G., Yang, Y., Huang, Z. H., & Kang, F. (2015). MnO-carbon hybrid nanofiber composites as superior anode materials for lithium-ion batteries. *Electrochimica Acta*, *170*, 164–170. https://doi.org/10.1016/j.electacta.2015.04.157. © 2015 Elsevier Ltd.

M30C electrode has the highest specific capacity among the anodes. The typical voltage profiles of the M30C electrode in the 1st, 2nd, 3rd, 10th, 25th, and 50th cycles were obtained and presented in Fig. 12.7B. The first cycle discharge and charge capacities were 1021 and 663 mAh/g, respectively. The initial Coulombic efficiency was 65%. The Coulombic efficiency increased rapidly to 97% in the second cycle and kept close to 100% afterward. The irreversible capacity loss in the first cycle was mainly attributed to the formation of a solid-electrolyte-interphase (SEI) layer at the surface of the anode. The reductive decomposition of the electrolyte and the confinement of some Li ions within the CNFs also took the roles. This argument is supported by the disappearance of the voltage plateau at 0.75–1.0 V in the second and later cycles (Ji et al., 2009). The M30C displayed good cycling stability with a reversible capacity of approximately 557 mAh/g being retained after 50 cycles. The capacity retention (84%) is higher than that of either the MnOx/PAN-derived CNFs (76%) (Please refer to Ji et al. (2009)) or the porous PI-derived CNFs (61%) (Nan et al., 2013). Anodes made from other composites including the CNF, M15C, M25C, M40C, and M50C delivered the lower reversible capacities of 221, 457, 501, 436, and 195 mAh/g, respectively. Overall, the MnO/CNF electrodes with low MnO loadings (i.e., CNF, M15C, M25C, and M30C) showed good capacity retention, whereas the M40C and M50C electrodes suffered from considerable capacity degradation (i.e., 42.4% and 72.1%). The different cycling performance could come from the varying uniformity of MnO distribution in the composites. The MnO nanoparticles in the M15C, M25C, and M30C composites were well-encapsulated by the CNF. Consequently, the CNF served as a robust matrix to well accommodate the large volume change caused by the MnO phase during the charge/discharge cycling. As for the M40C and the M50C composites with much higher MnO content than others, more MnO nanoparticles were exposed at the surface of the CNF. They lost the strain buffering protection offered by the CNF. The lithiation−delithiation cycling would result in the pulverization of the MnO nanoparticles. Consequently, the electrical connections to CNF were lost. Serious capacity degradation was observed. To obtain better insight into the cycling stability, the microstructures of M30C and M50C nanofiber composites after 50 cycles were characterized by SEM. As shown in Fig. 12.7C, the M30C nanofiber composite preserved the original interwoven network structure, and each individual nanofiber was well coated with a uniform thin SEI layer. This morphology is consistent with its good structural integrity during the cycling tests. On the contrary, the nanofibers in the M50C nanocomposite (see Fig. 12.7D) are covered by thick and nonuniform SEI films. Such microstructural features were generated from the pulverization of the exposed MnO nanoparticles.

Ji et al. (2012) introduced alpha-Fe_2O_3 nanoparticle-loaded CNF composites as stable and high-capacity anode materials for rechargeable LIBs. The nanofibers were fabricated via electrospinning the solution containing $FeCl_3$ and PAN in DMF solvent and carbonization in a protective atmosphere. Scanning electron microscopy, transmission electron microscopy, energy dispersive X-ray spectroscopy, X-ray photoelectron spectroscopy, XRD, and elemental analysis were used to observe the morphology and analyze the compositions of the alpha-Fe_2O_3-CNF composites. The produced alpha-Fe_2O_3 nanoparticles have an average size of 20 nm. These particles are uniformly aligned at the CNF surface. The resultant alpha-Fe_2O_3-CNF composites were made into anodes in rechargeable lithium half cells. The electrochemical performances of the cells were investigated. The alpha-Fe_2O_3-CNF composites demonstrated high reversible capacity, good capacity retention, and acceptable rate capability when tested as the anode materials for rechargeable LIBs.

Iron is an element with multiple valence values. Thus, there exist several types of iron oxides. Fe_2O_3 is just one of them, while Fe_3O_4 is another one. Wang et al. (2008) showed Fe_3O_4-loaded carbon-based nanofibers as anode materials for LIBs. Both pristine CNF and C/Fe_3O_4 composite nanofiber were prepared by electrospinning and subsequent carbonization processes. The composition and structures were characterized by various techniques such as Fourier transformation infrared spectroscopy, XRD, scanning, and transmission electron microscopy. The electrochemical properties were evaluated in coin-typed cells versus metallic lithium. After annealing at the temperatures range of 500°C−700°C, the carbon showed a disordered structure, while Fe_3O_4 exhibited a nanocrystalline structure with particle sizes ranging from 8.5 to 52 nm.

Compared with the pristine CNF, the composite C/Fe$_3$O$_4$ nanofiber obtained at 600°C exhibited much better electrochemical performances with a high reversible capacity of 1007 mAh/g at the 80th cycle and excellent rate capability. A beneficial pulverization phenomenon was revealed during the electrochemical cycling. The optimized C/Fe$_3$O$_4$ composite nanofiber may be considered as a promising anode material for high-performance LIBs.

There are reports on oxide nanofibers as anode materials. For example, Yu et al. (2020) used the PNb$_9$O$_{25}$ nanofiber as a high-voltage anode material for advanced lithium-ion batteries. In the work performed by Pakki et al. (2017), the electrochemical performance of lithium batteries was enhanced using the electrospun SiO$_2$ nanofiber as the binder-free anode. Hu et al. (2020) proposed grafting polyethyleneimine (PEI) on electrospun nanofibers to stabilize lithium metal anode for Li−S batteries. A functional ammoniated PAN nanofiber was found inhibiting Li dendrite formation and polysulfide shuttling. Branched PEI was attached to the electrospun PAN nanofiber mat via a chemical grafting to provide amino groups. In such a PAN with large polarity, the Li-ion distribution is uniform. The Li$_3$N-rich SEI layer containing spherical Li deposits showed a dendrite-free 3D structure. The coulombic efficiency of the resulting Li anode can be improved up to 98.8% with a low overpotential of 15 mV. Because PEI has strong chemical adsorption capability, polysulfide shutting was prevented, which promotes the capacity retention of sulfur.

12.3.2 Nanofiber cathodes

Organic semiconducting nanofibers have been used as cathode materials in lithium batteries. Cai et al. (2013) prepared polyindole nanofiber using an electrospinning technique for cathode material application in lithium-ion batteries. The polyindole polymer was synthesized by chemical oxidation of indole. The obtained polyindole was dissolved into acetonitrile for electrospinning. In electrospinning, the polyindole solution was filled into a glass syringe connected with a stainless-steel needle. The inner diameter of the needle is 0.35 mm. The syringe was placed in an automatic pump and polyindole solution was ejected at a rate of 0.8 mL/h. The voltages were in the range from 16 to 24 kV. The tip-to-collector distance was kept at 20 cm. The polyindole nanofibers were collected on a conducting carbon paper. The experiment was performed in an environmental chamber. The temperature was maintained at 25°C and the RH was kept at 35%. The battery consisted of a polyindole nanofiber cathode, lithium metal anode, glass filter separator, and 1 M LiPF$_6$ in ethylene carbonate (EC) and dimethyl carbonate (DMC) as the electrolyte. The volume ratio of EC to DMC is 1:1.

The surface morphology of the polyindole nanofiber cathode was examined by Cai et al. (2013) using scanning electron microscopy (SEM). A porous structure was observed. The diameter of the fibers ranged from 180 to 347 nm. Electrochemical characterization of the battery consisting of the Li anode and the polyindole nanofiber cathode revealed good cycling properties as well as fast charge and discharge characteristics. The battery achieved about 3.0 V electromotive force with a discharge capacity of 83 mAh/g which is about 99% of the theoretical value. The capacity faded away slowly during cycling. The discharge capacity decreased to 72% of the theoretical capacity after 500 cycles at a discharge current density of 40 mA/g. At a higher discharge current density like 200 and 400 mA/g, the battery still showed stable discharge capacity values of 79 and 70 mAh/g, respectively. The discharge capacity of the battery increased with the increase in temperature. The results confirmed that the polyindole nanofiber is a good cathode material for lithium-ion batteries. However, mass loss of polyindole leading to a decrease in capacity was found. As shown by Fig. 12.8 (Cai et al., 2013), the SEM image of polyindole nanofiber electrode after cycling, the nanofibrous structure was kept. However, the diameter of the indole nanofiber became bigger. A lot of etching pores were shown at the surface of the polyindole nanofibers. This should be due to the Li-ion transport in and out from the nanofibers during the charge/discharge period. The pore generation caused polyindole loss and led to a decrease in the capacity. Therefore, an improvement in the cathode capacity is needed for this battery.

In sodium batteries, organic semiconducting nanofiber cathode can also be found. Cai et al. (2017) adopted the poly(5-cyanoindole) nanofiber cathode for sodium battery applications. The purpose of their work was to make a sodium/electrospun poly (5-cyanoindole) nanofiber secondary battery with a comparable performance to the currently used lithium-ion battery. In their prototyped sodium battery, a piece of sodium foil was used to make the anode. The electrospun poly(5-cyanoindole) nanofiber served as the cathode material. In addition, NaPF$_6$ containing organic electrolyte solution was used as well in the battery. The performance of the sodium/electrospun poly(5-cyanoindole) nanofiber secondary battery was evaluated. The charge/discharge, rate performance, cycle stability, and electrochemical impedance tests

Figure 12.8 The scanning electron microscopic (SEM) image of polyindole nanofiber cathode after cycling. Polyindole nanofibers formed conductive networks.
Source: Reproduced with permission from Cai, Z. J., Shi, X. J., & Fan, Y. N. (2013). Electrochemical properties of electrospun polyindole nanofibers as a polymer electrode for lithium ion secondary battery. *Journal of Power Sources, 227*, 53–59. https://doi.org/10.1016/j.jpowsour.2012.10.081. © 2012 Elsevier B.V.

revealed that the charge and discharge of the sodium/electrospun poly(5-cyanoindole) nanofiber battery were fast. The cycle life is long, and its performance is approaching that of lithium-ion batteries. But it would be much less expensive than a Li battery.

Inorganic nanofibers as the cathode materials should offer better structural stability than organic cathode materials. Dimesso et al. (2013) investigated the LiCoPO$_4$−3D CNF composites as potential cathode materials for high-voltage batteries. Electrospun and carbonized 3D nanofiber mats were coated with olivine-structured lithium cobalt phosphate (LiCoPO$_4$) in a sol−gel process. The obtained material was used to make cathodes for LIBs. The experimental procedures included immersing the 3D nonwoven nanofiber mats in an aqueous solution containing lithium acetate, cobalt acetate, citric acid, and phosphoric acid at 80°C for 2 h. After air-drying, the coated mats were annealed at 730°C for 2−12 h in a nitrogen atmosphere. Crystalline deposits were uniformly distributed at the CNF surface. The active cathode material loaded on the 3D CNF composite mats reached as high as 300 wt.%. The discharge-specific capacity (measured at a discharge rate of 0.1 C and room temperature) achieved a maximum value of 46 mAh/g for the specimen annealed for 5 h. The two micrographs in Fig. 12.9 show the surface morphology of the composite nanofiber mat. Fig. 12.9A shows the SEM image of the CNF nonwoven mat. According to Dimesso et al. (2013), the nanofiber mat was prepared by electrospinning PAN polymer solution added with 15 wt.% carbon black (CB) followed by high-temperature carbonization. The sizes of the carbonized nanofibers were very uniform. Their diameters were in the range of 500−800 nm. These nanofibers formed a 3D nano-textured mat with high surface areas. The micrograph in Fig. 12.9B shows the formation of flower-like crystalline structures with sizes ranging from 2.0 to 50 μm. These crystals covered the whole surface of the CNF mats. The effect of the nanocrystals at the surface of the CNFs on the electrochemical performance of the cathode remains to be studied.

Yuan et al. (2017) made conductive hollow CNFs for holding elemental sulfur. The SEM images of the prepared hollow CNFs are shown in Fig. 12.10. TEM images showing the detailed hollow feature may be found in the similar hollow CNF cathode research work published by Yuan et al. (2016). Even though there is no information given by Yuan et al. (2016) about how to prepare the hollow CNF, hollow carbon can generally be made by hydrothermal synthesis (see the work in Brun et al. (2013)) or solvent exchange method (see earlier work by Chen et al. (2012)). Yuan et al. (2017) investigated the influences of preparation methods and sulfur content on the structure and properties of the CNFs. They found that the agglomeration of sulfur produced large-sized sulfur particles. Uniform distribution of sulfur with good contact with carbon can be achieved by the liquid phase precipitation at appropriate sulfur content. The hollow nanofibers formed a porous but mechanically stable three-dimensional framework. The hollow CNF framework-based cathode showed high conductivity, which enhanced the electrochemical performance of lithium sulfur battery (Yuan et al., 2017).

12.3.3 Nanofiber separators

Separators are key parts to ensure the safety of LIBs. A good separator can improve the performance of a battery. Polyolefin films are currently used as commercial

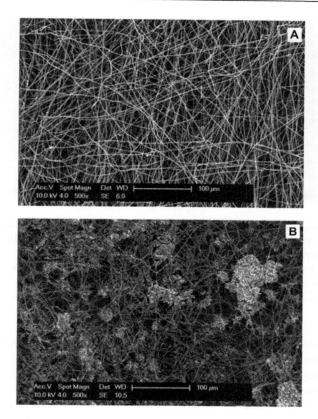

Figure 12.9 SEM images of (A) carbon nanofibers prepared by electrospinning of PAN + 15 wt.% carbon black (CB), and (B) LiCoPO$_4$ loaded composite carbon nanofibers. *PAN*, the polymer.
Source: Reproduced with permission from Dimesso, L., Spanheimer, C., Jaegermann, W., Zhang, Y., & Yarin, A. L. (2013). LiCoPO$_4$−3D carbon nanofiber composites as possible cathode materials for high voltage applications. *Electrochimica Acta*, 95, 38−42. https://doi.org/10.1016/j.electacta.2013.02.002. © 2013 Elsevier Ltd.

lithium-ion battery separators. The major limitation of polyolefin is its thermal instability, which could cause battery short circuits and fire risks. PI with good thermal stability was studied very early for battery separators (see Miao et al. (2013)). Composite nanofibers have been extensively studied for battery separator applications. The commonly used technology for making nanofiber separators is electrospinning. In the work performed by Li et al. (2017), a poly-m-phenylene isophthalamide (PMIA) nanofiber membrane containing SiO$_2$ nanoparticles was made by electrospinning. The morphology, crystallinity, thermal shrinkage, porosity and electrolyte uptake, and electrochemical performance of the SiO$_2$/PMIA nanofiber membranes were studied. The nanofiber membrane with 6 wt.% SiO$_2$ maintained thermal stability, higher porosity, and electrolyte uptake. It showed an ionic conductivity as high as 3.23×10^{-3} S/cm, which is

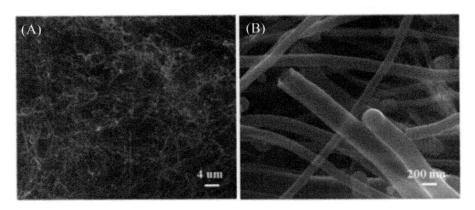

Figure 12.10 SEM images of hollow carbon fiber (HCF): (A) low-magnification; (B) high-magnification. The hollow nanofibers formed a porous but mechanically stable three-dimensional framework for holding sulfur.
Source: Reproduced from Yuan, Y., Fang, Z., & Liu, M. (2017). Sulfur/hollow carbon nanofiber composite as cathode material for lithium-sulfur batteries. *International Journal of Electrochemical Science*, 12(2), 1025−1033. https://doi.org/10.20964/2017.02.27. © 2017 The Authors.

much higher than that of the pure PMIA nanofiber membrane. The SiO_2/PMIA composite nanofiber membrane was used to make Li/LiCoO$_2$ cells. This cell exhibited high cycling stability with a capacity retention of 95% after 50 cycles. The SiO_2-loaded PMIA nanofiber membrane showed potential for separator application in high-temperature resistance LIBs.

Yanilmaz et al. (2014) made SiO_2/nylon 6,6 nanofiber membranes with different SiO_2 contents (0, 3, 6, 9, and 12 wt.%) by electrospinning. The prepared membranes were used to make separators in Li-ion batteries. The electrochemical performances of the separators were evaluated. It was found that the new nanofiber membrane separators showed high performance. They also have the advantages of enhanced mechanical properties and good thermal stability. They are better than microporous polyolefin membranes from the comparison of the electrochemical performance. For example, the SiO_2/nylon 6,6 nanofiber membranes have larger liquid electrolyte uptake, higher electrochemical oxidation limit, and lower lithium/membrane interfacial resistance. The SiO_2/nylon 6,6 nanofiber membranes were assembled into lithium/lithium cobalt oxide and lithium/lithium iron phosphate cells, respectively. Both cells showed high capacities and good cycling performance at room temperature.

An et al. (2014) reported their work on the enhancement of the thermal stability and electrolyte wetting of a porous polyethylene (PE) separator. An Al_2O_3 nanopowder layer and an electrospun poly-vinylidene fluoride (PVDF) nanofiber layer were coated on both sides of the PE separator. The Al_2O_3 layer provided excellent thermal stability; the area shrinkage is only 7.8% at 180°C. There was no meltdown up to 200°C. The electrolyte uptake of the multilayer separator was increased with the thickness of the nanofiber layer. As a result, the discharge capacity, rate capability, and cycle life of the LIBs containing the PVDF nanofiber layers were improved. The improvement can overcompensate for a loss of performance caused by the Al_2O_3 layer. This multilayer approach is highly effective in improving both the performance and safety of LIBs.

Chen et al. (2018) made the covalently-bonded poly(vinyl alcohol)-silica (PVA-SiO_2) composite nanofiber membranes via sol–gel electrospinning and investigated their physical and electrochemical properties for application as separators in LIBs. The PVA-SiO_2 membranes showed a unique three-dimensional interconnected porous structure and displayed a higher porosity of 73%, better electrolyte affinity, higher electrolyte uptake (405%), and lower thermal shrinkage compared to commercial polypropylene (PP) membranes. The batteries using PVA-SiO_2 composite nanofiber membranes as separators exhibited an enhanced ionic conductivity of 1.81 mS/cm. The cycling stability and C-rate performance were better than those of the batteries using PP separators. Such findings suggested that the PVA-SiO_2 composite nanofiber membrane could be a promising alternative to the commercial polyolefin separator for high-performance LIBs. Yanilmaz (2020) also examined electrospun PVA/SiO_2 nanofiber separator membranes for LIBs. The advantages of using the composite nanofiber membrane separator were given. SiO_2 was also added to the PAN polymer to make composite nanofiber through sol–gel electrospinning for improving the performance of the separators (see Yanilmaz et al. (2016)).

The quality of separator materials is one of the factors that determine the safety of LIBs. Traditional polyolefin separators have the limitations of low thermal stability and poor wettability. Developing new battery separator materials with high heat resistance and electrolyte wettability is needed. Wang et al. (2019) made a PI nanofiber matrix composite material with graphene oxide (GO) as the reinforcement. The nanofiber membrane with the three-dimensional porous structure was prepared by the electrospinning technology as shown in Fig. 12.11. Several chemicals were used for the processing, as illustrated in Fig. 12.11A. PMDA (97% purity, Dry at 120°C for 1 h), 4,4,-diaminodiphenyl ether (97% purity, Drying at 60°C for 0.5 h) powder, DMAc (99.5% purity) solvent, and single-layered GO were used. For battery making, a liquid electrolyte solution consisting of 1.0 M lithium hexafluorophosphate ($LiPF_6$) in a mixture with equal volume of EC, diethyl carbonate, and methyl ethylene carbonate was made.

Wang et al. (2019) showed the details of preparing the PI polymer by the two-step method, which is considered the simplest and most used method. In brief, the precursor solution of PAA was synthesis from the reaction of PMDA and ODA (see Fig. 12.11B, the first step reaction) in a molar ratio of PMDA/ODA = 1.02:1 in nitrogen atmosphere. The ODA was dissolved into the solvent (DMAc). Then PMDA powder was added at 40°C and stirred for 5 h. A homogeneous PAA solution was obtained at a concentration of 23 wt.%. The 0.2 wt.% GO/PAA solution was prepared by adding 0.5 wt.% GO/DMAc into the DAMc solution. After ultrasonication, a dark brown homogeneous solution was obtained. Electrospinning was carried out by setting the spinning voltages between 22 and 26 kV. The distance between the tip of the needle and the collector was about 22.5 cm. The ejection rate was 0.2 mL/h. After the

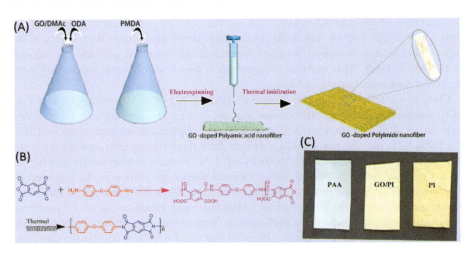

Figure 12.11 Schematic showing procedures for making PI nanofiber membrane (A), chemical equations for producing PI (B), and photographs of polyamic acid (PAA), GO/PI, and PI (C). The chemical reaction shows the mechanism of imidization.
Source: Reproduced with permission from Wang, L., Liu, F., Shao, W., Cui, S., Zhao, Y., Zhou, Y., & He, J. (2019). Graphite oxide dopping polyimide nanofiber membrane via electrospinning for high performance lithium-ion batteries. *Composites Communications*, *16*, 150−157. https://doi.org/10.1016/j.coco.2019.09.004. © 2019 Elsevier Ltd.

thickness of the nonwoven mats reached about 40 μm, the PAA and GO/PAA nanofibers were imidized at 80°C for 0.5 h, 120°C for 0.5 h, 200°C for 0.5 h, 250°C for 0.5 h, 300°C for 0.5 h, and 350°C for 1 h. The heating rate was 5°C/min. The imidizing experiments were carried out in flowing nitrogen. Fig. 12.11C shows the PAA, GO/PI, and PI, three nanofiber mats. SEM images and diameter distributions of the prepared nanofibers can be found in Fig. 12.12. Comparing the results in Fig. 12.12A and B, the diameter of PAA nanofiber is smaller than that of the PI. The results in Fig. 12.12C and D indicate that the GO/PI has smaller diameter than that of the GO/PAA nanofiber. The solvent evaporation at elevated temperature was considered as the main reason. The addition of GO significantly reduced the diameter of the PAA nanofiber. Wang et al. (2019) believed that the increased conductivity of the spinning solution due to the addition of GO caused such a result. Because the higher conductivity induced a stronger electric field during the electrospinning process, the polymer experienced larger stretching forces. This allowed the GO/PAA nanofiber to be finer than the PAA nanofiber without GO. Accordingly, after high temperature treatment, the GO/PI nanofiber is finer than the PI fiber. The higher magnification images as the inset in Fig. 12.12A−D clearly show the intra-fiber pores. The final product, GO/PI nanofiber separator, is smooth at macroscale, but with pores at micro- and nanoscales. The average diameter of GO/PI fiber is 194 nm. The porosity of the GO/PI fiber mat is about 68%, while for PI fiber mat this value is 45%. The small-sized pores within the GO/PI nanofiber mat are required for serving the proper function as the battery separator.

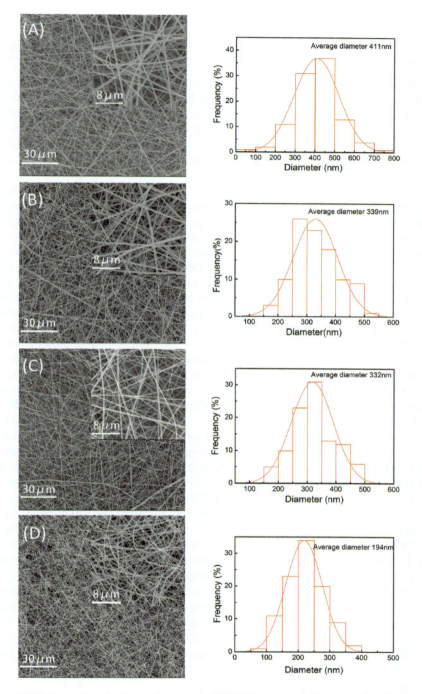

Figure 12.12 Scanning electron microscopic (SEM) images and diameter distributions of the prepared nanofibers. (A) PAA, (B) PI, (C) GO/PAA, and (D) GO/PI. *PI*, polyimide; *GO*, graphene oxide; *PAA*, the polymer.
Source: Reproduced with permission from Wang, L., Liu, F., Shao, W., Cui, S., Zhao, Y., Zhou, Y. & He, J. (2019). Graphite oxide dopping polyimide nanofiber membrane via electrospinning for high-performance lithium-ion batteries. *Composites Communications*, *16*, 150−157. https://doi.org/10.1016/j.coco.2019.09.004. © 2019 Elsevier Ltd.

Wang et al. (2019) also showed the thermal stability of the GO/PI nanofiber separator. No thermal shrinkage was found up to 200°C. A super high electrolyte absorption rate of 2200% was achieved. The ionic conductivity was as high as 2.04 mS/cm. Batteries assembled with the GO/PI composite nanofiber membrane showed outstanding cycling performance and C-rate discharge capability. It was concluded that the PI-based nanofiber separator could offer better performance than most of the existing separators.

The limitation of PI was also identified. PI nanofiber membrane prepared by electrospinning technique exhibits poor mechanical strength for the slippage of fibers, which limits its application into lithium-ion battery (LIB). Li et al. (2021) increased the mechanical strength of a PI nanofiber membrane by limiting the slippage among the fibers. Lithium polyacrylate (PAALi) as an excellent binder was used to promote the adhesions among the fibers. As a result, the PI nanofiber membrane treated by PAALi solution shows the mechanical strength of 16.1 MPa, higher than 5.0 MPa of the pristine PI membrane. The binder caused the originally loose and disordered PI nanofibers cross-linking with each other. This effectively prevented the slip of the fibers. Interestingly, the PAALi binder did not negatively impact on the thermal stability and electrochemical performance of the PI nanofiber separator. The $LiCoO_2$/Li cells using the modified PI nanofiber separators exhibited better cycling stability and rate performance than the cells made from the pristine PI separators. Therefore, the use of the PAALi binder is a promising approach to improving the mechanical strength of nanofiber membranes in high-power LIB.

Because commercial polyolefin separator cannot guarantee the safety of LIBs at high temperatures due to their poor thermal stability, various other types of nanofiber membranes with high thermal stability become the focus. In addition to the above-mentioned polymer nanofibers, aramid nanofiber composite separators for high performance Li-S batteries were reported by Emre et al. (2019). Fluoropolymers are more stable than other ones. Therefore, PVDF and its derived compounds were also studied for separator applications. As an example for using the PVDF separator, we can find from the work performed by Gao et al. (2022). Morphology optimization of the PVDF nanofiber separator for safe lithium-ion battery was the focus. The PVDF nanofiber membrane was made by electrospinning. The relationship between the electrochemical performance and nanofiber morphology was investigated. The PVDF changed from beads to nanofibers with the increase in the concentration of spinning solution. The PVDF nanofiber membrane prepared from the spinning solution with a concentration of 24 wt.% contains coarse nanofibers and fine nanofibers. The membrane demonstrated better electrochemical properties than other membranes for example PE membrane separator. For high-energy-density battery, it is much safe to use the PVDF nanofiber separator.

Composite or multilayered separators are intensively researched for improving the performances of batteries. A flexible SiO_2 nanofiber membrane combined with a poly(vinylidene fluoride-hexafluoropropylene) (PVDF-HFP) nanofiber membrane was prepared by Xu et al. (2020), using the electrospinning method. The mechanical strength of the SiO_2/PVDF-HFP composite nanofiber membrane solar protective films (SPF) is twice as high as that of the pure SiO_2 nanofiber membrane. At an elevated temperature of 200°C, no dimensional change was found in the SPF separators. Compared to commercial PE separators, SPF shows excellent thermal stability. SPF separators were tested using large area closed cells at 180°C. The porosity of SPF is 89.7%, which is two times higher than that of an ordinary PE separator. The liquid absorption rate of SPF is much higher than an ordinary PE separator and has reached 483%. Furthermore, the cycle and rate performance of LIBs prepared by SPF have been improved significantly. The excellent property of SPF makes it as a potential separator material for high-power batteries.

Fluoride coatings can be applied to the surface of commercially used PP battery separators. Alcoutlabi et al. (2013) used two fluoride compound nanofiber coatings to make composite membrane battery separators via electrospinning. Electrospun PVDF and polyvinylidene fluoride-co-chlorotrifluoroethylene (PVDF-co-CTFE) nanofibers were coated onto the PP microporous battery separators. The PVDF and PVDF-co-CTFE nanofiber coatings were carried out using single needle jet and needleless electrospinning methods. The nanofiber coating prepared by the needleless electrospinning method was found to have better adhesion to the microporous separator membrane than the nanofiber coating prepared by the single needle electrospinning. The PVDF and PVDF-co-CTFE nanofiber coatings increased the electrolyte uptake capacity in a secondary lithium-ion battery. The PVDF-co-CTFE copolymer nanofiber-coated microporous membrane showed higher electrolyte uptake capacity than the PVDF homopolymer-coated microporous membrane. It was further confirmed that the PVDF and PVDF-co-CTFE nanofiber coatings improved the adhesion of the porous microporous membrane to the battery electrode. It has been concluded by Alcoutlabi et al. (2013) that the nanofiber coatings prepared by the needleless electrospinning method have better adhesion properties and higher electrolyte uptake capacity than those made by single nozzle electrospinning. Similar works on electrospun PVDF-co-CTFE nanofiber coatings on various microporous membranes were presented by Lee et al. (2014). The results further confirmed that coating PVDF-co-CTFE nanofibers onto microporous membrane substrates should be a promising approach to obtaining new and high-performance separators for rechargeable LIBs.

For making composite nanofiber membrane separator, Yang et al. (2019) inserted CoS and Ketjen black (a unique electro-conductive CB) into CNF framework to make separators for high performance Li−S batteries. Li−S batteries have the theoretical energy density as high as 2600 Wh/kg. They are inexpensive and environmentally friendly. However, the detrimental polysulfides tend to build up in lithium sulfur batteries. The migration of such polysulfides causes serious problem of degrading their electrochemical performances. Besides polymer nanofibers, composite carbon-based separators are developed to solve this problem. In the study performed by Yang et al. (2019), the composite separator consisting of CNF, CoS, and ketjen black served as an effective polysulfide blocker. The cell with the CNF/CoS/ketjen black-modified separator and the high sulfur content (70 wt.%) cathode showed excellent stability with a capacity decay of 0.076% per cycle for more than 700 cycles under 1.0 C current density.

Many other materials are considered as the candidates for making separators in batteries. Li and Yin (2018) introduced the polyphenylene oxide-based nanofiber separator prepared by electrospinning method for LIBs. Huang et al. (2019) investigated making the lithium battery separators from sustainable sources. They studied the composite nanofiber membranes of bacterial cellulose/halloysite nanotubes as lithium-ion battery separators. In the work reported by Sun et al. (2022), polyetherketone (PEEK) nanofiber membrane was prepared. PEEK is classified as a semicrystalline high performance engineering thermoplastic material with the combination of thermal stability, chemical resistance, and excellent mechanical properties over a wide temperature range. It is considered as a novel material for high-performance lithium battery separator.

12.3.4 Nanfibers for Li−S batteries

Among various lithium batteries, the Li−S battery shows some advantages. The value of theoretical specific energy for Li−S is much higher than those of the commercial lithium ion (Li-ion) batteries, as mentioned by Liu et al. (2019). This allows the Li−S battery to have high potential for next generation energy storage applications, especially in the fields of portable electronics and electric vehicles. Nevertheless, the so called "shuttle effect" of the polysulfides and the damage of lithium dendrites are to be solved before the commercial application of Li−S battery. Electrospun nanofibers possess unique nanostructures and show some unique characteristics. They may help solve the two problems simultaneously. In the review given by Liu et al. (2019), various cathodes, separators, and interlayers of electrospun nanofiber materials were dealt with. Both fundamental research and technological development of electrospun nanofibers for Li−S batteries were introduced.

Although Li−S rechargeable batteries have high theoretical specific capacity, the frequent shuttle of soluble lithium polysulfides and severe Li corrosion have caused serious problems. Zhou et al. (2021) proposed to solve the problems by developing a bifunctional polyvinyl alcohol/poly(lithium acrylate) (C-PVA/PAA-Li) composite nanofiber separator. The nanofiber separator allowed rapid lithium-ion transport and ionic shielding of polysulfides. The C-PVA/PAA-Li composite nanofiber membrane is prepared through electrospinning followed by thermal crosslinking and in situ lithiation. The C-PVA/PAA-Li composite nanofiber membrane showed well-developed porous structures and high ionic conductivity. Consequently, the charge transfer resistance was reduced, and the growth of lithium dendrites was suppressed. The resulting Li−S batteries exhibit an ultra-low fading rate of 0.08% per cycle after 400 cycles at 0.2 C, and a capacity of 633 mAh/g at a high current density of 3 C. This study presents an inspiring and promising strategy to fabricate emerging dual-functional separators, which paves the pathway for the practical implementation of ultrastable and reliable Li−S battery systems.

According to Meng et al. (2021), the shuttle effect of polysulfides during charge and discharge processes is a critical problem that seriously limits the application of Li−S batteries. To solve the problem, a flexible nanofiber carbon paper was fabricated from poly(p-phenylene terephthalamide) (PPTA) nanofiber and CF as an effective interlayer for high-performance Li−S batteries. The nanofiber carbon paper possesses plenty of meso-/micropores and high heteroatom contents, which are beneficial for alleviating the shuttle effect of polysulfides through physical adsorption and chemical catalysis. The electrochemical reaction kinetics are improved remarkably. The primary discharge capacity of cells with PPTA/CF carbon paper achieves 1373 mAh/g at 0.1 C with a 96.2% utilization rate of active materials. The specific reversible discharge capacity maintains 755 mAh/g at 5°C, with a 99.7% coulomb efficiency. This flexible nanofiber carbon paper was considered as a promising candidate for application in high-performance lithium−sulfur batteries.

Designing functional interlayers is considered as an efficient method to inhibit the shuttle effect and improve the redox kinetics of lithium polysulfides. Zhao et al. (2021) introduced MoS_2 nanofiber (MSNF) interlayers in the Li−S batteries. The polysulfide migration slowed down due to the chemical adsorption between MoS_2 and polysulfide. The 3D nanofiber structure of MoS_2 provides fast rapid electron/ion transport and fast redox kinetics. Therefore, the Li−S batteries with MSNF interlayers have high specific capacity and excellent cycling stability. In the study performed by Xiang et al. (2022), a novel fluorine-containing emulsion and 3,4-ethylene dioxyethiophene (EDOT) codoped poly-m-phenyleneisophthalamide (PMIA) nanofiber membrane (EDOT/F-PMIA) was made by electrospinning to serve as the separator for Li-S battery. The multiscale EDOT/F-PMIA nanofiber membrane served as the matrix to fabricate gel polymer electrolyte (GPE). The existence of fluorine-containing emulsion and EDOT allowed the PMIA-based GPE to have excellent thermostability, eminent mechanical property, and well-distributed lithium-ions flux. The pore size of the nanofiber membrane decreased after adding the fluorine-containing emulsion and EDOT. S and O in EDOT with lone pair electrons could build connections with the lithium polysulfides. This inhibited the "shuttle effect" of lithium polysulfides by combining the physical confinement and chemical binding. Therefore, the Li−S battery assembled with the EDOT/F-PMIA separator exhibited excellent electrochemical performance, which delivered a high initial capacity of 851.9 mAh/g and maintained a discharge capacity of 641.1 mAh/g after 200 cycles with a capacity retention rate of 75.2% at 0.5 C.

In the study performed by Wang et al. (2018), the lithiated S, Li_2S, served as the cathode material. The electrospun TiO_2-loaded hollow CNFs (TiO_2-HCFs) were used as the conductive framework and the fiber membrane served as the protective barrier for Li_2S in the Li−S batteries. Both the electron and ionic conductivities were improved significantly by the TiO_2−HCFs. The nanofiber membrane as a strong physical barrier prevented the dissolution of those lithium polysulfides. The cell consisting of Li_2S/TiO_2-HCF composite showed a discharge capacity of 851 mAh/g(Li_2S) at 0.1 C. The TiO_2-HCFs/Li_2S/TiO_2-HCF composite cell delivered a relatively high specific capacity of 400 mAh/g(Li_2S) at 5 C. Huang et al. (2019) used the carpet-like TiO_2 nanofiber interlayer as an absorber for Li to improve the performance. Yuan et al. (2017) built the CNF conductive matrix for holding the elemental sulfur. The influences of preparation method and sulfur content on the structure and properties of the composite were investigated. It is found that the treatment process is inclined to produce large-sized sulfur particles accompanied by severe agglomeration. On the contrary, uniform distribution and compact contact can be established between sulfur and carbon for the composite based on liquid phase precipitation and appropriate sulfur content. The hollow nanofiber material with superior conductivity, plenty of pores, and high mechanical strength formed a stable three-dimensional framework, which enhanced the electrochemical performances of lithium sulfur batteries (Yuan et al., 2017).

Jiang et al. (2022) prepared a CoSe@NC nanofiber membrane by electrospinning and used as an independent functional interlayer for the Li-S batteries. The CoSe@NC nanofiber exhibited strong adsorption capability for polysulfides; The CoSe nanoparticles showed strong catalytic activity and accelerated the conversion of polysulfides. The adsorption-catalysis synergy of CoSe@NC enhanced the rate performance and cycling stability of Li–S batteries. The cell assembled with CoSe@NC functional interlayer delivers a high initial reversible discharge capacity of 1317 mAh/g at 0.1 C rate. After 200 cycles at 1.0 C, the capacity decay rate is only 0.16% per cycle. This work provides a way to promote the electrochemical performance of Li–S batteries.

12.3.5 Nanofibers in compliant and flexible batteries

Flexible and compliant batteries are the key components for powering various flexible electronic devices. In flexible batteries, electrodes with excellent mechanical durability and electrochemical performances are needed. Nanofibers make it possible to manufacture thin and flexible composite polymer electrolyte membranes. Watanabe et al. (2019) showed a very thin, flexible, and safe composite polymer electrolyte membrane consisting of a lithium-ion conductive nanofiber framework and polymer electrolyte matrix. Such a membrane was proposed as an alternative solid-state electrolyte for use in high-performance rechargeable all-solid-state lithium-ion batteries. Due to its unique structural integrity, the composite solid polymer electrolyte (SPE) possesses excellent mechanical stability, high ionic conductivity, and a large lithium-ion transference number. The ion conductivity reached 10^{-4} S/cm at room temperature and the lithium-ion transference number was 0.5. Good charge/discharge cycling behavior and moderate rate capability were found. A high-voltage multistacked battery with a solid polymer composite electrolyte was designed. The battery showed a doubled cell potential of 6.8 V. The applications in future batteries operating at high voltage and with high energy densities were proposed.

The design of electrodes with superior mechanical flexibility was shown by Xia et al. (2020). High mechanical flexibility is one of the key needs to develop energy storage devices with mechanical durability and excellent electrochemical performance. The compound tin selenide (SnSe)/C nanofiber flexible membrane was prepared by electrospinning and subsequent calcination. The SnSe/C nanofiber membrane allowed 180 degrees bending without any observable breakage, indicating the excellent flexibility of the nanofiber membrane. The SnSe nanoparticles were observed to distribute uniformly in the CNF and at the surface of the fiber. The content of SnSe reached 81.2 wt.%. The CNF framework as a conductive matrix increased the electrical conductivity of the composite. It also served as a buffer material that reduced the volume expansion during electrochemical reactions. When the

SnSe/C nanofibers were used for making the binder-free and current collector-free anode for lithium and sodium ion batteries, excellent electrochemical performances were obtained. The SnSe/C nanofiber anode delivered a stable discharge capacity of 405 mAh/g at 1000 mA/g after 500 cycles in LIBs and 290 mAh/g at 200 mA/g after 200 cycles in sodium-ion batteries. Therefore, the SnSe/C nanofiber is considered a promising anode material for flexible lithium-ion and sodium-ion batteries.

According to an earlier discussion in this chapter, electrospinning is the way to produce flexible electrodes. However, the nanofiber membranes without additives prepared by ordinary electrospinning methods showed low mechanical flexibility because the collapse of the fiber structure during heat treatment led to the compatibility of the fibers. Low adhesion of the interface between the polymer and transition metal ions was also observed. To solve the problem, Xia et al. (2021) introduced several coupling agents to enhance the thermal stability and strengths of nanofibers. For example, a porous $Sb_2S_3/TiO_2/C$ nanofiber membrane was prepared by using an appropriate amount of titanium (IV) isopropoxide as the coupling agent. The prepared nanofiber membranes maintained the- crease-free state even after being folded several times. The porous $Sb_2S_3/TiO_2/C$ nanofiber membranes can be cut into electrodes and directly assembled into lithium-ion half-cells or full-cells without any slurry coating. In lithium-ion half-cells, the porous $Sb_2S_3/TiO_2/C$ nanofiber membranes could cycle 800 times at a current density of 2000 mA/g. In lithium-ion full-cells, a high discharge capacity of 261.6 mAh/g can be obtained after 100 cycles when cycled at 50 mA/g. A practical application of the porous $Sb_2S_3/TiO_2/C$ nanofiber membrane in flexible LIBs was demonstrated by powering 16 red light-emitting diodes at the same time using one fully charged nanofiber full-cell. A flexible film electrode is fabricated by Liu, Wang et al. (2013) via growing polymorph TiO_2 nanosheets on electrospun CNF fabric. The hybrid electrode exhibits remarkable electrochemical performance with high reversible capacity, excellent rate capability, and ultralong cycle life for thousands of cycles, which makes it highly attractive for high-power flexible LIBs.

Flexible or stretchable lithium batteries are the required components in wearable devices and body-attached healthcare devices. Kwon et al. (2020) designed stretchable lithium batteries using nanofiber active materials, stretchable GPE, and wrinkle structure electrodes. A SnO_2/C nanofiber anode and a $LiFePO_4/C$ nanofiber cathode were integrated into the batteries because they contain the much-needed meso- and micropores for lithium-insertion and electrolyte containment. The poly(dimethylsiloxane) (PDMS) elastomer wrapping film was used to pack the stretchable full cell. A polymer gel electrolyte was sandwiched between the SnO_2/C and $LiFePO_4/C$ nanofiber electrodes. These electrodes with a wrinkled structure were stuck to the PDMS wrapping film by a polymer adhesive. The specific capacity of the stretchable battery was 128.3 mAh/g with a capacity retention of 92%. A 30% strain only caused a 6.3% decrease in the specific capacity. The energy densities were 458.8 Wh/kg in the strain-free state and 423.4 Wh/kg in the stretched state, respectively.

12.3.6 Miscellaneous topics on nanofibers for battery applications

There are many other research topics on nanofibers for battery applications. In this part, oxide nanofibers in batteries, sodium batteries using nanofibers, sustainable nanofibers, special nanofiber architectures, large-scale nanofiber production technologies, the advantages of CNF vs other nanofibers, etc. will be briefly discussed.

12.3.6.1 Oxide nanofibers for lithium batteries

In addition to using carbon-based nanofibers for batteries, transition metal oxide nanofibers were applied for battery applications. Yue et al. (2018) introduced CoO nanofiber decorated nickel foams as lithium dendrite-suppressing host skeletons for high-energy lithium metal batteries. In the work performed by Huang et al. (2019) and Huang and Wang (2019), TiO_2 nanofiber was used as the interlayer (absorber) for high-performance Li-S batteries. Bimetal oxide nanofibers, such as mesoporous nickel tungstate nanofiber, was investigated for application as anode material for LIBs (see Peng et al. (2017)). In the research performed by Wang et al. (2018), the lithiated S, Li_2S, served as the cathode material. The electrospun TiO_2-loaded hollow carbon nanofibers (TiO_2-HCFs) were used as the conductive framework and the fiber membrane served as the protective barrier for Li_2S in the Li−S batteries. Both the electron and ionic conductivities were improved significantly by the TiO_2-HCFs. The nanofiber membrane as a strong physical barrier prevented the dissolution of those lithium polysulfides. The cell consisting of Li_2S/TiO_2-HCF composite showed a discharge capacity of 851 mAh/g (Li_2S) at 0.1 C. The TiO_2-HCFs/Li_2S/TiO_2-HCF composite cell delivered a relatively high specific capacity of 400 mAh/g(Li_2S) at 5 C.

High-voltage solid lithium-metal batteries (HVSLMBs) are not only energy denser but also safer than state-of-the-art LIBs. However, the narrow electrochemical stability windows and low ionic conductivities of polymer solid electrolytes severely hinder the practical application of this type of battery. To simultaneously address these issues, complex oxide nanofibers were used as reinforcements for the battery applications. For example, a $Li_7La_3Zr_2O_{12}$ ceramic nanofiber was incorporated into composite polymer electrolytes for lithium metal batteries (Li et al., 2019). Liu et al. (2022) prepared the Janus-faced, perovskite nanofiber. This nanofiber served as the framework to reinforce composite electrolytes for HVSLMBs. Liu et al. (2022) fabricate a thin yet mechanically strong composite solid electrolyte by forming a layer of a reduction-tolerant polyethylene glycol diacrylate and oxidation-resistant PVDF within the perovskite nanofiber framework. The unique Janus-faced design enables stable and intimate contact with the lithium metal anode and high-voltage cathode. The perovskite inorganic electrolyte $Li_{0.33}La_{0.557}TiO_3$ (LLTO) nanofiber framework provided the membrane with a high ionic conductivity of 0.1 mS/cm at room temperature and excellent mechanical strength with a thickness of as low as 24 μm. When paired with a high-voltage $LiNi_{0.8}Co_{0.1}Mn_{0.1}O_2$ cathode, the solid lithium battery can deliver a reversible discharge capacity of as high as 176 mAh/g at 0.2 C at room temperature.

12.3.6.2 Nanofibers for sodium batteries

Some work on nanofibers for sodium battery fabrication can be found in the literature (see Cai et al. (2017)). Compared with sodium, the availability of lithium is very limited. There are about 43.6 million tons of Li on land sources. The land sources of lithium are about to run out because each year the demand for lithium from the electric vehicle industry is nearly 1.6 million tons per year by 2030. The annual consumption of Li is still increasing following a rapidly rising curve. It is predicted that the lithium available from land sources will be completely depleted by the year 2040. To solve this problem, it is necessary to replace lithium with other metals for battery manufacturing. Sodium is considered to have a similar property to lithium. As a result, the research on sodium rechargeable batteries has been an active field. Cai et al. (2017) investigated nanofibers for solid batteries. The purpose of their work was to make a sodium/electrospun poly(5-cyanoindole) nanofiber secondary battery with comparable performance to the currently used lithium-ion battery. In their prototyped sodium battery, a piece of sodium foil was used to make the anode. The electrospun poly(5-cyanoindole) nanofiber served as the cathode material. In addition, $NaPF_6$ containing organic electrolyte solution was used as well in the battery. The performance of the sodium/electrospun poly(5-cyanoindole) nanofiber secondary battery was evaluated. The charge/discharge, rate performance, cycle stability, and electrochemical impedance tests revealed that the charge and discharge of the sodium/electrospun poly(5-cyanoindole) nanofiber battery were fast. The cycle life is long, and its performance is approaching that of lithium-ion batteries. But it would be much less expensive than Li batteries.

12.3.6.3 Nanofibers extracted from biodegradable sources

Naturally grown polymers are biodegradable. They can be extracted from renewable or sustainable sources, which could cut the material costs for manufacturing batteries. One of the sustainable nanofibers, cellulose nanofibers, with the characteristics of 1D nanowires structure and rich in functional groups, has been modified and used for potential applications in Li–S batteries (see the work by Liu et al. (2022) and Liu and Zhang (2022)). Nanofibers from sustainable sources, such as bacterial cellulose nanofiber membranes, as high-performance separators for LIBs were also shown by Huang et al. (2020). As known, the electrical insulating properties of cellulose nanofibers prevent their applications as host materials in Li–S batteries. To increase the conductivity of the nanofibers, different modification strategies were used. In one of the approaches, tuning the charge states (positively or negatively charged) was conducted via attaching delaminated-MXene (d-MXene) to the nanofibers to form composites. Two different types of cellulose nanofibers (TEMPO-mediated oxidation cellulose nanofiber and cationic cellulose nanofiber) were attached with d-MXene through the directional assembly of ice templates to prepare the d-MXene/TEMPO-mediated oxidation cellulose nanofiber (MT) and d-MXene/cationic cellulose nanofiber (MC) composite aerogels, respectively. One of the MT nanofiber specimens, MT-4 aerogel (with 40 wt.% percentages of d-MXene in the mixed solution), as the cathode for Li-S battery showed excellent rate performance. The capacity was 848.1 mAh/g at the current density of 2 C. The MC-6 aerogel (with 60 wt.% percentages of d-MXene in the mixed solution) cathode exhibited an ultra-high capacity of 1573.7 mAh/g at 0.1 C. The MC-6 cathode also demonstrated excellent long-term cycling performance with a capacity retention rate of 96.3% after 200 cycles. The renewable cellulose nanofiber composite materials offered their advantages for making low-cost and high-performance Li−S batteries. Nanofibers produced by insects were also studied for battery applications. For example, Wu et al. (2019) carbonized the regenerated silk nanofiber and used it as reinforcement to make a multifunctional interlayer for high-performance Li−S batteries. The as-supplied silk sponge was dissolved in anhydrous formic acid with stirring for 24 h. The silk protein mass concentration in the formic acid was 12%. Polyethylene oxide (PEO) with an average molecular weight of 600,000 was added into the solution until its content reached 2 wt.% relative to the silk fibroin. Electrospinning at 18 kV was conducted to obtain the silk nanofiber collected on an aluminum (Al) foil. The aluminum foil was kept 18 cm from the tip of the metal needle. The electrospinning was carried out at room temperature with the RH ranging from 20% to 40%. The as-spun silk nanofiber film was heated at 250°C for 2.5 h in the Ar atmosphere and subsequently carbonized at 900°C for 2 h to produce the freestanding membranes as the interlayers in high-performance Li−S batteries.

12.3.6.4 Nanofibers with tuned structures

Nanofibers in special structures have found applications for lithium batteries. For example, a foamed nanofiber composite was made by Wang et al. (2015). This composite contains mesoporous carbon and silicon particles. The composite nanofiber was made for anode applications in LIBs. The porous composite nanofiber processing route including coelectrospinning and foaming can be seen from the sketch shown in Fig. 12.13A. In the integrated process, aluminum acetylacetonate (AACA) was used as the foaming agent in nanofibers consisting of PAN/silicon (Si) nanoparticles. The PAN/Si composite nanofibers were produced through the coelectrospinning process, and the mesopores in the nanofibers were generated in the foaming procedure during which AACA sublimated at 280°C. After the subsequent carbonization at 700°C in Ar, the obtained mesoporous carbon/silicon composite nanofiber mats were found very flexible as shown in Fig. 12.13B. The nanofiber mats were made into anodes for LIBs. The mesopores provided the buffering space to accommodate the high stresses and the large volume expansion within the silicon nanoparticles during lithiation. Therefore, the pulverization of silicon was mediated, which helped the new batteries achieve higher reversible capacity and better capacity retention than the existing batteries made of nonporous composite nanofibers and pristine CNFs.

A hierarchical structure containing polyamide 6 (PA6) nanofiber and PEO polymer was introduced by Gao et al. (2021). The high-performance all-solid-state polymer electrolyte with a fast conductivity pathway was made by the hierarchical structure.

Nanofibers and their composites for battery, fuel cell and solar cell applications 307

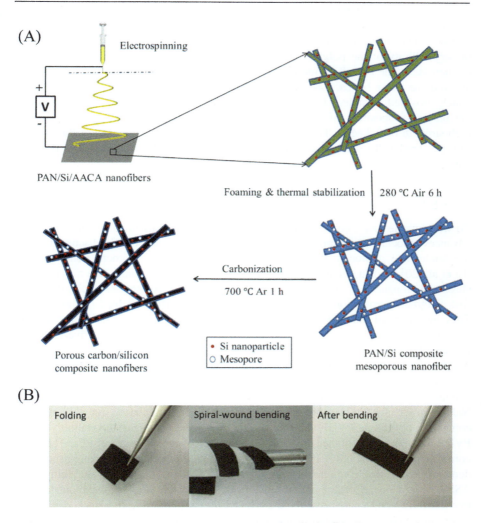

Figure 12.13 **Schematic showing the integrated coelectrospinning/foaming process for making the foamed composite nanofibers (A) and optical images of the mesoporous composite nanofibers under folding and winding, and the state after the mechanical deformation.** The strength and flexibility of the nanofiber mat are well shown by the photos in subfigure (B).
Source: Reproduced with permission from Wang, Y. X., Wen, X. F., Chen, J., & Wang, S. N. (2015). Foamed mesoporous carbon/silicon composite nanofiber anode for lithium ion batteries. *Journal of Power Sources*, *281*, 285–292. https://doi.org/10.1016/j.jpowsour.2015.01.184. © 2015 Elsevier B.V.

The utilization of all-solid-state electrolytes is an effective way to enhance the safety performance of lithium metal batteries. However, the low ionic conductivity and poor interface compatibility are the existing barriers to be removed in all-solid-state batteries. The composite electrolyte combining the electrospun PA6 nanofiber membrane with a hierarchical structure and the PEO polymer showed the potential for overcoming the barriers. The introduction of PA6 nanofiber membrane can effectively reduce the crystallinity of the polymer so that the ionic conductivity of the electrolyte can be increased. The branched fine fibers in the hierarchically structured PA6 membrane allowed the polar bonds, C=O and N—H, to be fully exposed at the fiber surface, which provided sufficient functional sites for lithium transport and helped regulate the uniform deposition of lithium. The hierarchical structure electrolyte showed a high strength of 9.2 MPa. This improved the safety and cycle stability of the battery. The prepared Li/Li symmetric battery was cycled stably for 1500 h under 0.3 mA/cm^2 at 60°C.

12.3.6.5 Other advances in nanofibers for batteries

To wrap up the nanofibers for battery applications, some miscellaneous topics on the advances associated with improving battery performances are briefly discussed. First, the binder-free approach allows low cost and reduced environment pollution (see (Li et al., 2014)). Liu, Huang, et al. (2013) showed making a binder-free Si nanoparticles@carbon nanofiber fabric for energy storage material. Another example of binder-free electrode preparation can be found in the work performed by Li et al. (2014). They incorporated multiple elements including Si, Ge, and Sn into CNFs for making Si/C, Ge/C, and Sn/C composite nanofiber anodes. SPEs have been actively studied for all-solid-state lithium-ion batteries (ASS-LIBs). Second, for the nanofiber electrolyte development, Nakazawa et al. (2021) illustrated the polymer composite electrolyte membranes with beta-crystalline-rich PVDF nanofibers. The enhanced lithium ionic conductivity was found in all-solid-state secondary batteries. Specifically, the work by Nakazawa et al. (2021) showed high-performance SPEs based on lithium salt-containing crystalline PVDF nanofibers. The higher the beta-phase crystallinity of the PVDF nanofibers, the higher the ionic conductivity of the composite SPEs. The ionic conductivity achieved 6.0×10^{-4} S/cm at 60°C. The PVDF nanofiber composite SPEs demonstrated high electrochemical stability and excellent mechanical durability. The ASS-LIB composed of the PVDF nanofiber composite SPEs showed outstanding rate capability and charge-discharge cycling behavior. Multilayered stacked batteries were fabricated. The potential of a quintuple cell reached 17.0 V. Batteries with even higher operation voltages and higher energy densities could readily be built. Electrospinning multiple functional CNFs for capacitors/batteries was reviewed by Wu et al. (2013). The applications of the spun CNF and nanocomposites in energy conversion systems were shown.

The electrospinning technique has been well known for its facile production of 1D continuous fibrous structures. The derived technologies include multiple jet spinning, sol—gel spinning, air bulb spinning, coaxial spinning, triaxial spinning, and needleless spinning. Electrospinning can be integrated into 3D printing facilities to generate controlled architectures. By controlling the spinning parameters, electrospinning can be run concurrently with electrospraying. With the advantage of simple setup and scalable capability using this technology, electrospun CNF has raised great interest in recent years due to its superior conductive potentiality, mechanical flexibility, and well-established fabrication method. This review addresses the various applications of CNF produced via electrospinning technique, mainly focusing on energy conversion systems, such as lithium-ion battery and supercapacitor. Different precursors, preparation conditions, and characterization methods are also reviewed to reveal the very nature of electrospun CNF because its overall performance depends upon its crystallinity, molecular orientation, fiber diameter, surface area, and porosity. Nanocomposites combining metal oxides with CNFs stand as another important topic in this review as the integration of both has been observed to achieve novel superiority, especially in the application of lithium-ion batteries and supercapacitors. To achieve the mass production level, needleless or syringeless spinning was proposed by Moon et al. (2019). Specifically, the mass production of electrospun CNF containing SiO_x for LIBs with enhanced capacity was shown. The electrospinning solution contained PAN and TEOS. After needleless spinning, the pyrolysis at 1000°C for 3 h was performed. The syringeless electrospinning system produced many composite nanofibers in a short time, and the obtained composite nanofibers exhibited uniform diameter and morphology. The composite nanofibers were converted into CNFs containing SiO_x during the pyrolysis. The obtained SiO_x-CNF mat exhibited higher charge/discharge capacity than the ordinary CNF mat. The composite nanofibers showed higher retention than the single crystalline silicon. They believed that the mass production of the SiO_x-CNFs from syringeless electrospinning should be promising in processing anodic materials for Li-ion batteries.

12.4 Nanofibers for solar cells

Nanofibers have been researched for a long time for solar cell applications due to their high surface areas, efficient charge carrier transport properties, and flexibility. Because the topic of nanofibers for solar cells is very broad and many papers on the associated research are available, it is not the intent of this chapter to provide an exhaustive review. Instead, the next two subsections just provide a brief discussion on nanofibers for PV solar cells and photothermal solar evaporator cells.

12.4.1 For photovoltaic solar cells

Organic solar cells have the advantages of being low-cost and easy to make. Their energy conversion efficiencies reach as high as those of their inorganic counterparts. Cha et al. (2021) prepared poly(3,4-ethylenedioxythiophene):poly(styrenesulfonate) (PEDOT:PSS)/ PVDF nanofiber-web-based transparent conducting electrodes (TCEs) for dye-sensitized PV textile applications. The PEDOT:PSS solution was mixed with dimethyl sulfoxide (DMSO) solvent, and the PEDOT:PSS/DMSO mixture was applied on the PVDF nanofiber web using a simple brush-painting technique to prepare ultrathin, lightweight, and highly TCEs. When the PVDF nanofiber web was treated with a 3:7 PEDOT:PSS (P3D7) and DMSO mixture, it showed a high transparency of 84% at a wavelength of 550 nm. The average sheet resistance was 1.5 kΩ/□. The figure of merit reached 1.04×10^{-4}. Preliminary results showed that the P3D7 TCE-based PV textile generated an average voltage of 73.2 mV and an average current of 0.44 mA/cm^2.

Because zinc oxide (ZnO) is one of the most used electron transport metal oxides in organic photovoltaic (OPV) cells, the addition of ZnO nanofibers into organic compounds can improve the electron transport behavior. As shown by Mohtaram et al. (2020), high-performance inverted OPV cells from electrospun ZnO nanofiber-based electron transport layers were better than their thin-film counterparts. ZnO nanofiber-based electron transporting layers with different nanofiber sizes were made by electrospinning. The electrospun ZnO nanofibers increased the short-circuit current density in the cells. The power conversion efficiency of the cells reached 7.6%. This improvement is due to the increase of the charge collection efficiency. The direct and continuous ZnO nanofiber pathway for electron extraction inside the solar cell structure helped increase the charge collection efficiency. Therefore, the application of 1D nanofiber-based charge extraction layers is a promising route for the development of high-performance OPV solar cells.

Tanveer et al. (2012) indicated that OPV solar cells consisted of electron donor polymers such as poly (3-hexylthiophene) (P3HT) and acceptor fullerenes such as (6,6)-phenyl C61 butyric acid methyl ester (PCBM). They were blended to form active layers. Ultra-fast electron transfer and large interfaces between donors and acceptors maximized the exciton dissociation. The absorption of light in the cells can be increased by increasing the thickness of the active layers. Nevertheless, a thick active layer slows down the charge carrier transport due to space charge effects and charge recombination. Increased series resistance and reduced efficiency in thick-layer cells were observed. Tanveer et al. (2012) reported that the efficiency of the PV device based on P3HT and PCBM bulk heterojunction can be improved by incorporating small-diameter electrospun ZnO nanofibers into the active layer. Fig. 12.14A shows the multilayered structure of the solar cell. A hole-blocking layer of ZnO was spin-coated on the cleaned ITO substrate. This ZnO thin film also served as the transition layer to promote the adhesion of the ZnO nanofibers to the substrate. Fig. 12.14B reveals the energy diagram of the materials for building the cell. The diameter, diffusion state, and melting condition of nanofibers can be well controlled by the calcination temperature. The thickness of the active layer was optimized for efficient PV solar cells by varying the electrospinning time. In a typical experiment, zinc acetate and 0.3 g PVP (with a molecular weight of 40,000) were dissolved in 3 mL 2-methoxyethanol. Then, 0.05 mL ethanolamine and 1.5 mL isopropanol were added. The solution was stirred for 14 h. After that, the prepared solution was loaded into a plastic syringe. The distance between the tip of the needle and the collector plate was kept at 10 cm during the electrospinning process. The flow rate was 0.35 mL/h. The voltage applied between the needle and the grounded metal collector plate was 7.5 kV. Three solar cell samples (Sample 1, Sample 2, and Sample 3) were

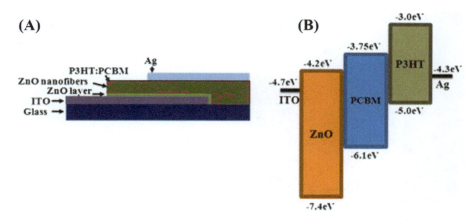

Figure 12.14 (A) Multilayered structure of the electrospun ZnO nanofibers/P3HT: PCBM photovoltaic solar cell; (B) energy diagram of the materials used in fabricating the solar cell. ZnO nanofibers/P3HT:PCBM is the active layer.
Source: Reproduced with permission from Tanveer, M., Habib, A., & Khan, M. B. (2012). Improved efficiency of organic/inorganic photovoltaic devices by electrospun ZnO nanofibers. *Materials Science and Engineering B: Solid-State Materials for Advanced Technology, 177* (13), 1144−1148. https://doi.org/10.1016/j.mseb.2012.05.025. © 2012 Elsevier B.V.

fabricated by electrospinning for 240, 300, and 360 s, respectively. The as-spun composite ZnO/PVP nanofibers were calcined at 450°C for 2.5 h. The calcination temperature and time were optimized to obtain a well-interconnected ZnO nanofiber without melting the nanofibers completely.

Tanveer et al. (2012) indicated that the ZnO nanofiber layer should not be too thick. The increase in the electrospinning time beyond the optimized value degraded the solar cell performance. The reason is that increasing the thickness of the layer caused increased series resistance. Consequently, increased traps and reduced blend infiltration through the nanofiber pores were found. The electrospinning time should be optimized to have a high active area for energy absorption and exciton dissociation. In the study performed by Tanveer et al. (2012), an optimum time of electrospinning of 300 s was recommended.

Tanveer et al. (2012) also presented the field emission SEM images of the three nanofiber-containing solar cell samples. The images in Fig. 12.15A−C reveal the cross sections of the three samples. The high magnification images in Fig. 12.15D−F show the ZnO nanofiber networks in the active layers of the three samples. The diameters of the calcined nanofibers range from 55 nm to 135 nm. The thickness of each cell's active layer was determined as 230 ± 23, 310 ± 20, and 405 ± 15 nm for Sample 1, 2, and 3, respectively. The thickness of the film without nanofibers was only 180 ± 25 nm.

There are various other nanofibers for solar cell applications. Those fibers are made from various materials via different techniques. Oxide nanofibers synthesized by thermal evaporation and vapor-phase transport approach have found applications

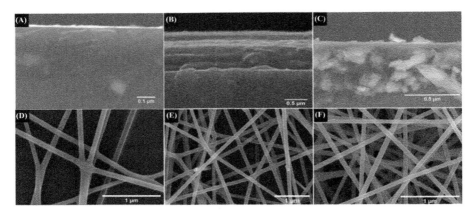

Figure 12.15 Field emission scanning electron microscopic (FE-SEM) images of various photovoltaic solar cells and the ZnO electrospun nanofibers. (A) cross-section of Sample 1 (ITO/HBL/ZnO nanofibers with 240 s electrospinning (ES) time/P3HT:PCBM/Ag), (B) cross-section of Sample 2 (ITO/HBL/ZnO nanofibers with 300 s ES time/P3HT:PCBM/Ag), (C) cross-section of Sample 3 (ITO/HBL/ZnO nanofibers with 360s ES time/P3HT:PCBM/Ag), (D) ZnO nanofibers network of Sample 1; (E) ZnO nanofibers network of Sample 2 and (F) ZnO nanofibers network of Sample 3. *ITO*, indium tin oxide; *HBL*, hole block layer.
Source: Reproduced with permission from Tanveer, M., Habib, A., & Khan, M. B. (2012). Improved efficiency of organic/inorganic photovoltaic devices by electrospun ZnO nanofibers. *Materials Science and Engineering B: Solid-State Materials for Advanced Technology*, 177(13), 1144–1148. https://doi.org/10.1016/j.mseb.2012.05.025. © 2012 Elsevier B.V.

as photoanodes in DSSCs. Electrospun metal-doped oxide nanofibers were fabricated into electrodes to improve the energy conversion efficiency of dye-sensitized cells. TiO_2 and CuO nanofibers are some of the examples. Like ZnO, the electrospun CuO nanofibers can serve as a blocking layer. Electrospun TiO_2 nanofibers were used very early in solar cells because of their high electron affinity and high electron mobility. In addition, nanofibers made from SnO_2, $SrTiO_3$, and $ZnSnO_4$ have also been studied as photoanode materials in DSSCs by various researchers. It must be pointed out that electrospun polymer blends were used as electrolytes in dye-sensitized solar cells to improve the performances of solar cells.

12.4.2 For photothermal solar cells

Numerous photothermal materials with high light absorption have been developed to fabricate advanced solar evaporator cells. But nanofibers for photothermal solar cells are still under research. In the paper published by Shi et al. (2018), a template synthesized silica-carbon-silica tri-layered nanofiber was shown high performance of harvesting solar energy and generating heat. The nanofiber sample showed a gray color in the dry state with an average solar light reflectance of 14.2%. It is believed that the high reflectance is caused by the strong subsurface scattering at the air/solid interface inside the highly porous structure. The wetted nanofiber changed color to deep black. A much smaller average reflectance of 3.40% was obtained for the wet sample. The subsurface scattering was found greatly diminished because the solid/air interfaces inside the dry membrane were replaced by solid/water interfaces in the wet sample. Shi et al. (2018) reported that by making the 3D structure to create multiple internal reflection of the incident light, extremely low reflectance can be achieved. For example, a 3D cup shaped nanofiber photothermal solar cell was prepared and the light reflectance of the cell was decreased significantly to 0.65%. This value is only 19% of that of the wetted 2D cell (3.40%). The reason for this is that more than 80% of the reflected light by the bottom of the 3D cup is absorbed by its wall.

Shi et al. (2018) also explained the logic for coating an additional silica layer on top of the carbon layer. It is mainly for adjusting the wettability. Without the outer silica layer, the carbon coated silica composite nanofiber is hydrophobic with a contact angle as high as 132 degrees. As a result, only a few water-wettable regions turned from gray to dark black color when the cell was placed under water. With the addition of silica outer, the tri-layered nanofiber became superhydrophilic with a contact angle of 0 degree. Water spread onto its surface within a short period of just 0.04 s. So, the gray-colored dry sample turned dark black immediately when contacted with water. The hydrophilicity of the nanofiber membrane is essential to ensure for the photothermal solar cell when used as a solar evaporator so that it can automatically draw water rapidly from sea of ocean source via capillary effects. In the photothermal solar cell made by Shi et al. (2018), the outer silica layer also acted as an antireflection layer, because it showed a moderate reflective index of $n = 1.5$ in the solar spectrum range. This value is higher than that of water ($n = 1.3$), but lower than that of amorphous carbon ($n = 2.3$). Consequently, the solar light reflectance of this coaxial nanofiber is extremely low.

12.5 Nanofibers for fuel cells

Composite nanofibers, especially those containing oxide nanoparticles, are highly sensitive to visible or ultraviolet light. In the work performed by Gan and Gan (2020), cobalt acetate was incorporated into PAN nanofibers through electrospinning as the cobalt oxide source. The PAN polymer-based composite nanofibers were oxidized and stabilized. Then, pyrolysis was conducted to produce a partially carbonized composite nanofiber containing cobalt oxide. The electrical and photonic properties of the composite nanofiber under visible light irradiation were carried out to evaluate the photoelectric behavior of the composite nanofiber. The p-type semiconducting behavior of the composite fiber was found by measuring the open circuit voltages of biophotofuel cells consisting of the photosensitive anode made from the composite nanofiber. The applications of the composite nanofiber for energy generation from glucose and other biomass were demonstrated.

During the nanofiber processing, a stainless-steel ejection needle was used as the spinneret for the electrospinning. The needle has a gage size of 20. PAN polymer with a molecular weight of 150, 000 was dissolved into DMF. Then, cobalt (II) acetate tetrahydrate was added. A precision syringe pump was used to inject the polymer/cobalt salt solution during electrospinning. The polymer composite fiber was preoxidized first. Then, it was put in hydrogen steam for pyrolysis during which the carbonized composite nanofiber was produced. In a typical experiment, two solutions (named I and II) were made for the nanofiber electrospinning. Solution I is the PAN polymer in DMF. Solution II represents the cobalt acetate in DMF. Solution I was made by dissolving 0.5 g PAN powder in 2.5 mL DMF under vigorous stirring for 30 min at 50°C. Solution II was obtained by adding 0.1 g cobalt acetate into 2.5 mL DMF. Then Solution II was poured into Solution I to form a uniform liquidous mixture for electrospinning. During the electrospinning, the rate of ejection was kept at 0.6 mL/h. The DC power supply was tuned to 12 kV so that the electric field intensity was kept at about 1.0 kV/cm. The fiber was deposited on a piece of soft tissue paper, which was rolled on a rotating metallic can. After air-drying, the nanofiber was cut into small pieces and oxidized in the air for 2 h. The oxidation was carried out at 250°C in a compact tube furnace. Finally, pyrolysis of the oxidized fiber at 500°C was performed in the same split tube furnace under hydrogen protection. The pyrolysis time is 2 h. The obtained nanofiber was attached to an aluminum plate to make the photoanode for the fuel cell. The configuration of the fuel cell is schematically shown in Fig. 12.16. The fuel cell consists of three electrodes: the composite nanofiber working electrode (W), the platinum counter electrode (C), and the Ag/AgCl reference electrode (R).

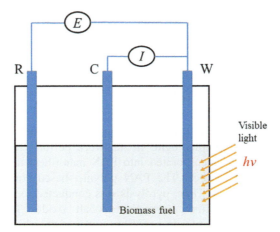

Figure 12.16 Schematic of the biophotofuel cell. W, working electrode; R, reference electrode; C, counter electrode.
Source: Reproduced with permission from Gan, Y. X. & Gan, J. B. (2020). Measuring the electrical and photonic properties of cobalt oxide-containing composite carbon fibers. *Journal of Composites Science*, 4(4). https://doi.org/10.3390/jcs4040156. © 2020 The Authors.

A SEM image of the biophotofuel cell anode in Fig. 12.17 reveals randomly aligned nanofibers. High porosity was found, but no bead formation was observed. Such a morphological feature is associated with the uniform distribution of the cobalt oxide in the CNFs. The chemical compositions of the nanofiber were analyzed by energy-dispersive X-ray diffraction spectroscopy. Mapping the selected areas was performed to generate the elemental information. The qualitative result for this fiber shows that C, O, and Co are the major elements in the fiber. The characteristic peaks can be listed as follows: For cobalt, the Kα peak is located at 6.93 keV. It is also found that the Lα (0.78 keV) peak can be seen from the spectrum shown. For oxygen, the Kα peak at 0.53 keV is shown. For carbon, the Kα peak is located at 0.28 keV.

The photoelectric behavior of the composite CF was evaluated using various solutions including sodium chloride, sugar, isopropyl, raspberry juice, and orange juice. The photoelectric response of the fiber electrode in a 5.0 wt.% sodium chloride aqueous solution revealed the p-type semiconducting behavior. The voltage data were recorded when the sample was exposed to sunlight followed by being blocked with a black screen. The first cycle is 15 s, corresponding to the visible light just being blocked ("OFF" state). Then the sunlight-blocking screen was removed, and the illumination lasted for another 15 s in the "ON" state, etc. It was clearly shown that the difference in voltage between the peak and valley ranged from 0.05 to 0.10 V when the light illumination was switched from the "OFF" to the "ON" state. The positive voltage shift at the instant sunlight illumination confirmed that the composite CF showed a general p-type behavior.

The trend of voltage variation with time for the composite CF specimen in a 5.0 wt. % sugar solution with and without visible light illumination was captured. When the sunlight was ON, the voltage generated by the specimen showed a 10-mV jump. When

Figure 12.17 SEM image of the cobalt oxide containing composite carbon fiber. The fiber was placed on aluminum foil to make the photoanode.
Source: Reproduced with permission from Gan, Y. X. & Gan, J. B. (2020). Measuring the electrical and photonic properties of cobalt oxide-containing composite carbon fibers. *Journal of Composites Science*, *4*(4). https://doi.org/10.3390/jcs4040156. © 2020 The Authors.

the light was OFF, the voltage dropped about 10 mV. A steep increase in the electrochemical potential was observed when the fiber specimen was excited by photon energy. Such a rapid response behavior is also good for building highly sensitive sensors for monitoring the existence of sugar. But the potential recovered very rapidly to the equilibrium around −0.487 V. This is because the sugar is a "hole-scavenger" as shown by the reaction formula in (Gan and Gan (2020)). The immediate recovery of the potential indicates that the hole-consumption is very fast. Therefore, the decomposition of sugar follows a fast kinetics of the reaction. The voltage data for raspberry juice, orange juice, and isopropyl (all at 5.0 wt.%) do not show any significant sensitivity to the ON and OFF of the visible light. This indicates that the hole-electron pair generation and transport in the CF are at almost the same rate. However, the differences in the values of equilibrium potential versus Ag/AgCl were found. For the composite fiber electrode in raspberry juice, the electrochemical potential was −0.7160 V. The equilibrium potential for the electrode in orange juice was −0.7704 V. For 5.0 wt.% isopropyl, this value was found at −0.7846 V. Orange juice and raspberry juice are biodegradable materials. Earlier studies on biomass energy harvesting under the photocatalysis of a titanium oxide nanotube electrode under ultraviolet (UV) light excitation were shown by Gan et al. (2011) and Gan et al. (2012). A switching behavior with the UV in ON and OFF states was observed by Gan et al. (2011, 2012), which is different from that of the cobalt oxide containing CNF anode as shown in Gan and Gan (2020).

Gan and Gan (2020) also examined the concentration effect on the equilibrium potential of the fuel cell. The cobalt oxide containing composite fiber electrode was inserted into isopropyl aqueous solutions with various concentrations of 5.0%, 10.0%, 15.0%, 20.0%, 25.0%, 35.0%, and 70.0% (all in weight percent). The time-dependent potential drift for each concentration case was investigated. No sensitivity to the ON and OFF visible light was observed for the composite fiber electrode in various isopropyl solutions within a wide range of concentrations. In addition, the measured equilibrium potential values based on the mean of the potentials were given. With the increase of the concentration of isopropyl in water, a slight increase in the equilibrium potential was found.

12.6 Concluding remarks

In this chapter, various nanofibers for batteries, solar cells, and fuel cells are summarized with an emphasis on nanofiber processing technologies. Nanofibers with enhanced performances provide many opportunities for the design and fabrication of various advanced energy conversion and storage devices and systems. Energy storage and conversion cells made by composite nanofibers are flexible, which is a much-needed property for wearable technology development.

The applications of nanofibers for batteries, PV solar cells, photothermal solar evaporator cells, and fuel cells have found success. Among the three important nanofiber processing technologies: electrospinning, template-assisted chemical deposition, and spray pyrolysis, the electrospinning technology is the most promising one due to its simplicity, versatility, and scalability. Nanofibers are very commonly used for battery separators, interlayers, and anodes. The applications for cathodes are much less studied.

In view of the research and development on nanofiber solar cells, both PV and photothermal solar cells containing nanofibers are successfully made. The photosensitive performances of various nanofibers, especially semiconducting oxide-loaded nanofibers are well demonstrated, and they found wide applications in solar energy harvesting. ZnO, TiO_2, CuO, CoO, SnO_2, MnO, $LiNbO_3$, and $SrTiO_3$ are some of the oxides showing strong performances.

The fabrication and characterization of flexible biophotofuel cell anodes using oxide nanoparticle-loaded composite CNFs are shown. Concerning the applications of nanofibers in biophotofuel cells, a case study on cobalt oxide-loaded composite CNF photoanode is presented. Cobalt oxide can be incorporated into the partially carbonized composite nanofiber via electrospinning followed by pyrolysis. The SEM analysis reveals the existence of major elements: C, Co, and O. The cobalt oxide containing CF demonstrates p-type behavior under visible light excitation. Positive potential shift has been found associated with the positively charged hole (h^+) formation in the photosensitive anode made by the composite nanofiber. The high sensitivity of the composite nanofiber to visible light allows for generating electricity from various biomass and detecting sugar in aqueous solutions. The photoelectric behavior of the electrode in various biomass solutions including raspberry juice and orange juice confirmed the power-generating capability of the nanofiber anode. The photosensitive activity in isopropyl solution is also confirmed through the open circuit potential measurement. With the increase of isopropyl concentration in water, the equilibrium potential of the electrode shifts toward the more positive value direction.

Acknowledgments

"This research was performed under an appointment to the U.S. Department of Homeland Security (DHS) Science & Technology (S&T) Directorate Office of University Programs Summer Research Team Program for Minority Serving Institutions, administered by the Oak Ridge Institute for Science and Education (ORISE) through an interagency agreement between the U.S. Department of Energy (DOE) and DHS. ORISE is managed by ORAU under DOE contract number DE-SC0014664. All opinions expressed in this work are the author's and do not necessarily reflect the policies and views of DHS, DOE, or ORAU/ORISE."

References

Alcoutlabi, M., Lee, H., Watson, J. V., & Zhang, X. (2013). Preparation and properties of nanofiber-coated composite membranes as battery separators via electrospinning. *Journal of Materials Science*, *48*(6), 2690−2700. Available from https://doi.org/10.1007/s10853-012-7064-0.

An, M. Y., Kim, H. T., & Chang, D. R. (2014). Multilayered separator based on porous polyethylene layer, Al 2O₃ layer, and electro-spun PVdF nanofiber layer for lithium batteries. *Journal of Solid State Electrochemistry, 18*(7), 1807–1814. Available from https://doi.org/10.1007/s10008-014-2412-4, http://link.springer-ny.com/link/service/journals/10008/index.htm.

Brun, N., Sakaushi, K., Yu, L., Giebeler, L., Eckert, J., & Titirici, M. M. (2013). Hydrothermal carbon-based nanostructured hollow spheres as electrode materials for high-power lithium-sulfur batteries. *Physical Chemistry Chemical Physics, 15*(16), 6080–6087. Available from https://doi.org/10.1039/c3cp50653c.

Cai, Z. J., Shi, X. J., & Fan, Y. N. (2013). Electrochemical properties of electrospun polyindole nanofibers as a polymer electrode for lithium ion secondary battery. *Journal of Power Sources, 227*, 53–59. Available from https://doi.org/10.1016/j.jpowsour.2012.10.081.

Cai, Z. J., Zhang, Q., Zhu, C., Song, X. Y., & Liu, Y. P. (2017). Development of a sodium/electrospun poly(5-cyanoindole) nanofiber secondary battery system with high performance. *Synthetic Metals, 231*, 15–18. Available from https://doi.org/10.1016/j.synthmet.2017.06.010, http://www.journals.elsevier.com/synthetic-metals/.

Cha, S., Lee, E., & Cho, G. (2021). Fabrication of poly(3,4-ethylenedioxythiophene):poly(styrenesulfonate)/poly(vinylidene fluoride) nanofiber-web-based transparent conducting electrodes for dye-sensitized photovoltaic textiles. *ACS Applied Materials and Interfaces, 13*(24), 28855–28863. Available from https://doi.org/10.1021/acsami.1c06081, http://pubs.acs.org/journal/aamick.

Chen, J. J., Zhang, Q., Shi, Y. N., Qin, L. L., Cao, Y., Zheng, M. S., & Dong, Q. F. (2012). A hierarchical architecture S/MWCNT nanomicrosphere with large pores for lithium sulfur batteries. *Physical Chemistry Chemical Physics, 14*(16), 5376–5382. Available from https://doi.org/10.1039/c2cp40141j.

Chen, S., Zhang, Z., Li, L., & Yuan, W. (2018). Covalently-bonded poly(vinyl alcohol)-silica composite nanofiber separator with enhanced wettability and thermal stability for lithium-ion battery. *Chemistry Select, 3*(47), 13365–13371. Available from https://doi.org/10.1002/slct.201802794, http://onlinelibrary.wiley.com/journal/10.1002/(ISSN)2365-6549.

Dhamodharan, P., Chen, J., & Manoharan, C. (2021). Fabrication of In doped ZnO thin films by spray pyrolysis as photoanode in DSSCs. *Surfaces and Interfaces, 23*100965. Available from https://doi.org/10.1016/j.surfin.2021.100965.

Dimesso, L., Spanheimer, C., Jaegermann, W., Zhang, Y., & Yarin, A. L. (2013). LiCoPO4 − 3D carbon nanofiber composites as possible cathode materials for high voltage applications. *Electrochimica Acta, 95*, 38–42. Available from https://doi.org/10.1016/j.electacta.2013.02.002.

Emre, A., & Gerber Kotov, N. (2019). Abstracts of Papers of the American Chemical Society, *257th National Meeting of the American-Chemical-Society (ACS)*, Orlando, FL 309, ACS Orlando Unpublished content Aramid nanofiber composite separators for high performance lithium-sulfur batteries, 257 (31.03.2019).

Gan, Y. X., Gan, B. J., Clark, E., Su, L., & Zhang, L. (2012). Converting environmentally hazardous materials into clean energy using a novel nanostructured photoelectrochemical fuel cell. *Materials Research Bulletin, 47*(9), 2380–2388. Available from https://doi.org/10.1016/j.materresbull.2012.05.049.

Gan, Y. X., Gan, B. J., & Su, L. (2011). Biophotofuel cell anode containing self-organized titanium dioxide nanotube array. *Materials Science and Engineering B: Solid-State Materials for Advanced Technology, 176*(15), 1197–1206. Available from https://doi.org/10.1016/j.mseb.2011.06.014.

Gan, Y. X., & Gan, J. B. (2020). Measuring the electrical and photonic properties of cobalt oxide-containing composite carbon fibers. *Journal of Composites Science, 4*(4). Available from https://doi.org/10.3390/jcs4040156, https://www.mdpi.com/2504-477X/4/4/156/pdf.

Gao, L., Li, J., Ju, J., Cheng, B., Kang, W., & Deng, N. (2021). High-performance all-solid-state polymer electrolyte with fast conductivity pathway formed by hierarchical structure polyamide 6 nanofiber for lithium metal battery. *Journal of Energy Chemistry, 54*, 644−654. Available from https://doi.org/10.1016/j.jechem.2020.06.035, http://elsevier.com/journals/journal-of-energy-chemistry/2095-4956.

Gao, X., Sheng, L., Xie, X., Yang, L., Bai, Y., Dong, H., Liu, G., Wang, T., Huang, X., & He, J. (2022). Morphology optimizing of polyvinylidene fluoride (PVDF) nanofiber separator for safe lithium-ion battery. *Journal of Applied Polymer Science, 139*(20). Available from https://doi.org/10.1002/app.52154, http://onlinelibrary.wiley.com/journal/10.1002/(ISSN)1097-4628.

Gong, X., Yang, J., Jiang, Y., & Mu, S. (2014). Application of electrospinning technique in power lithium-ion batteries. *Progress in Chemistry, 26*(1), 41−47. Available from https://doi.org/10.7536/PC130641.

Hu, M., Ma, Q., Yuan, Y., Pan, Y., Chen, M., Zhang, Y., & Long, D. (2020). Grafting polyethyleneimine on electrospun nanofiber separator to stabilize lithium metal anode for lithium sulfur batteries. *Chemical Engineering Journal, 388*. Available from https://doi.org/10.1016/j.cej.2020.124258, http://www.elsevier.com/inca/publications/store/6/0/1/2/7/3/index.htt.

Huang, C., Ji, H., Guo, B., Luo, L., Xu, W., Li, J., & Xu, J. (2019). Composite nanofiber membranes of bacterial cellulose/halloysite nanotubes as lithium ion battery separators. *Cellulose, 26*(11), 6669−6681. Available from https://doi.org/10.1007/s10570-019-02558-y, http://www.springer.com/journal/10570.

Huang, C., Ji, H., Yang, Y., Guo, B., Luo, L., Meng, Z., Fan, L., & Xu, J. (2020). TEMPO-oxidized bacterial cellulose nanofiber membranes as high-performance separators for lithium-ion batteries. *Carbohydrate Polymers, 230*115570. Available from https://doi.org/10.1016/j.carbpol.2019.115570.

Huang, P., & Wang, Y. (2019). Carpet-like TiO_2 nanofiber interlayer as advanced absorber for high performance Li−S batterie. *International Journal of Electrochemical Science, 14*(6), 5154−5160. Available from https://doi.org/10.20964/2019.06.59, http://www.electrochemsci.org/papers/vol14/140605154.pdf.

Ji, L., Medford, A. J., & Zhang, X. (2009). Porous carbon nanofibers loaded with manganese oxide particles: Formation mechanism and electrochemical performance as energy-storage materials. *Journal of Materials Chemistry, 19*(31), 5593. Available from https://doi.org/10.1039/b905755b.

Ji, L., Toprakci, O., Alcoutlabi, M., Yao, Y., Li, Y., Zhang, S., Guo, B., Lin, Z., & Zhang, X. (2012). α-Fe_2O_3 nanoparticle-loaded carbon nanofibers as stable and high-capacity anodes for rechargeable lithium-ion batteries. *ACS Applied Materials and Interfaces, 4*(5), 2672−2679. Available from https://doi.org/10.1021/am300333s.

Jiang, X., Zhang, S., Zou, B., Li, G., Yang, S., Zhao, Y., Lian, J., Li, H., & Ji, H. (2022). Electrospun CoSe@NC nanofiber membrane as an effective polysulfides adsorption-catalysis interlayer for Li−S batteries. *Chemical Engineering Journal, 430*131911. Available from https://doi.org/10.1016/j.cej.2021.131911.

Kwon, O. H., Oh, J. H., Gu, B., Jo, M. S., Oh, S. H., Kang, Y. C., Kim, J. K., Jeong, S. M., & Cho, J. S. (2020). Porous SnO_2/C Nanofiber Anodes and $LiFePO_4$/C nanofiber cathodes with a wrinkle structure for stretchable lithium polymer batteries with high electrochemical performance. *Advanced Science*, *7*(17). Available from https://doi.org/10.1002/advs.202001358, http://onlinelibrary.wiley.com/journal/10.1002/(ISSN)2198-3844.

Lee, B.-S. (2020). A review of recent advancements in electrospun anode materials to improve rechargeable lithium battery performance. *Polymers*, *12*(9), 2035. Available from https://doi.org/10.3390/polym12092035.

Lee, H., Alcoutlabi, M., Toprakci, O., Xu, G., Watson, J. V., & Zhang, X. (2014). Preparation and characterization of electrospun nanofiber-coated membrane separators for lithium-ion batteries. *Journal of Solid State Electrochemistry*, *18*(9), 2451−2458. Available from https://doi.org/10.1007/s10008-014-2501-4, http://link.springer-ny.com/link/service/journals/10008/index.htm.

Li, K. F., & Yin, X. Y. (2018). Polyphenylene oxide-based nanofiber separator prepared by electrospinning method for lithium-ion batteries. *Cailiao Gongcheng/Journal of Materials Engineering*, *46*(10), 120−126. Available from https://doi.org/10.11868/j.issn.1001-4381.2017.000464, http://jme.biam.ac.cn/CN/1001-4381/home.shtml.

Li, M., Sheng, L., Xu, R., Yang, Y., Bai, Y., Song, S., Liu, G., Wang, T., Huang, X., & He, J. (2021). Enhanced the mechanical strength of polyimide (PI) nanofiber separator via PAALi binder for lithium ion battery. *Composites Communications*, *24*. Available from https://doi.org/10.1016/j.coco.2020.100607, http://www.journals.elsevier.com/composites-communications.

Li, S., Chen, C., Fu, K., Xue, L., Zhao, C., Zhang, S., Hu, Y., Zhou, L., & Zhang, X. (2014). Comparison of Si/C, Ge/C and Sn/C composite nanofiber anodes used in advanced lithium-ion batteries. *Solid State Ionics*, *254*, 17−26. Available from https://doi.org/10.1016/j.ssi.2013.10.063.

Li, Y., Li, Q., & Tan, Z. (2019). A review of electrospun nanofiber-based separators for rechargeable lithium-ion batteries. *Journal of Power Sources*, *443*227262. Available from https://doi.org/10.1016/j.jpowsour.2019.227262.

Li, Y., Ma, X., Deng, N., Kang, W., Zhao, H., Li, Z., & Cheng, B. (2017). Electrospun SiO_2/PMIA nanofiber membranes with higher ionic conductivity for high temperature resistance lithium-ion batteries. *Fibers and Polymers*, *18*(2), 212−220. Available from https://doi.org/10.1007/s12221-017-6772-0, http://link.springer.com/journal/volumesAndIssues/12221.

Li, Y., Xu, G., Yao, Y., Xue, L., Zhang, S., Lu, Y., Toprakci, O., & Zhang, X. (2013). Improvement of cyclability of silicon-containing carbon nanofiber anodes for lithium-ion batteries by employing succinic anhydride as an electrolyte additive. *Journal of Solid State Electrochemistry*, *17*(5), 1393−1399. Available from https://doi.org/10.1007/s10008-013-2005-7.

Li, Y., Zhang, W., Dou, Q., Wong, K. W., & Ng, K. M. (2019). $Li_7La_3Zr_2O_{12}$ ceramic nanofiber-incorporated composite polymer electrolytes for lithium metal batteries. *Journal of Materials Chemistry A*, *7*(7), 3391−3398. Available from https://doi.org/10.1039/c8ta11449h, http://pubs.rsc.org/en/journals/journal/ta.

Liu, K., Wu, M., Jiang, H., Lin, Y., Xu, J., & Zhao, T. (2022). A Janus-faced, perovskite nanofiber framework reinforced composite electrolyte for high-voltage solid lithium-metal batteries. *Journal of Power Sources*, *526*. Available from https://doi.org/10.1016/j.jpowsour.2022.231172, https://www.journals.elsevier.com/journal-of-power-sources.

Liu, M., Deng, N., Ju, J., Fan, L., Wang, L., Li, Z., Zhao, H., Yang, G., Kang, W., Yan, J., & Cheng, B. (2019). A review: Electrospun nanofiber materials for lithium-sulfur batteries. *Advanced Functional Materials*, *29*(49)1905467. Available from https://doi.org/10.1002/adfm.201905467.

Liu, S., Wang, Z., Yu, C., Wu, H. B., Wang, G., Dong, Q., Qiu, J., Eychmüller, A., & Lou, X. W. (2013). A flexible TiO$_2$(B)-based battery electrode with superior power rate and ultralong cycle life. *Advanced Materials.*, *25*(25), 3462−3467. Available from https://doi.org/10.1002/adma.201300953.

Liu, Y., Huang, K., Fan, Y., Zhang, Q., Sun, F., Gao, T., Wang, Z., & Zhong, J. (2013). Binder-free Si nanoparticles@carbon nanofiber fabric as energy storage material. *Electrochimica Acta*, *102*, 246−251. Available from https://doi.org/10.1016/j.electacta.2013.04.021.

Liu, Y. E., & Zhang, M. G. (2022). Investigation of the effect of anion/cation-modified cellulose nanofibers/MXene composite aerogels on the high-performance lithium-sulfur batteries. *Ionics*, *28*(6), 2805−2815. Available from https://doi.org/10.1007/s11581-022-04498-3, http://www.springerlink.com/content/120106/.

Liu, Z., Zhang, X., Poyraz, S., Surwade, S. P., & Manohar, S. K. (2010). Oxidative template for conducting polymer nanoclips. *Journal of the American Chemical Society*, *132*(38), 13158−13159. Available from https://doi.org/10.1021/ja105966c.

Meng, L., Li, Y., Lin, Q., Long, J., Wang, Y., & Hu, J. (2021). Nitrogen and oxygen dual self-doped flexible PPTA nanofiber carbon paper as an effective interlayer for lithium-sulfur batteries. *ACS Applied Energy Materials*, *4*(8), 8592−8603. Available from https://doi.org/10.1021/acsaem.1c01780, http://pubs.acs.org/journal/aaemcq.

Miao, Y. E., Zhu, G. N., Hou, H., Xia, Y. Y., & Liu, T. (2013). Electrospun polyimide nanofiber-based nonwoven separators for lithium-ion batteries. *Journal of Power Sources*, *226*, 82−86. Available from https://doi.org/10.1016/j.jpowsour.2012.10.027.

Mohtaram, F., Borhani, S., Ahmadpour, M., Fojan, P., Behjat, A., Rubahn, H. G., & Madsen, M. (2020). Electrospun ZnO nanofiber interlayers for enhanced performance of organic photovoltaic devices. *Solar Energy*, *197*, 311−316. Available from https://doi.org/10.1016/j.solener.2019.12.079, http://www.elsevier.com/inca/publications/store/3/2/9/index.htt.

Moon, S., Yun, J., Lee, J. Y., Park, G., Kim, S. S., & Lee, K. J. (2019). Mass-production of electrospun carbon nanofiber containing SiO$_x$ for lithium-ion batteries with enhanced capacity. *Macromolecular Materials and Engineering*, *304*(3). Available from https://doi.org/10.1002/mame.201800564, http://onlinelibrary.wiley.com/journal/10.1002/(ISSN)1439-2054.

Nakazawa, S., Matsuda, Y., Ochiai, M., Inafune, Y., Yamato, M., Tanaka, M., & Kawakami, H. (2021). Enhancing Lithium ion conductivity and all-solid-state secondary battery performance in polymer composite electrolyte membranes with β-crystalline-rich poly(vinylidene fluoride) nanofibers. *Electrochimica Acta*, *394*139114. Available from https://doi.org/10.1016/j.electacta.2021.139114.

Nan, D., Wang, J. G., Huang, Z. H., Wang, L., Shen, W., & Kang, F. (2013). Highly porous carbon nanofibers from electrospun polyimide/SiO$_2$ hybrids as an improved anode for lithium-ion batteries. *Electrochemistry Communications*, *34*, 52−55. Available from https://doi.org/10.1016/j.elecom.2013.05.010.

Pakki, T., Sarma, S. S., Hebalkar, N. Y., Anandan, S., Mantravadi, K. M., & Rao, T. N. (2017). Enhanced electrochemical performance of electrospun SiO$_2$ nanofibers as binder-free anode. *Chemistry Letters*, *46*(7), 1007−1009. Available from https://doi.org/10.1246/cl.170080, http://www.journal.csj.jp/doi/pdf/10.1246/cl.170080.

Peng, T., Liu, C., Hou, X., Zhang, Z., Wang, C., Yan, H., Lu, Y., Liu, X., & Luo, Y. (2017). Control growth of mesoporous nickel tungstate nanofiber and its application as anode material for lithium-ion batteries. *Electrochimica Acta*, *224*, 460−467. Available from https://doi.org/10.1016/j.electacta.2016.11.154, http://www.journals.elsevier.com/electrochimica-acta/.

Qie, L., Chen, W. M., Wang, Z. H., Shao, Q. G., Li, X., Yuan, L. X., Hu, X. L., Zhang, W. X., & Huang, Y. H. (2012). Nitrogen-doped porous carbon nanofiber webs as anodes for lithium ion batteries with a superhigh capacity and rate capability. *Advanced Materials*, *24*(15), 2047−2050. Available from https://doi.org/10.1002/adma.201104634.

Shi, F., Chen, C., & Xu, Z. L. (2021). Recent advances on electrospun nanofiber materials for post-lithium ion batteries. *Advanced Fiber Materials*, *3*(5), 275−301. Available from https://doi.org/10.1007/s42765-021-00070-2, https://www.springer.com/journal/42765.

Shi, Y., Zhang, C., Li, R., Zhuo, S., Jin, Y., Shi, L., Hong, S., Chang, J., Ong, C., & Wang, P. (2018). Solar evaporator with controlled salt precipitation for zero liquid discharge desalination. *Environmental Science and Technology*, *52*(20), 11822−11830. Available from https://doi.org/10.1021/acs.est.8b03300, http://pubs.acs.org/journal/esthag.

Sun, Y., Chen, K., Zhang, C., Yu, H., Wang, X., Yang, D., Wang, J., Huang, G., Zhang, S., & Novel, A. (2022). Material for high-performance Li−O_2 battery separator: Polyetherketone nanofiber membrane. *Small*, *18*(21)2201470. Available from https://doi.org/10.1002/smll.202201470.

Tanveer, M., Habib, A., & Khan, M. B. (2012). Improved efficiency of organic/inorganic photovoltaic devices by electrospun ZnO nanofibers. *Materials Science and Engineering B: Solid-State Materials for Advanced Technology*, *177*(13), 1144−1148. Available from https://doi.org/10.1016/j.mseb.2012.05.025.

Wang, J. G., Yang, Y., Huang, Z. H., & Kang, F. (2015). MnO-carbon hybrid nanofiber composites as superior anode materials for lithium-ion batteries. *Electrochimica Acta*, *170*, 164−170. Available from https://doi.org/10.1016/j.electacta.2015.04.157, http://www.journals.elsevier.com/electrochimica-acta/.

Wang, L., Ding, C. X., Zhang, L. C., Xu, H. W., Zhang, D. W., Cheng, T., & Chen, C. H. (2010). A novel carbon-silicon composite nanofiber prepared via electrospinning as anode material for high energy-density lithium ion batteries. *Journal of Power Sources*, *195*(15), 5052−5056. Available from https://doi.org/10.1016/j.jpowsour.2010.01.088.

Wang, L., Liu, F., Shao, W., Cui, S., Zhao, Y., Zhou, Y., & He, J. (2019). Graphite oxide dopping polyimide nanofiber membrane via electrospinning for high performance lithium-ion batteries. *Composites Communications*, *16*, 150−157. Available from https://doi.org/10.1016/j.coco.2019.09.004, http://www.journals.elsevier.com/composites-communications.

Wang, L., Yu, Y., Chen, P. C., Zhang, D. W., & Chen, C. H. (2008). Electrospinning synthesis of C/Fe_3O_4 composite nanofibers and their application for high performance lithium-ion batteries. *Journal of Power Sources*, *183*(2), 717−723. Available from https://doi.org/10.1016/j.jpowsour.2008.05.079.

Wang, X., Bi, X., Wang, S., Zhang, Y., Du, H., & Lu, J. (2018). High-rate and long-term cycle stability of Li−S batteries enabled by Li_2S/TiO_2-impregnated hollow carbon nanofiber cathodes. *ACS Applied Materials and Interfaces*, *10*(19), 16552−16560. Available from https://doi.org/10.1021/acsami.8b03201, http://pubs.acs.org/journal/aamick.

Wang, Y. X., Wen, X. F., Chen, J., & Wang, S. N. (2015). Foamed mesoporous carbon/silicon composite nanofiber anode for lithium ion batteries. *Journal of Power Sources*, *281*, 285−292. Available from https://doi.org/10.1016/j.jpowsour.2015.01.184, http://www.journals.elsevier.com/journal-of-power-sources/.

Wang, Z., Xiong, X., Qie, L., & Huang, Y. (2013). High-performance lithium storage in nitrogen-enriched carbon nanofiber webs derived from polypyrrole. *Electrochimica Acta*, *106*, 320−326. Available from https://doi.org/10.1016/j.electacta.2013.05.088, http://www.journals.elsevier.com/electrochimica-acta/.

Watanabe, T., Inafune, Y., Tanaka, M., Mochizuki, Y., Matsumoto, F., & Kawakami, H. (2019). Development of all-solid-state battery based on lithium ion conductive polymer nanofiber framework. *Journal of Power Sources*, *423*, 255−262. Available from https://doi.org/10.1016/j.jpowsour.2019.03.066, https://www.journals.elsevier.com/journal-of-power-sources.

Wu, K., Hu, Y., Cheng, Z., Pan, P., Jiang, L., Mao, J., Ni, C., Gu, X., & Wang, Z. (2019). Carbonized regenerated silk nanofiber as multifunctional interlayer for high-performance lithium-sulfur batteries. *Journal of Membrane Science*, *592*. Available from https://doi.org/10.1016/j.memsci.2019.117349, http://www.elsevier.com/locate/memsci.

Wu, Y., Bobba, C. V. R., & Ramakrishna, S. (2013). Research and application of carbon nanofiber and nanocomposites via electrospinning technique in energy conversion systems. *Current Organic Chemistry*, *17*(13), 1411−1423. Available from https://doi.org/10.2174/1385272811317130008.

Xia, J., Yuan, Y., Yan, H., Liu, J., Zhang, Y., Liu, L., Zhang, S., Li, W., Yang, X., Shu, H., Wang, X., & Cao, G. (2020). Electrospun SnSe/C nanofibers as binder-free anode for lithium−ion and sodium-ion batteries. *Journal of Power Sources*, *449*. Available from https://doi.org/10.1016/j.jpowsour.2019.227559, https://www.journals.elsevier.com/journal-of-power-sources.

Xia, J., Zhang, X., Yang, Y. A., Wang, X., & Yao, J. N. (2021). Electrospinning fabrication of flexible, foldable, and twistable $Sb_2S_3/TiO_2/C$ nanofiber anode for lithium ion batteries. *Chemical Engineering Journal*, *413*127400. Available from https://doi.org/10.1016/j.cej.2020.127400.

Xiang, H., Liu, X., Deng, N., Cheng, B., & Kang, W. (2022). A novel EDOT/FCo-doped PMIA nanofiber membrane as separator for high-performance lithium-sulfur battery. *Chemistry − An Asian Journal*, *17*(20). Available from https://doi.org/10.1002/asia.202200669, http://onlinelibrary.wiley.com/journal/10.1002/(ISSN)1861-471X.

Xu, Y., Zhu, J. W., Fang, J. B., Li, X., Yu, M., & Long, Y. Z. (2020). Electrospun high-thermal-resistant inorganic composite nonwoven as lithium-ion battery separator. *Journal of Nanomaterials*. Available from https://doi.org/10.1155/2020/3879040, 2020, http://www.hindawi.com/journals/jnm/.

Yang, Y., Wang, S., Zhang, L., Deng, Y., Xu, H., Qin, X., & Chen, G. (2019). CoS-interposed and Ketjen black-embedded carbon nanofiber framework as a separator modulation for high performance Li−S batteries. *Chemical Engineering Journal*, *369*, 77−86. Available from https://doi.org/10.1016/j.cej.2019.03.034, http://www.elsevier.com/inca/publications/store/6/0/1/2/7/3/index.htt.

Yanilmaz, M. (2020). Evaluation of electrospun PVA/SiO_2 nanofiber separator membranes for lithium-ion batteries. *Journal of the Textile Institute*, *111*(3), 447−452. Available from https://doi.org/10.1080/00405000.2019.1642070, http://www.tandf.co.uk/journals/titles/00405000.asp.

Yanilmaz, M., Dirican, M., & Zhang, X. (2014). Evaluation of electrospun SiO_2/nylon 6,6 nanofiber membranes as a thermally-stable separator for lithium-ion batteries. *Electrochimica Acta*, *133*, 501−508. Available from https://doi.org/10.1016/j.electacta.2014.04.109.

Yanilmaz, M., Lu, Y., Zhu, J., & Zhang, X. (2016). Silica/polyacrylonitrile hybrid nanofiber membrane separators via sol-gel and electrospinning techniques for lithium-ion batteries. *Journal of Power Sources*, *313*, 205−212. Available from https://doi.org/10.1016/j.jpowsour.2016.02.089.

Yu, H., Zhang, J., Xia, M., Deng, C., Zhang, X., Zheng, R., Chen, S., Shu, J., & Wang, Z. B. (2020). PNb_9O_{25} nanofiber as a high-voltage anode material for advanced lithium ions batteries. *Journal of Materiomics*, 6(4), 781−787. Available from https://doi.org/10.1016/j.jmat.2020.07.003, https://www.journals.elsevier.com/journal-of-materiomics/.

Yuan, Y., Fang, Z., & Liu, M. (2017). Sulfur/hollow carbon nanofiber composite as cathode material for lithium-sulfur batteries. *International Journal of Electrochemical Science*, 12(2), 1025−1033. Available from https://doi.org/10.20964/2017.02.27, http://www.electrochemsci.org/papers/vol12/120201025.pdf.

Yuan, Y., Lu, H., Fang, Z., & Chen, B. (2016). Preparation and performance of sulfur−carbon composite based on hollow carbon nanofiber for lithium-sulfur batteries. *Ionics*, 22(9), 1509−1515. Available from https://doi.org/10.1007/s11581-016-1677-2, http://www.springerlink.com/content/120106/.

Yue, X. Y., Wang, W. W., Wang, Q. C., Meng, J. K., Zhang, Z. Q., Wu, X. J., Yang, X. Q., & Zhou, Y. N. (2018). CoO nanofiber decorated nickel foams as lithium dendrite suppressing host skeletons for high energy lithium metal batteries. *Energy Storage Materials*, 14, 335−344. Available from https://doi.org/10.1016/j.ensm.2018.05.017, http://www.journals.elsevier.com/energy-storage-materials/.

Zhang, P., Huang, L., Li, Y., Ren, X., Deng, L., & Yuan, Q. (2016). Si/Ni$_3$Si-encapulated carbon nanofiber composites as three-dimensional network structured anodes for lithium-ion batteries. *Electrochimica Acta*, 192, 385−391. Available from https://doi.org/10.1016/j.electacta.2016.01.223, http://www.journals.elsevier.com/electrochimica-acta/.

Zhang, X., Ji, L., Toprakci, O., Liang, Y., & Alcoutlabi, M. (2011). Electrospun nanofiber-based anodes, cathodes, and separators for advanced lithium-ion batteries. *Polymer Reviews*, 51(3), 239−264. Available from https://doi.org/10.1080/15583724.2011.593390.

Zhao, Q., Yang, S., Wang, H., & Zhang, H. (2021). Modifying polysulfide conversion and confinement by employing MoS$_2$ nanofiber interlayer for high electrochemical performance. *Ionics*, 27(6), 2615−2619. Available from https://doi.org/10.1007/s11581-021-04014-z, http://www.springerlink.com/content/120106/.

Zhao, Y., Yan, J., Yu, J., & Ding, B. (2022). Electrospun nanofiber electrodes for lithium-ion batteries. *Macromolecular Rapid Communications*2200740. Available from https://doi.org/10.1002/marc.202200740, 2022, http://onlinelibrary.wiley.com/journal/10.1002/(ISSN)1521-3927.

Zhou, C., Wang, J., Zhu, X., Chen, K., Ouyang, Y., Wu, Y., Miao, Y. E., & Liu, T. (2021). A dual-functional poly(vinyl alcohol)/poly(lithium acrylate) composite nanofiber separator for ionic shielding of polysulfides enables high-rate and ultra-stable Li−S batteries. *Nano Research*, 14(5), 1541−1550. Available from https://doi.org/10.1007/s12274-020-3213-y, http://www.springer.com/materials/nanotechnology/journal/12274.

Tribological applications/lubricant additive applications of nanofibers and their composites

13

Muhammad Ullah[1], Sidra Subhan[1], Muhammad Shakir[2], Ata Ur Rahman[1] and Muhammad Yaseen[1]
[1]Institute of Chemical Sciences, University of Peshawar, Peshawar, Khyber Pakhtunkhwa, Pakistan, [2]Institute of Space Technology, Islamabad, Pakistan

13.1 Introduction

Tribology is the science of lubrication, friction, wear, and other surface interactions in motion. Among other factors, lubrication is important in the field because it helps to reduce the wear and friction between the moving parts in a variety of sectors including transportation, manufacturing, and energy generation. For the fast mechanized system, proper lubrication of the mechanical parts is significantly important to lower friction and wear, which increases the energy efficiency of the system and also extends the lifespan of the equipment (Bobzin et al., 2011). Enormous research has been done on the making and utilization of quality additives, which can reduce the tension between the moving parts in a cost-effective and simply mechanized manner (Briscoe & Sinha, 2008). With the advancement in the commuting industry, considerable attention has been paid to the composites used as lubricant additives. For this purpose, materials possessing distinctive qualities, including large surface area, high aspect ratio, and superior mechanical capabilities, are widely experimented. Commonly used lubricant additives tend to improve the lubricant performance and increase the longevity of the mechanical parts (Gupta, 2017; Wang, 2022). Owing to their small size and high surface area-to-volume ratio, t nanofiber (NF)-based composites are more susceptible to be used as promising lubricant additives, which not only decrease the friction and wear but also minimize the accumulation of detrimental deposits on the mechanical surfaces (Huang, 2003; Rana et al., 2021; Zhang, 2016). Despite their usefulness, the application of NFs and their composites as lubricant additives is still in its growing stage, and further exploration is necessary to fully comprehend their characteristics and applications.

It is important to keep in mind that the tribological characteristics of NF lubricant additives heavily rely on the specific materials employed, the production techniques, and the application situations (Huang, 2017a). Hence, the performance of NFs can be further enhanced by the addition of friction modifiers and develop NF-reinforced composites. Moreover, NFs can also be employed in self-healing lubricants owing to their high surface area, which enables them to absorb more healing and recovery agents

(Karger-Kocsis & Kéki, 2017). These materials have the ability to replace the existing lubricant additives that provide up-to-date prospects for tribological applications.

This chapter will go into considerable depth about the qualities, benefits, and drawbacks of using NFs and their composites as lubricant additives in tribological applications. It will also go over recent findings and suggested future studies in this area.

13.2 Nanofibers and their composites

NFs are fiber-shaped, one-dimensional nanometer-scale materials with diameters ranging from tens to hundreds of nanometers, or simply these are the fibers with nanometer-scale diameters, as shown in Fig. 13.1. Polymers, ceramics, and metals are just a few of the numerous materials that may be used to construct different NF structures. They have special qualities that make them suitable for various applications due to their compact size and high surface area to volume ratio.

Moreover, high mechanical and thermal performance can be achieved by synergistically blending the NFs with other substances, such as metal nanoparticles (Briscoe & Sinha, 2008).

Some examples of NF composites are:

13.2.1 Polymer–metal composites

Combining metal nanoparticles with polymer-based NF composites can enhance the thermal stability and electrical conductivity of the pristine material. For example,

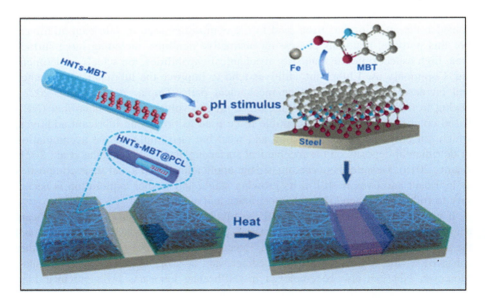

Figure 13.1 A typical nanofiber-based composite.

polystyrene reinforced with steel powder (polymer−metal), transition metal dichalcogenides (TMDs) in which molybdenum, tungsten, tantalum (Mo, W, and Ta), etc. are sandwiched in a bilayer of chalcogens with sulfur, selenium, and tellurium (S, Se, and Te) atoms and boron nitride (BN). More recently, Mxenes, a group of two-dimensional inorganic compounds that were originally identified in the field of materials research in 2011, composed of layers of transition metal carbides, nitrides, or carbonitrides that are only a few atoms thick (Prabhakar, 2018; Ronchi, 2023; Zahran & Marei, 2019).

13.2.2 Ceramic-polymer composites

By combining polymers and ceramic NFs, composite materials with increased mechanical durability and strength may be produced. They are often produced by incorporating ceramic NFs or particles into a polymer matrix, which produces a substance with special qualities that cannot be found in either ceramic or polymer materials by themselves. Ceramic-polymer composites can be utilized in tribological applications to enhance the wear and friction characteristics of lubricants. These composites then offer a firmer, smoother surface that lowers friction and wear by adding ceramic particles or NFs to the lubricant, thus improving the equipment. The following are some examples of ceramic-polymer composites utilized in tribological applications, which include alumina and polytetrafluoroethylene (PTFE) composites, polyimide (PI) and silicon carbide, boron carbide polyamide, and imide composites (Chao, 2019; Dziadek et al., 2017; Haidar et al., 2018; Wozniak, 2016; Zakaulla et al., 2020).

13.2.3 Metal-metal composites

By fusing metal NFs with other metals, composites with increased mechanical strength and wear resistance may be produced. Metal-metal composites are utilized in tribology to enhance the wear and friction characteristics of lubricants. These composites can offer a smoother, tougher surface that decreases wear and friction, resulting in better lubrication and longer machine life. For tribological applications, metal-metal composites can be made in a variety of ways; some of which are explored briefly as follows (Vencl, 2021).

13.2.3.1 Metal matrix composites

Metal matrix composites are substances made of a metal matrix that has been reinforced with one or more components, typically ceramic or intermetallic particles. In the field of tribology, which explores friction, wear, and lubrication, it has been found that metal matrix composites have shown better tribological characteristics than their unreinforced equivalents. These composite materials have a metal matrix strengthened by hard metal/ceramic particles. One such is the usage of silicon carbide-reinforced aluminum matrix composites, which have been proven to have better tribological and wear-resistant properties (Suryanarayanan et al., 2013).

Some of the advantages of metal matrix composites in the field of tribology include enhanced hardness, improved wear resistance, increased lubrication, enhanced load-bearing capability, and minimized friction and wear and tear. These advantages make the metal matrix composites captivating materials for implementation in a wide range of tribological applications, such as engine parts, bearings, gears, and wear-resistant coatings. It is important to note that the particular tribological characteristics of metal matrix composites can vary depending on the kind of reinforcement particles and the processing conditions applied to manufacture the composite (Sahoo & Das, 2011).

There are many methods of synthesizing these metal matrix composites, such as powder metallurgy in which the metal matrix and reinforcing powders are to be mixed. The mixture must be compressed into the desired shape followed by heating to a temperature near or above the melting point of the metal matrix. These metal matrix composites can also be prepared by an in situ casting method, which involves dispersing reinforcing particles in the liquid matrix metal before pouring the mixture into a mold followed by heating the composite after solidification in order to achieve the desired properties. Stir casting, where a mechanical stirrer is used to stir the reinforcement particles into the molten matrix metal followed by pouring of mixture into a mold to harden, and hot pressing approach, which involves applying high pressure and temperature to a mixture of matrix and reinforcing powders, are also employed for the synthesis of metal matrix composites (Mahmood et al., 2020; Meignanamoorthy & Ravichandran, 2018).

13.2.3.2 Metal fiber composites

Metal fiber composites are a sort of composite material in which the metal fibers are inserted into the metal matrix. High wear resistance, negligible friction, and exceptional thermal stability are just a few of the remarkable tribological characteristics that distinguish these composites among others. These are frequently employed as slide bearings and wear-resistant parts in a variety of industrial applications. To enhance the tribological properties of lubricants, metal fibers made of copper or aluminum can be added. By serving as a barrier between two moving surfaces, these fibers can lessen both wear and friction (Prasad & Asthana, 2004).

Such metal fiber composites can be prepared through the powder metallurgy method, which involves mixing of metal fibers with matrix powders followed by compaction and heating to form a homogeneous composite. In situ *casting* method, in which the metal fibers are scattered in a liquid metal matrix, molded, solidified, and heated to acquire desired properties, and hot-pressing method, in which the mixture of metal fibers and matrix powders is compressed at a high temperature and pressure, are also used. The selection of the preparation approach depends on the desired composite microstructure, the size and shape of the metal fibers, and the interaction between the matrix and the fibers (Akhlaghi et al., 2013).

13.2.3.3 Metal-lubricant composites

Metal-lubricant composites are a subclass of composite materials, which can be made by combining metal matrix with lubricant particles for reinforcement. As a barrier between two moving surfaces, these metal particles can enhance the lubricant's tribological qualities by lowering the wear and friction. Comparing metal-lubricant composites to traditional lubricants, they can offer a longer-lasting lubrication effect and also increase the lubricant's heat stability and oxidation resistance. For a range of tribological applications, metal-lubricant composites have shown outstanding potential as efficient and effective lubricant additives. In order to create a novel material with appealing self-lubricating qualities, solid lubricant ingredients, such as carbonous materials like graphite, molybdenum disulfide (MoS_2), and hexagonal boron nitride, are implanted into the metal matrices as reinforcements (Omrani, 2016).

The mixing of metal particles like that of aluminum (Al) metal matrix composites with reinforcing particles of boron carbide (B_4C), silicon carbide (SiC), alumina (Al_2O_3), and lubricating fluid into a single composite material is the basis for the mechanism of metal-lubricant composites as lubricant additives. The fluid produces a hydrodynamic lubrication effect, while the metal particles produce a solid lubrication effect. In the case of solid lubrication, the metal particles form a barrier between the surfaces that come into contact, reducing friction and shielding them from direct metal-to-metal contact. The mechanical components' lifespan is extended by this layer, which also contributes to lower wear. In the case of hydrodynamic lubrication, the lubricating fluid decreases friction by separating the surfaces in contact, lowering the friction coefficient, and supplying a steady stream of new lubricant. As dispersants, the metal particles in the composite maintain the lubricant fluid suspended in the material and stop its depletion. Because of this, metal lubricant composites have demonstrated significant promise as powerful and useful lubricant additives for a number of tribological applications (Qiu & Khonsari, 2011; Zhang et al., 2012).

The structure of metal lubricant composites is composed of metal particles and a lubricant fluid, which are combined into a single composite material. The metal particles are spread throughout the lubricating fluid and typically take the form of nanoparticles or microparticles. The qualities of the composite and its effectiveness as a lubricant additive can be impacted by the size, shape, and distribution of the metal particles. Metal lubricant composites can be divided into two main categories based on their microstructure, that is, homogeneous and heterogeneous. Homogeneous composites have a homogeneous structure with an even distribution of metal particles throughout the lubricant. While in heterogeneous composites, the metal particles are unevenly distributed and the composite has a more complex structure (Khelge, 2022; Sliney, 1982; Zhou, 2020a).

13.2.3.4 Metal-metal intermetallic

These are alloys comprised of two or more metals that possess a special set of characteristics not present in the constituent metals individually. Comparing these alloys to monometallic alloys, it is reported that they have better wear resistance and tribological properties. The nickel metal coated with graphite and aluminum powder (Ni-Gr/Al) on steel support was studied as Ni−Al metal−metal intermetallic, which shows an enhanced resistance to higher temperatures when used in lubricants. Such intermetallic improves the viscosity even at higher temperatures and pressures, along with the reduction in wear and tear of the sliding parts (Bandyopadhyay & Heer, 2018; Huang, 2017b).

13.2.4 Carbon nanofiber-based composites

To enhance the mechanical, electrical, and thermal characteristics of composites, carbon NFs are combined with a variety of matrix materials (including metals, ceramics, and polymers) (Zhou, 2020b). A variety of NF composites are constructed by adjusting the structure and content of the NFs and other elements utilized in the composite. The characteristics of these materials may be customized. Engine oil, gear oil, and lubricants for bearings and other moving components are just a few of the applications for which NFs-based lubricant additives have been explored (Guo, 2021) (Fig. 13.2).

13.3 Properties of nanofibers and their composites in tribology

Numerous distinctive qualities of NFs and their composites make them appropriate for use as lubricant additives. These properties include:

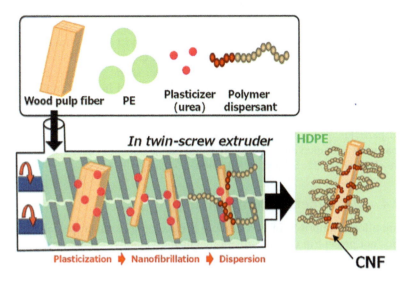

Figure 13.2 A typical carbon nanofiber and its composite.

13.3.1 High surface area

The large surface area to volume ratio of NFs and their composites improves their tribological characteristics. Owing to the high surface area to volume ratio, stresses are uniformly distributed across a greater surface and thus create more chances of lubrication among the mechanical parts, which in turn decreases the wear and friction. The increased surface area also makes it possible to incorporate nanoscale additives, such as graphene and other lubricating particles, which can enhance tribological performance even further. This makes tribological applications, such as bearings, gears, and coatings, more promising than the other additives (Xue, 2017).

13.3.2 High aspect ratio

The high aspect ratio describes the proportion of a specific nanoscale or microscale object, such as a fiber, particle, or platelet, to its length, width, or diameter. A high aspect ratio is seen as having a high surface area-to-volume ratio since it often denotes that the object is elongated or has a greater length relative to its width. The tribological performance of NFs and their composites is significantly influenced by the high aspect ratio of these materials. The mechanical strength and rigidity of the long, thin fibers further help to prevent deformation and cracking under heavy loads. For example, when NFs with a high aspect ratio are added to a lubricant, they can create a protective layer on the surfaces in contact, reducing friction and wear by acting as a physical barrier. The high aspect ratio also allows the NFs to penetrate into surface defects and cracks, helping to prevent further wear by reducing the contact area between surfaces. Additionally, the high aspect ratio of nanoparticles can improve their ability to be evenly dispersed throughout a lubricant, allowing them to be more effective in reducing friction and wear. As a result, tribological components composed of NFs and composites are more resilient and long-lasting, making them appropriate for usage in challenging situations. Furthermore, the high aspect ratio improves the capacity of NFs and composites to capture and hold lubricants, which can lessen wear and friction (Jung & Sodano, 2020).

13.3.3 Excellent mechanical properties

Due to their superior mechanical qualities, such as high strength, stiffness, and resilience, NFs and their composites can enhance the efficiency of the lubricant additives by establishing a protective barrier on the surfaces that are in contact. Owing to their high stiffness and strength, they can withstand heavy loads without deforming or cracking. This increases their toughness and longevity and allows them to be used in the challenging tribological conditions. NFs and composites' mechanical qualities also enable them to tolerate high stress and strain, which can lower wear and friction and increase tribological performance. They are also perfect to be used in high-performance tribological applications where weight reduction is a key factor because of their compact size and light weight (Sahoo, 2010).

13.3.4 Thermal stability

Due to their great thermal stability, NFs and their composites can prohibit lubricants from disintegrating at high temperatures, which can improve the effectiveness of the lubricant additives. The tribological performance of NFs and the composites made of them is strongly impacted by their heat stability. They are suitable for high-temperature use in tribological applications because they can keep their mechanical characteristics and integrity under extreme conditions. Their resistance to heat deterioration and oxidation, which can result in friction and wear in tribological systems, is further improved by their thermal stability. This makes tribological components built from NFs and composites more resilient and long-lasting, making them appropriate for use in challenging tribological situations. High thermal stability also improves their capacity to hold onto lubricants and prevent the creation of wear debris, which can consequently decrease wear and friction (Chen, 2019).

These characteristics have made NFs and their composites a very potential lubricant additive material as they can boost the efficiency of lubricant additives by offering a protective cover on the surfaces in contact and by lowering friction and wear.

13.3.5 Low density

NFs are lightweight and convenient to handle because of their low density. In tribology, the low density of NFs and their composites is a preferred characteristic because it results in lighter tribological components. This is crucial in high-performance tribological applications where weight reduction is an essential factor. Furthermore, the low density improves the capacity of NFs and composites to capture and hold lubricants, which can lessen wear and friction. As a result, composites and NF-based tribological components perform better tribologically and are more durable. Furthermore, because they can distribute stresses evenly and resist deformation and cracking, NFs and composites with low densities are better suited to sustain large loads (Li, 2010).

13.3.6 Biocompatibility

NFs can easily be designed and synthesized with different chemical groups, allowing them to be modified for specialized purposes. The biocompatibility of NFs and their composites is a crucial characteristic in tribology, especially in applications related to medicine and biology. They are safe to use in close proximity to live tissues because they are nontoxic and do not have negative effects on humans. Additionally, NFs and composites' resistance to biological degradation, such as corrosion and biofilm formation, which can result in friction and wear in tribological systems, is improved by their biocompatibility. This qualifies them for application in implants and medical devices where tribological performance and biocompatibility are crucial factors. The performance and endurance of medical and biological tribological systems may be enhanced by NFs and composites' improved capacity to integrate with living tissues due to their biocompatibility (Xu, 2015).

13.3.7 Cost-effectiveness

Some varieties of NFs and the composites they form can be manufactured in vast quantities for a comparatively inexpensive cost. The type of material used, the manufacturing process, and the intended use all influence the cost-effectiveness of NFs and their composites. NFs and their composites, on the other hand, are becoming more cost-effective as technology advances and production methods become more effective. The performance standards and the cost of substitute materials or technologies determine how cost-effective NFs and their composites are for a given application. The enhanced performance of NFs and their composites may, in some applications, exceed the greater price, resulting in overall cost savings over the course of the product or system. In other circumstances, the cost of manufacture or the accessibility of substitute materials may place more restrictions on the cost-effectiveness of NFs and the composites they produce (Zhu, 2016).

13.3.8 High thermal conductivity

NFs can be used in thermal management applications because some varieties, such as carbon NFs, have high thermal conductivity and thermal management applications. In tribology, NFs and their composites have high heat conductivity qualities, which have various benefits, such as:

13.3.8.1 Enhanced heat transfer

NFs and their composites with high thermal conductivity allow for better heat transfer, making them beneficial in tribological applications where heat management is essential. Due to their high surface area to volume ratio, which enables effective heat transfer at the nanoscale, NFs and their composites have been found to have improved heat transfer capabilities. Additionally, the special qualities of NFs and their composites can be modified for particular applications, providing enhanced thermal conductivity, lower thermal resistance, and other advantageous heat transfer characteristics. Using NFs and their composites for improved heat transfer has a number of benefits, one of which is their capacity to increase heat transfer efficiency while minimizing system size and weight. For example, adding NFs to heat exchangers can considerably speed up heat transfer, enhancing energy efficiency and lowering operational costs. In a number of applications, using insulation made of NFs can decrease heat loss and enhance thermal management. Due to the high thermal conductivity of carbon and the high surface area to volume ratio of NFs, carbon NF composites in particular have been shown to exhibit good thermal conductivity. This makes them ideal materials for use in a variety of heat transfer applications, such as heat exchangers, other industrial processes, and the thermal control of electronic equipment.

In order to make NFs and their composites feasible for use in high-temperature applications, these materials can be developed to have particular thermal characteristics, such as low thermal expansion, a high melting point, and high thermal stability. Also, a variety of geometries and combinations can utilize effective heat transfer to the flexibility and porosity of nanocomposites (Meng, 2022a, 2022b).

13.3.8.2 Energy efficiency and reduced temperature accumulation

The strong thermal conductivity in tribological applications decreases the temperature buildup, increases component longevity, improves heat transport, and lowers temperature generation, hence increasing the energy efficiency of the tribological systems. In a variety of applications, NFs and their composites have potentially increased energy efficiency and reduced temperature buildup. The high surface area to volume ratio of NFs and their composites, which permits effective heat transport and lowers temperature accumulation, is one of their main advantages. Improved energy efficiency and lower operating costs are possible, especially in situations where heat management is important. It is possible to construct NFs and the composites they form to have particular thermal characteristics, such as low thermal conductivity or low thermal expansion, which can help reduce temperature buildup and improve thermal management. For instance, incorporating insulation made of NFs in buildings and cars can lower heat loss and temperate swings, resulting in increased energy efficiency. It has been proven that NF-based lubricants and coatings greatly minimize wear and friction in a variety of tribological systems. NFs' high surface area to volume ratio enhances lubricant retention and dispersion, reducing friction and wear. As a result, machinery and equipment may use less energy, have a longer lifespan, and require less maintenance. NF-based thermal interface materials can also help with thermal management and lower temperature accumulation in tribological systems. The risk of overheating is decreased and energy efficiency is increased owing to the high thermal conductivity and ability to effectively transmit heat away from frictional contacts of these materials. Furthermore, adding carbon NFs to polymer-based composites can dramatically enhance those materials' mechanical and tribological characteristics, reducing wear and friction and increasing energy efficiency (Rasul, 2022; Shi, 2021; Zhang, 2021).

13.3.8.3 Reduced maintenance costs

NFs and their composites' enhanced thermal performance might lessen the need for routine maintenance and replacement, which can save money over time. Extended service life and lower maintenance costs can result from the use of NFs in lubricants and coatings to dramatically reduce friction and wear in tribological systems. NFs' high surface area to volume ratio enhances lubricant distribution and retention, reducing friction and wear on moving parts. Throughout the course of the equipment's lifetime, this may result in longer maintenance intervals and cheaper maintenance expenses. The durability and wear resistance of tribological components can be improved with NF-based composites. For instance, adding NFs to polymer-based composites can greatly enhance their mechanical and tribological characteristics, resulting in less wear and a longer useful life. This can cut maintenance costs by reducing the need for frequent repairs and component replacements for tribological systems. Moreover, the use of NF-based thermal interface materials in tribological systems can enhance thermal management and lower the danger of overheating, which can result in fewer equipment failures and maintenance expenses (Prabhakar, 2018; Werner, 2004).

13.3.8.4 Enhanced performance

NFs and their composites with high thermal conductivity can enhance the overall performance of tribological systems, resulting in higher production and efficiency. It has been shown that NF-based lubricants and coatings greatly minimize wear and friction in a variety of tribological systems. NFs' high surface area to volume ratio enhances lubricant distribution and retention, reducing friction and wear on moving parts. This may result in improved machinery and equipment performance and dependability. The mechanical and tribological qualities of the materials employed in tribological systems can also be enhanced by the use of NF-based composites. For instance, adding carbon NFs can dramatically enhance the mechanical and tribological characteristics of polymer-based composites, resulting in less wear and better performance. The thermal and electrical characteristics of tribological components can also be improved by the use of NF-based materials, which can further boost their performance. To enhance their tribological capabilities, NF-based materials can also be tailored with particular surface chemistries and morphologies. For instance, using functionalized NFs can increase their capacity to adhere to tribological components, resulting in improved durability and performance. Similar to this, the mechanical characteristics and wear resistance of tribological components can be enhanced by the employment of aligned NFs. Consequently, the performance of tribological parts and systems has the potential to be greatly improved by the use of NFs and their composites. The potential uses for NFs and their composites in improving tribological performance are projected to increase as technology develops and new production procedures are developed, as shown in Table 13.1 (Ibrahim, 2018; Lin et al., 2021).

13.4 Lubricants

Lubricants are compounds used to reduce wear and friction between surfaces that are moving relative to one another. They contribute to greater productivity, less wear and tear, and longer equipment and machinery lifespans. For example, oil, grease, and synthetic lubricants, as shown in Fig. 13.3.

Table 13.1 Various tribological applications of nanocomposites.

Matrix	Fillers		Size of filler (nm)		Optimum filler		Coefficient of friction		Specific wear rate ($\times 10^{-6}$ mm^3/Nm)	
	With lowest COF (wt.%)	With lowest wear (wt.%)	Without filler	With filler	Without filler	With filler	Without filler	With filler		
PTFE		ZnO	50	15	15	0.202	0.209	1125.3	13
PTFE		Al$_2$O$_3$	40	20	20	0.152	0.219	715	1.2
Epoxy		TiO$_2$	10	7	3	0.54	0.4	26	1.63
Epoxy		SiO$_2$	9	2.2 vol.%	2.2 vol.%	0.58	0.45	200	45
PEEK		ZrO$_2$	10	7.5	7.5	0.38	0.29	7.5	3.9
PEEK		SiO$_2$	<100	7.5	7.5	0.37	0.21	7.5	1.4
PET		Al$_2$O$_3$	38	2	2	0.32	0.3	17.4	9.5
BMI		SiC	<100	8	6	0.36	0.24	6.8	2.2
PS		MWCNT	10–20	1.5	1.5	0.42	0.31	130	8
PPS		CuO	30–50	10 vol.%	2 vol.%	0.43	0.34	32.4	7.4

Figure 13.3 Lubricant in sliding and movable parts.

13.4.1 Types of lubricants

There are numerous kinds of lubricants, and each has special qualities and traits that make it appropriate for a particular use. The selection of lubricant relies on the particular application and operating conditions. Each of these types of lubricants has special features and characteristics of its own. Among the most popular kinds of lubricants include:

13.4.1.1 Mineral oil-based lubricants

Lubricants derived from mineral oil are the most popular and are derived from crude oil. They are employed in a variety of fields, such as marine, industrial, and automotive. Mineral oil-based lubricants are constituted of mostly hydrocarbons and are formed from petroleum, as shown in Fig. 13.4. In addition to a variety of additives that can enhance their performance in particular applications, they often contain a blend of saturated and unsaturated hydrocarbons. Depending on the particular application and performance requirements, mineral oil-based lubricants might have different compositions. They typically consist of a basic oil that is enhanced with a range of additives, such as corrosion inhibitors, detergents, and antiwear compounds, to improve their performance attributes. The accessibility and affordability of mineral oil-based lubricants are two of their main benefits. They are a practical option for many industrial and automotive applications because they are reasonably simple to build and can be made in huge quantities. Mineral oil-based lubricants can, however, have some performance restrictions. For instance, they may degrade more rapidly in hot environments, which may result in decreased lubrication and greater wear on machinery. They are also often less stable than synthetic lubricants, which can result in shorter service life and increased maintenance expenses over time (Masjuki, 1999; Schneider, 2006).

Figure 13.4 Commonly used mineral oil-based lubricants.

13.4.1.2 Synthetic lubricants

Synthetic lubricants are chemically manufactured from a variety of base stocks and additives to provide high-performance lubricants. They are utilized in a variety of industrial, automotive, and aerospace applications and are designed to offer improved lubrication and protection in comparison to conventional mineral oil-based lubricants. These lubricants, which are comprised of chemical compounds, are intended to function better in high-performance applications and harsh temperatures than mineral oil-based lubricants. Synthetic ester, polyalphaolefin, polyglycol, and diester base stocks are some examples of synthetic lubricants as shown in Fig. 13.5. These base stocks are next blended with other performance-improving additives, such as antioxidants, detergents, and antiwear agents, to produce lubricants with particular qualities and performance traits. The ability of synthetic lubricants to work consistently throughout a wide range of operating temperatures is one of its fundamental benefits. This is because, in comparison to lubricants based on mineral oil, synthetic lubricants may be made to have a viscosity that is more stable throughout a wider temperature range. In high-temperature situations, in particular, this may lead to enhanced machinery protection. In comparison to mineral oil-based lubricants, synthetic lubricants are additionally typically more resistant to oxidation, thermal breakdown, and other types of degradation. This may lead to improved general performance, lower maintenance costs, and longer service life. Synthetic lubricants come in a variety of forms, each with its own set of benefits and drawbacks. Ester-based lubricants, polyglycol-based lubricants, and Polyalphaolefin-based lubricants are a few of the most often-used varieties (Hackler, 2021; Ray et al., 2012).

13.4.1.3 Biodegradable lubricants

Biodegradable lubricants are designed to be more environmentally friendly than conventional mineral oil-based lubricants because they are made from natural, renewable resources. They are typically produced using synthetic esters or vegetable oils. Vegetable oil-based lubricants, synthetic ester-based lubricants, and

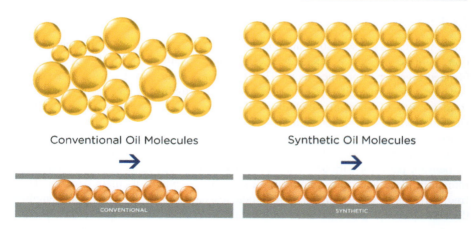

Figure 13.5 Comparison between conventional and synthetic lubricants.

Figure 13.6 Biodegradable lubricant in engine rotors.

other bio-based lubricants are only a few of the several varieties of biodegradable lubricants that are readily available. These lubricants are usually designed to have similar performance characteristics to traditional lubricants, such as corrosion resistance, extreme pressure performance, and antiwear protection. The lower environmental effect of biodegradable lubricants is one of their main benefits. They are less likely to harm the environment in the event of a spill or leak because they are made to decompose organically. Furthermore, a lot of biodegradable lubricants are produced using renewable resources, which can lessen the need for nonrenewable resources such as petroleum.

Biodegradable lubricants may, however, potentially have some drawbacks. They may not function as well in some applications and are often more expensive than conventional lubricants, as shown in Fig. 13.6. Additionally, depending on the formulation, the biodegradability of these lubricants can change and may be altered by variables including temperature, pressure, and other environmental conditions (Darminesh, 2017; Nagendramma & Kaul, 2012).

13.4.1.4 Greases

These lubricants are semisolid and made to stay intact, offering long-lasting lubrication. A base oil and a thickening agent, such as lithium or calcium soap, are blended together to produce them. The thickening agent is added to the base oil to enhance viscosity and stop it from flowing or dripping. It can be manufactured from a range of materials, including lithium, calcium, aluminum, or polyurea. In situations when a lubricant with a thicker consistency is required, like in bearings, gears, or other machinery parts, greases are often employed. They can also be utilized in situations when a lubricant needs to remain in place for a long time, like in suspension parts for cars or other nonroutinely serviced parts. Grease's capacity to remain in place and offer long-lasting lubrication is one of its main benefits. They frequently have water and other pollutant resistance as well, which can lengthen their lifespan and lower the need for maintenance. In order to improve their performance qualities, greases are frequently compounded with additives such as antiwear agents, rust inhibitors, and severe pressure additives. Nevertheless, applying greases may come with certain drawbacks. They can be trickier to apply and distribute uniformly because they are thicker and more viscous than other kinds of lubricants. In the case of a spill or leak, they may also be more challenging to remove or clean up, as shown in Fig. 13.7 (Lugt, 2009).

13.4.1.5 Air compressor lubricants

These lubricants are exclusively designed for use in air compressors, and they are intended to offer efficient lubrication and cooling. To help increase the service life of the compressor and its parts, these lubricants are normally designed to offer a high level of performance and protection. Mineral oil-based, synthetic, and semi-synthetic lubricants, as well as other varieties, are offered as lubricants for air compressors. The type of compressor, the working environment, and the manufacturer's recommendations will all affect the kind of lubricant that is most appropriate for a given application. Viscosity, oxidation resistance, antiwear protection, and corrosion resistance are a few of the crucial characteristics that are crucial for air

Figure 13.7 Grease as lubricant type commonly used in sliding contacts.

compressor lubricants, as shown in Fig. 13.8. The lubricant ought to be able to easily pass through the compressor's lubrication system and ought to be resistant to breakdown brought on by extreme temperatures and other external elements. In order to assist the compressor and its components to last longer, it should also offer suitable protection against wear and corrosion (Jacobson, 2004).

13.4.1.6 Gear lubricants

These lubricants are made especially for use in gearboxes, and they are designed to offer efficient lubrication and protection against wear and tear. To improve their performance qualities, these lubricants are frequently compounded with additives such as antiwear agents, rust inhibitors, and extreme pressure additives. The viscosity of gear lubricants, which is carefully chosen to offer the required level of lubrication and protection for the particular application, is one of their main characteristics. The lubricant's viscosity must be suitable for the gears' speed and load, as well as for the operating temperature and other external conditions. Mineral oil-based lubricants, synthetic lubricants, and semisynthetic lubricants are only a few of the formulations available for gear lubricants. The optimal sort of lubricant for a given application will vary depending on the type of gear, the working environment, and the manufacturer's recommendations. The capacity of high-quality gear lubricants to assist increase the service life of the gears and other parts of industrial machinery is one of its main benefits, as shown in Fig. 13.9. These lubricants can assist in decreasing the frequency of maintenance and repair tasks by providing proper lubrication and protection, which can ultimately help to save time and money (Höhn et al., 2001).

Figure 13.8 A typical air compressor lubricant.

Figure 13.9 Gear lubricant in gear boxes of vehicles.

13.4.1.7 Metalworking fluids

It has been investigated to employ NF-based lubricant additives in metalworking fluids (MWFs). For instance, studies have demonstrated that incorporating polyacrylonitrile NFs into MWF can enhance lubrication and reduce cutting tool wear. MWFs are lubricants employed in numerous industrial procedures, such as the cutting, grinding, and shaping of metal parts. They are made to lubricate, cool, and shield the tool and workpiece surfaces throughout these processes. MWFs are crucial in tribology because they lower wear, heat, and friction between the tool and workpiece surfaces (Osama, 2017).

There are various MWF types, each with unique characteristics and functions. Here are a few examples:

Mineral oil-based metalworking fluids

Mineral oil is used to make the most popular type of MWFs, known as mineral oil-based MWFs. They are employed in numerous operations, such as turning, drilling, and tapping. They are also utilized as lubricants in mechanical systems such as gears.

Synthetic metalworking fluids

These are employed in high-performance applications including hard turning, grinding, and high-speed machining because they are derived from synthetic oils or other synthetic fluids. Compared to MWFs based on mineral oil, they are more thermally and chemically stable.

Semisynthetic metalworking fluids

These are utilized in moderate-performance applications and combine mineral oils with synthetic fluids. They provide a balance between price and performance.

Water-based metalworking fluids

These are created using water together with a number of additives, such as biocides, emulsifiers, and corrosion inhibitors. They are used for many different tasks, such as grinding, honing, and polishing. Due to their simplicity of disposal and lack of mist or smoke production, they are also environmentally benign.

These are only a few examples of the potential uses for lubricant additives made from NFs. To fully realize their strength and to tailor their design for particular applications, more study is required.

13.4.1.8 Solid lubricants

An emerging field of study is the use of NF-reinforced composites as solid lubricants. The NFs can create a protective coating on the surfaces they come into touch with, which can lessen friction and wear due to their high surface area and aspect ratio. This is due to the NFs' high surface area and aspect ratio, which enable many fibers to be packed into a narrow area, creating a dense network of fibers that can create a protective layer on the surfaces in interaction, as shown in Fig. 13.10.

The production of gears is one practical application of solid lubricants based on NF-reinforced composites. The mechanical elements known as gears are used to transfer power and rotational motion between two shafts. Particularly in high-speed and high-load applications, hence gears are heavily loaded components that are susceptible to wear and collapse. In comparison to gears lubricated with conventional lubricants, solid lubricants based on NF-reinforced composites have been found to minimize both friction and wear tremendously (Friedrich, 2018).

Another application is in the study of friction, lubrication, and material wear or tribology. In sliding contact applications including bearings, gears, and camshafts, solid lubricants based on NF-reinforced composites have been utilized to lower friction and wear. In comparison to conventional lubricants, these lubricants have been found to have superior wear resistance and greater load-bearing ability (Rodriguez, 2016).

In conclusion, the use of NF-reinforced composites as solid lubricants is a rapidly expanding field of research that has shown significant promise in the study of tribology because it can lessen friction and wear by generating a thin protective

Figure 13.10 Solid lubricant in the form of bearing greases.

barrier on the interacting surfaces. These solid lubricants have been applied in tribology and gear applications in real life (Askarnia, 2020).

13.5 Nanofibers and their composites as lubricant additives

Chemical substances known as lubricant additives are added to base oils to improve their performance characteristics. More research is being done on the potential of NFs and their composites as lubricant additives, as shown in Fig. 13.11. As lubricant additives, NFs and their composites can provide numerous advantages, such as improved lubrication, enhanced wear resistance, increased thermal stability, increased load-bearing capability, compatibility with other materials, viscosity stability improvement, and improved chemical resistance, especially in neutralization of acids. The performance of lubricants and the longevity of mechanical parts could

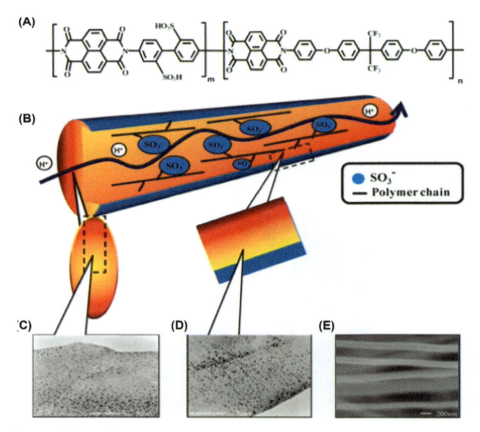

Figure 13.11 A typical polymeric composite as a lubricant additive.

both be greatly increased by the use of NFs and their composites as lubricant additives. Additionally, NFs and their composites can be modified and designed for particular applications, making them very adaptable and appealing for use in a variety of fields (Ibrahim, 2018; Liu, 2022; Sun, 2017).

13.5.1 Types of lubricant additives

To improve the characteristics of lubricants, a variety of lubricant additives can be utilized. The following are some of the most common types of lubricant additives:

13.5.1.1 Anti-wear additives

In the past few years, NF-reinforced composites have been explored as antiwear additives because of their distinctive qualities, such as high rigidity and strength. The abrasive forces that lead to wear can be resisted with the aid of these characteristics. Because of their high stiffness and strength, NFs can operate as a barrier between the abrasive particles and the surfaces they come into direct contact with, thereby minimizing wear. This layer of protection can be applied to the surfaces in concern, as shown in Fig. 13.12. The purpose of these lubricant additives is to diminish wear on mechanical surfaces. These additives frequently comprise metal-containing substances, including zinc dialkyldithiophosphates, tricresyl phosphate, calcium borate/cellulose acetate laurate, alpha-zirconium phosphate, and zinc borate (ZB), which provide a protective film on the metal's surface to lessen wear and friction (Dai, 2016; Dong & Hu, 1998; Friedrich & Schlarb, 2011; Guan et al., 2016; Yang, 2020).

For example, engine oil additives are one area where NF-reinforced composite-based antiwear additives have been successfully applied. Engine oils are designed to lubricate and guard against wear and strain on engine parts. Comparatively to conventional additives, engine oil additives based on NF-reinforced composites have been found to have higher wear resistance and greater load-bearing ability. This is due to the fact that the NFs' great rigidity and strength may develop a layer of defense on the surfaces they

Figure 13.12 A typical nanofiber antiwear lubricant additive.

come into touch with. This layer can operate as a barrier to stop abrasive particles from directly engaging with the surfaces, hence minimizing wear (Sarath, 2021).

Another illustration is in the study of friction, lubrication, and material wear, or tribology. NF-reinforced composite-based antiwear additives have been employed to lessen wear in sliding contact applications such as bearings, gears, and cams. This has been shown to achieve better wear resistance and increase load-bearing ability in comparison to conventional additives. The high density and stiffness of the NFs can provide a thick shield on the surfaces in contact, acting as a barrier to prevent the abrasive particles from directly interacting with the surfaces, significantly lowering wear and tear (Friedrich & Schlarb, 2011).

Conclusively, the use of NF-reinforced composites as antiwear additives is a developing topic of study that has shown significant promise in the study of tribology because it has the potential to minimize wear by creating a protective layer on the surfaces in contact. Abrasive particles may be blocked from directly interacting with surfaces and producing wear, thanks to the great strength and stiffness of the NFs. These antiwear compounds have really been put to use in tribology and automotive engine lubricating oil applications (Friedrich & Schlarb, 2011).

Examples of their usage in tribology applications include the use of these additives in bearings, gears, and cams to reduce wear and enhance wear resistance, as well as the use of NF-reinforced composites in engine lubricants to improve wear resistance and boost capacity for carrying loads. It should be highlighted that more study is required to completely comprehend the capabilities and restrictions of these compounds and to create practical processes for their production and incorporation into lubricants and other materials (Friedrich et al., 2012).

13.5.1.2 Antifoam additives

Antifoam additives are added to lubricants in order to avoid foam formation, which has a detrimental influence on lubrication and might result in mechanical damage. When air and oil are combined, foam can develop, which can reduce lubrication and raise the risk of mechanical damage by thinning the lubricant coating. In order to stop foam bubbles from expanding and having a detrimental effect on lubrication, antifoam additives break down and disperse the bubbles. Due to their capacity to disperse and disintegrate foam bubbles, silicones are frequently utilized as antifoam additives. Silicone-based compounds, polyglycols, mineral oil-based compounds, and fatty acid derivatives as antifoam additives are frequently added to lubricants in negligible quantities (usually 0.1%, 0.5%, 1%, 1.5%, and 2%) and are effective in preventing foam formation even in high-speed or high-pressure applications (Duncanson, 2003; Ifejika et al., 2022; Street & Goole, 1991; Tuszynski, 2014).

13.5.1.3 Extreme pressure additives

Extreme pressure additives function by creating a protective coating on the metal's surface, which lowers friction and wear. These lubricant additives are added to lubricants to enhance their performance at high loads and fast speeds, which are factors that thrust lubricants under extreme stress. Metal-to-metal contact at extreme pressure levels can increase friction and wear on mechanical surfaces. Compounds having sulfur, chlorine, and boron, such as sulfurized fats and oils, chlorinated paraffin, metal dithiophosphate, BN, boron carbide, and boron oxide, are the most prevalent forms of extreme pressure additive. These substances function by creating a protective coating that can endure heavy loads and fast speeds through chemical interaction with the metal surface. Extreme pressure additives can be utilized in a variety of applications, including gears, bearings, and other mechanical components that are subjected to high loads and high speeds. They are normally added to lubricants in minute amounts (usually about 0.5%—5%) (Baş & Karabacak, 2014; Canter 2007, 2016; Farng & Rudnick, 2003; Kim et al., 2010; Li, 2014; Shah et al., 2013).

13.5.1.4 Viscosity index improver (VII) additives

The viscosity index (VI), which is a measurement of a lubricant's capacity to retain its viscosity across a wide range of temperatures, is improved by the use of viscosity index improvers (VIIs), which are lubricant additives. The viscosity of the lubricant also varies with temperature, which can have a detrimental effect on lubrication and result in mechanical damage. VIIs function by preserving the lubricant's viscosity across a broad temperature range, ensuring that the lubricant can provide enough lubrication at all times. The most prominent kind of VIIs are polymeric compounds, which may raise the viscosity of the lubricant throughout a wide range of temperatures. Examples include polyisobutenes, polymethylmethacrylates, Polyalphaolefins, ethylene—propylene—diene monomer (EPDM), and phosphorus compounds. Engine oils, gear oils, and other lubricants that are exposed to large temperature fluctuations can all benefit from the addition of these polymers, which are normally added to lubricants in tiny amounts (typically between 0.5% and 3%) for a variety of purposes (Azizi, 2020; Cosimbescu, 2018; De Carvalho, 2010; Patil, 2021; Qu, 2014).

13.5.1.5 Detergents and dispersants

In order to keep lubricants clean and free of impurities, detergents and dispersants are two examples of lubricant additives. They function by preventing dirt, dust, and other particles from building up in the lubricant, which can have a detrimental influence on lubrication and result in mechanical damage. In order to assist remove impurities from lubricants, detergents are substances that are added to the lubricant. They normally function by adsorbing impurities on their surface, which keeps them from building up inside the lubricant. Sodium and calcium compounds, such as sodium and calcium alkylbenzene sulfonates, which are often used in transmission fluids and engine lubricants, are typical examples of detergents. Dispersants are substances that are added to lubricants to prevent impurities from settling out and to help keep them suspended in the lubricant. They generally function by forming a chemical connection with the pollutants and avoiding clumping. Polyisobutenes and polymethacrylates are typical dispersants that are frequently used in transmission fluids and engine oils. Engine oils, gear oils, and other lubricants that are subjected to high loads and high speeds can all benefit from the addition of detergents and dispersants, which are normally added to lubricants in modest amounts (typically about 1%–5%) (Nassar, 2017; Singh & Singh, 2012).

13.5.1.6 Friction modifiers

Due to their distinctive qualities, including their high surface area, high aspect ratio, and outstanding mechanical characteristics, NF-reinforced composites have recently been explored as friction modifiers. These characteristics make them appropriate for applications involving lubricant additives, particularly as friction modifiers (Ali, 2019).

To decrease friction and increase the effectiveness of mechanical systems, friction modifiers are added to lubricants. The NFs can create a protective coating on the surfaces they come into touch with, which helps lessen friction, thanks to their large surface area and aspect ratio. By serving as a barrier to avoid direct contact between the surfaces, these protective layers of NFs can lower the energy needed to move the surfaces against one another (Tang & Li, 2014).

For example, engine oil additives are one area where friction modifiers based on NF-reinforced composites have been employed in reality. Engine oils are designed to lubricate and guard against wear and strain on engine parts. It has been demonstrated that engine oil additives based on NF-reinforced composites have increased friction-reducing capabilities, leading to higher fuel economy, as shown in Fig. 13.13 (Moghadam, 2015). The study of friction, lubrication, and material wear is called tribology, and here is another area where it is evident. In sliding contact applications, such as bearings, gears, and cams, friction modifiers based on NF-reinforced composites have been utilized to lower friction. As a result of these compounds' increased friction-reducing abilities, performance and energy efficiency have both improved (Shahnazar et al., 2016). Hence, the use of NF-reinforced composites as friction modifiers is a developing topic of study that has shown significant promise in the study of tribology due to its potential to reduce friction by creating a protective layer on the surfaces in contact. These friction modifiers have been applied in tribology and engine oil applications in real life (Farsadi et al., 2017).

13.5.1.7 Self-healing lubricants

A new breed of lubricants called self-healing lubricants is intended to repair and restore worn-out surfaces. Due to its special qualities, including a large surface area, a high aspect ratio, and great mechanical characteristics, NF-reinforced composites have been researched as promising materials for self-healing lubricants (Dorri Moghadam, 2014).

The NFs can create a protective layer on the surfaces they come into touch with due to their large surface area and aspect ratio. The amount of wear and tear on the

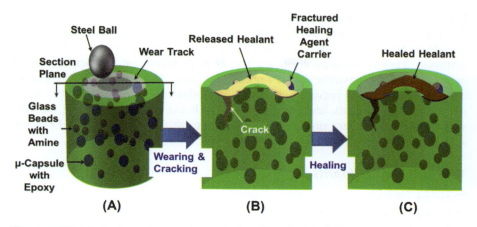

Figure 13.13 Mechanism of epoxy composite's self-healing during wear evaluation.

surfaces can be decreased by using this protective layer as a barrier to avoid direct contact between the surfaces. The large surface area and aspect ratio of the NFs enable them to swiftly repair and rebuild damaged surfaces when the protective layer is destroyed (Nosonovsky & Rohatgi, 2011).

Engine oil additives are a case study in the practical use of self-healing lubricants based on NF-reinforced composites. Engine oils are designed to lubricate and shield the moving parts of the engine from damage. The increased self-healing capabilities of engine oil additives based on NF-reinforced composites have been demonstrated to lead to higher engine performance and extended engine life (Lee, 2006). Another illustration is in the study of friction, lubrication, and material wear, or tribology. In sliding contact applications including bearings, gears, and cams, self-healing lubricants based on NF-reinforced composites have been utilized to repair and restore damaged surfaces. These lubricants have been demonstrated to have enhanced self-healing capabilities, which increase mechanical component performance and extend component life (Gu et al., 2013).

Conclusively, the utilization of NF-reinforced composites as self-healing lubricants is a developing topic of study with enormous promise. The NFs' high surface area, high aspect ratio, and outstanding mechanical qualities make them ideal for use in self-healing lubricants. These lubricants have been demonstrated to have enhanced self-healing capabilities, which increase mechanical component performance and extend component life. Examples of applications include tribological applications such as bearings, gears, and cams, as well as additives for motor oil (Sun, 2018).

13.5.1.8 Rust and corrosion inhibitors

These are lubricant additives added to lubricants to prevent rust and corrosion on metal surfaces. Typical examples of these additions are zinc dithiophosphates and other chemical substances that can help metal surfaces produce a protective coating.

Materials can be treated with rust inhibitors to stop or slow down corrosion that creates a barrier of protection on the metal's surface or by altering the environment to impede corrosion. Materials can be treated with rust inhibitors to stop or slow down corrosion. They function by creating a barrier of protection on the metal's surface or by altering the environment to impede corrosion. To increase the lifespan of metal components and prevent rust, corrosion inhibitors are typically used in items like paint, fuel, coolants, and other industrial or home chemicals. Among the substances that can prevent rust are phosphates, nitrites, benzotriazoles, and organic acids (Tang, 2019).

13.5.1.9 Antioxidants

These are lubricant additives that are added to lubricants to stop the lubricant from oxidizing and deteriorating due to heat. Phenols, amines, and other chemical substances that can prevent the lubricant from oxidation are frequently used in these additives. Antioxidant additives can make lubricants more stable, which lowers the production of dangerous oxidation products and lengthens the life of the lubricant. Antioxidant compounds can lessen the deterioration of moving parts, extending the lifespan of the component and lowering maintenance expenses. Antioxidant additives can enhance lubricants' lubricating capabilities, reducing wear and friction on moving parts. Antioxidant additives can boost the performance of lubricants, cutting down on energy use and operating expenses. Synthetic lubricants can benefit from the addition of antioxidant compounds for better performance and stability. Antioxidant additives can enhance lubricants' performance in high-temperature settings, which lowers the likelihood of component failure and extends the component lifespan (Cai, 2010; Shah et al., 2013).

13.5.1.10 Biodegradable lubricants

These are additives added to lubricants to increase their biodegradability. Typical examples of these additions include vegetable oils and other biodegradable natural substances. Biodegradable lubricants may be broken down into nontoxic byproducts, thus they do not affect the environment. Biodegradable lubricants have a lower propensity to contaminate soil or water, lowering the danger of environmental harm. Because biodegradable lubricants tend to be less toxic than conventional lubricants, there is a lower possibility that they would harm wildlife or humans. A variety of tribological applications are ideal for biodegradable lubricants because they can offer enough lubrication. Biodegradable lubricants are frequently compatible with renewable energy sources, such as wind and solar power, further supporting the use of renewable energy sources (Darminesh, 2017; Maleque et al., 2003).

13.5.1.11 Nanoadditives

Nanosized lubricant additives are added to lubricants to improve their properties by the use of nanoparticles like metal oxides, carbon nanotubes, and other nanostructures. Friction can be considerably reduced by nanosized additions like nanoparticles, nanotubes, and NFs, which improve tribological performance. Nanosized additives can enhance the tribological materials' resistance to wear, extending the life of individual components. Nanosized additions can boost tribological materials' ability to withstand loads, making them appropriate for usage in high-stress situations. Nanoscale additions can increase the surface toughness of tribological materials, increasing their resistance to abrasion and wear. The lubricating qualities of tribological materials can be improved by nanosized additions, which highly lower the friction and wear and increase the compatibility with various operating settings and situations, resulting in better performance and reliability (Xiao, 2022).

References

Akhlaghi, F., Eslami-Farsani, R., & Sabet, S. (2013). Synthesis and characteristics of continuous basalt fiber reinforced aluminum matrix composites. *Journal of Composite Materials, 47*(27), 3379–3388.

Ali, I., et al. (2019). Advances in carbon nanomaterials as lubricants modifiers. *Journal of Molecular Liquids, 279*, 251–266.

Askarnia, R., et al. (2020). Improvement of tribological, mechanical and chemical properties of Mg alloy (AZ91D) by electrophoretic deposition of alumina/GO coating. *Surface and Coatings Technology, 403*, 126410.

Azizi, S., et al. (2020). Performance improvement of EPDM and EPDM/silicone rubber composites using modified fumed silica, titanium dioxide and graphene additives. *Polymer Testing, 84*, 106281.

Bandyopadhyay, A., & Heer, B. (2018). Additive manufacturing of multi-material structures. *Materials Science and Engineering: R: Reports, 129*, 1–16.

Baş, H., & Karabacak, Y. E. (2014). Investigation of the effects of boron additives on the performance of engine oil. *Tribology Transactions, 57*(4), 740–748.

Bobzin, K., Bartels, T., & Mang, T. (2011). *Industrial tribology: Tribosystems, friction, wear and surface engineering, lubrication* (pp. 1–644). John Wiley & Sons.

Briscoe, B. J., & Sinha, S. K. (2008). *Tribological applications of polymers and their composites: Past, present and future prospects. Tribology and interface engineering series* (pp. 1–14). Elsevier.

Cai, M., et al. (2010). Imidazolium ionic liquids as antiwear and antioxidant additive in poly (ethylene glycol) for steel/steel contacts. *ACS Applied Materials & Interfaces, 2*(3), 870–876.

Canter, N. (2007). Special report: Trends in extreme pressure additives. *Tribology and Lubrication Technology, 63*(9), 10.

Canter, N. (2016). Status of medium-and long-chain chlorinated paraffins. *Tribology & Lubrication Technology, 72*(3), 34.

Chao, M., et al. (2019). Functionalized multiwalled carbon nanotube-reinforced polyimide composite films with enhanced mechanical and thermal properties. *International Journal of Polymer Science, 2019*.

Chen, B., et al. (2019). Tribological properties of epoxy lubricating composite coatings reinforced with core-shell structure of CNF/MoS2 hybrid. *Composites Part A: Applied Science and Manufacturing, 122*, 85–95.

Cosimbescu, L., et al. (2018). Low molecular weight polymethacrylates as multi-functional lubricant additives. *European Polymer Journal, 104*, 39–44.

Dai, W., et al. (2016). *Formation of anti-wear tribofilms via* α*-ZrP nanoplatelet as lubricant additives. Lubricants, 4*(3), 28.

Darminesh, S. P., et al. (2017). Recent development on biodegradable nanolubricant: A review. *International Communications in Heat and Mass Transfer, 86*, 159–165.

De Carvalho, M. J. S., et al. (2010). Lubricant viscosity and viscosity improver additive effects on diesel fuel economy. *Tribology international, 43*(12), 2298–2302.

Dong, J., & Hu, Z. (1998). A study of the anti-wear and friction-reducing properties of the lubricant additive, nanometer zinc borate. *Tribology international, 31*(5), 219–223.

Dorri Moghadam, A., et al. (2014). Functional metal matrix composites: Self-lubricating, self-healing, and nanocomposites-an outlook. *JOM, 66*, 872–881.

Duncanson, M. (2003). Effects of physical and chemical properties on foam in lubricating oils (C). *Tribology & Lubrication Technology, 59*(5), 9.

Dziadek, M., Stodolak-Zych, E., & Cholewa-Kowalska, K. (2017). Biodegradable ceramic-polymer composites for biomedical applications: A review. *Materials Science and Engineering: C, 71*, 1175–1191.

Farng, L. O., & Rudnick, L. (2003). Ashless antiwear and extreme-pressure additives. *Lubricant Additives: Chemistry and Application*, 223–258.

Farsadi, M., Bagheri, S., & Ismail, N. A. (2017). Nanocomposite of functionalized graphene and molybdenum disulfide as friction modifier additive for lubricant. *Journal of Molecular Liquids, 244*, 304–308.

Friedrich, K., & Schlarb, A. K. (2011). *Tribology of polymeric nanocomposites: Friction and wear of bulk materials and coatings*. Elsevier.

Friedrich, K., Almajid, A., & Chang, L. (2012). Modern polymer composites for friction and wear applications. *Polymer Journal, 34*(2). (18181724).

Friedrich, K. (2018). Polymer composites for tribological applications. *Advanced Industrial and Engineering Polymer Research, 1*(1), 3–39.

Gu, S., Zhang, Y., & Yan, B. (2013). *Solvent-free ionic molybdenum disulfide (MoS$_2$) nanofluids with self-healing lubricating behaviors. Materials Letters, 97*, 169–172.

Guan, B., Pochopien, B. A., & Wright, D. S. (2016). *The chemistry, mechanism and function of tricresyl phosphate (TCP) as an anti-wear lubricant additive. Lubrication Science (New York, N.Y.), 28*(5), 257–265.

Guo, L., et al. (2021). Recent advance in the fabrication of carbon nanofiber-based composite materials for wearable devices. *Nanotechnology, 32*(44), 442001.

Gupta, R. K., et al. (2017). Oil/water separation techniques: A review of recent progresses and future directions. *Journal of Materials Chemistry A, 5*(31), 16025–16058.

Hackler, R. A., et al. (2021). Synthetic lubricants derived from plastic waste and their tribological performance. *ChemSusChem, 14*(19), 4181–4189.

Haidar, D. R., Alam, K. I., & Burris, D. L. (2018). Tribological insensitivity of an ultralow-wear poly (etheretherketone)−polytetrafluoroethylene polymer blend to changes in environmental moisture. *The Journal of Physical Chemistry C, 122*(10), 5518–5524.

Höhn, B.-R., Michaelis, K., & Doleschel, A. (2001). Frictional behaviour of synthetic gear lubricants*, in. Tribology series* (pp. 759–768). Elsevier.

Huang, C., et al. (2017b). In-situ formation of Ni−Al intermetallics-coated graphite/Al composite in a cold-sprayed coating and its high temperature tribological behaviors. *Journal of Materials Science & Technology, 33*(6), 507–515.

Huang, G., et al. (2017a). Manganese-iron layered double hydroxide: A theranostic nanoplatform with pH-responsive MRI contrast enhancement and drug release. *Journal of Materials Chemistry B, 5*(20), 3629–3633.

Huang, Z.-M., et al. (2003). A review on polymer nanofibers by electrospinning and their applications in nanocomposites. *Composites Science and Technology*, *63*(15), 2223−2253.

Ibrahim, A. M. M., et al. (2018). Enhancing the tribological performance of epoxy composites utilizing carbon nano fibers additives for journal bearings. *Materials Research Express*, *6*(3), 035307.

Ifejika, V. E., Joel, O. F., & Aimikhe, V. J. (2022). Characterization of selected plant seed oils as anti-foam agents in natural gas treatment units. *Biomass Conversion and Biorefinery*, 1−9.

Jacobson, A. (2004). Air compressor lubricants: The next generation. *Tribology & Lubrication Technology*, *60*(7), 30.

Jung, J., & Sodano, H. A. (2020). Aramid nanofiber reinforced rubber compounds for the application of tire tread with high abrasion resistance and fuel saving efficiency. *ACS Applied Polymer Materials*, *2*(11), 4874−4884.

Karger-Kocsis, J., & Kéki, S. (2017). Review of progress in shape memory epoxies and their composites. *Polymers*, *10*(1), 34.

Khelge, S., et al. (2022). Effect of reinforcement particles on the mechanical and wear properties of aluminium alloy composites. *Materials Today: Proceedings*, *52*, 571−576.

Kim, B., Mourhatch, R., & Aswath, P. B. (2010). Properties of tribofilms formed with ashless dithiophosphate and zinc dialkyl dithiophosphate under extreme pressure conditions. *Wear*, *268*(3−4), 579−591.

Lee, S., et al. (2006). Self-healing behavior of a polyelectrolyte-based lubricant additive for aqueous lubrication of oxide materials. *Tribology Letters*, *24*, 217−223.

Li, B., et al. (2010). Effectual dispersion of carbon nanofibers in polyetherimide composites and their mechanical and tribological properties. *Polymer Engineering & Science*, *50*(10), 1914−1922.

Li, W., et al. (2014). Natural garlic oil as a high-performance, environmentally friendly, extreme pressure additive in lubricating oils. *ACS Sustainable Chemistry & Engineering*, *2*(4), 798−803.

Lin, Z., et al. (2021). Improved mechanical/tribological properties of polyimide/carbon fabric composites by in situ-grown polyaniline nanofibers. *Materials Chemistry and Physics*, *258*, 123972.

Liu, Y., et al. (2022). *Progresses on electrospun metal−organic frameworks nanofibers and their wastewater treatment applications.*. Materials Today Chemistry (Weinheim an der Bergstrasse, Germany), *25*, 100974.

Lugt, P. M. (2009). A review on grease lubrication in rolling bearings. *Tribology Transactions*, *52*(4), 470−480.

Mahmood, M. A., Popescu, A. C., & Mihailescu, I. N. (2020). Metal matrix composites synthesized by laser-melting deposition: A review. *Materials*, *13*(11), 2593.

Maleque, M., Masjuki, H., & Sapuan, S. (2003). Vegetable-based biodegradable lubricating oil additives. *Industrial lubrication and Tribology*, 137−143.

Masjuki, H., et al. (1999). Palm oil and mineral oil based lubricants—Their tribological and emission performance. *Tribology International*, *32*(6), 305−314.

Meignanamoorthy, M., & Ravichandran, M. (2018). Synthesis of metal matrix composites via powder metallurgy route: A review. *Mechanics and Mechanical Engineering*, *22*(1), 65−76.

Meng, Z., et al. (2022b). Carbon fiber modified by attapulgite for preparing ultra-high molecular weight polyethylene composite with enhanced thermal, mechanical, and tribological properties. *Polymers for Advanced Technologies*, *33*(12), 4142−4151.

Meng, Z., et al. (2022a). *Reinforced UHMWPE composites by grafting TiO$_2$ on ATP nanofibers for improving thermal and tribological properties. Tribology International, 172,* 107585.

Moghadam, A. D., et al. (2015). Mechanical and tribological properties of self-lubricating metal matrix nanocomposites reinforced by carbon nanotubes (CNTs) and graphene — A review. *Composites Part B: Engineering, 77,* 402—420.

Nagendramma, P., & Kaul, S. (2012). Development of ecofriendly/biodegradable lubricants: An overview. *Renewable and Sustainable Energy Reviews, 16*(1), 764—774.

Nassar, A. M., et al. (2017). Preparation and evaluation of the mixtures of sulfonate and phenate as lube oil additives. *International Journal of Industrial Chemistry, 8,* 383—395.

Nosonovsky, M., & Rohatgi, P. K. (2011). *Biomimetics in materials science: Self-healing, self-lubricating, and self-cleaning materials* (152). Springer Science & Business Media.

Omrani, E., et al. (2016). New emerging self-lubricating metal matrix composites for tribological applications. *Ecotribology: Research Developments,* 63—103.

Osama, M., et al. (2017). Recent developments and performance review of metal working fluids. *Tribology International, 114,* 389—401.

Patil, P. P., et al. (2021). Experimental analysis of tribological properties of polyisobutylene thickened oil in lubricated contacts. *Tribology International, 159,* 106983.

Prabhakar, K., et al. (2018). *A review of mechanical and tribological behaviour of polymer composite materials. IOP conference series: Materials science and engineering.* IOP Publishing.

Prasad, S., & Asthana, R. (2004). Aluminum metal-matrix composites for automotive applications: Tribological considerations. *Tribology Letters, 17*(3), 445—453.

Qiu, Y., & Khonsari, M. (2011). Experimental investigation of tribological performance of laser textured stainless steel rings. *Tribology International, 44*(5), 635—644.

Qu, J., et al. (2014). Comparison of an oil-miscible ionic liquid and ZDDP as a lubricant anti-wear additive. *Tribology International, 71,* 88—97.

Rana, A., Khan, I., & Saleh, T. A. (2021). Advances in carbon nanostructures and nanocellulose as additives for efficient drilling fluids: Trends and future perspective—A review. *Energy & Fuels, 35*(9), 7319—7339.

Rasul, M. G., et al. (2022). Improvement of the thermal conductivity and tribological properties of polyethylene by incorporating functionalized boron nitride nanosheets. *Tribology International, 165,* 107277.

Ray, S., Rao, P. V., & Choudary, N. V. (2012). Poly-α-olefin-based synthetic lubricants: A short review on various synthetic routes. *Lubrication Science, 24*(1), 23—44.

Rodriguez, V., et al. (2016). Reciprocating sliding wear behaviour of PEEK-based hybrid composites. *Wear, 362,* 161—169.

Ronchi, R. M., et al. (2023). Tribology of polymer-based nanocomposites reinforced with 2D materials. *Materials Today Communications,* 105397.

Sahoo, N. G., et al. (2010). Polymer nanocomposites based on functionalized carbon nanotubes. *Progress in Polymer Science, 35*(7), 837—867.

Sahoo, P., & Das, S. K. (2011). Tribology of electroless nickel coatings — A review. *Materials & Design, 32*(4), 1760—1775.

Sarath, P., et al. (2021). *Tribology of fiber reinforced polymer composites*: Effect *of fiber length, fiber orientation, and fiber size. Tribological Applications of Composite Materials,* 99—117.

Schneider, M. P. (2006). Plant-oil-based lubricants and hydraulic fluids. *Journal of the Science of Food and Agriculture, 86*(12), 1769—1780.

Shah, F. U., Glavatskih, S., & Antzutkin, O. N. (2013). Boron in tribology: From borates to ionic liquids. *Tribology Letters, 51*, 281–301.

Shahnazar, S., Bagheri, S., & Hamid, S. B. A. (2016). Enhancing lubricant properties by nanoparticle additives. *International Journal of Hydrogen Energy, 41*(4), 3153–3170.

Shi, Y., et al. (2021). In situ micro-fibrillization and post annealing to significantly improve the tribological properties of polyphenylene sulfide/polyamide 66/polytetrafluoroethylene composites. *Composites Part B: Engineering, 216*, 108841.

Singh, A. K., & Singh, R. K. (2012). A search for ecofriendly detergent/dispersant additives for vegetable-oil based lubricants. *Journal of Surfactants and Detergents, 15*(4), 399–409.

Sliney, H. E. (1982). Solid lubricant materials for high temperatures—A review. *Tribology International, 15*(5), 303–315.

Street, A., & Goole, N. H. (1991). Oleochemicals in lubricant additives. *Industrial Lubrication and Tribology, 43*(5), 3–8.

Sun, D., et al. (2018). Chemically and thermally stable isocyanate microcapsules having good self-healing and self-lubricating performances. *Chemical Engineering Journal, 346*, 289–297.

Sun, X., et al. (2017). Rheology, curing temperature and mechanical performance of oil well cement: Combined effect of cellulose nanofibers and graphene nano-platelets. *Materials & Design, 114*, 92–101.

Suryanarayanan, K., Praveen, R., & Raghuraman, S. (2013). Silicon carbide reinforced aluminium metal matrix composites for aerospace applications: A literature review. *International Journal of Innovative Research in Science, Engineering and Technology, 2*(11), 6336–6344.

Tang, Z., & Li, S. (2014). A review of recent developments of friction modifiers for liquid lubricants (2007–present). *Current Opinion in Solid State and Materials Science, 18*(3), 119–139.

Tang, Z. (2019). A review of corrosion inhibitors for rust preventative fluids. *Current Opinion in Solid State and Materials Science, 23*(4), 100759.

Tuszynski, W., et al. (2014). The potential of the application of biodegradable and non-toxic base oils for the formulation of gear oils—Model and component scuffing tests. *Lubrication Science, 26*(5), 327–346.

Vencl, A., et al. (2021). Production, microstructure and tribological properties of Zn-Al/Ti metal-metal composites reinforced with alumina nanoparticles. *International Journal of Metalcasting*, 1–10.

Wang, B., et al. (2022). Role of nano-sized materials as lubricant additives in friction and wear reduction: A review. *Wear, 490*, 204206.

Werner, P., et al. (2004). Tribological behaviour of carbon-nanofibre-reinforced poly (ether ether ketone). *Wear, 257*(9–10), 1006–1014.

Wozniak, A. I., et al. (2016). Thermal properties of polyimide composites with nanostructured silicon carbide. *Oriental Journal of Chemistry, 32*(6), 2967–2974.

Xiao, N., et al. (2022). Multidimensional nanoadditives in tribology. *Applied Materials Today, 29*, 101641.

Xu, S., et al. (2015). Mechanical properties, tribological behavior, and biocompatibility of high-density polyethylene/carbon nanofibers nanocomposites. *Journal of Composite Materials, 49*(12), 1503–1512.

Xue, M., et al. (2017). Preparation of $TiO_2/Ti_3C_2T_x$ hybrid nanocomposites and their tribological properties as base oil lubricant additives. *RSC Advances, 7*(8), 4312–4319.

Yang, M., et al. (2020). In-situ synthesis of calcium borate/cellulose acetate-laurate nanocomposite as efficient extreme pressure and anti-wear lubricant additives. *International Journal of Biological Macromolecules, 156*, 280–288.

Zahran, M., & Marei, A. H. (2019). Innovative natural polymer metal nanocomposites and their antimicrobial activity. *International Journal of Biological Macromolecules, 136*, 586–596.

Zakaulla, M., Parveen, F., & Ahmad, N. (2020). Artificial neural network based prediction on tribological properties of polycarbonate composites reinforced with graphene and boron carbide particle. *Materials Today: Proceedings, 26*, 296–304.

Zhang, B., et al. (2016). Recent advances in electrospun carbon nanofibers and their application in electrochemical energy storage. *Progress in Materials Science, 76*, 319–380.

Zhang, L., Xiao, J., & Zhou, K. (2012). Sliding wear behavior of silver−molybdenum disulfide composite. *Tribology Transactions, 55*(4), 473–480.

Zhang, N., et al. (2021). Significantly enhanced tribology and thermal management by dual-network graphene/epoxy composites. *Tribology International, 164*, 107239.

Zhou, H., et al. (2020a). *Investigation of mechanical and tribological performance of Ti_6Al_4V-based self-lubricating composites with different microporous channel parameters*. *Journal of Materials Engineering and Performance, 29*, 3995–4008.

Zhou, X., et al. (2020b). Production, structural design, functional control, and broad applications of carbon nanofiber-based nanomaterials: A comprehensive review. *Chemical Engineering Journal, 402*, 126189.

Zhu, Y., et al. (2016). Surface functionalized carbon nanofibers and their effect on the dispersion and tribological property of epoxy nanocomposites. *Journal Wuhan University of Technology, Materials Science Edition, 31*(6), 1219–1225.

Anticorrosive applications of nanofibers and their composites

Omar Dagdag[1], Rajesh Haldhar[2], Elyor Berdimurodov[3] and Hansang Kim[1]

[1]Department of Mechanical Engineering, Gachon University, Seongnam, Republic of Korea, [2]School of Chemical Engineering, Yeungnam University, Gyeongsan, Republic of Korea, [3]Faculty of Chemistry, National University of Uzbekistan, Tashkent, Uzbekistan

14.1 Introduction

Many metals including their alloys are seriously destroyed by corrosion. This may have a detrimental effect on the world's industrial and economic systems (Gianni et al., 2014; Muller et al., 2013). The majority of corrosion incidents occur when bare metal surfaces come into touch with their environment. The creation of efficient protective coatings to shield the metal substrate from hostile conditions (atmospheric, chemical, etc.) and stop the corrosive process is urgently required to solve this issue. Oil pipelines and reservoirs are some examples of specific circumstances where the metal surface must be coated with anticorrosive and antistatic coatings to stop corrosion of metallic structures, reduce the formation of static loads, and hasten the release of loads into the interior wall of oil tanks (Chen et al., 2008; Weng et al., 2013).

Organic coatings are frequently utilized to shield metallic components and slow down substrate corrosion (Chen et al., 2020). Comparatively speaking to other techniques, organic coatings are easier and more effective. They have a strong resistance to adhesion and great corrosion resistance, which can operate as a physical barrier between metal and a corrosive environment (Xiong et al., 2020).

The insertion of nanofillers may create a new interface with both the matrix, creating a new interface with subpar mechanical characteristics and poor compatibility (Shao et al., 2018). For instance, the inert wettability of carbon fibers would lead to limited coating dispersion and low interfacial adhesion strength, which would reduce the performance of the composite layer as a whole (Yue et al., 2017). Moreover, some nanofillers, such as graphene, are thought to be the most anticipated barrier filler. Yet, because of its electrical conductivity, it can serve as a cathode and quicken the pace of corrosion at exposed interfaces (Cui et al., 2017). Because of their high elasticity, superior mechanical resistance, and large specific surface area, nanofibers may improve the coating's mechanical qualities (Zhao et al., 2020).

14.2 Nanofiber composites and anticorrosive uses

Tetraaniline-based conductive nanofibers (TANF) were used to improve the corrosion resistance of water-based epoxy coatings on Q235 mild steel, according to Liu et al. Tetraaniline was self-assembled in a 1M HCl solution to create TANF. Fig. 14.1 depicts the standard synthetic pathway employed by TANF.

The effectiveness of TANF mostly on the anticorrosion capability of epoxy coating was examined using electrochemical impedance spectroscopic (EIS), polarizing curves (PDP), and scanning vibrating electrode technique. The findings demonstrated that the aqueous epoxy coating with 0.5 percent by weight of TANF outperformed the pure epoxy coating in terms of corrosion resistance. The analysis of rust layers by X-ray diffraction shows that the addition of electroactive TANI might enhance the creation of a passive metal oxide layer underneath the coating.

A simple way to concurrently add anticorrosion, antistatic, and antibacterial capabilities to water-based polyurethane (WPU) coatings without affecting their original qualities has been devised by Mirmohseni et al. (2020). To impart these qualities in WPU coatings, the polyaniline/reduced cationic graphene oxide (P-RGO$^+$) nanohybrid has already been created. Through in situ reduction and functionalization using DMF, reduction cationic graphene oxide (RGO$^+$) nanosheets have been produced. To create a binary nanohybrid (P-RGO$^+$), polyaniline nanofibers (PANI) were intercalated with cationic reduced graphene oxide nanosheets. The various polyaniline/cationic reduced

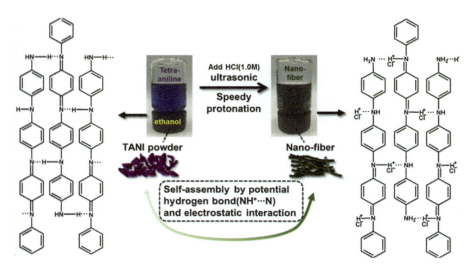

Figure 14.1 Schematic representation for preparing the tetraaniline-based conductive nanofiber (Liu et al., 2019).
Source: Reproduced with permission from Liu, T., Li, J., Li, X., Qiu, S., Ye, Y., Yang, F., et al. (2019). Effect of self-assembled tetraaniline nanofiber on the anticorrosion performance of waterborne epoxy coating, Progress in Organic Coatings, *128*, 137–147.

graphene oxide nanohybrids (P-RGO$^+$) seen in Fig. 14.2 were created using the in situ interfacial polymerization approach.

The findings of the Tafel as well as EIS diagrams demonstrated that changed coatings perform better in terms of preventing corrosion on steel substrates. Also, in the event of the nanohybrid, the coating's electrical surface strength was lowered to $9.8 \times 10 + 6\ \Omega/sq$, qualifying it as an antistatic covering.

Qiu et al. (2017) prepared sulfonated polyaniline (SPANI) by copolymerizing aniline and 2-aminobenzenesulfonic acid and applied as a corrosive inhibitor in the epoxy coating. SPANI was achieved by the following procedure illustrated in Fig. 14.3.

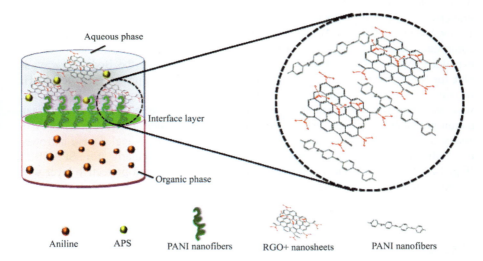

Figure 14.2 Polyaniline/reduced cationic graphene oxide binary nanohybrid synthesis diagram.
Source: Reproduced with permission from Mirmohseni, A., Azizi, M., & Dorraji, M. S. S. (2020). Cationic graphene oxide nanosheets intercalated with polyaniline nanofibers: A promising candidate for simultaneous anticorrosion, antistatic, and antibacterial applications, Progress in Organic Coatings, *139*, 105419.

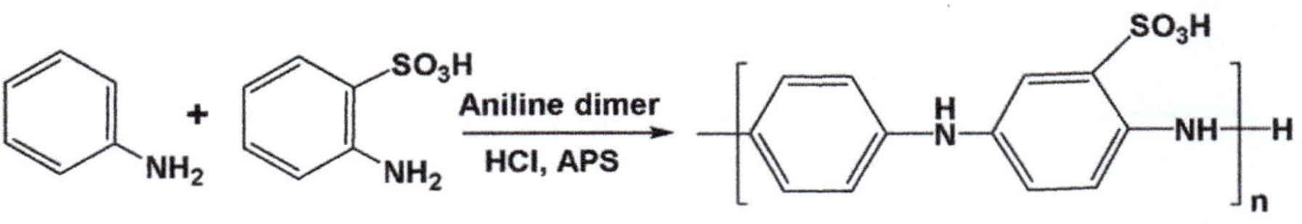

Figure 14.3 Copolymerization scheme for aniline and 2-aminobenzenesulfonic acid.
Source: Reproduced with permission from Qiu, S., Chen, C., Zheng, W., Li, W., Zhao, H., & Wang, L. (2017). Long-term corrosion protection of mild steel by epoxy coating containing self-doped polyaniline nanofiber, *Synthetic Metals*, *229*, 39–46.

In a 3.5% NaCl solution, the EIS and PDP investigated how well-produced composite coatings protected against corrosion. In contrast to blank epoxy coating, which loses its efficacy after 80 days of immersion, composite coatings with SPANI displayed strong protective performance with such a high impedance module during the 120-day immersion (Fig. 14.4).

The SPANI-0 Bode-modulus plot revealed a steadily decreasing $|Z|_{0.01Hz}$ value from 1.05×10^{10} Ωcm^2 (1 day) to 3.97×10^5 Ωcm^2 (80 days), suggesting that the corrosion resistance of the blank epoxy coating weakened as the corrosion media permeated through it (Fig. 14.4A). The initial $|Z|_{0.01Hz}$ for SPANI-0.5, SPANI-1, and SPANI-2 in the epoxy coating containing SPANI is 1.05×10^{11}, 2.37×10^{10}, and 1.40×10^{10} Ωcm^2, respectively (Fig. 14.4C, F, and G). The $|Z|_{0.01Hz}$ value of these coatings scarcely decreased by one or two orders of magnitude after 120 days of immersion, reaching 3.83×10^9 (SPANI-0.5), 1.03×10^9 (SPANI-1), and 1.26×10^8 Ωcm^2 (SPANI-2). Additionally, there were marginal improvements in the composite coating's $|Z|_{0.01Hz}$ value between days 80 and 100 of immersion, which was explained by the healing ability of SPANi. Fig. 14.4B shows the Bode-phase angle plot for SPANI-0, which showed that during immersion, the phase angle at 0.01 Hz increased to 0 degree. Additionally, the band in the low-frequency range (10−100 Hz) was associated with the reaction of metal corrosion and revealed that the metal foundation gradually lost its protection and experienced corrosion. However, the SPANI/epoxy coatings' Bode-phase angle plots revealed that the phase angles were narrowed to −90 degrees across a large frequency range, particularly for SPANI-0.5 (Fig. 14.4D). In this instance, the coating equated to an isolated protective layer with enormous resistance and little capacitance (Qiu et al., 2017).

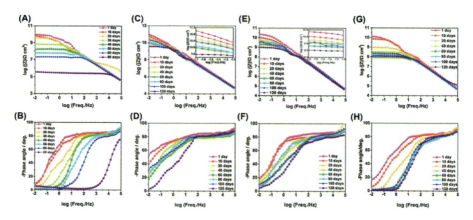

Figure 14.4 Bode plots for 120 days using 3.5% NaCl solution. (A and B) SPANI-0, (C and D) SPANI-0.5, (E anf F) SPANI-1, and (G and H) SPANI-2.
Source: Reproduced with permission from Qiu, S., Chen, C., Zheng, W., Li, W., Zhao, H., & Wang, L. (2017). Long-term corrosion protection of mild steel by epoxy coating containing self-doped polyaniline nanofiber, *Synthetic Metals*, 229, 39−46.

Fig. 14.5 illustrates the SPANI/epoxy coating's ability to resist corrosion. The epoxy coating's SPANI, a conductive polymer, provides both active as well as passive defense (Behzadnasab et al., 2013). By absorbing electrons from the metal's anodic dissolution, SPANI performs a cathodic reaction to change oxidized emeraldine base (EB) into reduced leucoemeraldine (LE) form. Its redox activity protects against the impact of isolating the anodic/cathodic reaction and hinders the anodic reaction's ability to obtain the OH$^-$ it requires. During the immersion time, the rising Fe ions (Fe^{2+} and Fe^{3+}) change into passive Fe$_2$O$_3$ as well as Fe$_3$O$_4$ as an oxide layer in response to the rising impedance module. Moreover, the self-oxidation of SPANI's LE form allows it to transform back into its EB form (Lu et al., 2003). Moreover, a variety of imine groups on the SPANI framework absorb Fe^{2+} and Fe^{3+} ions from corrosive regions to create chelated Fe-NH functional groups, which stabilize the metal potential in the passive zone. This reacts to the PDP curves' higher corrosion potential.

Wu et al. (2022) synthesized a new photosensitive prepolymer using urushiol epoxy acrylate (UEA). In order to perform the hardening of the UEA, photoinitiators of modified cellulose nanofiber photoinitiators (MCNFI) were prepared through the modification of cellulose nanofiber (CNF). The UEA synthesis process is illustrated in Fig. 14.6.

The process for preparing the MCNFI is presented in Fig. 14.7. First, 40 mL of anhydrous acetone and 0.5 g of CNF were combined. Once CNF was evenly distributed in acetone, this combination was sonicated using an ultrasonic machine for 30 minutes at room temperature. Then, a 100 mL round-bottomed flask containing the catalyst, acetone-dispersed CNF, 2.8 g of tolylene-2,4-diisocyanate (TDI), and 0.04 g of DBTDL was heated to 50°C in a water bath and agitated for 4 hours. As a

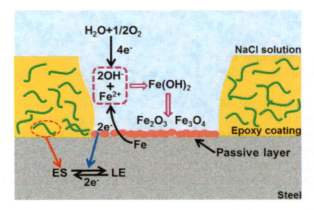

Figure 14.5 Diagrammatic representation of the corrosion process using sulfonated polyaniline epoxy coating.
Source: Reproduced with permission from Qiu, S., Chen, C., Zheng, W., Li, W., Zhao, H., & Wang, L. (2017). Long-term corrosion protection of mild steel by epoxy coating containing self-doped polyaniline nanofiber, Synthetic Metals, *229*, 39–46.

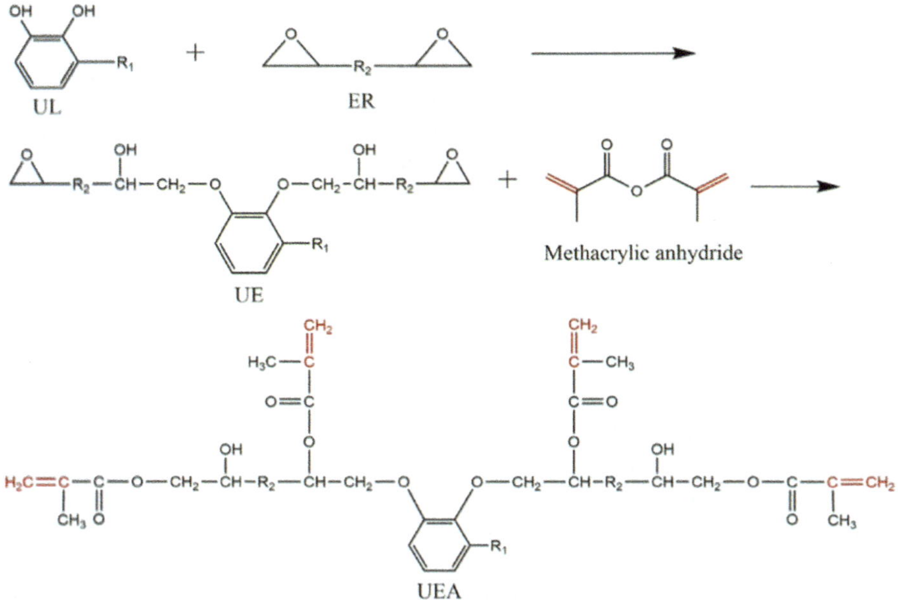

Figure 14.6 Synthesis process of urushiol epoxy acrylate.
Source: Reproduced with permission from Wu, H., Han, X., Zhao, W., Zhang, Q., Zhao, A., & Xia, J. (2022). Mechanical and electrochemical properties of UV-curable nanocellulose/urushiol epoxy acrylate anti-corrosive composite coatings, *Industrial Crops and Products*, *181*, 114805.

result, CNF-TDI, a TDI-modified CNF, was created. They were then given 1.9 g of photoinitiator 2959. The reaction system was held for 5 hours after being gradually heated to 60°C. To eliminate the unreacted materials shortly after the reaction, the products were centrifuged along repeatedly rinsed with acetone. The product was then dried for 24 hours at 50°C in a vacuum oven. The final result was the pure product MCNFI.

Fig. 14.8 illustrates the MCNFI/UEA development process. Different loading levels of MCNFI (4, 6, 8, and 10 wt.%) were dispersed in the prescribed quantity of UEA base component and stirred for 30 minutes. For comparison, a CNF/UEA sample containing unchanged CNF was also created by combining 3 wt.% 2959 and 0.8 wt.% CNF with UEA. Additionally, a pure UEA sample was similarly prepared by combining 3 wt.% 2959 with UEA. The last step included roll coating clean, dry tinplates and glass sheets with the processed MCNFI/UEA mixes. To accomplish full curing, the coated substrates were progressively exposed to UV radiation from a high-pressure mercury lamp with a wavelength of 365 nm and a power capacity of 2 kW for 60 seconds. The sample was 10 cm distant from the lamp envelope, while 98 mW/cm^2 of light per square centimeter was shining on the coated surface. The sample surface was heated to a temperature of 65°C–70°C

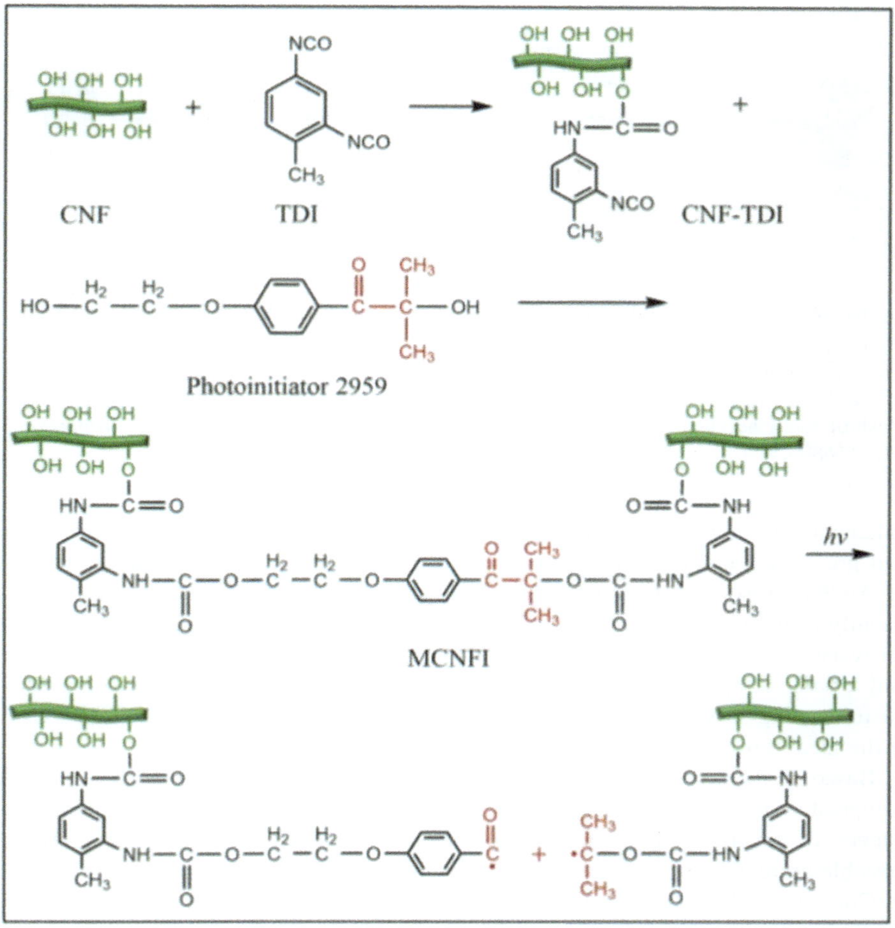

Figure 14.7 Preparation and decomposition of modified cellulose nanofiber photoinitiators. *Source*: Reproduced with permission from Wu, H., Han, X., Zhao, W., Zhang, Q., Zhao, A., & Xia, J. (2022). Mechanical and electrochemical properties of UV-curable nanocellulose/urushiol epoxy acrylate anti-corrosive composite coatings, *Industrial Crops and Products*, *181*, 114805.

during irradiation. Consequently, the dried MCNFI/UEA coatings were produced. The preparation of the CNF/UEA and pure UEA coatings was similar. The coating has a thickness of 605 μm. The polytetrafluoroethylene mold was filled with UEA composite resin, which was then dried in an oven at 80°C for four hours before being UV-cured.

The findings demonstrated that the MCNFI/UEA coating's examined qualities were superior to those of the pure UEA coating and the unaltered CNF reinforced UEA (CNF/UEA) coating. MCNFI has demonstrated a remarkable advantage in

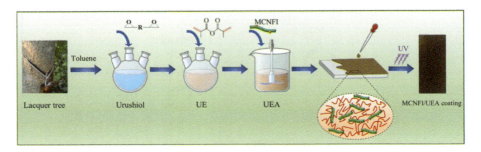

Figure 14.8 Modified cellulose nanofiber photoinitiators/urushiol epoxy acrylate compound coating preparation process.
Source: Reproduced with permission from Wu, H., Han, X., Zhao, W., Zhang, Q., Zhao, A., & Xia, J. (2022). Mechanical and electrochemical properties of UV-curable nanocellulose/urushiol epoxy acrylate anti-corrosive composite coatings, *Industrial Crops and Products*, *181*, 114805.

enhancing the anticorrosion capabilities of UEA coatings due to good dispersion and greater compatibility between it and the UEA matrix. Also, when the amount of MCNFI rose, the passivation of MCNFI/UEA initially increased and subsequently declined, as well as the MCNFI (8% by weight) /UEA sample had the highest resistance to corrosion. Low water absorption, a physical barrier, high adhesion, and a passivating action are just a few of the elements that contribute to the composite coating's exceptional corrosion resistance. In general, biobased MCNFI/UEA at the ideal ratio may be a strong contender for anticorrosion coatings.

Based on the aforementioned findings, it can be concluded that the MCNFI/UEA composite coating has higher anticorrosion effectiveness, which is likely the consequence of a variety of variables. Fig. 14.9 demonstrates the composite coating's possible corrosion-resistant mechanism.

Core-charged electrospun fibers were created by Ji et al. (2022) utilizing an oleic acid with an alkyd varnish resin core and a cellulose acetate shell. For the purpose of creating self-healing as well as pH-sensitive coatings for a steel substrate, the cores were synthesized and utilized in poly(dimethylsiloxane) (PDMS).

In order to create the self-healing coatings, hydrophobic fibers were added to the PDMS matrix. To do this, the PDMS was thoroughly mixed with its curing agent at a 10:1 ratio before being centrifuged to eliminate bubbles. Electrostatic spinning was used to cover the Q235 carbon steel substrates with fibers. The prepared fiber/steel substrate was then coated with PDMS utilizing a spin coater at 800 rpm for 20 seconds as well as 1500 rpm for 10 seconds. The hydrophobic fiber coatings were dried at 80°C for an hour after being placed in a vacuum oven for 5 minutes to eliminate bubbles. The items were then moved and allowed to naturally dry at room temperature. The coatings' combined thickness was around 150 μm. In order to examine the self-healing capabilities of the coatings, fissures were created on their surface using a surgical knife blade. Fig. 14.10 provides a schematic illustration of how coatings are created.

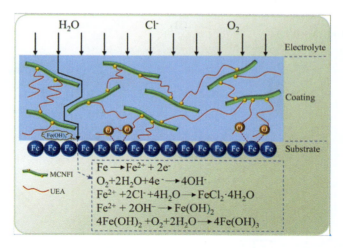

Figure 14.9 Diagrammatic representation of the corrosion resistance mechanism by modified cellulose nanofiber photoinitiators/urushiol epoxy acrylate composite coatings on tin supports.
Source: Reproduced with permission from Wu, H., Han, X., Zhao, W., Zhang, Q., Zhao, A., & Xia, J. (2022). Mechanical and electrochemical properties of UV-curable nanocellulose/urushiol epoxy acrylate anti-corrosive composite coatings, Industrial Crops and Products, *181*, 114805.

The self-healing performance of scratched coatings is studied by EIS measurements, which is an effective and reliable method for investigating the corrosion resistance of samples. The Bode and Nyquist plots of the scratched coatings tested by wet/dry cyclic corrosion tests (immersed in 3.5 wt.% NaCl solution (pH = 11.8)) are presented in Fig. 14.11A–F.

The fiber-PDMS composite coating's $|Z|_{0.01Hz}$ rises with immersion duration, reaching a maximum value of 89.67 k/Ωcm² after 12 days of immersion, as shown in Fig. 14.11B. After 15 days of immersion, this sample then experiences a small downward trend, and the $|Z|_{0.01Hz}$ value is 65.80 kcm². As a result, the highest self-healing rate of the fiber-PDMS composite coating is 4.90 k/Ωcm²/d. This outcome can be linked to the fibers' release of oleic acid (OA) and alkyd varnish (AVR) during the wet/dry cyclic corrosion testing, which coats the steel substrate's surface with a shielding coating. The wet/dry cyclic corrosion tests reveal a declining trend in the impedance modulus of the blank PDMS coating. The blank PDMS coating's $|Z|_{0.01Hz}$ values are 4.76 and 2.38 k/Ωcm² after 1 and 15 days, respectively, demonstrating the remarkable self-healing protective capability of the fiber-PDMS composite coating. Additionally, the emergence of the self-healing layer in the scratched region causes the phase angle of the fiber-PDMS coating at high frequency to be much greater than that of the blank PDMS coating. Fig. 14.11G shows that the impedance of the fiber-PDMS composite coating progressively increases while that of the blank PDMS coating gradually drops, demonstrating the better anticorrosion and self-healing ability of fiber-PDMS composite coatings.

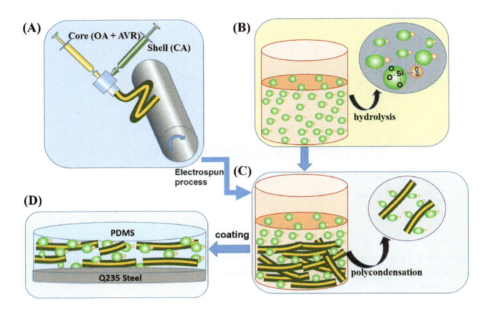

Figure 14.10 Process schematic for preparing self-healing coatings: (A) synthesis of CA@(OA + AVR) core-shell electrospun fibers, (B) preparation of hydrophobic solution, (C) preparing hydrophobic fiber materials, and (D) apply the composite fiber coating to the Q235 steel surface.
Source: Reproduced with permission from Ji, X., Wang, W., Zhao, X., Wang, L., Ma, F., & Wang, Y., et al. (2022). Poly (dimethyl siloxane) anti-corrosion coating with wide pH-responsive and self-healing performance based on core − shell nanofiber containers, *Journal of Materials Science & Technology*, *101*, 128−145.

According to Wang et al. (2019), a simple procedure for creating conductive thermoplastic polyurethane/carbon nanotubes/polydimethylsiloxane (TPU/CNT/PDMS)-based superhydrophobic deformation sensors has been proposed. These materials were created by ultrasonically inducing CNT decoration on the surface of electrospun TPU nanofibers, followed by PDMS modification. The schematic preparation method for TPU, CNT, and PDMS nanofibers is shown in Fig. 14.12A−C. The conductor network is created by CNTs that are evenly dispersed throughout the nanofiber surface and have a hierarchical structure. The superhydrophobicity and, therefore, the anticorrosion feature of the nanofiber composite are provided by the low surface energy PDMS layer. Corrosive alkali and salt could be repelled by the composite of superhydrophobic nanofibers, and the membrane continued to be super hydrophobic even under heavy pressure.

By electrospinning polyvinylidene fluoride nanofibers filled with MBT and IBU, Piao et al. (2022) created self-healing composite-coated nanofibers. Fig. 14.13 displays the preparation schematic for the PFP@MBT-IBU nanofiber membrane.

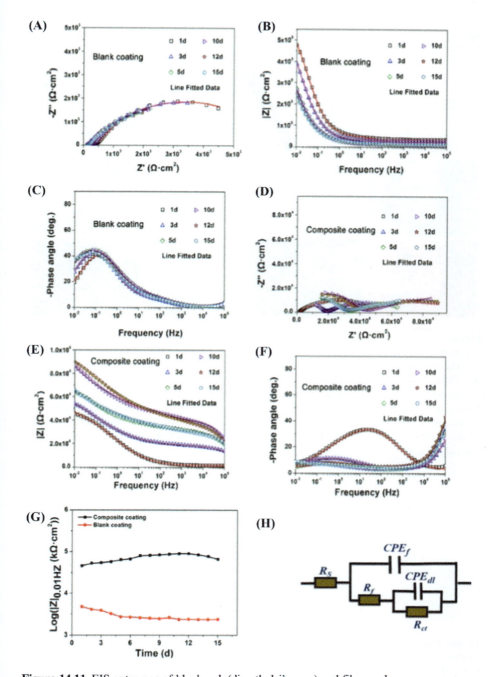

Figure 14.11 EIS outcomes of blank poly(dimethylsiloxane) and fiber- poly (dimethylsiloxane) composite coatings tested by wet/dry cyclic corrosion tests (pH = 11.8): (A–F) Nyquist, Bode and phase; (G) the values of $|Z|_{0.01Hz}$.
Source: Reproduced with permission from Ji, X., Wang, W., Zhao, X., Wang, L., Ma, F., & Wang, Y., et al. (2022). Poly (dimethyl siloxane) anti-corrosion coating with wide pH-responsive and self-healing performance based on core − shell nanofiber containers, Journal of Materials Science & Technology, *101*, 128−145.

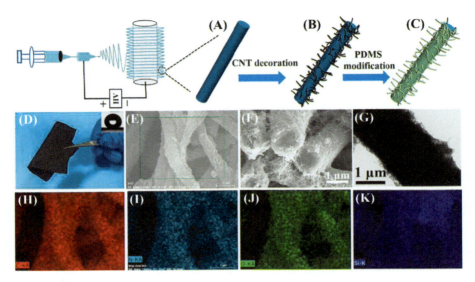

Figure 14.12 (A−C) Schematic illustration for preparing TPU/CNT/PDMS nanofiber composites. (D) Photo of TPU/CNT/PDMS nanofiber composite with good flexibility. (E) The SEM image on the nanofiber composite surface. (F) cross sectional SEM image and (G) TEM picture of TPU/CNT/PDMS nanofiber composite. (H−K) Digitization of cartographic images of the green zone in (E) for C, N, O, and Si, respectively.
Source: Reproduced with permission from Wang, L.,Chen, Y., Lin, L., Wang, H., Huang, X., & Xue, H., et al. (2019). Highly stretchable, anti-corrosive and wearable strain sensors based on the PDMS/CNTs decorated elastomer nanofiber composite, Chemical Engineering Journal, *362*, 89−98.

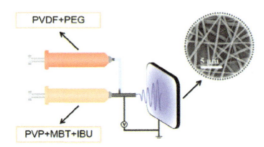

Figure 14.13 Preparation scheme of PFP@MBT-IBU electrospun nanofibers.
Source: Reproduced with permission from Piao, J., Wang, W., Cao, L., Qin, X., Wang, T., & Chen, S. (2022). Self-healing performance and long-term corrosive resistance of Polyvinylidene fluoride nanofiber alkyd coating, *Composites Communications*, *36*, 101404.

Fig. 14.14 depicts typical electrochemical test outcomes for scratched coatings in a pH = 4, 3.5 wt.% NaCl solution. The OCP can assess the long-term anticorrosion qualities immediately. The scratched coatings' OCP values in Fig. 14.14A,

Anticorrosive applications of nanofibers and their composites 369

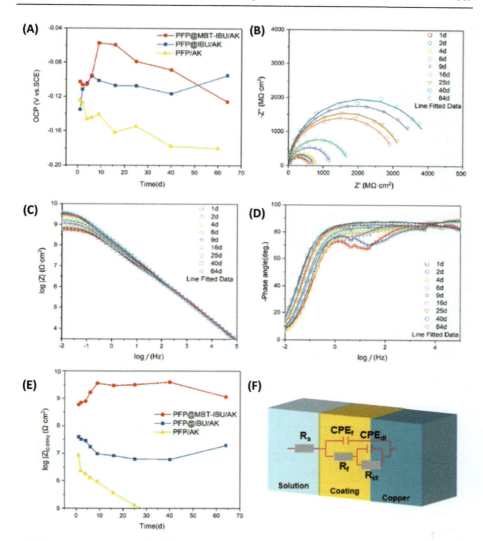

Figure 14.14 Electrochemistry of 3.5% NaCl coatings (pH = 4): (A) OCP results of the coatings; (B–D) Nyquist and Bode plots of Cu coated with PFP@MBT-IBU/AK coating; (E) $|Z|_f = 0.01$Hz of the scratched coatings.
Source: Reproduced with permission from Piao, J., Wang, W., Cao, L., Qin, X., Wang, T., & Chen, S. (2022). Self-healing performance and long-term corrosive resistance of Polyvinylidene fluoride nanofiber alkyd coating, Composites Communications, 36, 101404.

from high to low, are PFP@MBT-IBU/AK, PFP@IBU/AK, and PFP/AK coating. It shows that adding PFP nanofibers improves the coating's capacity for protection. Coatings are given self-healing properties by the inclusion of corrosion inhibitors, and the corrosion resistance alongside the self-healing properties of the coating further improved due to the synergistic action of several corrosion inhibitors. Fig. 14.14B−D displays the Nyquist along with Bode spectra of PFP@MBT-IBU/AK scratched coating in 3.5 wt.% NaCl solution (pH = 4), which may be used to study the corrosion protection property and self-healing ability of the coatings

According to Yang et al. (2018), corrosion resistance of coatings formed on the substrate Q235 was prepared using a zinc-rich epoxy composite coating comprising a doped conductive SPANi nanofiber (Fig. 14.15).

The Nyquist plot showed decreasing capacitive arc radii for the neat zinc-rich epoxy corrosion coating (Fig. 14.16A) throughout the whole immersion, indicating a deterioration in the coating's capacity to resist corrosion. The capacitive arc radii substantially decreased after 12 days of immersion, as shown in Fig. 14.16A. This was due to the porous nature of the zinc-rich corrosion coating, which reduced the amount of interaction between zinc particles and finally resulted in a clear weakening of cathodic protection. After 8−10 days of immersion, the capacitive loop radii for three composite coatings significantly increased, indicating that the aggressive medium had only barely penetrated the zinc layers before the oxide (such as ZnO or $Zn(OH)_2$) was created to replace the minute imperfections in the zinc-rich layer. The passivation product generated by zinc oxide/hydroxide and interlayer SPANi caused the radii of Nyquist plots at day 8 to increase beyond those of day 1 in particular for 0.5 and 1.0 wt.% samples.

A strong SPANi zinc coating's corrosion process is shown in Fig. 14.17. The organic layer of the surface's nanofiber served as a barrier to stop the penetration of the corrosion medium during the early stages of corrosion (Wang & Tan, 2006). Also, the metal substrate has been protected by the remarkable redox characteristics

Figure 14.15 Structural formulations and SEM images of water-soluble SPANi nanofibers and Zn-rich composite coating preparation/SPANi.
Source: Reproduced with permission from Yang, F., Liu, T., Li, J., Qiu, S., & Zhao, H. (2018). Anticorrosive behavior of a zinc-rich epoxy coating containing sulfonated polyaniline in 3.5% NaCl solution, *RSC Advances*, 8, 13237−13247.

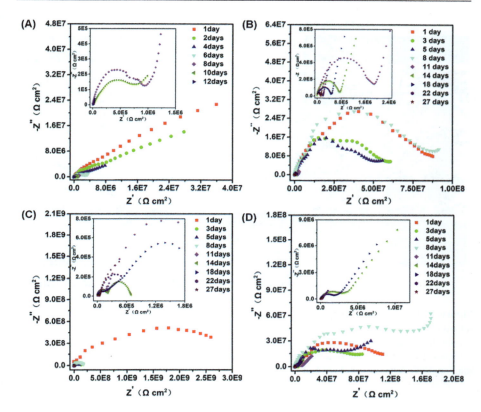

Figure 14.16 Nyquist plots of (A) pure Zn-rich coating, (B) 0.5%, (C) 1.0%, and (D) 2.0% SPANi Zn-rich coatings during 27 days of immersion in 3.5% NaCl solution.
Source: Reproduced with permission from Yang, F., Liu, T., Li, J., Qiu, S., & Zhao, H. (2018). Anticorrosive behavior of a zinc-rich epoxy coating containing sulfonated polyaniline in 3.5% NaCl solution, *RSC Advances*, 8, 13237−13247.

of the SPANi nanofiber by creating a passive coating. More specifically, SPANi's cathodic reaction also occurred. By gaining dissociative electrons from the anodic dissolution of Zn powder into the iron coating and substrate, SPANi's emerald salt form was converted to the reduced form of LE.

According to Saikia and Karak (2019), the multi-walled carbon nanotube (MWCNT)-polyaniline nanofiber-carbon dot (CD) nanohybrid being created utilizing in situ aniline polymerization when MWCNT and CD were present. The multi-wire proportional chambers (MWPC) nanohybrid was synthesized using aniline in situ polymerization in the presence of MWCNT and CDs. A likely mechanistic pathway to this process is described in Fig. 14.18.

The nanocomposite was prepared using an ex situ polymerization technique. In this case, the ricin oil-based SME sorbitol and monoglyceride were used as a matrix and the MWPC was used as a reinforcing agent. The trajectory of nanocomposite

Figure 14.17 Corrosion mechanism proposed for the SPANi zinc-rich coating in 3.5% NaCl solution.
Source: Reproduced with permission from Yang, F., Liu, T., Li, J., Qiu, S., & Zhao, H. (2018). Anticorrosive behavior of a zinc-rich epoxy coating containing sulfonated polyaniline in 3.5% NaCl solution, RSC Advances, *8*, 13237–13247.

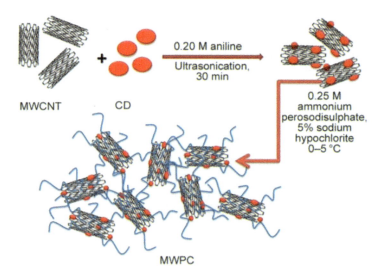

Figure 14.18 Possible schematic route for MWPC nanohybrid preparation.
Source: Reproduced with permission from Saikia, A. & Karak, N. (2019). Fabrication of renewable resource based hyperbranched epoxy nanocomposites with MWCNT-polyaniline nanofiber-carbon dot nanohybrid as tough anticorrosive materials, eXPRESS Polymer Letters, *13*.

Anticorrosive applications of nanofibers and their composites

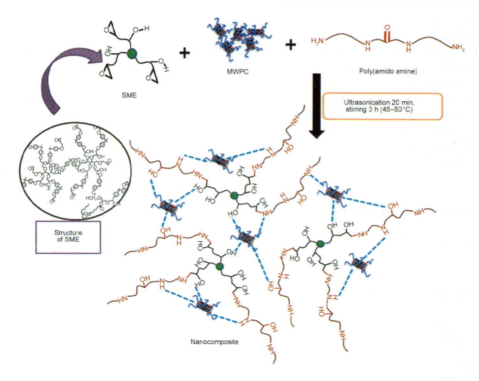

Figure 14.19 Diagram of the ex situ production of nanocomposites (Saikia & Karak, 2019).

formation is presented in Fig. 14.19. The corrosion performance of hardened nanocomposites was investigated on soft steel sheets in 3.5% NaCl solution by the PDP method. According to the study, nanocomposites with the greatest concentration of nanohybrids outperformed thermosets in terms of corrosion resistance (corrosion rate of 4.62.104 mpy). The hyperbranched, steady dispersion resin of the nanocomposite-based nanohybrid was, therefore, shown to have potential for use as a high-performance anticorrosive material.

14.3 Conclusion

This chapter provides a brief review of nanofibers for use in coatings, including natural and synthetic fibers. It also discusses composite coatings, various types of coating fibers and their uses, and the impact of varying fiber aspect ratios on coating qualities. In general, anticorrosive applications of composite nanofibers and nanofillers exhibit enhanced physical behavior even with low nanofiller concentration, owing to their nanoscale diameters, which provide a substantial contact zone between nanofillers along with polymeric matrices' elevated aspect ratio.

Additionally, the enhanced barrier performance of nanoclay structures allows for less gas penetration and improved anticorrosive properties. Anticorrosive nanofiller composites exhibit superior inhibition against corrosion on metallic surfaces when tested using a variety of standard electrochemical anticorrosion equipment, including impedance spectroscopy, corrosion potential, polarization resistance, and corrosion current. According to studies on free-standing coatings of "as-prepared" composite nanofibers coatings, which show that O_2 along with H_2O permeability is increased, the better anticorrosion behavior of composite nanofibers coatings over pristine polymers is due to the even dispersion of nanofillers sheets in the polymeric matrix. This increases the diffusion tortuosity pathway for H_2O and O_2 molecules.

Author contribution statement

Equal contribution by all the authors.

Abbreviations

TANF	Conducting nanofiber
XDR	X-ray diffraction
EIS	Electrochemical impedance spectroscopic
PDP	Polarizing curves
SVET	Scanning vibrating electrode technique
WPU	waterborne polyurethane
DMF	Dimethylformamide
PANI	Polyaniline
SPANI	Sulfonated polyaniline
UEA	Urushiol epoxy acrylate
MCNFI	Modified cellulose nanofiber photoinitiators
EDS	Energy dispersive spectrometer
PVDF	Polyvinylidene fluoride
SPANi	Conducting sulfonated polyaniline
MWPC	Multiwire proportional chambers
TDI	Tolylene-2,4-diisocyanate
OA	Oleic acid
AVR	Alkyd varnish
MWCNT	Multi-walled carbon nanotube
CD	carbon dot
EB	oxidized emeraldine base

References

Behzadnasab, M., Mirabedini, S., & Esfandeh, M. (2013). Corrosion protection of steel by epoxy nanocomposite coatings containing various combinations of clay and nanoparticulate zirconia. *Corrosion Science*, 75, 134–141.

Chen, B., Jin, Z., Pei, X., Li, J., Liu, C., & Zhao, H. (2020). Liquid-like secondary amine containing POSS as reactive nanoadditive to improve the anticorrosion performance of epoxy coating. *Materials Letters, 261*, 127103.

Chen, Z.-H., Tang, Y., Yu, F., Chen, J.-H., & Chen, H.-H. (2008). Preparation of light color antistatic and anticorrosive waterborne epoxy coating for oil tanks. *Journal of Coatings Technology and Research, 5*, 259−269.

Cui, C., Lim, A. T. O., & Huang, J. (2017). A cautionary note on graphene anti-corrosion coatings. *Nature Nanotechnology, 12*, 834−835.

Gianni, L., Gigante, G. E., Cavallini, M., & Adriaens, A. (2014). Corrosion of bronzes by extended wetting with single versus mixed acidic pollutants. *Materials, 7*, 3353−3370.

Ji, X., Wang, W., Zhao, X., Wang, L., Ma, F., Wang, Y., et al. (2022). Poly (dimethyl siloxane) anti-corrosion coating with wide pH-responsive and self-healing performance based on core-shell nanofiber containers. *Journal of Materials Science & Technology, 101*, 128−145.

Liu, T., Li, J., Li, X., Qiu, S., Ye, Y., Yang, F., et al. (2019). Effect of self-assembled tetraaniline nanofiber on the anticorrosion performance of waterborne epoxy coating. *Progress in Organic Coatings, 128*, 137−147.

Lu, J., Liu, N., Wang, X., Li, J., Jing, X., & Wang, F. (2003). Mechanism and life study on polyaniline anti-corrosion coating. *Synthetic Metals, 135*, 237−238.

Mirmohseni, A., Azizi, M., & Dorraji, M. S. S. (2020). Cationic graphene oxide nanosheets intercalated with polyaniline nanofibers: A promising candidate for simultaneous anticorrosion, antistatic, and antibacterial applications. *Progress in Organic Coatings, 139*, 105419.

Muller, J., Laïk, B., & Guillot, I. (2013). α-CuSn bronzes in sulphate medium: Influence of the tin content on corrosion processes. *Corrosion Science, 77*, 46−51.

Piao, J., Wang, W., Cao, L., Qin, X., Wang, T., & Chen, S. (2022). Self-healing performance and long-term corrosive resistance of polyvinylidene fluoride nanofiber alkyd coating. *Composites Communications, 36*, 101404.

Qiu, S., Chen, C., Zheng, W., Li, W., Zhao, H., & Wang, L. (2017). Long-term corrosion protection of mild steel by epoxy coating containing self-doped polyaniline nanofiber. *Synthetic Metals, 229*, 39−46.

Saikia, A., & Karak, N. (2019). Fabrication of renewable resource based hyperbranched epoxy nanocomposites with MWCNT-polyaniline nanofiber-carbon dot nanohybrid as tough anticorrosive materials. *eXPRESS Polymer Letters, 13*.

Shao, Z.-B., Zhang, M.-X., Li, Y., Han, Y., Ren, L., & Deng, C. (2018). A novel multifunctional polymeric curing agent: Synthesis, characterization, and its epoxy resin with simultaneous excellent flame retardance and transparency. *Chemical Engineering Journal, 345*, 471−482.

Wang, L., Chen, Y., Lin, L., Wang, H., Huang, X., Xue, H., et al. (2019). Highly stretchable, anti-corrosive and wearable strain sensors based on the PDMS/CNTs decorated elastomer nanofiber composite. *Chemical Engineering Journal, 362*, 89−98.

Wang, T., & Tan, Y.-J. (2006). Understanding electrodeposition of polyaniline coatings for corrosion prevention applications using the wire beam electrode method. *Corrosion Science, 48*, 2274−2290.

Weng, C. J., Chen, Y. L., Jhuo, Y. S., Yi-Li, L., & Yeh, J. M. (2013). Advanced antistatic/ anticorrosion coatings prepared from polystyrene composites incorporating dodecylbenzenesulfonic acid-doped SiO_2@ polyaniline core−shell microspheres. *Polymer International, 62*, 774−782.

Wu, H., Han, X., Zhao, W., Zhang, Q., Zhao, A., & Xia, J. (2022). Mechanical and electrochemical properties of UV-curable nanocellulose/urushiol epoxy acrylate anti-corrosive composite coatings. *Industrial Crops and Products, 181*, 114805.

Xiong, H., Qi, F., Zhao, N., Yuan, H., Wan, P., Liao, B., et al. (2020). Effect of organically modified sepiolite as inorganic nanofiller on the anti-corrosion resistance of epoxy coating. *Materials Letters, 260*, 126941.

Yang, F., Liu, T., Li, J., Qiu, S., & Zhao, H. (2018). Anticorrosive behavior of a zinc-rich epoxy coating containing sulfonated polyaniline in 3.5% NaCl solution. *RSC Advances, 8*, 13237–13247.

Yue, Z., Vakili, A., & Duran, M. P. (2017). Surface treatments of solvated mesophase pitch-based carbon fibers. *Journal of Materials Science, 52*, 10250–10260.

Zhao, X., Yuan, S., Jin, Z., Zhu, Q., Zheng, M., Jiang, Q., et al. (2020). Fabrication of composite coatings with core-shell nanofibers and their mechanical properties, anticorrosive performance, and mechanism in seawater. *Progress in Organic Coatings, 149*, 105893.

Gas adsorption and storage of nanofibers and their composites

Nancy Elizabeth Davila-Guzman[1], Margarita Loredo-Cancino[1], Sandra Pioquinto-Garcia[1] and Alan A. Rico-Barragán[2]
[1]School of Chemistry, Autonomous University of Nuevo Leon, San Nicolas de los Garza, Nuevo Leon, Mexico, [2]Department of Environmental Engineering, National Technological of Mexico/Higher Technological Institute of Misantla, Misantla, Veracruz, Mexico

15.1 Introduction

Gas adsorption is a key industry separation process for the purification or separation of gaseous mixtures due to its easy operation, high performance, and low operational cost (Pérez-Pellitero & Pirngruber, 2020). Generally, gas adsorption in continuous systems is carried out in packed columns with porous materials to separate a mixture. The porous materials, known as adsorbents, interact with the gas molecules by physical or chemical means to retain them on their surface by adsorption. Several adsorbent materials, such as zeolite, carbon-based materials, silica gel, polymeric resins, inorganic materials, and composite materials, have been employed to separate mixtures in the industry (Morris & Wheatley, 2008).

The choice of the adsorbent material for a specific gas separation relies on the physicochemical properties of the mixture components, the surface chemistry and textural properties of the adsorbent material, and the operational conditions (temperature, pressure, initial concentration of the mixture, and the desired degree of separation, among others). Thus it is not an easy task to properly select an adsorbent material for gas separation and purification.

Due to the increasing demand for efficient and sustainable energy systems, the development of new adsorbent materials has sped up. In this sense, gas adsorption and storage in nanofibers and their composites have emerged as a promising approach for addressing the pressing challenges of energy sustainability and environmental conservation. The unique properties and structural characteristics of nanofibers provide remarkable opportunities to enhance gas adsorption capacities and storage densities for diverse applications, from clean energy systems to environmental remediation.

This book chapter delves into the fascinating area of gas adsorption and storage in nanofibers and their composites. We look at the performance features and adsorption mechanisms that make these materials ideal for gas storage applications. By examining the interplay between nanofiber architectures, surface chemistries, and gas adsorption phenomena, we aim to comprehensively understand the mechanisms driving gas storage in these advanced materials.

Polymeric Nanofibers and their Composites. DOI: https://doi.org/10.1016/B978-0-443-14128-7.00015-8
© 2025 Elsevier Ltd. All rights are reserved, including those for text and data mining, AI training, and similar technologies.

The chapter covers several gas species, including hydrogen, carbon dioxide, sulfur, VOCs, and other important gases for clean energy systems and environmental applications. Furthermore, the advancements in developing nanofiber composites for enhanced gas storage are discussed. The synergistic effects that can be achieved by combining the unique features of nanofibers with metal-organic frameworks (MOFs), carbon-based adsorbents, or zeolites are also discussed.

Overall, this book chapter aims to provide a valuable resource for researchers, scientists, and engineers interested in exploring the potential of nanofibers and their composites for efficient gas storage, offering insights into this rapidly evolving research area and encouraging additional study and development for real-world applications.

15.2 Carbon dioxide capture

Carbon dioxide (CO_2) adsorption on nanofibers and their composites is an area of active research that addresses the challenges of CO_2 capture and storage, an essential strategy for mitigating climate change. Nanofibers, with their high surface area and porous structure, offer promising opportunities for efficient CO_2 adsorption. Here are some critical aspects of CO_2 and its adsorption on nanofibers and their composites.

Carbon dioxide is a triatomic molecule with a molecular weight of 44 g/mol. It is a colorless gas at room temperature and pressure. It presents a linear structure in which each carbon-oxygen bond has a length of 1.16 Å and comprises a σ bond and a π bond. It has a kinetic diameter of 3.30 Å, a zero-dipole moment, and a quadrupole moment of 0.00134 cm². Organizations such as the IPCC and the National Aeronautics Administration have found a strong relationship between climate change and the increase in greenhouse gas (GHG) emissions due to active infrared interactions presenting these gases (North, 2015). The Earth's atmosphere is transparent to the visible light from the sun, falling on its surface and reemitted as infrared radiation. The main components of the Earth's atmosphere (oxygen and nitrogen) are transparent to infrared radiation. However, carbon dioxide and other atmospheric gases such as water vapor, methane, and nitrous oxide adsorb some of this radiation, trapping it within the Earth's atmosphere (Trogler, 1995).

Although CO_2 has a lower global warming potential than other GHGs, as shown in Table 15.1, it also ranks first in anthropogenic gas emissions. These emissions are primarily caused by the combustion of fossil fuels and natural gas (Intergovernmental Panel on Climate Change, 2015; US EPA, 2008), having a significant increase in their concentration in the atmosphere in the last 50 years, going from 280 to 418 ppm, and having an impact on proportionally in the average temperature of the Earth's surface.

The increased concentration of CO_2 would cause key risks for physical, biological, and human systems. Some of the consequences of climate change are disruption of livelihoods due to weather variations, sea level rise, coastal flooding, periods of

Table 15.1 Global warming potential of greenhouse gases (Intergovernmental Panel on Climate Change, 2015).

Greenhouse gas	Global warming potential	Concentration (ppm)	Permanence in the atmosphere (years)
Carbon dioxide (CO_2)	1	400	5−200
Methane (CH_4)	25	0.180	12
Nitrogen dioxide (N_2O)	298	0.324	114
Hydrofluorocarbons (HFCs)	124−14,800	0.132	260

extreme heat, ocean acidification, loss of ecosystems, food and water insecurity, health problems, and propagation of vectors, such as mosquitoes, among others (Intergovernmental Panel on Climate Change, 2015).

Despite all the efforts aimed at reducing CO_2 emissions, billions of tons are still discharged into the environment, which is why the IPCC has proposed carbon capture and utilization technologies as a necessary tool to curb emissions from significant stationary sources, such as thermoelectric plants, cement plants, refineries, steel, and iron industry plants, among others.

One of the most widely used methods currently for CO_2 capture is adsorption, based on the intramolecular interaction between the surface of the adsorbent material and the adsorbate (CO_2). Adsorption is considered one of the methods applicable to postcombustion with lower energy requirements and greater versatility when used in different pressure and temperature ranges (Chiang & Juang, 2017). However, this method depends to a large extent on the adsorbent material, which is why it is essential to develop materials that present high adsorption capacity and speed, selectivity, easy regeneration, low cost, and high chemical and thermal stability (Plaza et al., 2010).

In developing adsorbent materials, it is critical to obtain solids whose surface morphological and chemical structure present a high adsorption capacity without losing selectivity due to the multicomponent nature of the postcombustion streams (N_2, CO_2). Because adsorption is a surface phenomenon, controlling the contact area per unit mass of the adsorbent and its distribution and shape is imperative. The most relevant textural properties for adsorbent materials in CO_2 adsorption are specific area, volume, and micropore diameter (Li, Yang, et al., 2022; Rehman & Park, 2018; Singh et al., 2018; Sreńscek-Nazzal and Kiełbasa, 2019; Yuan et al., 2019).

Nanofibers, one of the most critical one-dimensional nanomaterials, are expected to have meager resistance for gas transport and high-speed kinetics; therefore they have excellent CO_2 capture and storage potential. They can be made from various materials, such as carbon-based materials (carbon nanotubes [CNTs], graphene, and activated carbon [AC]), MOFs, and polymers (polyacrylonitrile [PAN] and polyamide [PA], among others). Each material has its unique properties that influence CO_2 adsorption capacity and selectivity. The surface of nanofibers can be modified or functionalized to enhance their CO_2 adsorption performance. Functional groups such as amine ($-NH_2$), hydroxyl ($-OH$), or carboxyl ($-COOH$) can be introduced to increase the affinity of nanofibers for CO_2 molecules. Also, nanofibers can be incorporated into composite structures to enhance their CO_2 adsorption properties. For example, nanofibers can be dispersed within a polymer matrix or coated with a thin layer of metal oxide to create a hierarchical structure combining both materials' advantages.

Carbon dioxide adsorption on nanofibers primarily occurs through physical adsorption. Physical adsorption involves weak intermolecular forces, such as van der Waals interactions, while chemical adsorption involves the formation of covalent bonds between CO_2 and the surface functional groups. The adsorption capacity of nanofibers refers to the amount of CO_2 that can be adsorbed per unit mass or unit surface area of the material. Selectivity refers to the preference of nanofibers for CO_2 over other gases present in a mixture, such as nitrogen or methane. Both capacity and selectivity are crucial for efficient CO_2 capture. After CO_2 adsorption, the nanofibers need to be regenerated for reuse. Regeneration methods can include temperature swing adsorption (TSA) or pressure swing adsorption (PSA) processes, where the adsorbed CO_2 is desorbed by changing temperature or pressure conditions.

Nanofiber materials have been explored for CO_2 adsorption due to their high surface area, porous structure, and customizable properties. Porous carbon materials have gained significant preference for CO_2 adsorption owing to their exceptional porosity, extensive surface area, surface chemistry, and thermal stability. Here are some commonly studied nanofiber materials for CO_2 adsorption, with a particular remark on AC.

15.2.1 Activated carbon nanofibers

Carbonaceous materials have excelled as adsorbents due to their variety of precursors, synthesis flexibility, low cost, and simplicity. An adsorbent material's specific area corresponds to the pores' internal area and the adsorbent particles' external area. AC has a porous structure made up of macropores (diameter > 50 nm), mesopores (diameter between 50 and 2 nm), and micropores (diameter < 2 nm), where most of the specific area is found in the micropores. The adsorption capacity of CO_2 at atmospheric pressure in ACs behaves directly proportional to the increase in their specific area (Singh et al., 2017).

The textural characteristics of the material, such as the diameter and volume of the micropores, also affect both the adsorption capacity and its selectivity due to the exclusion principle. Theoretical studies on determining the ideal pore size in laminar-type structures have shown that diameters between 0.57 and 0.72 nm are ideal for CO_2 adsorption and storage (Heuchel et al., 1999). This range of sizes presents the minimum energy potential in the system. Likewise, they present a vast space for the movement of the carbon dioxide molecule and its interaction with the surface functional groups, as well as a higher resistance of larger molecules such as nitrogen. This ideal range of micropores has been experimentally corroborated for both activated and functionalized carbons with different heteroatoms (Rehman & Park, 2018) and unmodified types (Presser et al., 2011).

Heteroatomic doping is a frequently used method to modify the chemical properties of simple carbons. Literature reviews have compiled studies on ACs in which the inclusion of heteroatoms, such as oxygen, sulfur, phosphorus, and nitrogen to the carbon matrix, results in a greater adsorption capacity and selectivity compared to its nonmodified counterpart (Saha & Kienbaum, 2019). The increase in adsorption capacity is mainly associated with increased electrostatic interactions with acid-type gases such as CO_2.

Recently, ACs containing nitrogen in their surface structure have been shown to improve CO_2 adsorption on nonpolar or nonacidic gases, such as N_2 or CH_4 (Saha et al., 2017). Incorporating nitrogen generates greater basicity in the carbon matrix, helping attract the acid gas CO_2 more efficiently. Theoretical studies have shown that pyridine and pyrrole nitrogenous groups have a shorter bond junction distance and a more significant adsorption energy difference than N_2, obtaining a high selectivity (Lim et al., 2016).

Analyzing these studies, the critical stage in synthesizing AC by chemical activation focuses on controlling the textural and elemental properties at the time of the activation stage. High impregnation and temperature ratios increase the surface area and the diameter of the micropores but reduce the amount of final nitrogen in the material. Although nitrogen-doped CAs present promising characteristics for CO_2 adsorption, they also present a disadvantage like most currently synthesized materials, which is a particle size in the order of microns [95% smaller than 150 μm (Cooney, 1998)].

Powder adsorbents represent a fundamental problem when integrating such materials in a fixed bed to perform the separation efficiently due to large pressure drops or dustiness. The structuring of materials is presented as an alternative to solve this problem. Structured adsorbent materials can help control kinetic and engineering factors associated with separation processes, including pressure drop, enthalpy effects of adsorption, external heat integration (for TSA), performance, mechanical stability, optimization of gas flow routes to reaction sites, mass, and energy transfer, among others (Ruiz-Morales et al., 2017; Stephen et al., 2018).

Recently, the fabrication of oxyfluorinated activated electrospun carbon nanofibers (CNFs) for CO_2 capture and the adsorption mechanism was reported (Bai et al., 2011). The presence of nonpolar CO_2 molecules near carbon pores is influenced by the semiionic interaction of oxygen groups, which possess lone-pair electrons. This interaction leads to the attraction of electrons in the CO_2 molecules. This phenomenon could serve as a mechanism for enhancing the capacity of CO_2 storage. Ultimately, the CO_2 gas can be stored within the microporous pores of carbon material. Additionally, some residual CO_2 molecules within the carbon pores can be influenced by the effects of oxygen groups, such as their grabbing effects through semiionic interaction. As a result, the efficiency of CO_2 storage is significantly enhanced.

A structural transformation mechanism has been developed to help understand the CO_2 adsorption mechanism on PAN-based CNFs (PAN-based CNFs); CO_2

molecules can be adsorbed in three potential locations: between the graphitic layers, at the edges of the graphitic layers, or on the surface of crystallites (Park et al., 2021). Based on analysis using selected area electron diffraction, they propose that most CO_2 adsorption occurs on the surface of crystallites and within the spaces between the crystallites, creating ultramicropores.

The properties of nanofibers depend strongly on the type of oxide incorporated into their structure. Different types of oxides can have a range of effects on the characteristics and functionality of nanofibers. For instance, one of the challenges in synthesizing activated carbon nanofibers (ACNFs) is achieving highly porous carbons with exceptional flexibility. To address this, SnO_2 nanoparticles have been introduced into the carbon matrix to enhance the flexibility of CNFs by acting as a plasticizer for connecting single fiber cracks (Ali et al., 2020). The incorporation of SnO_2 nanoparticles also led to increased surface area and pore volume in the CNFs, creating more oxygen vacancies that can capture CO_2 molecules. The resulting material exhibited a CO_2 uptake of 2.6 mmol/g. Notably, CO_2 capture capacities exceeding 1 mmol/g are significant in potentially reducing the cost of CO_2 sequestration. The sample also offered excellent cyclic stability (up to 20 cycles) and remarkable CO_2/N_2 selectivity of 20. On the other hand, the incorporation of metal crystallites such as Co_3O_4 into CNF membranes resulted in enhanced porosity and a significant increase in surface area, reaching 483 m^2/g (Iqbal et al., 2016). Although there were slight changes in the structure of the resulting membrane, it exhibited an exceptional CO_2-capturing capacity of 5.4 mmol/g at 1 bar and room temperature. The CNF membranes were carbonized at 850°C under a nitrogen atmosphere showing a three-dimensional fibrous structure. Therefore careful selection of the oxide component is critical in tailoring the specific characteristics and capabilities of nanofibers for targeted applications.

Another approach to increasing the CO_2 adsorption capacity of adsorbents is the content of nitrogen functional groups, and the Pearson correlation analysis can be helpful in investigating the influence of these functional groups on the CO_2 adsorption capacities of nanofibers (Chiang et al., 2021). The presence of nitrogen groups, mainly pyridine-N oxides, is closely associated with the nanofiber's performance. Additionally, high cyclability can be achieved without significant degradation.

The effect of incorporating nitrogen groups into the carbon framework using a high nitrogen-containing precursor, such as polypyrrole (PPy), in a single step has been investigated (Li et al., 2016). The obtained N-doped porous CNF webs exhibited CO_2 adsorption capacities of 4.42 mmol/g at 25°C, and the adsorption capacities are nearly identical in three adsorption/desorption cycles. Moreover, the material can be used simultaneously as a catalyst for CO_2 conversion to epoxides to synthesize cyclic carbonates.

Polyacrylonitrile (PAN) has been employed as a nitrogen-rich precursor for carbon during the electrospinning process, followed by carbonization (Zainab et al., 2020). The resulting material exhibited remarkable characteristics, including high selectivity for CO_2 gas (selectivity factor of 20) and superior CO_2 adsorption performance (3.11 mmol/g at 25°C). Notably, the synthesized adsorbent demonstrated long-term stability as there was no noticeable change in mass even after undergoing

50 cycles of CO_2 adsorption and desorption. The fabricated samples contained various nitrogen species, including pyridinic nitrogen, pyrrolic nitrogen, and quaternary nitrogen. Additionally, most of the pores in the material were micropores, with a significant presence of finely tuned micropores measuring approximately 0.71 nm. Notably, this pore size was nearly double the kinetic diameter of a CO_2 gas molecule (0.33 nm), which is considered the optimal range for effective gas adsorption performance. These findings were further supported by Monte Carlo simulation, validating the significance of the pore size distribution for efficient gas adsorption.

The adsorption mechanism of CO_2 onto electrospun PAN-based CNFs containing pyridine-type N, quaternary N, and amine groups has been proposed (Chiang et al., 2020). At low CO_2 partial pressure, micropores and polar nitrogen functionalities on the pore walls or surface defect sites contribute to the attraction of CO_2 molecules and facilitate their adsorption through interactions. The nitrogen functionalities in the adsorbent, acting as Lewis bases, enhance these interactions by their electron-donor properties, as they possess lone electron pairs. Simultaneously, the micropore confinement further increases the adsorption potential of CO_2, resulting in higher CO_2 uptake. At higher CO_2 partial pressure, the role of ultramicropores is replaced by the overall surface area dominance in CO_2 adsorption. Importantly, there is no formation of chemical bonds between CO_2 molecules and the adsorbents, ensuring the reversibility of CO_2 adsorption on the fiber.

15.2.2 Carbon nanotubes

CNTs are cylindrical carbon structures with excellent mechanical and electrical properties. Their high aspect ratio and large specific surface area make them suitable for CO_2 adsorption. Functionalization of CNTs with amine groups further enhances their CO_2 adsorption capacity. Amine-functionalized CNTs have been studied to enhance the performance of electrospun CNFs (Iqbal et al., 2017). Incorporating amine-functionalized CNTs resulted in significantly increased porosity, improved tolerance of the CNF membrane against flexural loads, and enhanced adsorption capacity compared to a pristine CNF membrane. The NH_2 groups in the amine-functionalized CNTs exhibited a high affinity for nonpolar CO_2 molecules, improving adsorption capacity. Nitrogen groups led to a more significant amount of CO_2 gas being stored in the slit pores of the synthesized membrane. The resulting membrane demonstrated exceptional mechanical strength and exhibited an impressive CO_2 uptake of 6.3 mmol/g at 1.0 bar and 298K. Additionally, it displayed a remarkable CO_2/N_2 selectivity of 78.

15.2.3 Graphene and graphene oxide

Graphene is a single layer of carbon atoms arranged in a hexagonal lattice. Graphene oxide (GO), derived from graphene, contains oxygen-containing functional groups. Both graphene and GO nanofibers exhibit high CO_2 adsorption capacity due to their large surface area and interaction with CO_2 molecules through van der Waals forces. Porous graphite nanofibers can be created by employing a KOH etching technique within the 700°C–1000°C temperature range (Meng & Park, 2010). Through the analysis of CO_2 adsorption isotherms, the material treated at 900°C exhibited the most substantial BET surface area, measuring 567 m^2/g, along with an impressive CO_2 adsorption capacity of 59.2 mg/g. AC/graphene nanocomposites (gACNFs) have been successfully produced through electrospinning (Othman et al., 2021). Adding graphene to the ACNF resulted in a noticeable enhancement in the CO_2 adsorption capacity. Specifically, the CO_2 adsorption capacity of the ACNF increased from 1.51 to 3.4 mmol/g at 25°C and 3 bar, indicating a significant improvement in the adsorption performance of the nanocomposites.

15.2.4 Metal-organic frameworks

MOFs are crystalline materials composed of metal ions or clusters coordinated with organic ligands. MOFs exhibit high porosity and can be tailored for specific CO_2 adsorption applications by selecting appropriate metal ions and ligands. Their tunable structure allows for high CO_2 adsorption capacity and selectivity. However, limitations arose as the MOF particles are embedded within the polymer nanofibers produced through electrospinning, which creates an additional barrier for gas diffusion. Consequently, the gas uptake and access to MOFs are reduced. To address this drawback, a two-step procedure involving the preparation of electrospun fibers doped with MOF nanocrystals (e.g., zeolite-imidazolate framework-8 or ZIF-8) followed by a second growth process has been proposed (Wu et al., 2012). Utilizing the MOF crystal-embedded electrospun fibrous mats as the seeding layer and framework for the second growth, they successfully obtained a MOF membrane predominantly composed of MOF crystals (Fig. 15.1). When subjected to a 1:1

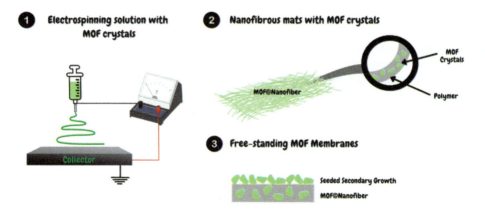

Figure 15.1 Schematic illustration of free-standing metal-organic framework membranes. *Source*: Adapted from Wu, Y. N., Li, F., Liu, H., Zhu, W., Teng, M., Jiang, Y., Li, W., Xu, D., He, D., Hannam, P., & Li, G. (2012). Electrospun fibrous mats as skeletons to produce free-standing MOF membranes, *Journal of Materials Chemistry*, 22, 16971–16978. https://doi.org/10.1039/C2JM32570E.

mixture of N_2/CO_2, the ZIF-8 membrane displayed a preferential affinity for CO_2; the average separation factor of N_2/CO_2 reached 2.4.

15.2.5 Polymer nanofibers

Various polymers can be electrospun into nanofiber mats for CO_2 adsorption. PAN, PA, and polyethylenimine (PEI) are commonly used polymer materials. PEI, which contains a single amine group per carbon atom, shows excellent potential for adsorbing and capturing carbon dioxide. However, its effectiveness is hindered by its high viscosity resulting from its branched structure and high molecular weight (Liu et al., 2018). As a result, carbon dioxide adsorption is considerably slower in polyethylene amine. PAN possesses high mechanical strength attributed to its molecular structure. Additionally, it exhibits excellent chemical resistance, sunlight resistance, and low moisture susceptibility (Shaki, 2023).

It is important to note that the surface area of polymer nanofibers is generally lower than certain inorganic materials or AC, which can have specific surface areas in the range of thousands of m^2/g. Therefore studies on polymer nanofibers are only expected if the surface area is increased by the activation process (i.e., turning polymer nanofibers onto AC), improving basicity by surface modification and preparing hybrid composites. Higher CO_2 uptakes ($>300\%$) have been reached after steam activation of polymer nanofibers (Heo et al., 2019).

Functional groups such as amine or carboxyl can be introduced to enhance CO_2 adsorption capacity. For instance, the impregnation with polyethyleneimine of a composite membrane of PA-6/CNTs (Zainab et al., 2017). The nanofibrous structure of the membrane primarily consists of amide groups, providing an advantage over other polymers in terms of CO_2 capture. Amine-functionalized CNTs were chosen as the carbonaceous material due to their high affinity and large storage capacity for CO_2. The resulting composite membranes exhibited a maximum CO_2 capture capacity of 51 mg per gram of composite membrane, comparable to other reported CO_2-capturing membranes based on organic polymers. Furthermore, the composite membranes demonstrated consistent performance over 12 adsorption and desorption cycles, indicating their stability and reliability.

15.2.6 Ionic liquid

Ionic liquid (ILs) exhibit favorable CO_2 selectivity, but their liquid state at room temperature necessitates their immobilization onto suitable support materials. In this sense, the electrospinning of a solution containing poly(ionic liquid) has been successfully demonstrated (Chen & Elabd, 2009). While the CO_2 capture capabilities of poly(ionic liquid) nanofibers have yet to be documented, these innovative nanofibrous materials are anticipated to hold great promise as sorbents and membranes for CO_2 separation.

It is important to note that the selection of nanofiber materials depends on factors such as adsorption capacity, selectivity, stability, cost, and scalability. Researchers continue to explore novel nanofiber materials and optimize their properties to achieve efficient CO_2 adsorption for practical applications in carbon capture and storage systems.

Finally, it is worth noting that while research in this field is progressing rapidly, practical implementation and scalability of nanofiber-based CO_2 adsorption systems are still under development. Nonetheless, these materials hold promise for addressing the challenges associated with CO_2 capture and contributing to climate change mitigation.

15.3 Volatile organic compound adsorption for indoor air pollution control

The term "indoor air pollution" describes the presence of dangerous contaminants in the air inside buildings and other enclosed areas. When inhaled over a prolonged length of time, these toxins can harm human health (Casas et al., 2023); indoor air pollution causes nearly 2.5 million premature deaths annually (Rajkumar et al., 2019). VOCs are significant pollutants found in indoor environments and toxic to human health. The VOC emissions are produced using thinners, paints, adhesives, and cigarette smoke (Lin et al., 2022; Salthammer, 2004; Wang et al., 2021).

Adsorption is frequently used to eliminate dangerous contaminants from enclosed environments, including VOCs (Choung et al., 2001) and specific gases (Azhagapillai et al., 2020; Belarbi et al., 2019; Wang et al., 2015). The adsorption technique is frequently used in various air purification systems, such as controlling industrial pollutants, improving indoor air quality, and controlling vehicle emissions. AC is one of the most widely used adsorbents for indoor air pollution control (da Costa Lopes et al., 2015); other VOC adsorbents are zeolites, silica gels, and molecular sieves (Lefevere et al., 2018; Megías-Sayago et al., 2019). Adsorbent materials must present properties, such as surface area, pore size, and chemical characteristics, which permit the removal of specific indoor air pollutants.

Another option for material adsorbent of VOCs in indoor areas is nanofibers (Buyukada-Kesici et al., 2021). Some key points of the application of nanofibers in VOC adsorption are the surface area and porosity; the last one permitted the diffusion of VOCs into the composite, increasing the accessibility of the adsorption sites and facilitating efficient adsorption. Applied nanofibers based on different amounts of SiO_2 aerogel (SA) and PAN (polyacrylonitrile) have been evaluated for VOC adsorption (25°C) (Yu et al., 2020). The nanofiber of SA (PAN + SA 71.4% and PAN + SA 100%) presented higher VOC adsorption capacity than AC. Even though AC surface area (1248.50 m^2/g) was higher compared with PAN + SA 100% (289.30 m^2/g), the value of xylene adsorption capacity was 15% lower. Meanwhile, using chloroform in the adsorption experiment increases the difference to 39%. The relevant point is that PAN + SA 100% showed a pore size (12.47 nm) and a pore volume (1.037 cm^3/g) larger compared to AC (3.46 nm and 0.81 cm^3/g). In this sense, not only surface area is an important characteristic in adsorbent material, but also pore size and pore volume helps in VOCs adsorption using nanofibers (PAN + SA 100%). VOC adsorption not only occurred on the nanofibers' surface but also within the interior of the PAN/SIO$_2$ composite, where the additionally adsorption sites are available.

The chemical structures and structural effects of VOCs can help describe nanofibers' adsorption process. For example, the interaction between aromatic compounds on the surface of electrospun fly ash/polyurethane (PU) nanofiber can occur by forming an π-complex (Kim et al., 2013). The compound with higher binding energy increases the VOC-nanofiber interaction due to its aromatic ring acting as an electron donor (Awad et al., 2021).

The nanofibers can be fabricated from various materials, including polymers, carbon, metals, and metal oxides. Polystyrene (PS), acrylic resin, and PS-AR nanofibers have been tested versus VOC adsorption (Chu et al., 2015). Nanofibers formed by nonpolar substance presented affinity to benzene hydrocarbons; in this case, the similar chemical behavior of the polymer backbone and the adsorbate favors the adsorption. This versatility enables the selection of nanofibers with properties suited to the target adsorbate or adsorption process of VOCs in enclosed environments.

Nanofibers' characteristics, such as their surface chemistry and shape, can be modified to suit specific adsorption applications. It is possible to add functional groups to the nanofiber surface to improve selectivity or increase adsorption capacity for adsorbates; for example, TiO$_2$ nanofibers were modified by Yb^{3+} ions for toluene removal (Liao et al., 2019). The data of the VOC removal experiment fitted with pseudo-second-order kinetics models ($R^2 > 0.99$), indicating that both chemical and physical adsorption can be attributed to the toluene adsorption mechanisms on the nanofibers. Another example is a study where PAN-Cellulose nanocrystal (CNC) nanofibers were evaluated in the presence of methyl ethyl ketone (MEK). The addition of CNC in the composite enables internal mass transfer due to the decreased average size of nanofibers. The small size and the high aspect ratio of nanofibers promote faster adsorption kinetics compared to bulk materials. The short diffusion pathways within the nanofiber structure allow the rapidly transporting adsorbates to the adsorption sites, resulting in quicker adsorption kinetics, as the example explained before. The poly (L-lactic acid) (PLLA) and methyl-β-cyclodextrin (CD) were mixed to form a nanocomposite for toluene removal. PLLA/CD nanofibers presented higher adsorption capacity; creating a toluene inclusion complex in CD could explain the result (Wanwong et al., 2023).

Applications of nanofibers in VOC adsorption are still in progress to validate various factors such as temperature, humidity, VOC concentrations, and time. However, recent studies presented nanofibers with a VOC adsorption capacity 150 times higher than a commercial mask, thus highlighting the opportunity to use nanofibers in VOC adsorption (Kadam et al., 2021).

15.4 Gas adsorption for fuel purification

As the demand for energy increases due to the increase in population and the use of technology for services, transportation, and other daily activities, it is necessary to find methods that make fossil fuels or biofuels more efficient. One of the strategies to achieve this is by removing the unwanted compounds that lower the calorific value and spoil the combustion system, thus ensuring more excellent performance and, above all, protecting the equipment, machinery, or artifact fed by the fuel in question. Some unwanted compounds in fuels are H_2S, water, VOCs, and CO_2, among others. Reduction techniques of different natures have been developed, including adsorption, absorption, cryogenics, and membranes. However, the most common is adsorption.

Adsorption has been used to reduce the concentration of sulfur compounds in some processes, such as coal gas production. Coal gas can be used as fuel, but sulfur impurities poison catalysts and corrode equipment (Feng et al., 2023). One of the most efficient techniques used for coal gas desulfurization is using metallic compounds or oxides (Fe_2O_3, ZnO) that react with H_2S. Using this desulfurization technique entails some problems with the active components in which the structure deteriorates or is plugged. For this reason, porous structures have been developed to support the active components and allow for better distribution (Liu et al., 2021; Wang et al., 2021; Zhang et al., 2022). One of the structures studied as a support is the electrospun nanofiber, and it has been demonstrated that it offers a large surface area-to-volume ratio and a greater number of sites for the active components (Feng et al., 2023). For instance, aerogel nanofibers obtained from phenolic resins with different proportions of cetyltrimethylammonium bromide (CTAB) have been evaluated for SO_2 adsorption. There is a strong correlation between the surface area and SO_2 adsorption capacity. The aerogel nanofibers presented a swelling mechanism by which the SO_2 molecules were introduced into the nanofibers favoring the desorption process as well. In addition, high SO_2 selectivity was obtained when the nanofiber was exposed to SO_2/CO_2 mixture (Li et al., 2022).

Fossil fuels, including natural gas, have contributed significantly to supplying global energy demands. Hydrogen sulfide (H_2S), an extremely poisonous and corrosive gas, is one of the contaminants that natural gas may include. Commercial PAN fibers functionalized with Cu by an amino-ligand reaction have efficiently removed sulfur compounds from the gas phase even in wet conditions and in the presence of CO_2. The main sulfur removal mechanism proposed is based on coordination bonds between H_2S and Cu, mainly by the complexation-ligand type interactions that would give rise to an N−Cu−S structure (Liu et al., 2023). Similarly, boron carbon nitride nanofiber doped with Cu formed coordination bonds with sulfur compounds present in a model fuel made of n-octane by the interactions of the empty 4s orbitals of the Cu atoms and the lone electrons of the S atoms (Zhang, Ran, et al., 2023).

Chemical adsorption of sulfur compounds has been reported for Cu/Cu_xO supported in electrospun CNFs. The activation of the CNFs at 190°C with oxygen oxidized part of the Cu, resulting in the chemical interaction with the sulfur compounds favoring the adsorption of hydrogen sulfide. In contrast, physical adsorption is the main adsorption mechanism responsible for the removal of sulfur compounds in electrospun CNFs without Cu/Cu_xO (Bajaj et al., 2018). The chemical reaction between the sulfur compounds and copper occurs as follows:

$$Cu + H_2S \rightarrow CuS + H_2$$

$$Cu_xO + H_2S \rightarrow Cu_xS + H_2O$$

According to the above chemical reactions, Cu and Cu_xO function as active sites for H_2S adsorption-producing Cu_xS compounds (Bajaj et al., 2018). Similar behavior has been reported for the adsorption of sulfur compounds by metal oxides such as Cu, Fe, and Zn. However, better performance in the adsorption capacity of sulfurous compounds can be achieved when metal oxides are immobilized in nanofibers (Feng et al., 2020). For instance, CuO and Cu_xO nanofibers (Fig. 15.2) have been employed in the design of H_2S dosimeters because the continuous formation of CuS and Cu_xS is related to the reduction in electrical resistance. In addition, it is possible to detect between 5 and 10 ppm of H_2S for 100 cycles, which is promising and attractive for the use of CuO and CuxO nanofibers as H_2S sensors (Werner et al., 2021). In this same sense, we can talk about other H_2S sensors that use nanofibers as support, varying the active component, for example, SnO_2 (Phuoc et al., 2020), CuO/ZnO (Hsu et al., 2021), and others, in which there can be a sulfuration-desulfurization reaction, a reduction-oxidation reaction, or both.

The strong chemical attraction that metal oxides have for H_2S molecules is crucial in increasing the effectiveness of desulfurization procedures. For instance, the H_2S adsorption capacity of PU at breakthrough was increased 2.5 folds when low concentrations of CNTs/ferrite composite (0.7%) were incorporated at the PU nanofiber. Interestingly, the incorporation of Ni and Zn into ferrite nanoparticles decreased the performance of the PU nanofiber. It seems that the iron content

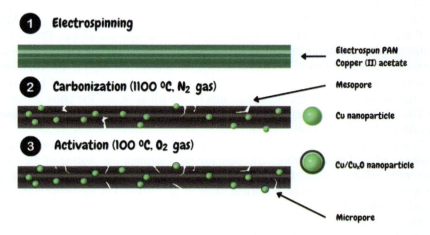

Figure 15.2 Schematic illustration of the preparation of activated carbon nanofiber modified with Cu/Cu_xO nanoparticles.
Source: Adapted from Y. Feng, Y., Lu, J., Wang, J., Mi, J., Zhang, M., Ge, M., Li, Y., Zhang, Z., & Wang, W. (2020). Desulfurization sorbents for green and clean coal utilization and downstream toxics reduction: A review and perspectives, *Journal of Cleaner Production*, *273*, 123080. https://doi.org/10.1016/j.jclepro.2020.123080.

decreased with the inclusion of nickel and zinc in the nanofibers, causing a reduction in the interaction between the metal and the H_2S. Thus an adsorption mechanism is suggested based on the reaction of H_2S with iron ions (Fe^{3+}), forming sulfides or sulfates through their catalytic oxidation (Maddah et al., 2020).

Iron-based carbon nanofibrous composites carbonized by microwaves were also reported, demonstrating nucleation and growth of zero-valent iron on nanofibers. Compared to the conventional carbonization technique, this technique favored porosity and high sulfur adsorption capacity and promoted energy savings in generating adsorbent materials (Zhang, Ru, et al., 2023).

Molecularly imprinted polybenzimidazole nanofibers produced by the electrospinning technique proved to be efficient for the removal of sulfonated compounds from fuels. DFT analysis indicated that $\pi-\pi$ interactions and hydrogen bonding between the oxygen of the sulfonated compounds and the $-NH$ groups of polybenzimidazole are the main sulfur adsorption mechanisms, which are confirmed by an effective and rapid regeneration by Soxhlet extraction with methanol/acetonitrile (Ogunlaja et al., 2014).

In addition to sulfur compounds, removing other impurities from fossil fuels and biofuels is necessary. Siloxane is an undesirable biogas compound due to its adverse effects on combustion systems. However, few efforts to remove siloxane compounds from biogas have been made using nanofibers. For instance, a hollow fiber membrane contactor (premade from Zhejiang Dongda Environment Engineering Co.) was used for the adsorption of volatile methyl siloxane as an interface between different absorbents (Selexol, tetradecane, polyethylene glycol dimethyl ether, and propylene carbonate) and the gas with siloxanes. This approach effectively prevented overflow, foaming, and entrainment (Wang et al., 2023; Wang et al., 2023).

Polyacrylonitrile (PAN) nanofibers have served as support media for MOF DUT-4 particles for the removal of D4 siloxane from the gas phase. The coaxial electrospinning technique enhanced the surface distribution of the DUT-4 particles, increasing the adsorption of D4 siloxane molecules and reducing the environmental impacts of the nanofiber synthesis. The adsorption mechanism occurred by a complex interaction between the methyl groups of the siloxane and the polar groups of the DUT-4 in the nanofibers, which implies a complex regeneration of the adsorbent (Pioquinto-García et al., 2023).

As support media, nanofibers have also been tested to increase the production of cellulosic ethanol by removing cellobiose products through the coating of the nanofiber with β-Glucosidase. In addition, these fibers can be highly recyclable by changing to a magnetic-type nanofiber (Lee et al., 2010). On the other hand, the use of carbonized and oxidized nanofibers for ethanol adsorption has also been reported, in which it was observed that acid oxidation increases oxygen functional groups (Bai et al., 2014). Nanofibers used as membranes have also been tested in the separation of ethanol/water mixtures to promote biofuel production. A polymer poly-(vinylidene fluoride) (PVDF), spin-coated with dihydroxy polydimethylsiloxane (PDMS) membrane, was synthesized to generate a corrugated membrane (Li, Pan, et al., 2022). This corrugated membrane outperformed certain other PDMS membranes, demonstrating improved membrane flux and exhibiting potential for more efficient separation processes.

15.5 Hydrogen storage

Hydrogen storage on nanofibers is an area of research aimed at developing efficient and effective methods for storing hydrogen gas. Hydrogen is a promising clean energy carrier due to its high energy content and the potential for producing zero-emission power when used in fuel cells. However, one of the main challenges in utilizing hydrogen as a fuel is its low density and the need for safe and compact storage methods. The hydrogen storage capacities of various materials can vary widely, depending on the specific material, its structure, and the conditions under which hydrogen is stored, but a 1−8 wt.% storage capacity is expected on various materials, excluding the controversial super hydrogen sorption in graphite nanofibers (73 wt.%) (Krishnankutty et al., 1997; Nechaev et al., 2014; Rzepka et al., 2005; Simonyan & Johnson, 2002). Remarkable experimental findings recently published, showcasing the accumulation of approximately 20−30 wt.% of "reversible" hydrogen and about 7−10 wt.% of "irreversible" hydrogen (Nechaev et al., 2022).

Nanofibers, fibers with diameters on the nanometer scale, have garnered attention for hydrogen storage due to their unique properties. These materials can provide a high surface area and offer many active sites for hydrogen adsorption. Additionally, nanofibers can be tailored to specific structures, compositions, and surface properties, influencing their hydrogen storage capabilities. There are several approaches to achieving hydrogen storage on nanofibers.

The physical adsorption mechanism involves van der Waals forces, including London dispersion forces and dipole−dipole interactions, between the hydrogen molecules and the surface of the storage material. These forces are relatively weak compared to chemical bonds; therefore the main advantage of physical adsorption for hydrogen storage is its reversibility, which allows for the storage and subsequent release of hydrogen on demand. Materials with high surface areas and well-defined pores, such as AC or specific MOFs (Ren et al., 2015), are often used for maximizing hydrogen adsorption capacity. However, the storage capacities are typically lower than other storage methods, and the adsorption and desorption kinetics can be relatively slow. Ongoing research aims to enhance the hydrogen storage capacity and kinetics of physical adsorption by exploring novel materials, optimizing their surface properties, and developing strategies such as nanostructuring and hybrid systems. For example, after doping porous CNFs with lithium and fluorine, the hydrogen storage capacity increased 24 times concerning the pure porous CNFs (Fig. 15.3) (Chen et al., 2021).

Nanofibers can be modified to have reactive functional groups, such as metal hydride-forming species or catalysts, which can chemically bond with hydrogen atoms. This approach involves the formation of strong covalent or ionic bonds between hydrogen and the surface of the nanofibers. Nickel is commonly used as a catalyst or catalyst support material in hydrogen adsorption processes for several reasons: high hydrogen adsorption capacity, favorable thermodynamics, catalytic activity, abundance and cost-effectiveness, and compatibility with other materials. In this regard, porous nickel nanofibers (NFs) have been synthesized to catalyze

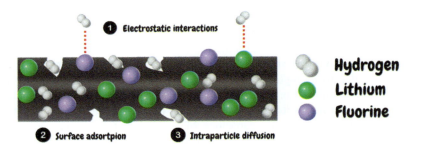

Figure 15.3 Schematic illustration of the adsorption mechanism of hydrogen onto porous carbon nanofibers with lithium and fluorine.
Source: Adapted from Chen, X., Xue, Z., Niu, K., Liu, X., Wei Lv, Zhang, B., Li, Z., Zeng, H., Ren, Y., Wu, Y., & Zhang, Y. (2021). Li—fluorine codoped electrospun carbon nanofibers for enhanced hydrogen storage, *RSC Advances*, *11*, 4053—4061. https://doi.org/10.1039/D0RA06500E.

MgH$_2$ (Chen et al., 2015). Despite being an appealing material for solid hydrogen storage, one of the significant limitations of this material is its thermodynamic stability and sluggish kinetics, which hinder its widespread practical use. On the other hand, nanostructured nickel catalyst aggregated in CNFs revealed a remarkable catalytic effect on improving the hydrogen storage properties of MgH$_2$, with capacities of as high as 6.12 wt.%.

Nanofibers can create a nanoconfinement effect, where hydrogen molecules are trapped within the small spaces or pores in the fiber structure. This confinement can increase the density of hydrogen and enhance its storage capacity. It has been demonstrated that hydrogen uptake onto amorphous carbon nanofibers depends on the narrow microporosity of the material and, at room temperature, has a much better storage capacity than any other carbon material (Kunowsky et al., 2012). Ammonia borane (AB) (NH$_3$BH$_3$), often abbreviated as AB, is a compound that has received significant attention for its potential as a hydrogen storage material. It is a solid-state substance that can release hydrogen gas when it undergoes thermal or chemical decomposition. One of the advantages of AB is its high hydrogen content, with approximately 19.6% by weight, which means that a relatively small amount of AB can store a significant amount of hydrogen.

Additionally, AB is relatively stable at room temperature and atmospheric pressure, making it easier to handle and transport than gaseous or liquid hydrogen. AB can release hydrogen through different processes, such as thermal decomposition, that is, heating AB to temperatures typically above 100°C to release hydrogen gas. The thermal decomposition process results in the formation of gaseous hydrogen (H$_2$), borazine (B$_3$N$_3$H$_6$), and boron nitride (BN) as by-products. The synergetic nanoconfinement effect of AB supported in nanofiber structures on dehydrogenation temperature has been studied for the removal of unwanted by-products of AB, achieving promising results (Alipour et al., 2014).

Another mechanism for hydrogen storage is the electrochemical method that reversibly stores hydrogen using electrochemical reactions. Hydrogen is typically stored in a material known as an electrode or an active material. During the charging or hydrogen uptake phase, an electric current is applied to the system, causing the electrode to undergo a reduction reaction. In this process, hydrogen ions or atoms from the electrolyte are adsorbed onto the electrode surface, leading to hydrogen storage within the electrode material. The applied electric current is reversed during the discharging or hydrogen release phase, initiating an oxidation reaction that releases hydrogen ions or atoms from the electrode back into the electrolyte, making the stored hydrogen available for use. Although electrochemical hydrogen storage offers several advantages, there are challenges associated with electrochemical hydrogen storage. That is, the development of suitable electrode materials with high hydrogen storage capacity, good reversibility, and long-term stability is crucial. Hierarchical nanofibers have been synthesized and evaluated for hydrogen storage to overcome some of these challenges. Mesoporous CNFs showing a hierarchical nanostructure, consisting of the open microporous channels connected with large mesopores and micropores, presented a large number of active sites for hydrogen storage and reduced the volume change during the charge−discharge cycling (Xing et al., 2014). The incorporation of nickel and ceria nanoparticles on hierarchically porous carbon micronanofiber material enhanced electrochemical hydrogen storage capacity and cyclic stability resulting from multiple factors (George et al., 2021). These included the high surface area provided by the AC microfiber substrate, graphitic CNFs' presence, and nickel's catalytic activity. Furthermore, introducing ceria nanoparticles created variations in charge carriers within the material, further contributing to its improved performance in electrochemical hydrogen storage.

Various types of nanofibers have been researched for hydrogen storage due to their unique properties and promise for increased hydrogen adsorption. Among the important types of nanofibers currently under investigation for hydrogen storage are:

15.5.1 Carbon nanofibers

AC is a material that has been extensively studied for hydrogen storage applications. AC has several characteristics that make it suitable for hydrogen storage: a high surface area that allows for a more significant number of active sites where hydrogen molecules can physically adsorb, tunable porosity, and reversible adsorption, making it suitable for storage and subsequent use, among others. Despite the advantages of AC, there are still challenges to address. The gravimetric and volumetric hydrogen storage capacities of AC are relatively modest compared to some other materials (Chung et al., 2005; Marella & Tomaselli, 2006), such as metal hydrides or complex metal hydrides, although values as high as 6.5 wt.% have been reported (Browning et al., 2002). Additionally, hydrogen adsorption and desorption kinetics on AC can be relatively slow. Ongoing research focuses on improving the hydrogen storage performance of AC through advancements in material synthesis, functionalization techniques, and understanding the fundamental interactions between hydrogen and carbon surfaces. Other strategies have been explored to enhance the hydrogen storage capacity of AC, such as doping and functionalization with metals, incorporating heteroatoms, or nanostructuring (Kim et al., 2008a; 2008b; Xia et al., 2013), which provide additional adsorption sites and can offer higher binding energies for hydrogen.

Plant-derived natural fibers such as cotton filament have been transformed into graphitic carbon through carbonization or pyrolysis methods, with or without catalysts (Mukherjee, 2021). This process preserves the inherent pores of the fibers, resulting in carbonaceous structures. CNFs obtained from plant-based precursors yield a specific structure that is very difficult to synthesize otherwise and is exceptionally economical.

15.5.2 Metal oxide nanofibers

Researchers have explored various metal oxide nanofibers for hydrogen storage, including materials such as titanium oxide (TiO_2), zinc oxide (ZnO), iron oxide (Fe_2O_3), and cerium oxide (CeO_2), among others. These metal oxide nanofibers can be synthesized using electrospinning, sol-gel methods, or vapor-phase deposition techniques, allowing for precise control over their properties. For instance, TiO_2 and $LiTi_2O_4$ nanofibers calcined at 450°C showed high hydrogen storages of 1.11 and 0.74 wt.% despite their low surface areas of 49.4 and 50.2 m^2/g, respectively, due to high adsorption potential energy (Jo & Jo, 2012). In another study, doping ZnO nanofibers with Al enhanced the hydrogen storage capabilities due to the catalytic characteristics of the doped materials (Yaakob et al., 2012).

15.5.3 Polymer nanofibers

Polymer nanofibers can be engineered with specific compositions, structures, and morphologies, enabling the tailoring of their properties for optimal hydrogen storage. The manipulation of factors such as polymer type, molecular weight, and fiber diameter can influence hydrogen adsorption capacities and kinetics. Also, polymer nanofibers are lightweight and flexible, making them suitable for various hydrogen storage applications, including portable devices and lightweight vehicles. Their mechanical flexibility allows for easy integration into different form factors, enabling efficient storage and release of hydrogen. More importantly, polymer nanofibers can be combined with other materials, such as metal nanoparticles or carbon-based materials, to create hybrid structures. These hybrids take advantage of the unique properties of both components, resulting in improved hydrogen storage performance. For example, incorporating metal nanoparticles can enhance the catalytic properties of the polymer nanofibers, facilitating hydrogen adsorption and desorption reactions. Despite their potential, challenges remain in utilizing polymer nanofibers for hydrogen storage. These include optimizing the nanofibers' porosity and surface area, improving the materials' stability under cycling conditions, and enhancing the kinetics of hydrogen adsorption and desorption. As a result, alternative materials have received more attention, and few works on pure polymer nanofibers have been published (Phani et al., 2014).

Ongoing research focuses on developing novel nanofiber materials, exploring advanced fabrication techniques, and investigating innovative approaches such as nanostructuring and hybrid systems to improve the hydrogen storage performance of polymer nanofibers. These advancements aim to increase the hydrogen storage capacity, improve kinetics, and enhance the overall efficiency of hydrogen storage systems using polymer nanofibers.

It is important to note that hydrogen storage remains a complex and challenging problem. While nanofibers show promise, issues remain, such as the reversibility of hydrogen adsorption and desorption, storage capacity, temperature and pressure requirements, and overall system efficiency. Further research and development are necessary to overcome these challenges and enable practical and safe hydrogen storage on nanofibers for real-world applications.

15.6 Conclusions

Nanofibers have garnered a lot of interest in gas adsorption and storage applications. They contain distinguishing characteristics such as high surface area, variable pore size, and outstanding adsorption capabilities, making them appealing choices for a wide range of applications. In the context of carbon dioxide (CO_2) capture, nanofibers provide an effective platform for adsorbing and sequestering CO_2 emissions. Functionalized nanofibers, such as amine-functionalized CNFs or MOF nanofibers, can selectively adsorb CO_2 from gas streams, assisting in the reduction of GHG emissions. Adsorption capacity is high for nanofiber-based adsorbents, which are routinely functionalized with specific chemical groups or metal nanoparticles.

Gas adsorption on nanofibers also appears to be interesting for fuel purification operations. Nanofiber-based adsorbents can remove impurities and contaminants from fuels, enhancing their quality and performance. For hydrogen storage applications, nanofibers, with their vast surface area and unique properties, maybe a potential solution for the development of clean and sustainable energy systems. For example, CNFs and metal hydride-functionalized nanofibers have been investigated for hydrogen storage to enhance hydrogen uptake, release kinetics, and storage capacity.

Notwithstanding these notable advancements, there are still hurdles to be addressed and overcome before nanofiber technology can be considered mature. To illustrate, there is a requirement for more in-depth research on the performance of nanofibers under conditions that closely resemble real-world scenarios, such as low gas concentrations and varying humidity levels. Furthermore, despite its essential role in facilitating the upscaling of the adsorption processes, the available literature on the adsorption dynamics of nanofibers is limited. Nevertheless, we envision that continued research and development in the field of nanofibers for gas adsorption and storage holds enormous promise for resolving environmental problems, increasing air quality, expanding clean energy technologies, and contributing to a more sustainable future.

References

Ali, N., Babar, A. A., Zhang, Y., Iqbal, N., Wang, X., Yu, J., & Ding, B. (2020). Porous, flexible, and core-shell structured carbon nanofibers hybridized by tin oxide nanoparticles for efficient carbon dioxide capture. *Journal of Colloid and Interface Science, 560,* 379–387. Available from https://doi.org/10.1016/J.JCIS.2019.100.034.

Alipour, J., Shoushtari, A. M., & Kaflou, A. (2014). Electrospun PMMA/AB nanofiber composites for hydrogen storage applications. *E-Polymers, 14,* 305–311. Available from https://doi.org/10.1515/EPOLY-2014-0071/MACHINEREADABLECITATION/RIS.

Awad, R., Haghighat Mamaghani, A., Boluk, Y., & Hashisho, Z. (2021). Synthesis and characterization of electrospun PAN-based activated carbon nanofibers reinforced with cellulose nanocrystals for adsorption of VOCs. *Chemical Engineering Journal, 410,* 128412. Available from https://doi.org/10.1016/J.CEJ.2021.128412.

Azhagapillai, P., Vijayanathan Pillai, V., Al Shoaibi, A., & Chandrasekar, S. (2020). Selective adsorption of benzene, toluene, and m-xylene on sulfonated carbons. *Fuel, 280,* 118667. Available from https://doi.org/10.1016/J.FUEL.2020.118667.

Bai, B.-C., Kim, J.-G., Im, J.-S., Jung, S.-C., & Lee, Y.-S. (2011). Influence of oxyfluorination on activated carbon nanofibers for CO_2 storage. *Carbon Letters, 12,* 236–242. Available from https://doi.org/10.5714/cl.2011.12.40.236.

Bai, Y., Huang, Z.-H., & Kang, F. (2014). Surface oxidation of activated electrospun carbon nanofibers and their adsorption performance for benzene, butanone and ethanol. *Colloids and Surfaces A: Physicochemical and Engineering Aspects, 443,* 66–71. Available from https://doi.org/10.1016/j.colsurfa.2013.100.057.

Bajaj, B., Joh, H.-I., Jo, S. M., Park, J. H., Yi, K. B., & Lee, S. (2018). Enhanced reactive H2S adsorption using carbon nanofibers supported with Cu/Cu_xO nanoparticles. *Applied Surface Science, 429,* 253–257. Available from https://doi.org/10.1016/j.apsusc.2017.060.280.

Belarbi, H., Gonzales, P., Basta, A., & Trens, P. (2019). Comparison of the benzene sorption properties of metal organic frameworks: Influence of the textural properties. *Environmental Science: Processes & Impacts, 21,* 407–412. Available from https://doi.org/10.1039/C8EM00481A.

Browning, D. J., Gerrard, M. L., Lakeman, J. B., Mellor, I. M., Mortimer, R. J., & Turpin, M. C. (2002). Studies into the storage of hydrogen in carbon nanofibers: Proposal of a possible reaction mechanism. *Nano Letters, 2,* 201–205. Available from https://doi.org/10.1021/NL015576G/ASSET/IMAGES/MEDIUM/NL015576GN00001.GIF.

Buyukada-Kesici, E., Gezmis-Yavuz, E., Aydin, D., Cansoy, C. E., Alp, K., & Koseoglu-Imer, D. Y. (2021). Design and fabrication of nano-engineered electrospun filter media with cellulose nanocrystal for toluene adsorption from indoor air. *Materials Science and Engineering B, 264,* 114953. Available from https://doi.org/10.1016/J.MSEB.2020.114953.

Casas, L., Dumas, O., & Le Moual, N. (2023). Indoor air and respiratory health: Volatile organic compounds and cleaning products. *Asthma in the 21st Century: New Research Advances,* 135–150. Available from https://doi.org/10.1016/B978-0-323-85419-1.00002-5.

Chen, H., & Elabd, Y. A. (2009). Polymerized ionic liquids: Solution properties and electrospinning. *Macromolecules, 42,* 3368–3373. Available from https://doi.org/10.1021/MA802347T/ASSET/IMAGES/MEDIUM/MA-2008-02347T_0009.GIF.

Chen, J., Xia, G., Guo, Z., Huang, Z., Liu, H., & Yu, X. (2015). Porous Ni nanofibers with enhanced catalytic effect on the hydrogen storage performance of MgH_2. *Journal of*

Materials Chemistry A, *3*, 15843—15848. Available from https://doi.org/10.1039/C5TA03721B.

Chen, X., Xue, Z., Niu, K., Liu, X., Wei Lv., Zhang, B., Li, Z., Zeng, H., Ren, Y., Wu, Y., & Zhang, Y. (2021). Li—fluorine codoped electrospun carbon nanofibers for enhanced hydrogen storage. *RSC Advances*, *11*, 4053—4061. Available from https://doi.org/10.1039/D0RA06500E.

Chiang, Y.-C., Huang, C.-C., Chin, W.-T., Hsieh, W.-H., Sheu, J.-S., & Ragulskis, M. (2021). Carbon dioxide adsorption on carbon nanofibers with different porous structures. *Applied Sciences*, *11*, 7724. Available from https://doi.org/10.3390/APP11167724.

Chiang, Y. C., & Juang, R. S. (2017). Surface modifications of carbonaceous materials for carbon dioxide adsorption: A review. *Journal of the Taiwan Institute of Chemical Engineers*, *71*, 214—234. Available from https://doi.org/10.1016/j.jtice.2016.120.014.

Chiang, Y. C., Wu, C. Y., & Chen, Y. J. (2020). Effects of activation on the properties of electrospun carbon nanofibers and their adsorption performance for carbon dioxide. *Separation and Purification Technology*, *233*, 116040. Available from https://doi.org/10.1016/J.SEPPUR.2019.116040.

Choung, J. H., Lee, Y. W., Choi, D. K., & Kim, S. H. (2001). Adsorption equilibria of toluene on polymeric adsorbents. *Journal of Chemical & Engineering Data*, *46*, 954—958. Available from https://doi.org/10.1021/JE000282I.

Chu, L., Deng, S., Zhao, R., Zhang, Z., Li, C., & Kang, X. (2015). Adsorption/desorption performance of volatile organic compounds on electrospun nanofibers. *RSC Advances*, *5*, 102625—102632. Available from https://doi.org/10.1039/C5RA22597C.

Chung, H. J., Lee, D. W., Jo, S. M., Kim, D. Y., & Lee, W. S. (2005). Electrospun poly (vinylidene fluoride)-based carbon nanofibers for hydrogen storage. *Materials Research Society Symposium — Proceedings*, *837*, 77—82. Available from https://doi.org/10.1557/PROC-837-N3.15/METRICS.

Cooney, D. O. (1998). *Adsorption design for wastewater treatment*. CRC Press. Available from https://books.google.com/books?id = jS3BVK1T3iIC&pgis = 1, accessed December 29, 2015.

da Costa Lopes, A. S., de Carvalho, S. M. L., do Socorro Barros Brasil, D., de Alcantara Mendes, R., & Lima, M. O. (2015). Surface modification of commercial activated carbon (CAG) for the adsorption of benzene and toluene. *American Journal of Analytical Chemistry*, *6*, 528—538. Available from https://doi.org/10.4236/AJAC.2015.66051.

Feng, Y., Lu, J., Wang, J., Mi, J., Zhang, M., Ge, M., Li, Y., Zhang, Z., & Wang, W. (2020). Desulfurization sorbents for green and clean coal utilization and downstream toxics reduction: A review and perspectives. *Journal of Cleaner Production*, *273*, 123080. Available from https://doi.org/10.1016/j.jclepro.2020.123080.

Feng, Y., Zhang, M., Sun, Y., Cao, C., Wang, J., Ge, M., Cai, W., Mi, J., & Lai, Y. (2023). Porous carbon nanofibers supported Zn@MnO sorbents with high dispersion and loading content for hot coal gas desulfurization. *Chemical Engineering Journal*, *464*, 142590. Available from https://doi.org/10.1016/j.cej.2023.142590.

George, J. K., Yadav, A., & Verma, N. (2021). Electrochemical hydrogen storage behavior of Ni-Ceria impregnated carbon micro-nanofibers. *International Journal of Hydrogen Energy*, *46*, 2491—2502. Available from https://doi.org/10.1016/J.IJHYDENE.2020.100.077.

Heo, Y. J., Zhang, Y., Rhee, K. Y., & Park, S. J. (2019). Synthesis of PAN/PVDF nanofiber composites-based carbon adsorbents for CO_2 capture. *Composites Part B: Engineering*, *156*, 95—99. Available from https://doi.org/10.1016/J.COMPOSITESB.2018.080.057.

Heuchel, M., Davies, G. M., Buss, E., & Seaton, N. A. (1999). Adsorption of carbon dioxide and methane and their mixtures on an activated carbon: Simulation and experiment. *Langmuir*, *15*, 8695−8705. Available from https://doi.org/10.1021/LA9904298.

Hsu, K.-C., Fang, T.-H., Hsiao, Y.-J., & Li, Z.-J. (2021). Rapid detection of low concentrations of H_2S using CuO-doped ZnO nanofibers. *Journal of Alloys and Compounds*, *852*, 157014. Available from https://doi.org/10.1016/j.jallcom.2020.157014.

Intergovernmental Panel on Climate Change. (2015). *Climate change 2014: Mitigation of climate change*, Cambridge University Press, https://doi.org/10.1017/cbo9781107415416.

Iqbal, N., Wang, X., Ge, J., Yu, J., Kim, H. Y., Al-Deyab, S. S., El-Newehy, M., & Ding, B. (2016). Cobalt oxide nanoparticles embedded in flexible carbon nanofibers: Attractive material for supercapacitor electrodes and CO_2 adsorption. *RSC Advances*, *6*, 52171−52179. Available from https://doi.org/10.1039/C6RA06077C.

Iqbal, N., Wang, X., Yu, J., & Ding, B. (2017). Robust and flexible carbon nanofibers doped with amine functionalized carbon nanotubes for efficient CO_2 capture. *Advanced Sustainable Systems*, *1*, 1600028. Available from https://doi.org/10.1002/ADSU. 201600028.

Jo, S. M., & Jo, S. M. (2012). Electrospun nanofibrous materials and their hydrogen storage. *Hydrogen Storage*. Available from https://doi.org/10.5772/50521.

Kadam, V., Truong, Y. B., Schutz, J., Kyratzis, I. L., Padhye, R., & Wang, L. (2021). Gelatin/β−cyclodextrin bio-nanofibers as respiratory filter media for filtration of aerosols and volatile organic compounds at low air resistance. *Journal of Hazardous Materials*, *403*, 123841. Available from https://doi.org/10.1016/J.JHAZMAT.2020.123841.

Kim, B. J., Lee, Y. S., & Park, S. J. (2008a). A study on the hydrogen storage capacity of Ni-plated porous carbon nanofibers. *International Journal of Hydrogen Energy*, *33*, 4112−4115. Available from https://doi.org/10.1016/J.IJHYDENE.2008.050.077.

Kim, B. J., Lee, Y. S., & Park, S. J. (2008b). Preparation of platinum-decorated porous graphite nanofibers, and their hydrogen storage behaviors. *Journal of Colloid and Interface Science*, *318*, 530−533. Available from https://doi.org/10.1016/J.JCIS.2007.100.018.

Kim, H. J., Pant, H. R., Choi, N. J., & Kim, C. S. (2013). Composite electrospun fly ash/polyurethane fibers for absorption of volatile organic compounds from air. *Chemical Engineering Journal*, *230*, 244−250. Available from https://doi.org/10.1016/J.CEJ.2013.060.090.

Krishnankutty, N., Park, C., Rodriguez, N. M., & Baker, R. T. K. (1997). The effect of copper on the structural characteristics of carbon filaments produced from iron catalyzed decomposition of ethylene. *Catalysis Today*, *37*, 295−307. Available from https://doi.org/10.1016/S0920-5861(97)00019-9.

Kunowsky, M., Marco-Lozar, J. P., Oya, A., & Linares-Solano, A. (2012). Hydrogen storage in CO_2-activated amorphous nanofibers and their monoliths. *Carbon*, *50*, 1407−1416. Available from https://doi.org/10.1016/J.CARBON.2011.110.013.

Lee, S.-M., Jin, L. H., Kim, J. H., Han, S. O., Bin Na, H., Hyeon, T., Koo, Y.-M., Kim, J., & Lee, J.-H. (2010). β-Glucosidase coating on polymer nanofibers for improved cellulosic ethanol production. *Bioprocess and Biosystems Engineering*, *33*, 141−147. Available from https://doi.org/10.1007/s00449-009-0386-x.

Lefevere, J., Mullens, S., & Meynen, V. (2018). The impact of formulation and 3D-printing on the catalytic properties of ZSM-5 zeolite. *Chemical Engineering Journal*, *349*, 260−268. Available from https://doi.org/10.1016/j.cej.2018.050.058.

Li, D., Yang, J., Zhao, Y., Yuan, H., & Chen, Y. (2022). Ultra-highly porous carbon from Wasted soybean residue with tailored porosity and doped structure as renewable multipurpose absorbent for efficient CO_2, toluene and water vapor capture. *Journal of*

Cleaner Production, *337*, 130283. Available from https://doi.org/10.1016/J.JCLEPRO.2021.130283.

Li, J., Pan, Y., Ji, W., Zhu, H., Liu, G., Zhang, G., & Jin, W. (2022). High-flux corrugated PDMS composite membrane fabricated by using nanofiber substrate. *Journal of Membrane Science*, *647*, 120336. Available from https://doi.org/10.1016/j.memsci.2022.120336.

Li, Y., Zou, B., Hu, C., & Cao, M. (2016). Nitrogen-doped porous carbon nanofiber webs for efficient CO_2 capture and conversion. *Carbon*, *99*, 79−89. Available from https://doi.org/10.1016/J.CARBON.2015.110.074.

Li, Z.-M., Zhu, S.-X., Mao, F.-F., Zhou, Y., Zhu, W., & Tao, D.-J. (2022). CTAB-controlled synthesis of phenolic resin-based nanofiber aerogels for highly efficient and reversible SO_2 capture. *Chemical Engineering Journal*, *431*, 133715. Available from https://doi.org/10.1016/j.cej.2021.133715.

Liao, F., Chu, L. F., Guo, C. X., Guo, Y. J., Ke, Q. F., & Guo, Y. P. (2019). Ytterbium doped TiO_2 nanofibers on activated carbon fibers enhances adsorption and photocatalytic activities for toluene removal. *ChemistrySelect*, *4*, 9222−9231. Available from https://doi.org/10.1002/SLCT.201902002.

Lim, G., Lee, K. B., & Ham, H. C. (2016). Effect of N-containing functional groups on CO_2 adsorption of carbonaceous materials: A density functional theory approach. *Journal of Physical Chemistry C*, *120*, 8087−8095. Available from https://doi.org/10.1021/ACS.JPCC.5B12090/SUPPL_FILE/JP5B12090_SI_001.PDF.

Lin, K. H., Tsai, J. H., Cheng, C. C., & Chiang, H. L. (2022). Emission of volatile organic compounds from consumer products. *Aerosol and Air Quality Research*, *22*, 220250. Available from https://doi.org/10.4209/AAQR.220250.

Liu, F., Kuang, Y., Wang, S., Chen, S., & Fu, W. (2018). Preparation and characterization of molecularly imprinted solid amine adsorbent for CO_2 adsorption. *New Journal of Chemistry*, *42*, 10016−10023. Available from https://doi.org/10.1039/C8NJ00686E.

Liu, Q., Liu, B., Liu, Q., Guo, S., & Wu, X. (2021). Probing mesoporous character, desulfurization capability and kinetic mechanism of synergistic stabilizing sorbent $Ca_xCu_yMn_zO_t$/MAS-9 in hot coal gas. *Journal of Colloid and Interface Science*, *587*, 743−754. Available from https://doi.org/10.1016/j.jcis.2020.110.034.

Liu, Z., Sun, G., Chen, Z., Ma, Y., Qiu, K., Li, M., & Ni, B.-J. (2023). Anchoring Cu-N active sites on functionalized polyacrylonitrile fibers for highly selective H_2S/CO_2 separation. *Journal of Hazardous Materials*, *450*, 131084. Available from https://doi.org/10.1016/j.jhazmat.2023.131084.

Maddah, B., Yavaripour, A., Ramedani, S. H., Hosseni, H., & Hasanzadeh, M. (2020). Electrospun PU nanofiber composites based on carbon nanotubes decorated with nickel-zinc ferrite particles as an adsorbent for removal of hydrogen sulfide from air. *Environmental Science and Pollution Research*, *27*, 35515−35525. Available from https://doi.org/10.1007/s11356-020-09324-9.

Marella, M., & Tomaselli, M. (2006). Synthesis of carbon nanofibers and measurements of hydrogen storage. *Carbon*, *44*, 1404−1413. Available from https://doi.org/10.1016/J.CARBON.2005.110.020.

Megías-Sayago, C., Bingre, R., Huang, L., Lutzweiler, G., Wang, Q., & Louis, B. (2019). CO_2 adsorption capacities in zeolites and layered double hydroxide materials. *Frontiers in Chemistry*, *7*, 466568. Available from https://doi.org/10.3389/FCHEM.2019.00551/BIBTEX.

Meng, L. Y., & Park, S. J. (2010). Effect of heat treatment on CO_2 adsorption of KOH-activated graphite nanofibers. *Journal of Colloid and Interface Science*, *352*, 498−503. Available from https://doi.org/10.1016/J.JCIS.2010.080.048.

Morris, R. E., & Wheatley, P. S. (2008). Gas storage in nanoporous materials. *Angewandte Chemie International Edition*, *47*, 4966−4981. Available from https://doi.org/10.1002/ANIE.200703934.

Mukherjee, B. (2021). Carbon nanofiber for hydrogen storage. *Carbon Nanofibers: Fundamentals and Applications*, 175−209. Available from https://doi.org/10.1002/9781119769149.CH7.

Nechaev, Y. S., Yürüm, A., Tekin, A., Yavuz, N. K., Yürüm, Y., Veziroglu, T. N., Nechaev, Y. S., Yürüm, A., Tekin, A., Yavuz, N. K., Yürüm, Y., & Veziroglu, T. N. (2014). Fundamental open questions on engineering of "super" hydrogen sorption in graphite nanofibers: Relevance for clean energy applications. *American Journal of Analytical Chemistry*, *5*, 1151−1165. Available from https://doi.org/10.4236/AJAC.2014.516122.

Nechaev, Y. S., Denisov, E. A., Cheretaeva, A. O., Shurygina, N. A., Kostikova, E. K., Öchsner, A., & Davydov, S. Y. (2022). On the problem of "super" storage of hydrogen in graphite Nanofibers. *C*, *8*, 23. Available from https://doi.org/10.3390/C8020023.

North, M. (2015). What is CO_2? Thermodynamics, basic reactions and physical chemistry. *Carbon dioxide utilisation. Closing carbon cycle* (First Ed., pp. 3−17). Elsevier. Available from https://doi.org/10.1016/B978-0-444-62746-9.00001-3.

Ogunlaja, A. S., Du Sautoy, C., Torto, N., & Tshentu, Z. R. (2014). Design, fabrication and evaluation of intelligent sulfone-selective polybenzimidazole nanofibers. *Talanta*, *126*, 61−72. Available from https://doi.org/10.1016/j.talanta.2014.030.035.

Othman, F. E. C., Yusof, N., Samitsu, S., Abdullah, N., Hamid, M. F., Nagai, K., Abidin, M. N. Z., Azali, M. A., Ismail, A. F., Jaafar, J., Aziz, F., & Salleh, W. N. W. (2021). Activated carbon nanofibers incorporated metal oxides for CO_2 adsorption: Effects of different type of metal oxides. *Journal of CO_2 Utilization*, *45*, 101434. Available from https://doi.org/10.1016/J.JCOU.2021.101434.

Park, J., Kretzschmar, A., Selmert, V., Camara, O., Kungl, H., Tempel, H., Basak, S., & Eichel, R. A. (2021). Structural study of polyacrylonitrile-based carbon nanofibers for understanding gas adsorption. *ACS Applied Materials & Interfaces*, *13*, 46665−46670. Available from https://doi.org/10.1021/ACSAMI.1C13541/ASSET/IMAGES/LARGE/AM1C13541_0005.JPEG.

Pérez-Pellitero, J., & Pirngruber, G. D. (2020). Industrial zeolite applications for gas adsorption and separation processes. *Structure and Bonding*, *184*, 195−225. Available from https://doi.org/10.1007/430_2020_75/COVER.

Phani, A., De BrittoMT, R., Srinivasan, S., & Stefanakos, L. (2014). *Polyaniline nanofibers obtained by electrospin process for hydrogen storage applications*. Available from http://www.ripublication.com, accessed: June 29, 2023.

Phuoc, P. H., Hung, C. M., Van Toan, N., Van Duy, N., Hoa, N. D., & Van Hieu, N. (2020). One-step fabrication of SnO_2 porous nanofiber gas sensors for sub-ppm H_2S detection. *Sensors and Actuators A: Physical*, *303*, 111722. Available from https://doi.org/10.1016/j.sna.2019.111722.

Pioquinto-García, S., Álvarez, J. R., Rico-Barragán, A. A., Giraudet, S., Rosas-Martínez, J. M., Loredo-Cancino, M., Soto-Regalado, E., Ovando-Medina, V. M., Cordero, T., Rodríguez-Mirasol, J., & Dávila-Guzmán, N. E. (2023). Electrospun Al-MOF fibers as D4 siloxane adsorbent: Synthesis, environmental impacts, and adsorption behavior. *Microporous and Mesoporous Materials*, *348*, 112327. Available from https://doi.org/10.1016/j.micromeso.2022.112327.

Plaza, M. G., García, S., Rubiera, F., Pis, J. J., & Pevida, C. (2010). Post-combustion CO_2 capture with a commercial activated carbon: Comparison of different regeneration

strategies. *Chemical Engineering Journal, 163*, 41−47. Available from https://doi.org/10.1016/j.cej.2010.070.030.
Presser, V., McDonough, J., Yeon, S. H., & Gogotsi, Y. (2011). Effect of pore size on carbon dioxide sorption by carbide derived carbon. *Energy & Environmental Science, 4*, 3059−3066. Available from https://doi.org/10.1039/C1EE01176F.
Rajkumar, S., Young, B. N., Clark, M. L., Benka-Coker, M. L., Bachand, A. M., Brook, R. D., Nelson, T. L., Volckens, J., Reynolds, S. J., L'Orange, C., Good, N., Koehler, K., Africano, S., Osorto Pinel, A. B., & Peel, J. L. (2019). Household air pollution from biomass-burning cookstoves and metabolic syndrome, blood lipid concentrations, and waist circumference in Honduran women: A cross-sectional study. *Environmental Research, 170*, 46−55. Available from https://doi.org/10.1016/J.ENVRES.2018.120.010.
Rehman, A., & Park, S. J. (2018). Comparative study of activation methods to design nitrogen-doped ultra-microporous carbons as efficient contenders for CO_2 capture. *Chemical Engineering Journal, 352*, 539−548. Available from https://doi.org/10.1016/J.CEJ.2018.070.046.
Ren, J., Musyoka, N. M., Annamalai, P., Langmi, H. W., North, B. C., & Mathe, M. (2015). Electrospun MOF nanofibers as hydrogen storage media. *International Journal of Hydrogen Energy, 40*, 9382−9387. Available from https://doi.org/10.1016/j.ijhydene.2015.050.088.
Ruiz-Morales, J. C., Tarancón, A., Canales-Vázquez, J., Méndez-Ramos, J., Hernández-Afonso, L., Acosta-Mora, P., Marín Rueda, J. R., & Fernández-González, R. (2017). Three dimensional printing of components and functional devices for energy and environmental applications. *Energy & Environmental Science, 10*, 846−859. Available from https://doi.org/10.1039/C6EE03526D.
Rzepka, M., Bauer, E., Reichenauer, G., Schliermann, T., Bernhardt, B., Bohmhammel, K., Henneberg, E., Knoll, U., Maneck, H. E., & Braue, W. (2005). Hydrogen storage capacity of catalytically grown carbon Nanofibers. *The Journal of Physical Chemistry B, 109*, 14979−14989. Available from https://doi.org/10.1021/JP051371A.
Saha, D., & Kienbaum, M. J. (2019). Role of oxygen, nitrogen and sulfur functionalities on the surface of nanoporous carbons in CO_2 adsorption: A critical review. *Microporous and Mesoporous Materials, 287*, 29−55. Available from https://doi.org/10.1016/J.MICROMESO.2019.050.051.
Saha, D., Van Bramer, S. E., Orkoulas, G., Ho, H. C., Chen, J., & Henley, D. K. (2017). CO_2 capture in lignin-derived and nitrogen-doped hierarchical porous carbons. *Carbon, 121*, 257−266. Available from https://doi.org/10.1016/J.CARBON.2017.050.088.
Salthammer, T. (2004). Emissions of volatile organic compounds from products and materials in indoor environments. *The handbook of environmental chemistry, 4*, 37−71. Available from https://doi.org/10.1007/B94830/COVER.
Shaki, H. (2023). The use of electrospun nanofibers for absorption and separation of carbon dioxide: A review. *Journal of Industrial Textiles, 53*. Available from https://doi.org/10.1177/15280837231160290, https://doi.org/10.1177/15280837231160290, 152808372311602.
Simonyan, V. V., & Johnson, J. K. (2002). Hydrogen storage in carbon nanotubes and graphitic nanofibers. *Journal of Alloys and Compounds, 330−332*, 659−665. Available from https://doi.org/10.1016/S0925-8388(01)01664-4.
Singh, G., Kim, I. Y., Lakhi, K. S., Srivastava, P., Naidu, R., & Vinu, A. (2017). Single step synthesis of activated bio-carbons with a high surface area and their excellent CO_2

adsorption capacity. *Carbon*, *116*, 448−455. Available from https://doi.org/10.1016/J.CARBON.2017.020.015.

Singh, J., Bhunia, H., & Basu, S. (2018). Synthesis of porous carbon monolith adsorbents for carbon dioxide capture: Breakthrough adsorption study. *Journal of Taiwan Institute of Chemical Engineers*, *89*, 140−150. Available from https://doi.org/10.1016/J.JTICE.2018.040.031.

Sreńscek-Nazzal, J., & Kiełbasa, K. (2019). Advances in modification of commercial activated carbon for enhancement of CO_2 capture. *Applied Surface Science*, *494*, 137−151. Available from https://doi.org/10.1016/J.APSUSC.2019.070.108.

Stephen, D., Sinha, A., Kalyanaraman, J., Zhang, F., Realff, M. J., & Lively, R. P. (2018). Critical comparison of structured contactors for adsorption-based gas separations. *Annual Review of Chemical and Biomolecular Engineering*, *9*, 129−152. Available from https://doi.org/10.1146/ANNUREV-CHEMBIOENG-060817-084120, https://doi.org/10.1146/Annurev-Chembioeng-060817-084120.

Trogler, W. C. (1995). The environmental chemistry of trace atmospheric gases. *Journal of Chemical Education*, *72*, 973−976. Available from https://doi.org/10.1021/ED072P973.

US EPA. (2008). Inventory of U.S. greenhouse gas emissions and sinks, *Environmental Protection*, 1990−2009. https://www.epa.gov/ghgemissions/inventory-us-greenhouse-gas-emissions-and-sinks-1990-2017 (accessed: June 28, 2023).

Wang, G., Dou, B., Zhang, Z., Wang, J., Liu, H., & Hao, Z. (2015). Adsorption of benzene, cyclohexane and hexane on ordered mesoporous carbon. *Journal of Environmental Sciences*, *30*, 65−73. Available from https://doi.org/10.1016/J.JES.2014.100.015.

Wang, J., Ke, C., Jia, X., Ma, C., Liu, X., Qiao, W., & Ling, L. (2021). Polyethyleneimine-functionalized mesoporous carbon nanosheets as metal-free catalysts for the selective oxidation of H_2S at room temperature. *Applied Catalysis B: Environment and Energy*, *283*, 119650. Available from https://doi.org/10.1016/j.apcatb.2020.119650.

Wang, J., Liu, L., Wang, L., Lu, J., & Li, Y. (2023a). Volatile methyl siloxane separation from biogas using hollow fiber membrane contactor with polyethylene glycol dimethyl ether: A numerical and experimental study. *Process Safety and Environmental Protection*, *171*, 250−259. Available from https://doi.org/10.1016/j.psep.2023.010.023.

Wang, J., Wang, L., & Li, Y. (2023b). Absorbent screening using a hollow fiber membrane contactor for removing volatile methyl siloxanes from biogas. *Journal of Cleaner Production*, *404*, 136865. Available from https://doi.org/10.1016/j.jclepro.2023.136865.

Wang, Y., Wang, H., Tan, Y., Liu, J., Wang, K., Ji, W., Sun, L., Yu, X., Zhao, J., Xu, B., & Xiong, J. (2021). Measurement of the key parameters of VOC emissions from wooden furniture, and the impact of temperature. *Atmospheric Environment*, *259*, 118510. Available from https://doi.org/10.1016/J.ATMOSENV.2021.118510.

Wanwong, S., Sangkhun, W., & Jiamboonsri, P. (2023). Electrospun cyclodextrin/poly(l-lactic acid) nanofibers for efficient air filter: Their PM and VOC removal efficiency and triboelectric outputs. *Polymers (Basel)*, *15*, 722. Available from https://doi.org/10.3390/POLYM15030722/S1.

Werner, S., Seitz, C., Beck, G., Hennemann, J., & Smarsly, B. M. (2021). Porous SiO_2 nanofibers loaded with CuO nanoparticles for the dosimetric detection of H_2S. *ACS Applied Nano Materials*, *4*, 5004−5013. Available from https://doi.org/10.1021/acsanm.1c00518.

Wu, Y. N., Li, F., Liu, H., Zhu, W., Teng, M., Jiang, Y., Li, W., Xu, D., He, D., Hannam, P., & Li, G. (2012). Electrospun fibrous mats as skeletons to produce free-standing MOF membranes. *Journal of Materials Chemistry*, *22*, 16971−16978. Available from https://doi.org/10.1039/C2JM32570E.

Xia, G., Li, D., Chen, X., Tan, Y., Tang, Z., Guo, Z., Liu, H., Liu, Z., & Yu, X. (2013). Carbon-coated Li$_3$N nanofibers for advanced hydrogen storage. *Advanced Materials, 25*, 6238−6244. Available from https://doi.org/10.1002/ADMA.201301927.

Xing, Y., Fang, B., Bonakdarpour, A., Zhang, S., & Wilkinson, D. P. (2014). Facile fabrication of mesoporous carbon nanofibers with unique hierarchical nanoarchitecture for electrochemical hydrogen storage. *International Journal of Hydrogen Energy, 39*, 7859−7867. Available from https://doi.org/10.1016/J.IJHYDENE.2014.030.106.

Yaakob, Z., Jafar Khadem, D., Shahgaldi, S., Wan Daud, W. R., & Tasirin, S. M. (2012). The role of Al and Mg in the hydrogen storage of electrospun ZnO nanofibers. *International Journal of Hydrogen Energy, 37*, 8388−8394. Available from https://doi.org/10.1016/J.IJHYDENE.2012.020.092.

Yu, Y., Ma, Q., Bin Zhang, J., & Bin Liu, G. (2020). Electrospun SiO$_2$ aerogel/polyacrylonitrile composited nanofibers with enhanced adsorption performance of volatile organic compounds. *Applied Surface Science, 512*, 145697. Available from https://doi.org/10.1016/J.APSUSC.2020.145697.

Yuan, H., Chen, J., Li, D., Chen, H., & Chen, Y. (2019). 5 Ultramicropore-rich renewable porous carbon from biomass tar with excellent adsorption capacity and selectivity for CO$_2$ capture. *Chemical Engineering Journal, 373*, 171−178. Available from https://doi.org/10.1016/J.CEJ.2019.040.206.

Zainab, G., Iqbal, N., Babar, A. A., Huang, C., Wang, X., Yu, J., & Ding, B. (2017). Freestanding, spider-web-like polyamide/carbon nanotube composite nanofibrous membrane impregnated with polyethyleneimine for CO$_2$ capture. *Composites Communications, 6*, 41−47. Available from https://doi.org/10.1016/J.COCO.2017.090.001.

Zainab, G., Babar, A. A., Ali, N., Aboalhassan, A. A., Wang, X., Yu, J., & Ding, B. (2020). Electrospun carbon nanofibers with multi-aperture/opening porous hierarchical structure for efficient CO$_2$ adsorption. *Journal of Colloid and Interface Science, 561*, 659−667. Available from https://doi.org/10.1016/J.JCIS.2019.110.041.

Zhang, X., Ru, Z., Sun, Y., Zhang, M., Wang, J., Ge, M., Liu, H., Wu, S., Cao, C., Ren, X., Mi, J., & Feng, Y. (2022). Recent advances in applications for air pollutants purification and perspectives of electrospun nanofibers. *Journal of Cleaner Production, 378*, 134567. Available from https://doi.org/10.1016/j.jclepro.2022.134567.

Zhang, X., Ru, Z., Wang, T., Feng, W., Zhang, M., Wang, J., Mi, J., Ge, M., & Feng, Y. (2023). Insights to the microwave effect in formation and performance promotion of iron-based carbon nanofibrous composites for H2S removal. *Composites Communications, 37*, 101468. Available from https://doi.org/10.1016/j.coco.2022.101468.

Zhang, Y., Ran, H., Liu, X., Zhang, X., Yin, J., Zhang, J., He, J., Li, H., & Li, H. (2023). Cu-doped BCN nanofibers for highly selective adsorption desulfurization through S-Cu coordination and π−π interaction. *Separation and Purification Technology, 318*, 123963. Available from https://doi.org/10.1016/j.seppur.2023.123963.

Catalysis and electrocatalysis application of nanofibers and their composites

Elyor Berdimurodov[1,2,3], Khasan Berdimuradov[4,5], Ashish Kumar[6], Ilyos Eliboev[7,8], Nodira Eshmamatova[7], Bakhtiyor Borikhonov[9] and Sardorbek Otajonov[7]

[1]Chemical & Materials Engineering, New Uzbekistan University, Tashkent, Uzbekistan, [2]University of Tashkent for Applied Sciences, Tashkent, Uzbekistan, [3]Department of Physical Chemistry, National University of Uzbekistan, Tashkent, Uzbekistan, [4]Faculty of Industrial Viticulture and Food Production Technology, Shahrisabz Branch of Tashkent Institute of Chemical Technology, Shahrisabz, Uzbekistan, [5]Physics and Chemistry, Western Caspian University, Baku, Azerbaijan, [6]Department of Science and Technology, NCE, Bihar Engineering University, Government of Bihar, Patna, Bihar, India, [7]Faculty of Chemistry, National University of Uzbekistan, Tashkent, Uzbekistan, [8]Chemistry, Uzbek-Finnish Pedagogical Institute, Samarqand, Uzbekistan, [9]Faculty of Chemistry and Biology, Karshi State University, Karshi City, Uzbekistan

16.1 Introduction

Nanofiber composites have attracted attention in recent years due to their unique physical and chemical properties, including high surface area, high aspect ratio, and excellent mechanical properties. These properties make them promising materials for catalytic and electrocatalytic applications. In this chapter, we will provide an overview of catalysis and electrocatalysis, as well as the application of nanofiber composites in these fields (Liu et al., 2021; Wang et al., 2019).

Catalysis and electrocatalysis are fundamental processes that play a critical role in various fields, including energy conversion, chemical manufacturing, and environmental remediation. The development of efficient and sustainable catalysts and electrocatalysts is, therefore, essential to address the growing demand for clean energy technologies (Gui et al., 2020a; Zhang et al., 2019).

Nanofiber composites, with their unique physical and chemical properties, have emerged as promising materials for catalytic and electrocatalytic applications. These materials have high surface area, tailorable selectivity, and excellent electrical conductivity. They can be synthesized from a variety of materials, including metals, metal oxides, and carbon-based materials, using various techniques such as electrospinning, template-assisted synthesis, and chemical vapor deposition (CVD) (Gui et al., 2020b; Tong et al., 2019).

The application of nanofiber composites in catalysis and electrocatalysis has shown great promise in various fields, including fuel cells, metal−air batteries, water splitting, and environmental remediation. The synergistic effect of combining catalysis and electrocatalysis in the form of nanofiber composites has opened up new avenues for developing highly efficient and selective catalysts (Xiang et al., 2023; Zhang et al., 2022).

This chapter aims to provide a comprehensive overview of the application of nanofiber composites in catalysis and electrocatalysis. The book will cover the synthesis methods of nanofibers, the types of nanofibers used as catalysts and electrocatalysts, their advantages in catalysis and electrocatalysis, and their application in various fields. The book will also discuss the challenges and future prospects of using nanofiber composites in catalysis and electrocatalysis. The application of nanofiber composites in catalysis and electrocatalysis has gained significant attention in recent years due to their unique physical and chemical properties. Nanofiber composites have a high surface area-to-volume ratio, which provides a larger number of active sites for catalytic reactions and enhances electrocatalytic activity. Additionally, their unique physical and chemical properties enable the tailoring of the catalysts' selectivity and activity, making them highly efficient and selective catalysts and electrocatalysts (Gong et al., 2019; Li et al., 2023; Yang et al., 2020). Electrospinning (Fig. 16.1) is a useful method for synthesizing ordered mesoporous inorganic fibers, including silicates, ZSM-5, beta zeolite, SBA, faujasite, MOFs, and ZIFs. Other techniques, such as growing mesoporous material on carbon and polymeric electrospun fibers, have also been developed. In the inverse procedure,

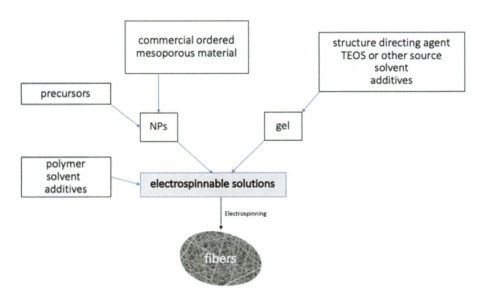

Figure 16.1 Standard method for the production of arranged porous inorganic threads by electrospinning (Guerrero-Pérez, 2022). *NPs*, Nanoparticles; *TEOS*, tetraethyl orthosilicate.

electrospun mesoporous inorganic nanofibers can be used as a template for creating mesoporous carbon material through CVD and matrix removal with an appropriate solvent. These methods have potential applications in catalysis, energy storage, and biomedical engineering.

The use of nanofiber composites in catalysis and electrocatalysis has shown great potential in various applications, including energy conversion and storage, chemical manufacturing, and environmental remediation. For example, nanofiber composites have been used as catalysts in fuel cells and metal−air batteries, which are promising technologies for clean and sustainable energy. They have also been used in water splitting, a process that involves the conversion of water into hydrogen and oxygen using renewable energy sources, as well as in the removal of pollutants from wastewater. The combination of catalysis and electrocatalysis in the form of nanofiber composites has opened up new avenues for developing highly efficient and selective catalysts. By incorporating metal nanoparticles into nanofiber composites, the catalytic activity and selectivity can be enhanced due to the synergistic effect between the metal nanoparticles and the nanofiber matrix. Despite the promising results, there are still challenges that need to be addressed in the development of nanofiber composites for catalysis and electrocatalysis. These challenges include the optimization of the synthesis methods, the control of the properties of the nanofiber composites, and the understanding of the underlying mechanisms of the catalytic and electrocatalytic reactions. Overall, this chapter will be a valuable resource for researchers, scientists, and engineers who are interested in the development of efficient and sustainable catalysts and electrocatalysts using nanofiber composites (Berdimurodov et al., 2021; Nie et al., 2021; Peng et al., 2020) (Fig. 16.2).

16.2 Nanofiber synthesis methods

Nanofibers have emerged as promising materials in the field of catalysis and electrocatalysis due to their unique properties such as high surface area, small size, and tunable porosity. The synthesis of nanofibers with controlled morphology and composition is essential for achieving high catalytic activity and selectivity. In this chapter, we will discuss the various synthesis methods of nanofibers and their composites for catalysis and electrocatalysis applications.

Electrospinning is a widely used method for the synthesis of nanofibers. It involves the use of an electric field to draw a polymer solution or melt from a syringe tip to a collector. The electric field induces a charge on the surface of the solution, which leads to the formation of a Taylor cone. When the electrostatic repulsion overcomes the surface tension, the polymer jet is ejected from the tip of the cone and is collected on a grounded substrate. The resulting nanofibers can be further processed by calcination, reduction, or functionalization to obtain the desired properties for catalysis and electrocatalysis applications (Cao et al., 2020; Xie et al., 2021; Zhang & Wang, 2021).

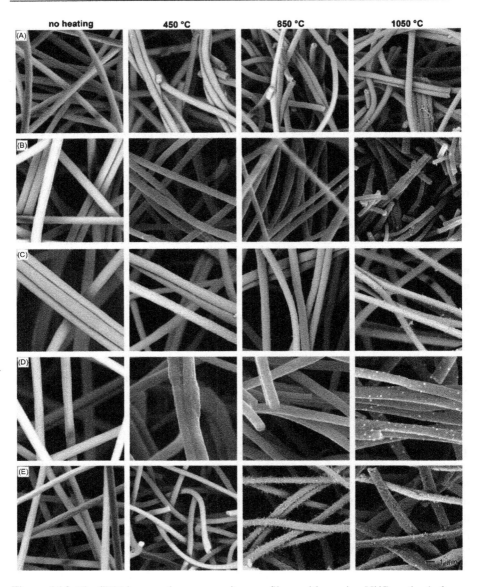

Figure 16.2 The SEM images show composite nanofibers with varying Ni/Co ratios before and after heating at different temperatures: 450°C, 850°C, and 1050°C. The first column depicts the images before heating, while the second, third, and fourth columns show the images after heating at 450°C, 850°C, and 1050°C, respectively. The images labeled (A) depict fibers consisting of only Ni, (B) show a 1:0.5 ratio of Ni to Co, (C) show a 1:1 ratio, (D) show a 1:2 ratio, and (E) show fibers consisting of only Co. The scale bar of 1 μm in the image labeled (E), which was heated to 1050°C, applies to all images (Schossig et al., 2023).

Electrospinning is a popular method for the synthesis of nanofibers because it is a simple and versatile technique that can produce nanofibers from a wide range of materials, including polymers, metals, and metal oxides. The resulting nanofibers can have a wide range of morphologies, including straight, twisted, and branched structures, depending on the processing conditions. Electrospinning can also be combined with other techniques, such as templating and functionalization, to further enhance the properties of the resulting nanofibers (Ali et al., 2020; Zhang et al., 2020).

The formation of BOS nanofibers (Fig. 16.3) during electrospinning is primarily influenced by the properties of the polymer solution and the process parameters. A balance between the solution's viscosity and surface tension is crucial for the formation of BOS structures. Low viscosity and high surface tension can lead to the formation of beads due to the dominance of surface tension in the jet's instability, while high viscosity and low surface tension favor the formation of smooth fibers (Li et al., 2017).

Template synthesis involves the use of a template to guide the growth of nanofibers. The template can be a sacrificial material or a self-assembled structure. In sacrificial template synthesis, the template is removed after the growth of nanofibers, leaving behind a porous nanofiber structure. For example, a porous polymer template can be used to synthesize metal oxide nanofibers by electrospinning a polymer solution containing metal precursors, followed by calcination to remove the polymer template. In self-assembled template synthesis, the nanofibers are grown on a self-assembled structure such as a surfactant micelle or a block copolymer. The resulting nanofibers can have a controlled morphology and composition, which is

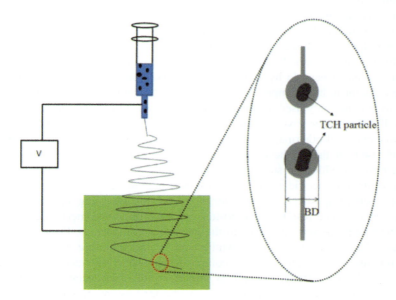

Figure 16.3 A diagrammatic representation depicting nanofibers in the form of beads strung together, enclosing drug particles (Li et al., 2017).

important for catalysis and electrocatalysis applications (Li et al., 2022; Liu et al., 2020; Lu et al., 2021).

This technique involves the use of a template to guide the growth of nanofibers. The template can be a sacrificial material or a self-assembled structure, as I mentioned earlier. Sacrificial template synthesis is particularly useful for the synthesis of porous nanofiber structures, which can be used for catalytic reactions that require high surface area and accessibility of reactants to the catalyst. Self-assembled template synthesis, on the other hand, can produce nanofibers with controlled morphology and composition, which can be important for achieving specific catalytic properties (Li et al., 2019; Wang et al., 2021).

Vapor-phase synthesis involves the use of a gaseous precursor to grow nanofibers. The precursor is typically decomposed in a reactor, and the resulting atoms or molecules are deposited on a substrate to form nanofibers. CVD is a common vapor-phase synthesis method used for the synthesis of carbon nanofibers. In CVD, a hydrocarbon gas is decomposed on a substrate at high temperature to form carbon nanofibers. The resulting nanofibers can have a high degree of crystallinity and can be functionalized for catalysis and electrocatalysis applications. This technique involves the use of a gaseous precursor to grow nanofibers. CVD is a common vapor-phase synthesis method used for the synthesis of carbon nanofibers, as I mentioned earlier. CVD can also be used for the synthesis of other materials, such as metal oxides and nitrides. Vapor-phase synthesis can produce nanofibers with high crystallinity and purity, which can be important for achieving specific catalytic properties (Verma et al., 2021; Wang et al., 2021).

Solution-based synthesis involves the use of a solution to grow nanofibers. The solution can contain precursors that are chemically transformed into nanofibers or can act as a solvent for the growth of nanofibers. In sol−gel synthesis, a precursor solution is hydrolyzed and condensed to form a gel, which is then dried and calcined to form nanofibers. Sol−gel synthesis is commonly used for the synthesis of metal oxide nanofibers. In hydrothermal synthesis, a precursor solution is heated in a sealed reactor at high temperature and pressure to form nanofibers. Hydrothermal synthesis is commonly used for the synthesis of metal sulfide and metal oxide nanofibers. This technique involves the use of a solution to grow nanofibers. Sol−gel synthesis and hydrothermal synthesis are two common solution-based synthesis methods used for the synthesis of metal oxide nanofibers. These methods can produce nanofibers with high surface area and tunable pore size, which can be important for achieving specific catalytic properties (Lin et al., 2021; Wang et al., 2021).

Nanofiber composites can be synthesized by incorporating nanofibers into a matrix material. The resulting composites can have improved mechanical, thermal, and electrical properties compared to the individual components. For example, carbon nanofiber composites can be synthesized by incorporating carbon nanofibers into a polymer matrix. The resulting composites can have high electrical conductivity and mechanical strength, which is important for electrocatalysis applications. Metal oxide nanofiber composites can be synthesized by incorporating metal oxide nanofibers into a mesoporous silica matrix. The resulting composites can have high surface area and tunable pore size, which is important for catalysis applications.

Nanofiber composites can be synthesized by incorporating nanofibers into a matrix material, such as a polymer or a mesoporous silica matrix. The resulting composites can have improved properties, such as increased mechanical strength, electrical conductivity, and surface area, which can be important for achieving specific catalytic properties.

Therefore, the synthesis of nanofibers and their composites is essential for achieving high catalytic activity and selectivity in catalysis and electrocatalysis applications. Electrospinning, template synthesis, vapor-phase synthesis, and solution-based synthesis are common methods used for the synthesis of nanofibers. Nanofiber composites can be synthesized by incorporating nanofibers into a matrix material, which can improve the properties of the resulting composite (Cao et al., 2021; Lei et al., 2020).

16.3 Catalysis application of nanofiber composites

Nanofiber composites have emerged as a promising class of materials for catalysis applications due to their high surface area, tunable pore size, and improved mechanical, thermal, and electrical properties. In this chapter, we will discuss the various applications of nanofiber composites in catalysis and electrocatalysis.

Nanofiber composites have been extensively studied for catalytic applications such as heterogeneous catalysis, photocatalysis, and enzymatic catalysis. Heterogeneous catalysis involves the use of a solid catalyst to catalyze a chemical reaction in a liquid or gas phase. Nanofiber composites can be used as catalyst supports due to their high surface area and tunable pore size, which can improve the accessibility of reactants to the catalyst. For example, metal oxide nanofiber composites have been used as catalyst supports for the oxidation of organic compounds, such as benzene and toluene, due to their high surface area and thermal stability (Dong et al., 2021; Li et al., 2022).

In this study (Wang et al., 2020) (Fig. 16.4), a promising electrocatalyst for both oxygen reduction reaction (ORR) and oxygen evolution reaction (OER) was introduced. The electrocatalyst consists of nanostructured FeNi alloy nanoparticles embedded in N-doped carbon nanotubes (CNTs) entwined with porous carbon fibers (FeNi/N-CPCF). The FeNi/N-CPCF-950, with its hierarchically porous structure, bamboo-like CNTs, and strong synergistic coupling between FeNi alloys and N-doped carbon species, exhibits a half-wave potential of 0.867 V for ORR and a low operating potential of 1.585 V at 10 mA/cm^2 for OER in 0.1 M KOH. This performance surpasses that of commercial Pt/C and RuO$_2$. Furthermore, this bifunctional catalyst enables the zinc−air batteries built in-house to achieve a high energy efficiency of 61.5%, a small charge-discharge voltage gap of 0.764 V, and excellent cycling performance (640 h, 960 cycles) at 10 mA/cm^2 under ambient conditions (Wang et al., 2020).

Nanofiber composites have been extensively studied as catalyst supports for heterogeneous catalysis. Heterogeneous catalysis involves the use of a solid catalyst to

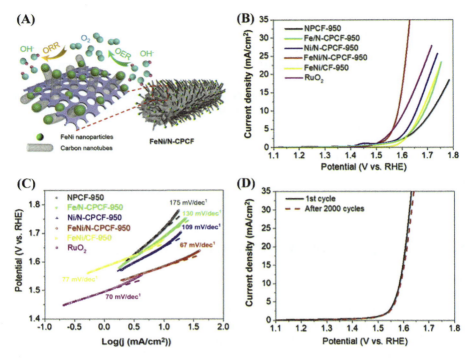

Figure 16.4 (A) A schematic illustration showing the structure of the FeNi/N-CPCF membrane. (B) Linear sweep voltammetry (LSV) curves for various catalysts in a 0.1 M KOH electrolyte, recorded at a 5 mV/s scan rate and including IR correction. (C) Tafel plots derived from the respective LSV curves. (D) Oxygen evolution reaction (OER) polarization curves for the FeNi/N-CPCF-950 catalyst before and after 2000 cycles of cyclic voltammetry (CV) scanning (Wang et al., 2020).

catalyze a chemical reaction in a liquid or gas phase. The high surface area and tunable pore size of nanofiber composites make them attractive as catalyst supports because they can improve the accessibility of reactants to the catalyst. For example, metal oxide nanofiber composites have been used as catalyst supports for the oxidation of organic compounds, such as benzene and toluene, due to their high surface area and thermal stability. The performance of a catalyst is often influenced by its surface area, pore size distribution, and active site density. Nanofiber composites can be tailored to meet specific catalytic requirements by adjusting the properties of the nanofibers and the matrix material. For example, the incorporation of metal nanoparticles into the nanofiber composites can increase the active site density and improve catalytic performance (Bian & Sun, 2021; Li et al., 2021; Sankar et al., 2020; Zhu et al., 2020).

The Tafel slope provides insights into the mechanism and kinetics of the hydrogen evolution reaction (HER). In alkaline media, the HER mechanism follows either the Volmer–Heyrovsky or Volmer–Tafel reaction. A Tafel slope (b) of 120,

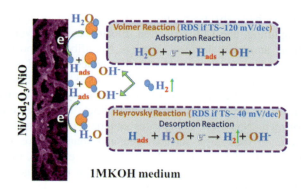

Figure 16.5 Hydrogen evolution reaction mechanism on the Ni/Gd$_2$O$_3$/NiO NF surface (El-Maghrabi et al., 2021).

40, or 30 mV/dec indicates that the Volmer, Heyrovsky, or Tafel reaction, respectively, is the rate-determining step (RDS) (Fig. 16.5) (El-Maghrabi et al., 2021).

The Volmer reaction involves the electroreduction of water molecules with electrons on the cathode's active sites (*H$_2$O + e$^-$ → *H$_{ads}$ + OH$^-$).

The Heyrovsky reaction describes electrochemical hydrogen desorption (*H$_{ads}$ + H$_2$O + e$^-$ → H$_2$↑ + OH$^-$).

The Tafel reaction involves chemical desorption (2H$_{ads}$ ↔ H^2↑).

For Ni/Gd$_2$O$_3$/NiO nanofiber (NF) surfaces, the HER process with a Tafel slope (b) of 45 mV/dec follows the Volmer–Heyrovsky mechanism. The RDS is governed by electrochemical desorption, which includes water molecule discharge and hydrogen desorption from the catalyst surface.

Nanofiber composites have also been studied as photocatalysts for the degradation of organic compounds under UV irradiation. Photocatalysis involves the use of a catalyst to initiate a chemical reaction using light energy. The high surface area and ability to absorb light energy of nanofiber composites make them attractive as photocatalysts. For example, TiO$_2$ nanofiber composites have been used as photocatalysts for the degradation of organic compounds, such as phenol and methyl orange, under UV irradiation.

The performance of a photocatalyst is often influenced by its structure, band gap, and surface area. For example, the incorporation of metal nanoparticles into the nanofiber composites can increase light absorption and improve photocatalytic performance (Dong et al., 2021; Gui et al., 2020a; Li et al., 2019).

Nanofiber composites have also been studied as enzyme immobilization supports for enzymatic catalysis. Enzymatic catalysis involves the use of enzymes to catalyze a chemical reaction. The high surface area and biocompatibility of nanofiber composites make them attractive as enzyme immobilization supports. For example, cellulose nanofiber composites have been used as enzyme immobilization supports for the hydrolysis of cellulose. The performance of an enzymatic catalyst is often influenced by the stability and activity of the enzyme. For example, the incorporation of functional groups into the nanofiber composites can improve enzyme binding and activity.

Nanofiber composites have also been studied for electrocatalytic applications such as fuel cells, water splitting, and CO_2 reduction. Electrocatalysis involves the use of a catalyst to facilitate an electrochemical reaction. Nanofiber composites can be used as electrocatalysts due to their high electrical conductivity and surface area. For example, carbon nanofiber composites have been used as electrocatalysts for the ORR in fuel cells due to their high electrical conductivity and mechanical strength. Water splitting involves the use of a catalyst to facilitate the separation of water into hydrogen and oxygen. Nanofiber composites can be used as catalyst supports for water splitting due to their high surface area and tunable pore size, which can improve the accessibility of reactants to the catalyst. For example, metal oxide nanofiber composites have been used as catalyst supports for the oxygen evolution reaction in water splitting (Lin et al., 2021; Wang et al., 2021; Xiang et al., 2023).

Nanofiber composites have also been studied for electrocatalytic applications such as fuel cells, water splitting, and CO_2 reduction. Electrocatalysis involves the use of a catalyst to facilitate an electrochemical reaction. Nanofiber composites can be used as electrocatalysts due to their high electrical conductivity and surface area. For example, carbon nanofiber composites have been used as electrocatalysts for the ORR in fuel cells due to their high electrical conductivity and mechanical strength (Bian & Sun, 2021; Li et al., 2023; Lin et al., 2021).

Fuel cells are electrochemical devices that convert the chemical energy of a fuel into electrical energy. Nanofiber composites have been studied as electrocatalysts for the ORR in fuel cells. The ORR is a key reaction in the cathode of a fuel cell, and it involves the reduction of oxygen to water. The high electrical conductivity and surface area of nanofiber composites make them attractive as electrocatalysts for the ORR.

The performance of an electrocatalyst in a fuel cell is often influenced by its activity, stability, and selectivity. For example, the incorporation of metal nanoparticles into the nanofiber composites can increase the active site density and improve electrocatalytic performance (Nie et al., 2021; Sankar et al., 2020; Verma et al., 2021).

Water splitting is a promising technology for the production of hydrogen as a clean and sustainable energy source. Nanofiber composites have been studied as catalyst supports for water splitting. Water splitting involves the use of a catalyst to facilitate the separation of water into hydrogen and oxygen. The high surface area and tunable pore size of nanofiber composites make them attractive as catalyst supports for water splitting. The performance of a catalyst in water splitting is often influenced by its activity, stability, and selectivity. For example, the incorporation of metal oxide nanoparticles into the nanofiber composites can increase the active site density and improve catalytic performance (El-Maghrabi et al., 2021; Wang et al., 2019, 2021).

Developing efficient, stable, and cost-effective electrocatalysts for oxygen and hydrogen evolution reactions (OER and HER) is a critical challenge in achieving overall water splitting. In this study (Fig. 16.6) (Xie et al., 2022), a hierarchically structured CoP/carbon nanofibers (CNFs) composite was successfully synthesized, and its potential as a high-efficiency bifunctional electrocatalyst for overall water splitting was assessed. The synergetic effect of two-dimensional (2D) CoP

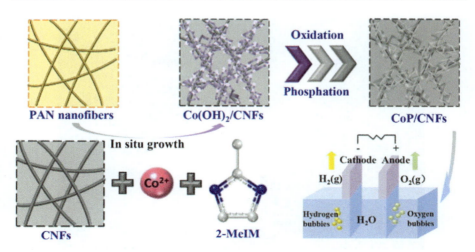

Figure 16.6 A hierarchically structured CoP/CNFs composite, consisting of 1D carbon nanofibers and 2D CoP nanosheets, was successfully fabricated and investigated as an efficient bifunctional electrocatalyst for overall water splitting (Xie et al., 2022).

nanosheets and one-dimensional (1D) CNFs in the CoP/CNFs composite provided abundant active sites and facilitated rapid electron and mass transport pathways, significantly enhancing electrocatalytic performance. The optimized CoP/CNFs achieved a current density of 10 mA/cm^2 at low overpotentials of 325 mV for OER and 225 mV for HER. For overall water splitting, CoP/CNFs reached a low potential of 1.65 V at 10 mA/cm^2. The straightforward approach presented in this work can aid in designing and developing multifunctional nonnoble metal catalysts for various energy applications (Xie et al., 2022).

CO_2 reduction is a promising technology for the conversion of CO_2 into value-added chemicals such as methane, methanol, and formic acid. Nanofiber composites have been studied as catalyst supports for CO_2 reduction. CO_2 reduction involves the use of a catalyst to convert CO_2 into value-added chemicals. The high surface area and tunable pore size of nanofiber composites make them attractive as catalyst supports for CO_2 reduction. The performance of a catalyst in CO_2 reduction is often influenced by its activity, stability, and selectivity. For example, the incorporation of metal oxide nanoparticles into the nanofiber composites can increase the active site density and improve catalytic performance (Lu et al., 2021; Peng et al., 2020; Xiang et al., 2023).

In this study (Gong et al., 2022) (Fig. 16.7), it was reported the development of porous nitrogen-doped carbon nanofibers (NCPF) with highly dispersed Ni and molybdenum phosphide nanoparticles (Ni-MoP@NCPF) for photocatalytic CO_2 reduction. The porous carbon nanofibers' high conductivity and open ends enhance charge/mass transfer and CO_2 adsorption. The Ni species are highly dispersed and stabilized by coordinating Ni−N bonds with pyrrolic nitrogen in NCPF. The Ni-MoP@NCPF achieves a CO product selectivity of up to 98.95% with a rate of

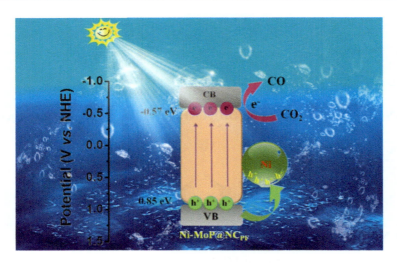

Figure 16.7 The mechanism of Ni-MoP@NCPF as a photocatalyst for CO_2 reduction (Gong et al., 2022).

953.33 µmol/g/h under visible light irradiation, which is 9.37 times faster than Ni-free MoP@NCPF. These findings offer new insights into photocatalytic CO_2 reduction using cost-effective transition metal compounds through combined morphology, composition, and heterointerface engineering. The mechanism for photocatalytic CO_2 reduction on Ni-MoP@NCPF is based on the band structure analysis, which indicates that the conduction and valence band positions are suitable for reducing CO_2 to CO. Additionally, the shifted XPS peaks of Ni 2p and Mo 3d in Ni-MoP@NCPF suggest valence electron transfer from Ni to MoP across their heterointerfaces, creating an internal electrical field that separates photogenerated electron-hole pairs, impeding recombination. This enables efficient photocatalytic performance for CO_2 reduction to CO (Gong et al., 2022).

Nanofiber composites have a wide range of applications in electrocatalysis due to their high electrical conductivity, surface area, and tunable pore size. Nanofiber composites can be used as electrocatalysts for fuel cells, water splitting, and CO_2 reduction. The performance of the electrocatalyst can be tailored by adjusting the properties of the nanofibers and the matrix material.

16.4 Future suggestions

While nanofiber composites have shown promising applications in catalysis and electrocatalysis, there is still a lot of room for improvement and further research. Here are some future suggestions:

- Development of new nanofiber composites: Researchers can explore new combinations of materials to develop nanofiber composites with unique properties. For example,

combining metal nanoparticles with 2D materials such as graphene or transition metal dichalcogenides could lead to novel electrocatalysts with high activity and selectivity.
- Optimization of synthesis methods: The synthesis of nanofiber composites can be further optimized to improve their properties and reduce costs. For example, electrospinning can be used to fabricate nanofiber composites with precise control over the fiber diameter and pore size.
- Scaling up production: Currently, the production of nanofiber composites is limited to laboratory-scale. Efforts should be made to scale up the production of these materials for industrial applications.
- Understanding the mechanisms of synergy: While the synergistic effect of nanofiber composites in catalysis and electrocatalysis is well-established, the mechanisms behind this effect are not completely understood. Further research is needed to fully understand the interactions between different materials in these composites and how they contribute to their enhanced performance.
- Real-world testing: To fully realize the potential of nanofiber composites in catalysis and electrocatalysis, more testing under real-world conditions is needed. This will help to identify any challenges or limitations that may arise in practical applications.
- Multifunctional nanofiber composites: Researchers are exploring the development of multifunctional nanofiber composites that can perform multiple tasks simultaneously. For example, nanofiber composites that can catalyze reactions while also serving as a support for electrodes in fuel cells or batteries could reduce the overall weight and complexity of these systems.
- Integration with renewable energy: The growing demand for renewable energy sources has created new opportunities for the use of nanofiber composites in catalysis and electrocatalysis. For example, nanofiber composites could be used to develop more efficient and cost-effective electrocatalysts for energy storage and conversion devices such as batteries, fuel cells, and electrolyzers.
- Artificial photosynthesis: Researchers are exploring the use of nanofiber composites in artificial photosynthesis, a process that mimics the natural process of photosynthesis to produce fuels from sunlight, water, and carbon dioxide. Nanofiber composites could be used as electrocatalysts for the reduction of carbon dioxide and water to produce renewable fuels such as hydrogen and methanol.

Overall, the future of nanofiber composites in catalysis and electrocatalysis looks promising, and continued research and development in this area will lead to new and improved applications for these materials.

16.5 Conclusion

In conclusion, nanofiber composites have a wide range of applications in catalysis and electrocatalysis due to their unique properties such as high surface area, tunable pore size, and the ability to incorporate different types of materials. The synergistic effect of nanofiber composites arises from the combination of different materials with complementary properties, which can improve the catalytic or electrocatalytic activity, stability, and selectivity of the composite. Nanofiber composites have shown promising applications in various fields such as environmental remediation, energy conversion, chemical synthesis, electrochemical energy conversion, carbon dioxide reduction, water splitting, and biomedical applications. Future research could focus on developing new nanofiber composites, optimizing synthesis methods, scaling up production, understanding the mechanisms of synergy, real-world testing, developing multifunctional nanofiber composites, integrating with renewable energy, artificial photosynthesis, nanofiber-based sensors, and biomedical applications. Overall, the future of nanofiber composites in catalysis and electrocatalysis looks promising, and continued research and development in this area will lead to new and improved applications for these materials. Looking to the future, further research could focus on developing new nanofiber composites, optimizing synthesis methods, scaling up production, and understanding the mechanisms of synergy. Additionally, real-world testing, developing multifunctional nanofiber composites, integrating with renewable energy, artificial photosynthesis, nanofiber-based sensors, and biomedical applications are areas where future research could bring significant advancements. Overall, the future of nanofiber composites in catalysis and electrocatalysis looks promising, and continued research and development in this area will lead to new and improved applications for these materials.

Reference

Ali, Z., Mehmood, M., Ahmed, J., & Nizam, M. N. (2020). Synthesis of graphitic nanofibers and carbon nanotubes by catalytic chemical vapor deposition method on nickel chloride alcogel for high oxygen evolution reaction activity in alkaline media. *Nano-Structures & Nano-Objects, 24*, 100574.

Berdimurodov, E., Kholikov, A., Akbarov, K., & Guo, L. (2021). Experimental and theoretical assessment of new and eco−friendly thioglycoluril derivative as an effective corrosion inhibitor of St2 steel in the aggressive hydrochloric acid with sulfate ions. *Journal of Molecular Liquids, 335*, 116168.

Bian, J., & Sun, C. (2021). NiCoFeP nanofibers as an efficient electrocatalyst for oxygen evolution reaction and zinc−air batteries. *Advanced Energy and Sustainability Research, 2*(6), 2000104.

Cao, X., Deng, J., & Pan, K. (2020). Electrospinning Janus type CoO_x/C nanofibers as electrocatalysts for oxygen reduction reaction. *Advanced Fiber Materials, 2*, 85−92.

Cao, X., Wang, T., & Jiao, L. (2021). Transition-metal (Fe, Co, and Ni)-based nanofiber electrocatalysts for water splitting. Advanced Fiber. *Materials, 1-19*, 33.

Dong, K., Liang, J., Wang, Y., Xu, Z., Liu, Q., Luo, Y., et al. (2021). Honeycomb carbon nanofibers: a superhydrophilic O_2-entrapping electrocatalyst enables ultrahigh mass activity for the two-electron oxygen reduction reaction. *Angewandte Chemie International Edition, 60*(19), 10583−10587.

El-Maghrabi, H. H., Nada, A. A., Bekheet, M. F., Roualdes, S., Riedel, W., Iatsunskyi, I., et al. (2021). Coaxial nanofibers of nickel/gadolinium oxide/nickel oxide as highly effective electrocatalysts for hydrogen evolution reaction. *Journal of Colloid and Interface Science, 587*, 457−466.

Gong, S., Hou, M., Niu, Y., Teng, X., Liu, X., Xu, M., et al. (2022). Molybdenum phosphide coupled with highly dispersed nickel confined in porous carbon nanofibers for enhanced photocatalytic CO_2 reduction. *Chemical Engineering Journal, 427*, 131717.

Gong, T., Qi, R., Liu, X., Li, H., & Zhang, Y. (2019). N, F-codoped microporous carbon nanofibers as efficient metal-free electrocatalysts for ORR. *Nano-Micro Letters, 11*, 1−11.

Guerrero-Pérez, M. O. (2022). Research Progress on the Applications of Electrospun Nanofibers in Catalysis. *Catalysts [Internet], 12*(1), 13.

Gui, L., Liu, Y., Zhang, J., He, B., Wang, Q., & Zhao, L. (2020a). In situ exsolved Co nanoparticles coupled on $LiCoO_2$ nanofibers to induce oxygen electrocatalysis for rechargeable Zn−air batteries. *Journal of Materials Chemistry A, 8*(38), 19946−19953.

Gui, L., Huang, Z., Ai, D., He, B., Zhou, W., Sun, J., et al. (2020b). Integrated ultrafine $Co_{0.85}Se$ in carbon nanofibers: an efficient and robust bifunctional catalyst for oxygen electrocatalysis. *Chemistry−A European Journal, 26*(18), 4063−4069.

Lei, Y., Wang, Q., Peng, S., Ramakrishna, S., Zhang, D., & Zhou, K. (2020). Electrospun inorganic nanofibers for oxygen electrocatalysis: design, fabrication, and progress. *Advanced Energy Materials, 10*(45), 1902115.

Li, C., Wu, M., & Liu, R. (2019). High-performance bifunctional oxygen electrocatalysts for zinc-air batteries over mesoporous Fe/Co-NC nanofibers with embedding FeCo alloy nanoparticles. *Applied Catalysis B: Environmental, 244*, 150−158.

Li, H., Sun, Y., Wang, J., Liu, Y., & Li, C. (2022). Nanoflower-branch LDHs and CoNi alloy derived from electrospun carbon nanofibers for efficient oxygen electrocatalysis in microbial fuel cells. *Applied Catalysis B: Environmental, 307*, 121136.

Li, S., Guo, H., He, S., Yang, H., Liu, K., Duan, G., et al. (2022). Advanced electrospun nanofibers as bifunctional electrocatalysts for flexible metal-air (O_2) batteries: Opportunities and challenges. *Materials & Design*, 110406.

Li, T., Ding, X., Tian, L., Hu, J., Yang, X., & Ramakrishna, S. (2017). The control of beads diameter of bead-on-string electrospun nanofibers and the corresponding release behaviors of embedded drugs. *Materials Science and Engineering: C, 74*, 471−477.

Li, T., Lu, T., Li, X., Xu, L., Zhang, Y., Tian, Z., et al. (2021). Atomically dispersed Mo sites anchored on multichannel carbon nanofibers toward superior electrocatalytic hydrogen evolution. *ACS nano, 15*(12), 20032−20041.

Li, T., Lu, T., Zhong, H., Xi, S., Zhang, M., Pang, H., et al. (2023). Atomically dispersed $V-O_2N_3$ sites with axial V—O coordination on multichannel carbon nanofibers achieving superior electrocatalytic oxygen evolution in acidic media. *Advanced Energy Materials, 13*(3), 2203274.

Lin, H., Xie, J., Zhang, Z., Wang, S., & Chen, D. (2021). Perovskite nanoparticles@ N-doped carbon nanofibers as robust and efficient oxygen electrocatalysts for Zn-air batteries. *Journal of Colloid and Interface Science, 581*, 374−384.

Liu, C., Zuo, P., Jin, Y., Zong, X., Li, D., & Xiong, Y. (2020). Defect-enriched carbon nanofibers encapsulating NiCo oxide for efficient oxygen electrocatalysis and rechargeable Zn-air batteries. *Journal of Power Sources, 473*, 228604.

Liu, M., He, Y., & Zhang, J. (2021). Co_3Fe_7 nanoparticles encapsulated in porous nitrogen-doped carbon nanofibers as bifunctional electrocatalysts for rechargeable zinc−air batteries. *Materials Chemistry Frontiers, 5*(17), 6559−6567.

Lu, Q., Wu, H., Zheng, X., Chen, Y., Rogach, A. L., Han, X., et al. (2021). Encapsulating cobalt nanoparticles in interconnected n-doped hollow carbon nanofibers with enriched Co—N—C moiety for enhanced oxygen electrocatalysis in Zn-Air Batteries. *Advanced Science, 8*(20), 2101438.

Nie, G., Zhang, Z., Wang, T., Wang, C., & Kou, Z. (2021). Electrospun one-dimensional electrocatalysts for oxygen reduction reaction: insights into structure−activity relationship. *ACS Applied Materials & Interfaces, 13*(32), 37961−37978.

Peng, W., Wang, Y., Yang, X., Mao, L., Jin, J., Yang, S., et al. (2020). Co9S8 nanoparticles embedded in multiple doped and electrospun hollow carbon nanofibers as bifunctional oxygen electrocatalysts for rechargeable zinc-air battery. *Applied Catalysis B: Environmental*, *268*, 118437.

Sankar, S. S., Karthick, K., Sangeetha, K., Karmakar, A., & Kundu, S. (2020). Polymeric nanofibers containing CoNi-based zeolitic imidazolate framework nanoparticles for electrocatalytic water oxidation. *ACS Applied Nano Materials*, *3*(5), 4274–4282.

Schossig, J., Gandotra, A., Arizapana, K., Weber, D., Wildy, M., Wei, W., et al. (2023). CO_2 to value-added chemicals: synthesis and performance of mono- and bimetallic nickel–cobalt nanofiber. *Catalysts. Catalysts [Internet]*, *13*(6), 17.

Tong, J., Li, Y., Bo, L., Li, W., Li, T., Zhang, Q., et al. (2019). CoP/N-doped carbon hollow spheres anchored on electrospinning core–shell N-doped carbon nanofibers as efficient electrocatalysts for water splitting. *ACS Sustainable Chemistry & Engineering*, *7*(20), 17432–17442.

Verma, D. K., Kazi, M., Alqahtani, M. S., Syed, R., Berdimurodov, E., Kaya, S., et al. (2021). N–hydroxybenzothioamide derivatives as green and efficient corrosion inhibitors for mild steel: Experimental, DFT and MC simulation approach. *Journal of Molecular Structure*, *1241*, 130648.

Wang, P., Zhang, X., Wei, Y., & Yang, P. (2019). Ni/NiO nanoparticles embedded inporous graphite nanofibers towards enhanced electrocatalytic performance. *International Journal of Hydrogen Energy*, *44*(36), 19792–19804.

Wang, X., Zhang, X., Fu, G., & Tang, Y. (2021). Recent progress of electrospun porous carbon-based nanofibers for oxygen electrocatalysis. *Materials Today Energy*, *22*, 100850.

Wang, Y., Zhao, M., Hou, C., Chen, W., Li, S., Ren, R., et al. (2021). Efficient degradation of perfluorooctanoic acid by solar photo-electro-Fenton like system fabricated by MOFs/carbon nanofibers composite membrane. *Chemical Engineering Journal*, *414*, 128940.

Wang, Z., Ang, J., Liu, J., Ma, X. Y. D., Kong, J., Zhang, Y., et al. (2020). FeNi alloys encapsulated in N-doped CNTs-tangled porous carbon fibers as highly efficient and durable bifunctional oxygen electrocatalyst for rechargeable zinc-air battery. *Applied Catalysis B: Environmental*, *263*, 118344.

Wang, Z. M., Liu, P., Cao, Y. P., Ye, F., Xu, C., & Du, X. Z. (2021). Characterization and electrocatalytic properties of electrospun Pt-IrO_2 nanofiber catalysts for oxygen evolution reaction. *International Journal of Energy Research*, *45*(4), 5841–5851.

Xiang, F., Zhao, X., Yang, J., Li, N., Gong, W., Liu, Y., et al. (2023). Enhanced Selectivity in the Electroproduction of H_2O_2 via F/S Dual-Doping in Metal-Free Nanofibers. *Advanced Materials*, *35*(7), 2208533.

Xie, Q., Wang, Z., Lin, L., Shu, Y., Zhang, J., Li, C., et al. (2021). Nanoscaled and atomic ruthenium electrocatalysts confined inside super-hydrophilic carbon nanofibers for efficient hydrogen evolution reaction. *Small (Weinheim an der Bergstrasse, Germany)*, *17*(38), 2102160.

Xie, X.-Q., Liu, J., Gu, C., Li, J., Zhao, Y., & Liu, C.-S. (2022). Hierarchical structured CoP nanosheets/carbon nanofibers bifunctional eletrocatalyst for high-efficient overall water splitting. *Journal of Energy Chemistry*, *64*, 503–510.

Yang, P., Zhao, H., Yang, Y., Zhao, P., Zhao, X., & Yang, L. (2020). Fabrication of N, P-codoped Mo_2C/carbon nanofibers via electrospinning as electrocatalyst for hydrogen evolution reaction. *ES Materials & Manufacturing*, *7*(6), 34–39.

Zhang, B., Wu, Z., Shao, W., Gao, Y., Wang, W., Ma, T., et al. (2022). Interfacial atom-substitution engineered transition-metal hydroxide nanofibers with high-valence Fe for efficient electrochemical water oxidation. *Angewandte Chemie International Edition*, *61*(13), e202115331.

Zhang, W., Chu, J., Li, S., Li, Y., & Li, L. (2020). CoN$_x$C active sites-rich three-dimensional porous carbon nanofibers network derived from bacterial cellulose and bimetal-ZIFs as efficient multifunctional electrocatalyst for rechargeable Zn−air batteries. *Journal of Energy Chemistry*, *51*, 323−332.

Zhang, X., & Wang, L. (2021). Research progress of carbon nanofiber-based precious-metal-free oxygen reaction catalysts synthesized by electrospinning for Zn-Air batteries. *Journal of Power Sources*, *507*, 230280.

Zhang, Y.-Q., Tao, H.-B., Chen, Z., Li, M., Sun, Y.-F., Hua, B., et al. (2019). In situ grown cobalt phosphide (CoP) on perovskite nanofibers as an optimized trifunctional electrocatalyst for Zn−air batteries and overall water splitting. *Journal of Materials Chemistry A*, *7*(46), 26607−26617.

Zhu, Y., Bu, L., Shao, Q., & Huang, X. (2020). Structurally ordered Pt3Sn nanofibers with highlighted antipoisoning property as efficient ethanol oxidation electrocatalysts. *ACS Catalysis*, *10*(5), 3455−3461.

Sensors application of nanofibers and their composites

17

Shveta Sharma[1], Richika Ganjoo[2], Elyor Berdimurodov[3,4], Alok Kumar[5] and Ashish Kumar[6]

[1]Department of Chemistry, Government College Una, Affiliated to Himachal Pradesh University, Una, Himachal Pradesh, India, [2]Department of Chemistry, School of Chemical Engineering and Physical Sciences, Lovely Professional University, Phagwara, Punjab, India, [3]Faculty of Chemistry, National University of Uzbekistan, Tashkent, Uzbekistan, [4]Akfa University, Tashkent, Uzbekistan, [5]Department of Mechanical Engineering, NCE, Bihar Engineering University, Department of Science, Technology and Technical Education, Nalanda, Bihar, India, [6]Department of Chemistry, NCE, Bihar Engineering University, Department of Science, Technology and Technical Education, Nalanda, Bihar, India

17.1 Introduction

As a result of advances in nanotechnology and material science, a wide variety of zero-dimensional to three-dimensional (3D) materials have been developed specifically for use in sensor applications. One-dimensional (1D) nanomaterials (NMs) showed great promise for the future, as it is common knowledge that 1D structures have the potential to offer shorter routes for the transmission of electrons and to promote the penetration of electrolyte along the axis of longitudinal NFs and nanowires, leading to an overall improvement in the sensing performance (Hahm, 2016; Li et al., 2011; Wang et al., 2006, 2017, 2018, 2019; Xia et al., 2018; Zhao et al., 2017). There has been a significant amount of development in the realm of chemical and biological sensors as a direct result of the urgent need for more affordable, user-friendly, and accurate detection methods (Aussawasathien et al., 2005).

Researchers have already worked on a number of different materials, which can be used for sensing. Depending on the need, the materials are used in form of NMs, for example, carbon nanotubes, also known as carbon nanotubes (CNTs), are one of the most common types of 1D NMs. Due to their one-of-a-kind mechanical, electrical, and magnetic capabilities, CNTs have been employed in the past to fabricate a variety of high-performance sensors and biosensors. CNTs are particularly attractive candidates for the fabrication of chemical and biological sensors with high sensitivity and selectivity because they have a large surface area and a strong adsorption ability toward diverse chemicals and biomolecules (Barsan et al., 2015; Meyyappan, 2016; Wang et al., 2019). In recent years, determining the existence of heavy metal ions and the quantity of those ions has garnered steadily rising degrees of attention as a result of developing concerns about human and environmental health (Amendola et al., 2006; Rurack & Resch-Genger, 2002). Therefore, it is very necessary to construct sensitive and selective fluorescent chromogenic probes made of chelating ligands in order to detect heavy transition metal (HTM) cations in biological and environmental sensing devices. Various derivatives of colorimetric with 1,8-naphthalimide in it (as they were stable in light and were giving high yield) were successfully used in fluorescent HTM detection (Liu & Tian, 2005; Wang et al., 2013; Zou & Tian, 2010). Researchers have also used metal oxide NFs for the detection of toxic gases. Semiconductor metal oxide gas sensors have garnered a lot of interest recently owing to the many benefits that come with using them, including their quick reaction time, cheap cost, straightforward construction, and high compatibility. Because of their great performance, low cost, and good gas-sensing capabilities, semiconductor metal oxide materials are employed extensively in a broad variety of applications. Conducting polymers have also been used as they have π electrons, which makes them good conductors of electricity and very sensitive over a large range. Conducting polymers and their composites provide a larger usefulness in the application of strain and pressure sensors, particularly in providing a better figure of merits, such as enhanced responsiveness, detecting span, resilience, and robustness (Veeramuthu et al., 2021). There have been ongoing efforts and attempts committed to the development of extremely efficient optoelectronic devices, and significant breakthroughs have been produced that effectively demonstrate how to utilize the durability of polymer composites (Huang et al., 2019). Also, as a result of developments in a variety of disciplines, including various processes of determining which disease or condition explains a person's symptoms and signs and environmental protection, there is now a greater need for analytical instruments that are able to facilitate the identification and monitoring of certain chemicals. This requirement may be satisfied by sensors and biosensors, which allow for very selective detection with high sensitivity of numerous molecules in the samples that are analyzed. Researchers have taken a particular interest in nanotechnology, which has led to the creation of sensors that are more sensitive and have improved performance. NMs, which also include NFs, have features that are not obtained in bulk materials. As a result, they have proven to be an intriguing and desirable supplement to traditional methods of sensor construction. The use of NMs paves the way for the modification of conventional methods, which ultimately results in an increased number of substances loaded per unit weight, decreased sample size, and the potential for the formation of small designs (Agasti et al., 2010).

Monitoring of chemicals in the fields of diagnostics and therapeutics is essential for the avoidance, detection, and treatment of illnesses. These compounds include hormones, medicines, and protein markers. The glucometer, which is a sensor that detects the concentration of glucose in biological fluids (often blood), continues to be one of the most widely used biosensors in diagnostics due to the fact that it is necessary for those who are afflicted with diabetes (Chen et al., 2018). Even though metals are present in everyday life and in the business world, and are crucial for many cellular processes, such as the activity of enzymes, some of them also pose a risk to the environment and health. Because of this, sensors that are able to detect trace amounts of hazardous heavy metal ions are required (Chowdhury et al., 2018).

17.2 Nanofibers in sensors

Wang et al. (2019) reviewed various past research to identify the usage of CNTs in the formation of sensors. In addition to CNTs, carbon NFs (CNFs) have also received a significant amount of attention owing to the fact that they possess exceptional chemical and physical characteristics and have a structure that is comparable to that of CNTs and fullerenes (Chen et al., 2017; Ning et al., 2016). CNFs have exclusively both graphite structures consisting of two-dimensional conjugated sp2 carbon atoms at both the planes, exhibiting significant possibilities of alterations on the top or integration to create CNF-based NMs, which were then used in various fields such as sensors (Li et al., 2015; Magana et al., 2016). CNFs have a large total surface area of material per unit of mass, a limited number of imperfections, a high width-to-hight ratio, low density, high elastic modulus per mass density of material and strength, high conductivity, etc. As a result, they have extensive scope to be applied in various sectors such as energy storage and sensing. This is because CNFs have these characteristics (Al-Saleh & Sundararaj, 2011; Chen et al., 2017; Llobet, 2013; Tiwari et al., 2016). Nowdays, the most common approaches to the production of CNFs are mentioned in Fig. 17.1

In recent years, as a result of the fast development of technology for nanofabrication, an increasing number of NMs formed from carbon is employed in sensing applications for identifying and detecting a variety of materials. Pure CNFs, CNFs loaded with metal NPs, CNFs loaded with metal oxides, CNFs loaded with metal alloys, and other forms of CNFs loaded with other sorts of loaded materials are the five categories of carbon-based NFs that may be utilized as sensors (Chung, 2012; Marín-Barroso et al., 2019; Matlock-Colangelo & Baeumner, 2012). CNFs are used in various types of sensors such as gas sensors, pressure sensors, detection of small molecules, and sensors for bio-molecules. For example, 1D CNFs made of graphitic

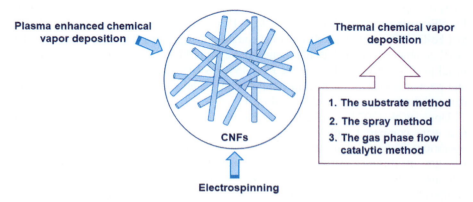

Figure 17.1 Various ways for synthesizing carbon nanofibers.
Source: From Wang, Zhuqing, Wu, Shasha, Wang, Jian, Yu, Along & Wei, Gang. (2019). Carbon nanofiber-based functional nanomaterials for sensor applications. *Nanomaterials*, 9 (7). https://doi.org/10.3390/nano9071045.

nano rolls were prepared by Li et al. (2012). The graphitic CNFs as prepared exhibit sensitivity to various gases, such as hydrogen, methane, and ethanol gases, at room temperature. Zhu et al. (2011) used a polymer framework as a host to manufacturing polymeric nanocomposites with CNFs in it and thus prepared sensor was utilized as the strain sensor. CNFs can also be used for detecting small molecules. For example, Liu et al. worked on a sensor made up of Ni and CNF and used it successfully for glucose detection (Liu et al., 2009). CNFs are able to offer direct electron transfer and stabilize enzyme activity. In addition to having an enormous surface area and many different active areas, which may be the grounds for the adsorption of proteins and enzymes, CNFs also have the potential to adsorb proteins and enzymes. This is because CNFs can provide the grounds for the adsorption of proteins and enzymes and, therefore, can be used for preparing the biosensors (Arkan et al., 2017; Eissa et al., 2018; Rizwan et al., 2018; Zhang et al., 2019). For example, Gupta et al. used CNFs to prepare a nanoelectrode arrangement and confirmed that the concentration of the C-reactive protein was proportional to a reduction in current and an increased amount of resistance (Gupta et al., 2014).

Liang et al. (2017) prepared chemosensors based on fluorescent electrospun (ES) NFs with sensitivity toward magnetism, temperature, and mercury ions. The reagents used included poly(N-isopropylacrylamide)-co-(N-methylolacrylamide)-co-(acrylic acid) (poly (NIPAAm-co-NMA-co-AA)), fluorescent 1-benzoyl-3-[2-(2-allyl-1,3-dioxo-2,3-dihydro-1Hbenzo[de]isoquinolin-6-ylamino)- ethyl]-thiourea (BNPTU), Fe_3O_4 NPs, and a single-capillary spinneret. First, the fluorescent BNPTU was synthesized. Then, as shown in Fig. 17.2A, the polymer was synthesized. In Fig. 17.2B, the NFs were subjected to a chemical treatment to boost the stability of the compounds in water. Fluorescence emission from BNPTU is very selective and depends on the presence of Hg^{2+}. When BNPTU is used to detect Hg^{2+}, the color of its fluorescence emission shifts from green to blue, and the thermos-responsive magnetic characteristics of NIPAAm and Fe_3O_4 NPs, respectively, are shown to be present (Fig. 17.2C). Two types of copolymers, P1 and P2, were synthesized. By employing both the copolymers, ES NFs were prepared. After that, Fe_3O_4 NPs were also prepared, and finally, P1, P2, and Fe_3O_4 NPs were blended. The final composites were used for sensing purposes. P1 or P2 combined with 10% BNPTU and 5% Fe_3O_4 NP ES NFs, which are able to detect metal ions. Results revealed that the P2%−5% ES NFs displayed substantial blue shifts and were able to detect an extremely dilute concentration of Hg^{2+}. Temperature dependency and magnetism were also confirmed.

Dong et al. (2022) synthesized $ZnSnO_3$ NFs and $ZnSnO_3$/ZnO composite NFs and were compared with each other for their gas sensitivity and temperature of operation. To know about the gas sensitization, both the NFs were studied for ethanol gas at different temperatures demonstrated in Fig. 17.3. $ZnSnO_3$ gave a response factor of 13.4, which increased with the increased temperature. For the composite, the response factor was 19.6 at 225°C. Composite was also showing a better response to ethanol sensing.

Mohamad et al. (2017) manufactured chemically modified polyaniline and graphene composite NFs by using aniline monomer and graphene in the presence of a solution containing poly(methyl vinyl ether-alt-maleic acid) (PMVEA). This resulted

Sensors application of nanofibers and their composites 427

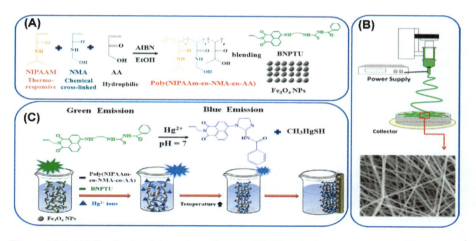

Figure 17.2 (A) Synthesis of the polymeric unit. (B) Chemical treatment to nanofibers (C) detection of Hg^{2+}.
Source: From Liang, F. C., Luo, Y. L., Kuo, C. C., Chen, B. Y., Cho, C. J., Lin, F. J., Yu, Y. Y., & Borsali, R. (2017). Novel magnet and thermoresponsive chemosensory electrospinning fluorescent nanofibers and their sensing capability for metal ions. *Polymers*, *9*(4). https://doi.org/10.3390/polym9040136.

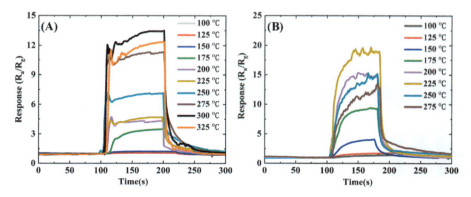

Figure 17.3 Effect of temperature on $ZnSnO_3$ NFs and $ZnSnO_3/ZnO$ composite.
Source: From Dong, S., Jin, X., Wei, J., & Wu, H. (2022). Electrospun $ZnSnO_3/ZnO$ composite nanofibers and its ethanol-sensitive properties. *Metals*, *12*(2). https://doi.org/10.3390/met12020196.

in the production of polyaniline (PANI)/graphene (GP) NFs. PANI/GP NFs were then utilized as DNA sensors. The identification of *Mycobacterium tuberculosis* was accomplished by using a particular sequence of DNA, which has been shown to be reliable and accurate (Wang et al., 2009). The incorporation of PANI into GP framework NFs not only improved the fixation of the DNA probe but also increased the signal intensification. First of all, the PANI/GP NFs were prepared, and after that, cyclic voltammetry was used to manufacture the sensor (Fig. 17.4). Surface morphology was studied by using a scanning electron microscope (SEM) of PANI NFs containing an amount of PMVEA. With an increase in PMVEA, there was a corresponding rise in the NFs diameter, and also extra stability was provided to PANI NFs. The results showed a strong electrochemical activity of PANI/GP composite NFs owing to their high sensitivity and specificity with synthesized DNA. The suggested DNA biosensor has demonstrated tremendous promise for the detection of Mycobacterium tuberculosis.

Wolf et al. (2015) prepared optical chemical sensors by employing NFs. First, the NFs were prepared by electrospinning. It was shown that the reaction time of opto-chemical sensors may be significantly improved by two times when the actual sensing layers are realized as a highly-porous 3D membrane composed of ES polymer NFs rather than a compact polymer film. This is because ES polymer NFs have a greater surface area per unit volume than compact polymer films do. However, performance attributes of the sensors, such as their pinpointing, exactness, and responsiveness, remain maintained throughout the process. Electrospinning is a reasonably simple, dependable, and scalable method for the processing of polymer solutions into NFs. This method also requires relatively inexpensive equipment. Because these nanostructured sensors have the same optical characteristics as their predecessors, they are capable of performing as an upgraded substitute in the same kinds of typical optoelectronic sensor setups. Chinnappan et al. (2017) reviewed

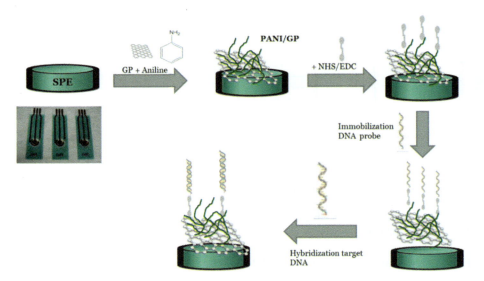

Figure 17.4 Manufacturing of DNA biosensors.
Source: From Mohamad, F., Mat Zaid, M., Abdullah, J., Zawawi, R., Lim, H., Sulaiman, Y., & Abdul Rahman, N. (2017). Synthesis and characterization of polyaniline/graphene composite nanofiber and its application as an electrochemical DNA biosensor for the detection of mycobacterium tuberculosis. *Sensors*, *17*(12). https://doi.org/10.3390/s17122789.

various past research focusing on the latest breakthroughs in the area of smart and electronic textiles, and it pays particular attention to the ES NF-based materials employed in wearable and flexible electronics, sensors, and energy storage. The vast majority of sensors that make use of ES fibers are constructed in the form of a nonwoven mesh, which does not need any adjustments to the conventional arrangement. When compared to cast film sensors made of the same material, the performance of sensors built from NFs is often superior in terms of sensitivity. Targeted applications and sensing approaches are often dependent, not on the shape, but on the material that is chosen. In order to build a sensor, there has to be a shift in the properties or traits of the material when it is exposed to the target, and there also needs to be some mechanism to assess the change in those characteristics or qualities (Castano & Flatau, 2016; Khoshaman, 2011). In comparison to traditional electrical sensors, polymer-based textile sensors provide a number of significant benefits. They can conform to any surface or shape thanks to the compliant substrates, which also make them portable in the event that any kind of close-range monitoring is implemented (Gao et al., 2010). Direct writing of poly(vinylidene fluoride) NFs onto a prestrained polydimethylsiloxane substrate was demonstrated by Duan and coworkers (Ding et al., 2015; Duan et al., 2014) using a technique called kinetically controlled mechano-electrospinning (MES). Results revealed that electrospinning is a long-standing technology that has been documented in published works for more than seventy years; nowadays, it is the technique that is most suited for the production of construction of continuous NFs. Although electrospinning is a flexible and significant process for synthesizing 1D nanostructures, there are still several challenges that need to be solved before these materials can fulfill their promise in terms of industrialization and practical applications. The manufacturing rate is limited, and it has proved challenging to create consistent NFs with diameters of less than 50 nm. In addition, the recent developments in electrospinning methods led to the conclusion that it is an effective instrument. This is shown by the fact that these techniques are making progress. It will be an extremely important contributor to the growth of the wearable and flexible electronics industry. Talwar et al. (2014) tried to formulate polyaniline NFs with different amounts of Zinc oxide (ZnO), and various analyses were performed by employing spectrochemical analysis and field emission scanning electron microscopy (FESEM) technique. Thereafter, the gas sensor was formed, by using polyaniline powder and m-cresol. X-ray diffraction (XRD) study of ZnO and different samples of polyaniline with varying concentrations of ZnO were analyzed, and it was found that the crystalline quality of PANI powder was greatly improved when ZnO nanoparticles were added to it during the polymerization process. The ultra violet (UV)/visible spectra showed that the characteristic peaks were there in the ZnO sample, but when added to the PANI sample, there was no characteristic peak. FESEM was used to conduct structural analysis on ZnO, PANI, and PANI synthesized with varying amounts of ZnO nanoparticles. The analysis was done on all three materials. The FESEM analysis of commercial ZnO powder revealed that they consisted of nanoparticles of ZnO. PANI NFs were generated when ZnO nanoparticles were added to the polyaniline sample (S0), which exhibited a conglomerated nanostructure of PANI. Results revealed that because the doping of PANI may be regulated by acid and base reactions, it is particularly effective in the detection of a wide variety of gases. After being exposed to ammonia, polyaniline gas sensors undergo deprotonation, which results in a shift in conductivity from the conductive emeraldine salt form to the insulating emeraldine base form (Yin & Yang, 2011). The morphology of the synthesized samples provides insight into the rapid reaction time of the polyaniline NFs that were used in the study. Polyaniline NFs have a rapid reaction time because of their vast surface area and their very tiny diameters. This is because diffusion into cylinders with such a small diameter occurs very quickly. Different PANI samples were studied for ammonia gas sensing, and it was confirmed that sensing properties were increased with the increased amount of ZnO. PANI NFs with 30% ZnO showed better results. The sensor response became less sensitive when the amount of ZnO powder used in the synthesis was increased further. The varying shape of the NFs and, therefore, their surface areas are to blame for the inconsistent responses produced by the sensors. Wang et al. (2006) investigated polyvinyl acetate-tungsten oxide NFs and tungsten oxide ceramic NFs. The NFs were further characterized by SEM and transmission electron microscopy (TEM), and their elements were analyzed by XRD. The study showed that during the calcination process, the temperature had a significant impact on both the shape and structure of ceramic tungsten oxide NFs. The sensing performance of the material may be strongly impacted by the material microstructure, which can be modified by the preparation processes. At an operating temperature of 350°C, pure WO_3 NFs exhibited a very rapid reaction toward NH_3 at very low concentrations. The orthorhombic phase WO_3 NFs showed a high sensitivity to ammonia across a broad range of concentrations, which suggests that the NFs may have potential uses as a sensor material in the sectors of homeland security, environmental protection, and health care.

17.3 Future outlook

Out of various techniques, electrospinning is an age-old technology that has been documented in published works for more than seventy years; nowadays, it is the technique that is most suited for the production of construction of NFs that are continuous. Electrospinning is a flexible and significant process for synthesizing 1D nanostructures; nevertheless, before these materials can fulfill their potential for industrialization and practical applications, there are still several difficulties that need to be solved. It has been challenging to generate consistent NFs with diameters of less than 50 nm, and the manufacturing pace has been slow. In the not-too-distant future, each of the difficulties that are connected with this technology will need to be resolved via the process of research, discovery, and development of the electrospinning method in conjunction with other production techniques. Although there has been a lot of research done over the last several years on forming and using CNFs and materials made up of CNFs, we believe that additional work may be done on the following elements of the topic. To begin, fresh approaches to the CNF synthesis process might be explored. Further investigation might be put into modifying CNFs to be used in biosensors, materials that inhibit the growth of bacteria, bone tissue engineering, and other similar endeavors. In the future, however, the most significant obstacle that the innovative sensors will need to overcome is the possibility of achieving economic viability. There has not been much work that has effectively transitioned from the proof-of-principle scale to the commercialization level. Although NFs may be made in vast quantities, the manufacturing process for the sensors themselves is often difficult and involves a number of stages. In addition, the stability of such systems over the long term has, for the most part, not been characterized. The final aim for such sensors has not yet been reached, which is to build sensors that are not only cost-effective but also sensitive and selective.

17.4 Conclusion

NM-based sensors are gaining popularity nowadays. When NFs are used, the qualities of many different things have changed. The fact that the characteristics of materials on the micro and nano sizes are distinct from those of bulk materials is one of the primary arguments in favor of the use of NF-based systems. Commonly used NFs include carbon nanofibers (CNFs), CNTs, natural and synthetic polymers, and inorganic materials. Further, the diagnosis and treatment of disease have been and will continue to be the primary focus of healthcare for all individuals, and NFs are becoming more applicable in a wide range of research fields, such as immune sensing and disease detection applications. Because of NFs, it is possible to expand the detecting area of sensors, which, in many cases, leads to increased selectivity or lower detection limits for analytes. NFs are making great strides in disease monitoring and diagnosis, particularly in the cases of malaria and cancer. Based on previous research on CNF-based functional NMs, it is possible to draw the conclusion that CNFs serve vital roles in the creation of diverse sensors, which can work in different fields such as gas detection, strain measurement, and small and large molecule detection. The CNFs, regardless of the sort of porosity they contained, were able to increase the electrode materials' expanse of surface. The sensing ability was significantly improved by the modification of CNFs with a variety of NPs, polymers, and biomolecules. The sensing sensitivity of the fabricated sensors was increased by the interactions that were there in the fibers and solvents used. Because of their great mechanical strength and chemical inertness, CNF-based sensors often demonstrate good stability and selectivity to target molecules. This is owing to the fact that CNFs are made of carbon NFs. Optical sensors made up of polymer NFs gave two times better sensing as compared to plane polymeric sensors. The performance of the polymer NF sensors was very pinpointing and exact, and maintained responsiveness remained throughout the process.

References

Agasti, S. S., Rana, S., Park, M. H., Kim, C. K., You, C. C., & Rotello, V. M. (2010). Nanoparticles for detection and diagnosis. *Advanced Drug Delivery Reviews*, *62*(3), 316−328. Available from https://doi.org/10.1016/j.addr.2009.11.004.

Al-Saleh, M. H., & Sundararaj, U. (2011). Review of the mechanical properties of carbon nanofiber/polymer composites. *Composites Part A: Applied Science and Manufacturing*, *42*(12), 2126−2142. Available from https://doi.org/10.1016/j.compositesa.2011.08.005.

Amendola, V., Fabbrizzi, L., Foti, F., Licchelli, M., Mangano, C., Pallavicini, P., Poggi, A., Sacchi, D., & Taglietti, A. (2006). Light-emitting molecular devices based on transition metals. *Coordination Chemistry Reviews*, *250*(3−4), 273−299. Available from https://doi.org/10.1016/j.ccr.2005.04.022.

Arkan, E., Paimard, G., & Moradi, K. (2017). A novel electrochemical sensor based on electrospun TiO_2 nanoparticles/carbon nanofibers for determination of Idarubicin in biological samples. *Journal of Electroanalytical Chemistry*, *801*, 480−487. Available from https://doi.org/10.1016/j.jelechem.2017.08.034.

Aussawasathien, D., Dong, J.-H., & Dai, L. (2005). Electrospun polymer nanofiber sensors. *Synthetic Metals*, *154*(1−3), 37−40. Available from https://doi.org/10.1016/j.synthmet.2005.07.018.

Barsan, M. M., Ghica, M. E., & Brett, C. M. A. (2015). Electrochemical sensors and biosensors based on redox polymer/carbon nanotube modified electrodes: A review. *Analytica Chimica Acta*, *881*, 1−23. Available from https://doi.org/10.1016/j.aca.2015.02.059, http://www.journals.elsevier.com/analytica-chimica-acta/.

Castano, L. M., & Flatau, A. B. (2016). Smart textile transducers: Design, techniques, and applications. *Industrial applications for intelligent polymers and coatings* (pp. 121−146). United States: Springer International Publishing. Available from http://doi.org/10.1007/978-3-319-26893-4, https://doi.org/10.1007/978-3-319-26893-4_6.

Chen, L., Hwang, E., & Zhang, J. (2018). Fluorescent nanobiosensors for sensing glucose. *Sensors*, *18*(5). Available from https://doi.org/10.3390/s18051440.

Chen, L. F., Lu, Y., Yu, L., & Lou, X. W. (2017). Designed formation of hollow particle-based nitrogen-doped carbon nanofibers for high-performance supercapacitors. *Energy and Environmental Science*, *10*(8), 1777−1783. Available from https://doi.org/10.1039/c7ee00488e, http://www.rsc.org/Publishing/Journals/EE/About.asp.

Chinnappan, A., Baskar, C., Baskar, S., Ratheesh, G., & Ramakrishna, S. (2017). An overview of electrospun nanofibers and their application in energy storage, sensors and wearable/flexible electronics. *Journal of Materials Chemistry C*, *5*(48), 12657−12673. Available from https://doi.org/10.1039/c7tc03058d.

Chowdhury, S., Rooj, B., Dutta, A., & Mandal, U. (2018). Review on recent advances in metal ions sensing using different fluorescent probes. *Journal of Fluorescence*, *28*(4), 999−1021. Available from https://doi.org/10.1007/s10895-018-2263-y.

Chung, D. D. L. (2012). Carbon materials for structural self-sensing, electromagnetic shielding and thermal interfacing. *Carbon*, *50*(9), 3342−3353. Available from https://doi.org/10.1016/j.carbon.2012.01.031.

Ding, Y., Duan, Y., & Huang, Y. A. (2015). Electrohydrodynamically printed, flexible energy harvester using in situ poled piezoelectric nanofibers. *Energy Technology*, *3*(4), 351−358. Available from https://doi.org/10.1002/ente.201402148.

Dong, S., Jin, X., Wei, J., & Wu, H. (2022). Electrospun $ZnSnO_3$/ZnO composite nanofibers and its ethanol-sensitive properties. *Metals*, *12*(2). Available from https://doi.org/10.3390/met12020196.

Duan, Y., Huang, Y., Yin, Z., Bu, N., & Dong, W. (2014). Non-wrinkled, highly stretchable piezoelectric devices by electrohydrodynamic direct-writing. *Nanoscale*, *6*(6), 3289−3295. Available from https://doi.org/10.1039/c3nr06007a, http://www.rsc.org/publishing/journals/NR/Index.asp.

Eissa, S., Alshehri, N., Abduljabbar, M., Rahman, A. M. A., Dasouki, M., Nizami, I. Y., Al-Muhaizea, M. A., & Zourob, M. (2018). Carbon nanofiber-based multiplexed immunosensor for the detection of survival motor neuron 1, cystic fibrosis transmembrane conductance regulator and Duchenne Muscular Dystrophy proteins. *Biosensors and Bioelectronics*, *117*, 84−90. Available from https://doi.org/10.1016/j.bios.2018.05.048, http://www.elsevier.com/locate/bios.

Gao, Y., Zheng, Y., Diao, S., Toh, W. D., Ang, C. W., Je, M., & Heng, C. H. (2010). Low-power ultrawideband wireless telemetry transceiver for medical sensor applications. *IEEE Transactions on Biomedical Engineering*, *58*(3), 768−772.

Gupta, R. K., Periyakaruppan, A., Meyyappan, M., & Koehne, J. E. (2014). Label-free detection of C-reactive protein using a carbon nanofiber based biosensor. *Biosensors and Bioelectronics*, *59*, 112−119. Available from https://doi.org/10.1016/j.bios.2014.03.027, http://www.elsevier.com/locate/bios.

Hahm, J. I. (2016). Fundamental properties of one-dimensional zinc oxide nanomaterials and implementations in various detection modes of enhanced biosensing. *Annual Review of Physical Chemistry*, *67*, 691−717. Available from https://doi.org/10.1146/annurev-physchem-031215-010949, http://www.annualreviews.org/journal/physchem.

Huang, M. Y., Veeramuthu, L., Kuo, C. C., Liao, Y. C., Jiang, D. H., Liang, F. C., Yan, Z. L., Borsali, R., & Chueh, C. C. (2019). Improving performance of Cs-based perovskite light-emitting diodes by dual additives consisting of polar polymer and n-type small molecule. *Organic Electronics*, *67*, 294−301. Available from https://doi.org/10.1016/j.orgel.2018.12.042, http://www.elsevier.com/locate/orgel.

Khoshaman, A.H. (2011). Application of electrospun thin films for supra-molecule based gas sensing.

Li, W., Zhang, F., Dou, Y., Wu, Z., Liu, H., Qian, X., Gu, D., Xia, Y., Tu, B., & Zhao, D. (2011). A self-template strategy for the synthesis of mesoporous carbon nanofibers as advanced supercapacitor electrodes. *Advanced Energy Materials*, *1*(3), 382−386. Available from https://doi.org/10.1002/aenm.201000096.

Li, W., Zhang, L. S., Wang, Q., Yu, Y., Chen, Z., Cao, C. Y., & Song, W. G. (2012). Low-cost synthesis of graphitic carbon nanofibers as excellent room temperature sensors for explosive gases. *Journal of Materials Chemistry*, *22*(30), 15342−15347. Available from https://doi.org/10.1039/c2jm32031b, http://www.rsc.org/Publishing/Journals/jm/index.asp.

Li, Y., Zhang, M., Zhang, X., Xie, G., Su, Z., & Wei, G. (2015). Nanoporous carbon nanofibers decorated with platinum nanoparticles for non-enzymatic electrochemical sensing of H_2O_2. *Nanomaterials*, *5*(4), 1891−1905. Available from https://doi.org/10.3390/nano5041891.

Liang, F. C., Luo, Y. L., Kuo, C. C., Chen, B. Y., Cho, C. J., Lin, F. J., Yu, Y. Y., & Borsali, R. (2017). Novel magnet and thermoresponsive chemosensory electrospinning fluorescent nanofibers and their sensing capability for metal ions. *Polymers*, *9*(4). Available from https://doi.org/10.3390/polym9040136, http://www.mdpi.com/2073-4360/9/4/136/pdf.

Liu, B., & Tian, H. (2005). A selective fluorescent ratiometric chemodosimeter for mercury ion. *Chemical Communications*, *25*, 3156−3158. Available from https://doi.org/10.1039/b501913c, http://pubs.rsc.org/en/journals/journal/cc.

Liu, Y., Teng, H., Hou, H., & You, T. (2009). Nonenzymatic glucose sensor based on renewable electrospun Ni nanoparticle-loaded carbon nanofiber paste electrode. *Biosensors and Bioelectronics*, *24*(11), 3329−3334. Available from https://doi.org/10.1016/j.bios.2009.04.032.

Llobet, E. (2013). Gas sensors using carbon nanomaterials: A review. *Sensors and Actuators B: Chemical*, *179*, 32−45. Available from https://doi.org/10.1016/j.snb.2012.11.014.

Magana, J. R., Kolen'Ko, Y. V., Deepak, F. L., Solans, C., Shrestha, R. G., Hill, J. P., Ariga, K., Shrestha, L. K., & Rodriguez-Abreu, C. (2016). From chromonic self-assembly to hollow carbon nanofibers: efficient materials in supercapacitor and vapor-sensing applications. *ACS Applied Materials and Interfaces*, *8*(45), 31231−31238. Available from https://doi.org/10.1021/acsami.6b09819, http://pubs.acs.org/journal/aamick.

Marín-Barroso, E., Messina, G. A., Bertolino, F. A., Raba, J., & Pereira, S. V. (2019). Electrochemical immunosensor modified with carbon nanofibers coupled to a paper platform for the determination of gliadins in food samples. *Analytical Methods*, *11*(16), 2170−2178. Available from https://doi.org/10.1039/c9ay00255c, http://pubs.rsc.org/en/journals/journal/ay.

Matlock-Colangelo, L., & Baeumner, A. J. (2012). Recent progress in the design of nanofiber-based biosensing devices. *Lab on a Chip*, *12*(15), 2612−2620. Available from https://doi.org/10.1039/c2lc21240d, http://pubs.rsc.org/en/journals/journal/lc.

Meyyappan, M. (2016). Carbon nanotube-based chemical sensors. *Small (Weinheim an der Bergstrasse, Germany)*, *12*(16), 2118−2129. Available from https://doi.org/10.1002/smll.201502555.

Mohamad, F., Zaid, M. M., Abdullah, J., Zawawi, R., Lim, H., Sulaiman, Y., & Rahman, N. A. (2017). Synthesis and characterization of polyaniline/graphene composite nanofiber and its application as an electrochemical DNA biosensor for the detection of mycobacterium tuberculosis. *Sensors*, *17*(12). Available from https://doi.org/10.3390/s17122789.

Ning, P., Duan, X., Ju, X., Lin, X., Tong, X., Pan, X., Wang, T., & Li, Q. (2016). Facile synthesis of carbon nanofibers/MnO_2 nanosheets as high-performance electrodes for asymmetric supercapacitors. *Electrochimica Acta*, *210*, 754−761. Available from https://doi.org/10.1016/j.electacta.2016.05.214.

Rizwan, M., Koh, D., Booth, M. A., & Ahmed, M. U. (2018). Combining a gold nanoparticle-polyethylene glycol nanocomposite and carbon nanofiber electrodes to develop a highly sensitive salivary secretory immunoglobulin A immunosensor. *Sensors and Actuators, B: Chemical*, *255*, 557−563. Available from https://doi.org/10.1016/j.snb.2017.08.079.

Rurack, K., & Resch-Genger, U. (2002). Rigidization, preorientation and electronic decoupling—The 'magic triangle' for the design of highly efficient fluorescent sensors and switches. *Chemical Society Reviews*, *31*(2), 116−127. Available from https://doi.org/10.1039/b100604p.

Talwar, V., Singh, O., & Singh, R. C. (2014). ZnO assisted polyaniline nanofibers and its application as ammonia gas sensor. *Sensors and Actuators, B: Chemical*, *191*, 276−282. Available from https://doi.org/10.1016/j.snb.2013.09.106.

Tiwari, J. N., Vij, V., Kemp, K. C., & Kim, K. S. (2016). Engineered carbon-nanomaterial-based electrochemical sensors for biomolecules. *ACS Nano*, *10*(1), 46−80. Available from https://doi.org/10.1021/acsnano.5b05690, http://pubs.acs.org/journal/ancac3.

Veeramuthu, L., Venkatesan, M., Benas, J. S., Cho, C. J., Lee, C. C., Lieu, F. K., Lin, J. H., Lee, R. H., & Kuo, C. C. (2021). Recent progress in conducting polymer composite/nanofiber-based strain and pressure sensors. *Polymers*, *13*(24). Available from https://doi.org/10.3390/polym13244281, https://www.mdpi.com/2073-4360/13/24/4281/pdf.

Wang, Y., Ji, X., Huang, X., Yang, P. I., & Gouma, M. (2006). Dudley Fabrication and characterization of polycrystalline WO_3 nanofibers and their application for ammonia sensing. *Journal of Physical Chemistry B*, *110*(47), 23777−23782. Available from https://doi.org/10.1021/jp0635819.

Wang, G., Ji, Y., Huang, X., Yang, X., Gouma, P. I., & Dudley, M. (2006). Fabrication and characterization of polycrystalline WO_3 nanofibers and their application for ammonia sensing. *Journal of Physical Chemistry B*, *110*(47), 23777−23782. Available from https://doi.org/10.1021/jp0635819, http://pubs.acs.org/journal/jpcbfk.

Wang, H., Hao, Q., Yang, X., Lu, L., & Wang, X. (2009). Graphene oxide doped polyaniline for supercapacitors. *Electrochemistry Communications*, *11*(6), 1158−1161. Available from https://doi.org/10.1016/j.elecom.2009.03.036.

Wang, L., Wu, A., & Wei, G. (2018). Graphene-based aptasensors: From molecule-interface interactions to sensor design and biomedical diagnostics. *Analyst*, *143*(7), 1526−1543. Available from https://doi.org/10.1039/c8an00081f, http://pubs.rsc.org/en/journals/journal/an.

Wang, L., Zhang, Y., Wu, A., & Wei, G. (2017). Designed graphene-peptide nanocomposites for biosensor applications: A review. *Analytica Chimica Acta*, *985*, 24−40. Available from https://doi.org/10.1016/j.aca.2017.06.054.

Wang, W., Wen, Q., Zhang, Y., Fei, X., Li, Y., Yang, Q., & Xu, X. (2013). Simple naphthalimide-based fluorescent sensor for highly sensitive and selective detection of Cd^{2+} and Cu^{2+} in aqueous solution and living cells. *Dalton Transactions*, *42*(5), 1827−1833. Available from https://doi.org/10.1039/c2dt32279j.

Wang, Z., Wu, S., Wang, J., Yu, A., & Wei, G. (2019). Carbon nanofiber-based functional nanomaterials for sensor applications. *Nanomaterials*, *9*(7). Available from https://doi.org/10.3390/nano9071045.

Wolf, C., Tscherner, M., & Köstler, S. (2015). Ultra-fast opto-chemical sensors by using electrospun nanofibers as sensing layers. *Sensors and Actuators B: Chemical*, *209*, 1064−1069. Available from https://doi.org/10.1016/j.snb.2014.11.070.

Xia, Y., Li, R., Chen, R., Wang, J., & Xiang, L. (2018). 3D architectured graphene/metal oxide hybrids for gas sensors: A review. *Sensors*, *18*(5). Available from https://doi.org/10.3390/s18051456.

Yin, H., & Yang, J. (2011). Synthesis of high-performance one-dimensional polyaniline nanostructures using dodecylbenzene sulfonic acid as soft template. *Materials Letters*, *65*(5), 850−853. Available from https://doi.org/10.1016/j.matlet.2010.12.031.

Zhang, T., Xu, H., Xu, Z., Gu, Y., Yan, X., Liu, H., Lu, N., Zhang, S., Zhang, Z., & Yang, M. (2019). A bioinspired antifouling zwitterionic interface based on reduced graphene oxide carbon nanofibers: electrochemical aptasensing of adenosine triphosphate. *Microchimica Acta*, *186*(4). Available from https://doi.org/10.1007/s00604-019-3343-7.

Zhao, Q., Zhao, M., Qiu, J., Lai, W.-Y., Pang, H., & Huang, W. (2017). One dimensional silver-based nanomaterials: Preparations and electrochemical applications. *Small (Weinheim an der Bergstrasse, Germany)*, *13*(38). Available from https://doi.org/10.1002/smll.201701091.

Zhu, J., Wei, S., Ryu, J., & Guo, Z. (2011). Strain-sensing elastomer/carbon nanofiber "metacomposites.". *The Journal of Physical Chemistry C*, *115*(27), 13215−13222. Available from https://doi.org/10.1021/jp202999c.

Zou, Q., & Tian, H. (2010). Chemodosimeters for mercury(II) and methylmercury(I) based on 2,1,3-benzothiadiazole. *Sensors and Actuators B: Chemical*, *149*(1), 20−27. Available from https://doi.org/10.1016/j.snb.2010.06.040.

Water splitting application of nanofibers and their composites

18

Abhinay Thakur[1], Valentine Chikaodili Anadebe[2] and Ashish Kumar[3]
[1]Department of Chemistry, School of Chemical Engineering and Physical Sciences, Lovely Professional University, Phagwara, Punjab, India, [2]Department of Chemical Engineering, Alex Ekwueme Federal University Ndufu Alike, Abakakili, Ebonyi State, Nigeria, [3]Department of Science and Technology, NCE, Bihar Engineering University, Government of Bihar, Patna, Bihar, India

18.1 Introduction

In recent years, the global demand for clean and sustainable energy solutions has surged, driven by concerns over climate change and environmental degradation. This trend is evidenced by substantial investments in renewable energy, which reached a staggering $303.5 billion in 2020 according to the UN Environment Program (Eswaran et al., 2023; Ge et al., 2019; Nabgan et al., 2024; Zheng et al., 2022). Despite the challenges posed by the COVID-19 pandemic, this investment reflects a growing recognition of the imperative to transition away from polluting fossil fuels toward cleaner alternatives. The rapid expansion of solar photovoltaic (PV) installations further underscores this shift, with global capacity hitting a record high of 139 gigawatts (GW) in the same year, as reported by the International Energy Agency (Abdelhamid, 2023; Balazadeh Meresht et al., 2023; Pinto et al., 2021). Moreover, the burgeoning electric vehicle market exemplifies a move toward sustainable transportation, with global sales soaring by 43% in 2020 according to BloombergNEF (Borawski, 2022; Wang & Ru, 2019). Concurrently, wind power capacity has experienced substantial growth, with the Global Wind Energy Council reporting an addition of 53 GW in 2020 alone, reaching a total capacity of 743 GW (Wang & Yin, 2022; Wang et al., 2021). Additionally, the rise of green hydrogen as an energy carrier, with over 30 GW of electrolyzer capacity announced by early 2021 according to the International Renewable Energy Agency, highlights the increasing diversification of clean energy sources. As traditional methods of energy production continue to pose significant challenges in terms of pollution and resource depletion, the need for alternative, renewable energy sources has become increasingly evident. Among these sources, hydrogen has emerged as a promising candidate due to its abundance and potential as a clean fuel. However, the widespread adoption of hydrogen as an energy carrier hinges on the development of efficient and economically viable methods for its production. One such method is water splitting, a process that involves breaking water molecules into hydrogen and oxygen using renewable energy sources such as solar or wind power (Balitskii et al., 2021; Waqas et al., 2021).

Traditional methods of water splitting, such as electrolysis, often rely on expensive catalysts or require large amounts of energy, limiting their scalability and affordability. These limitations have spurred significant interest in exploring alternative approaches that can overcome these challenges and make water splitting economically viable on a large scale. Nanotechnology has emerged as a promising avenue for revolutionizing water-splitting processes, offering unparalleled advantages in terms of catalysis and energy conversion (Pardhi et al., 2022; Vodovozov et al., 2022). Among the various nanomaterials being investigated for this purpose, nanofibers have garnered significant attention due to their unique properties at the nanoscale.

Nanofibers are one-dimensional nanostructures with diameters typically ranging from a few nanometers to several micrometers. They can be synthesized from a variety of materials, including polymers, metals, metal oxides, and carbon-based materials, using techniques such as electrospinning, template synthesis, and chemical vapor deposition. One of the key advantages of nanofibers lies in their high surface area-to-volume ratio, which allows for enhanced interaction with reactants and improved catalytic activity. Additionally, nanofibers exhibit tunable morphology and excellent mechanical properties, making them ideal candidates for a wide range of applications, including catalysis, sensing, filtration, and energy storage. In the context of water splitting, nanofibers offer several advantages over traditional catalysts and electrode materials (Ammar et al., 2023; Ma et al., 2022; Zhang et al., 2021). Their high surface area allows for efficient utilization of active sites, leading to enhanced catalytic activity and reaction rates. Moreover, their tunable morphology enables precise control over properties such as porosity, surface chemistry, and electronic structure, further optimizing their performance in water-splitting reactions. Additionally, nanofibers can be easily integrated into composite materials with other nanomaterials or support matrices, allowing for the creation of multifunctional materials with synergistic properties. Several examples demonstrate the wide-ranging applications and global impact of nanofibers across various industries, highlighting their importance in addressing critical challenges and advancing technological innovations worldwide.

- Catalysis: Nanofibers are extensively utilized as catalyst supports or active catalysts due to their high surface area-to-volume ratio and tunable surface properties. For instance, in the automotive industry, nanofiber-based catalysts are employed in catalytic converters to facilitate the conversion of harmful emissions into less harmful substances, contributing to reduced air pollution on a global scale (Kindrachuk et al., 2021; Martyushev et al., 2023; Najjar et al., 2022; Van Tran et al., 2021).
- Sensing: Nanofibers are widely employed in various sensing applications, including environmental monitoring, healthcare diagnostics, and food safety. For example, nanofiber-based sensors can detect trace amounts of pollutants in water or air, enabling early detection and mitigation of environmental hazards, thereby safeguarding public health and ecosystems worldwide.
- Filtration: Nanofibers are utilized in filtration systems for air and water purification due to their high surface area and small pore size (Li et al., 2020; Österle & Dmitriev, 2016; Yaqoob et al., 2021). These filters can effectively remove particulate matter, bacteria, viruses, and other contaminants, contributing to improved air and water quality in both industrial and domestic settings across the globe.

Water splitting application of nanofibers and their composites   439

- Energy storage: Nanofibers play a crucial role in advancing energy storage technologies, such as batteries, supercapacitors, and fuel cells. By providing high surface area electrodes and facilitating rapid ion transport, nanofiber-based materials enable the development of high-performance energy storage devices, which are essential for the transition to renewable energy sources and the electrification of transportation on a global scale.
- Tissue engineering: In the field of regenerative medicine, nanofibers are utilized as scaffolds for tissue engineering applications. These scaffolds mimic the extracellular matrix, providing a conducive environment for cell growth, proliferation, and differentiation (Baena et al., 2015; Ilie & Cristescu, 2022; Marlinda et al., 2023). Nanofiber-based tissue engineering constructs hold immense potential for repairing and regenerating damaged tissues and organs, addressing critical healthcare challenges worldwide.
- Protective clothing: Nanofibers are incorporated into protective clothing and personal protective equipment to enhance their performance. For instance, nanofiber-based membranes can provide barrier properties against hazardous chemicals, pathogens, and UV radiation, ensuring the safety and well-being of workers in industries such as healthcare, chemical processing, and construction worldwide. In an experiment, Huang et al. (Huang et al., 2020) focused on synthesizing a highly efficient bifunctional catalyst, namely, nickel phosphide doped with reduced graphene oxide nanosheets supported on nickel foam (Ni$_2$P/rGO/NF). The catalyst was prepared through a hydrothermal and calcination process tailored for both hydrogen evolution reaction (HER) and oxygen evolution reaction (OER). Electrochemical tests were conducted using a parallel two-electrode electrolyzer setup (Ni$_2$P/rGO/NF‖Ni$_2$P/rGO/NF). Fig. 18.1 illustrates the comprehensive process of water splitting occurring on the Ni$_2$P/rGO/NF electrode. This map provides a visual representation of the electrochemical reactions involved in both the HER and the OER on the catalyst surface. Remarkably, Ni$_2$P/rGO/NF‖Ni$_2$P/rGO/NF demonstrates a low voltage requirement of 1.676 V to drive 10 mA/cm^2, comparable to Pt/C/NF‖IrO$_2$/NF (1.502 V). Moreover, it maintains a stable current density for at least 30 hours, indicating promising commercial viability due to its robust activity and stability.

18.2 Fundamentals of nanofibers and composite

Similarly, Zhang et al. (2023) explored a novel GO/Rh-SrTiO$_3$ nanocomposite synthesized via a facile hydrothermal method. The optimized GO/Rh-SrTiO$_3$ nanocomposite

Figure 18.1 Diagram illustrating the complete process of water splitting occurring on the Ni2P/rGO/NF electrode.
Source: From Huang, J., Li, F., Liu, B., & Zhang, P. (2020). Ni$_2$P/rGO/NF nanosheets as a bifunctional high-performance electrocatalyst for water splitting. *Materials*, *13*(3). https://doi.org/10.3390/ma13030744.

demonstrated outstanding photocatalytic performance for overall water splitting, achieving an H_2 evolution rate of 55.83 μmol/g/h and an O_2 production rate of 23.26 μmol/g/h under visible light ($\lambda \geq 420$ nm). This represented a significant improvement over single-phased $SrTiO_3$. The photocatalytic performance of the synthesized photocatalysts was evaluated through overall water splitting for the production of H_2 and O_2. Pure $SrTiO_3$ (STO) exhibited no overall water splitting under visible light ($\lambda \geq 420$ nm), as depicted in Fig. 18.2A, attributed to its limited ability to absorb and utilize UV light exclusively. In contrast, the GO/STO heterostructure enabled overall water splitting under visible light. Remarkably, the leading performer, GO/Rh-SrTiO₃ nanocomposite, demonstrated exceptional photocatalytic overall water-splitting performance, yielding an H_2 evolution rate of 55.83 μmol/g/h and an O_2 production rate of 23.26 μmol/g/h under visible light ($\lambda \geq 420$ nm). This underscores the role of heterostructure construction and Rh metal active site modification in enhancing visible light utilization and the photocatalytic efficiency of STO. Fig. 18.2B illustrates that the H_2 and O_2 production rates over the xGO/Rh-STO samples initially increase and then decrease with increasing GO content. This trend is attributed to the excessive GO reducing the exposed active sites of STO, consequently decreasing the overall water-splitting rate. Notably, the 7.2% GO/Rh-STO photocatalyst exhibits the highest H_2 production rate (55.83 μmol/g/h) and O_2 production rate (23.26 μmol/g/h) under visible light ($\lambda \geq 420$ nm). Additionally, Fig. 18.3 provides the mechanism behind the process.

The motivation behind this chapter lies in the desire to comprehensively explore and elucidate the potential of nanofibers and their composites in advancing water-splitting technologies. By addressing specific objectives, such as reviewing synthesis techniques, characterizing properties, investigating composite formation, evaluating catalytic activity and electrochemical performance, examining applications, and discussing recent developments and future directions, this chapter aims to provide valuable insights for researchers, engineers, and policymakers involved in the

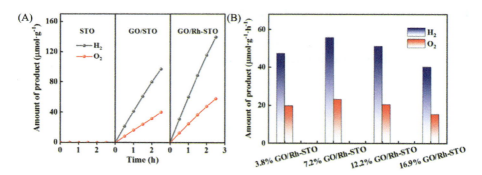

Figure 18.2 Time course of H_2 and O_2 evolution over different catalysts: STO, GO/STO, GO/Rh − STO, and the photocatalytic activity of xGO/Rh − STO (x = 3.8%, 7.2%, 12.2%, or 16.9%) under visible light ($\lambda \geq 420$ nm) irradiation. *STO*, SrTiO₃.
Source: From Zhang, S., Jiang, E., Wu, J., Liu, Z., Yan, Y., Huo, P., & Yan, Y. (2023). Visible-light-driven GO/Rh-SrTiO₃ photocatalyst for efficient overall water splitting. *Catalysts*, *13*(5). https://doi.org/10.3390/catal13050851.

Water splitting application of nanofibers and their composites 441

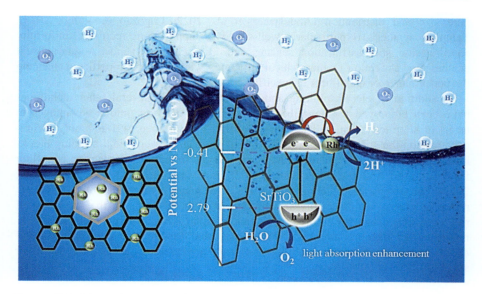

Figure 18.3 Schematic representation illustrating the potential photocatalytic degradation mechanism by nanofibers.
Source: From Zhang, S., Jiang, E., Wu, J., Liu, Z., Yan, Y., Huo, P., & Yan, Y. (2023). Visible-light-driven GO/Rh-SrTiO$_3$ photocatalyst for efficient overall water splitting. *Catalysts*, *13*(5). https://doi.org/10.3390/catal13050851.

development and deployment of sustainable energy solutions. Through a detailed examination of nanofibers and their applications in water splitting, this chapter seeks to inspire innovation and contribute to the ongoing efforts toward a clean and renewable energy future.

18.2.1 Nanofiber synthesis techniques

Nanofibers, owing to their exceptional properties and diverse applications, can be synthesized using a variety of techniques, each with its own set of advantages and limitations. Among these methods, electrospinning stands out as one of the most widely utilized approaches. Electrospinning involves the application of a high voltage to a polymer solution or melt, resulting in the formation of ultrafine fibers through the stretching and elongation of the polymer jet (Borawski, 2021; Casati & Vedani, 2014; Nayl et al., 2022). This technique offers precise control over fiber morphology and composition, allowing for the fabrication of nanofibers with tailored properties. Moreover, electrospinning is compatible with a wide range of polymers and can be adapted to suit various applications, including tissue engineering, drug delivery, filtration, and energy storage. Another popular method for nanofiber synthesis is template synthesis, where nanofibers are formed within nonporous templates through processes such as sol-gel synthesis or chemical vapor deposition. In this approach, the template serves as a scaffold for the growth of nanofibers, allowing for precise control over pore size and distribution (Abdelazeez et al., 2022; Logozzo & Valigi, 2023). Template synthesis is particularly well-suited for applications requiring uniformity and precision, such as filtration, sensing, and drug delivery systems. Additionally, it enables the fabrication of hierarchical structures with tailored properties, enhancing the performance of nanofibers in specific applications.

In addition to electrospinning and template synthesis, several other techniques have been employed for nanofiber synthesis, each offering unique advantages in terms of simplicity, scalability, and versatility. For instance, self-assembly techniques rely on the spontaneous organization of molecular or colloidal building blocks into nanofiber structures, offering a simple and cost-effective approach to nanofiber fabrication (Al-Abduljabbar & Farooq, 2023; Jamesh et al., 2022; Wang, Lv, et al., 2023). Phase separation techniques, on the other hand, involve the controlled separation of phases within a polymer solution or melt, resulting in the formation of nanofiber networks with tailored properties. Similarly, hydrothermal synthesis techniques utilize high-pressure, high-temperature conditions to induce the formation of nanofibers from precursor materials, offering a versatile approach to nanofiber synthesis.

18.2.2 Properties of nanofibers

Nanofibers represent a class of materials with remarkable properties that set them apart from bulk materials, rendering them highly sought-after for an array of applications across diverse fields. One of their standout features is their exceptionally high surface area-to-volume ratio. This means that nanofibers possess a vast amount of surface area relative to their volume, offering a plethora of active sites for interactions with surrounding molecules. This characteristic is particularly advantageous in catalysis, where the efficacy of a catalyst often hinges on its surface area and reactivity (Montani et al., 2020; Stejskal, 2022; Zhou, Jia, et al., 2023). Nanofibers excel in this regard, serving as efficient catalyst supports or active sites for chemical reactions. Their large surface area enhances the exposure of catalytic sites, leading to faster reaction rates and improved overall efficiency, which is crucial for various industrial processes and environmental remediation efforts. Moreover, nanofibers boast tunable morphology, allowing for precise control over their structural characteristics such as diameter, length, and alignment. This tunability empowers researchers to tailor nanofibers to meet the specific requirements of different applications. For instance, in tissue engineering, the alignment of nanofibers plays a pivotal role in guiding cell behavior. By precisely controlling the alignment of nanofibers within scaffolds, researchers can mimic the natural extracellular matrix and provide cues that direct cell adhesion, proliferation, and differentiation (Do et al., 2020; Ji et al., 2022; Kadja et al., 2023). This capability holds immense promise for regenerative medicine, where engineered tissues and organs are created to replace or repair damaged tissues in the body. Furthermore, nanofibers exhibit exceptional mechanical properties, including high strength, flexibility, and toughness. Despite their diminutive size, nanofibers possess remarkable mechanical robustness, making them suitable for applications requiring durability and resilience. For instance, infiltration systems and nanofiber membranes demonstrate superior mechanical strength and flexibility, enabling efficient capture and removal of contaminants from air and water streams. Additionally, in the realm of electronics, nanofibers can be integrated into flexible and stretchable electronic devices, allowing for the development of wearable sensors, flexible displays, and other innovative technologies.

18.2.3 Composite formation and characteristics

Composite materials, incorporating nanofibers dispersed within a matrix or embedded in another material, represent a class of advanced materials with synergistic properties that surpass those of individual components (Christoforidou et al., 2023; Song et al., 2023). These composites are engineered to combine the unique characteristics of nanofibers with the desirable properties of the matrix material, resulting in materials with tailored properties suitable for a wide range of applications. The formation of nanofiber composites can be achieved through several fabrication methods, including solution blending, melt mixing, in situ polymerization, and layer-by-layer assembly. These methods allow for the integration of nanofibers with different materials, such as polymers, metals, ceramics, and carbon-based materials, facilitating the creation of multifunctional materials with enhanced performance and functionality. The characteristics of nanofiber composites are influenced by various factors, including the type of nanofibers, matrix material, and fabrication method, as well as the morphology, orientation, and distribution of nanofibers within the composite (Cui et al., 2022). For instance, nanofiber-reinforced polymer composites exhibit improved mechanical properties, such as stiffness, strength, and toughness, compared to neat polymers. This enhancement is attributed to the high aspect ratio and exceptional mechanical strength of nanofibers, which act as reinforcements within the polymer matrix, resisting deformation and fracture under applied loads (Han et al., 2022; Su et al., 2020; Sun et al., 2023). By controlling parameters such as nanofiber content, orientation, and dispersion, researchers can tailor the mechanical properties of composite materials to meet specific requirements for structural applications in aerospace, automotive, construction, and sports equipment.

Similarly, nanofiber-based catalyst composites demonstrate enhanced catalytic activity and selectivity, making them promising materials for applications in environmental remediation, chemical synthesis, and energy conversion. Nanofibers, with their high surface area and reactive surfaces, serve as excellent supports or active sites for catalytic materials, facilitating efficient mass transport and catalytic reactions. By incorporating nanofibers into catalyst composites, researchers can enhance catalytic performance, increase reaction rates, and improve product selectivity, leading to more sustainable and cost-effective processes for various industrial applications (La et al., 2023; Nyamai & Phaahlamohlaka, 2023; Zhou, Zhang, et al., 2023). Additionally, the tunable properties of nanofibers, such as surface chemistry and morphology, enable the design of catalyst composites with tailored properties for specific reactions and applications, further expanding their utility in catalysis. Furthermore, nanofiber composites offer a versatile platform for tailoring material properties to meet specific application requirements across various technological domains. For example, in the field of energy storage and conversion, nanofiber-based composites are being investigated for use in batteries, supercapacitors, fuel cells, and solar cells. Nanofibers can improve the performance of these devices by enhancing charge transport, increasing surface area for electrochemical reactions, and providing mechanical stability and flexibility. Similarly, in biomedical applications, nanofiber composites are utilized for tissue engineering, drug delivery, and medical implants. The high surface area and porosity of nanofibers enable efficient drug loading and release, while their biocompatibility and bioactivity promote cell adhesion, proliferation, and tissue regeneration.

18.3 Nanofibers and composites in water splitting

18.3.1 Role of nanofibers in water splitting

Nanofibers are playing an increasingly crucial role in the advancement of water-splitting technologies, driven by their exceptional properties that enhance catalytic activity and electrochemical performance (Aguilar-Ferrer et al., 2022; Kumbhar et al., 2019). These nanoscale materials offer a range of unique characteristics that make them highly effective catalysts and electrode materials in water-splitting processes. This essay elaborates on the significant role of nanofibers in water splitting, focusing on their high surface area-to-volume ratio, tunable morphology, and effectiveness as catalyst supports or active sites. One of the primary reasons nanofibers are valuable in water splitting is their high surface area-to-volume ratio (De Alvarenga et al., 2020). This characteristic enables them to interact more efficiently with water molecules and catalytic species, enhancing the adsorption of reactants onto their surfaces. In water-splitting reactions, such as electrolysis or photocatalysis, the availability of active sites for catalytic reactions is crucial for promoting the desired chemical transformations. Nanofibers, with their vast surface area, provide ample sites for reactant adsorption, facilitating the catalytic reactions necessary for water splitting. This increased surface area allows for more reactive sites per unit volume, thereby improving the efficiency of the water-splitting process. Furthermore, the tunable morphology of nanofibers enables precise control over their surface properties, such as surface chemistry and roughness. By manipulating the fabrication parameters during synthesis, researchers can tailor the surface characteristics of nanofibers to enhance their catalytic activity. For instance, adjusting the composition of the precursor solution or modifying the electrospinning parameters can alter the surface chemistry of nanofibers, making them more reactive toward specific reactants or catalytic species involved in water splitting (Ponnalagar et al., 2023; Wang, Ricote, et al., 2023). Additionally, controlling the morphology, such as the diameter, length, and alignment of nanofibers, allows for optimization of their catalytic performance. Nanofibers with well-defined morphologies exhibit improved mass transport and accessibility of active sites, leading to enhanced catalytic activity in water-splitting reactions. Some global examples highlighting the crucial role of nanofibers in advancing water-splitting technologies:

- Research at Massachusetts Institute of Technology (MIT): Scientists at the MIT have developed nanofiber-based catalysts for water splitting using earth-abundant materials. These catalysts exhibit high efficiency and stability, offering promising prospects for large-scale water electrolysis as a means of producing clean hydrogen fuel.
- Hydrogen production in Japan: Japan, a global leader in hydrogen technology, has been actively researching nanofiber-based materials for water splitting. Collaborative efforts between academia and industry have resulted in the development of efficient nanofiber catalysts that enhance the performance of water electrolysis systems, contributing to Japan's vision of a hydrogen-based society (Liu, Xu, et al., 2023; Wang et al., 2017).
- European Union projects: Several research projects funded by the European Union focus on leveraging nanofiber-based materials for water splitting. For example, the "NanoHydroChem" project aims to develop innovative nanofiber catalysts for efficient water electrolysis, addressing the challenges associated with current electrolysis technologies and promoting the transition to renewable hydrogen production.
- South Korean initiatives: South Korea has invested significantly in nanofiber research for water-splitting applications. Researchers in South Korean institutions such as KAIST (Korea Advanced Institute of Science and Technology) and KIST (Korea Institute of Science and Technology) are exploring novel nanofiber-based catalysts and electrodes to improve the performance and durability of water electrolysis systems, supporting the country's efforts toward sustainable energy production (Balaji et al., 2024; Cieluch et al., 2024; Xin et al., 2021).
- Collaborative efforts in Australia: Collaborative research projects between universities and industry partners in Australia are focused on developing nanofiber-based materials for water splitting. These initiatives aim to address the technical challenges associated with water electrolysis and pave the way for cost-effective and environmentally friendly hydrogen production in the country.

18.3.2 Recent developments and advances of nanofibers in water splitting

Moreover, nanofibers can serve as effective catalyst supports or active sites, promoting heterogeneous catalysis and charge transfer processes during electrochemical reactions. The high surface area and porous structure of nanofibers make them ideal substrates for anchoring catalytic species, such as metal nanoparticles or metal oxides, which serve as active sites for catalytic reactions in water splitting (Niu et al., 2020; Šimko & Lukáš, 2016; Wan et al., 2021). These catalytic species can be immobilized onto the surface of nanofibers through various deposition techniques, such as impregnation, chemical vapor deposition, or electrochemical deposition. Once anchored, these catalytic species facilitate the adsorption and activation of reactants, accelerating the rate of water-splitting reactions. Additionally, nanofibers themselves can exhibit intrinsic catalytic activity, particularly in the case of carbon-based nanofibers or certain metal oxide nanofibers, further enhancing their effectiveness as catalysts in water-splitting processes.

18.3.3 Synergistic effects on composite materials

The integration of nanofibers into composite materials represents a significant advancement in water-splitting technologies, offering synergistic effects that surpass those of individual components. By combining nanofibers with other materials, such as polymers, metals, ceramics, or carbon-based materials, researchers can create composite materials with tailored properties optimized for water-splitting applications (Jaleh et al., 2023; Magrini et al., 2021; Xiao et al., 2023). This essay discusses and elaborates on the synergistic effects of nanofiber composites in water splitting, focusing on their enhanced charge transport, catalytic activity, and long-term stability. One of the key advantages of integrating nanofibers into composite materials is the enhancement of charge transport and conductivity, particularly in electrochemical water-splitting processes. Nanofibers embedded within a conductive matrix, such as graphene or carbon nanotubes, can serve as pathways for electron transport, facilitating efficient charge transfer during electrochemical reactions. Graphene, with its excellent conductivity and high surface area, is particularly well-suited as a matrix material for nanofiber composites in water-splitting applications (Ahmed et al., 2021; Gebreslase et al., 2022). When combined with nanofibers, graphene forms a conductive network that promotes rapid electron transfer between catalytic sites, thereby improving the efficiency of electrochemical water splitting. Similarly, carbon nanotubes offer superior electrical conductivity and mechanical strength, making them ideal candidates for enhancing charge transport in nanofiber composites.

Moreover, nanofibers can serve as supports or scaffolds for other active materials, such as metal nanoparticles or metal oxides, providing a hierarchical structure that maximizes catalytic activity and stability (Gowrisankar et al., 2021; Khan et al., 2022). Metal nanoparticles, such as platinum, ruthenium, or nickel, are widely used as catalysts in water-splitting reactions due to their high catalytic activity. By anchoring these metal nanoparticles onto the surface of nanofibers, researchers can create composite materials with enhanced catalytic performance. The high surface area and porous structure of nanofibers provide ample sites for anchoring metal nanoparticles, ensuring uniform distribution and accessibility of catalytic sites. Additionally, the hierarchical structure of nanofiber composites allows for synergistic interactions between the nanofiber support and the catalytic species, further enhancing catalytic activity and stability. Furthermore, nanofiber composites offer improved long-term stability compared to individual components, making them highly promising for practical applications in water splitting. The integration of nanofibers with matrix materials provides mechanical reinforcement and protection against degradation, resulting in composite materials with enhanced durability and resilience. For instance, polymer matrices, such as polyvinyl alcohol or polyacrylonitrile, can encapsulate nanofibers, providing a protective barrier against environmental factors and chemical reactions (Garkal et al., 2022; Rosmini et al., 2022). Similarly, ceramic matrices, such as titanium dioxide or zirconium oxide, offer excellent chemical and thermal stability, ensuring the long-term performance of nanofiber composites in harsh operating conditions. By combining the unique properties of nanofibers with the stability of matrix materials, researchers can develop composite materials with superior performance and longevity in water-splitting applications.

18.3.4 Catalytic activity enhancement

Nanofiber-based catalyst composites represent a significant advancement in water-splitting technologies, offering enhanced catalytic activity, selectivity, and durability. Their effectiveness stems from several key factors, with their high surface area, tunable surface properties, and hierarchical structures playing crucial roles in driving superior performance in water-splitting applications. Firstly, the large surface area of nanofibers is instrumental in facilitating efficient catalytic reactions for water splitting (Cuenca et al., 2021; Liu, Zhang, et al., 2023; Momeni et al., 2024). Nanofibers inherently possess a high surface area-to-volume ratio, providing numerous active sites for catalytic reactions to take place. This expansive surface area enables increased interactions between water molecules and catalytic species, facilitating adsorption and reaction at the nanofiber surface. Consequently, nanofibers serve as efficient platforms for catalyzing various steps involved in water splitting, including the dissociation of water molecules and the subsequent release of hydrogen and oxygen gas. Furthermore, the tunable surface properties of nanofibers offer precise control over catalyst-substrate interactions and reaction kinetics. Through methods such as functionalization or doping, researchers can modify the surface chemistry of nanofibers to tailor their catalytic activity and selectivity. For instance, introducing specific functional groups or dopants onto nanofiber surfaces enhances their affinity toward certain reactants or catalytic species, thereby promoting desired chemical transformations while suppressing unwanted side reactions. Additionally, controlling the surface roughness or morphology of nanofibers allows for the optimization of catalyst-substrate interactions, leading to improved reaction kinetics and overall catalytic performance in water-splitting reactions (Gawel et al., 2022; Liu et al., 2020; Rusdi et al., 2022).

Moreover, the hierarchical structures present in nanofiber-based catalyst composites further contribute to their superior performance. These structures, which involve the integration of nanofibers with other catalytic materials or support matrices, create synergistic effects that enhance catalytic activity and stability (Chahal et al., 2024; Koskin et al., 2020; Yun et al., 2023). For example, nanofibers decorated with metal nanoparticles exhibit enhanced catalytic activity for water splitting due to synergistic interactions between the metal nanoparticles and the nanofiber support. This hierarchical architecture maximizes the utilization of active sites and promotes efficient charge transfer processes during electrochemical reactions, resulting in improved overall performance of the catalyst composite. Furthermore, the hierarchical structures of nanofiber composites, where nanofibers are integrated with other catalytic materials, create synergistic effects that enhance catalytic activity and stability (Sharma & Abhinay Thakur, 2023; Sharma et al., 2023; Thakur & Kumar, 2023a, 2023b; Thakur et al., 2021; Thakur, Sharma, et al., 2022). Nanofibers decorated with metal nanoparticles, for example, exhibit enhanced catalytic activity for water splitting due to the synergistic effects between the metal nanoparticles and the nanofiber support. The presence of metal nanoparticles on the surface of nanofibers provides additional active sites for catalytic reactions, while the nanofiber support offers mechanical stability and protection against agglomeration or deactivation of the metal nanoparticles (Maafa et al., 2021; Zhang et al., 2018). This hierarchical structure ensures efficient mass transport of reactants and products, facilitating rapid reaction rates and improving overall efficiency in water splitting. Moreover, nanofiber-based catalyst composites offer significant improvements in durability compared to conventional catalysts, making them promising materials for efficient and sustainable water-splitting technologies. The mechanical robustness and chemical stability of nanofibers provide long-term support for catalytic materials, ensuring their performance and functionality over extended periods of operation (Bao et al., 2018; Macdonald et al., 2018; Yuanzheng et al., 2021). Additionally, the hierarchical structures of nanofiber composites enhance their resistance to deactivation mechanisms, such as poisoning or leaching of active species, leading to prolonged catalyst lifetimes and reduced maintenance requirements. As a result, nanofiber-based catalyst composites offer a viable solution for addressing the durability challenges associated with water-splitting technologies, paving the way for the development of efficient and sustainable methods for producing hydrogen fuel from water.

18.3.5 Electrochemical performance improvement

Nanofiber-based electrode materials have emerged as promising candidates for enhancing the electrochemical performance of water-splitting reactions, owing to their unique properties such as high conductivity, large surface area, and tailored surface properties. In this discussion, we will elaborate on how nanofibers, when embedded within conductive matrices such as carbon-based materials or metal substrates, facilitate rapid charge transfer and electron transport, thereby enhancing the efficiency of electrochemical water splitting (Agboola & Shakir, 2022; Azizi et al., 2023). We will also explore how the large surface area of nanofibers provides ample active sites for electrochemical reactions, allowing for efficient adsorption and desorption of reactants and reaction intermediates. Additionally, we will discuss how the tailored surface properties of nanofibers enable precise control over electrode—electrolyte interfaces, reducing interfacial resistance and improving overall electrochemical performance. Overall, nanofiber-based electrode materials offer significant improvements in electrocatalytic activity, stability, and durability, making them highly promising for practical applications in water-splitting technologies (Kotb & Velev, 2023; Manda et al., 2022). One of the primary advantages of nanofiber-based electrode materials lies in their high conductivity, which is essential for efficient charge transfer and electron transport during electrochemical reactions. Nanofibers, when embedded within conductive matrices such as carbon-based materials or metal substrates, provide pathways for rapid electron transfer between the electrode and the electrolyte. Carbon-based materials, such as graphene or carbon nanotubes, are particularly well-suited as matrix materials for nanofiber composites due to their excellent electrical conductivity and mechanical strength. When combined with nanofibers, these carbon-based matrices form a conductive network that promotes rapid electron transfer, leading to enhanced electrochemical performance in water-splitting reactions. Similarly, metal substrates, such as gold or platinum, offer high electrical conductivity and catalytic activity, further facilitating electron transport and improving the efficiency of water-splitting processes.

Moreover, the large surface area of nanofibers provides ample active sites for electrochemical reactions, allowing for efficient adsorption and desorption of reactants and reaction intermediates. Nanofibers possess a high surface area-to-volume ratio, resulting in a greater number of active sites available for catalytic reactions at the electrode–electrolyte interface. This increased surface area enhances the accessibility of reactants to active sites, thereby improving the efficiency of electrochemical water splitting. Additionally, the porous structure of nanofibers enables rapid diffusion of reactants and products throughout the electrode material, further enhancing mass transport and reaction kinetics (Cimesa & Moustafa, 2022; Lau et al., 2016; Santangelo, 2019). As a result, nanofiber-based electrode materials exhibit enhanced electrochemical performance, including higher reaction rates and improved Faradaic efficiency, compared to conventional electrode materials. Furthermore, the tailored surface properties of nanofibers enable precise control over electrode–electrolyte interfaces, reducing interfacial resistance and improving overall electrochemical performance. By modifying the surface chemistry or morphology of nanofibers through surface functionalization or doping, researchers can tailor the electrode–electrolyte interface to optimize electrochemical reactions (Dhonchak & Agnihotri, 2023; Kaya, Lgaz, et al., 2023; Thakur & Kumar, 2023c; Thakur, Kumar, Singh, 2023; Thakur, Sharma, et al., 2022). For instance, introducing specific functional groups or dopants onto the surface of nanofibers can enhance their affinity toward certain reactants or reaction intermediates, thereby promoting desired chemical transformations and suppressing side reactions. Additionally, controlling the surface roughness or porosity of nanofibers allows for the optimization of electrode–electrolyte interactions, leading to reduced interfacial resistance and improved charge transfer kinetics. As a result, nanofiber-based electrode materials offer enhanced electrochemical performance, including lower overpotentials, higher current densities, and improved stability, making them highly promising for practical applications in water-splitting technologies.

18.4 Recent developments and advances of nanofibers in water splitting

One notable recent development in the field of nanofibers for water splitting is the advancement in synthesis techniques, allowing for precise control over nanofiber morphology, composition, and structure. Traditional methods such as electrospinning have been refined to produce nanofibers with improved uniformity, orientation, and alignment, which are crucial for optimizing catalytic performance (Abdel Hameed et al., 2022; Kim et al., 2018). Additionally, novel synthesis approaches, such as template-assisted synthesis, self-assembly, and vapor-phase deposition, have emerged to create nanofibers with tailored properties, enabling researchers to design materials specifically optimized for water-splitting applications. These advancements in synthesis techniques have led to the development of nanofibers with enhanced catalytic activity, stability, and durability, paving the way for more efficient and sustainable water-splitting technologies. Another recent trend in nanofiber-based materials for water splitting is the design of composite structures incorporating nanofibers with other functional materials, such as metals, metal oxides, polymers, and carbon-based materials. These nanofiber composites offer synergistic effects that surpass those of individual components, leading to enhanced catalytic performance and electrochemical activity. For instance, nanofibers decorated with metal nanoparticles exhibit improved catalytic activity for water splitting, owing to the synergistic interactions between the metal nanoparticles and the nanofiber support (Ali et al., 2023; Huang et al., 2024). Similarly, nanofiber composites incorporating carbon-based materials, such as graphene or carbon nanotubes, demonstrate enhanced charge transport and conductivity, resulting in improved electrochemical water-splitting efficiency. Furthermore, the integration of nanofibers into polymer matrices enhances mechanical stability and flexibility, making the composites suitable for a wide range of practical applications in water-splitting technologies. For instance, in an experiment, Woo et al. (2020) presented a novel approach for fabricating electrocatalysts aimed at improving the efficiency of HER and overall water splitting. The method involved the combination of electrospinning and pyrolysis to create carbon nanofibers (CNFs) embedded with transition-metal-based nanoparticles, specifically Co-CeO$_2$. These Co-CeO$_2$ nanoparticle-embedded CNFs, denoted as Co-CeO$_2$@CNF, exhibit exceptional electrocatalytic performance for HER, with an impressively low overpotential of 92 mV and a Tafel slope of 54 mV/decade. Additionally, to facilitate the OER, we integrate Ni$_2$Fe catalyst with high activity into the CNFs, termed Ni$_2$Fe@CNF. The electrochemical properties of these novel electrocatalysts, Co-CeO$_2$@CNF and Ni$_2$Fe@CNF, were comprehensively evaluated in an alkaline electrolyzer for overall water splitting. Remarkably, using Co-CeO$_2$@CNF as the HER electrocatalyst and Ni$_2$Fe@CNF as the OER electrocatalyst, they achieved a remarkable overall water-splitting current density of 10 mA/cm^2 with an applied voltage of only 1.587 V. Notably, this voltage was significantly lower compared to that required by conventional electrolyzers equipped with Pt/C and IrO$_2$ electrodes, which typically require 400 mV. Furthermore, the electrocatalysts demonstrated exceptional long-term durability, sustaining over 70 hours of continuous overall water splitting. This durability can be attributed to the conformal incorporation of nanoparticles into the CNFs, ensuring robust performance over extended operation periods.

Shaikh et al. (2021) introduced a novel approach for synthesizing a Co$_3$O$_4$@TiO$_2$ composite through a two-step liquid-phase deposition method. Initially, ZIF-67 was synthesized and employed as a template for the subsequent composite synthesis, leveraging the combined photocatalytic properties of Co$_3$O$_4$ and TiO$_2$. The resulting composite was engineered to offer superior properties compared to its individual components, with the aim of enhancing electrochemical water splitting efficiency. To characterize the synthesized Co$_3$O$_4$@TiO$_2$ composite, various analytical techniques including powder X-ray diffraction (PXRD), Brunauer–Emmet–Teller (BET) analysis for surface area determination, atomic force microscopy (AFM), scanning electron microscopy (SEM), and transmission electron microscopy (TEM) are employed. Notably, linear sweep voltammetry (LSV) results at a scan rate of 1 mV/s demonstrate that the incorporation of metal oxides within other metal oxides results in higher current density and lower onset potential compared to pure metal oxides, highlighting the synergistic effects within the composite. In Fig. 18.4A, the PXRD patterns of the Co_3O_4@TiO_2 composite are depicted. The characteristic peaks of the composite are observed at 31, 37, 45, 48, 50, 59, and 65 degrees, corresponding to the 101, 110, and 220 facets, which represent the diffraction peaks of Co_3O_4 and TiO_2. These peaks exhibit sharp and well-defined characteristics, indicative of the crystalline nature of the composite material. Additionally, BET analysis was conducted to evaluate the surface area of the Co_3O_4@TiO_2 composite, as illustrated in Fig. 18.4B. The obtained surface area of the material was measured to be 391 m^2/g, which is notably high. This significant surface area holds potential importance in determining the catalytic performance of the material. In comparison, the surface areas of Co_3O_4 and TiO_2 individually were found to be 132 and 197 m^2/g, respectively. Fig. 18.4C presents the AFM image of the Co_3O_4@TiO_2 composite, revealing the semispherical morphology of the particles with an estimated particle size ranging from 20 to 40 nm. Furthermore, Fig. 18.4D displays the SEM image of the composite, providing visual confirmation of the morphology and shape of the particles. Moreover, TEM

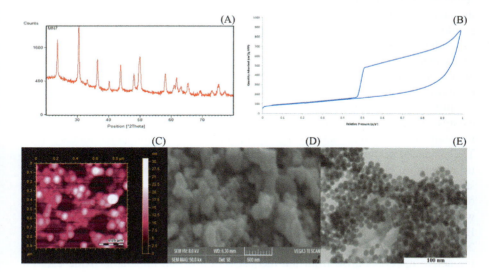

Figure 18.4 Various characterization techniques applied to the Co_3O_4@TiO_2 composite: (A) X-ray diffraction, (B) Brunauer–Emmet–Teller, (C) atomic force microscopy, (D) scanning electron microscopy, and (E) transmission electron microscopy analyses.
Source: From Shaikh, Z. A., Laghari, A. A., Litvishko, O., Litvishko, V., Kalmykova, T., & Meynkhard, A. (2021). Liquid-phase deposition synthesis of zif-67-derived synthesis of CO_3O_4 @TiO_2 composite for efficient electrochemical water splitting. *Metals*, *11*(3), 1–10. https://doi.org/10.3390/met11030420.

analysis was conducted to examine the inner morphology of the composite material. The TEM results, shown in Fig. 18.4E, highlight the nanoscale size of the particles and illustrate the combination of two distinct types of metal oxides within a single structure, exhibiting a high degree of crystallinity.

Yang et al. (2023) developed bifunctional electrocatalysts, Co nanoparticles embedded in N-doped carbon nanotubes/graphitic nanosheets (Co@NCNTs/NG), using a facile high-temperature pyrolysis method. They systematically investigated the influence of the cobalt precursor type (cobalt oxide or cobalt nitrate) on the morphology and particle size of the synthesized products. Their results revealed that the pyrolysis product obtained using cobalt oxide as the precursor (Co@NCNTs/NG-1) exhibited smaller particle size and higher specific surface area compared to the product obtained using cobalt nitrate (Co@NCNTs/NG-2). Notably, Co@NCNTs/NG-1 demonstrated significantly lower potentials of -0.222 V versus reversible hydrogen electrode (RHE) for HER and 1.547 V versus RHE for OER at a benchmark current density of 10 mA/cm^2 compared to Co@NCNTs/NG-2, indicating higher bifunctional catalytic activities. Furthermore, they evaluated the performance of a water-splitting device utilizing Co@NCNTs/NG-1 as both anode and cathode. The device achieved a potential of 1.92 V to reach 10 mA/cm^2 with exceptional stability over 100 hours of operation. This study presents a facile pyrolysis strategy for the development of highly efficient and stable bifunctional electrocatalysts for water-splitting applications, contributing to the advancement of sustainable energy technologies. Similarly, Fan et al. (2023) employed electrospinning followed by postheat treatment at varying temperatures to synthesize $CoMoO_4$ nanofibers. Their investigation revealed that $CoMoO_4$ nanofibers exhibit remarkable activity for the HER, requiring only an overpotential of 80 mV to achieve a current density of 10 mA/cm^2. Additionally, these nanofibers demonstrate impressive performance for the OER in a 1 M KOH electrolyte, despite the higher energy demand associated with this process. Furthermore, the $CoMoO_4$ catalyst displayed exceptional stability in multiple cyclic voltammetry cycles and maintained its catalytic activity for up to 80 hours in chronopotentiometry tests. The enhanced performance of $CoMoO_4$ nanofibers can be attributed to the synergistic interaction between different metallic elements, which surpasses the activity of single oxide counterparts. As depicted in Fig. 18.5, $CoMoO_4$ was found to persist on both sides of the reaction, highlighting the catalyst's stability. Notably, there were no substantial alterations observed in the samples during the HER test. However, during the OER test, the formation of the CoOOH substance (PDF 26−0480) was detected at the Co-active site. These findings suggest that the genuine active materials arise from the reconstruction of the bimetallic oxide catalyst, accentuating the synergistic coupling between different metal elements, and thereby enhancing the electrochemical activity of the catalyst. This study underscores the potential of bimetallic oxides, such as $CoMoO_4$, in advancing energy production through electrochemical water splitting.

Khan et al. (2021) explored a novel approach to enhance the photocatalytic activity of nickel titanate nanofibers ($NiTiO_3$ NFs) by incorporating them with acetic acid-treated exfoliated and sintered sheets of graphitic carbon nitride (AAs-

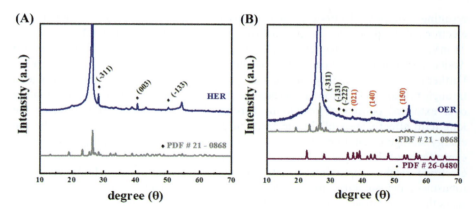

Figure 18.5 X-ray diffraction patterns of CMO-650 following the stability test: (A) posthydrogen evolution reaction and (B) postoxygen evolution reaction.
Source: From Fan, J., Chang, X., Li, L., & Zhang, M. (2024). Synthesis of CoMoO$_4$ nanofibers by electrospinning as efficient electrocatalyst for overall water splitting. *Molecules*, *29*(1). https://doi.org/10.3390/molecules29010007.

gC$_3$N$_4$). The porous AAs-gC$_3$N$_4$ sheets, synthesized through acetic acid-treated exfoliation followed by sintering, offer a substantially enlarged surface area and an abundance of catalytically active sites compared to bulk gC$_3$N$_4$. The hybrid photocatalysts were synthesized through a two-step process, starting with the production of NiTiO$_3$ NFs with a diameter of 360 nm using electrospinning. Subsequently, the NiTiO$_3$ NFs were sensitized with exfoliated gC$_3$N$_4$ sheets via a sonication process. By adjusting the weight ratio of NiTiO$_3$ fibers to porous AAs-gC$_3$N$_4$, it was determined that NiTiO$_3$ NFs containing 40 wt.% of porous AAs-gC$_3$N$_4$ exhibited optimal activity in terms of methylene blue removal and H$_2$ evolution. Under visible light irradiation for 60 minutes, the hybrid photocatalyst demonstrated exceptional performance, achieving the removal of 97% of methylene blue molecules, surpassing pristine AAs-gC$_3$N$_4$, NiTiO$_3$ NFs, and bulk gC$_3$N$_4$. Moreover, the optimal hybrid structure exhibited superior H$_2$ evolution performance compared to pure AAs-gC$_3$N$_4$, NiTiO$_3$ NFs, and bulk gC$_3$N$_4$. The enhanced photocatalytic activity of the hybrid sample is attributed to the synergistic interaction between the holey AAs-gC$_3$N$_4$ nanosheets and NiTiO$_3$ NFs, which prolongs the lifetime of photogenerated charge carriers and reduces recombination losses. Additionally, the higher BET surface area and the presence of catalytically active sites further contribute to the improved photocatalytic performance. Overall, this study presents a novel strategy for fabricating highly efficient photocatalysts with tunable operating windows and enhanced charge separation, offering promising prospects for a wide range of photocatalytic applications. Additionally, Zhang et al. (2019) presented a novel approach utilizing nanostructured CoFe$_2$O$_4$, which holds significant promise for addressing pressing global energy concerns. The synthesis method involved the ultra-fast electrodeposition of CoFe$_2$O$_4$ onto conductive substrates, offering an efficient and cost-effective means to fabricate high-performance energy devices. The resulting CoFe$_2$O$_4$ electrodes, when deposited on Ni-foam, exhibited remarkable electrocatalytic properties, with a low overpotential of 270 mV and a Tafel slope of 31 mV/decade. These characteristics indicated enhanced conductivity and superior device performance compared to conventionally dip-coated CoFe$_2$O$_4$ electrodes. Furthermore, the durability of the catalytic electrode was thoroughly evaluated through bending and chronoamperometry studies, demonstrating excellent stability over extended periods of operation. This resilience was crucial for ensuring the long-term effectiveness of the electrode in practical applications. Additionally, the energy storage capabilities of CoFe$_2$O$_4$ were investigated, revealing a high specific capacitance of 768 F/g at a current density of 0.5 A/g. Even after 10,000 cycles, the electrode retains approximately 80% of its initial capacitance, highlighting its robustness and suitability for long-term energy storage applications. Fig. 18.6A presents the polarization curves obtained from the LSV test conducted on various electrolyzer cells. Notably, the CoFe$_2$O$_4$//Pt−C electrolyzer configuration exhibited remarkable efficiency, requiring a significantly low potential of 1.56 V to achieve a current density of 10 mA/cm^2. In comparison, benchmark electrolyzer configurations, including IrO$_2$//Pt−C, IrO$_2$//IrO$_2$, and Pt−C//Pt−C, necessitated higher potentials of 1.57, 1.77, and 1.64 V, respectively. Furthermore, the chronoamperometry test results, as illustrated in Fig. 18.6B, demonstrate the exceptional stability of the CoFe$_2$O$_4$//Pt−C electrolyzer system over an extended period of more than 17 hours. This sustained performance underscores the robustness and reliability of the CoFe$_2$O$_4$//Pt−C configuration, indicating its suitability for prolonged operation in practical applications.

Moreover, recent advancements in nanofiber-based materials have led to the exploration of novel architectures and functionalities for water-splitting applications. For example, hierarchical nanofiber structures with controlled porosity and morphology have been developed to optimize mass transport and reaction kinetics, leading to

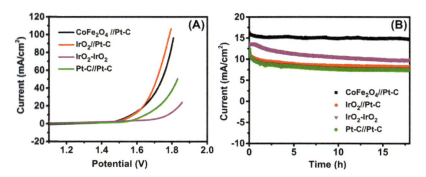

Figure 18.6 Depiction of the polarization (A) and chronoamperometry (B) plots for the overall water splitting process of a two-electrode electrolyzer.
Source: From Zhang, C., Bhoyate, S., Zhao, C., Kahol, P. K., Kostoglou, N., Mitterer, C., Hinder, S. J., Baker, M. A., Constantinides, G., Polychronopoulou, K., Rebholz, C., & Gupta, R. K. (2019). Electrodeposited nanostructured CoFe$_2$O$_4$ for overall water splitting and supercapacitor applications. *Catalysts*, *9*(2). https://doi.org/10.3390/catal9020176.

improved overall efficiency in water-splitting reactions. Additionally, functionalized nanofibers with tailored surface chemistry and reactivity have been investigated to enhance selectivity and specificity in catalytic processes. Furthermore, the integration of nanofibers into flexible and scalable electrode designs has enabled the development of portable and integrated water-splitting devices for on-demand hydrogen production. These novel architectures and functionalities open up new avenues for the design and optimization of nanofiber-based materials for water splitting, promising further advancements in the field. In addition to materials development, recent research efforts have focused on understanding the fundamental mechanisms underlying nanofiber-mediated water-splitting reactions. Advanced characterization techniques, such as electron microscopy, spectroscopy, and surface analysis, have been employed to elucidate the structural, chemical, and electronic properties of nanofibers and their interfaces with catalytic species (Arifa Farzana et al., 2023; Kaya, Thakur, et al., 2023; Kumar & Thakur, 2020; Thakur & Kumar, 2024; Thakur, Kumar, et al., 2022; Thakur, Kumar, Kaya, et al., 2023; Thakur, Savaş, et al., 2023; Verma et al., 2023). Computational modeling and simulation studies have also provided insights into the kinetics and thermodynamics of water-splitting reactions on nanofiber surfaces, guiding the rational design and optimization of nanofiber-based materials for improved performance. Furthermore, interdisciplinary research collaborations between materials scientists, chemists, physicists, and engineers have facilitated the translation of fundamental discoveries into practical applications, accelerating the development of next-generation nanofiber-based water-splitting technologies.

18.5 Challenges and future prospects

18.5.1 Technical challenges and limitations

Despite significant progress in nanofiber-based materials for water splitting, several technical challenges and limitations remain, hindering their widespread implementation and commercialization. These challenges encompass various aspects, including material synthesis, catalytic activity, stability, and scalability.

- Material synthesis challenges:
    One of the primary challenges in nanofiber-based water-splitting technologies lies in the synthesis of nanofibers with precisely controlled morphology, composition, and structure. Traditional synthesis methods, such as electrospinning, often require complex processing conditions and may result in nonuniform or inconsistent nanofiber structures (Ng et al., 2024; Zou et al., 2021). Achieving high throughput and reproducibility while maintaining desired properties poses a significant challenge, particularly for large-scale production. Furthermore, the integration of nanofibers into composite structures with other functional materials presents additional synthesis challenges. Ensuring homogeneous dispersion and strong interfacial interactions between nanofibers and matrix materials is crucial for optimizing catalytic performance and stability. However, achieving uniform dispersion and interfacial bonding across large areas or volumes remains a formidable challenge, limiting the scalability and practicality of nanofiber-based water-splitting technologies.

- Catalytic activity and stability challenges:
  While nanofiber-based catalyst composites demonstrate enhanced catalytic activity for water splitting, challenges remain in achieving optimal performance, selectivity, and long-term stability. The choice of catalytic materials, their loading density, and their distribution on the nanofiber surface can significantly impact catalytic activity and stability (Coy et al., 2021; Min et al., 2021; Wang, chieh Chiu, et al., 2024). However, maintaining high catalytic activity while ensuring stability under harsh operating conditions, such as high temperatures, corrosive environments, and prolonged operation, poses a major challenge. Moreover, understanding and mitigating degradation mechanisms, such as metal leaching, nanoparticle aggregation, and surface poisoning, is essential for improving the stability and durability of nanofiber-based catalyst composites. Developing strategies to enhance catalyst stability, prevent degradation, and mitigate surface fouling without compromising catalytic activity is critical for the practical implementation of nanofiber-based water-splitting technologies.
- Scalability and cost challenges:
  Another significant challenge facing nanofiber-based water-splitting technologies is scalability and cost-effectiveness (Lu et al., 2019; Zhu et al., 2017). While nanofibers offer unique properties and performance advantages, scaling up production processes to meet industrial-scale demands remains a major hurdle. Issues such as batch-to-batch variability, equipment scalability, raw material costs, and energy consumption need to be addressed to enable cost-effective large-scale production of nanofiber-based materials. Additionally, the cost of raw materials, synthesis precursors, and postprocessing steps can contribute to the overall cost of nanofiber-based water-splitting technologies. Developing cost-effective synthesis routes, utilizing abundant and sustainable resources, and optimizing manufacturing processes are essential for reducing production costs and enhancing the economic viability of nanofiber-based water-splitting technologies.

18.5.2 Strategies for optimization

To address the technical challenges and limitations associated with nanofiber-based water-splitting technologies, various optimization strategies can be pursued. These strategies encompass material design, synthesis techniques, catalytic engineering, and system integration, aiming to enhance performance, stability, scalability, and cost-effectiveness.

- Material design and synthesis optimization:
  Efforts to optimize nanofiber-based materials for water splitting should focus on precise control over morphology, composition, and structure to maximize catalytic activity, stability, and scalability (Bandaru et al., 2023; Díaz-Abad et al., 2022; Milikić et al., 2022). Advanced synthesis techniques, such as template-assisted synthesis, self-assembly, and vapor-phase deposition, offer opportunities to tailor nanofiber properties with improved uniformity, orientation, and alignment. Designing hierarchical structures and composite architectures with optimized interfacial interactions can further enhance catalytic performance and stability. Moreover, exploring novel materials, such as two-dimensional (2D) materials, metal-organic frameworks (MOFs), and hybrid nanocomposites, holds promise for improving water-splitting efficiency and durability. These materials offer unique properties, such as high surface area, tunable porosity, and enhanced charge transport, which can be leveraged to overcome existing limitations and optimize performance in water-splitting applications.

- Catalytic engineering and surface modification:

    Catalytic engineering plays a crucial role in optimizing the performance and stability of nanofiber-based catalyst composites for water splitting. Tailoring the composition, size, shape, and distribution of catalytic species on nanofiber surfaces can significantly impact catalytic activity, selectivity, and durability. Strategies such as atomic layer deposition (ALD), chemical vapor deposition (CVD), and electrochemical deposition enable precise control over catalyst loading density and distribution, enhancing catalytic performance while minimizing degradation (Gao et al., 2024; Wang, Chen, et al., 2024). Furthermore, surface modification techniques, such as functionalization, doping, and alloying, offer opportunities to tailor nanofiber surface chemistry and reactivity for improved catalytic activity and stability. Introducing specific functional groups, dopants, or coatings onto nanofiber surfaces can enhance catalytic selectivity, suppress side reactions, and mitigate degradation mechanisms, leading to enhanced overall performance in water-splitting reactions.
- System integration and process optimization:

In addition to material and catalytic optimization, system integration and process optimization are essential for enhancing the efficiency, scalability, and cost-effectiveness of nanofiber-based water-splitting technologies. Developing integrated device architectures, such as flow-through reactors, membrane-electrode assemblies (MEAs), and photoelectrochemical cells, can improve mass transport, reactant utilization, and overall system performance. Furthermore, optimizing process parameters, such as temperature, pressure, pH, and electrolyte composition, can enhance reaction kinetics, maximize product yields, and minimize energy consumption. Implementing advanced characterization techniques, process monitoring systems, and feedback control algorithms can enable real-time optimization and fine-tuning of water-splitting processes, leading to improved efficiency and reliability.

18.5.3 Opportunities for further research

Despite the challenges and limitations, nanofiber-based water-splitting technologies offer exciting opportunities for further research and development. Key areas for future exploration include advanced materials design, catalytic engineering, system integration, and sustainable manufacturing.

- Advanced materials design and characterization:

    Advancing materials design and characterization techniques is essential for developing next-generation nanofiber-based materials with enhanced performance and functionality. Exploring novel materials, such as 2D materials, perovskites, and MOFs, offers opportunities to overcome existing limitations and unlock new possibilities in water-splitting applications. Additionally, developing advanced characterization techniques, such as in situ spectroscopy, operando microscopy, and computational modeling, can provide deeper insights into the structure-property relationships of nanofiber-based materials, guiding the rational design and optimization of materials for water splitting.
- Catalytic engineering and surface science:

    Further research in catalytic engineering and surface science is needed to develop tailored catalysts with improved activity, selectivity, and stability for water-splitting applications. Investigating new catalytic materials, nanostructures, and surface modifications can help optimize catalytic performance and mitigate degradation mechanisms (Badmus et al., 2021; Cai et al., 2024). Additionally, understanding the fundamental mechanisms underlying catalytic reactions on nanofiber surfaces, such as adsorption, activation, and desorption processes, is crucial for designing efficient and durable catalysts for water splitting.

- System integration and process optimization:

  Exploring innovative system integration strategies and process optimization techniques can enhance the efficiency, scalability, and cost-effectiveness of nanofiber-based water-splitting technologies. Developing advanced reactor designs, electrode configurations, and system architectures can improve mass transport, reactant utilization, and overall system performance. Moreover, implementing advanced process monitoring and control systems can enable real-time optimization of water-splitting processes, leading to improved efficiency and reliability.

18.6 Conclusion

Addressing sustainability and scalability concerns is essential for the widespread adoption and commercialization of nanofiber-based water-splitting technologies. Developing sustainable manufacturing processes, utilizing renewable resources, and minimizing environmental impacts are critical for ensuring the long-term viability and societal acceptance of these technologies. Additionally, exploring cost-effective production methods, optimizing resource utilization, and leveraging economies of scale can help reduce production costs and enhance the economic competitiveness of nanofiber-based water-splitting technologies.

Nanofiber-based materials present an exciting avenue for propelling water-splitting technologies forward and tackling the pressing global issues surrounding clean energy production. These materials offer a unique advantage through their ability to precisely manipulate morphology, composition, and structure, resulting in enhanced catalytic activity and electrochemical performance. This makes them highly promising candidates for facilitating efficient and sustainable water-splitting processes. Despite the significant strides made in this field, there are still formidable challenges to overcome. Issues such as synthesizing nanofibers with consistent properties, optimizing catalytic activity and stability, ensuring scalability, and achieving cost-effectiveness remain pressing concerns. Nonetheless, with continued dedication to research and development, nanofiber-based water-splitting technologies hold immense potential for ushering in a more sustainable energy future. By focusing on strategies to optimize performance, identifying new avenues for exploration in research, and finding solutions to overcome barriers to commercialization, nanofiber-based materials can indeed play a pivotal role in driving the transition toward a cleaner and greener world. Through collaboration, innovation, and perseverance, nanofiber-based materials can revolutionize the landscape of clean energy production and contribute significantly to mitigating environmental challenges on a global scale.

References

Abdel Hameed, R. M., Al-Enizi, A. M., Karim, A., & Yousef, A. (2022). Free-standing and binder-free nickel polymeric nanofiber membrane for electrocatalytic oxidation of ethanol in alkaline solution. *Journal of Materials Research and Technology, 21*, 4591–4606. Available from https://doi.org/10.1016/j.jmrt.2022.110.054.

Abdelazeez, A. A. A., Hadia, N. M. A., Mourad, A. H. I., El-Fatah, G. A., Shaban, M., Ahmed, A. M., Alzaid, M., Cherupurakal, N., & Rabia, M. (2022). Effect of Au plasmonic material on poly M-toluidine for photoelectrochemical hydrogen generation from sewage water. *Polymers (Basel), 14*. Available from https://doi.org/10.3390/polym14040768.

Abdelhamid, H. N. (2023). An introductory review on advanced multifunctional materials. *Heliyon, 9*, e18060. Available from https://doi.org/10.1016/j.heliyon.2023.e18060.

Agboola, P. O., & Shakir, I. (2022). Facile fabrication of SnO_2/MoS_2/rGO ternary composite for solar light-mediated photocatalysis for water remediation. *Journal of Materials Research and Technology, 18*, 4303–4313. Available from https://doi.org/10.1016/j.jmrt.2022.040.109.

Aguilar-Ferrer, D., Szewczyk, J., & Coy, E. (2022). Recent developments in polydopamine-based photocatalytic nanocomposites for energy production: Physico-chemical properties and perspectives. *Catalysis Today, 397–399*, 316–349. Available from https://doi.org/10.1016/j.cattod.2021.080.016.

Ahmed, M. K., Zayed, M. A., El-dek, S. I., Hady, M. A., El Sherbiny, D. H., & Uskoković, V. (2021). Nanofibrous ε-polycaprolactone scaffolds containing Ag-doped magnetite nanoparticles: Physicochemical characterization and biological testing for wound dressing applications in vitro and in vivo. *Bioactive Materials, 6*, 2070–2088. Available from https://doi.org/10.1016/j.bioactmat.2020.120.026.

Al-Abduljabbar, A., & Farooq, I. (2023). Electrospun polymer nanofibers: Processing, properties, and applications. *Polymers (Basel), 15*. Available from https://doi.org/10.3390/polym15010065.

Ali, A., Liang, F., Feng, H., Tang, M., Jalil Shah, S., Ahmad, F., Ji, X., Kang Shen, P., & Zhu, J. (2023). Gram-scale production of in-situ generated iron carbide nanoparticles encapsulated via nitrogen and phosphorous co-doped bamboo-like carbon nanotubes for oxygen evolution reaction. *Materials Science for Energy Technologies, 6*, 301–309. Available from https://doi.org/10.1016/j.mset.2023.010.004.

De Alvarenga, G., Hryniewicz, B. M., Jasper, I., Silva, R. J., Klobukoski, V., Costa, F. S., Cervantes, T. N. M., Amaral, C. D. B., Schneider, J. T., Bach-Toledo, L., Peralta-Zamora, P., Valerio, T. L., Soares, F., Silva, B. J. G., & Vidotti, M. (2020). Recent trends of micro and nanostructured conducting polymers in health and environmental applications. *Journal of Electroanalytical Chemistry, 879*. Available from https://doi.org/10.1016/j.jelechem.2020.114754.

Ammar, Z., Ibrahim, H., Adly, M., Sarris, I., & Mehanny, S. (2023). Influence of natural fiber content on the frictional material of brake pads—A review. *Journal of Composites Science, 7*. Available from https://doi.org/10.3390/jcs7020072.

Arifa Farzana, B., Mujafarkani, N., Thakur, A., Kumar, A., Mushira Banu, A., & Shifana, M. (2023). Evaluating (p-semidine-guanidine-formaldehyde) terpolymer resin efficiency as anti-corrosive agent for mild steel in 1M H_2SO_4: An experimental and computational approach. *Inorganic Chemistry Communications, 158*111572. Available from https://doi.org/10.1016/j.inoche.2023.111572.

Azizi, H., Koocheki, A., & Ghorani, B. (2023). Structural elucidation of gluten/zein nanofibers prepared by electrospinning process: Focus on the effect of zein on properties of nanofibers. *Polymer Testing*, *128*108231. Available from https://doi.org/10.1016/j.polymertesting.2023.108231.

Badmus, M., Liu, J., Wang, N., Radacsi, N., & Zhao, Y. (2021). Hierarchically electrospun nanofibers and their applications: A review. *Nano Materials Science*, *3*, 213–232. Available from https://doi.org/10.1016/j.nanoms.2020.110.003.

Baena, J. C., Wu, J., & Peng, Z. (2015). Wear performance of UHMWPE and reinforced UHMWPE composites in arthroplasty applications: A review. *Lubricants*, *3*, 413–436. Available from https://doi.org/10.3390/lubricants3020413.

Balaji, K. V., Shirvanimoghaddam, K., & Naebe, M. (2024). Multifunctional basalt fiber polymer composites enabled by carbon nanotubes and graphene. *Composites Part B: Engineering*, *268*111070. Available from https://doi.org/10.1016/j.compositesb.2023.111070.

Balazadeh Meresht, N., Moghadasi, S., Munshi, S., Shahbakhti, M., & McTaggart-Cowan, G. (2023). Advances in vehicle and powertrain efficiency of long-haul commercial vehicles: A review. *Energies*, *16*. Available from https://doi.org/10.3390/en16196809.

Balitskii, A., Kolesnikov, V., Abramek, K. F., Balitskii, O., Eliasz, J., Havrylyuk, M., Ivaskevych, L., & Kolesnikova, I. (2021). Influence of hydrogen-containing fuels and environmentally friendly lubricating coolant on nitrogen steels' wear resistance for spark ignition engine pistons and rings kit gasket set. *Energies*, *14*. Available from https://doi.org/10.3390/en14227583.

Bandaru, A. K., Khan, A. N., Durmaz, T., Alagirusamy, R., & O'Higgins, R. M. (2023). Improved mechanical properties of multi-layered PTFE composites through hybridisation. *Construction and Building Materials*, *374*130921. Available from https://doi.org/10.1016/j.conbuildmat.2023.130921.

Bao, J., Xie, J., Lei, F., Wang, Z., Liu, W., Xu, L., Guan, M., Zhao, Y., & Li, H. (2018). Two-dimensional Mn-Co LDH/graphene composite towards high-performance water splitting. *Catalysts*, *8*. Available from https://doi.org/10.3390/catal8090350.

Borawski, A. (2021). Impact of operating time on selected tribological properties of the friction material in the brake pads of passenger cars. *Materials (Basel)*, *14*, 1–13. Available from https://doi.org/10.3390/ma14040884.

Borawski, A. (2022). Testing passenger car brake pad exploitation time's impact on the values of the coefficient of friction and abrasive wear rate using a pin-on-disc method. *Materials (Basel)*, *15*. Available from https://doi.org/10.3390/ma15061991.

Cai, H., Zhang, P., Li, B., Zhu, Y., Zhang, Z., & Guo, W. (2024). High-entropy oxides for energy-related electrocatalysis. *Materials Today Catalysis*, *4*100039. Available from https://doi.org/10.1016/j.mtcata.2024.100039.

Casati, R., & Vedani, M. (2014). Metal matrix composites reinforced by nano-particles—A review. *Metals (Basel)*, *4*, 65–83. Available from https://doi.org/10.3390/met4010065.

Chahal, R., Dalal, Y., Dahiya, S., Punia, R., Maan, A. S., Singh, K., & Ohlan, A. (2024). Insitu assembly of Fe_3O_4@$FeNi_3$ spherical mesoporous nanoparticles embedded on 2D reduced graphene oxide (RGO) layers as protective barrier for EMI pollution. *Applied Surface Science Advances*, *19*100545. Available from https://doi.org/10.1016/j.apsadv.2023.100545.

Christoforidou, A., Verleg, R., & Pavlovic, M. (2023). Static, fatigue and hygroscopic performance of steel-reinforced resins under various temperatures. *Construction and Building Materials*, *403*133079. Available from https://doi.org/10.1016/j.conbuildmat.2023.133079.

Cieluch, M., Düerkop, D., Kazamer, N., Wirkert, F., Podleschny, P., Rost, U., Schmiemann, A., & Brodmann, M. (2024). Manufacturing and investigation of MEAs for PEMWE based on glass fibre reinforced PFSA/ssPS composite membranes and catalyst-coated substrates prepared via catalyst electrodeposition. *International Journal of Hydrogen Energy, 52*, 521−533. Available from https://doi.org/10.1016/j.ijhydene.2023.070.310.

Cimesa, M., & Moustafa, M. A. (2022). Experimental characterization and analytical assessment of compressive behavior of carbon nanofibers enhanced UHPC. *Case Studies in Construction Materials, 17*, e01487. Available from https://doi.org/10.1016/j.cscm.2022.e01487.

Coy, E., Siuzdak, K., Grądzka-Kurzaj, I., Sayegh, S., Weber, M., Ziółek, M., Bechelany, M., & Iatsunskyi, I. (2021). Exploring the effect of BN and B-N bridges on the photocatalytic performance of semiconductor heterojunctions: Enhancing carrier transfer mechanism. *Applied Materials Today, 24*. Available from https://doi.org/10.1016/j.apmt.2021.101095.

Cuenca, E., Mezzena, A., & Ferrara, L. (2021). Synergy between crystalline admixtures and nano-constituents in enhancing autogenous healing capacity of cementitious composites under cracking and healing cycles in aggressive waters. *Construction and Building Materials, 266*121447. Available from https://doi.org/10.1016/j.conbuildmat.2020.121447.

Cui, Y. p, Shang, Y. r, Shi, R. x, Che, Q. d, & Wang, J. p (2022). Pt-decorated $NiWO_4/WO_3$ heterostructure nanotubes for highly selective sensing of acetone. *Nonferrous Metals Society of China (English Edition), 32*, 1981−1993. Available from https://doi.org/10.1016/S1003-6326(22)65924-7.

Dhonchak, C., & Agnihotri, N. (2023). Computational insights in the spectrophotometrically 4H-chromen-4-one complex using DFT method. *Biointerface Research in Applied Chemistry, 13*, 357.

Do, H. H., Nguyen, D. L. T., Nguyen, X. C., Le, T. H., Nguyen, T. P., Trinh, Q. T., Ahn, S. H., Vo, D. V. N., Kim, S. Y., & Van Le, Q. (2020). Recent progress in TiO_2-based photocatalysts for hydrogen evolution reaction: A review. *Arabian Journal of Chemistry, 13*, 3653−3671. Available from https://doi.org/10.1016/j.arabjc.2019.120.012.

Díaz-Abad, S., Fernández-Mancebo, S., Rodrigo, M. A., & Lobato, J. (2022). Enhancement of SO_2 high temperature depolarized electrolysis by means of graphene oxide composite polybenzimidazole membranes. *Journal of Cleaner Production, 363*. Available from https://doi.org/10.1016/j.jclepro.2022.132372.

Eswaran, M., Rahimi, S., Pandit, S., Chokkiah, B., & Mijakovic, I. (2023). A flexible multifunctional electrode based on conducting PANI/Pd composite for non-enzymatic glucose sensor and direct alcohol fuel cell applications. *Fuel, 345*128182. Available from https://doi.org/10.1016/j.fuel.2023.128182.

Fan, J., Chang, X., Li, L., & Zhang, M. (2023). Synthesis of $CoMoO_4$ nanofibers by electrospinning as efficient electrocatalyst for overall water splitting. *Molecules (Basel, Switzerland), 29*, 7. Available from https://doi.org/10.3390/molecules29010007.

Gao, Y., Li, G., Chen, W., Shi, X., Gong, C., Shao, Q., & Liu, Y. (2024). Graphene oxide coated fly ash for reinforcing dynamic tensile behaviours of cementitious composites. *Construction and Building Materials, 411*134289. Available from https://doi.org/10.1016/j.conbuildmat.2023.134289.

Garkal, A., Bangar, P., & Mehta, T. (2022). Thin-film nanofibers for treatment of age-related macular degeneration. *OpenNano, 8*100098. Available from https://doi.org/10.1016/j.onano.2022.100098.

Gawel, K., Wenner, S., Jafariesfad, N., Torsæter, M., & Justnes, H. (2022). Portland cement hydration in the vicinity of electrically polarized conductive surfaces. *Cement and Concrete Composites, 134*104792. Available from https://doi.org/10.1016/j.cemconcomp.2022.104792.

Ge, J., Zhang, Y., Heo, Y. J., & Park, S. J. (2019). Advanced design and synthesis of composite photocatalysts for the remediation of wastewater: A review. *Catalysts*. Available from https://doi.org/10.3390/catal9020122.

Gebreslase, G. A., Martínez-Huerta, M. V., Sebastián, D., & Lázaro, M. J. (2022). Transformation of $CoFe_2O_4$ spinel structure into active and robust CoFe alloy/N-doped carbon electrocatalyst for oxygen evolution reaction. *Journal of Colloid and Interface Science*, *625*, 70–82. Available from https://doi.org/10.1016/j.jcis.2022.060.005.

Gowrisankar, A., Sherryn, A. L., & Selvaraju, T. (2021). In situ integrated 2D reduced graphene oxide nanosheets with MoSSe for hydrogen evolution reaction and supercapacitor application. *Applied Surface Science Advances*, *3* 100054. Available from https://doi.org/10.1016/j.apsadv.2020.100054.

Han, N., Wang, S., Rana, A. K., Asif, S., Klemeš, J. J., Bokhari, A., Long, J., Thakur, V. K., & Zhao, X. (2022). Rational design of boron nitride with different dimensionalities for sustainable applications. *Renewable and Sustainable Energy Reviews*, *170*. Available from https://doi.org/10.1016/j.rser.2022.112910.

Huang, J., Li, F., Liu, B., & Zhang, P. (2020). Ni_2P/rGO/NF nanosheets as a bifunctional high-performance electrocatalyst for water splitting. *Materials (Basel)*, *13*, 1–10. Available from https://doi.org/10.3390/ma13030744.

Huang, T., Liu, W., Liu, Y., Hou, Q., Chen, S., Li, R., & Liu, H. (2024). Structural trade-off regulation of composite aerogels via "island-bridge" design for advanced nickel-iron batteries. *Next Energy*, *2* 100076. Available from https://doi.org/10.1016/j.nxener.2023.100076.

Ilie, F., & Cristescu, A. C. (2022). Tribological behavior of friction materials of a disk-brake pad braking system affected by structural changes—A review. *Materials (Basel)*, *15*. Available from https://doi.org/10.3390/ma15144745.

Jaleh, B., Moradi, A., Eslamipanah, M., Khazalpour, S., Tahzibi, H., Azizian, S., & Gawande, M. B. (2023). Laser-assisted synthesis of Au NPs on MgO/chitosan: Applications in electrochemical hydrogen storage. *Journal of Magnesium and Alloys*, *11*, 2072–2083. Available from https://doi.org/10.1016/j.jma.2023.050.003.

Jamesh, M. I., Akila, A., Sudha, D., Gnana Priya, K., Sivaprakash, V., & Revathi, A. (2022). Fabrication of earth-abundant electrocatalysts based on green-chemistry approaches to achieve efficient alkaline water splitting—A teview. *Sustainability*, *14*. Available from https://doi.org/10.3390/su142416359.

Ji, Y., Song, W., Xu, L., Yu, D. G., & Bligh, S. W. A. (2022). A review on electrospun poly (amino acid) nanofibers and their applications of hemostasis and wound healing. *Biomolecules*, *12*. Available from https://doi.org/10.3390/biom12060794.

Kadja, G. T. M., Mualliful Ilmi, M., Mardiana, S., Khalil, M., Sagita, F., Culsum, N. T. U., & Fajar, A. T. N. (2023). Recent advances of carbon nanotubes as electrocatalyst for insitu hydrogen production and CO_2 conversion to fuels. *Results in Chemistry*, *6* 101037. Available from https://doi.org/10.1016/j.rechem.2023.101037.

Kaya, S., Lgaz, H., Thakkur, A., Kumar, A., Özbakır, D., Karakuş, N., & Ben Ahmed, S. (2023). Molecular insights into the corrosion inhibition mechanism of omeprazole and tinidazole: A theoretical investigation. *Molecular Simulation*, 1–15. Available from https://doi.org/10.1080/08927022.2023.2256888.

Kaya, S., Thakur, A., & Kumar, A. (2023). The role of in silico/DFT investigations in analyzing dye molecules for enhanced solar cell efficiency and reduced toxicity. *Journal of Molecular Graphics & Modelling*, *124*. Available from https://doi.org/10.1016/j.jmgm.2023.108536.

Khan, H., Kang, S., & Lee, C. S. (2021). Evaluation of efficient and noble-metal-free nitio3 nanofibers sensitized with porous gc3 n4 sheets for photocatalytic applications. *Catalysts*, *11*, 1−16. Available from https://doi.org/10.3390/catal11030385.

Khan, I., Lee, J. H., Park, J., & Wooh, S. (2022). Nano/micro-structural engineering of Nafion membranes for advanced electrochemical applications. *Journal of Saudi Chemical Society*, *26*101511. Available from https://doi.org/10.1016/j.jscs.2022.101511.

Kim, S., Nguyen, N. T., & Bark, C. W. (2018). Ferroelectric materials: A novel pathway for efficient solar water splitting. *Applied Sciences*, *8*. Available from https://doi.org/10.3390/app8091526.

Kindrachuk, M., Volchenko, D., Balitskii, A., Abramek, K. F., Volchenko, M., Balitskii, O., Skrypnyk, V., Zhuravlev, D., Yurchuk, A., & Kolesnikov, V. (2021). Wear resistance of spark ignition engine piston rings in hydrogen-containing environments. *Energies*, *14*. Available from https://doi.org/10.3390/en14164801.

Koskin, A. P., Larichev, Y. V., Mishakov, I. V., Mel'gunov, M. S., & Vedyagin, A. A. (2020). Synthesis and characterization of carbon nanomaterials functionalized by direct treatment with sulfonating agents. *Microporous and Mesoporous Materials*, *299*110130. Available from https://doi.org/10.1016/j.micromeso.2020.110130.

Kotb, Y., & Velev, O. D. (2023). Hierarchically reinforced biopolymer composite films as multifunctional plastics substitute. *Cell Reports Physical Science*, *4*101732. Available from https://doi.org/10.1016/j.xcrp.2023.101732.

Kumar, A., & Thakur, A. (2020). *Encapsulated nanoparticles in organic polymers for corrosion inhibition*. Elsevier Inc. Available from https://doi.org/10.1016/B978-0-12-819359-4.00018-0.

Kumbhar, V. S., Lee, H., Lee, J., & Lee, K. (2019). Recent advances in water-splitting electrocatalysts based on manganese oxide. *Carbon Resources Conversion*, *2*, 242−255. Available from https://doi.org/10.1016/j.crcon.2019.110.003.

La, D. D., Dang, T. D., Le, P. C., Bui, X. T., Chang, S. W., Chung, W. J., Kim, S. C., & Nguyen, D. D. (2023). Self-assembly of monomeric porphyrin molecules into nanostructures: Self-assembly pathways and applications for sensing and environmental treatment. *Environmental Technology & Innovation*, *29*103019. Available from https://doi.org/10.1016/j.eti.2023.103019.

Lau, J., Dey, G., & Licht, S. (2016). Thermodynamic assessment of CO_2 to carbon nanofiber transformation for carbon sequestration in a combined cycle gas or a coal power plant. *Energy Conversion and Management*, *122*, 400−410. Available from https://doi.org/10.1016/j.enconman.2016.060.007.

Li, W., Yang, X., Wang, S., Xiao, J., & Hou, Q. (2020). Comprehensive analysis on the performance and material of automobile brake discs. *Metals (Basel)*, *10*. Available from https://doi.org/10.3390/met10030377.

Liu, C., Zhang, X., Zhao, L., Hui, L., & Liu, D. (2023). Multilayer amnion-PCL nanofibrous membrane loaded with celecoxib exerts a therapeutic effect against tendon adhesion by improving the inflammatory microenvironment. *Heliyon*, *9*, e23214. Available from https://doi.org/10.1016/j.heliyon.2023.e23214.

Liu, F., Wang, L., Li, D., Liu, Q., & Deng, B. (2020). Preparation and characterization of novel thin film composite nanofiltration membrane with PVDF tree-like nanofiber membrane as composite scaffold. *Materials & Design*, *196*109101. Available from https://doi.org/10.1016/j.matdes.2020.109101.

Liu, G., Xu, Y., Yang, T., & Jiang, L. (2023). Recent advances in electrocatalysts for seawater splitting. *Nano Materials Science*, *5*, 101−116. Available from https://doi.org/10.1016/j.nanoms.2020.120.003.

Logozzo, S., & Valigi, M. C. (2023). Experimental wear characterization and durability enhancement of an aeronautic braking system. *Applied Sciences*, *13*. Available from https://doi.org/10.3390/app13137646.

Lu, F., Wang, J., Chang, Z., & Zeng, J. (2019). Uniform deposition of Ag nanoparticles on ZnO nanorod arrays grown on polyimide/Ag nanofibers by electrospinning, hydrothermal, and photoreduction processes. *Materials & Design*, *181*108069. Available from https://doi.org/10.1016/j.matdes.2019.108069.

Ma, L., Ding, S., Zhang, C., Zhang, M., & Shi, H. (2022). Study on the wear performance of brake materials for high-speed railway with intermittent braking under low-temperature environment conditions. *Materials (Basel)*, *15*. Available from https://doi.org/10.3390/ma15248763.

Maafa, I. M., Al-Enizi, A. M., Abutaleb, A., Zouli, N. I., Ubaidullah, M., Shaikh, S. F., Al-Abdrabalnabi, M. A., & Yousef, A. (2021). One-pot preparation of CdO/ZnO core/shell nanofibers: An efficient photocatalyst. *Alexandria Engineering Journal*, *60*, 1819−1826. Available from https://doi.org/10.1016/j.aej.2020.110.030.

Macdonald, T. J., Ambroz, F., Batmunkh, M., Li, Y., Kim, D., Contini, C., Poduval, R., Liu, H., Shapter, J. G., Papakonstantinou, I., & Parkin, I. P. (2018). TiO_2 nanofiber photoelectrochemical cells loaded with sub-12 nm AuNPs: Size dependent performance evaluation. *Materials Today Energy*, *9*, 254−263. Available from https://doi.org/10.1016/j.mtener.2018.060.005.

Magrini, T., Bouville, F., & Studart, A. R. (2021). Transparent materials with stiff and tough hierarchical structures. *Open Ceramics*, *6*. Available from https://doi.org/10.1016/j.oceram.2021.100109.

Manda, A. A., Drmosh, Q. A., Elsayed, K. A., Al-Alotaibi, A. L., Olanrewaju Alade, I., Onaizi, S. A., Dafalla, H. D. M., & Elhassan, A. (2022). Highly efficient UV-visible absorption of TiO_2/Y_2O_3 nanocomposite prepared by nanosecond pulsed laser ablation technique. *Arabian Journal of Chemistry*, *15*104004. Available from https://doi.org/10.1016/j.arabjc.2022.104004.

Marlinda, A. R., Thien, G. S. H., Shahid, M., Ling, T. Y., Hashem, A., Chan, K. Y., & Johan, M. R. (2023). Graphene as a lubricant additive for reducing friction and wear in its liquid-based form. *Lubricants*, *11*. Available from https://doi.org/10.3390/lubricants11010029.

Martyushev, N. V., Malozyomov, B. V., Khalikov, I. H., Kukartsev, V. A., Kukartsev, V. V., Tynchenko, V. S., Tynchenko, Y. A., & Qi, M. (2023). Review of methods for improving the energy efficiency of electrified ground transport by optimizing battery consumption. *Energies*, *16*. Available from https://doi.org/10.3390/en16020729.

Milikić, J., Tapia, A., Stamenović, U., Vodnik, V., Otoničar, M., Škapin, S., Santos, D. M. F., & Šljukić, B. (2022). High-performance metal (Au,Cu)−polypyrrole nanocomposites for electrochemical borohydride oxidation in fuel cell applications. *International Journal of Hydrogen Energy*, *47*, 36990−37001. Available from https://doi.org/10.1016/j.ijhydene.2022.080.229.

Min, G., Pullanchiyodan, A., Dahiya, A. S., Hosseini, E. S., Xu, Y., Mulvihill, D. M., & Dahiya, R. (2021). Ferroelectric-assisted high-performance triboelectric nanogenerators based on electrospun P(VDF-TrFE) composite nanofibers with barium titanate nanofillers. *Nano Energy*, *90*106600. Available from https://doi.org/10.1016/j.nanoen.2021.106600.

Momeni, P., Nourisefat, M., Farzaneh, A., Shahrousvand, M., & Abdi, M. H. (2024). The engineering, drug release, and in vitro evaluations of the PLLA/HPC/*Calendula officinalis* electrospun nanofibers optimized by response surface methodology. *Heliyon*, *10*, e23218. Available from https://doi.org/10.1016/j.heliyon.2023.e23218.

Montani, M., Vitaliti, D., Capitani, R., & Annicchiarico, C. (2020). Performance review of three car integrated abs types: Development of a tire independent wheel speed control. *Energies, 13*. Available from https://doi.org/10.3390/en13236183.

Nabgan, W., Nabgan, B., Jalil, A. A., Ikram, M., Hussain, I., Bahari, M. B., Tran, T. V., Alhassan, M., Owgi, A. H. K., Parashuram, L., Nordin, A. H., & Medina, F. (2024). A bibliometric examination and state-of-the-art overview of hydrogen generation from photoelectrochemical water splitting. *International Journal of Hydrogen Energy, 52*, 358−380. Available from https://doi.org/10.1016/j.ijhydene.2023.050.162.

Najjar, I. R., Sadoun, A. M., Fathy, A., Abdallah, A. W., Elaziz, M. A., & Elmahdy, M. (2022). Prediction of tribological properties of alumina-coated, silver-reinforced copper nanocomposites using long short-term model combined with golden jackal optimization. *Lubricants, 10*. Available from https://doi.org/10.3390/lubricants10110277.

Nayl, A., Abd-Elhamid, A., Awwad, N., Abdelgawad, M., Wu, J., Mo, X., Gomha, S., Aly, A., & Brase, S. (2022). Recent progress and potential biomedical applications of electrospun nanofibers in regeneration of tissues and organs. *Polymers (Basel), 14*. Available from https://doi.org/10.3390/polym14081508.

Ng, C. H., Mistoh, M. A., Teo, S. H., Galassi, A., Taufiq-Yap, Y. H., Siambun, N. J., Foo, J., Sipaut, C. S., Seay, J., & Janaun, J. (2024). The roles of carbonaceous wastes for catalysis, energy, and environmental remediation. *Catalysis Communications, 187*106845. Available from https://doi.org/10.1016/j.catcom.2024.106845.

Niu, Q., Ola, O., Chen, B., Zhu, Y., Xia, Y., & Ma, G. (2020). Metal sulfide nanoparticles anchored N, S co-doped porous carbon nanofibers as highly efficient bifunctional electrocatalysts for oxygen reduction/evolution reactions. *International Journal of Electrochemical Science, 15*, 4869−4883. Available from https://doi.org/10.20964/2020.06.28.

Nyamai, N., & Phaahlamohlaka, T. (2023). Significantly advanced hydrogen production via water splitting over Zn/TiO$_2$/CNFs and Cu/TiO$_2$/CNFs nanocomposites. *Journal of Industrial and Engineering Chemistry*. Available from https://doi.org/10.1016/j.jiec.2023.120.061.

Pardhi, S., Chakraborty, S., Tran, D. D., El Baghdadi, M., Wilkins, S., & Hegazy, O. (2022). A review of fuel cell powertrains for long-haul heavy-duty vehicles: Technology, hydrogen, energy and thermal management solutions. *Energies, 15*, 1−55. Available from https://doi.org/10.3390/en15249557.

Pinto, R. L. M., Gutiérrez, J. C. H., Pereira, R. B. D., de Faria, P. E., & Rubio, J. C. C. (2021). Influence of contact plateaus characteristics formed on the surface of brake friction materials in braking performance through experimental tests. *Materials (Basel), 14*. Available from https://doi.org/10.3390/ma14174931.

Ponnalagar, D., Hang, D. R., Islam, S. E., Te Liang, C., & Chou, M. M. C. (2023). Recent progress in two-dimensional Nb$_2$C MXene for applications in energy storage and conversion. *Materials & Design, 231*112046. Available from https://doi.org/10.1016/j.matdes.2023.112046.

Rosmini, C., Tsoncheva, T., Kovatcheva, D., Velinov, N., Kolev, H., Karashanova, D., Dimitrov, M., Tsyntsarski, B., Sebastián, D., & Lázaro, M. J. (2022). Mesoporous Ce−Fe−Ni nanocomposites encapsulated in carbon-nanofibers: Synthesis, characterization and catalytic behavior in oxygen evolution reaction. *Carbon, 196*, 186−202. Available from https://doi.org/10.1016/j.carbon.2022.040.036.

Rusdi, R. A. A., Halim, N. A., Nurazzi, N. M., Abidin, Z. H. Z., Abdullah, N., Ros, F. C., Ahmad, N., & Azmi, A. F. M. (2022). The effect of layering structures on mechanical and thermal properties of hybrid bacterial cellulose/Kevlar reinforced epoxy composites. *Heliyon, 8*, e09442. Available from https://doi.org/10.1016/j.heliyon.2022.e09442.

Santangelo, S. (2019). Electrospun nanomaterials for energy applications: Recent advances. *Applied Sciences*. Available from https://doi.org/10.3390/app9061049.

Shaikh, Z. A., Laghari, A. A., Litvishko, O., Litvishko, V., Kalmykova, T., & Meynkhard, A. (2021). Liquid-phase deposition synthesis of zif-67-derived synthesis of CO_3O_4@TiO_2 composite for efficient electrochemical water splitting. *Metals (Basel)*, *11*, 1−10. Available from https://doi.org/10.3390/met11030420.

Sharma, D., Abhinay Thakur, M. K. S., Sharma, M. K., Jakhar, K., Kumar, A., Sharma, A. K., & Om, H. (2023). Synthesis, electrochemical, morphological, computational and corrosion inhibition studies of 3-(5-naphthalen-2-yl-[1,3,4]oxadiazol-2-yl)-pyridine against mild steel in 1M HCl. *Asian Journal of Chemistry*, *35*, 1079−1088. Available from https://doi.org/10.14233/ajchem.2023.27711.

Sharma, D., Thakur, A., Kumar, M., Sharma, R., Kumar, S., & Om, H. (2023). Effective corrosion inhibition of mild steel using novel 1, 3, 4-oxadiazole-pyridine hybrids: Synthesis, electrochemical, morphological, and computational insights. *Environmental Research*, *234*116555. Available from https://doi.org/10.1016/j.envres.2023.116555.

Song, Z., Chen, Y., Ren, N., & Duan, X. (2023). Recent advances in the fixed-electrode capacitive deionization (CDI): Innovations in electrode materials and applications. *Environmental Functional Materials*. Available from https://doi.org/10.1016/j.efmat.2023.110.001.

Stejskal, J. (2022). Recent advances in the removal of organic dyes from aqueous media with conducting polymers, polyaniline and polypyrrole, and their composites. *Polymers (Basel)*, *14*. Available from https://doi.org/10.3390/polym14194243.

Su, R., Xie, C., Alhassan, S. I., Huang, S., Chen, R., Xiang, S., Wang, Z., & Huang, L. (2020). Oxygen reduction reaction in the field of water environment for application of nanomaterials. *Nanomaterials*, *10*, 1−34. Available from https://doi.org/10.3390/nano10091719.

Sun, H., Qin, P., Liang, Y., Yang, Y., Zhang, J., Guo, J., Hu, X., Jiang, Y., Zhou, Y., Luo, L., & Wu, Z. (2023). Sonochemically assisted the synthesis and catalytic application of bismuth-based photocatalyst: A mini review. *Ultrasonics Sonochemistry*, *100*106600. Available from https://doi.org/10.1016/j.ultsonch.2023.106600.

Thakur, A., Kaya, S., & Kumar, A. (2021). Recent innovations in nano container-based self-healing coatings in the construction industry. *Current Nanoscience*, *18*, 203−216. Available from https://doi.org/10.2174/1573413717666210216120741.

Thakur, A., & Kumar, A. (2023a). Exploring the potential of ionic liquid-based electrochemical biosensors for real-time biomolecule monitoring in pharmaceutical applications: From lab to life. *Results in Engineering*, *20*101533. Available from https://doi.org/10.1016/j.rineng.2023.101533.

Thakur, A., & Kumar, A. (2023b). Ecotoxicity analysis and risk assessment of nanomaterials for the environmental remediation. *Macromolecular Symposia*, *410*, 1−23. Available from https://doi.org/10.1002/masy.202100438.

Thakur, A., & Kumar, A. (2023c). Recent trends in nanostructured carbon-based electrochemical sensors for the detection and remediation of persistent toxic substances in real-time analysis. *Materials Research Express*, *10*. Available from https://doi.org/10.1088/2053-1591/acbd1a.

Thakur, A., & Kumar, A. (2024). Unraveling the multifaceted mechanisms and untapped potential of activated carbon in remediation of emerging pollutants: A comprehensive review and critical appraisal of advanced techniques. *Chemosphere*, *346*140608. Available from https://doi.org/10.1016/j.chemosphere.2023.140608.

Thakur, A., Kumar, A., Kaya, S., Benhiba, F., & Sharma, S. (2023). Electrochemical and computational investigations of the *Thysanolaena latifolia* leaves extract: An eco-benign solution for the corrosion mitigation of mild steel. *Results in Chemistry*, *6*101147. Available from https://doi.org/10.1016/j.rechem.2023.101147.

Thakur, A., Kumar, A., Sharma, S., Ganjoo, R., & Assad, H. (2022). Computational and experimental studies on the efficiency of *Sonchus arvensis* as green corrosion inhibitor for mild steel in 0.5 M HCl solution. *Materials Today: Proceedings*, *66*, 609–621. Available from https://doi.org/10.1016/j.matpr.2022.060.479.

Thakur, A., Kumar, A., & Singh, A. (2023). Adsorptive removal of heavy metals, dyes, and pharmaceuticals: Carbon-based nanomaterials in focus. *Carbon*, *217*118621. Available from https://doi.org/10.1016/j.carbon.2023.118621.

Thakur, A., Savaş, K., & Kumar, A. (2023). Recent trends in the characterization and application progress of nano-modified coatings in corrosion mitigation of metals and alloys. *Applied Sciences*, *13*, 730. Available from https://doi.org/10.3390/app13020730.

Thakur, A., Sharma, S., Ganjoo, R., Assad, H., & Kumar, A. (2022). Anti-corrosive potential of the sustainable corrosion inhibitors based on biomass waste: A review on preceding and perspective research. *Journal of Physics: Conference Series*, *2267*012079. Available from https://doi.org/10.1088/1742-6596/2267/1/012079.

Van Tran, V., Nu, T. T. V., Jung, H. R., & Chang, M. (2021). Advanced photocatalysts based on conducting polymer/metal oxide composites for environmental applications. *Polymers (Basel)*, *13*. Available from https://doi.org/10.3390/polym13183031.

Verma, C., Thakur, A., Ganjoo, R., Sharma, S., Assad, H., Kumar, A., Quraishi, M. A., & Alfantazi, A. (2023). Coordination bonding and corrosion inhibition potential of nitrogen-rich heterocycles: Azoles and triazines as specific examples. *Coordination Chemistry Reviews*, *488*215177. Available from https://doi.org/10.1016/j.ccr.2023.215177.

Vodovozov, V., Raud, Z., & Petlenkov, E. (2022). Review of energy challenges and horizons of hydrogen city buses. *Energies*, *15*. Available from https://doi.org/10.3390/en15196945.

Wan, K., Li, Y., Wang, Y., & Wei, G. (2021). Recent advance in the fabrication of 2D and 3D metal carbides-based nanomaterials for energy and environmental applications. *Nanomaterials*, *11*, 1–34. Available from https://doi.org/10.3390/nano11010246.

Wang, J., Ong, W. L., Ho, J. H., & Ho, G. W. (2017). Inorganic-organic hybrid membranes for photocatalytic hydrogen generation and volatile organic compound degradation. *Procedia Engineering*, *215*, 202–210. Available from https://doi.org/10.1016/j.proeng.2017.110.010.

Wang, N., & Yin, Z. (2022). The influence of mullite shape and amount on the tribological properties of non-asbestos brake friction composites. *Lubricants*, *10*. Available from https://doi.org/10.3390/lubricants10090220.

Wang, P., Lv, H., Cao, X., Liu, Y., & Yu, D. G. (2023). recent progress of the preparation and application of electrospun porous nanofibers. *Polymers (Basel)*, *15*. Available from https://doi.org/10.3390/polym15040921.

Wang, Q., Ricote, S., & Chen, M. (2023). Oxygen electrodes for protonic ceramic cells. *Electrochimica Acta*, *446*142101. Available from https://doi.org/10.1016/j.electacta.2023.142101.

Wang, S., Chen, S., Sun, J., Liu, Z., He, D., & Xu, S. (2024). Effects of rare earth oxides on the mechanical and tribological properties of phenolic-based hybrid nanocomposites. *Polymers (Basel)*, *16*. Available from https://doi.org/10.3390/polym16010131.

Wang, S., chieh Chiu, H., & Demopoulos, G. P. (2024). Tetragonal phase-free crystallization of highly conductive nanoscale cubic garnet ($Li_{6.1}Al_{0.3}La_3Zr_2O_{12}$) for all-solid-state lithium-metal batteries. *Journal of Power Sources, 595*234061. Available from https://doi.org/10.1016/j.jpowsour.2024.234061.

Wang, X., & Ru, H. (2019). Effect of lubricating phase on microstructure and properties of Cu-Fe friction materials. *Materials (Basel), 12*. Available from https://doi.org/10.3390/ma12020313.

Wang, Z., Shuai, S., Li, Z., & Yu, W. (2021). A review of energy loss reduction technologies for internal combustion engines to improve brake thermal efficiency. *Energies, 14*, 1−18. Available from https://doi.org/10.3390/en14206656.

Waqas, M., Zahid, R., Bhutta, M. U., Khan, Z. A., & Saeed, A. (2021). A review of friction performance of lubricants with nano additives. *Materials (Basel), 14*. Available from https://doi.org/10.3390/ma14216310.

Woo, S., Lee, J., Lee, D. S., Kim, J. K., & Byungkwon, L. (2020). Electrospun carbon nanofibers with embedded co-ceria nanoparticles for efficient hydrogen evolution and overall water splitting. *Materials (Basel), 13*. Available from https://doi.org/10.3390/ma13040856.

Xiao, Z., Xiao, X., Kong, L. B., Dong, H., Li, X., Sun, X., He, B., Ruan, S., & Zhai, J. (2023). MXenes and MXene-based composites for energy conversion and storage applications. *Journal of Materials Chemistry, 9*, 1067−1112. Available from https://doi.org/10.1016/j.jmat.2023.040.013.

Xin, W., Lin, C., Fu, L., Kong, X. Y., Yang, L., Qian, Y., Zhu, C., Zhang, Q., Jiang, L., & Wen, L. (2021). Nacre-like mechanically robust heterojunction for lithium-ion extraction. *Matter, 4*, 737−754. Available from https://doi.org/10.1016/j.matt.2020.120.003.

Yang, W., Li, H., Li, P., Xie, L., Liu, Y., Cao, Z., Tian, C., Wang, C. A., & Xie, Z. (2023). Facile synthesis of Co nanoparticles embedded in N-doped carbon nanotubes/graphitic nanosheets as bifunctional electrocatalysts for electrocatalytic water splitting. *Molecules (Basel, Switzerland), 28*. Available from https://doi.org/10.3390/molecules28186709.

Yaqoob, H., Teoh, Y. H., Sher, F., Farooq, M. U., Jamil, M. A., Kausar, Z., Sabah, N. U., Shah, M. F., Rehman, H. Z. U., & Rehman, A. U. (2021). Potential of waste cooking oil biodiesel as renewable fuel in combustion engines: A review. *Energies, 14*. Available from https://doi.org/10.3390/en14092565.

Yuanzheng, L., Youqi, W., Jiang, H., & Buyin, L. (2021). Nanofiber enhanced graphene−elastomer with unique biomimetic hierarchical structure for energy storage and pressure sensing. *Materials & Design, 203*109612. Available from https://doi.org/10.1016/j.matdes.2021.109612.

Yun, J., Zhou, C., Guo, B., Wang, F., Zhou, Y., Ma, Z., & Qin, J. (2023). Mechanically strong and multifunctional nano-nickel aerogels based epoxy composites for ultra-high electromagnetic interference shielding and thermal management. *Journal of Materials Research and Technology, 24*, 9644−9656. Available from https://doi.org/10.1016/j.jmrt.2023.050.193.

Zhang, C., Bhoyate, S., Zhao, C., Kahol, P. K., Kostoglou, N., Mitterer, C., Hinder, S. J., Baker, M. A., Constantinides, G., Polychronopoulou, K., Rebholz, C., & Gupta, R. K. (2019). Electrodeposited nanostructured $CoFe_2O_4$ for overall water splitting and supercapacitor applications. *Catalysts, 9*. Available from https://doi.org/10.3390/catal9020176.

Zhang, S., Jiang, E., Wu, J., Liu, Z., Yan, Y., Huo, P., & Yan, Y. (2023). Visible-light-driven $GO/Rh-SrTiO_3$ photocatalyst for efficient overall water splitting. *Catalysts, 13*, 1−14. Available from https://doi.org/10.3390/catal13050851.

Zhang, Y., Heo, Y. J., Lee, J. W., Lee, J. H., Bajgai, J., Lee, K. J., & Park, S. J. (2018). Photocatalytic hydrogen evolution via water splitting: A short review. *Catalysts, 8.* Available from https://doi.org/10.3390/catal8120655.

Zhang, Z., Zhou, H., Yao, P., Fan, K., Liu, Y., Zhao, L., Xiao, Y., Gong, T., & Deng, M. (2021). Effect of fe and cr on the macro/micro tribological behaviours of copper-based composites. *Materials (Basel), 14.* Available from https://doi.org/10.3390/ma14123417.

Zheng, C., Ma, Z., Yu, L., Wang, X., Zheng, L., & Zhu, L. (2022). Effect of silicon carbide nanoparticles on the friction-wear properties of copper-based friction discs. *Materials (Basel), 15.* Available from https://doi.org/10.3390/ma15020587.

Zhou, J., Zhang, L. n, Liu, B. b, Xu, C. x, & Liu, H. (2023). Preparation of hollow core-shelled MnCoSex/MnO@nitrogen-doped carbon composite by multiple interfaces coupling and its electrochemical properties. *Transactions of Nonferrous Metals Society of China (English Edition), 33,* 2471–2482. Available from https://doi.org/10.1016/S1003-6326(23)66274-0.

Zhou, Z., Jia, Y., Wang, Q., Jiang, Z., Xiao, J., & Guo, L. (2023). Recent progress on molybdenum carbide-based catalysts for hydrogen evolution: A review. *Sustainabilty, 15.* Available from https://doi.org/10.3390/su151914556.

Zhu, L., Nuo Peh, C. K., Gao, M., & Ho, G. W. (2017). Hierarchical heterostructure of TiO_2 nanosheets on CuO nanowires for enhanced photocatalytic performance. *Procedia Engineering, 215,* 180–187. Available from https://doi.org/10.1016/j.proeng.2017.110.007.

Zou, X., Yang, Y., Chen, H., Shi, X. L., Song, S., & Chen, Z. G. (2021). Hierarchical meso/macro-porous TiO_2/graphitic carbon nitride nanofibers with enhanced hydrogen evolution. *Materials & Design, 202*109542. Available from https://doi.org/10.1016/j.matdes.2021.109542.

Österle, W., & Dmitriev, A. I. (2016). The role of solid lubricants for brake friction materials. *Lubricants, 4,* 1–22. Available from https://doi.org/10.3390/lubricants4010005.

Šimko, M., & Lukáš, D. (2016). Mathematical modeling of a whipping instability of an electrically charged liquid jet. *Applied Mathematical Modelling, 40,* 9565–9583. Available from https://doi.org/10.1016/j.apm.2016.060.018.

Water, air, and soil purification from the application of nanofibers and their composites

19

Ainun Zulfikar[1,2], Marita Wulandari[1,3], Abdul Halim[4] and Bimastyaji Surya Ramadan[5]

[1]Graduate School of Science and Technology, University of Tsukuba, Tsukuba, Ibaraki, Japan, [2]Department of Materials and Metalurgical Engineering, Institut Teknologi Kalimantan, Balikpapan, Indonesia, [3]Department of Environmental Engineering, Institut Teknologi Kalimantan, Balikpapan, Indonesia, [4]Department of Chemical Engineering, Universitas International Semen Indonesia, Gresik, Indonesia, [5]Department of Environmental Engineering, Faculty of Engineering, Environmental Sustainability Research Group (ENSI-RG), Universitas Diponegoro, Semarang, Indonesia

Abbreviations

ACF	activated carbon fiber
CA	cellulose acetate
CNF	carbon nanofiber
CNT	carbon nanotubes
EKR-PRB	electrokinetic remediation − permeable reactive barrier
EOF	electroosmosis flow
ENFM	electrospun nanofiber membrane
EKG	electrokinetic geosynthetic
HVAC	heating, ventilation, and air conditioning
HPEI	hyperbranched polyethylenimine
NF	nanofiber
PAN	polyacrylonitrile
PE	polyethylene
PLA	polylactic acid
PCL	polycaprolactone
PANN PRB	polyacrylonitrile nanofiber permeable reactive barrier
PM	particulate matter
PP	polypropylene
PPE	personal protection equipment
PVA	polyvinyl alcohol
PVP	polyvinyl pyrrolidone

19.1 Introduction

Pollution in the air is a growing concern worldwide, as it harms human health and claims the lives of thousands of people each year. Particulate matter (PM) and gaseous pollutants found in polluted air can be the root cause of a wide range of adverse health effects, including but not limited to asthma, skin irritation, high blood pressure, cancer, congenital disabilities, and respiratory and cardiovascular diseases (Lv et al., 2019). The nature of the air pollutants and their exposure level play a role in the severity of the health risks. Covid-19 has infected more than 54 billion people and has been responsible for more than 6 million deaths worldwide, making air purification technology more important presently (Babaahmadi et al., 2021; Leung et al., 2020). The treatment of wastewater is another critical area of focus. About a quarter of the total population of the world does not have access to water that is both clean and safe to drink (HMTShirazi et al., 2022). The contamination of the soil is another significant environmental problem such as microplastics (Uwamungu et al., 2022) and oil (Michael-Igolima et al., 2022).

It is necessary to have strict environmental guidelines to prevent the contamination of clean water, air, and soil resources and to ensure that people can still access them. Industries such as food, oil, gas, transportation, pharmaceuticals, electronics, biotechnology, petrochemicals, paints, and coatings significantly contribute to environmental issues such as air quality, wastewater contamination, and land pollution. The effective handling and treatment of industrial wastes are necessary to block all pollution and ensure a consistently clean environment. Furthermore, air and water are the main transport options for particulate contaminants, which can also impact to the soil. The term "purifying" is not commonly used in soil; "remediation" is the appropriate term frequently used in the soil industry. The term "soil remediation" also means cleaning or removing contaminated soil from contaminants (Wang, Hou, et al., 2021).

Removing particulate and biological contaminants is essential for air and water purification processes because these contaminants can cause fouling of the membranes used in reverse osmosis and reduce throughput. Because of the relatively large pores in the filters, fibrous filters, commonly used in filtration devices for water and air, have a low efficiency for removing submicron nanoparticles. Electrospinning is produced in nanofiber membranes with a small fiber diameter, a controllable porous structure, a high specific surface area, good internal connectivity, and steerable morphology. These features make nanofiber membranes effective screens for ultrafine particles.

In addition, nanofiber membranes can potentially incorporate functional or active chemistry at the nanoscale. Nanofiber membranes have a high surface area to volume ratio, nanosized pores, and high porosity, improving the efficiency of conventional materials used for filtration, adsorption, and separating particulate materials. Electrospinning is the most effective method for producing web-like nonwoven ultrafine fibers from various types of polymers, including microfibers ($>\mu1$ m) or nanofibers (<1000 nm) (Fahimirad et al., 2021). The electrospinning process makes it very simple to incorporate bioactive, antimicrobial, and antiviral agents into nanofiber structures. This is an emerging area of research. The specific surface of electrospun nanofiber membranes comes from the external surface, which is advantageous for regeneration. Nanofiber is used in soil remediation as an adsorption material in electrokinetic (EK) remediation combined with a permeable reactive barrier. In addition, nanofiber membranes are simple to scale up, even a few meters in length, and simple to recycle and replace (Wang, Hou, et al., 2021). This chapter will discuss nanofibers' role in the techniques used to purify water, air, and soil. The review also covered research in nanofibers because of their application—the advantages and disadvantages—which were also covered in the last part of this section.

19.2 Water purification application of nanofibers and their composites

Due to their high surface area-to-volume ratio, small pore size, and good mechanical and chemical capabilities, nanofibers and their composites have emerged as potential materials for different water purification applications. These materials can be utilized in various water purification processes, such as filtration, adsorption, and catalysis, to remove impurities and effectively enhance water quality.

Wastewater comprises several types of pollutants, from dissolved pollutants, suspended solids or colloids, and relatively large sediment. Large solid sediment can be separated by sieving or sedimentation, while dissolved and suspended solids need more advanced separation. Nanofiber (carbon nanofiber (CNF)) is used as a water treatment coagulant, adsorbent, and membrane filtration. Nanofiber filtration membranes are one of the most common applications of nanofibers in water purification. Due to their small hole size and high porosity, nanofiber membranes can remove bacteria, viruses, and other particles from water. In addition, the nanofiber membranes can be functionalized with various substances, including graphene oxide, carbon nanotubes (CNTs), and metal nanoparticles, to improve their selectivity and performance. Graphene oxide-coated nanofiber membranes, for instance, have been demonstrated to efficiently remove heavy metal ions from water (El-Aswar et al., 2022; HMTShirazi et al., 2022). Fig. 19.1 demonstrates the most current and essential uses of nanofiber in wastewater treatment to remove contaminants such as pesticides, pharmaceutical and personal care heavy metals, nutrients, and radioactive elements.

As a natural polymer, cellulose provides a sustainable and environmentally friendly offer and is potentially used to substitute petroleum-based polymers (Halim, Xu, et al., 2020). Cellulose nanofiber can be made from several bioresources, and several processes provide many options for wastewater treatment. Even the properties of cellulose nanofiber are affected by its resource and fabrication process. However, the common feature of cellulose nanofiber is its large surface area and high negative surface charge (Halim et al., 2019). The review related

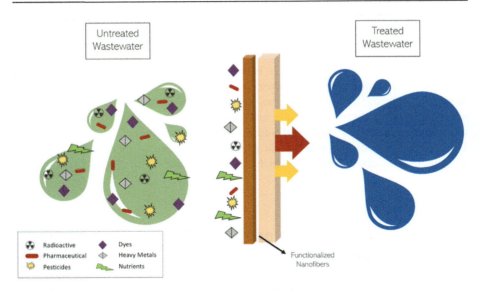

Figure 19.1 Schematic illustration of the capacity of functionalized nanofibers to remove some pollutants substances in water.

to the properties of cellulose nanofiber and its properties can be found in a previous review by the Cranston group (Vanderfleet & Cranston, 2021).

Relatively small and stable particles are commonly found in wastewater as a colloid. The stability of the small floc was accomplished by electrostatically and sterically stabilizing. The surface charge density of the flocculant creates electrostatic repulsion, while the chemical structure, such as the hydrophobicity of the flocculant, creates steric repulsion. Flocculants are introduced to promote the collision between colloid particles. The flocculants will destabilize the particles and promote coagulation. The high negative zeta potential of cellulose nanofiber neutralizes the colloid electrostatically, which enables coagulation (Mohammed et al., 2018; Suopajärvi et al., 2013). The long fiber of cellulose nanofiber bridges the particles to form larger flocks. Balea et al. (2019) reported the application of CNF to clarify ink wastewater from the flexographic printing industry. CNF and polyacrylamide combination generates 100% turbidity reduction (Balea et al., 2019). Silylated cellulose nanofiber decreases the turbidity of chalcopyrite and pyrite as high as 90%–99% (Coelho Braga de Carvalho et al., 2021). However, many current types of research are still on a laboratory scale and need further steps on a prototype or commercial scale.

Furthermore, nanofibers and their composites can be utilized as adsorbents to remove contaminants from water. Due to the nanofibers' extremely high surface area to volume ratio, they can effectively absorb contaminants. In addition, the nanofibers can be functionalized with various substances, including activated carbon, zeolites, and metal oxides, to increase their adsorption capacity and selectivity. For instance, nanofibers impregnated with activated carbon have been proven to effectively remove organic contaminants from water. As an adsorbent, the high surface area of cellulose nanofiber and surface charge are advantages. Cellulose nanofiber has potential applications for heavy metals, dyes, polyelectrolytes, biomolecules, pharmaceuticals, or pesticide adsorbents. Positively charged pollutants attract negatively charged cellulose nanofiber. A high surface area as high as 482 m^2/g optimizes the adsorption capacity (Sehaqui et al., 2011). Cellulose nanofiber-based adsorbents are in the form of several types, such as cellulose acetate (CA) nanofibers (Hamad et al., 2020), bacteria, cellulose/chitosan composite (Li et al., 2020), cellulose nanofiber membrane (Choi et al., 2020; Zhang et al., 2020), cellulose nanofiber aerogel (Ji et al., 2020), carboxymethyl cellulose nanofibril (Yu et al., 2020), and sugarcane bagasse cellulose (Sankararamakrishnan et al., 2020). Thiol-functionalized cellulose nanofiber membrane shows that the adsorption capacity of Cu(II), Cd(II), and Pb(II) ions were 49.0, 45.9, and 22.0 mg/g, respectively (Choi et al., 2020). Pomelo waste was also reported as a source of cellulose nanofiber. Pomelo waste-derived CNF shows high carboxylate content and high adsorption capacity of malachite green (MG) and Cu(II). The adsorption capacity of MG and Cu(II) is 530 mg/g to Mg and 74.2 mg/g to Cu(II) (Tang et al., 2020). Cellulose nanofiber incorporated with nano Fe(0)-FeS adsorbs anionic congo red dye and cationic methylene blue with adsorption capacities of 111.1 and 200.0 mg/g, respectively (Sankararamakrishnan et al., 2020). Isogai group used cellulose nanofiber as a backbone for radiative substances absorption (Vipin et al., 2016). Cellulose nanofiber absorbs radiative cesium. The combined radioactivity of Cs-134 and Cs-137 decreased from 27,146 Bq/kg ($n = 5$) to 14,816 Bq/kg ($n = 5$) at farmland before and after restoration.

As absorbent, cellulose nanofiber is fabricated to be aerogel through vacuum freeze-drying to absorb oil or organic pollutants in oily wastewater. The oil will be absorbed and retained within its porous structure (Sayyed et al., 2021). Aerogel is usually functionalized or coated with hydrophobic chemicals to turn the surface wettability from hydrophilic to hydrophobic. Otherwise, nanoparticles are added to increase the hydrophobicity (Halim et al., 2023). The Polydimethylsiloxane (PDMS) coated CNF aerogel shows a separation efficiency of 99.9% and a flux of 145 L/h/g. Anisotropic cellulose nanofiber-chitosan aerogel fabricated via freeze drying offers an oil adsorption capacity of 82−253 g/g (Zhang et al., 2021). Magnetic hydrophobic polyvinyl alcohol (PVA)/CNF aerogels prepared by vacuum freeze drying have an oil absorption capacity of 136 g/g (Xu et al., 2018). Silylated cellulose nanofibers and silica nanoparticles composite aerogel show more than 99% separation efficiency and a high flux of 1910 ± 60 L/m^2/h during the separation of surfactant-stabilized water-in-oil emulsions (Zhou et al., 2018).

Cellulose nanofiber shows a suitable self-cleaning and antifouling property. Cellulose nanofibers absorb many water molecules, which are very hydrophilic and oleophobic. The properties mimic the fish scale that keeps fish clean even in a muddy environment (Halim et al., 2022). Chemically modified cellulose nanofiber shows a higher contact angle than mechanically cellulose nanofiber (Halim, Lin, et al., 2020). During chemical production, the hydroxyl group converts to the carboxylic group, which has stronger hydrophilicity (Cheng et al., 2017). CNF-coated cellulose sponge as oil-water separation shows a high flux of 3.73×10^3 L/m^2/h and 166 L/m^2/h for mechanical cellulose nanofiber and TEMPO-oxidized cellulose nanofiber, respectively. Both have separation efficiency higher than 99% only by gravitational force alone (Halim et al., 2019). Cellulose nanofiber prepared by electrospinning reaches a flux of 34,300.6 L/m^2/h for the oil-water mixture and 2503.7 L/m^2/h for surfactant stabilized oil-water emulsion, respectively. The separation efficiency is more than 98.3% (Shu et al., 2020). CNF-coated stainless steel rapidly separates the oil−water mixture with a flux of 139556 ± 3733 L/m^2/h without external force (Yin et al., 2022).

19.3 Air purification application of nanofibers and their composites

Nanofiber-based air filters are being developed to combat global air pollution. These filters effectively remove micro and nanoscale particles, volatile organic compounds, and nitrogen oxides. Plasma treatment and chemical functionalization can improve nanofiber-based air filter efficiency. Materials, fiber shape, and filter structure can optimize these filters' performance (Roegiers & Denys, 2021).

Electrospinning is a cost-effective, scalable method for making nanofibers for air filters. Nanofiber materials and crosslinking agents can improve the strength and stiffness of nanofiber-based air filters. Despite its potential benefits, nanofiber-based air filters need further research to optimize their scalability, durability, cost-effectiveness, and performance in varied applications and operating circumstances (Gough et al., 2021). Residential, commercial, and industrial nanofiber-based air filters have been tested. However, more research is needed to determine their pros and cons. Air pollution mitigation must include emissions reduction and urban design. Developing and optimizing nanofiber-based air filters could improve air quality and public health (Deng et al., 2021).

Air filters vary in effectiveness and capability. Some home and commercial filters collect only big particles. In hospitals, laboratories, and other air-quality-sensitive environments, other filters capture particulates such as pollen, animal hair, and smoke. Air filtering systems can be freestanding or integrated into building heating, ventilation, and air conditioning systems. To function appropriately, the filter must be maintained or cleaned. Indoor air quality relies on air filtration, especially in places where people spend a lot of time. Air filtration reduces the incidence of respiratory illnesses and improves health by absorbing pollutants and allergens (Bian et al., 2023; Rana et al., 2023).

19.3.1 Nanofibers in air filtration

Air filtration removes particles, contaminants, and pollutants by passing them through a filter. The air filter is essential for a broad range of industrial sectors, from home care to offices, hospitals, and factories, where clean air is fundamental for comfort, health, and productivity. The air filter traps particles and pollutants to produce clean air. The filter could be made from fibers with designated pore size, fiber diameter, and many more depending on size and mechanism to trap the air contaminants. The basic concept of filtration terms is removal efficiency, pressure drop, and quality factors (QF) (Bian et al., 2023).

Removal efficiency can be defined as particle capture ability and following this formula.

$$E(\%) = \frac{(C_1 - C_2)}{C_1} \times 100\% \qquad (19.1)$$

Where C is particle concentration in either mass ($\mu g/m^3$) or number ($\#/m^3$), and C_1 and C_2 are the particles upstream and downstream

Pressure drop is used to evaluate the permeability of the air filter, which can be defined as:

$$\Delta P = P_1 - P_2 \qquad (19.2)$$

Where P_1 and P_2 are pressure in the upstream and downstream (Pa)

$$QF = \frac{-\ln(1-E)}{\Delta P} \qquad (19.3)$$

QF are the parameters that evaluate the overall filtration performance of the filter. In the formula of QF consider pressure drop (ΔP) and removal efficiency (E).

The capture mechanism of air filtration also gives the knowledge to help develop air filters better. Air filtration usually works with controlled fluid dynamics, which could capture particles in the fiber surface (Barhate & Ramakrishna, 2007) and bond through van der Waals forces (Jordan, 1954). The primary mechanism for fiber filters is illustrated in Fig. 19.2, such as interception, inertial impaction, Brownian diffusion, and electrostatic effects.

Based on the concept and work principle of air filtration, fiber plays a vital role in this phenomenon. Microfiber and nanofiber for the air filter can develop a better air filtration process. Nanoscale fiber utilization means that the advantages of new mechanical, chemical, and physical properties are meaningful in the air filter application (Kadam et al., 2018). Fiber diameters are known to have an impact on filter efficiency. Under relatively similar conditions, 300 nm fibers can obtain 99% capture efficiency for 300 nm particle size (Kadam et al., 2018) compared to fibers with 1000 nm diameters, which only reach 48.21% (Liu, Hsu, et al., 2015; Liu, Zhang, et al., 2015). In another case, reducing the diameters of Polyacrylonitrile (PAN) fibers from 1000 to 200 nm significantly increases the filter efficiency from 48.21% to 98.11% (Liu, Hsu, et al., 2015; Liu, Zhang, et al., 2015). This result illustrates that nanofiber as an air filter also offers exciting alternatives. Nanofibers have been shown to have higher filtration efficiency than micro-scale fiber under specific conditions (Al-Attabi et al., 2019; Hinds, 1999).

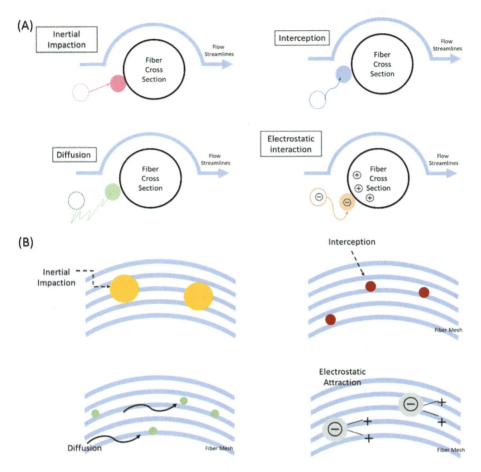

Figure 19.2 Illustration of the main capturing particulate matter or volatile organic compounds (VOCs) mechanism for fiber filters in different perspectives: (A) Fiber cross section (B) Fiber Mesh.

The filter with large and small pore size distribution has higher air filtration efficiency than the filter with uniform small pore size. However, nanofibers also give advantages such as small pore size distribution, low basis weight and thickness, and high permeability as an outstanding air filter (Al-Attabi et al., 2019). Moreover, nanofibers, the main topic of air filter materials, have several parameters influencing the filter's performance. Fiber's diameters, surface area, and pore size could be several parameters that affect the filter's performance. Tiny fiber's diameter also provides a larger surface area and higher specific surface area added by an interconnecting structure that incorporates surface functionalities across the fibers and exposes active sites. Higher surface area also can obtain troughs by forming rough fibers and grafting particles in the fiber surface (Wan et al., 2014; Wang & Pan, 2015).

19.3.2 Nanofibers materials for air filters

Nanofiber-based air filters provide various advantages over conventional air filters, such as greater filtration effectiveness, a smaller pressure drop, and a longer filter life. Typical nanofiber construction materials include polymers, ceramics, and metals. In the past, polymers derived from fossil fuels, such as polyimide (PI) (Zhang et al., 2016), PAN (Naragund & Panda, 2022; Vinh & Kim, 2016), polyethylene (PE), and polypropylene (PP), were a preferred raw material for manufacturing commercial air filters. In order to make air filters more eco-friendly and biodegradable while enhancing their active functional groups, significant research is being conducted on biopolymer-based alternatives (Rana et al., 2023). The diagram in Fig. 19.3 represents the classification of nanofiber materials in air filter applications.

Because the severe acute respiratory syndrome coronavirus 2 (SARS-CoV-2) can spread by respiratory droplets (Tang et al., 2020), degradable nanofibers for air filtration are vital. The coronavirus has infected over 54 billion people and caused over 6 million deaths globally (Leung et al., 2020). People around the world are mandated to put on personal protection equipment such as masks, which has the potential to generate a large amount of waste. In addition, urban air pollution costs in developing nations reach 2 percent of the gross domestic product (Lewis & Edwards, 2016). Current breakthroughs in biodegradable materials to make more eco-friendly masks for future demands are being explored (Babaahmadi et al., 2021). Fig. 19.3 classified degradable raw materials of fiber air filters. Biopolymers such as cellulose (Deng et al., 2021; Rana et al., 2023), alginate (Deng et al., 2022; Wang et al., 2020), chitosan (Choi et al., 2021; Han et al., 2022; Sun et al., 2020), silk fibroin (Gao et al., 2018; Wang, Cui, et al., 2021), and gelatin (Arican et al., 2022) are trying to be utilized as an air filter. Researchers need an environmentally friendly air filter immediately, and biopolymers are a promising candidate.

The most popular and promising candidate shortly is cellulose and its derivatives. The cellulose-based air filter in some work shows excellent performance, and some researchers combine cellulose derivatives to make it better. Balgis's group combined polyvinylpyrrolidone (PVP) with cellulose, creating dual-size and

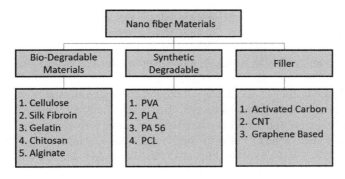

Figure 19.3 Classification of nanofibers materials in air filter application.

multistructure composite nanofiber membranes with varying precursor concentrations. This research also showed that PVP concentration affected composite nanofiber membrane morphology. Polymers integrate nanofiber membranes, indicating their importance. The cellulose/PVP nanofiber composite membrane demonstrated good air filtration ability (QF = 0.117 Pa^{-1}), which was promising. Nevertheless, the lack of active functional group in cellulose, such as carboxyl functional groups will limit the polymer mixing option (Balgis et al., 2017). Due to its solubility, biodegradability, and group activity, CA, the most famous cellulose derivative is appropriate for electrospun nanofiber membranes. The research showed that the acetylation of cellulose nanofibers increases their interface with the polymer solvent during electrospinning, making CA possible for nanofiber membrane formation (Chattopadhyay et al., 2016).

Other materials such as bio-based choline alginate material have excellent antimicrobial activity and PM removal performance (99.69% PM2.5 removal efficiency, 99.89% PM10 removal efficiency, pressure drop below 2 Pa, and 2.8882 QF) (Wang et al., 2020). Furthermore, after 120 min, the number of *Escherichia coli* still alive on the pure nylon, nylon/chitosan, and chitosan-dipped nylon-6 nanofibrous filter media went down to 8.4%, 7.1%, and 2.8%, respectively (Sun et al., 2020). Moreover, silk fibroin's characteristics improve filtration. Due to the richness of functional groups, silk fibroin fibrous membranes had a 99.99% filtering efficiency and 75 Pa air resistance, resulting in a higher quality factor (Gao et al., 2018). Gelatin as biomaterials also achieved 95% filtration efficiency in average fiber diameters 232−778 nm. It shows that gelatin might be a promising candidate for N95 respiratory filters in the future (Arican et al., 2022).

The other class is a degradable synthetic polymer. Some of these polymers come from natural resources, while others are synthesized from various organic compounds, such as Polylactic Acid (PLA) (Sun et al., 2022), PVA (Givehchi et al., 2016), Polyamide-56 (PA56) (Liu, Zhang, et al., 2015), and Polycaprolactone (PCL) (Abuabed & Pallipparambil Varghese, 2019; Rao et al., 2017). Due to their wide application range and production capability, PLA nanofibers as the air filters are promising candidates in the next era.

In their study on PLA nanofiber production, Sun et al. reported using PLA and tetrabutylammonium chloride concentrations of 8 and 5 wt.%, respectively, resulting in branched PLA nanofibers. The air filter had a small pore size (0.70 μm) and high porosity (92.3%), resulting in high PM0.3 removal efficiency (99.95%) and low air resistance (79.67 Pa). Branched T-PLA-5 nanofibers exhibited excellent filtration when combined with cellulose wood pulp paper. The exciting result was that the filtration efficiency remained stable above 85% for PM0.3 (32 L/min) after 5000 backflushing (Sun et al., 2022). Moreover, PVA air filters also show an excellent result when applied as multilayer filters by stacking thin nanofibrous to increase the filter QF (Givehchi et al., 2016). Polyamide-56 nanofiber research performs some attractive results; PA-56 NFN membranes exhibit high filtration efficiency (99.995%), low-pressure drop (111 Pa), large dust holding capacity (49 g/m^2), and dust-cleaning regeneration ability (Liu, Zhang, et al., 2015).

Moreover, developing nanofibers as air filters also involves carbon-based material as filler. Activated carbon (Son et al., 2020), CNT (Li et al., 2014), and graphene (Stanford et al., 2019) can be used to produce more active sites in fiber for better removal efficiency. Furthermore, carbon-based materials also help remove contaminants, such as NO_x (Kim & Kim, 2022) and HONO (Yoo et al., 2015), and enhance in specific situations. The other techniques are fiber treatment and fiber composite making. This way also could be the alternative for boosting air filter performance. Several modifications have been made like the Janus fiber system (Park et al., 2021; Wang et al., 2023), fiber treatment to enhance antibacterial (Han et al., 2022; Son et al., 2020; Wang et al., 2020), degradability (Bian et al., 2023), and moisture resistance (Liu et al., 2021).

Table 19.1 summarizes the advantages and references of air filter materials work from many group research worldwide.

Table 19.1 Summarize of nanofiber's air filter work.

Composite/ fiber	Preparation method	Advantages	References
Cellulose nanofibers (CNF)	Softwood cellulose by the TEMPO oxidation process	Transparent, high gas barrier, promising for flexible transparent device application	Fukuzumi et al. (2009)
CNF with a high crystallinity degree	Household heat-press to dry wet pulp into filter paper.	Excellent performed and could be applied for air filter virus removal	Metreveli et al. (2014)
Silk fibroin	Water solvent electrospinning	Green air filter Fabrication, Improves filtration efficiency (reach 99.99% for 0.3−10 micron particles), Rich functional group	Wang, Cui et al. (2021)
PA-56	Polyamide 56- NFN membranes	Two-dimensional ultra-thin air filter (20 nm), high removal efficiency 99.95%, low-pressure drop 111 Pa, and large holding capacity 49 g/m2	Liu, Zhang et al. (2015)
PVP/CNF	Dual-size PVP/CNF composite nanofibers by one-step electrospinning	Mixed sizes nanofibers allow better permeability and particle capture	Balgis et al. (2017)
Cholinium alginate	Foam dipping into alginate solution to make membranes	Antimicrobial activity, high removal efficiency 99.89%, PM10 pressure drop ultra-low below 2 Pa.	Wang et al. (2020)

19.4 Soil purification applications of nanofibers and their composites

Due to the proliferation of numerous industrial technologies, a portion of the soil is now contaminated with various contaminants (Rajendran et al., 2022; Zahedifar, 2021). Contaminated soil can be treated by several techniques, including physical, chemical, biological treatment, and thermal processes (Effendi et al., 2019, 2022; Mazarji et al., 2021; Michael-Igolima et al., 2022; Ramadan et al., 2018). Nanomaterials have been extensively used in all fields of science and technology, including environmental remediation (Trujillo-Reyes et al., 2014). Such materials can be connected to functional groups that can attach to specific molecules, such as contaminants for environmental cleanup, via covalent bonding. Nanomaterials' size, shape, porosity, and chemical composition can be advantageous for pollutant cleanup (Edgar et al., 2023; Guerra et al., 2018). Metal oxide, carbonaceous nanomaterials, nano polymer, and semiconductor materials are all examples of nanomaterial classes that have been produced. Some items, such as nanofibers and nanomaterials, are now being produced (Zahedifar, 2021).

Various instrumental techniques explored synthesized nanofiber's structural, physicochemical, and mechanical behavior. The remediation of contaminated soil using nanotechnology and nanomaterials has emerged as a very efficient, quick, and promising method compared to conventional materials. High specific surface area and reactivity of nanomaterials facilitate rapid interactions and high efficiency, as most interactions during the remediation process occur at the interface between nanoparticles and contaminants. Moreover, the smaller particle size of nanomaterials enhances their penetration into the soil matrix and delivery into pollutant sites (Alidokht et al., 2021; Zahedifar, 2021). Table 19.2 summarizes several research works on the use of nanofibers in soil purification and remediation. However, the evaluation of current soil remediation using nanofiber has yet to be presented. This review examines the various applications of nanofiber in treating soil contaminated by some pollutants.

19.4.1 Role of nanofiber in remediation by using electrokinetic technique

EK and PRB have enhanced the reduction of heavy metals from polluted soil (Li et al., 2022; Peng et al., 2015). Some nanomaterials, such as nanofiber, may be a suitable adsorption medium for metal ion remediation in a PRB system (Liang et al., 2013; Peng et al., 2015; Wang, Hou, et al., 2021). Peng et al. (2015) determined that the Electrokinetic-PAN nanofiber permeable reactive barrier (EK/PANNR) system could effectively remove heavy metals (Zn^{2+}, Fe^{2+}, Ca^{2+}) from polluted soil. The EK/PANNRB system efficiently removed metal ions from moderately contaminated soil. EK/PANNRB remediation was more effective than EK without PRB. PANNPRB-containing systems had increased current and ionic mobility, indicating improved metal ion electromigration and removal efficiency.

Table 19.2 A summary of the recent studies on the application of nanofiber for the remediation/ purification of contaminated soil.

Focus of the Study	Pollutant concentration	Remediation Procedure	Efficiency	References
Evaluation of the efficacy of the combination of (EKR)-PRB and (PAN/HPEI ENFM) in the cleanup of soil polluted with Cr (VI).	Cr (mg/kg) = 810	In this investigation, synthetic Cr (VI) soil was employed. The experimental setup comprises the soil cell, a PRB cell, two electrode compartments, and a DC power source. 2 g of PAN/HPEI ENFM was placed within the PRB cell. ICP-EOS discovered the entire Cr.	The Cr (VI) immobilization capacity of PAN/HPEI ENFM material in Cr (VI) contaminated soil may reach 72% without affecting the original pH environment of the soil.	Wang, Hou, et al. (2021)
To successfully remove metal ions from polluted soils, an electrospun PANN membrane	Zn^{2+}, Fe^{2+}, Ca^{2+} (mg/L) 5, 10, 100.	The EK machine mainly comprised a glass column with chambers for electrodes on both sides. A piece of PANN membrane connected two layers of 8 cm-wide filter paper between the end of the column and the electrode chamber. The adsorbent and filterable media in PRB was filter paper with a membrane. This system was operational for 312 h.	Removal with EK only, the maximum recovery of Zn^{2+}, Fe^{2+}, and Ca^{2} at 50 V was 83.2%, 82.7%, and 96.3%, whereas with PANNPRB, at 25 V, it was 74.6% and 94.3%, respectively.	Peng et al. (2015)

(*Continued*)

Table 19.2 (Continued)

Focus of the Study	Pollutant concentration	Remediation Procedure	Efficiency	References
ACF and CNFs are used to help plants absorb Cr (VI)	Cr concentration = 50 ± 0.23 mg/kg and 100 ± 0.33 mg/kg	CNFs were created using chemical vapor deposition (CVD) on the Cu-ACF samples. Tests were performed on the ability of the various therapies to adsorb Cr. All three stress groups—Cr(0), Cr(50), and Cr(100)—were used for the test, and their respective concentrations of 0, 50, and 100 mg of Cr(VI) per L of Ms medium were used in each group.	To simulate the removal of Cr (VI) from the soil by adsorption over ACF and the improved uptake of Cu by plants using CNFs as the micronutrient carrier, it is adequate to combine ACF with Cu-CNF. The therapy effectively prevented Cr-induced phytotoxicity-induced plant cell damage by promoting xylem and phloem development.	Kumar et al. (2020)
To develop a phytoremediation procedure for handling coco peat and nanofiber membranes from small-scale mining contaminated with heavy metals.	Cu Concentration = 636.32 mg/kg	The idea of choosing phytoremediation planting factors such as plot height and plant distance was inspired by root length, dry fern weight, plant height, and heavy metal uptake. After proposing the fern species and planting settings, the overall phytoremediation strategy for mine wastes was established.	Pityrogramma calomelanos can be used for the phytoremediation of wastes polluted with the aforementioned heavy metals. Based on the findings, the percent uptake of As and Cu by P. calomelanos was 0.16% and 0.01%, respectively, and the translocation factor of As and Cu was 6.78 and 0.04, respectively.	Win et al. (2020)

Electromigration transported extra metal ions to the electrodes and withdrew them from the soil during EK and PANN membrane adsorption and complexation.

In another study, a PRB was made from aminated electrospun nanofiber membrane PAN electrospun nanofiber membrane -(PAN/HPEI ENFM), and an EK can be used to clean up soil that is contaminated with Cr (VI). The porous electrospun nanofiber membrane is made out of nanofibers (ENFM). Polymer solutions were stretched at high voltage to make these nanofibers (Fan et al., 2019; Zhang et al., 2019). The experimental results indicated that this membrane has a high capacity for Cr (VI) adsorption and excellent reusability. In the test of contaminated soil by the Cr (VI) experiment, the PAN/ hyperbranched polyethylenimine- aminated electrospun nanofiber membrane PAN/HPEI ENFM material was able to immobilize Cr (VI) at a rate of 72.6% without changing the pH of the soil. Because of this, it has been shown that PAN/HPEI ENFM could be used as a PRB material to clean up heavy metal-contaminated soil (Wang, Hou, et al., 2021).

On the other hand, Behrouzinia et al. (2022) used electrokinetic geosynthetics (EKG) to reduce the Cu concentration in contaminated water and soil. To create the distinctive EKG, which can take on a 2D or 3D shape, conductive elements are introduced to geosynthetics or ordinary polymers via knitting, weaving, needle punching, extrusion, or laminating processes (Lamont-Black et al., 2015). Additionally, composite materials can be used to create EKG electrodes. Carbon fibers and composite nanofiber can also create a new EKG electrode. The combined application of EK consolidation and remediation, made possible by properties such as an effective contact surface with the soil profile and a porous structure containing nanofiller Cu removing agents, was the most significant advantage of the new EKG electrodes over conventional electrodes in this study. This EKG electrode worked well as a drainage route for removing copper, resulting in removal efficiencies of over 90%. Fig. 19.4 shows the schematic mechanism of nanofiber application in contaminated soil using EKR combined with PRB.

In addition to reducing pollutant concentrations in polluted soil, nanofiber can be utilized as a soil amendment (Zare et al., 2021). The findings indicate that applying NF reduced soil loss more than it reduced runoff volume. The primary reason for its effects is the remarkable capacity of nanofiber amendment to take in water, thereby increasing the amount of water that can be stored in the soil profile. It can also enhance the adherence of soil particles, which reduces the amount of soil that is lost. Conversely, nanofibers may form a microstructure on the soil's surface that resembles a net, thereby reducing soil loss.

19.4.2 Role of nanofibers in phytoremediation

Phytoremediation is regarded as an effective and eco-friendly alternative for removing heavy metal pollution from water or soil by utilizing different species of macrophytes, also known as hyperaccumulator plants, through various methods. Four mechanisms play essential roles in the phytoremediation process, including phytoextraction, rhizofiltration, phytovolatilization, and phytostabilization (Tan et al., 2023; Win et al., 2020)

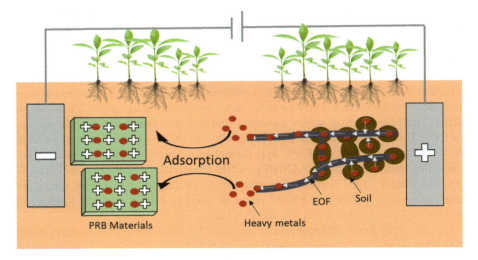

Figure 19.4 Schematic illustration of electrokinetic remediation combined with permeable reactive barrier (with nanofiber addition).

Some research utilized nanofiber for soil remediation. The method for reducing Cr (VI) in soil was conducted by Kumar et al. (2020). They used chickpea (*Cicer arietinum*) plants to minimize the uptake of Cr (VI) from the soil by utilizing microporous activated carbon microfibers (ACF). At the same time, CNFs formed on an ACF substrate efficiently carry Cu from the soil to the plant's root, stem, and leaf. Several adsorbents, such as nanomaterial and activated carbon, are widely employed as aqueous heavy metals, such as Cr (VI) (Kumar et al., 2020; Prajapati & Verma, 2017; Prajapati et al., 2016; Sharma et al., 2019). Recent research has demonstrated that nanosized CNFs produced on an ACF substrate are not only efficient at transporting themselves into plants through the xylem but are also effective in transporting micronutrients to the plant shoots and roots (Gupta et al., 2019; Kumar & Verma, 2018; Kumar et al., 2020). Based on this investigation, the results show that a combination of ACF and Cu-CNF can be utilized to simultaneously improve plant uptake of Cu, using CNFs as the micronutrient carrier, and scavenge Cr (VI) from the soil via adsorption over ACF. From this result, it is necessary to do additional research to test the efficacy of the suggested materials and approach in the existing Cr-contaminated soils. The schematic mechanism of plant uptake of heavy metals using CNF can be seen in Fig. 19.5.

Win et al. (2020) studied the phytoremediation of the mixture of nanofiber membrane, coco peat, and mine tailing using Pityrogramma *calomelanos*. This study mostly elucidated the issue of heavy metals and mine tailings in coco peat and nanofiber membranes produced in regions where small-scale gold mining occurs and contains significant As and Cu. This work aimed to develop a phytoremediation protocol for treating mine tailings and heavy metal-polluted coco peat and nanofiber membranes from small-scale mining. To do this, the results of phytoremediation trials were utilized, as determined by the percentage of As and Cu uptake by ferns

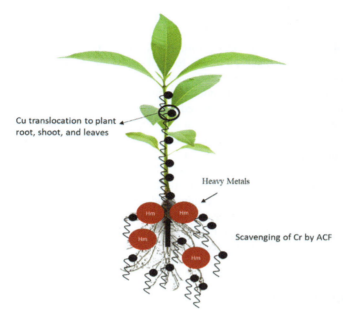

Figure 19.5 Nanofiber for the remediation of contaminated soil with phytoremediation.

and the percentage of As and Cu removal from the tailings mixture. A test conducted after five months revealed that the concentration of heavy metals in the tailing mixture exceeded the standard limit. It can be concluded that coco peat and nanofiber composite were used as adsorbents to clean up mining waste.

19.5 Future direction

Nanofibers are excellent for enhancing the efficacy of water and air filters. Meanwhile, for soil remediation, nanofiber membranes give a different approach. In soil remediation, nanofibers act as adsorption media for supporting soil remediation technology. Nanofiber for water and air filter performance exploits the advantages of nanoscale fiber properties and its possibility to be developed beyond the limit of traditional nanotechnology. Nanofibers are superior in surface area with or without the richness of active sites (Wan et al., 2014). Active sites in nanofibers help enhance the capture and adsorption of pollutants in water, air, and soil purification. This potential could be obtained with chemical modification, fabrication technique, and combined materials (composite) of the fiber. Fiber-making methods and new materials expansion for nanofibers also significantly impact better performance in the future. The sustainability and eco-friendly nanofibers trend is still attractive (Babaahmadi et al., 2021; Bian et al., 2023) to reduce the secondary pollutant. The main concept of nanofibers as purifying media is to collect and capture pollutants, and it will be a contrast to the goal if the filter could be a potential contaminant itself. Biodegradable materials such as protein-based polymers (Bian et al., 2023; Gao et al., 2018), tannic acid, and other polysaccharide polymers (Deng et al., 2021; Sun et al., 2020; Wang et al., 2020) are worth to investigate more due to their functional group richness. Likewise, exploring fiber-making methods and posttreatment is also interesting. Further, for improving performance, research about combining materials to make a nanocomposite in nanofibers form like carbon-based nanomaterials CNT (Li et al., 2014), graphene (Kim & Kim, 2022; Stanford et al., 2019), activated carbon (Son et al., 2020) as filler, Janus fiber system (Park et al., 2021; Wang et al., 2023), blending polymers, grafting (Wan et al., 2014; Wang & Pan, 2015), essential oil addition (Son et al., 2020) and other combining method could be pushed for a broader perspective. Research in more expansive areas could be enhanced with unique features such as antibacterial, degradability, self-cleaning, and moisture resistance. Moreover, the application of nanofibers for water, air, and soil purification needs further research due to some potential materials still waiting to be utilized as an alternative to nanofibers to get a higher quality of this technology.

19.6 Conclusion

Nanofibers have a potentially positive future in the application of water, air, and soil decontamination. The most challenging topics in each scientific area have their own unique difficulties. Nanofibers play an essential part for membranes and substrates in the treatment and purification of water, which helps to speed up the process. On the other hand, nanofibers are an essential component of the next-generation technology that is used in air filters to adsorb PM and volatile organic pollutants. Nanoscale fibers used in the process of air purification have many different alternative materials that can enhance the performance of the nanofibers. Consequently, the behavior of nanofibers in soil remediation is distinct from that of water and air filters. In most cases, nanofibers are used in soil remediation as a supporting device to augment the effect of established technologies such as EK and PRB.

Conflict of interest

The authors declare that they have no conflict of interest.

References

Abuabed, A. S., & Pallipparambil Varghese, B. (2019). Aligned electrospun polycaprolactone nanofiber matrix as a functional air filter. In A. Adibi, S.-Y. Lin, & A. Scherer (Eds.), *Photonic and phononic properties of engineered nanostructures IX* (p. 83). San Francisco, United States: SPIE. Available from https://doi.org/10.1117/12.2514891.

Al-Attabi, R., Morsi, Y. S., Schütz, J. A., & Dumée, L. F. (2019). Electrospun membranes for airborne contaminants capture. In A. Barhoum, M. Bechelany, & A. S. H. Makhlouf (Eds.), *Handbook of nanofibers* (pp. 961–978). Cham: Springer International Publishing. Available from https://doi.org/10.1007/978-3-319-53655-2_37.

Alidokht, L., Anastopoulos, I., Ntarlagiannis, D., Soupios, P., Tawabini, B., Kalderis, D., & Khataee, A. (2021). Recent advances in the application of nanomaterials for the remediation of arsenic-contaminated water and soil. *Journal of Environmental Chemical Engineering*, 9, 105533. Available from https://doi.org/10.1016/j.jece.2021.105533.

Arican, F., Uzuner-Demir, A., Polat, O., Sancakli, A., & Ismar, E. (2022). Fabrication of gelatin nanofiber webs via centrifugal spinning for N95 respiratory filters. *Bulletin of Materials Science*, 45, 93. Available from https://doi.org/10.1007/s12034-022-02668-7.

Babaahmadi, V., Amid, H., Naeimirad, M., & Ramakrishna, S. (2021). Biodegradable and multifunctional surgical face masks: A brief review on demands during COVID-19 pandemic, recent developments, and future perspectives. *Science of The Total Environment*, 798, 149233. Available from https://doi.org/10.1016/j.scitotenv.2021.149233.

Balea, A., Monte, M. C., Fuente, E., Sanchez-Salvador, J. L., Blanco, A., & Negro, C. (2019). Cellulose nanofibers and chitosan to remove flexographic inks from wastewaters. *Environmental Science: Water Research & Technology*, 5, 1558–1567. Available from https://doi.org/10.1039/C9EW00434C.

Balgis, R., Murata, H., Goi, Y., Ogi, T., Okuyama, K., & Bao, L. (2017). Synthesis of dual-size cellulose–polyvinylpyrrolidone nanofiber composites via one-step electrospinning method for high-performance air filter. *Langmuir: the ACS Journal of Surfaces and Colloids*, 33, 6127–6134. Available from https://doi.org/10.1021/acs.langmuir.7b01193.

Barhate, R., & Ramakrishna, S. (2007). Nanofibrous filtering media: Filtration problems and solutions from tiny materials. *Journal of Membrane Science*, 296, 1–8. Available from https://doi.org/10.1016/j.memsci.2007.03.038.

Behrouzinia, S., Ahmadi, H., Abbasi, N., & Javadi, A. A. (2022). Experimental investigation on a combination of soil electrokinetic consolidation and remediation of drained water using composite nanofiber-based electrodes. *Science of The Total Environment*, 836, 155562. Available from https://doi.org/10.1016/j.scitotenv.2022.155562.

Bian, Y., Zhang, C., Wang, H., & Cao, Q. (2023). Degradable nanofiber for eco-friendly air filtration: Progress and perspectives. *Separation and Purification Technology*, 306, 122642. Available from https://doi.org/10.1016/j.seppur.2022.122642.

Chattopadhyay, S., Hatton, T. A., & Rutledge, G. C. (2016). Aerosol filtration using electrospun cellulose acetate fibers. *Journal of Materials Science*, 51, 204–217. Available from https://doi.org/10.1007/s10853-015-9286-4.

Cheng, G., Liao, M., Zhao, D., & Zhou, J. (2017). Molecular understanding on the underwater oleophobicity of self-assembled monolayers: Zwitterionic versus nonionic. *Langmuir: The ACS Journal of Surfaces and Colloids*, 33, 1732–1741. Available from https://doi.org/10.1021/acs.langmuir.6b03988.

Choi, H. Y., Bae, J. H., Hasegawa, Y., An, S., Kim, I. S., Lee, H., & Kim, M. (2020). Thiol-functionalized cellulose nanofiber membranes for the effective adsorption of heavy metal ions in water. *Carbohydrate Polymers*, 234, 115881. Available from https://doi.org/10.1016/j.carbpol.2020.115881.

Choi, S., Jeon, H., Jang, M., Kim, H., Shin, G., Koo, J. M., Lee, M., Sung, H. K., Eom, Y., Yang, H., Jegal, J., Park, J., Oh, D. X., & Hwang, S. Y. (2021). Biodegradable, efficient, and breathable multi-use face mask filter. *Advanced Science*, 8, 2003155. Available from https://doi.org/10.1002/advs.202003155.

Coelho Braga de Carvalho, A. L., Ludovici, F., Goldmann, D., Silva, A. C., & Liimatainen, H. (2021). Silylated thiol-containing cellulose nanofibers as a bio-based flocculation agent for ultrafine mineral particles of chalcopyrite and pyrite. *Journal of Sustainable Metallurgy*, 7, 1506−1522. Available from https://doi.org/10.1007/s40831-021-00439-y.

Deng, Y., Lu, T., Cui, J., Keshari Samal, S., Xiong, R., & Huang, C. (2021). Bio-based electrospun nanofiber as building blocks for a novel eco-friendly air filtration membrane: A review. *Separation and Purification Technology*, 277, 119623. Available from https://doi.org/10.1016/j.seppur.2021.119623.

Deng, Y., Lu, T., Cui, J., Ma, W., Qu, Q., Zhang, X., Zhang, Y., Zhu, M., Xiong, R., & Huang, C. (2022). Morphology engineering processed nanofibrous membranes with secondary structure for high-performance air filtration. *Separation and Purification Technology*, 294, 121093. Available from https://doi.org/10.1016/j.seppur.2022.121093.

Edgar, V.-N., Hermes, P.-H., Denisse, V.-G. J., Andrea, P.-M., Roberto, S.-C. C., Ileana, V.-R., Ingle, A. P., & Fabián, F.-L. (2023). 11 − Nanotechnology for environmental remediation: A sustainable approach. In A. P. Ingle (Ed.), *Nanotechnology in agriculture and agroecosystems, micro and nano technologies* (pp. 297−346). Elsevier. Available from https://doi.org/10.1016/B978-0-323-99446-0.00008-8.

Effendi, A. J., Ramadan, B. S., & Helmy, Q. (2022). Enhanced remediation of hydrocarbons contaminated soil using electrokinetic soil flushing − Landfarming processes. *Bioresource Technology Reports*, 17, 100959. Available from https://doi.org/10.1016/j.biteb.2022.100959.

Effendi, A. J., Wulandari, M., & Setiadi, T. (2019). Ultrasonic application in contaminated soil remediation. *Current Opinion in Environmental Science & Health, Environmental Impact Assessment: Green technologies for environmental remediation*, 12, 66−71. Available from https://doi.org/10.1016/j.coesh.2019.09.009.

El-Aswar, E. I., Ramadan, H., Elkik, H., & Taha, A. G. (2022). A comprehensive review on preparation, functionalization and recent applications of nanofiber membranes in wastewater treatment. *Journal of Environmental Management*, 301, 113908. Available from https://doi.org/10.1016/j.jenvman.2021.113908.

Fahimirad, S., Fahimirad, Z., & Sillanpää, M. (2021). Efficient removal of water bacteria and viruses using electrospun nanofibers. *Science of The Total Environment*, 751, 141673. Available from https://doi.org/10.1016/j.scitotenv.2020.141673.

Fan, J.-P., Luo, J.-J., Zhang, X.-H., Zhen, B., Dong, C.-Y., Li, Y.-C., Shen, J., Cheng, Y.-T., & Chen, H.-P. (2019). A novel electrospun β-CD/CS/PVA nanofiber membrane for simultaneous and rapid removal of organic micropollutants and heavy metal ions from water. *Chemical Engineering Journal*, 378, 122232. Available from https://doi.org/10.1016/j.cej.2019.122232.

Fukuzumi, H., Saito, T., Iwata, T., Kumamoto, Y., & Isogai, A. (2009). Transparent and high gas barrier films of cellulose nanofibers prepared by tempo-mediated oxidation. *Biomacromolecules*, 10, 162−165. Available from https://doi.org/10.1021/bm801065u.

Gao, X., Gou, J., Zhang, L., Duan, S., & Li, C. (2018). A silk fibroin based green nano-filter for air filtration. *RSC Advances*, 8, 8181−8189. Available from https://doi.org/10.1039/C7RA12879G.

Givehchi, R., Li, Q., & Tan, Z. (2016). Quality factors of PVA nanofibrous filters for airborne particles in the size range of 10−125 nm. *Fuel*, 181, 1273−1280. Available from https://doi.org/10.1016/j.fuel.2015.12.010.

Gough, C. R., Callaway, K., Spencer, E., Leisy, K., Jiang, G., Yang, S., & Hu, X. (2021). Biopolymer-based filtration materials. *ACS Omega*, 6, 11804−11812. Available from https://doi.org/10.1021/acsomega.1c00791.

Guerra, F. D., Attia, M. F., Whitehead, D. C., & Alexis, F. (2018). Nanotechnology for Environmental remediation: Materials and applications. *Molecules (Basel, Switzerland)*, *23*, 1760. Available from https://doi.org/10.3390/molecules23071760.

Gupta, G. S., Kumar, A., & Verma, N. (2019). Bacterial homoserine lactones as a nanocomposite fertilizer and defense regulator for chickpeas. *Environmental Science: Nano*, *6*, 1246−1258. Available from https://doi.org/10.1039/C9EN00199A.

Halim, A., Ernawati, L., Ismayati, M., Martak, F., & Enomae, T. (2022). Bioinspired cellulose-based membranes in oily wastewater treatment. *Frontiers of Environmental Science & Engineering*, *16*, 94. Available from https://doi.org/10.1007/s11783-021-1515-2.

Halim, A., Gabriel, A. A., Ismayati, M., Rayhan, P. L. N., & Azizah, U. (2023). Expanded polystyrene waste valorization as a superhydrophobic membrane for oil spill remediation. *Waste and Biomass Valorization Accepted*, *14*, 2025−2036. Available from https://doi.org/10.1007/s12649-022-01976-7.

Halim, A., Lin, K.-H., & Enomae, T. (2020). Biomimicking properties of cellulose nanofiber under ethanol/water mixture. *Scientific Reports*, *10*, 21070. Available from https://doi.org/10.1038/s41598-020-78100-z.

Halim, A., Xu, Y., & Enomae, T. (2020). Fabrication of cellulose sponge: Effects of drying process and cellulose nanofiber deposition on the physical strength. *ASEAN Journal of Chemical Engineering*, *20*, 1−10. Available from https://doi.org/10.22146/ajche.51313.

Halim, A., Xu, Y., Lin, K. H., Kobayashi, M., Kajiyama, M., & Enomae, T. (2019). Fabrication of cellulose nanofiber-deposited cellulose sponge as an oil-water separation membrane. *Separation and Purification Technology*, *224*, 322−331. Available from https://doi.org/10.1016/j.seppur.2019.05.005.

Hamad, A. A., Hassouna, M. S., Shalaby, T. I., Elkady, M. F., Abd Elkawi, M. A., & Hamad, H. A. (2020). Electrospun cellulose acetate nanofiber incorporated with hydroxyapatite for removal of heavy metals. *International Journal of Biological Macromolecules*, *151*, 1299−1313. Available from https://doi.org/10.1016/j.ijbiomac.2019.10.176.

Han, M.-C., Cai, S.-Z., Wang, J., & He, H.-W. (2022). Single-side superhydrophobicity in Si_3N_4-doped and SiO_2-treated polypropylene nonwoven webs with antibacterial activity. *Polymers*, *14*, 2952. Available from https://doi.org/10.3390/polym14142952.

Hinds, W. C. (1999). *Aerosol technology: Properties, behavior, and measurement of airborne particles* (2nd ed.). New York: Wiley.

HMTShirazi, R., Mohammadi, T., Asadi, A. A., & Tofighy, M. A. (2022). Electrospun nanofiber affinity membranes for water treatment applications: A review. *Journal of Water Process Engineering*, *47*, 102795. Available from https://doi.org/10.1016/j.jwpe.2022.102795.

Ji, Y., Wen, Y., Wang, Z., Zhang, S., & Guo, M. (2020). Eco-friendly fabrication of a cost-effective cellulose nanofiber-based aerogel for multifunctional applications in Cu(II) and organic pollutants removal. *Journal of Cleaner Production*, *255*, 120276. Available from https://doi.org/10.1016/j.jclepro.2020.120276.

Jordan, D. W. (1954). The adhesion of dust particles. *British Journal of Applied Physics*, *5*, S194−S197. Available from https://doi.org/10.1088/0508-3443/5/S3/363.

Kadam, V. V., Wang, L., & Padhye, R. (2018). Electrospun nanofibre materials to filter air pollutants − A review. *Journal of Industrial Textiles*, *47*, 2253−2280. Available from https://doi.org/10.1177/1528083716676812.

Kim, Y.-S., & Kim, Y.-H. (2022). Removal of NO_x from graphene based photocatalyst ceramic filter. *Applied Chemistry for Engineering*, *33*, 600−605. Available from https://doi.org/10.14478/ACE.2022.1104.

Kumar, A., Gahoi, P., & Verma, N. (2020). Simultaneous scavenging of Cr(VI) from soil and facilitation of nutrient uptake in plant using a mixture of carbon microfibers and nanofibers. *Chemosphere*, *239*, 124760. Available from https://doi.org/10.1016/j.chemosphere.2019.124760.

Kumar, A., & Verma, N. (2018). Wet air oxidation of aqueous dichlorvos pesticide over catalytic copper-carbon nanofiberous beads. *Chemical Engineering Journal*, *351*, 428−440. Available from https://doi.org/10.1016/j.cej.2018.06.058.

Lamont-Black, J., Jones, C. J. F. P., & White, C. (2015). Electrokinetic geosynthetic dewatering of nuclear contaminated waste. *Geotextiles and Geomembranes*, *43*, 359−362. Available from https://doi.org/10.1016/j.geotexmem.2015.04.005.

Leung, N. H. L., Chu, D. K. W., Shiu, E. Y. C., Chan, K.-H., McDevitt, J. J., Hau, B. J. P., Yen, H.-L., Li, Y., Ip, D. K. M., Peiris, J. S. M., Seto, W.-H., Leung, G. M., Milton, D. K., & Cowling, B. J. (2020). Respiratory virus shedding in exhaled breath and efficacy of face masks. *Nature Medicine*, *26*, 676−680. Available from https://doi.org/10.1038/s41591-020-0843-2.

Lewis, A., & Edwards, P. (2016). Validate personal air-pollution sensors. *Nature*, *535*, 29−31. Available from https://doi.org/10.1038/535029a.

Li, D., Tian, X., Wang, Z., Guan, Z., Li, X., Qiao, H., Ke, H., Luo, L., & Wei, Q. (2020). Multifunctional adsorbent based on metal-organic framework modified bacterial cellulose/chitosan composite aerogel for high efficient removal of heavy metal ion and organic pollutant. *Chemical Engineering Journal*, *383*, 123127. Available from https://doi.org/10.1016/j.cej.2019.123127.

Li, P., Wang, C., Zhang, Y., & Wei, F. (2014). Air filtration in the free molecular flow regime: A review of high-efficiency particulate air filters based on carbon nanotubes. *Small (Weinheim an der Bergstrasse, Germany)*, *10*, 4543−4561. Available from https://doi.org/10.1002/smll.201401553.

Li, Y., Shao, M., Huang, M., Sang, W., Zheng, S., Jiang, N., & Gao, Y. (2022). Enhanced remediation of heavy metals contaminated soils with EK-PRB using β-CD/hydrothermal biochar by waste cotton as reactive barrier. *Chemosphere*, *286*, 131470. Available from https://doi.org/10.1016/j.chemosphere.2021.131470.

Liang, W., Liu, N., Dong, Z., Liu, L., Mai, J. D., Lee, G.-B., & Li, W. J. (2013). Simultaneous separation and concentration of micro- and nano-particles by optically induced electrokinetics. *Sensors and Actuators A: Physical*, *193*, 103−111. Available from https://doi.org/10.1016/j.sna.2013.01.020.

Liu, B., Zhang, S., Wang, X., Yu, J., & Ding, B. (2015). Efficient and reusable polyamide-56 nanofiber/nets membrane with bimodal structures for air filtration. *Journal of Colloid and Interface Science*, *457*, 203−211. Available from https://doi.org/10.1016/j.jcis.2015.07.019.

Liu, C., Hsu, P.-C., Lee, H.-W., Ye, M., Zheng, G., Liu, N., Li, W., & Cui, Y. (2015). Transparent air filter for high-efficiency PM 2.5 capture. *Nature Communications*, *6*, 6205. Available from https://doi.org/10.1038/ncomms7205.

Liu, T., Cai, C., Ma, R., Deng, Y., Tu, L., Fan, Y., & Lu, D. (2021). Super-hydrophobic cellulose nanofiber air filter with highly efficient filtration and humidity resistance. *ACS Applied Materials & Interfaces*, *13*, 24032−24041. Available from https://doi.org/10.1021/acsami.1c04258.

Lv, D., Wang, R., Tang, G., Mou, Z., Lei, J., Han, J., De Smedt, S., Xiong, R., & Huang, C. (2019). Ecofriendly electrospun membranes loaded with visible-light-responding nanoparticles for multifunctional usages: Highly efficient air filtration, dye scavenging, and bactericidal activity. *ACS Applied Materials & Interfaces, 11*, 12880−12889. Available from https://doi.org/10.1021/acsami.9b01508.

Mazarji, M., Minkina, T., Sushkova, S., Mandzhieva, S., Bidhendi, G. N., Barakhov, A., & Bhatnagar, A. (2021). Effect of nanomaterials on remediation of polycyclic aromatic hydrocarbons-contaminated soils: A review. *Journal of Environmental Management, 284*, 112023. Available from https://doi.org/10.1016/j.jenvman.2021.112023.

Metreveli, G., Wågberg, L., Emmoth, E., Belák, S., Strømme, M., & Mihranyan, A. (2014). A size-exclusion nanocellulose filter paper for virus removal. *Advanced Healthcare Materials, 3*, 1546−1550. Available from https://doi.org/10.1002/adhm.201300641.

Michael-Igolima, U., Abbey, S. J., & Ifelebuegu, A. O. (2022). A systematic review on the effectiveness of remediation methods for oil contaminated soils. *Environmental Advances, 9*, 100319. Available from https://doi.org/10.1016/j.envadv.2022.100319.

Mohammed, N., Grishkewich, N., & Tam, K. C. (2018). Cellulose nanomaterials: Promising sustainable nanomaterials for application in water/wastewater treatment processes. *Environmental Science: Nano, 5*, 623−658. Available from https://doi.org/10.1039/C7EN01029J.

Naragund, V. S., & Panda, P. K. (2022). Electrospun polyacrylonitrile nanofiber membranes for air filtration application. *International Journal of Environmental Science and Technology, 19*, 10233−10244. Available from https://doi.org/10.1007/s13762-021-03705-4.

Park, S., Koo, H. Y., Yu, C., & Choi, W. S. (2021). A novel approach to designing air filters: Ubiquitous material-based Janus air filter modules with hydrophilic and hydrophobic parts. *Chemical Engineering Journal, 410*, 128302. Available from https://doi.org/10.1016/j.cej.2020.128302.

Peng, L., Chen, X., Zhang, Y., Du, Y., Huang, M., & Wang, J. (2015). Remediation of metal contamination by electrokinetics coupled with electrospun polyacrylonitrile nanofiber membrane. *Process Safety and Environmental Protection, 98*, 1−10. Available from https://doi.org/10.1016/j.psep.2015.06.003.

Prajapati, Y. N., Bhaduri, B., Joshi, H. C., Srivastava, A., & Verma, N. (2016). Aqueous phase adsorption of different sized molecules on activated carbon fibers: Effect of textural properties. *Chemosphere, 155*, 62−69. Available from https://doi.org/10.1016/j.chemosphere.2016.04.040.

Prajapati, Y. N., & Verma, N. (2017). Adsorptive desulfurization of diesel oil using nickel nanoparticle-doped activated carbon beads with/without carbon nanofibers: Effects of adsorbate size and adsorbent texture. *Fuel C, 186*−194. Available from https://doi.org/10.1016/j.fuel.2016.10.044.

Rajendran, S., Priya, T. A. K., Khoo, K. S., Hoang, T. K. A., Ng, H.-S., Munawaroh, H. S. H., Karaman, C., Orooji, Y., & Show, P. L. (2022). A critical review on various remediation approaches for heavy metal contaminants removal from contaminated soils. *Chemosphere, 287*, 132369. Available from https://doi.org/10.1016/j.chemosphere.2021.132369.

Ramadan, B. S., Sari, G. L., Rosmalina, R. T., Effendi, A. J., & Hadrah. (2018). An overview of electrokinetic soil flushing and its effect on bioremediation of hydrocarbon contaminated soil. *Journal of Environmental Management, 218*, 309−321. Available from https://doi.org/10.1016/j.jenvman.2018.04.065.

Rana, A. K., Mostafavi, E., Alsanie, W. F., Siwal, S. S., & Thakur, V. K. (2023). Cellulose-based materials for air purification: A review. *Industrial Crops and Products, 194*, 116331. Available from https://doi.org/10.1016/j.indcrop.2023.116331.

Rao, C., Gu, F., Zhao, P., Sharmin, N., Gu, H., & Fu, J. (2017). Capturing PM 2.5 emissions from 3D printing via nanofiber-based air filter. *Scientific Reports*, *7*, 10366. Available from https://doi.org/10.1038/s41598-017-10995-7.

Roegiers, J., & Denys, S. (2021). Development of a novel type activated carbon fiber filter for indoor air purification. *Chemical Engineering Journal*, *417*, 128109. Available from https://doi.org/10.1016/j.cej.2020.128109.

Sankararamakrishnan, N., Singh, N., & Srivastava, I. (2020). Hierarchical nano Fe(0)@FeS doped cellulose nanofibres derived from agrowaste − Potential bionanocomposite for treatment of organic dyes. *International Journal of Biological Macromolecules*, *151*, 713−722. Available from https://doi.org/10.1016/j.ijbiomac.2020.02.155.

Sayyed, A. J., Pinjari, D. V., Sonawane, S. H., Bhanvase, B. A., Sheikh, J., & Sillanpää, M. (2021). Cellulose-based nanomaterials for water and wastewater treatments: A review. *Journal of Environmental Chemical Engineering*, *9*, 106626. Available from https://doi.org/10.1016/j.jece.2021.106626.

Sehaqui, H., Zhou, Q., Ikkala, O., & Berglund, L. A. (2011). Strong and tough cellulose nanopaper with high specific surface area and porosity. *Biomacromolecules*, *12*, 3638−3644. Available from https://doi.org/10.1021/bm2008907.

Sharma, M., Joshi, M., Nigam, S., Shree, S., Avasthi, D. K., Adelung, R., Srivastava, S. K., & Mishra, Y. K. (2019). ZnO tetrapods and activated carbon based hybrid composite: Adsorbents for enhanced decontamination of hexavalent chromium from aqueous solution. *Chemical Engineering Journal*, *358*, 540−551. Available from https://doi.org/10.1016/j.cej.2018.10.031.

Shu, D., Xi, P., Cheng, B., Wang, Y., Yang, L., Wang, X., & Yan, X. (2020). One-step electrospinning cellulose nanofibers with superhydrophilicity and superoleophobicity underwater for high-efficiency oil-water separation. *International Journal of Biological Macromolecules*, *162*, 1536−1545. Available from https://doi.org/10.1016/j.ijbiomac.2020.07.175.

Son, B. C., Park, C. H., & Kim, C. S. (2020). Fabrication of antimicrobial nanofiber air filter using activated carbon and cinnamon essential oil. *Journal of Nanoscience and Nanotechnology*, *20*, 4376−4380. Available from https://doi.org/10.1166/jnn.2020.17597.

Stanford, M. G., Li, J. T., Chen, Y., McHugh, E. A., Liopo, A., Xiao, H., & Tour, J. M. (2019). Self-sterilizing laser-induced graphene bacterial air filter. *ACS Nano*, *13*, 11912−11920. Available from https://doi.org/10.1021/acsnano.9b05983.

Sun, N., Shao, W., Zheng, J., Zhang, Y., Li, J., Liu, S., Wang, K., Niu, J., Li, B., Gao, Y., Liu, F., Jiang, H., & He, J. (2022). Fabrication of fully degradable branched poly (lactic acid) nanofiber membranes for high-efficiency filter paper materials. *Journal of Applied Polymer Science*, *139*. Available from https://doi.org/10.1002/app.53186.

Sun, Z., Yue, Y., He, W., Jiang, F., Lin, C.-H., Pui., David, Y. H., Liang, Y., & Wang, J. (2020). The antibacterial performance of positively charged and chitosan dipped air filter media. *Building and Environment*, *180*, 107020. Available from https://doi.org/10.1016/j.buildenv.2020.107020.

Suopajärvi, T., Liimatainen, H., Hormi, O., & Niinimäki, J. (2013). Coagulation−flocculation treatment of municipal wastewater based on anionized nanocelluloses. *Chemical Engineering Journal*, *231*, 59−67. Available from https://doi.org/10.1016/j.cej.2013.07.010.

Tan, H. W., Pang, Y. L., Lim, S., & Chong, W. C. (2023). A state-of-the-art of phytoremediation approach for sustainable management of heavy metals recovery. *Environmental*

Technology & Innovation, *30*, 103043. Available from https://doi.org/10.1016/j.eti.2023.103043.

Tang, F., Yu, H., Yassin Hussain Abdalkarim, S., Sun, J., Fan, X., Li, Y., Zhou, Y., & Chiu Tam, K. (2020). Green acid-free hydrolysis of wasted pomelo peel to produce carboxylated cellulose nanofibers with super absorption/flocculation ability for environmental remediation materials. *Chemical Engineering Journal*, *395*, 125070. Available from https://doi.org/10.1016/j.cej.2020.125070.

Trujillo-Reyes, J., Peralta-Videa, J. R., & Gardea-Torresdey, J. L. (2014). Supported and unsupported nanomaterials for water and soil remediation: Are they a useful solution for worldwide pollution? *Journal of Hazardous Materials*, *280*, 487−503. Available from https://doi.org/10.1016/j.jhazmat.2014.08.029.

Uwamungu, J. Y., Wang, Y., Shi, G., Pan, S., Wang, Z., Wang, L., & Yang, S. (2022). Microplastic contamination in soil agro-ecosystems: A review. *Environmental Advances*, *9*, 100273. Available from https://doi.org/10.1016/j.envadv.2022.100273.

Vanderfleet, O. M., & Cranston, E. D. (2021). Production routes to tailor the performance of cellulose nanocrystals. *Nature Reviews Materials*, *6*, 124−144. Available from https://doi.org/10.1038/s41578-020-00239-y.

Vinh, N., & Kim, H.-M. (2016). Electrospinning fabrication and performance evaluation of polyacrylonitrile nanofiber for air filter applications. *Applied Sciences*, *6*, 235. Available from https://doi.org/10.3390/app6090235.

Vipin, A. K., Fugetsu, B., Sakata, I., Isogai, A., Endo, M., Li, M., & Dresselhaus, M. S. (2016). Cellulose nanofiber backboned Prussian blue nanoparticles as powerful adsorbents for the selective elimination of radioactive cesium. *Scientific Reports*, *6*. Available from https://doi.org/10.1038/srep37009.

Wan, H., Wang, N., Yang, J., Si, Y., Chen, K., Ding, B., Sun, G., El-Newehy, M., Al-Deyab, S. S., & Yu, J. (2014). Hierarchically structured polysulfone/titania fibrous membranes with enhanced air filtration performance. *Journal of Colloid and Interface Science*, *417*, 18−26. Available from https://doi.org/10.1016/j.jcis.2013.11.009.

Wang, J., Hou, L., Yao, Z., Jiang, Y., Xi, B., Ni, S., & Zhang, L. (2021). Aminated electrospun nanofiber membrane as permeable reactive barrier material for effective in-situ Cr (VI) contaminated soil remediation. *Chemical Engineering Journal*, *406*, 126822. Available from https://doi.org/10.1016/j.cej.2020.126822.

Wang, M.-L., Yu, D.-G., & Bligh, S. W. A. (2023). Progress in preparing electrospun Janus fibers and their applications. *Applied Materials Today*, *31*, 101766. Available from https://doi.org/10.1016/j.apmt.2023.101766.

Wang, Y., Yuan, W., Zhang, L., Zhang, Z., Zhang, G., Wang, S., He, L., & Tao, G. (2020). Bio-based antimicrobial ionic materials fully composed of natural products for elevated air purification. *Advanced Sustainable Systems*, *4*, 2000046. Available from https://doi.org/10.1002/adsu.202000046.

Wang, Z., Cui, Y., Feng, Y., Guan, L., Dong, M., Liu, Z., & Liu, L. (2021). A versatile Silk Fibroin based filtration membrane with enhanced mechanical property, disinfection and biodegradability. *Chemical Engineering Journal*, *426*, 131947. Available from https://doi.org/10.1016/j.cej.2021.131947.

Wang, Z., & Pan, Z. (2015). Preparation of hierarchical structured nano-sized/porous poly (lactic acid) composite fibrous membranes for air filtration. *Applied Surface Science*, *356*, 1168−1179. Available from https://doi.org/10.1016/j.apsusc.2015.08.211.

Win, Z. C., Diaz, L. J. L., Perez, T. R., & Nakasaki, K. (2020). Phytoremediation of heavy metal contaminated wastes from small-scale gold mining using *Pityrogramma calomelanos*. *E3S Web of Conferences*, *148*, 05007. Available from https://doi.org/10.1051/e3sconf/202014805007.

Xu, Z., Jiang, X., Zhou, H., & Li, J. (2018). Preparation of magnetic hydrophobic polyvinyl alcohol (PVA)−cellulose nanofiber (CNF) aerogels as effective oil absorbents. *Cellulose*, *25*, 1217−1227. Available from https://doi.org/10.1007/s10570-017-1619-9.

Yin, X., Wu, J., Zhao, H., Zhou, L., He, T., Fan, Y., Chen, L., Wang, K., & He, Y. (2022). A microgel-structured cellulose nanofibril coating with robust antifouling performance for highly efficient oil/water and immiscible organic solvent separation. *Colloids and Surfaces A: Physicochemical and Engineering Aspects*, *647*, 128875. Available from https://doi.org/10.1016/j.colsurfa.2022.128875.

Yoo, J. Y., Park, C. J., Kim, K. Y., Son, Y.-S., Kang, C.-M., Wolfson, J. M., Jung, I.-H., Lee, S.-J., & Koutrakis, P. (2015). Development of an activated carbon filter to remove NO_2 and HONO in indoor air. *Journal of Hazardous Materials*, *289*, 184−189. Available from https://doi.org/10.1016/j.jhazmat.2015.02.038.

Yu, H., Hong, H.-J., Kim, S. M., Ko, H. C., & Jeong, H. S. (2020). Mechanically enhanced graphene oxide/carboxymethyl cellulose nanofibril composite fiber as a scalable adsorbent for heavy metal removal. *Carbohydrate Polymers*, *240*, 116348. Available from https://doi.org/10.1016/j.carbpol.2020.116348.

Zahedifar, M. (2021). Chapter 1 − Nanomaterials in soil remediation: An introduction. In A. Amrane, D. Mohan, T. A. Nguyen, A. A. Assadi, & G. Yasin (Eds.), *Nanomaterials for soil remediation* (pp. 3−12). Elsevier. Available from https://doi.org/10.1016/B978-0-12-822891-3.00001-3.

Zare, S., Sadeghi, S. H. R., & Khosravani, A. (2021). Controllability of soil and water loss in small plots using nanofiber amendment produced from recycled old paperboard containers. *Soil and Tillage Research*, *209*, 104949. Available from https://doi.org/10.1016/j.still.2021.104949.

Zhang, D., Xu, W., Cai, J., Cheng, S.-Y., & Ding, W.-P. (2020). Citric acid-incorporated cellulose nanofibrous mats as food materials-based biosorbent for removal of hexavalent chromium from aqueous solutions. *International Journal of Biological Macromolecules*, *149*, 459−466. Available from https://doi.org/10.1016/j.ijbiomac.2020.01.199.

Zhang, M., Jiang, S., Han, F., Li, M., Wang, N., & Liu, L. (2021). Anisotropic cellulose nanofiber/chitosan aerogel with thermal management and oil absorption properties. *Carbohydrate Polymers*, *264*, 118033. Available from https://doi.org/10.1016/j.carbpol.2021.118033.

Zhang, R., Liu, C., Hsu, P.-C., Zhang, C., Liu, N., Zhang, J., Lee, H. R., Lu, Y., Qiu, Y., Chu, S., & Cui, Y. (2016). Nanofiber air filters with high-temperature stability for efficient $PM_{2.5}$ removal from the pollution sources. *Nano Letters*, *16*, 3642−3649. Available from https://doi.org/10.1021/acs.nanolett.6b00771.

Zhang, S., Shi, Q., Christodoulatos, C., Korfiatis, G., & Meng, X. (2019). Adsorptive filtration of lead by electrospun PVA/PAA nanofiber membranes in a fixed-bed column. *Chemical Engineering Journal*, *370*, 1262−1273. Available from https://doi.org/10.1016/j.cej.2019.03.294.

Zhou, S., You, T., Zhang, X., & Xu, F. (2018). Superhydrophobic cellulose nanofiber-assembled aerogels for highly efficient water-in-oil emulsions separation. *ACS Applied Nano Materials*, *1*, 2095−2103. Available from https://doi.org/10.1021/acsanm.8b00079.

Electronics application of nanofibers and their composites

Manoj Kumar Banjare[1], Kamalakanta Behera[2], Ramesh Kumar Banjare[3], Mamta Tandon[1], Siddharth Pandey[4] and Kallol K. Ghosh[5]

[1]MATS School of Sciences, MATS University, Raipur, Chhattisgarh, India, [2]Department of Chemistry, University of Allahabad, Prayagraj, Uttar Pradesh, India, [3]Department of Chemistry(MSET), MATS University, Raipur, Chhattisgarh, India, [4]Department of Chemistry, Indian Institute of Technology Delhi, New Delhi, India, [5]School of Studies in Chemistry, Pt. Ravishankar Shukla University, Raipur, Chhattisgarh, India

20.1 Introduction

Recent years have seen an increase in interest in flexible and stretchable electronics due to their potential characteristics, including thinness, lightness, flexibility, stretchability, conformability, and compatibility (Dagdeviren et al., 2016). It has taken a tremendous amount of work to find intrinsically soft materials, flexible/stretchable architectures, and/or to increase the performance of soft electronics (Rajan et al., 2018). Due to their high specific surface area and superior mechanical and electrical properties, several sophisticated nanomaterials, including graphene, carbon nanotubes (CNT), nanoparticles, nanowires (NW)/nanofibers, and nanomembranes, have been developed to manufacture soft electronics (Cho et al., 2022; Son & Bao, 2018). There are numerous ways to construct electronics with nanomaterials, including wavy, restrain, composite, and transfer.

To generate one-dimensional nanofibers with intriguing mechanical and electrical properties, such as high porosity, surface area, conductivity, transparency, and ultrahigh flexibility, electrospinning offers a low-cost, effective, and large-scale approach (Babuab et al., 2023). Additionally, the development of thin, soft, lightweight, breathable, and conformable electronic devices depends on a variety of successful nanofiber assemblies. Additionally, electrospinning is the best method for creating substrate materials for flexible/stretchable electronic devices that directly interface with tissues, organs, or cells due to the applicability of biocompatible polymers (Xue et al., 2019).

Nanoparticle production is a relatively recent development in science. Nanotechnology developed quickly during the 1990s, giving rise to new terms and concepts such as nanoelectronics, nanomaterials, and nanobiology (Bayda et al., 2020). "There's Plenty of Room at the Bottom" was the title of the talk delivered by Professor Richard Feynman at the California Institute of Technology on December 29, 1959, during the American Physical Society meeting (Laucht et al., 2021).

This book chapter seeks to educate readers on the current state of electronic nanofiber application research and to offer ideas for several intriguing advances. Modern electronic applications are covered in detail, including those involving flexible/stretchable conductors, transparent electrodes, strain sensors, pressure sensors, energy harvesting and storage devices, transistors, and optoelectronics (Zhao et al., 2019). Finally, a forecast of potential outcomes is offered.

20.2 Terminology for nanotechnology

20.2.1 Nano

The Greek word "Nannos," which means "extremely short man," is where the prefix gets its name (Tabassum, 2020).

20.2.2 Nanoscale

Sizes typically vary from 1 to 100 nm.

20.2.3 Nanoparticle

All three exterior dimensions of a nanoobject are measured in the nanometric space (1–100 nm) (Jeevanandam et al., 2018).

20.2.4 Nanofiber

Nanoobject have three much greater exterior dimensions compared to their two smaller nanoscale counterparts (Ross et al., 2016).

20.2.5 Nanocomposite

Multiphase structure with at least one phase having at least one nanometrically small dimension (Baig et al., 2021).

20.2.6 Nanomaterials

"A natural, incidental, or manufactured material containing particles, in an unbound state, as an aggregate or as an agglomerate, where for 50% or more of the particles in the number size distribution, one or more external dimensions is in the size range of 1–100 nm," according to the definition of nanomaterials (Tabassum, 2020).

20.2.7 Nanoscience

Nanoscience is defined as "the study, discovery, and comprehension of matter at the nanoscale, where size- and structure-dependent features and phenomena, separate from those connected to individual atoms or molecules or with bulk materials, can develop" (Zhao et al., 2019).

20.2.8 Nanotechnology

A new field of engineering called nanotechnology uses techniques from nanoscience to produce things that are useful, marketable, and profitable. Norio Taniguchi, a Japanese physicist, coined the word "nanotechnology" in 1974 to describe the advancement of mechanical processing and material precision (Baig et al., 2021).

20.3 The evolution of nanofiber technologies over time

It is challenging to separate the history of nanofiber technology from the overall development of nanoscience and nanotechnology. Nanofiber production processes have a long history that spans over four centuries of inventions and outcomes. This section provides an overview of the history of innovations in nanofiber technology, with a focus on the growth of electrostatic production and fiber drawing.

Electrospinning history and development of nanofiber technology up to 2008 in chronological order:

William Gilbert, a physicist, conducted an experiment in which a round water drop on a dry surface deflected into a cone shape when it approached an electrically charged amber (Gilbert & Wright, 1967).

Bose (1744): An article by Georg Mathias Bose titled "Electricity - its discovery and evolution, with poetic drawings" was published (Bose, 1744).

Rayleigh and others (1878a, 1878b, 1879 and 1882): Physicist Lord Rayleigh studied the unstable states of electrically charged liquid droplets. He observed that the liquid was released in tiny jets after the surface tension and electrostatic force reached equilibrium. He wrote several insightful works on topics such as the equilibrium of liquid conducting masses charged with electricity, the instability of jets, and the impact of electricity on colliding waterdrops (Rayleigh, 1878a, 1878b, 1879, 1882).

Louis Schwabe and others developed several techniques for spinning silk and producing synthetic fibers (Kauffman, 1993).

Boys (1887): Charles Vernon Boys reported the procedure in a study on the production of nanofibers in 1988 (Boys, 1887).

The electrostatic atomization of conductive fluid under an applied voltage was observed by Lord Raleigh and others in 1897 (Cloupeau & Prunet-Foch, 1989).

John Francis Cooley, an American inventor, received the first patent for electrospinning in Cooley. Morton invented a less complex low-throughput device in the same year (Rayleigh, 1879; Zeleny, 1917).

Discharge of Electricity from Pointed Conductors Differing in Size, published by Zeleny (1907).

Burton and Wiegand (1912) investigated how electricity affected water drop streams (Burton & Wiegand, 1912).

The efforts to mathematically represent the behavior of fluids under electrostatic forces started at that point with the publication of Zeleny's (1914a, 1914b, 1917, 1920) study on the behavior of fluid droplets at the end of metal capillaries (Zeleny 1914, 1915, 1917, 1920).

The distortion and shattering of waterdrops in intense electric fields were studied by Macky in 1931 (Macky, 1931).

The first patent for the experimental manufacturing of nanofibers was published by Formhals in 1934 (Anton, 1934).

The electric moments of molecules in liquids were studied by Onsager in 1936 (Onsager, 1936).

Electrospun fibers were created by N.D. Rozenblum and I.V. Petryanov-Sokolov in 1938, and they later transformed them into filter materials (Tucker et al., 2013).

In the Union of Soviet Socialist Republics (USSR), electrospun air filter materials known as "Petryanov filters" were used commercially (Tucker et al., 2013).

English (1948) studied a water drop's corona (English, 1948).

John Zeleny, the inventor, studied the electrospray process phenomena in 1914 (Guan & Cole, 2016).

Hollow graphitic carbon fibers—Created in 1952 by Radushkevich and Lukyanovich, (Mostofizadeh et al., 2011).

"There's Plenty of Room at the Bottom - An Invitation to Enter a New Field of Physics" was the title of a speech given by Richard P. Feynman in 1959 at the California Institute of Technology at the American Physical Society's annual meeting (Caltech). This was the first article about nanotechnology to be published (Feynman, 1959).

Geoffrey Ingram Taylor (1964, 1969) developed the theoretical underpinnings of electrospinning during this period. He looked at the fluid droplet's Taylor cone-shaped formation when an electric field was present. In conducting fluids, he continued to collaborate with J.R. Melcher (1969) to create the leaky dielectric model (Melcher & Taylor, 1969; Taylor, 1964, 1966, 1969).

In 1966 Professor Harold L. Simons published a patent for a device that creates the lightest, ultrathin nanofiber fabrics in a variety of patterns (Simons, 1966).

Creation of nanoscale carbon fibers through the development of a chemical vapor deposition method (Oberlin et al., 1976).

In 1977 Martin and Cockshott suggested using electrospun fibers as a treatment for wounds (Martin & Cockshott, 1977).

A study of electrospun fibers as an implanted (vascular) graft was done by Annis and colleagues in 1978. In the 1980s, Donaldson Co. Larrondo (1981) and John Manley electrospun continuous filaments of quickly crystallizing polymers from their melts to establish the production of electrospun polymeric nanofibers at an industrial scale (Larrondo & St John Manley, 1981a, 1981b, 1981c).

The first US patent for hollow carbon fibers was granted (Tennent, 1987).

Multiwalled CNTs were discovered by Sumio Iijima in 1991 (Iijima, 1991).

Single-walled CNTs have been found (McEuen, 2000).

The electrospinning procedure for creating electrospun fibers from various polymers with a range of cross-sectional morphologies and size ranges between Reneker and Doshi's analysis describes wavelengths between and 5000 nm (1955) (Doshi & Reneker, 1995).

Wang et al. (2001) created inorganic fibers using electrospinning (Wang et al., 2001).

A method known as core-shell electrospinning has been created (Sun et al., 2003).

Prof. Seeram Ramakrishna and coauthors published "A review on polymer nanofibers by electrospinning and their applications in nanocomposites" (Huang et al., 2003).

Electrospun continuous yarn's initial publication in a scientific journal (Smit et al., 2005).

Authorized by: Inventor

"An Introduction to Electrospinning and Nanofibers" is the first book written by Professor Seeram Ramakrishna et al. (2005) (Scheme 20.1).

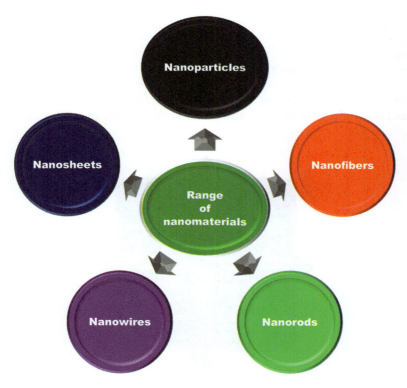

Scheme 20.1 Classification of nanomaterials based on range.

20.4 Nanofiber and nanofibrous material types

The advancements in nanoscience and nanotechnology during the past two decades have produced a variety of nanomaterials, including

1. Nanoparticles,
2. Nanofibers,
3. Nanorods,
4. Nanowires, and
5. Nanosheets.

Gleiter proposed the first classification scheme for nanomaterials in 1995 (Gleiter, 2000). He categorized the nanoparticles according to their chemical makeup and crystalline shapes. The dimensionality of the nanomaterials was not taken into account; hence the Gleiter scheme was not finished. Utilizing the dimensions of the nanostructure itself and its constituent parts, several types of nanostructured materials are evaluated and categorized.

A new classification system for nanomaterials was published in 2007 by Pokropivny and Skorokhod (Pokropivny & Skorokhod, 2007), that is,

1. Zero dimensional
2. One dimensional
3. Two dimensional
4. Three dimensional (Scheme 20.2).

Numerous different types of nanofiber materials have been documented as of late, and it is anticipated that new nanofiber and nanofibrous material varieties may

Scheme 20.2 Classification of nanomaterials based on dimension.

Scheme 20.3 Classification of nanomaterials based on morphology.

emerge in the future. Materials made of nanofibers and nanofibrils are often categorized according (Lundahl et al., 2017) to

1. **Composition**, for example,
    a. Polymers,
    b. Metals,
    c. Metal Oxides,
    d. Ceramics,
    e. Carbon, and
    f. Hybrid
2. **Size**, for example, diameter, length, and pore size
3. **Morphology**
    a. Nonporous,
    b. Mesoporous,
    c. Hollow,
    d. Core-Shell,
    e. Biocomponent, and
    f. Multicomponent (Scheme 20.3).

20.5 Classification of nanofibers based on their chemical composition

Numerous fibrous materials, including metals, ceramics, metal oxides, and even natural and synthetic polymers, have been electrospun into homogeneous fibers with carefully controlled diameters, compositions, and morphologies (Anusiya & Jaiganesh, 2022). Nanofiber materials are categorized into four primary groups according to their chemical composition:

1. Carbon-based nanofibers,
2. Inorganic-based nanofibers,
3. Organic-based nanofibers, and
4. Composite-based nanofibers (Scheme 20.4).

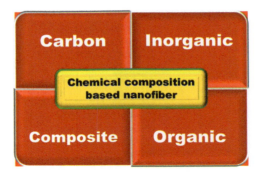

Scheme 20.4 Classification of nanofibers based on chemical composition.

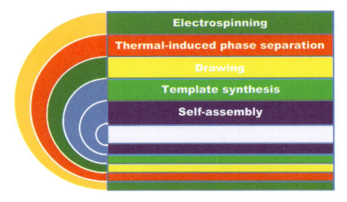

Scheme 20.5 Synthesis method of nanofibers.

20.6 Synthesis methods of nanofibers

1. Electrospinning
2. Thermal-induced phase separation
3. Drawing
4. Template synthesis
5. Self-assembly (Scheme 20.5).

20.7 Application of electronic-based nanofibers and their composites

20.7.1 Electronics in textiles

This development is a significant new idea that opens up a wide range of multifunctional, wearable e-textiles for applications such as detecting and monitoring bodily functions, delivering communication capabilities, data transmission, controlling personal environments, and many more (Chen et al., 2022). Knitting, weaving, embroidering, coating/laminating, printing, dip coating, and atomic layer deposition are common methods for producing e-textiles. Commercial conductive threads are used in weaving, stitching, and embroidery. Traditional conductive threads have excellent conductivity but are expensive, heavy, and mechanically fragile.

20.7.2 Sensing in textiles

Electrospun nanofibers are a desirable material for sensors because of their high surface area to volume ratio. Due to their versatility in material selection and simplicity in adding active agents, electrospun fibers have been studied for usage in numerous sensor applications. Depending on the material and its qualities, electrospinning has been used to achieve optical change, electrical resistivity, electrochemical sensing, and acoustic waves as sensor techniques. The majority of electrospun fiber sensors are nonwoven meshes, which may be manufactured using the existing setup. Electrospun fiber sensors frequently perform better than cast film sensors made of the same material. The material that is chosen, rather than the form, usually determines targeted applications and sensing approaches (Jin & Bai, 2022).

20.7.3 Energy storage

20.7.3.1 Lithium-ion batteries

In lithium-ion batteries (LIBs), graphite is typically used as the anode, and layered $LiCoO_2$ is typically used as the cathode as an intercalation host for $Li+$. These two electrodes are separated by a porous permeable membrane that only permits $Li+$ ions and prevents a short circuit caused by electrodes coming into contact. During charging and discharging, Li ions are intercalated and deintercalated between graphite and $LiCoO_2$ through the electrolyte (Ozawa, 1994).

20.7.3.2 Supercapacitors

Supercapacitors (SCs) have drawn attention to energy storage devices because of their high power density, moderate energy density, long cycling life, and low maintenance requirements. Electric double-layer capacitors (EDLCs), one of two forms of SCs, rely on reversible ion absorption in an electrolyte onto the working materials. High specific surface area electrode materials are necessary. The second is pseudocapacitors (PCs), which use a quick and reversible Faradaic reaction to store charges. Pseudo-capacitive electrode materials, such as transition-metal oxides and conductive polymers, have greater electrochemical capacitances and energy densities compared to carbon electrode materials for electrochemical double-layer capacitors and can, therefore, meet the demands of high-performance SCs (Reece et al., 2020).

20.7.3.3 Hydrogen storage

A well-known energy source that makes a great fuel is hydrogen. Because water vapor makes up the majority of the exhaust gases produced by hydrogen-powered cars, it is a possible replacement for conventional energy sources. Hydrogen has a chemical energy density that is at least three times greater than other chemical fuels (142 kg). The safe, effective, and cost-effective storage of hydrogen is a major technological barrier to its widespread usage as a sustainable energy carrier. Finding systems that can satisfy thermodynamic, kinetic, and cycle stabilities for hydrogen storage applications is still a significant difficulty despite decades of investigation (Sordakis et al., 2018; Xue et al., 2019).

20.7.4 Strategies for electrospun nanofibers in soft electronics

Electrospinning has been used to directly or indirectly create a variety of nanomaterials, including organic polymer nanofibers, carbon nanofibers, metal nanofibers, ceramic nanofibers, and inorganic hybrid nanofibers. Their use in soft electronics, including transparent electrodes, conductors, transistors, optoelectronics, sensors, and energy devices, has grown steadily. To this end, various techniques have been used to design electrospun nanofibers as active and passive components for flexible/stretchable electronic devices49, including functional substrates, templates, and precursors, friction layers for nanogenerators, electrolyte separators for energy storage devices, and sensing layers for chemical sensors (Xue et al., 2019).

20.7.4.1 Single semiconducting nanofiber

Due to their high carrier mobility and exceptional mechanical flexibility, single electrospun semiconducting nanofibers have been employed as channel materials for flexible/stretchable transistors. One-dimensional nanostructures made of conjugated polymers show the following promising advantages over traditional film-type semiconducting materials: (1) a large contact area with a dielectric layer, (2) inherent flexibility, (3) the ability to engineer wrinkled or buckled structures for improved stretchability, (4) easier patentability via direct printing, and (5) efficient and simple scale-up fabrication (Chwee & Limb, 2017).

20.7.4.2 Entangled porous nanofiber sheet

Electrical components are typically deposited onto elastomeric substrates or embedded into elastomeric matrixes to create flexible/stretchable electronic devices. Polydimethylsiloxane (PDMS), polyimide (PI), polyethylene terephthalate (PET), Ecoflex, parylene, latex rubber, paper, and so on, are the principal elastomeric materials used for soft electronics. These polymeric substrates, however, have mechanical incompatibilities with frequently utilized electrical materials such as metals, metallic nanomaterials, and carbon-based materials (Medeiros et al., 2016).

20.7.4.3 Aligned nanofiber array

During the electrospinning process, the flow rate and collector conditions can be easily adjusted to produce well-aligned nanofibers (Medeiros et al., 2016). Aligned nanofibers made from conjugated polymers have also been employed as channel materials in addition to the single semiconducting nanofibers previously discussed in "Single semiconducting nanofiber." According to a study, the number of aligned P3HT electrospun nanofibers linking the source (S) and drain (D) electrodes can influence the maximum concurrent values (Rahman & Netravali, 2016).

20.7.4.4 Temporary template

Electrospun polymer nanofiber mats are frequently used as temporary templates when fabricating transparent electrodes (Choi et al., 2009). The porous mesh architectures of the conductive metal layers are kept after the removal of nanofibers by annealing or dissolving in solution. By altering the shape of the electrospun nanofiber sheet, characteristics such as the transparency and sheet resistance of the transparent electrode may be fine-tuned.

20.7.4.5 Reinforcement

Many materials based on nanotubes or NW have been widely used as reinforcement materials in composites to improve mechanical performance (Nurazzi et al., 2021). These materials were inspired by the reinforcing wires in concrete. Even though an elastomer substrate's thickness is on the order of a few micrometers, electrospun nanofibers can transform it into a skin-like substrate.

20.7.5 Yarn

Due to their potential applications in wearable screens, smart fabrics, and portable electronics, e-textiles are particularly significant for wearable electronics. Although other spinning techniques (dry, force, melting, and wet spinning) have been developed to create nano- and microfibers for e-textile applications, they are not appropriate for mass production (Hufenus et al., 2020). To this goal, conductive membranes made of electrospun nanofibers have been wrapped to create a variety of three-dimensional porous microfibers or yarn-shaped flexible/stretchable conductors. Graphene, silver nanoparticles (AgNPs), CNTs, and other electrically conducting substances are frequently included in entangled nanofibers. With a high maximum strain ($>50\%$) and strong conductivity, the as-prepared yarns are typically employed as highly sensitive strain and pressure sensors.

20.7.6 Applications of electrospun nanofibers in soft electronics

The impressive qualities of electrospun nanofibers in one-dimensional nanostructures, namely, the porosity, flexibility, conductivity, transparency, and diverse fibrous morphologies, are crucial for designing high-performance soft electronics. Several recently developed flexible/stretchable electronic devices, including conductors, transparent electrodes, strain sensors, pressure sensors, nanogenerators, and support structures, are discussed and outlined in the following sections (Xue et al., 2017).

20.7.6.1 Conductors

The development of the next generation of wearable electronics and soft robotics requires the use of flexible and/or stretchy conductors (He et al., 2020).

20.7.6.2 Nontransparent conductors

A flexible and/or stretchy conductor's primary characteristic is that it maintains high conductivity under specific mechanical deformations (Wang et al., 2018). Numerous flexible/stretchable conductive films or yarns have been created as a result of the high flexibility of the porous nanofiber-networked structure created by electrospinning.

20.7.6.3 Transparent electrodes

Transparent heaters, photovoltaics, SCs, light-emitting diodes, and batteries are just a few examples of soft electronic devices that use flexible transparent electrodes, which combine excellent electrical conductivity, optical transparency, and mechanical flexibility (Wang et al., 2021). Transparent conductive oxides are used as conducting materials all the time, but their use in flexible/stretchable electronics is limited by their brittleness and the high-pressure vacuum processes needed during fabrication.

20.7.6.4 Sensors

Electrospun nanofibers have been used to create a variety of very sensitive flexible/stretchable sensors due to their large surface area and ultra flexibility (Aliheidari et al., 2019). This section reviews the two most researched electrospun nanofiber sensors, strain sensors, and pressure sensors.

20.7.6.5 Pressure sensors

To effectively assess external stimuli on curved and dynamic surfaces such as human skin and natural tissues, the determination of tiny normal pressures is essential. Due to their porous architecture and great elasticity, electrospun nanofiber mats offer a high tolerance for repeated external pressing. The use of electrospun nanofibers has thus led to the development of numerous wearable pressure sensors (Xu et al., 2022). There are four different operating methods for these sensors: capacitive, resistive, triboelectric, and piezoelectric.

20.7.7 Energy harvesting and storage devices

Implementing independent wearable electronic platforms requires wearable energy-generating and storage technologies. In this section, we discuss the creation of soft energy storage devices such as SCs and harvesting devices based on electrospun nanofibers, such as nanogenerators and batteries (He et al., 2021).

20.7.7.1 Nanogenerators

The rapid depletion of fossil fuels and environmental concerns have led to an increase in demand for renewable energy. To address this issue, mechanical energy harvesting devices that can transform small-scale energy from vibrations and motions into electrical power have been created (He et al., 2021).

20.7.7.2 Supercapacitors

Electrochemical capacitors, also known as SCs, are useful energy storage devices that have quick charging/discharging capabilities, high power and energy densities, long-term cycling, and secure operation. A SC is made up of two electrodes and an ion-permeable electrolyte layer or separator placed between them. A SC must sustain excellent performance while being repeatedly mechanically deformed to be included in a soft electronic system (Kim et al., 2021). Following this objective, the design of very flexible/stretchable capacitive materials for electrodes is the primary focus of the development of flexible/stretchable SCs.

20.7.7.3 Transistors

The charge transport characteristics in a device are controlled by a field-effect transistor, which is typically thought of as an electrical switch (Di et al., 2007). It is made up of three electrodes (source, drain, and gate), an active channel made of a semiconductor, and a thin layer of dielectric acting as an insulator between the channel and gate electrodes. Charge carriers flow from the source to the drain when a voltage is applied at the gate electrode, creating a source-drain current.

20.8 Applications in energy devices of electrospun NFs

20.8.1 Energy harvesting and conversion devices
20.8.1.1 Solar cells

There have been three stages in the development of solar energy technology: monocrystalline and polycrystalline silicon solar cells, amorphous silicon thin film solar cells, and third-generation solar cells, which refer to novel concepts in high conversion efficiency solar cells such as dye-sensitized solar cells (DSSCs) and hybrid solar cells (Husain et al., 2018). Even though solar technology has come a long way, work still has to be done to significantly increase photovoltaic cells'

Dye-sensitized solar cells

1. Photoanode

With the use of a photosensitizing dye, DSSCs may directly convert light into electricity (Sharma et al., 2018). Three main parts make up a standard DSSC: an electrolyte, a counter electrode (CE), and a photoanode. The photosensitizing dye is deposited on the porous semiconductor film that is coated over the clear conducting glass that makes up a photoanode in most cases. After the photosensitizer in DSSCs absorbs photons, the photoelectrons from the photosensitizer first enter the semiconductor's conducting band before being gathered on the photoanode. To create current, the photoelectrons are then transferred to the CE via the external circuit. Maximum light absorption and effective charge transfer have an impact on the total photoelectric conversion efficiency in this process.

2. Counter Electrodes

The CE in DSSCs participates in electron transmission and collection, and its catalytic activity alters the internal series resistance of the component, changing the fill factor (Wu et al., 2017).

3. DSSCs Electrolytes

The practical application of DSSCs is constrained by the leakage and volatilization of liquid electrolytes, even though the power conversion efficiency of DSSCs based on liquid electrolytes has exceeded 11% (Iftikhar et al., 2019).

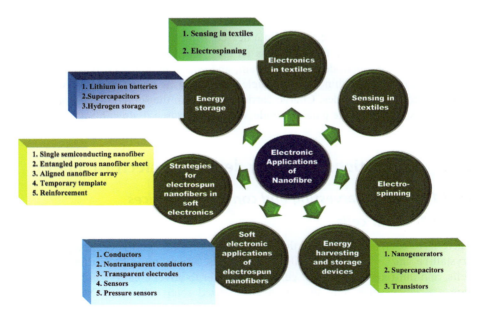

Scheme 20.6 Electronic application of nanofibers.

20.8.1.2 Fuel cells

The redox interaction between the anode and cathode in fuel cells allows chemical energy to be transformed into electrical energy. However, unlike batteries, fuel cells store the reactants externally rather than inside, allowing for continuous energy production as long as there is a steady supply of fuel (commonly hydrogen and methanol) (Winter & Brodd, 2004). Due to their high power density and low operating temperature, direct methanol fuel cells (DMFCs) and proton exchange membrane (PEM) fuel cells (PEMFCs) have been identified as two fuel cell types that have the potential to be industrialized. In comparison to PEMFC, DMFC has additional benefits, including no need for hydrogen preparation, storage, shipping, or security issues, direct fuel flow without the need for external reforming, simple construction, quick response times, and ease of operation.

1. **Electrospun Catalyst-Supports**
   The use of a pricy Pt catalyst in DMFCs restricts the commercial potential of these materials. Exploring new materials or materials that support Pt catalyst as an electrocatalyst is crucial to solving this issue (Guerrero-Pérez, 2022).
2. **Electrospun Electrolyte Membrane**
   Another crucial element in fuel cells is the PEM. The fuel crossover and proton conductivity are two elements that have an impact on the effectiveness of energy conversion. Because of its distinct chemical composition, Nafion has traditionally been employed as a commercial polymer electrolyte membrane with relatively strong proton conductivity (Guerrero-Pérez, 2022).

20.8.1.3 Mechanical energy harvesters

Nanogenerators

Piezoelectric nanomaterials are typically used to create mechanical nanogenerators, which can take small mechanical vibrations from the surrounding environment and transform them into electricity to power micro/nanodevices (Wang, 2012). The three primary types of architectures used for these nanogenerators are film-based, nanowire-based, and nanofiber-based (Wang, 2012). The preparation of nanofiber-based generators benefits specifically from the electrospinning technology.

20.8.1.4 Solar-driven hydrogen generation

Hydrogen generators—photocatalysts

In contrast to hydrogen energy, which is effective, clean, and friendly to the environment, solar energy is unrenewable and environmentally favorable (Oh et al., 2022). Photocatalytic water splitting for hydrogen production is a potential solution to the energy issue to maximize the use of solar energy.

20.8.2 Energy storage devices

20.8.2.1 Rechargeable lithium-ion batteries

Due to their high energy density and prolonged cycle life, LIBs have been used in a wide range of products, including mobile phones, laptop computers, camcorders, digital cameras, and many more commercial and military applications (Goodenough & Park, 2013). Anode (often graphite), an organic electrolyte with a carbonate structure, and a cathode make up the majority of LIBs (generally $LiCoO_2$).

Cathode materials

Due to its high specific capacity, high voltage, and extended cycle life, $LiCoO_2$ is a typical cathode material used in LIBs. Due to the close diffusion distance, the electrospun nanostructured $LiCoO_2$ fiber electrode showed improved Li ion and electron conductivity (Dai et al., 2016). The electrode as prepared may demonstrate excellent power density and good rate capabilities.

Anode materials

Due to several benefits, including inexpensive cost, easy availability, and extended cycle life, carbon NFs have been the most widely used anode materials for LIBs. Carbon NFs do have several disadvantages too, namely, their generally limited specific capacity and rate capability. This is because, in comparison to carbon NFs without PAN, C/PAN NFs generated using electrospinning and thermal treatments displayed a greater reversible capacity, which was explained by their highly disordered structure and flaws (Huang et al., 2021).

Separator materials

A separator is a crucial part of LIBs and has a big impact on how well the battery performs. However, there are several drawbacks to commercial separators based on microporous membranes, including low porosity, inadequate thermal stability, and poor wettability in liquid electrolytes (Song et al., 2021).

20.8.2.2 Supercapacitors

Due to their high power performance, lengthy cycle life, and low maintenance requirements, SCs are regarded as one of the most promising new energy storage devices in many fields, including transportation, electricity, communications, defense, consumer electronics, and other applications. SCs can be divided into PCs and electrical double-layer capacitors based on various energy storage techniques (EDLCs) (Dhandapani et al., 2022).

20.8.2.3 Hydrogen storage

Since water is the only byproduct of the energy conversion process, hydrogen energy, one of the significant new energy sources of the 21st century, has been seen as an appropriate substitute fuel. Finding materials that are acceptable for large-scale hydrogen storage while also meeting the demands of automotive applications in terms of weight, volume, and efficiency is currently a challenge due to their low cost, durability, and safety (Tashie-Lewis & Nnabuife, 2021).

20.8.3 Other applications

Nanofibers have also been used in other applications, including flexible/stretchable heaters, photodetectors, chemical sensors, optically encoded sensors, solar cells, and organic LEDs. For instance, exploiting the photoelectric action of electrospun ZnO/CdO (cadmium oxide) nanofibers, a fully transparent and flexible ultraviolet-visible photodetector has been created. ZnO and CdO nanoparticles were fused into separate nanofibers to create ZnO/CdO nanofibers. Wearable chemical sensing devices can utilize the developed fibrous membranes as chemiresistors (Lee et al., 2022; Wu et al., 2020; Yoon et al., 2022).

Due to their high surface area to volume ratio, flexibility in surface functions, variable pore size, high interconnected porosity, and exceptional mechanical qualities, nanofibers exhibit distinctive physical, chemical, and biological capabilities (stiffness and tensile strength) (Huang et al., 2003; Teixeira et al., 2020). In a variety of fields, including regenerative medicine and tissue engineering, medication delivery, sensors, energy production and storage, filtration, catalysis, textile, defense, and security, nanofibers have drawn extraordinary interest and a wide range of uses. Electrospinning is one of the few fabrication methods that can create nanofibers of various materials in a variety of fibrous configurations.

Nanofibers with modified surfaces that may communicate with the user or environment are referred to as smart textiles. With current electronics' speed and computing power combined with fabrics' flexibility, wearability, and continuous nature, flexible electronics and textiles have the potential to combine the best features of each technology. The fundamental component of garment devices and a requirement for creating self-powered e-textiles, wearable textiles have integrated energy storage systems, according to recent research publications on the subject (Sadi & Kumpikaitė, 2022; Zhang et al., 2021).

20.9 Conclusions

Due to their high surface area to volume ratio, flexibility in surface functions, variable pore size, high interconnected porosity, and exceptional mechanical qualities, nanofibers exhibit distinctive physical, chemical, and biological capabilities (stiffness and tensile strength). Nanofibers have generated a great deal of attention and have a wide variety of uses in many different industries, including textile electronics, textile sensing, and textile energy storage (i.e., LIBs, SCs, hydrogen storage). Methods for using electrospun nanofibers in soft electronics are particularly notable. Soft electrospun nanofibers have a unique ability to create nanofibers of various materials in different fibrous assemblies, making them useful in soft electronics, energy harvesting and storage devices, and other energy applications.

Electrospinning is very appealing to both academia and business because of its relatively easy setup and high output rate. Despite the advancement of new techniques for synthesizing nanofibers and the emergence of new uses for them, there are still obstacles in the way of the industrial-scale commercial production of high-quality nanofibers. The majority of the new uses for nanofibers are still in the proof-of-concept phase. More emphasis should be placed on in vivo investigations because the majority of nanofiber biomedical applications are currently carried out in vitro.

Authors contribution

The manuscript was written with the contributions of all authors. All authors have approved the final version of the manuscript.

Conflict of interest

All authors declare no competing financial interest.

Acknowledgments

The authors are grateful to HOD MATS School of Sciences, MATS University, Raipur, Chhattisgarh.

References

Aliheidari, N., Aliahmad, N., Agarwal, M., & Dalir, H. (2019). Electrospun nanofibers for label-free sensor applications. *Sensors*, *19*(16), 3587.

Anton, F. (1934). Process and apparatus for preparing artificial threads. In: Google patents.

Anusiya, G., & Jaiganesh, R. (2022). A review on fabrication methods of nanofibers and a special focus on application of cellulose nanofibers. *Carbohydrate Polymer Technologies and Applications*, *4*, 100262.

Babuab, A., Aazemab, I., Waldenab, R., Bairagic, S., Mulvihillc, D. M., & Pillaia, S. C. (2023). Electrospun nanofiber based TENGs for wearable electronics and self-powered sensing. *Chemical Engineering Journal*, *452*(1), 139060.

Baig, N., Kammakakam, I., & Falath, W. (2021). Nanomaterials: A review of synthesis methods, properties, recent progress, and challenges. *Materials Advances*, *2*, 1821–1871.

Bayda, S., Adeel, M., Tuccinardi, T., Cordani, M., & Rizzolio, F. (2020). The history of nanoscience and nanotechnology: From chemical–physical applications to nanomedicine. *Molecules (Basel, Switzerland)*, *25*(1), 112.

Bose, G.-M. (1744). *Die Electricität nach ihrer Entdeckung und Fortgang mit poetischer Feder entworffen*. Wittenberg: Joh. Joachim Ahlfelden.

Boys, C. V. (1887). On the production, properties, and some suggested uses of the finest threads. *Proceedings of the Physical Society of London, 9*(1), 8.

Burton, E., & Wiegand, W. (1912). *Effect of electricity on streams of water drops*. Toronto: University Library.

Chen, G., Xiao, X., Zhao, X., Tat, T., Bick, M., & Chen, J. (2022). Electronic textiles for wearable point-of-care systems. *Chemical Reviews, 122*(3), 3259−3291.

Cho, K. W., Sunwoo, S.-H., Hong, Y. J., Koo, J. H., Kim, J. H., Baik, S., Hyeon, T., & Kim, D.-H. (2022). Soft bioelectronics based on nanomaterials. *Chemical Reviews, 122*(5), 5068−5143.

Choi, S.-H., Ankonina, G., Youn, D.-Y., Oh, S.-G., Hong, J.-M., Rothschild, A., & Kim, I.-D. (2009). Hollow ZnO nanofibers fabricated using electrospun polymer templates and their electronic transport properties. *ACS Nano, 3*(9), 2623−2631.

Chwee, K., & Limb, T. (2017). Nanofiber technology: Current status and emerging developments. *Progress in Polymer Science, 70*, 1−17.

Cloupeau, M., & Prunet-Foch, B. (1989). Electrostatic spraying of liquids in cone-jet mode. *Journal of Electrostatics, 22*(2), 135−159.

Dagdeviren, C., Ozlem, P. J., Kwi-Il, L. T., Keon, P., Leee, J., Shifg, Y., Huanghijk, Y., & Rogerslmnop, J. A. (2016). Recent progress in flexible and stretchable piezoelectric devices for mechanical energy harvesting, sensing and actuation. *Extreme Mechanics Letters, 9*(1), 269−281.

Dai, X., Zhou, A., Xu, J., Lu, Y., Wang, L., Fan, C., & Li, J. (2016). Extending the high-voltage capacity of $LiCoO_2$ cathode by direct coating of the composite electrode with Li_2CO_3 via magnetron sputtering. *Journal of Physical Chemistry C, 120*(1), 422−430.

Dhandapani, E., Thangarasu, S., Ramesh, S., Ramesh, K., Vasudevan, R., & Duraisamy, N. (2022). Recent development and prospective of carbonaceous material, conducting polymer and their composite electrode materials for supercapacitor—A review. *Journal of Energy Storage, 52*, 104937.

Di, C.-a, Yu, G., Liu, Y., & Zhu, D. (2007). High-performance organic field-effect transistors: Molecular design, device fabrication, and physical properties. *The Journal of Physical Chemistry. B, 111*(51), 14083−14096.

Doshi, J., & Reneker, D. H. (1995). Electrospinning process and applications of electrospun fibers. *Journal of Electrostatics, 35*(2−3), 151−160.

English, W. (1948). Corona from a water drop. *Physical Review, 74*(2), 179.

Feynman, R. P. (1959). There's plenty of room at the bottom. *Miniaturization*, 282−296.

Gilbert, W., & Wright, E. (1967). De magnete, magneticisque corporibus, et de magno magnete tellure: Physiologia noua, plurimis & argumentis, & experimentis demonstrata. *Royal College of Physicians, London, 1600*.

Gleiter, H. (2000). Nanostructured materials: Basic concepts and microstructure. *Acta Materialia, 48*(1), 1−29.

Goodenough, J. B., & Park, K.-S. (2013). The Li-ion rechargeable battery: A perspective. *Journal of the American Chemical Society, 135*(4), 1167−1176.

Guan, B., & Cole, R. B. (2016) (Submitted for publication). The background to electrospray. In *The Encyclopedia of Mass Spectrometry Volume 9: Historical Perspectives, Part A: The Development of Mass Spectrometry*, 132−140.

Guerrero-Pérez, M. O. (2022). Research progress on the applications of electrospun nanofibers in catalysis. *Catalysts, 12*(1), 9.

He, J., Zhang, Y., Zhou, R., Meng, L., Chen, T., Mai, W., & Pan, C. (2020). Recent advances of wearable and flexible piezoresistivity pressure sensor devices and its future prospects. *Journal of Materiomics, 6*(1), 86−101.

He, W., Fu, X., Zhang, D., Zhang, Q., Zhu, K., Yuan, Z., & Ma, R. (2021). Recent progress of flexible/wearable self-charging power units based on triboelectric nanogenerators. *Nano Energy, 84*, 105880.

Huang, A., Ma, Y., Peng, J., Li, L., Chou, S.-l, Ramakrishna, S., & Peng, S. (2021). Tailoring the structure of silicon-based materials for lithium-ion batteries via electrospinning technology. *eScience, 1*(2), 141−162.

Huang, Z.-M., Zhang, Y.-Z., Kotaki, M., & Ramakrishna, S. (2003). A review on polymer nanofibers by electrospinning and their applications in nanocomposites. *Composites Science and Technology, 63*(15), 2223−2253.

Hufenus, R., Yan, Y., Dauner, M., & Kikutani, T. (2020). Melt-spun fibers for textile applications. *Materials, 13*(19), 4298.

Husain, A. A. F., Hasan, W. Z. W., Shafie, S., Hamidon, M. N., & Pandey, S. S. (2018). A review of transparent solar photovoltaic technologies. *Renewable and Sustainable Energy Reviews, 94*, 779−791.

Iftikhar, H., Sonai, G. G., Hashmi, S. G., Nogueira, A. F., & Lund, P. D. (2019). Progress on electrolytes development in dye-sensitized solar cells. *Materials, 12*(12), 1998.

Iijima, S. (1991). Helical microtubules of graphitic carbon. *Nature, 354*(6348), 56.

Jeevanandam, J., Barhoum, A., Chan, Y. S., Dufresne, A., & Danquah, M. K. (2018). Review on nanoparticles and nanostructured materials: History, sources, toxicity and regulations. *Beilstein Journal of Nanotechnology, 9*, 1050−1074.

Jin, C., & Bai, Z. (2022). MXene-based textile sensors for wearable applications. *ACS Sensors, 7*(4), 929−950.

Kauffman, G. B. (1993). Rayon: The first semi-synthetic fiber product. *Journal of Chemical Education, 70*(11), 887.

Kim, J., Kim, J. W., Kim, S., Keum, K., Park, J., Jeong, Y. R., Jin, S. W., & Ha, J. S. (2021). Stretchable, self-healable, and photodegradable supercapacitor based on a polyelectrolyte crosslinked via dynamic host-guest interaction. *Chemical Engineering Journal, 422*, 130121.

Larrondo, L., & St John Manley, R. (1981a). Electrostatic fiber spinning from polymer melts. I. Experimental observations on fiber formation and properties. *Journal of Polymer Science: Polymer Physics, 19*(6), 909−920.

Larrondo, L., & St John Manley, R. (1981b). Electrostatic fiber spinning from polymer melts. II. Examination of the flow field in an electrically driven jet. *Journal of Polymer Science: Polymer Physics, 19*(6), 921−932.

Larrondo, L., & St John Manley, R. (1981c). Electrostatic fiber spinning from polymer melts. III. Electrostatic deformation of a pendant drop of polymer melt. *Journal of Polymer Science: Polymer Physics, 19*(6), 933−940.

Laucht, A., Hohls, F., Ubbelohde, N., Gonzalez-Zalba, M. F., Reilly, D. J., Stobbe, S., Schröder, T., Scarlino, P., Koski, J. V., & Dzurak, A. (2021). Roadmap on quantum nanotechnologies. *Nanotechnology, 32*(16), 162003.

Lee, G., Zarei, M., Wei, Q., Zhu, Y., & Lee, S. G. (2022). Surface wrinkling for flexible and stretchable sensors. *Small (Weinheim an der Bergstrasse, Germany), 18*(42), 2203491.

Lundahl, M. J., Klar, V., Wang, L., Ago, M., & Rojas, O. J. (2017). Spinning of cellulose nanofibrils into filaments: A review. *Industrial & Engineering Chemistry Research, 56*(1), 8−19.

Macky, W. (1931). Some investigations on the deformation and breaking of water drops in strong electric fields. *Proceedings of the Royal Society of London. Series A, Containing Papers of a Mathematical and Physical Character, 133*(822), 565−587.

Martin, G. E., & Cockshott, I. D. (1977). Fibrillar product of electrostatically spun organic material. In: Google patents.

McEuen, P. L. (2000). Single-wall carbon nanotubes. *Physics World*, *13*(6), 31.
Medeiros, E. L. G., Braz, A. L., Porto, I. J., Menner, A., Bismarck, A., Boccaccini, A. R., Lepry, W. C., Nazhat, S. N., Medeiros, E. S., & Blaker, J. J. (2016). Porous bioactive nanofibers via cryogenic solution blow spinning and their formation into 3D macroporous scaffolds. *ACS Biomaterials Science & Engineering*, *2*(9), 1442–1449.
Melcher, J., & Taylor, G. (1969). Electrohydrodynamics: A review of the role of interfacial shear stresses. *Annual Review of Fluid Mechanics*, *1*(1), 111–146.
Mostofizadeh, A., Li, Y., Song, B., & Huang, Y. (2011). Synthesis, properties, and applications of low- dimensional carbon-related nanomaterials. *Journal of Nanomaterials*, *2011*, 16.
Nurazzi, N. M., Sabaruddin, F. A., Harussani, M. M., Kamarudin, S. H., Rayung, M., Asyraf, M. R. M., Aisyah, H. A., Norrrahim, M. N. F., Ilyas, R. A., Abdullah, N., Zainudin, E. S., Sapuan, S. M., & Khalina, A. (2021). Mechanical performance and applications of CNTs reinforced polymer composites—A review. *Nanomaterials (Basel)*, *11*(9), 2186.
Oberlin, A., Endo, M., & Koyama, T. (1976). Filamentous growth of carbon through benzene decomposition. *Journal of Crystal Growth*, *32*(3), 335–349.
Oh, V. B.-Y., Ng, S.-F., & Ong, W.-J. (2022). Is photocatalytic hydrogen production sustainable? – Assessing the potential environmental enhancement of photocatalytic technology against steam methane reforming and electrocatalysis. *Journal of Cleaner Production*, *379*(2), 134673.
Onsager, L. (1936). Electric moments of molecules in liquids. *Journal of the American Chemical Society*, *58*(8), 1486–1493.
Ozawa, K. (1994). Lithium-ion rechargeable batteries with $LiCoO_2$ and carbon electrodes: The $LiCoO_2$/C system. *Solid State Ionics*, *69*(3–4), 212–221.
Pokropivny, V. V., & Skorokhod, V. V. (2007). Classification of nanostructures by dimensionality and concept of surface forms engineering in nanomaterial science. *Materials Science and Engineering: C*, *27*(5–8), 990–993.
Rahman, M. M., & Netravali, A. N. (2016). Aligned bacterial cellulose arrays as "green" nanofibers for composite materials. *ACS Macro Letters*, *5*(9), 1070–1074.
Rajan, K., Garofalo, E., & Chiolerio, A. (2018). Wearable intrinsically soft, stretchable, flexible devices for memories and computing. *Sensors*, *18*(2), 367.
Ramakrishna, S., Fujihara, K., Teo, W., Lim, T.-C., & Ma, Z. (2005). *An introduction to electrospinning and nanofibers*. Singapura: World Scientific Publishing Company.
Rayleigh, L. (1878a). On the instability of jets. *Proceedings of the London Mathematical Society*, *1*(1), 4–13.
Rayleigh, L. (1878b). The influence of electricity on colliding water drops. *Proceedings of the Royal Society of London*, *28*, 405–409.
Rayleigh, L. (1879). On the capillary phenomena of jets. *Proceedings of the Royal Society of London*, *29*, 71–97.
Rayleigh, L. (1882). On the equilibrium of liquid conducting masses charged with electricity. *London, Edinburgh and Dublin Philosophical Magazine and Journal of Science*, *14*, 87.
Reece, R., Lekakou, C., & Smith, P. A. (2020). A high-performance structural supercapacitor. *ACS Applied Materials & Interfaces*, *12*(23), 25683–25692.
Ross, M. B., Mirkin, C. A., & Schatz, G. C. (2016). Optical properties of one-, two-, and three-dimensional arrays of plasmonic nanostructures. *Journal of Physical Chemistry C*, *120*(2), 816–830.
Sadi, M. S., & Kumpikaitė, E. (2022). Advances in the robustness of wearable electronic textiles: Strategies, stability, washability and perspective. *Nanomaterials*, *12*(12), 2039.
Sharma, K., Sharma, V., & Sharma, S. S. (2018). Dye-sensitized solar cells: Fundamentals and current status. *Nanoscale Research Letters*, *13*.

Simons, H. L. (1966). Process and apparatus for producing patterned non-woven fabrics. In: Google patents.
Smit, E., Bűttner, U., & Sanderson, R. D. (2005). Continuous yarns from electrospun fibers. *Polymer*, *46*(8), 2419–2423.
Son, D., & Bao, Z. (2018). Nanomaterials in skin-inspired electronics: Toward soft and robust skin-like electronic nanosystems. *ACS Nano*, *12*(12), 11731–11739.
Song, Y., Sheng, L., Wang, L., Xu, H., & He, X. (2021). From separator to membrane: Separators can function more in lithium ion batteries. *Electrochemistry Communications*, *124*, 106948.
Sordakis, K., Tang, C., Vogt, L. K., Junge, H., Dyson, P. J., Beller, M., & Laurenczy, G. (2018). Homogeneous catalysis for sustainable hydrogen storage in formic acid and alcohols. *Chemical Reviews*, *118*(2), 372–433.
Sun, Z., Zussman, E., Yarin, A. L., Wendorff, J. H., & Greiner, A. (2003). Compound core–shell polymer nanofibers by co-electrospinning. *Advanced Materials*, *15*(22), 1929–1932.
Tabassum, N. (2020). An empirical exploration of the nanotechnology. *International Journal of Advanced Research*, *8*(7), 885–915.
Tashie-Lewis, B. C., & Nnabuife, S. G. (2021). Hydrogen production, distribution, storage and power conversion in a hydrogen economy - A technology review. *Chemical Engineering Journal Advances*, *8*, 100172.
Taylor, G. (1964). Disintegration of water drops in an electric field. *Proceedings of the Royal Society of London. Series A, Mathematical, Physical and Engineering Sciences*, 383–397.
Taylor, G. (1966). The force exerted by an electric field on a long cylindrical conductor. *Proceedings of the Royal Society of London. Series A. Mathematical, Physical and Engineering Sciences*, 145–158.
Taylor, G. (1969). Electrically driven jets. *Proceedings of the Royal Society of London. Series A. Mathematical, Physical and Engineering Sciences*, 453–475.
Teixeira, M. A., Amorim, M. T. P., & Felgueiras, H. P. (2020). Poly(vinyl alcohol)-based nanofibrous electrospun scaffolds for tissue engineering applications. *Polymers*, *12*(1), 7.
Tennent, H. G. (1987). Carbon fibrils, method for producing same and compositions containing same. In: Google patents.
Tucker N., Hofman K., & Tazzaq H. (2013). A history of electrospinning 1600–1995. GB-06385.
Wang, J., Lin, M.-F., Park, S., & Lee, P. S. (2018). Deformable conductors for human–machine interface. *Materials Today*, *21*(5), 508–526.
Wang, T., Lu, K., Xu, Z., Lin, Z., Ning, H., Qiu, T., Yan, Z., Zheng, H., Yao, R., & Peng, J. (2021). Recent developments in flexible transparent electrode. *Crystals*, *11*(5), 511.
Wang, X. (2012). Piezoelectric nanogenerators—Harvesting ambient mechanical energy at the nanometer scale. *Nano Energy*, *1*(1), 13–24.
Wang, Y., Serrano, S., & Santiago-Aviles, J. J. (2001). Electrostatic synthesis and characterization of Pb (Zr x Ti 1-x)O$_3$ Micro/nano-fibers. *MRS Online Proceedings Library*, 702.
Winter, M., & Brodd, R. J. (2004). What are batteries, fuel cells, and supercapacitors? *Chemical Reviews*, *104*(10), 4245–4270.
Wu, J., Lan, Z., Lin, J., Huang, M., Huang, Y., Fan, L., Luo, G., Lin, Y., Xie, Y., & Wei, Y. (2017). Counter electrodes in dye-sensitized solar cells. *Chemical Society Reviews*, *46*, 5975–6023.
Wu, Z., Cheng, T., & Wang, Z. L. (2020). Self-powered sensors and systems based on nanogenerators. *Sensors*, *20*(10), 2925.
Xu, T., Ji, G., Li, H., Li, J., Chen, Z., Awuye, D. E., & Huang, J. (2022). Preparation and applications of electrospun nanofibers for wearable biosensors. *Biosensors (Basel)*, *12*(3), 177.

Xue, J., Wu, T., Dai, Y., & Xia, Y. (2019). Electrospinning and electrospun nanofibers: Methods, materials, and applications. *Chemical Reviews*, *119*(8), 5298−5415.

Xue, J., Xie, J., Liu, W., & Xia, Y. (2017). Electrospun nanofibers: New concepts, materials, and applications. *Accounts of Chemical Research*, *50*(8), 1976−1987.

Yoon, Y., Truong, P. L., Lee, D., & Ko, S. H. (2022). Metal-oxide nanomaterials synthesis and applications in flexible and wearable sensors. *ACS Nanoscience Au*, *2*(2), 64−92.

Zeleny, J. (1907). The discharge of electricity from pointed conductors differing in size. *Physical Review Letters*, *25*(5), 305.

Zeleny, J. (1914). The electrical discharge from liquid points, and a hydrostatic method of measuring the electric intensity at their surfaces. *Physical Review*, *3*(2), 69.

Zeleny, J. (1915). On the conditions of instability of electrified drops, with applications to the electrical discharge from liquid points. *Proceedings of the Cambridge Philosophical Society*, *18*, 71.

Zeleny, J. (1917). Instability of electrified liquid surfaces. *Physical Review*, *10*(1), 1.

Zeleny, J. (1920). Electrical discharges from pointed conductors. *Physical Review*, *16*(2), 102.

Zhang, Y., Wang, H., Lu, H., Li, S., & Zhang, Y. (2021). Electronic fibers and textiles: Recent progress and perspective. *iScience*, *24*(7), 23, 102716.

Zhao, Y., Kim, A., Wan, G., & Tee, B. C. K. (2019). Design and applications of stretchable and self-healable conductors for soft electronics. *Nano Convergence*, *6*(25).

Additive application of nanofibers and their composites for enhanced performance

21

Sirsendu Sengupta[1,2,], Surya Sarkar[2,3,4,*] and Priyabrata Banerjee[2,3]*
[1]Department of Chemistry, The university of Burdwan, Burdwan, West Bengal, India, [2]Electric Mobility and Tribology Research Group, CSIR-Central Mechanical Engineering Research Institute, Durgapur, West Bengal, India, [3]Academy of Scientific and Innovative Research (AcSIR), Ghaziabad, Uttar Pradesh, India, [4]Department of Physics, Durgapur Women's College, Durgapur, West Bengal, India

21.1 Introduction

Nanofiber is a fiber generated from polymers and its range of diameter is 1—1000 nm. Natural and synthetic polymers are used for generating a large molecular weight, have a small number of substituted chains, and can straightforwardly and systematically dissolve in definite solutions and under desirable circumstances (Ewaldz & Brettmann, 2019; Hassiba et al., 2016). Nanofibers have several applications, such as tissue engineering, drug delivery, air filtration, utilization as microsensors, treatment of wastewater, lithium-ion batteries, photocatalytics, fuel cells, and nanostructured piezoelectric substances because of their tunable physicochemical properties, such as higher surface area to volume ratio (approximately 1000 times larger than microfiber), lightweight, flexibility, tiny pore size (<3 μm), and small diameter (300 — 1000 nm) (Huang et al., 2015; Lou et al., 2017, 2019, 2020; Pham et al., 2006; Yildiz et al., 2019; Zhang et al., 2008). Matrices of nanofiber composite play an important role in the development of several engineering biomaterials for a variety of biomedical uses because they act as tremendous tunable properties such as mechanical, biological, and physical properties as desired (Lakshmi Priya et al., 2017). The structure of a native extracellular matrix is very similar to nanofibers because of its thermal conductivity, maximum surface-to-volume ratio, and efficacy of loading. Nanofiber composite arrangement gives various different properties such as biocompatibility, responsiveness stimuli, and tissue engineering (Nune et al., 2013). Thereafter, the surrounding matrix interface bonding with the nanofiber composites becomes a key factor in deciding the strength of the bond, interaction type, and densities of dislocation. Surface charge, size, shape, and composition of strengthening ingredients are extremely controlled by the connections in the nanofiber matrix (Sawicka & Gouma, 2006). The preparation methods of various nanofibers include electrospinning, emulsion spinning, melt spinning, and solution spinning. Contrasted to the other three methods, the

---

*Contributed equally to this chapter.

Table 21.1 Evaluating structure measuring or chemical composition methods.

Techniques	Range	Mechanism	Measured	Samples
Fourier Transform Infrared Radiation (FTIR)	4000–400 cm^{-1}	Infrared radiation	Light absorption or emission solid	Liquid or gas
UV – vis	200–1100 nm	UV light	Absorption or reflectance	Mainly gas, liquid; and solid that absorb UV
XRF		Fluorescent X-ray	The energy of the photon	Ceramic substances, Metals, glass, and building substances
XRD	∼1%	X-ray	Molecular as well as atomic structure	Crystal
XPS	Parts per thousand/ 1000 ppm	X-rays	Kinetic energy and number of electrons in solid	liquid or gas
XAFS	∼5%	Photo absorption of an X-ray	Emission of electron	Solid, liquid, or gas
ICP		Inductively coupled plasma to ionize	Color	Metals and several nonmetals

most common and user-friendly method is electrospinning. The dissimilarity in the methods, benefits, drawbacks, and branches of this process are recorded in Table 21.1.

21.2 Additive application of nanofibers and their composites for enhanced performance

21.2.1 Oil–water separation

Enlargement of urbanization, as well as industries, requires a rapid increase in sodium-ion technology to meet the growing demand for petroleum products. It should be extensively used for building up the utilization, treatment, and shipping of crude oil besides its derivatives (Pintor et al., 2016; Sarbatly et al., 2016). On the other hand, the massive discharge of oily hygienic sewage, plus repeated oil-spill accidents, and industrial wastewater continuously pollute the world (Banks, 2003; Díez et al., 2007; Peng & Guo, 2016; Peterson et al., 2003). Thereafter, Peterson et al. (2003) noticed that seep-out oils in water primarily contain the various hydrocarbons of petroleum and several poisonous substances that are responsible for toxic effects on existing things in the water, and appreciably pose threats to the human body (Peng & Guo, 2016). On the contrary, ships, automobiles, and even airplanes are affected due to the presence of water in petroleum oil products. Water presence in oil also damages the oil pipe, disturbs lubrication key parameters for the desired oil, engraves the cylinder, and also damages the engine efficiency (Lu et al., 2009). Therefore, water separate from oil is very important and highly advantageous. The method of separating water from oil plays an important role in removing both oily contaminants and part of wastewater from the targeted fuel−water mixture. Consequently, Nordvik et al. in 1996 built up a new method with the help of viscosity to distinct the water after oil in a water−oil mixture (Nordvik et al., 1996). Normally, the separation of water from oil depends on various conditions because, in various cases, its environment is different, such as water mixed with different types of oil, including emulsion formed between oil−water mixtures and oil slicks, etc. Demonstrating the procedure of creating various oil/water mixtures, the straightforward separation processes, such as skimming, separation of gravity, and flotation of air, are very helpful for the treatment of oil slicks as well as unbalanced emulsions. However, the above procedures are not enough to distinguish the stable emulsions due to lower processing efficacy and high operation cost. Therefore, the stability of oil/water emulsions can be disrupted due to the presence of chemical dispersants as well as biological agents, although secondary pollution is initiated due to lack of functional treatment. Therefore, other techniques are immediately required for oil−water separation. Newly designed techniques for the handling of oil−water blends are used by means of membrane separation between fibrous and porous absorbent materials. The benefit of this technique is the separation of water from oil is very effortless way and cost-effective. In the last decades, various porous materials, such as inorganic minerals, natural polymeric materials, and organic synthetic fibers, have been used to build up the technique for oil sorbent purposes. Along with the above absorbent material, adsorbents made from organic synthetic fibers such as mats prepared from polypropylene (PP) fibers, should be extensively used in the adsorption of oil slicks due to their hydrophobic nature, scalable fabrication, and open cell arrangement. Thereafter, the adsorption capacity (<30 g/g) of the above oil adsorption technique is very low, and reusable capacity is also very poor owing to low porosity, large diameter, and unstable porous skeleton. In distinction, multifunctional fibrous membranes used to separate water from the oil−water mixture are extensively useable materials due to the comparatively high separation efficacy and trouble-free operational procedure. According to the filtrates, generally, the above membrane separation technique can be categorized into three varieties, namely, "water-removing," "oil-removing," and "smart membrane separation." However, when it was applied practically, two major problems arose: a decrease in the separation efficacy and difficulty in maintaining permeation flux. Additionally, perfection is also required for an antifouling property of the targeted membranes. In recent times, with the introduction of nanotechnology, a variety of nanostructured substances with various constitutions and frameworks (*such as* nanowires, nanofibers, nanotubes, and nanoparticles) have been fruitfully grown for water separation from the oil−water mixture by tuning the surface properties and building hierarchical porous moieties. Thereafter, the above nanofiber materials are tuned to get large surface area, high porosity, wettability, multiporous structures, and large aspect ratio through distinctive physical, chemical, and mechanical phenomena. These properties should be examined to get excellent efficacy for the separation of water from oil−water mixture. Furthermore, for the improvement of nanofibrous substances, the most flexible and effective technique was the electrospinning technique. This technique was great attention because of its excellent potential to manage the porous structure of fibrous substances along with influencing the surface phenomenon of independent fibers by tuning skeleton. As a result, based on electrospinning technologies, different kinds of oil sorbents, as well as oil−water filters, should be introduced for the separation of water from oil−water mixture, which shows noticeable superiority in contrast with conventional fibrous membranes. Electrospun nanofibrous substances used for oil−water separation, including membranes, mats, and aerogels are discussed in this chapter. Furthermore, for the fabrication of nanofibrous substances, the electrospinning technique was used because of its versatility and simplicity approach. The electrospinning technique gives more attention because of its tunable large surface area, huge porosity, wettability, multiporous structures, and greater aspect ratio. These distinctive physical, chemical, and mechanical phenomena should be measured to get excellent efficacy for the separation of water from an oil−water mixture. Based on the above properties, different electrospun nanofibrous oil/water separation methods have an excellent sorption rate and high ability to separate water from the oil−water mixtures. These methods play a crucial role in realistic applications such as oil cleansing and oily wastewater management.

21.2.2 Volatile organic compounds

Nowadays volatile organic compounds (VOCs) have drawn significant awareness due to their harmful effects on the environment and can dangerously affect the quality of air. Furthermore, VOCs can be treated as a significant hazard to the environment and human health due to their content of toxic substances such as benzene and toluene. These chemicals can cause various painful reactions in the human body and affect people's physical condition (Vallecillos et al., 2019). Thereafter, different level concentrations of VOCs were challenged by currently available methods due to their low-level detection limit. The complexity of the method is further aggravated through another obstructing gas added in huge concentration to the VOCs. Subsequently, the preparation of samples before the concentration of VOCs at a gaseous state, including solid-sorbent enrichment techniques and whole-air collection methods, was the most common and important technique. However, the latter produced more significant results, and the adsorbents are essential (Li et al., 2012). Thereafter, in a few years, a quantity of fresh substances, especially nanostructured substances, should be examined and used as sorbent materials for adsorbing various gaseous organic substances (Li et al., 2004; Liu et al., 2015). Furthermore, the adsorbent capacity of CNTs can be modified due to the increasing adsorbent capacity by the tunable surface area, distribution of pore size, and uniform diameters of mesoporous materials. CNTs consist of a graphene layer rolled coaxially into cylinders of nanometric diameter and, furthermore, categorized into single-walled and multiwalled carbon nanotubes. The latter presents additional concentric cylinders of carbon. Thereafter, with comparatively bulky penetrate volumes along with very thin desorption bandwidth, the CNTs exhibit excellent encouraging adsorption plus desorption value (Karwa & Mitra, 2006; Li et al., 2004). Modification of surface chemistry is a very sensitive and challenging task for the research community. Chu et al. investigate the adsorption/desorption potential efficacy of VOC materials on homemade electrospun nanofiber (Chu et al., 2016). Thereafter, they analyzed its selectivity, adsorption/desorption efficacy, adsorption equilibrium time, and regeneration ability. The surface properties of the synthesized nanofiber molecules were examined with the help of TEM and SEM analyzing methods. Furthermore, in this examination, the adsorption efficiency of different VOCs, including ketone, hydrocarbon, ester, aromatic alcohol, and chlorinated substances, was compared against various sorbent materials. Thereafter, from this investigation, the authors found that the interactivity with sorbent plus target molecules is powerful on behalf of nanofiber materials. Similarly, the desorption efficacy of targeted VOC compounds with dissimilar sorbents was compared in this investigation by the thermal desorption and solvent desorption sequences from each of the tested objects to contrast the desorption properties of the targeted sorbents. Furthermore, the above work examined the adsorption efficacy of domestic electrospun nanofibers on behalf of VOCs, including their accuracy, adsorption/desorption efficacy, equilibrium time of adsorption, renewability, and the effect of temperature. Therefore, electrospun nanofibers play a key role and are used as an efficient adsorbent on behalf of the VOCs.

21.3 Energy conversion and storage

21.3.1 Solid oxide fuel cell

Fabrication of polymer/ceramic nanofiber materials includes several methods, of which the most important and well-organized method is electrospinning. Furthermore, the size of the electrospun fiber was very tiny thus it shows excellent properties, which makes it suitable for several vital applications such as solid oxide fuel cells (SOFCs). Thereafter, with the help of electrochemical reactions, SOFCs convert fuels directly into electricity, making them high-temperature fuel cells. Furthermore, when contrast to several already existing energy production methods,

SOFCs play an important role because of their excellent efficacy value. In the present scenario, researchers are generally paying attention to developing various strategies and techniques for growing the potential value of SOFCs and cost-effectiveness. The potential ability of a SOFC can be remarkably tuned by modifying the microstructure and the composition of the cathode. After that, in recent times, the enhancement of intermediate temperature (IT) and potential ability of SOFCs can be done by the cathode and anode fabrication of electrospun nanofibers. The electrospinning technique is a very old technique, which is broadly applied in biomedical and filtration procedures. The electrospinning technique is the most profitable as well as fruitful technique to produce nanolevel fibers (Orera et al., 2014). In the present scenario, clean energy production has been categorized as the main field of research areas. Therefore, the production of energy plays a key role due to its environmental impact value, as it contributes to pollution also. For the prevention of this problem, the most important and vital thing is to develop and produce sustainable energy. In past decades, several numbers of technologies have been produced viz. carbon-free energy and SOFCs. In the past decade, the creation of SOFC has happened. After that, researchers are trying to increase its potential value by tuning several parameters. It has been observed that, with decreasing the operational temperature, the ability of its potentiality is increased (Aruna et al., 2017). The commodity of SOFC materials has been a challanging task for the economic reason due to surpassing the conventional energy production technology. Pintauro et al. primarily reported research in 2007 on the topic of electrospinning and SOFC materials (Pintauro et al., 2007). SOFCs are energy conversion systems that convert fuel directly to energy with the help of electrochemical reactions. Thereafter, the fuel cell is important because it avoids the combustion procedure, thus it will release the least amount of carbon compared to others. Therefore, SOFCs exhibit excellent efficacy, and when cogeneration occurs, their efficacy value increases. The main benefit of SOFCs is their potential applicability in various fields such as hydrogen and hydrocarbon industries due to their fuel flexibility. A SOFC contains basically three components such as an anode, a cathode, and an electrolyte. A fuel and an oxidant are established in the anode and cathode correspondingly. Furthermore, the electrochemical reactions occur in a solid electrolyte system, which will exhibit various chemical potential values. The electrolyte is selected in such a way that it permits only oxide ions to pass through it while obstructing the electrons. Therefore, an external circuit is used to tap the electricity. After that, the above technology found some drawbacks in terms of operational temperature. Tsipis et al. exhibit that the conductivity of electrodes in terms of electronic and ionic conductivity electrodes showed the most preferable value due to their range of temperature, $1000°C-1200°C$ (Tsipis & Kharton, 2011). Furthermore, the advancement of the behavior of the SOFC system can be performed by minimizing the operational temperature of the cathode and anode without any deformation of its performance. The above procedure is the fundamental proposal of the IT of SOFCs. IT of the SOFCs operates the temperature range in between $600°C-800°C$. Thereafter, it plays an important role due to its improvement in the following key features: flexibility to utilize economical metals, quick startup and shutdown time, and minimized reaction time of cell components (Gao et al., 2013). Over the past few years, nickel (Ni) has been an important element in the preparation of SOFC anode. Furthermore, with the help of PVP/ethanol solution mixed with yttrium and zirconia, Ni-coated YSZ and 8YSZ nanofibers can be produced (Li et al., 2011). After that, for the preparation of nanofiber, the most potential and capable method is electrospinning. When operated appropriately, it can create an enormous impact on SOFC electrode manufacturing, which is presently controlled by powder-based ceramics. Electrospun nanofiber methods play an important role because of their uninterrupted porosity, comparatively more reaction-efficient sites, profitability as a production method, and notable enhancement of the SOFC potential by nearly doubling the power density and polarization.

21.3.2 Solar cells

In the 21st century, the most flexible energy in the world is electric energy. Furthermore, various recyclable resources such as solar, airstream, and tidal energy are used to develop and enhance methods for accomplishing the targeted electrical energy. The most important and usable technology is photovoltaic (PV) because it easily converts solar light directly into electrical energy. Therefore, solar power is remarked as an infinite source of energy. At the earliest time, Regan et al. reported using the fabrication of a dye-sensitized solar cell (DSSC) to obtain electricity with the help of the potential of nanotechnology (O'Regan & Grätzel, 1991). DSSC method is an "artificial photosynthesis" procedure that uses an electrolyte, a metal oxide film of nanoparticulate used as the electrode, and a dye sensitizer placed in the middle of conducting glasses. Furthermore, the free electron is deposited at the targeted electrode, which is inserted into the metal oxide nanoparticulate. The efficacy value increases if the collection of the electrons in the electrode is increased. In recent trends, researchers have focused on developing inexpensive solar energy conversion systems. For the achievement of the above-targeted system, researchers have prepared organic PV devices, which consist of mixtures of conjugated polymers and nanostructures of inorganic molecules. After that, for potential charge separation of photo-originated excitons, organic cells can be used for the creation of heterojunction of donor and acceptor constituents to produce a huge internal surface area. On the other hand, due to the large folded value and irregular topology of donor—acceptor interface, above devices exhibit unproductive charge transport capacity. Furthermore, for the enhancement of the transport process, annealing could be introduced, but there should be some limitations because it may not be suitable for the utilization of other polymers because their glass transition temperature is extremely low. Consequently, for the improvement of capable and flexible solar cells, different processing methods are required that can produce the same morphological yield without any thermal treatment (Berson et al., 2007). Thereafter, to get a higher energy potential, such a device must contain a polymer with inorganic composite or encapsulated functional materials (Lo et al., 2007). Electrospun nanofibers have various vital applications in various fields.

21.3.3 Sodium batteries

A variety of sophisticated energy storage and conversion devices are developed to improve the error-less reading of climate change and enhance energy security. In the present scenario, for the preparation of large-scale energy storage applications, sodium-ion batteries (SIBs) have been getting tremendous attention due to abundant resources and the inexpensiveness of sodium. On the other hand, SIBs contain those electrode materials that are affected by more crucial volume expansion and lethargic electrochemical kinetics, causing inferior cycling durability owing to the bigger ion radius (Na^+: 1.02 Å vs Li^+: 0.76 Å). After that, the energy density of commercial LIBs is relatively higher than SIBs because of greater standard electrochemical potential value (Na^+/Na: −2.71 V and Li^+/Li: −3.04 V vs SHE), which obstructs its further improvement (Hwang et al., 2017; Yabuuchi et al., 2014). After that, researchers selected appropriate anode materials on the basis of their excellent specific capacity and long cycling life. Unlike commercial graphite, carbon nanofibers (CNFs) show comparatively greater sodium storage capability due to several reasons. First of all, CNFs contain large interlayer distances and dense voids due to the amorphous nature of carbon, which may also be affected by distorted graphite layers and irregularly distributed graphite micro-crystallites. Secondly, CNFs exhibit a large surface-to-volume ratio and a small transport distance of ions due to their typical one-dimensional structure. Thirdly, it can stock up sodium through intercalation into disordered graphite interlayers, reversible plating/striping of sodium cluster in closed micropores, and storage capacity of sodium (Mukherjee et al., 2019; Zhang et al., 2020). Fourthly, unadulterated CNFs have a specific capacity that can be improved by the proper design and tunability of the targeted structure of CNFs. Furthermore, to get better electronic conductivity and surface area, CNFs can be doped with heteroatoms, therefore, CNFs increase surface area, making them better obstacles for the electrolyte and facilitating the transport of Na^+ ion. Thereafter, 1-D CNFs are accepted as potential anode substances for SIBs due to the above reasons (Tian et al., 2018; Wang et al., 2017). Thereafter, Chen et al. synthesized CNFs with the help of an uncomplicated electrospinning procedure as anode materials for SIBs. Subsequently, developed CNFs exhibit potential electrochemical behavior, including an excellent reversible capability (233 mAh/g at 0.05 A/g) and tremendous cycle stability (capacity retention ratio of 97.7% after 200 cycles) because graphite sheets have greater interlayer distance and form weak ordered turbostratic structure (Chen et al., 2014). Zhao et al. examined the effect of carbonization temperature on the capacity of sodium storage due to the fulvic acid-based electrospun structure of CNFs (Zhao et al., 2016). On the other hand, the prerequisite of huge-scale energy storage systems (ESSs) cannot be achieved by the sodium storage ability of pure CNF substances. Therefore, to achieve the above performance, researchers designed several schemes including building a unique morphology, doping the heteroatoms into the targeted skeleton, and combining them with other carbon substances (Yue et al., 2020). Additionally, CNFs are used as excellent potential anodes for SIBs because of their outstanding stability of the structure, greater interlayer distance, and high electronic conductivity. First, researchers are reviewing that due to the porous structure and doping of heteroatom into the targeted skeleton, the electrochemical properties of CNFs will increase.

21.3.4 Other applications

Li et al. reported the preparation of photocatalytic performance of a composite of carbon nanotubes loaded on poly nanofibers and TiO$_2$ nanoparticles, and they performed FT-IR, XRD, XPS, FE-SEM, and UV-visible spectroscopy (Li et al., 2020). The FE-SEM images of the nanofibers and composites are shown in Fig. 21.1. It is clear that f-CNTs are long, narrow tubes with diameters in the range of several tens of nanometers. The PMAA fibers have diameters in the range of 300–400 nm. On the other side, CNTs/PMMA show rough surfaces, and CNT was not observed. Again the electrospun TiO$_2$/PMMA exhibits a rough surface with many holes for the formation of TiO$_2$ nanoparticles. The nanocomposite of TiO$_2$ with CNT and PMAA nanofibers depicts a rougher surface with many particle aggregates. The exchange of acid-treated CNTs in the position of CNT results in a smoother surface.

Han et al. described self-standing and the binder-free application of polyamide-derived CNFs and manganese dioxide for supercapacitor electrode applications (Han et al., 2020). They performed the XRD, TEM, SEM, TGA analysis, and cyclic

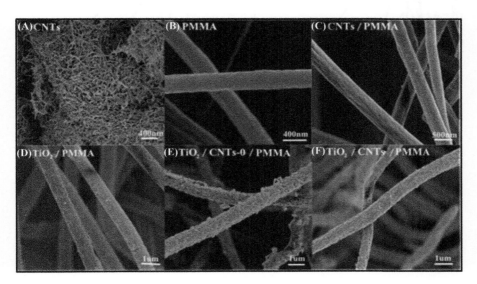

Figure 21.1 FE-SEM images of (A) f-CNTs, (B) PMMA nanofibers, (C) CNTs/PMMA, (D) TiO$_2$/PMMA, (E) TiO$_2$/CNTs-0/PMMA and (F) TiO$_2$/CNTs/PMMA.
Source: From Li, Y., Wang, Z., Zhao, H., & Yang, M. (2020). Composite of TiO$_2$ nanoparticles and carbon nanotubes loaded on poly(methyl methacrylate) nanofibers: Preparation and photocatalytic performance. *Synthetic Metals*, 269, 116529. https://doi.org/10.1016/j.synthmet.2020.116529.

voltammetry study for the electrochemical performance of the binder-free materials of supercapacitor electrode. The charge–discharge test of CNF/MnO$_2$ composites was carried out in the potential range of 0.0–0.8 V at different current densities, as shown in Fig. 21.2. The charging–discharging time is 1300 sec for PI/PVP derived CNF and the time increases to 2900 sec for CNF/MnO$_2$–10 composite, which indicates excellent super capacitive behavior. On another side, the charging–discharging times of CNF/MnO$_2$–30 and CNF/MnO$_2$–60 were 1700 and 1400 sec, respectively, which revealed the decrease of specific capacitance values. Thus CNF/MnO$_2$–10 composite has better performance for binder-free materials of supercapacitor electrodes.

Kim et al. highlighted the versatile behavior of TEMPO-oxidized nanofibers for stable silicon anode on lithium-ion batteries and performed XPS, SEM, and cyclic

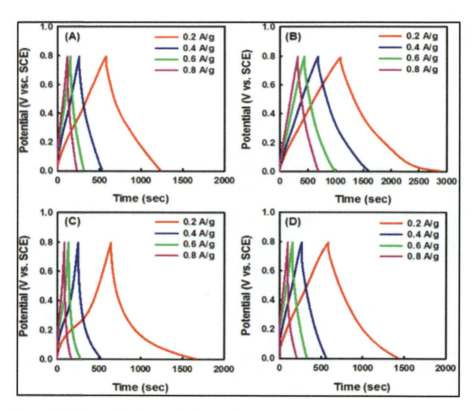

Figure 21.2 Curves of galvanostatic charge-discharge (A) PI/PVP-derived CNF, (B) CNF/MnO$_2$–10, (C) CNF/MnO$_2$–30, and (D) CNF/MnO$_2$–60 clicking.
Source: From Han, N. K., Choi, Y. C., Park, D. U., Ryu, J. H., & Jeong, Y. G. (2020). Core-shell type composites based on polyimide-derived carbon nanofibers and manganese dioxide for self-standing and binder-free supercapacitor electrode applications. *Composites Science and Technology*, *196*, 108212. https://doi.org/10.1016/j.compscitech.2020.108212.

voltammetry dispersion ability test (Kim et al., 2021). The depth of indentation increases in plain CMC, TOCNF-CMC with a ratio (1:14), and TOCNF-CMC with a ratio (5:10) due to the effect of strengthening effect of TOCNF in the binder of CMC, which is depicted in the Fig. 21.3. The lowest depth changes in the highest ratio of TOCNF-CMC at the highest load 0.2 mN indicate better structural integrity against stress. The values of reduced elastic modulus for the three different films are 3.15, 8.94, and 11.2 GPa, and their hardness values are 0.192, 0.405, and 0.716 GPa, respectively, which implies that the small amount in addition to TOCNF with CMC improves the mechanical properties of the binder in a better way. Thus TOCNF can act on both sides, firstly as the reinforcing fiber and secondly as an assistant binder for lithium-ion batteries, and hence increase the effective stress distribution and mechanical stability of the electrode.

Xu et al. reported the behavior of MnO/CNF composite for high-performance supercapacitors and they performed the XRD, XPS, cyclic voltammetry, and high angular dark field image study (Xu et al., 2021). The HAADF image of CNF and MnO/CNF composite material are shown in Fig. 21.4. The elemental mapping from EDS images of different elements C, Mn, O and N are shown where the element C is distrubuted all over and the distribution of Mn and O elements coincides in the position with the nanaoparticles of MnO. The three elements C, N, and O are uniformly distributed over the CNF and the interesting fact is the contents of N and O elements seem low in the composite material. The ratio of N/C and O/C decreases after excluding MnO and the resistance was reduced for the composite material. Because the thickness of the two materials remains the same, it is stated that the conductivity has been increased for the depletion of N and O elements.

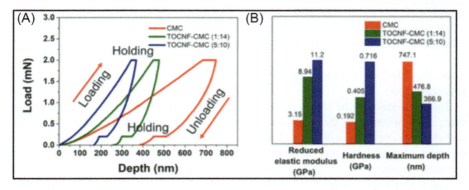

Figure 21.3 (A) Profiles of nano-indentation of CMC, TOCNF-CMC with ratio (1:14), TOCNF-CMC with ratio (5:10). (B) Parameters of nano-indentation of CMC, TOCNF-CMC with ratio (1:14), TOCNF-CMC with ratio (5:10).
Source: From Kim, J. M., Cho, Y., Guccini, V., Hahn, M., Yan, B., Salazar-Alvarez, G., & Piao, Y. (2021). TEMPO-oxidized cellulose nanofibers as versatile additives for highly stable silicon anode in lithium-ion batteries. *Electrochimica Acta*, *369*, 137708. https://doi.org/10.1016/j.electacta.2020.137708.

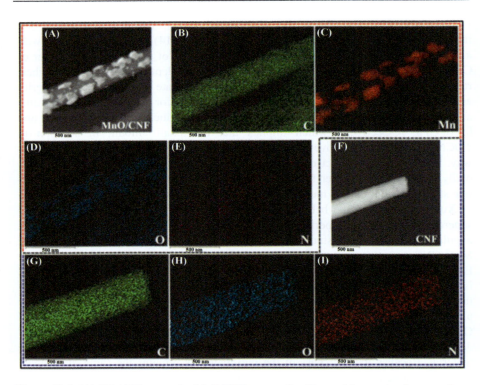

Figure 21.4 (A) HAADFimages for MnO/CNF composite. Elemental mapping images of (B) C, (C) Mn, (D) O, and (E) N elements in MnO/CNF composite. (F) HAADF image of CNF. Elemental mapping images of (G) C, (H) O, and (I) N elements in CNF.
Source: From Xu, W., Liu, L., & Weng, W. (2021). High-performance supercapacitor based on MnO/carbon nanofiber composite in extended potential windows. *Electrochimica Acta*, *370*, 137713. https://doi.org/10.1016/j.electacta.2021.137713.

Thus the MnO/CNF composite has excellent potential for becoming a flexible supercapacitor.

Jin et al. studied the thermal conduction behaviors of reduced graphene oxide-aramid nanofibers and they performed the XPS, SEM, Raman spectroscopy, FTIR, and thermal conduction studies (Jin et al., 2021). Aramid nanofibers possess high mechanical, isolating, and temperature resistance properties. Thus the addition of reduced graphene oxide to the aramid improves the thermal conduction performance than the pure aramid nanofibers. The thermal conductivities of the different composites of Go and rGO are shown in Fig. 21.5. This is clear from the study that conductivity increases with the increases of the GO or rGO contents in the composite material. The higher loading of rGO than GO loading increases the higher thermal conductivity. The 40% of rGO can enhance the value of the thermal conductivity of ANF by about 1250% and make it applicable in thermal interface materials and flexible electronics.

Additive application of nanofibers and their composites for enhanced performance   533

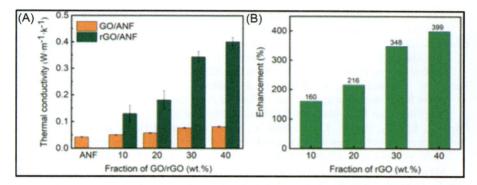

Figure 21.5 (A) Thermal conductivities of the GO/ANF and rGO/ANF films (B) enhancements of the thermal conductivities of the rGO/ANF films.
Source: From Jin, Z., E, S., Luo, Z., Ning, D., Huang, J., Ma, Q., Jia, F., & Lu, Z. (2021). Investigations on the thermal conduction behaviors of reduced graphene oxide/aramid nanofibers composites. *Diamond and Related Materials*, *116*, 108422. https://doi.org/10.1016/j.diamond.2021.108422.

21.4 Conclusions and future perspectives

This is clear that the use of composites of nanofibers has a bright outlook and the research with state-of-the-art technology exhibits major efficiency in microscale applications. The main attention of this chapter is to give attention to discuss the efficient properties and applications of nanofibers. Available literature generally discusses the common characteristics of nanofibers, but this chapter aims to review excellent potential applications and efficient properties of nanofibers in the fields, such as separation of water from oil−water mixture and storage of energy. Furthermore, the safety of the environment and neaten is also done with the help of nanofiber-based substances in terms of photocatalysis, oil/water separation, and removal of microorganisms, which has been talked about in this chapter. After that, this chapter also demonstrates the application of nanofiber substances in the storage of energy devices such as lithium-ion batteries, fuel cells, solar cells, and supercapacitors. Though there are some questions about the research methodology, the procedure of nanofiber interleaving is still commercially usable for unique damage resistance. In summary, the main aim of the above chapter is to be delivered in terms of as a prompt and crispy way for the research community in the field of advanced materials, mostly focused on the functionalized properties and their potential applications.

References

Aruna, S. T., Balaji, L. S., Kumar, S. S., & Prakash, B. S. (2017). Electrospinning in solid oxide fuel cells − A review. *Renewable and Sustainable Energy Reviews.*, *67*,

673−682. Available from https://doi.org/10.1016/j.rser.2016.09.003, https://www.journals.elsevier.com/renewable-and-sustainable-energy-reviews.

Banks, S. (2003). SeaWiFS satellite monitoring of oil spill impact on primary production in the Galápagos Marine Reserve. *Marine Pollution Bulletin*, *47*(7−8), 325−330. Available from https://doi.org/10.1016/S0025-326X(03)00162-0.

Berson, S., De Bettignies, R., Bailly, S., & Guillerez, S. (2007). Poly(3-hexylthiophene) fibers for photovoltaic applications. *Advanced Functional Materials*, *17*(8), 1377−1384. Available from https://doi.org/10.1002/adfm.200600922.

Chen, T., Liu, Y., Pan, L., Lu, T., Yao, Y., Sun, Z., Chua, D. H. C., & Chen, Q. (2014). Electrospun carbon nanofibers as anode materials for sodium ion batteries with excellent cycle performance. *Journal of Materials Chemistry A*, *2*(12), 4117. Available from https://doi.org/10.1039/c3ta14806h.

Chu, L., Deng, S., Zhao, R., Deng, J., Kang, X., & Zhang, Y. (2016). Comparison of adsorption/desorption of volatile organic compounds (VOCs) on electrospun nanofibers with tenax TA for potential application in sampling. *PLoS One*, *11*(10), e0163388. Available from https://doi.org/10.1371/journal.pone.0163388.

Díez, S., Jover, E., Bayona, J. M., & Albaigés, J. (2007). Prestige oil spill. III. Fate of a heavy oil in the marine environment. *Environmental Science and Technology*, *41*(9), 3075−3082. Available from https://doi.org/10.1021/es0629559.

Ewaldz, E., & Brettmann, B. (2019). Molecular interactions in electrospinning: From polymer mixtures to supramolecular assemblies. *ACS Applied Polymer Materials*, *1*(3), 298−308. Available from https://doi.org/10.1021/acsapm.8b00073, http://pubs.acs.org/journal/aapmcd.

Gao, C. L., Liu, Y. J., & Kumar, R. V. (2013). Synthesis and characterization of $La_{0.6}Sr_{0.4}Co_{0.2}Fe_{0.8}O_3$ nanofiber/$Ce_{0.9}Gd_{0.1}O_2$ nanoparticle composite as cathode material for intermediate temperature solid oxide fuel cells. *ECS Transactions*, *57*. Available from https://doi.org/10.1149/05701.2009ecst, http://ecst.ecsdl.org/.

Han, N. K., Choi, Y. C., Park, D. U., Ryu, J. H., & Jeong, Y. G. (2020). Core-shell type composites based on polyimide-derived carbon nanofibers and manganese dioxide for self-standing and binder-free supercapacitor electrode applications. *Composites Science and Technology*, *196*, 108212. Available from https://doi.org/10.1016/j.compscitech.2020.108212.

Hassiba, A. J., El Zowalaty, M. E., Nasrallah, G. K., Webster, T. J., Luyt, A. S., Abdullah, A. M., & Elzatahry, A. A. (2016). Review of recent research on biomedical applications of electrospun polymer nanofibers for improved wound healing. *Nanomedicine: Nanotechnology, Biology, and Medicine*, *11*(6), 715−737. Available from https://doi.org/10.2217/nnm.15.211, http://www.futuremedicine.com/loi/nnm.

Huang, T., Wang, C., Yu, H., Wang, H., Zhang, Q., & Zhu, M. (2015). Human walking-driven wearable all-fiber triboelectric nanogenerator containing electrospun polyvinylidene fluoride piezoelectric nanofibers. *Nano Energy*, *14*, 226−235. Available from https://doi.org/10.1016/j.nanoen.2015.01.038.

Hwang, J. Y., Myung, S. T., & Sun, Y. K. (2017). Sodium-ion batteries: Present and future. *Chemical Society Reviews*, *46*(12), 3529−3614. Available from https://doi.org/10.1039/c6cs00776g, http://pubs.rsc.org/en/journals/journal/cs.

Jin, Z., E, S., Luo, Z., Ning, D., Huang, J., Ma, Q., Jia, F., & Lu, Z. (2021). Investigations on the thermal conduction behaviors of reduced graphene oxide/aramid nanofibers composites. *Diamond and Related Materials*, *116*, 108422. Available from https://doi.org/10.1016/j.diamond.2021.108422.

Karwa, M., & Mitra, S. (2006). Gas chromatography on self-assembled, single-walled carbon nanotubes. *Analytical Chemistry*, *78*(6), 2064−2070. Available from https://doi.org/10.1021/ac052115x.

Kim, J. M., Cho, Y., Guccini, V., Hahn, M., Yan, B., Salazar-Alvarez, G., & Piao, Y. (2021). TEMPO-oxidized cellulose nanofibers as versatile additives for highly stable silicon anode in lithium-ion batteries. *Electrochimica Acta, 369*, 137708. Available from https://doi.org/10.1016/j.electacta.2020.137708.

Lakshmi Priya, M., Rana, D., Bhatt, A., & Ramalingam, M. (2017). *Nanofiber composites in gene delivery. Nanofiber Composites for Biomedical Applications* (pp. 253−274). India: Elsevier Inc. Available from http://www.sciencedirect.com/science/book/9780081001738, http://doi.org/10.1016/B978-0-08-100173-8.00031-4.

Li, L., Zhang, P., Liu, R., & Guo, S. M. (2011). Preparation of fibrous Ni-coated-YSZ anodes for solid oxide fuel cells. *Journal of Power Sources, 196*(3), 1242−1247. Available from https://doi.org/10.1016/j.jpowsour.2010.08.038.

Li, M., Biswas, S., Nantz, M. H., Higashi, R. M., & Fu, X. A. (2012). Preconcentration and analysis of trace volatile carbonyl compounds. *Analytical Chemistry, 84*(3), 1288−1293. Available from https://doi.org/10.1021/ac2021757.

Li, Q. L., Yuan, D. X., & Lin, Q. M. (2004). Evaluation of multi-walled carbon nanotubes as an adsorbent for trapping volatile organic compounds from environmental samples. *Journal of Chromatography A, 1026*(1-2), 283−288. Available from https://doi.org/10.1016/j.chroma.2003.10.109, http://www.elsevier.com/locate/chroma.

Li, Y., Wang, Z., Zhao, H., & Yang, M. (2020). Composite of TiO_2 nanoparticles and carbon nanotubes loaded on poly(methyl methacrylate) nanofibers: Preparation and photocatalytic performance. *Synthetic Metals, 269*, 116529. Available from https://doi.org/10.1016/j.synthmet.2020.116529.

Liu, G.-qiang, Wan, M.-xi, Huang, Z.-hong, & Kang, F.-yu (2015). Preparation of graphene/metal-organic composites and their adsorption performance for benzene and ethanol. *New Carbon Materials, 30*(6), 566−571. Available from https://doi.org/10.1016/s1872-5805(15)60205-0.

Lo, M. Y., Zhen, C., Lauters, M., Jabbour, G. E., & Sellinger, A. (2007). Organic-inorganic hybrids based on pyrene functionalized octavinylsilsesquioxane cores for application in OLEDs. *Journal of the American Chemical Society, 129*(18), 5808−5809. Available from https://doi.org/10.1021/ja070471m.

Lou, L., Wang, J., Lee, Y. J., & Ramkumar, S. S. (2019). Visible light photocatalytic functional TiO_2/PVDF nanofibers for dye pollutant degradation. *Particle & Particle Systems Characterization, 36*(9), 1900091. Available from https://doi.org/10.1002/ppsc.201900091.

Lou, L., Kendall, R. J., Smith, E., & Ramkumar, S. S. (2020). Functional PVDF/rGO/TiO_2 nanofiber webs for the removal of oil from water. *Polymer, 186*, 122028. Available from https://doi.org/10.1016/j.polymer.2019.122028.

Lou, L. H., Qin, X. H., & Zhang, H. (2017). Preparation and study of low-resistance polyacrylonitrile nano membranes for gas filtration. *Textile Research Journal, 87*(2), 208−215. Available from https://doi.org/10.1177/0040517515627171, https://journals.sagepub.com/home/TRJ.

Lu, X., Wang, C., & Wei, Y. (2009). One-dimensional composite nanomaterials: Synthesis by electrospinning and their applications. *Small (Weinheim an der Bergstrasse, Germany), 5*(21), 2349−2370. Available from https://doi.org/10.1002/smll.200900445, http://www3.interscience.wiley.com/cgi-bin/fulltext/122603475/PDFSTART, China.

Mukherjee, S., Mujib, S. B., Soares, D., & Singh, G. (2019). Electrode materials for high-performance sodium-ion batteries. *Materials, 12*(12), 1952. Available from https://doi.org/10.3390/ma12121952.

Nordvik, A. B., Simmons, J. L., Bitting, K. R., Lewis, A., & Strøm-Kristiansen, T. (1996). Oil and water separation in marine oil spill clean-up operations. *Spill Science and*

Technology Bulletin, *3*(3), 107−122. Available from https://doi.org/10.1016/S1353-2561 (96)00021-7.

Nune, M., Kumaraswamy, P., Krishnan, U. M., & Sethuraman, S. (2013). Self-assembling peptide nanofibrous scaffolds for tissue engineering: Novel approaches and strategies for effective functional regeneration. *Current Protein and Peptide Science*, *14*(1), 70−84. Available from https://doi.org/10.2174/1389203711314010010.

Orera, V. M., Laguna-Bercero, M. A., & Larrea, A. (2014). Fabrication methods and performance in fuel cell and steam electrolysis operation modes of small tubular solid oxide fuel cells: A review. *Frontiers in Energy Research*, *2*. Available from https://doi.org/10.3389/fenrg.2014.00022, http://journal.frontiersin.org/article/10.3389/fenrg.2014.00022/full.

O'Regan, B., & Grätzel, M. (1991). A low-cost, high-efficiency solar cell based on dye-sensitized colloidal TiO$_2$ films. *Nature*, *353*(6346), 737−740. Available from https://doi.org/10.1038/353737a0, http://www.nature.com/nature/index.html.

Peng, Y., & Guo, Z. (2016). Recent advances in biomimetic thin membranes applied in emulsified oil/water separation. *Journal of Materials Chemistry A*, *4*(41), 15749−15770. Available from https://doi.org/10.1039/c6ta06922c, http://pubs.rsc.org/en/journals/journalissues/ta.

Peterson, C. H., Rice, S. D., Short, J. W., Esler, D., Bodkin, J. L., Ballachey, B. E., & Irons, D. B. (2003). Long-term ecosystem response to the exxon valdez oil spill. *Science (New York, N.Y.)*, *302*(5653), 2082−2086. Available from https://doi.org/10.1126/science.1084282.

Pham, Q. P., Sharma, U., & Mikos, A. G. (2006). Electrospinning of polymeric nanofibers for tissue engineering applications: A review. *Tissue Engineering*, *12*(5), 1197−1211. Available from https://doi.org/10.1089/ten.2006.12.1197.

Pintauro, P. N., Mather, P., Arnoult, O., Choi, J., Wycisk, R., & Lee, K. M. (2007). Composite membranes for hydrogen/air PEM fuel cells. *ECS Transactions*, *11*(1), 79−87. Available from https://doi.org/10.1149/1.2780917.

Pintor, A. M. A., Vilar, V. J. P., Botelho, C. M. S., & Boaventura, R. A. R. (2016). Oil and grease removal from wastewaters: Sorption treatment as an alternative to state-of-the-art technologies. A critical review. *Chemical Engineering Journal*, *297*, 229−255. Available from https://doi.org/10.1016/j.cej.2016.03.121, http://www.elsevier.com/inca/publications/store/6/0/1/2/7/3/index.htt.

Sarbatly, R., Krishnaiah, D., & Kamin, Z. (2016). A review of polymer nanofibres by electrospinning and their application in oil-water separation for cleaning up marine oil spills. *Marine Pollution Bulletin*, *106*(1−2), 8−16. Available from https://doi.org/10.1016/j.marpolbul.2016.03.037, http://www.elsevier.com/locate/marpolbul.

Sawicka, K. M., & Gouma, P. (2006). Electrospun composite nanofibers for functional applications. *Journal of Nanoparticle Research*, *8*(6), 769−781. Available from https://doi.org/10.1007/s11051-005-9026-9.

Tian, P., Tang, L., Teng, K. S., & Lau, S. P. (2018). Graphene quantum dots from chemistry to applications. *Materials Today Chemistry*, *10*, 221−258. Available from https://doi.org/10.1016/j.mtchem.2018.09.007, https://www.journals.elsevier.com/materials-today-chemistry/.

Tsipis, E. V., & Kharton, V. V. (2011). Electrode materials and reaction mechanisms in solid oxide fuel cells: A brief review. III. Recent trends and selected methodological aspects. *Journal of Solid State Electrochemistry*, *15*(5), 1007−1040. Available from https://doi.org/10.1007/s10008-011-1341-8.

Vallecillos, L., Espallargas, E., Allo, R., Marcé, R. M., & Borrull, F. (2019). Passive sampling of volatile organic compounds in industrial atmospheres: Uptake rate determinations and application. *Science of the Total Environment, 666,* 235−244. Available from https://doi.org/10.1016/j.scitotenv.2019.02.213, http://www.elsevier.com/locate/scitotenv.

Wang, M., Yang, Y., Yang, Z., Gu, L., Chen, Q., & Yu, Y. (2017). Sodium-ion batteries: Improving the rate capability of 3D interconnected carbon nanofibers thin film by boron, nitrogen dual-doping. *Advanced Science, 4*(4), 1600468. Available from https://doi.org/10.1002/advs.201600468.

Xu, W., Liu, L., & Weng, W. (2021). High-performance supercapacitor based on MnO/carbon nanofiber composite in extended potential windows. *Electrochimica Acta, 370,* 137713. Available from https://doi.org/10.1016/j.electacta.2021.137713.

Yabuuchi, N., Kubota, K., Dahbi, M., & Komaba, S. (2014). Research development on sodium-ion batteries. *Chemical Reviews, 114*(23), 11636−11682. Available from https://doi.org/10.1021/cr500192f, http://pubs.acs.org/journal/chreay.

Yildiz, O., Dirican, M., Fang, X., Fu, K., Jia, H., Stano, K., Zhang, X., & Bradford, P. D. (2019). Hybrid carbon nanotube fabrics with sacrificial nanofibers for flexible high performance lithium-ion battery anodes. *Journal of the Electrochemical Society, 166*(4), A473−A479. Available from https://doi.org/10.1149/2.0821902jes, http://jes.ecsdl.org/content/166/4/A473.full.pdf + html.

Yue, L., Yue, L., Zhao, H., Wu, Z., Liang, J., Lu, S., Chen, G., Gao, S., Zhong, B., Guo, X., & Sun, X. (2020). Recent advances in electrospun one-dimensional carbon nanofiber structures/heterostructures as anode materials for sodium ion batteries. *Journal of Materials Chemistry A, 8*(23), 11493−11510. Available from https://doi.org/10.1039/d0ta03963b, http://pubs.rsc.org/en/journals/journal/ta.

Zhang, H., Huang, Y., Ming, H., Cao, G., Zhang, W., Ming, J., & Chen, R. (2020). Recent advances in nanostructured carbon for sodium-ion batteries. *Journal of Materials Chemistry A, 8*(4), 1604−1630. Available from https://doi.org/10.1039/c9ta09984k, http://pubs.rsc.org/en/journals/journal/ta.

Zhang, Y., He, X., Li, J., Miao, Z., & Huang, F. (2008). Fabrication and ethanol-sensing properties of micro gas sensor based on electrospun SnO_2 nanofibers. *Sensors and Actuators B: Chemical, 132*(1), 67−73. Available from https://doi.org/10.1016/j.snb.2008.01.006.

Zhao, P. Y., Zhang, J., Li, Q., & Wang, C. Y. (2016). Electrochemical performance of fulvic acid-based electrospun hard carbon nanofibers as promising anodes for sodium-ion batteries. *Journal of Power Sources, 334,* 170−178. Available from https://doi.org/10.1016/j.jpowsour.2016.10.029.

Nanofibers and their composites for supercapacitor applications

22

Ishita Ishita, Shriram Radhakanth, Pradeep Kumar Sow and Richa Singhal
Department of Chemical Engineering, BITS Pilani, Goa Campus, Zuarinagar, Goa, India

22.1 Introduction

The rapid depletion of fossil fuel reserves, alarming greenhouse gas emissions, and rising pollution levels are significant techno-environmental concerns due to the ever-increasing dependence on fossil fuels. Consequently, developing environmentally sustainable energy generation and storage technologies is critical. Batteries, fuel cells, and supercapacitors (SCs) are electrochemical energy conversion and storage devices that can cater to the rising demand for sustainable and renewable energy systems. SCs, also known as ultracapacitors, are energy storage devices that have gained significant attention recently for their high power density, fast charging/discharging, and longer lifespan. The history of SCs can be traced back to the early 1900s when scientists first discovered the phenomenon of electrostatic capacitance. However, it was not until the late 20th century that advances in materials science and manufacturing techniques allowed for the development of commercial SCs (Viswanathan, 2017). Fig. 22.1 shows the Ragone plot, which compares the energy density and power density of various energy storage devices. It can be seen that SCs offer lower energy density than batteries but have significantly higher power density, thereby making them suitable for applications where energy is required to be delivered at a high rate. Unlike traditional batteries, which store energy in a chemical form, SCs store energy via electrostatic charge separation at the electrode/electrolyte interface (Poonam et al., 2019). This allows them to store and release energy at relatively high rates. Because of their high-power density and longer lifespan, they can supplement or even replace batteries in certain applications. For example, SCs can supplement the batteries in the grid storage systems, where they can help balance the supply and demand of electricity. In electric vehicles, SCs can provide a high-power density required for short-term acceleration and a sudden energy requirement for braking. SCs can be designed to be lightweight, small in size, and flexible, making them useful in portable devices. The global efforts and policies to reduce carbon emissions have made the development of renewable and sustainable energy storage devices, such as SCs, more crucial now than ever (Zhang et al., 2009).

The choice of electrode material has a significant impact on the performance of a SC. Materials with high surface area and good electrical conductivity are crucial

Polymeric Nanofibers and their Composites. DOI: https://doi.org/10.1016/B978-0-443-14128-7.00022-5
© 2025 Elsevier Ltd. All rights are reserved, including those for text and data mining, AI training, and similar technologies.

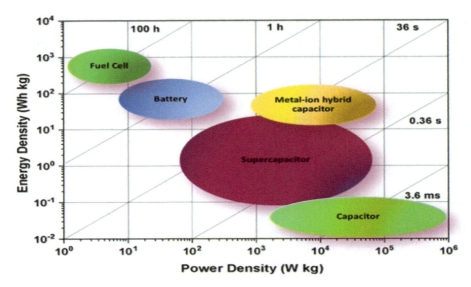

Figure 22.1 The Ragone plot displaying the comparison of the energy density and power density of various energy storage devices.

for attaining high performance. Among the various nanostructured materials explored, nanofibers are one of the most promising classes of materials used as electrodes in SCs. Nanofibers with their one-dimensional (1D) architecture exhibit high surface area and effective ion diffusion/electron transfer pathways. The nanofibrous materials are also used as separators in a SC due to their porous structure. This chapter aims to provide a comprehensive overview of the SC operation and the different strategies that are being explored with the use of nanofibrous materials toward improving the device's performance. This chapter discusses the fundamentals of SCs, including their components, the charge storage mechanisms involved, the performance metrics of SCs, and the application of nanofibers and their composites as promising materials in SCs.

22.1.1 Supercapacitor and its components

SCs are essentially capacitors that can deliver significantly higher energy and capacitance than traditional capacitors for a device of comparable dimensions. This is attributed to the notably higher surface area of the electrode and a thinner dielectric, resulting in $\sim 10^4 - 10^5$ times higher capacitance than traditional capacitors. SCs are typically composed of two electrodes, each coupled with a current collector and separated by an electrolyte-soaked separator, as shown in Fig. 22.2. During charging, the potential is applied across the electrodes causing the ions (negative and positive) to move from the electrolyte bulk toward the surface of the oppositely charged electrode. During discharging, the charges diffuse back from the electrode

Nanofibers and their composites for supercapacitor applications 541

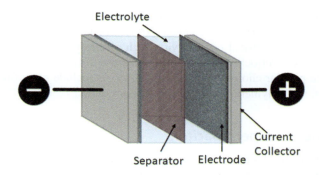

Figure 22.2 Schematic showing the different components of a supercapacitor.

surface to the bulk of the electrolyte, while electrons flow through the external circuit, providing electric current.

Electrodes are a critical component in SCs, and their properties determine the performance of the SC device. The selection of electrode material and structure is vital in SC design. A good SC electrode exhibits good electrical conductivity, high surface area, good electrochemical stability, high mechanical strength, and superior electrolyte wettability. In addition, they should also be economical and environmentally safe for commercial applicability. Carbon nanostructures, such as activated carbon, graphite, carbon aerogels, carbon nanotubes (CNTs), carbon nanofibers (CNFs), and transition metal oxide-based materials have been widely used as electrodes for SCs (Zhang et al., 2009; Zhang, Kang, et al., 2016).

Electrolytes used in SCs can be classified into various types, including aqueous, organic, ionic liquids (ILs), solid-state, and redox-active electrolytes. The desirable properties of an electrolyte include high ionic conductivity, high chemical and electrochemical stability, operation over a wide voltage window, low viscosity, low volatility, low toxicity, high thermal stability, and low cost (Zhang & Pan, 2015). The nature of the electrolyte, including its concentration, the ion type and size, the solvent used, the ion-solvent interaction, the electrolyte-electrode material interaction, and the potential window, can significantly affect the performance of the SC (Pal et al., 2019). The operating potential of a SC cell is determined by the electrochemically stable potential window of the electrolyte, which may be further limited by the stability of the electrodes in the presence of the respective electrolyte (Pal et al., 2019; Poonam et al., 2019). While aqueous electrolytes (such as H_2SO_4, KOH, Na_2SO_4, etc.) have a shorter operating potential window of around 1.0–1.3 V, in contrast, organic electrolytes (including acetonitrile, and propylene carbonate) can operate in the potential window of up to 2.5–2.7 V. The ILs such as imidazolium, pyrrolidinium, and quaternary ammonium salts like tetraalkylammonium and cyclic amines have significantly wider operating potential windows up to 4.0 V and higher (Pal et al., 2019).

A separator acts as a barrier between two electrodes in a SC. The primary role of the separator membrane is to allow ions to freely pass through it while preventing a short circuit. The power density of the SC is impacted by the ionic mobility and conductivity through the separator, making the selection of the separator material crucial (Poonam et al., 2019; Viswanathan, 2017). The factors for selecting a separator for SCs are low electrical conductivity, high electrolyte absorbency, high ionic conductivity, good electrochemical stability in the electrolyte used, and high thermal stability. Some popular choices for separator materials include glass fiber, cellulose, ceramic fibers, and polymeric film materials (Editor, n.d.).

22.1.2 Types of supercapacitors

SCs can be broadly classified into electric double-layer capacitors (EDLCs) and pseudocapacitors based on their charge storage mechanism (see Fig. 22.3). EDLCs store charges electrostatically. The electrochemical species are adsorbed on the interface between the electrode and the electrolyte, forming an electric double-layer. The electric double-layer formation is a fast process and responds instantaneously to a voltage variation, as the charging/discharging process does not require any charge transfer across the electrode/electrolyte interface. Here, the charge storage capability is majorly dependent on the accessible surface area of the electrode. The surface area available for the charge storage is dictated by the compatibility between the pore size of the electrode and ion size as well as the wettability of the electrode by the electrolyte. Carbonaceous materials, such as activated carbon, CNTs, CNFs, and graphene-based electrodes, are widely used EDLC electrodes because they have a high surface area with tunable pore size distribution.

Pseudocapacitors utilize fast and reversible Faradaic charge transfer processes (redox reactions) for charge storage. The term "pseudocapacitor" arises from the fact that the charge storage is fast and reversible like that of a capacitor but involves battery-like charge transfer across the electrode/electrolyte interface. Heteroatoms, transitional metal oxides, conducting polymers (CPs), etc. are prime examples of materials that impart pseudocapacitance (Jiang & Liu, 2019).

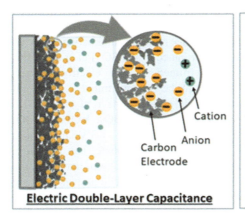
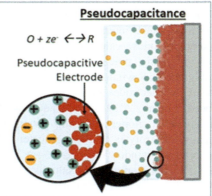

Figure 22.3 An illustration depicting the charge-storage mechanism of an electric double-layer capacitor and a pseudocapacitor.

The pseudocapacitance demonstrated by an active material can be of two types, intrinsic or extrinsic pseudocapacitance. If the pseudocapacitance arises due to the nanostructuring of the material, it is termed extrinsic pseudocapacitance, while if it is not dependent on the particle size and dimensions, it is termed intrinsic pseudocapacitance (Fleischmann et al., 2020). The various mechanisms contributing to charge storage in pseudocapacitors include a change in oxidation state via redox reactions for transition metal oxides, reversible doping and de-doping for CPs, adsorption pseudocapacitance (e.g., H^+ on Pt), and intercalation pseudocapacitance caused by rapid ion intercalation on the redox-active electrode without any crystallographic phase change (Shao et al., 2018).

Asymmetric and hybrid SCs are advanced SCs that can provide higher energy/power densities. Asymmetric SCs use two dissimilar electrode materials as their positive and negative electrodes. Such a device combines a pseudocapacitive electrode (as the cathode) and an EDLC-type electrode (as the anode) to offer a wide potential window, thereby enhancing the energy density. Conversely, hybrid SCs are also asymmetric devices that combine battery-type electrodes, that is, metal-ion and an EDLC electrode, to obtain enhanced charge storage. Here, metal ions are intercalated into the EDLC electrode during charging. Such devices exhibit the potential to bridge the gap between SCs and batteries. Numerous research studies are being carried out to integrate different electrode materials to obtain higher SC performance, that is, higher energy & power densities. The attempt to bring together different charge storage mechanisms in various configurations is one of the prominent strategies to boost the performance of SCs.

22.2 Performance metrics of supercapacitor

Some of the commonly used key performance metrics for evaluating the operation of an SC include capacitance, energy density, power density, cycle life, and self-discharge. For a traditional capacitor comprising of two electrode plates of area "A" separated by a dielectric medium of thickness "d" and relative permittivity "ε_r," the capacitance is given by:

$$C = \frac{\varepsilon_r \varepsilon_o A}{d} \tag{22.1}$$

where ε_o is the permittivity of free space. As we can see, the capacitance is directly proportional to the electrode area, enhancing the surface area can lead to higher charge storage and thereby higher capacitance. The SCs leverage on this fact and utilize electrodes with significantly larger surface areas than traditional capacitors leading to a huge increase in the capacitance. The amount of charge stored in a SC under a given potential is known as its capacitance and is given by:

$$C = \frac{\Delta Q}{\Delta V} \tag{22.2}$$

where C is the capacitance, Q is the charge, and V is the applied potential. The capacitance is further normalized by either mass (gravimetric capacitance), area (areal capacitance), or volume (volumetric capacitance) to define an intrinsic specific capacitance. For normalized electrode capacitance, the mass, area, or volume of the electrode can be used, while for the total SC cell capacitance, the whole cell mass or volume can be used. The charge storage capacity of SCs is commonly described in terms of specific capacitance, which is also referred to as the gravimetric capacitance and expressed in F/g. Different electrochemical techniques such as cyclic voltammetry (CV) and galvanostatic charge/discharge (GCD) are utilized to calculate the specific capacitance of the electrode materials or cell. The specific capacitance calculated from the CV in a three-electrode configuration is given by:

$$C = \frac{\int IdV}{m \times \nu \times \Delta V} \tag{22.3}$$

where the numerator is the area under the CV curve, ν is the scan rate (in V/s), m is the mass of the working electrode (in grams), and ΔV is the operation potential window. From GCD curves, the specific capacitance is given by the following equation:

$$C = \frac{I\Delta t}{m \times \Delta V} \tag{22.4}$$

where, I is the discharge current and Δt is the discharge time (in seconds).

In a two-electrode configuration, the overall capacitance (C_T) of the SC cell is given by:

$$\frac{1}{C_T} = \frac{1}{C_1} + \frac{1}{C_2} \tag{22.5}$$

where C_1 and C_2 are the capacitances of the two electrodes. For a symmetric SC with identical electrodes on either side, $C_1 = C_2$. Hence, the total capacitance is half of the electrode's capacitance. In an asymmetric SC where two dissimilar electrodes are used, it is necessary to balance the charge between the positive and negative electrodes for optimum performance. The mass of each electrode is decided based on the charge balance between the two electrodes, as given by the following equations:

$$Q_{\text{electrode}} = C_{\text{electrode}} \times m \times \Delta V \tag{22.6}$$

For, $Q^+ = Q^-$

$$\frac{m_+}{m_-} = \frac{C_{\text{electrode}-} \times \Delta V_-}{C_{\text{electrode}+} \times \Delta V_+} \tag{22.7}$$

Energy density and power density are the performance parameters that determine the end application of energy storage devices. The amount of energy that may be stored or released from a SC normalized by its mass or volume is called its energy density. It can be reported in terms of Wh/kg or Wh/L. The following relation can be used to compute energy density (E):

$$E = \int Q dV \tag{22.8}$$

For a SC with linear charge/discharge relationships, the above equation simplifies to:

$$E = \frac{1}{2}CV^2 \tag{22.9}$$

where C is the capacitance and V is the potential window. Here, the energy density is directly proportional to the square of the voltage window. Therefore, increasing the operating potential window can significantly enhance the energy density of the SC.

Power density (P) is defined as the amount of energy delivered per unit of time. It is expressed in W/kg and calculated using the following equation:

$$P = \frac{E}{\Delta t} \tag{22.10}$$

where Δt is the discharge time. The maximum power delivered is heavily influenced by the resistance offered by the internal components and given as:

$$P = \frac{V^2_{charged}}{4R_S} \tag{22.11}$$

where, $V_{charged}$ is the charging potential and R_S is the equivalent series resistance (ESR) of the SC. The ESR of a device depends on the contribution of different cell components such as the inherent resistance of the electrode material and current collector, the contact resistance between the electrode and the current collector, and the ionic resistance offered by the electrolyte. Electrochemical impedance spectroscopy and the IR drop in the GCD technique can be used to determine the ESR.

In addition to the above performance metrics, the stability of the SC is crucial for its practical application, which can be evaluated based on the cycling stability and the self-discharge behavior. To determine cycle life, the device is charged and discharged repeatedly for a finite number of cycles (10,000 or more), and the evolution of capacitance with the cycle number is studied. High capacitance retention implies good cycling stability of the SC cell. Besides cycling stability, another important parameter is the self-discharge behavior of the SC. The drop in potential of a charged SC in an open circuit condition characterizes the self-discharge.

Self-discharge can affect different types of SCs differently. It has been reported that EDLCs, which rely on surface charge adsorption for their operation, experience more significant self-discharge compared to pseudocapacitors based on the Faradaic charge storage mechanism due to the faster charge transfer kinetics (Shang et al., 2023). Moreover, SCs with redox-active electrolytes (also referred to as active electrolyte-enhanced SCs) demonstrate even faster self-discharge due to the shuttling of ions across the separator. Broadly, the self-discharge can occur through three pathways, leakage current, Faradaic reaction, and redistribution of charges (Liu et al., 2021). Leakage current can be caused by either a parasitic reaction occurring in the vicinity of the electrode−electrolyte interface, an internal short-circuit, or a combination of both. Faradaic reactions can be caused by functional groups on the electrode surface and/or redox mediators that are present in a redox-active electrolyte. These reactive species can undergo redox reactions to cause self-discharge and, therefore, voltage loss. Charge redistribution is caused when the electrode surface close to the current collector reaches the desired potential, while the inner pores/active material does not. Subsequently, in an open circuit, the charge species on the external electrode surface (near the current collector) could migrate inside the electrode surface, causing the redistribution and, consequently, a decline in the potential. Various techniques to combat self-discharge have been reported in the literature (Lee et al., 2019; Liu et al., 2021). Short-circuiting of the electrodes can be prevented by providing complete cell insulation and tight sealing between the electrodes. Shuttling of active species (such as soluble redox ions) between the electrodes can be curbed by using a suitable ion exchange membrane. The diffusion-limited charge redistribution can be minimized by reducing electrolyte volume or charging the cell either by holding the potential for a long period of time or using a lower current.

22.3 Nanofibers in supercapacitors

The primary component dictating the SC performance is the electrode and its characteristics. Nanofibers are 1D nanostructured materials with unique properties such as high aspect ratio, large surface area, good mechanical strength, and efficient electron transport, making them an excellent candidate for application in energy storage devices. Nanofibers have further been distinguished due to the facile incorporation of functional groups, porous structures, defects, and other active materials. These characteristics of nanofibers make them excellent contenders to be used as electrodes and/or separators in SCs.

22.3.1 Synthesis of nanofibers

Among various techniques used for synthesizing nanofibers, electrospinning is a more popular and facile technique that produces nanofibers with diameters less than 1 μm. It is a versatile technique that allows for easy tuning of structural (morphology and porosity) as well as chemical characteristics (incorporation of additional materials, functional groups, etc.) of nanofibers via optimization of setup geometry, precursor solution, and process parameters (Mao et al., 2013). Electrospinning can also provide free-standing nanofiber mats that can be used as electrodes without any binders, thereby eliminating additional processing steps, reducing the weight of the electrode, and avoiding nonconductive sites provided by the binder.

An electrospinning setup (Fig. 22.4) comprises a spinneret/needle connected to a syringe pump that ejects and controls the flow rate of the polymer spinning solution, a high-voltage power supply for the electrification of the polymer solution, and a collector/target to collect the generated fibers. The syringe pump at a predetermined flow rate pushes the polymer solution out of the needle. A pendant-shaped droplet is formed at the tip of the needle, held together by the surface tension force. Once the voltage is supplied, there is an electrostatic repulsion between the charged species of the droplet causing shear stresses (Baji et al., 2010). Due to this repulsion, the droplet distorts to form a cone known as the Taylor cone (Baji et al., 2010; Xue, Wu, et al., 2019). As the voltage increases, the surface tension is surpassed by the electrostatic forces resulting in a jet from the end of the Taylor cone. The jet extends out, initially, in a straight line but then is subjected to "bending instabilities" resulting in an intense "whipping" motion, which leads to the elongation of the jet (Xue, Wu, et al., 2019). Simultaneously, the solvent in the precursor solution evaporates, resulting in the solidification of the fibers, which eventually get deposited on the collector in the form of a nonwoven nanofiber mat. A huge number of synthetic and natural polymers have been electrospun successfully to date. A sufficiently high molecular weight of the polymer is required to realize the chain entanglement for forming fibers.

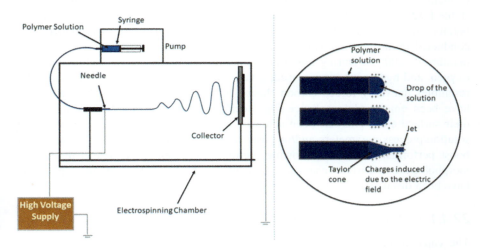

Figure 22.4 Depiction of an electrospinning setup with the inset showing the Taylor cone formation.

22.4 Nanofibers as electrodes

Electrospun nanofibers are promising electrode materials for SCs owing to their unique and superior properties, such as high aspect ratio, high surface area, tunable morphology, efficient electron/charge transfer, mechanical flexibility, and versatility to incorporate additional functionalities. This section discusses nanofiber-based electrode materials categorized into four main groups: CNFs, CNF-based composites, metal oxide nanofibers, and CP nanofibers.

22.4.1 Carbon nanofibers

CNFs are widely chosen as electrodes as they have excellent electrical conductivity, good chemical stability, and tunable morphology (Zhang, Kang, et al., 2016). Additionally, CNFs can be fabricated as free-standing structures, eliminating the need for binders while synthesizing electrodes for SCs. CNFs are synthesized by electrospinning polymer precursor solution, followed by pyrolysis. Various polymer precursors have been utilized for producing CNFs, such as polyacrylonitrile (PAN), polyimide (PI), polyvinyl alcohol (PVA), polyvinyl pyrrolidone (PVP), and pitch. Among these, PAN is the most popular choice as the polymer precursor to produce CNFs because it has a high carbon yield, >50%, and results in fibers with superior mechanical strength. To produce PAN-based CNFs, the electrospun PAN nanofibers are subjected to a two-step heat treatment process to convert polymeric nanofibers to CNFs: stabilization at moderate temperatures in the air followed by carbonization at high temperatures in an inert atmosphere. Oxidative stabilization is an important step that ensures a ladder-like structure due to various reactions arising at this stage such as cyclization, dehydrogenation, aromatization, oxidation, and crosslinking (Rahaman et al., 2007). The ladder-like structure is important to obtain thermally stable PAN fibers for the high-temperature carbonization process. The typical temperature range for stabilization lies between 200°C and 300°C. Carbonization is conducted in an inert atmosphere (usually with flowing N_2/Ar gas) at a high temperature ($\geq$800°C). During this process, noncarbon heteroatoms (such as nitrogen, oxygen, and hydrogen) are removed, giving the CNFs a "turbostratic" structure that resembles unevenly folded carbon sheets (Saha & Schatz, 2012). Many strategies have been employed to enhance the specific capacitance of CNFs by enhancing their surface area. This has been achieved by either modifying the electrospinning solution/process parameters and/or postprocessing to alter the CNF morphology to boost performance. In the following sections, two main strategies to modify the morphology and surface chemistry of CNFs to enhance the specific capacitance have been discussed, that is, pore engineering and heteroatom doping.

22.4.1.1 Pore engineering

The capacitance of an electrode is directly proportional to the electrochemically accessible electrode surface area. Therefore, it is essential to find means to increase the surface area of CNFs. Electrospun CNFs have a nonwoven structure with inherent macropores as a courtesy of interfiber spacing, which acts as diffusion pathways for ions. Further addition of pores on the CNFs can lead to an increase in the surface area. Pores also function as reservoirs for electrolytes and shorten the ion diffusion pathway. The optimal design of pore architecture on CNFs is crucial as that will not only determine the enhancement in the surface area and the associated active charge storage sites but also govern the rate of ion diffusion, thereby controlling the rate capability. The creation of hierarchical pores on the CNFs is desirable as that would enable faster ion transport from macro- to meso- to micropores. The introduction of pores must be done optimally. Even though micropores vastly improve the surface area, if the micropore size is incompatible with the electrolyte ion size, the specific capacitance would not increase significantly because the ion would not be able to access that pore. The addition of mesopores will ensure faster ion transport as well as contribute to the surface area. Therefore, an optimum balance would be required between micro-, meso- and macro-pores to achieve high charge storage along with a faster rate of charge/discharge. Various methods can be used to create pores on CNFs, as discussed below.

Activation of carbon nanofibers

CNFs can be activated physically, chemically, or by a combination of both. While physical activation utilizes carbon dioxide and steam as physical activation agents, chemical activation uses chemical activation agents such as potassium hydroxide, sodium hydroxide, phosphoric acid, zinc chloride, etc. The activation process of CNFs can be conducted either in conjunction with the heat treatment or postheat-treatment, that is, CNFs are first synthesized and then subjected to activation. These activation processes help in increasing porosity and surface area, as well as lead to the addition of functional groups on the electrode surface. Kim et al. (2019) investigated the effect of using nitrogen (N_2), steam (H_2O), and carbon dioxide (CO_2) as activating agents on CNFs. It was found that steam activation resulted in the smallest fiber diameter, followed by CO_2 and N_2 activation, with controlled pore sizes in the range of 0.64 to 0.81 nm. However, CO_2-activated CNF (at 900°C) displayed the highest specific capacitance. It was also found that CNFs with average pore sizes similar to that of the solvated ion sizes of the electrolyte provided the highest normalized capacitance. KOH activation of CNFs is a popular method to increase the surface area via chemical etching to produce micropores. Generally, in this process, CNFs are immersed in KOH for a predefined time, followed by heat treatment at a high temperature in an inert environment, and then washed with HCl and distilled water to neutralize the pH to remove any residual activating agent. Lawrence et al. (2015) reported that upon activation via KOH, the surface area of the CNFs increased from 1218 m^2/g to 2285 m^2/g, which led to a higher specific capacitance of activated porous CNFs of 144 F/g at 5 mV/s. As an alternative to the multistep approach for CNF activation, a small quantity of the activation agent, such as phosphoric acid (H_3PO_4), could be added directly to the electrospinning solution. For example, Zhi et al. (2014) reported a higher surface area of 709 m^2/g on adding H_3PO_4 to the electrospinning solution and achieved a specific capacitance of 156 F/g.

Addition of a sacrificial polymer (soft template method)

Another approach to enhance the surface area and incorporate porosity within CNFs is to add a sacrificial material, usually a polymer, directly to the electrospinning solution. Such sacrificial material decomposes or disintegrates during the heat-treatment process, thereby leaving behind empty spaces or pores within the CNFs. Various sacrificial materials have been reported, such as poly(methyl methacrylate) (PMMA), polyvinylpyrrolidone (PVP), Nafion, etc. The optimal performance with sacrificial material depends on its concentration (with respect to the carbon precursor polymer), molecular weight, compatibility with the carbon precursor polymer, and decomposition temperature. Liu et al. (2016) studied the effect of increasing PVP concentration in the electrospinning solution on the synthesized CNFs. It was found that increasing the amount of sacrificial polymer (PVP) led to an increase in the specific surface area with an increase in mesopores as well as micropores. The increase in mesopore contribution improved the wettability of the electrode as well as aided in the transportation of ions, which further led to a high SC performance with a specific capacitance of 200 F/g. Nafion as a sacrificial material to form porous CNFs with a surface area higher than 1500 m^2/g has been extensively studied by Tran and Kalra (2013a, 2013b). They found that with increased loading of Nafion in the PAN electrospinning solution, the total solid concentration in the solution must also increase as Nafion aggregates and reduces chain entanglement as well as extensional viscosity. They also studied the effects of increased Nafion concentration in PAN solution up to a ratio of 80:20 (Nafion: PAN). It was found that the cumulative pore volume and average pore size increased with increasing Nafion concentration. This larger pore size in the 80:20 (Nafion: PAN) CNF facilitated ion diffusion and demonstrated the highest specific capacitance of 210 F/g. Lai and Lo (2015) studied PMMA as the sacrificial polymer and found that varying the molecular weight of PMMA led to different nanostructured CNFs. It was observed that PMMA with the lowest molecular weight led to porous CNFs, while with higher molecular weight, the CNFs were hollow. The reason behind this structural change is the phase separation between PAN and PMMA due to their incompatible solubility parameter, wherein PMMA with lower molecular weight disperses like droplets in the PAN matrix, while PMMA with higher molecular weight aggregates and stretches into a channel during electrospinning leading to porous CNFs (for PMMA with lower molecular weight) and hollow CNFs (for PMMA with higher molecular weight). The highest specific capacitance (210 F/g) was obtained for the PMMA with the lowest molecular weight. Polymers such as PMMA and PVP that have low decomposition temperatures (often less than the stabilization temperature) can cause a lower amount of mesopore formation and less graphitization. Because mesopores function as diffusion channels for the electrolyte ions, the reduction in the mesopore concentration is detrimental to the SC performance. Thereby, high decomposition temperature polymers are preferred as sacrificial materials, as they lead to hierarchical (meso-microporous structure) as well as robust porosity. Wang et al. (2018) reported the use of a high-decomposition temperature thermoplastic such as polysulfone (PSF) as the sacrificial material. Using

20% PSF as the sacrificial polymer in the PAN spinning solution led to CNFs with higher surface area, increased mesopores, a higher degree of graphitization, and interconnected fibers, resulting in a specific capacitance of 289 F/g. High-amylose starch (HAS), another high melting point polymer, was compared with PVP and PMMA as sacrificial polymers by Wang, Wang, et al. (2019) and found to yield a high surface area of 809 m^2/g with 59% micropore content that had a corresponding specific capacitance of 282 F/g. Although most of the studies in the literature have used virgin and valuable polymers as sacrificial materials that disintegrate during pyrolysis, it is rather wasteful, enhances the cost of the process, and renders it nonenvironmentally-friendly. Alternatively, Ishita and Singhal (2020) reported using a waste material, polystyrene (PS) foam (from packing waste, "Styrofoam"), as the sacrificial material. Adding waste PS foam to the PAN precursor led to the segregation of PS in the PAN matrix due to incompatible solubility parameters. This segregation during electrospinning stretched out to form PS channels distributed throughout PAN fibers. Upon heat treatment, PS disintegrates and creates porous, multichannel CNF with a high specific capacitance of 271.6 F/g.

Addition of fillers and other pore-inducing materials (hard template method)

Besides the use of sacrificial polymers to create pores, other organic and inorganic materials can also be used to induce porosity. It follows a similar synthesis strategy as that of the sacrificial polymers., that is, the addition of fillers and/or other pore-inducing materials in the electrospinning polymer solution followed by its removal after the carbonization step. Lee et al. (2015) studied the variation of silica (SiO$_2$) in the electrospinning solution and related morphology of CNFs to the concentration of SiO$_2$ in the polymer solution. In this study, the CNFs were synthesized with increasing amounts of SiO$_2$ and chemically activated by KOH postcarbonization. It was found that by increasing the amount of SiO$_2$, the surface area increased along with the specific capacitance until a point beyond which the fibrous structure transformed into a bead-like structure due to SiO$_2$ agglomeration in the PAN matrix, resulting in a surface area lower than the CNF with no SiO$_2$. The CNF synthesized with 50 wt.% SiO$_2$ gave the highest specific capacitance of 197 F/g. One of the crucial parameters to realize uniform pore distribution within CNFs is to ensure that the filler material is homogeneously distributed throughout the electrospinning precursor solution. The aggregation of the filler material may induce nonhomogeneity in the pore distribution that may further result in the electrode's nonuniform mechanical and electrochemical properties. Zhang, Jiang, et al. (2016) reported using nano-CaCO$_3$, a carbonic salt, as a hard template to induce porosity in CNFs. In order to increase its dispersion in the carbon precursor solution, THF was added along with DMF as the solvent in the spinning solution. The resulting hierarchically porous CNF gave a specific capacitance of 251 F/g, which establishes the use of the bi-solvent method. Taking a step further, the synthesis of a porous hollow CNF via coaxial electrospinning of a polymer solution containing an organic silicate compound (tetraethyl orthosilicate [TEOS]) for SC electrode was reported by Xiao et al. (2020). A mixture of the carbon precursor and TEOS was the sheath solution, while paraffin oil was the core solution. Once the nanofibers were spun, they were first subjected to heat treatment to form hollow CNFs, which removed the paraffin oil, and then treated with HF to remove the silica and obtain hollow, porous CNFs, which gave a specific capacitance of 261 F/g. While newer approaches are being reported for creating porous CNFs, future works can exploit the combination of the above techniques and tailor-design pore sizes for the specific electrolytes to obtain CNFs with high surface area and designed porosity.

22.4.1.2 Heteroatom doping of carbon nanofibers

Heteroatoms in organic chemistry are noncarbon and nonhydrogen atoms. The addition of heteroatoms, such as oxygen and nitrogen, to carbonaceous material enables heterogeneity on the carbon surface, which can alter the electrical conductivity along with the wettability of the substrate while imparting pseudocapacitive properties through the incorporation of redox-active functional groups (Luo et al., 2021). Heteroatom doping can be achieved by adding a heteroatom doping precursor in the carbon precursor polymer solution or by a postprocess activation treatment. The presence of oxygen in the form of hydroxyl, carbonyl, and/or carboxyl functional groups on the carbon surface has proven to enhance wettability and specific capacitance with an aqueous electrolyte. This can be achieved by either pyrolysis of oxygen-containing precursors such as biomass, chemical activation by KOH or NaOH, or via treatment with chemical reagents such as HNO_3 and H_2O_2 (Luo et al., 2021). Alternatively, the addition of salt to the polymer precursor can also result in the heteroatom functionalization of CNF. Singhal and Kalra (2015) reported the addition of common salt to the electrospinning solution, which was then pyrolyzed and acid-treated; the resulting CNFs possessed enhanced oxygen content in the form of carboxyl groups, which in turn also helped in the assimilation of sulfur functional groups during the acid-treatment. The presence of the heteroatoms (O and S) led to a high specific capacitance of 204 F/g even with a low surface area CNF (only 24 m^2/g), attributed to the redox reactions resulting from these functional groups.

Nitrogen doping in the carbon electrodes has been reported to increase pseudocapacitance via redox reactions of pyrrolic-N and pyridinic-N as well as faster electron transport and improved conductivity contributed by quaternary-N and pyridinic-N-oxide (Lyu et al., 2019). Nitrogen integration on the carbon surface can also increase the wettability of the electrode. Most studies have used nitrogen-containing precursors to successfully dope a carbon material with nitrogen. For example, PVP is a nitrogen-rich polymer precursor that also increases the porosity of the electrode by acting as a sacrificial material. Li et al. (2017) reported a high specific capacitance of 198 F/g by utilizing PVP as a precursor for synthesizing porous CNFs with high cyclability performance. Such high performance is attributed to nitrogen doping and higher surface area obtained due to the addition of PVP. In another study by Cheng et al. (2015), PVP was used as a nitrogen precursor along with terephthalic acid in the electrospinning solution resulting in nitrogen-doped, flexible cross-linked CNFs, which showed a specific capacitance of 175 F/g. Further, combining the excellent properties of nitrogen with fluorine, which has a high electronegativity, has been reported to result in enhanced wettability, rate capability, and cycle life. Na et al. (2017) treated porous CNFs with nitrogen (during heat treatment) and octafluorocyclobutane (C_4F_8) via vacuum plasma treatment to produce nitrogen and fluorine-doped porous CNFs, which gave a specific capacitance of 252.6 F/g attributed to the porous structure and heteroatom doping.

22.4.2 Carbon nanofiber-based composites

CNF-based composites entail CNFs as the substrate along with other electro-active (pseudocapacitive) materials, such as transition metal-oxides, metal-organic frameworks (MOFs), CPs, etc. The utilization of transition metal-based electrodes has been widely reported due to their high theoretical specific capacity. However, they suffer from poor electrical conductivity, poor ion diffusion, and metal instability, which hinder their practical application. Thus, these materials are often incorporated into a conductive matrix such as CNF to address the above issues. The integration of pseudocapacitive materials on CNFs to form composites provides enhanced SC performance via combining Faradaic and nonFaradaic charge storage, improved electrical conductivity, effective dispersion of active material resulting in enhanced accessibility to active sites, improved stability of the composite electrode, and more effective transport of ions/electrons. There are two main strategies employed to synthesize the CNF composites: The first technique involves the in situ addition of the metal oxide precursor to the spinning solution, while the other technique utilizes the ex situ addition of metal oxide nanoparticles to the CNFs. The ex situ addition can be conducted either after electrospinning or after carbonization, that is, directly on CNFs. For the former case, the electrospun nanofibers are treated with a metal-oxide precursor solution followed by a heat-treatment step, where the polymer converts to carbon and oxidative conversion of the metal oxide precursor to the metal oxide takes place. The two widely employed CNF composites involving metal oxides and MOFs have been discussed below.

22.4.2.1 Metal oxide-carbon nanofiber composites

Several transition metal-oxides, such as MnO_2, Co_3O_4, and $NiCo_2O_4$, embedded in CNFs as composites have been reported in the literature. Among these, manganese oxides are a popular choice of pseudocapacitive electrodes in SCs due to their multiple oxidation states (-3 to $+7$), high theoretical capacitance, easy availability, and environmental benignity (Uke et al., 2017). Du et al. (2014) reported the freestanding and flexible MnO_x/CNFs as an electrode for SC, synthesized using manganese acetate as the precursor added along with PAN in the electrospinning solution. MnO_x/CNFs displayed a high specific capacitance of 211 F/g, which is attributed to the greater packing density and more thorough exploitation of the electro-active MnO_x sites due to the well-distributed morphology. The general energy storage mechanism for MnO_2-based SCs involves rapid and reversible redox reactions between the +3 and +4 oxidation states of manganese and is given by Eqs. (22.12) and (22.13) (Chen et al., 2017):

$$(MnO_2)_{surface} + C^{n+} + ne^- \rightleftharpoons \left(MnO_2^{n-} C^{n+}\right)_{surface} \tag{22.12}$$

$$MnO_2 + C^+ + e^- \rightleftharpoons MnOOC \tag{22.13}$$

where C^+ is an electrolyte cation.

Another report by Radhakanth and Singhal (2023) studied the effect of carbonization time on the MnO-embedded CNFs. It was found that the optimal two-step carbonization provided only MnO nanoparticles with a +2-oxidation state embedded throughout the CNF, while in other cases, mixed Mn_xO_y forms were obtained. A high specific capacitance of 246 F/g was achieved using 1 M H_2SO_4 as the electrolyte with excellent cycling stability of 97.5% after 10,000 cycles. Several studies have also reported the combination of different transition metal oxide nanoparticles incorporated into the CNF mat to obtain a mixed metal-CNF composite that exploits the performance enhancement contributed by the constituting metal oxides (Mishra et al., 2022; Wei et al., 2016). Kurtan et al. (2020) added cobalt and nickel acetylacetonate to the PAN polymer solution. The solution was electrospun, and the obtained nanofiber mat was then subjected to heat treatment to obtain CoNi-CNFs. The CoNi-CNF electrodes in 1 M KOH electrolyte exhibited a specific capacitance of 132 F/g, and 85.3% of the capacitance was retained after 10,000 cycles. Recently, Li et al. (2022) reported the ex situ incorporation of spinel oxide $NiCo_2O_4$ on hollow carbon nanofibers (HCNFs) prepared using cobalt nitrate, nickel nitrate, and poly(acrylonitrile-co-acrylamide) as precursors. The $NiCo_2O_4$/HCNFs were used in a flexible solid-state SC and achieved excellent results with a specific capacitance of 1864 F/g at 1 A/g and a capacitance retention of 91.7% after 5000 cycles. In another study, Kim et al. (2018) electroplated the CNFs derived from PAN with silver using a KS-700 solution. An asymmetric SC device was assembled with CNF/Ag composite and achieved an areal capacitance of 30.6 mF/cm^2 at 0.1 mA/cm^2 in 6 M KOH electrolyte with ~100% capacitance retention after 10,000 cycles.

22.4.2.2 Metal-organic framework-carbon nanofiber composites

MOFs are crystalline materials composed of metal ions or clusters connected by organic ligands to form porous structures. Thus, MOF is essentially made of two components, inorganic metal salt and the organic linker material. The MOFs by themselves have a high surface area and tuable pore sizes, making them useful in various applications (Salunkhe et al., 2017; Xue, Zheng, et al., 2019). The synergistic effect of the MOFs, along with CNFs, has been exploited to further improve the performance of the SCs. Based on the mode by which the MOF and CNF are combined, they can be classified into three broad categories, MOF-in-fiber, MOF-on-fiber, and MOF-seed-fiber (Dou et al., 2020; Salunkhe et al., 2017).

The MOF-in-fiber method involves the incorporation of MOF in situ in the CNF by mixing it with the electrospinning solution. The MOF nanoparticles embedded nanofibers are thus obtained. Lu et al. (2021) synthesized ZIF-67 nanocubes by centrifuging a mixture of cobalt nitrate with cetyltrimethylammonium bromide and DI water. These nanocubes were then dissolved in a polymer solution and electrospun. The electrospun nanofiber mat was then calcined in air for 2 hours at 400°C, and thus the ZIF-67-embedded CNF mat was obtained. The synthesized material showed a high specific capacitance of 970 F/g in 2 M KOH. Yao et al. (2018) reported the synthesis of a bimetallic coordination MOF, ZIF-C-0.05, using zinc nitrate and cobalt nitrate as precursors with 2-methylimidazole and methanol. The ZIF-C-0.05 was added to the polymer solution of PAN and PVP in DMF and electrospun. The mat was subjected to heat treatment to obtain the N-doped graphitic porous carbon nanofibers (NGHPCF). The NGHPCF achieved a high specific capacitance of 326 F/g with high capacitance retention at higher current densities. In this work, Co/Zn metal coordination content and the pyrolysis temperature were varied to control the specific surface area, N content, and the degree of graphitization of NGHPCF material.

The MOF-on-fiber approach involves direct-solvo or hydrothermal growth of MOF on the polymer nanofiber by immersing it in the MOF precursor solution. This leads to the growth of the MOF nanoparticles on the surface of the nanofibers constituting the nanofiber mat. Shin and Shin (2021) synthesized the Ni-MOF@CNF by placing the presynthesized CNF into the MOF precursor solution, followed by hydrothermal treatment. The Ni-MOF@CNF was further carbonized at 400°C for 1 hour in a N_2 atmosphere to obtain the NiO/C@CNF. The NiO/C@CNF showed improved electrical conductivity and had the distinct advantage of self-standing and a high electron transfer capability. It demonstrated the performance of 742 F/g in a 3 M KOH electrolyte. In another study, Tian et al. (2020) synthesized bimetallic MOF-embedded CNFs to take advantage of the synergistic effect of different transition metals to enhance the overall performance of the asymmetric SC. In this study, preoxidized PAN nanofibers (PPNFs) were added to a hydrothermal reactor containing the precursor solution for Co-Ni MOF, leading to PPNF@Co-Ni-MOF. This was used as the anode in an asymmetric SC, while the cathode was the CNF activated with KOH and coordinated with graphene oxide (GO), along with the PVA-KOH gel electrolyte. It demonstrated a performance of 263 F/g at 2 A/g.

The MOF-seed-fibers are synthesized by electrospinning of the polymer with one of the MOF precursors (inorganic metal salt or the organic linker) that acts as a seed for the growth of MOF. The nanofiber mat is further subjected to the solvothermal reaction in the autoclave with the other precursor of the MOF. This technique leads to more MOF loading with MOFs present both on the inside and on the surface of the fiber. Samuel et al. (2019) reported hierarchical Mn@ZIF-8 dodecahedral nanoparticles decorated on CNFs via electrospinning of the MOF precursor, 2-methylimidazole, along with PAN solution. The obtained nanofibers were soaked in methanol along with manganese nitrate and zinc acetate, followed by stabilization and carbonization. It achieved a specific capacitance of 501 F/g in a 6 M KOH electrolytic solution. In another study, Tian et al. (2021) prepared C-Co@MOFs, which showed a high specific capacitance of 1201.6 F/g. The Co nanoparticles modified CNFs were first prepared, followed by solvothermal deposition of Ni-MOFs. The Co nanoparticles present in CNFs assisted in the growth of Ni-MOFs on the surface of Co-CNFs.

22.4.3 Metal oxide nanofibers

Metal oxides are promising electrode materials because of their redox activity and intrinsic capability for energy storage. Designing them as 1D materials can impart additional properties such as a substantial increase in surface area, high resistance to stress change, and enhanced electronic and ionic transport, making them an excellent choice as electrodes in SCs. Such carbon-free 1D metal oxide nanofibers are synthesized by electrospinning the metal oxide precursor along with a polymer (such as PVA and PVP), followed by heat treatment under an air atmosphere (Liao et al., 2016). The polymer usually decomposes during the heat treatment process giving rise to metal oxide nanofibers. The carbon-free metal oxide nanofibers can be classified into various categories based on the combination and the nature/type of the metal oxide nanoparticles as unitary metal oxides, spinel-type metal oxides, perovskites, and metal oxide–metal oxide composites (Liang et al., 2020). The commonly used metals for the unitary metal oxides (i.e. single metal oxide) and the spinel metal oxides (i.e., two transition metal oxides with general formula AB_2O_4) are Mn, Ni, Co, and Fe. The perovskite materials, with the general formula of ABX_3 (where A and B represent cations, while X is an anion), combine two or more metal oxides with high melting points and have a heavier metals such as La and Sr. The metal oxide–metal oxide composites are the ones in which one of the metal oxide nanoparticles is grown or deposited on another metal oxide nanoparticle. Examples of each of the four types of metal oxide nanofibers are further discussed below.

22.4.3.1 Unitary metal oxides

The unitary metal oxide nanofibers consist of only one metal oxide. Kundu and Liu (2015) synthesized Ni nanofibers by electrospinning nickel acetate with citric acid and PVP in DI water directly on Ni foam. It was further subjected to heat treatment at 500°C in N_2 atmosphere for 2 hours. This was used as a free-standing, binder-free electrode and displayed a high specific capacitance of 737 F/g in 2 M KOH solution. The synthesis of flexible Fe_2O_3 and V_2O_5 nanofibers by electrospinning followed by calcination used in an asymmetric solid-state SC with Na_2SO_4 gel electrolyte was reported by Jiang et al. (2018). In their study, the specific capacitance of 71.5 F/g along with an average energy density of 32.2 Wh/kg and power density of 128.7 W/kg was achieved. In the study by Teli et al. (2020), the synthesis of MnO nanofibers was achieved by electrospinning a solution of manganese acetate, PVP, double distilled water, and ethanol, which was then annealed at 300°C for three hours. High performance of 526 F/g in 1 M Na_2SO_4 electrolyte solution and 96% retention after 1000 cycles were achieved. This was mainly attributed to the high specific surface area of the synthesized MnO nanofibers.

22.4.3.2 Spinel metal oxides

The spinel oxides have a general formula AB_2O_4, where A and B are transition metal oxides such as Mn, Ni, Co, and Fe. Peng et al. (2015) synthesized and characterized nanofibers obtained by electrospinning PAN and PVP together in DMF along with two inorganic metal salts. The $NiCo_2O_4$ nanofibers achieved a high specific capacitance of 1756 F/g at 1 A/g in 2 M KOH with about 87% capacitance retention after 10,000 cycles at 5 A/g. Bhagwan et al. (2018) electrospun a solution of zinc acetate, manganese acetate, ethanol, acetic acid, and PVP and synthesized $ZnMn_2O_4$ nanofibers with a high specific surface area of 30 m^2/g. The $ZnMn_2O_4$ nanofibers were tested in a symmetrical SC with 1 M Na_2SO_4 electrolyte and achieved 240 F/g at 1 A/g.

22.4.3.3 Perovskite metal oxides

Ma et al. (2019) studied the effect of Sr substitution in the perovskite-type $LaMn_{0.9}Ni_{0.1}O_3$ nanofibers. The nitrate salts of lanthanum, strontium, manganese, and nickel were added in stoichiometric amounts to a solution of PVP in DMF and electrospun. The obtained nanofiber mat was then annealed at 650°C for 2 hours in the air. The $La_{0.85}Sr_{0.15}Mn_{0.9}Ni_{0.1}O_3$ nanofibers achieved a high specific capacitance of 114 F/g at 0.13 A/g in 1 M Na_2SO_4 electrolyte. Another study by Wang, Lin, et al. (2019) synthesized $La_xSr_{1-x}FeO_3$ nanofibers using a very similar approach. The nitrate salts of lanthanum, strontium, and iron were mixed in PVP-DMF solution and electrospun, followed by annealing at 650°C for 2 hours in the air. The $La_{0.7}Sr_{0.3}FeO_3$ nanofibers showed a very high specific capacitance of 523 F/g at 1 A/g in 1 M Na_2SO_4 electrolyte with a good capacitance retention of 83.8% after 5000 cycles at 20 A/g in a wide potential window from -1 to $+1$ V.

22.4.3.4 Metal oxide–metal oxide composite

The different metal oxide nanostructures, such as nanosheets, can be deposited or grown on top of the metal oxide nanofibers to obtain high performance by the synergistic effect of both the structure and the metal oxide present. Dong et al. (2018) synthesized the hierarchical tubular yolk-shell-like $NiMoO_4$ nanosheets on Co_3O_4 nanofibers. First, the Co_3O_4/C fibers were synthesized by electrospinning PVP and cobalt nitrate in DMF and then annealed at 800°C in an N_2 atmosphere. The resulting Co_3O_4/C fibers were dispersed in a solution of nickel nitrate, sodium molybdate, and water, autoclaved at 160°C for 12 hours, and then filtered. The filtrate was washed, dried, and then calcined in air at 450°C for 1 hour to obtain the $Co_3O_4@NiMoO_4$ material. This exhibited a high specific capacitance of 913 F/g at 10 A/g in 3 M KOH electrolyte with an 89.9% retention after 3000 cycles. A unique 1D/2D architecture with MnO_2 nanosheets on TiO_2 nanofibers was described by Da Silva et al. (2019). Polyethylene oxide, acetonitrile, acetic acid, and titanium (IV) butoxide were electrospun and calcined at 550°C for 3 hours. The obtained TiO_2 nanofibers were added to a solution of $KMnO_4$ and water and autoclaved at 160°C for 12 hours. The mixture was cooled, filtered, and washed several times and then vacuum dried at 60°C for 12 hours. A high specific capacitance of 525 F/g at 0.25 A/g in 1 M Na$_2$SO$_4$ electrolyte with a good capacitance retention of 84% after 2000 cycles at 8 A/g. In another study by Gong et al. (2016), the heterogenous 1-D RuO$_2$@Co$_3$O$_4$ nanofibers were synthesized and investigated. Ruthenium chloride, cobalt nitrate, ethanol, DMF, and PVP were mixed and electrospun. The obtained nanofiber mat was vacuum-dried for 24 hours at 70°C and then annealed at 500°C for 2 hours. The RuO$_2$@Co$_3$O$_4$ showed a very high specific capacitance of 1103.6 F/g at 10 A/g in 2 M KOH solution and a good capacitance retention of 88% after 5000 cycles.

22.4.4 Conducting polymer nanofibers

CPs, as the name suggests, are electrically conductive organic polymers. These have been used widely as electrode material in SCs owing to their high electrical conductivity in the doped state, high storage capacity and reversibility, and tunable redox activity via chemical modification (Wang et al., 2012). The charge storage mechanism here is pseudocapacitive based on doping and de-doping of polymers. The redox reactions in CPs occur throughout the polymer bulk and not just the surface, thereby providing a higher number of redox-active sites per unit volume, higher structural stability, and reversibility during charging/discharging. Commonly used CPs are polyaniline (PANI), poly(3,4-ethylene dioxythiophene) (PEDOT), polypyrrole (PPy), polythiophene (PTh), etc. Electrical conductivity can be induced in these polymers by doping them via oxidation or reduction reactions. The "p-doping" results from the oxidation of the polymer backbone producing positively charged polymers, while the "n-doping" results in negatively charged polymers due to the reduction of the polymeric repeating units. Under high mass loadings, CPs can suffer from a lower accessible surface area due to dense packing, thereby directly impeding the participation of the redox sites of CP in the charge storage process, resulting in a lower specific capacitance. Essentially, increasing the surface area of the CP can lead to better utilization of redox-active sites. Therefore, electrospun CP nanofibers exhibit better performance compared to other forms due to their higher surface area, shorter diffusion lengths, and better ion-conduction pathways. Miao et al. (2013) utilized a porous/hollow CP nanofibrous (hollow PANI nanofibers) structure to further boost the surface-to-volume ratio that facilitated the use of the entire electrode surface for charge storage, overcoming the shortcomings of a densely packed structure. In their study, hollow fibers were fabricated using a poly (amic acid) (PAA) core and a PANI sheath. PAA was first electrospun to form nanofibers, which were then subjected to in-situ polymerization of aniline to form a PANI sheath. This core-and-sheath nanofiber was further treated with N, N-dimethylacetamide to remove PAA, thus, forming hollow PANI nanofibers. These hollow nanofibers gave a specific capacitance of 601 F/g with an H$_2$SO$_4$ electrolyte. However, the performance retention was only 62% after 500 cycles due to the expansion and contraction of PANI during continued charge/discharge, causing mechanical failure of the hollow fibers. Some of the prominent challenges associated with the usage of CPs as electrodes in SCs include the requirement of specific electrolytes, hindered doping/de-doping, swelling, and shrinking of CPs during charge/discharge, leading to mechanical degradation of the electrode (Meng et al., 2017). The addition of metal oxides to CPs not only enhances the charge storage but also functions as a protective layer for the CPs to resist any distortion due to charging/discharging. For example, in a study conducted by Subramani and Rajiv (2022), nanostructured CP, poly(3-methyl thiophene) (PMT), and RuO_2 nanoparticles were added to poly(ethylene oxide) (PEO), a biocompatible nonionic and porous polymer used as polymer substrate and electrospun to form PMT/PEO/RuO_2 nanofibers, which gave a specific capacitance of 685 F/g and retained 92% of its performance after 5000 cycles. RuO_2 is an electronic-protonic conductor that helps in overcoming the distortion in PMT during charging/discharging. Using a carbon-based material in conjunction with a CP is another way to address some of the above issues, including enhanced mechanical stability, increased ion mobility, and better performance retention during cycling studies (Fong et al., 2017). Anand et al. (2020) reported PANI-coated CNFs synthesized via in situ polymerization of aniline on CNF. These PANI-coated CNFs showed a specific capacitance of 493.7 F/g with 90.5% performance retention after 5000 cycles. Different sacrificial materials like TEOS have been reported by Yanilmaz et al. (2019) to produce porous CNFs with higher surface area followed by in situ aniline polymerization, which led to 96% performance retention. Another approach for the deposition of CP is electrodeposition. Mohd Abdah et al. (2017) utilized electrodeposition to coat PEDOT onto GO-incorporated PVA nanofibers. Because PVA is a non-CP, the mixing of GO provided conductivity to the fibers. Electrodeposition of PANI can be done via potentiodynamic, potentiostatic, or galvanostatic techniques, out of which galvanostatic electro-polymerization was demonstrated to provide the most uniform coating on CNFs (for both nonporous and porous CNFs) (Tran et al., 2015).

22.5 Nanofibers as separators

In a SC cell, the electrodes undergoing the electrochemical interactions need to be physically separated to prevent direct contact and short-circuiting while allowing the transport of ionic species between them. This functionality is achieved by employing a separator between the electrodes. The separator is a thin and nonconducting layer that stops the two opposing electrodes from coming into direct contact. A good separator should be able to retain the electrolyte in order to act as a reservoir of the electrochemical species and aid in ion transport. It should be a good dielectric material with high chemical and thermal stability and high mechanical strength. For wearable applications, flexibility is an additional requirement of the separator. Separators also play a crucial role in slowing and/or preventing self-discharge. Polymeric materials are a prevalent choice for use as separators due to their high chemical resistance, good mechanical strength, low cost, and ease of processing. Some of the commonly employed polymers that have been used to fabricate separators include polypropylene, polyethylene, polyvinylidene fluoride

(PVDF), poly(vinylidene fluoride-co-hexafluoropropylene) (PVDF-HFP), and polyethylene oxide (Ahankari et al., 2022). Electrospinning is a commonly used approach for synthesizing separator membranes with controlled porosity and pore sizes. It can be used to obtain nonwoven free-standing nanofiber mats that can serve directly as separators in SCs. Further, composite membranes with additional nanoparticles or with two or more polymer blends can be synthesized. The electrospun nanofiber-based separators possess high porosity and thereby high electrolyte uptake along with faster ionic diffusion compared to those fabricated using solution-casting or phase-inversion methods (Li et al., 2023).

PVDF has been a popular choice for nanofiber synthesis for separators. It was demonstrated that with a 20 wt.% concentration, it is thermally stable up to 450°C with good electrochemical stability over a 2.5 V potential range (Arthi et al., 2022). Much like the strategy used to make in situ porous CNFs via sacrificial material utilization, porous nanofibers have been developed with higher surface area for good electrolyte uptake. In an investigation by He et al. (2017), PVP was used along with PVDF in the spinning solution, which led to a phase separation that created pockets of PVP in the nanofibers. Removal of PVP was easily achieved by dissolving it in deionized water, leading to highly porous nanofibers with a high surface area of 111.5 m^2/g with good electrochemical stability in an organic electrolyte. PVDF nanofibers have demonstrated excellent compatibility with organic electrolytes. However, their hydrophobic nature leads to poor compatibility with aqueous electrolytes. This concern was addressed by Buxton et al. (2022), where an anionic surfactant, sodium dodecyl sulfate, was added to make the PVDF nanofibers superhydrophilic for aqueous electrolytes. During the electrospinning process, the stretching of the PVDF solution led to polarized nanofiber formation. Sodium dodecyl sulfate further increased the β-phase crystalline phase in the nanofibers amplifying the polarity of PVDF, thus, demonstrating piezoelectric properties. The resulting directionally polarized piezoelectric nanofibers with high dielectric constant and porosity displayed lower self-discharge and are promising materials for use as separators in SCs. In another study by Cao et al. (2022), a PAN separator was grown in situ epitaxially between two PANI/CNT fiber electrodes via the electrospinning technique. Here, a single fiber, when cut into a circular cross-section, showed four concentric circles: the one in the middle comprised PANI/CNT, followed by PAN separator, another layer of PANI/CNT, and everything wrapped in an encapsulation (PAN). This "coaxial" fiber SC demonstrated excellent folding resistance and 93% capacitance retention upon 1000 cycles when being held at 180 degrees.

22.6 Conclusions and future outlook

Owing to the increasing electrical energy storage demands, there is a huge market for devices such as SCs, which promise high-power density, rapid charge/discharge, and long cycle life, in various applications ranging from large-scale energy storage to portable devices. Primarily, SCs can store and release energy very quickly as compared to conventional batteries. Furthermore, the SCs and batteries can complement one another, and using them together can deliver a host of benefits that are difficult for one to offer independently. One of the essential requirements for designing high-performance SCs is to have functionalizable and tunable electrodes with high surface area, where the electrospun nanofibers can be a promising solution. Electrospun nanofibers, especially carbonized nanofibers, have been extensively studied and used in the development of SCs due to their high specific surface area, high porosity, and excellent electrical conductivity. Additionally, the CNF-based electrode can be made free-standing, eliminating the need for binder material and the associated problems such as inhomogeneity, low conductivity, and dependence on the coating technique. The free-standing nature and the flexibility of the nanofibrous electrodes also open up the possibilities of using flexible SCs in applications such as wearable and portable electronics. Additionally, the nanofibrous electrodes are not limited to CNFs, and nanofibrous electrodes can be fabricated from various materials, including metal oxides, conductive polymers, and composites. While the potential of generating a host of modifications and functionalities on the electrospun nanofibers for the application in SCs has been proven in the literature, the versatility of the electrospinning process opens up a wide range of new possibilities. We can expect a vast majority of the innovations in nanofibrous electrode synthesis from engineering the fibers' pore structure and tuning the nanofibers' surface/bulk chemistry. Such materials would be able to maximize the synergistic effect arising from porous architecture and surface/bulk chemistry. Nevertheless, electrospinning-based nanofiber synthesis is not without challenges that must be overcome to realize their practical applications. Some of the major challenges associated with the electrospinning-based approach include the low throughput of the process, thereby limiting the scalability, high synthesis cost, and the need for specific solvents, to name a few. Furthermore, for the practical application of SCs, it is necessary to boost their performance in terms of higher energy density, improve their cycle life and stability, and lower self-discharge. In addition to its critical role as the electrode, the nanofiber's role as a separator for SC applications cannot be undermined. Specifically, the nanofiber mats as separators with porous architecture influence the ion transport pathways and also act as a reservoir for the electrolyte. In summary, the use of high surface area nanofibers in SC applications offers significant advantages and has the potential to significantly improve their performance. Overall, the continued development and optimization of nanofibrous electrodes and separator material could make the SCs a more viable and attractive energy storage option for a range of applications.

References

Ahankari, S., Lasrado, D., & Subramaniam, R. (2022). Advances in materials and fabrication of separators in supercapacitors. *Materials Advances*, *3*(3), 1472–1496. Available from https://doi.org/10.1039/d1ma00599e, http://www.rsc.org/journals-books-databases/about-journals/materials-advances/.

Anand, S., Ahmad, M. W., Ali Al Saidi, A. K., Yang, D. J., & Choudhury, A. (2020). Polyaniline nanofiber decorated carbon nanofiber hybrid mat for flexible electrochemical supercapacitor. *Materials Chemistry and Physics, 254*. Available from https://doi.org/10.1016/j.matchemphys.2020.123480, http://www.journals.elsevier.com/materials-chemistry-and-physics.

Arthi, R., Jaikumar, V., & Muralidharan, P. (2022). Development of electrospun PVdF polymer membrane as separator for supercapacitor applications. *Energy Sources, Part A: Recovery, Utilization and Environmental Effects, 44*(1), 2294−2308. Available from https://doi.org/10.1080/15567036.2019.1649746, http://www.tandf.co.uk/journals/titles/15567036.asp.

Baji, A., Mai, Y. W., Wong, S. C., Abtahi, M., & Chen, P. (2010). Electrospinning of polymer nanofibers: Effects on oriented morphology, structures and tensile properties. *Composites Science and Technology, 70*(5), 703−718. Available from https://doi.org/10.1016/j.compscitech.2010.010.010, http://www.journals.elsevier.com/composites-science-and-technology/.

Bhagwan, J., Kumar, N., Yadav, K. L., & Sharma, Y. (2018). Probing the electrical properties and energy storage performance of electrospun $ZnMn_2O_4$ nanofibers. *Solid State Ionics, 321*, 75−82. Available from https://doi.org/10.1016/j.ssi.2018.040.007, http://www.journals.elsevier.com/solid-state-ionics/.

Buxton, W. G., King, S. G., & Stolojan, V. (2022). Suppression of self-discharge in aqueous supercapacitor devices incorporating highly polar nanofiber separators. *Energy and Environmental Materials, 6*(3), e12363. Available from https://doi.org/10.1002/eem2.12363, http://onlinelibrary.wiley.com/journal/25750356.

Cao, Y., Zhang, H., Zhang, Y., Yang, Z., Liu, D., Fu, H., Zhang, Y., Liu, M., & Li, Q. (2022). Epitaxial nanofiber separator enabling folding-resistant coaxial fiber-supercapacitor module. *Energy Storage Materials, 49*, 102−110. Available from https://doi.org/10.1016/j.ensm.2022.030.011, http://www.journals.elsevier.com/energy-storage-materials/.

Chen, H., Zeng, S., Chen, M., Zhang, Y., & Li, Q. (2017). A new insight into the rechargeable mechanism of manganese dioxide based symmetric supercapacitors. *RSC Advances, 7*(14), 8561−8566. Available from https://doi.org/10.1039/c6ra28040d, http://pubs.rsc.org/en/journals/journalissues.

Cheng, Y., Huang, L., Xiao, X., Yao, B., Yuan, L., Li, T., Hu, Z., Wang, B., Wan, J., & Zhou, J. (2015). Flexible and cross-linked N-doped carbon nanofiber network for high performance freestanding supercapacitor electrode. *Nano Energy, 15*, 66−74. Available from https://doi.org/10.1016/j.nanoen.2015.040.007, http://www.journals.elsevier.com/nano-energy/.

Da Silva, E. P., Rubira, A. F., Ferreira, O. P., Silva, R., & Muniz, E. C. (2019). In situ growth of manganese oxide nanosheets over titanium dioxide nanofibers and their performance as active material for supercapacitor. *Journal of Colloid and Interface Science, 555*, 373−382. Available from https://doi.org/10.1016/j.jcis.2019.070.064, http://www.elsevier.com/inca/publications/store/6/2/2/8/6/1/index.htt.

Dong, T., Li, M., Wang, P., & Yang, P. (2018). Synthesis of hierarchical tube-like yolk-shell Co_3O_4@$NiMoO_4$ for enhanced supercapacitor performance. *International Journal of Hydrogen Energy, 43*(31), 14569−14577. Available from https://doi.org/10.1016/j.ijhydene.2018.060.067, http://www.journals.elsevier.com/international-journal-of-hydrogen-energy/.

Dou, Y., Zhang, W., & Kaiser, A. (2020). Electrospinning of metal−organic frameworks for energy and environmental applications. *Advanced Science, 7*.

Du, Y., Zhao, X., Huang, Z., Li, Y., & Zhang, Q. (2014). Freestanding composite electrodes of MnO$_x$ embedded carbon nanofibers for high-performance supercapacitors. *RSC Advances*, *4*(74), 39087−39094. Available from https://doi.org/10.1039/c4ra06301e, http://pubs.rsc.org/en/journals/journalissues.

Editor. (n.d.). Series in Materials Science 300. *Handbook of Nanocomposite Supercapacitor Materials I Characteristics*.

Fleischmann, S., Mitchell, J. B., Wang, R., Zhan, C., Jiang, D. E., Presser, V., & Augustyn, V. (2020). Pseudocapacitance: From fundamental understanding to high power energy storage materials. *Chemical Reviews*, *120*(14), 6738−6782. Available from https://doi.org/10.1021/acs.chemrev.0c00170, http://pubs.acs.org/journal/chreay.

Fong, K. D., Wang, T., & Smoukov, S. K. (2017). Multidimensional performance optimization of conducting polymer-based supercapacitor electrodes. *Sustainable Energy & Fuels*, *1*(9), 1857−1874. Available from https://doi.org/10.1039/C7SE00339K.

Gong, D., Zhu, J., & Lu, B. (2016). RuO$_2$@Co$_3$O$_4$ heterogeneous nanofibers: A high-performance electrode material for supercapacitors. *RSC Advances*, *6*(54), 49173−49178. Available from https://doi.org/10.1039/C6RA04884F.

He, T., Jia, R., Lang, X., Wu, X., & Wang, Y. (2017). Preparation and electrochemical performance of PVdF ultrafine porous fiber separator-cum-electrolyte for supercapacitor. *Journal of the Electrochemical Society*, *164*(13), E379−E384. Available from https://doi.org/10.1149/2.0631713jes, http://jes.ecsdl.org/content/by/year.

Ishita, I., & Singhal, R. (2020). Porous multi-channel carbon nanofiber electrodes using discarded polystyrene foam as sacrificial material for high-performance supercapacitors. *Journal of Applied Electrochemistry*, *50*(8), 809−820. Available from https://doi.org/10.1007/s10800-020-01433-0, https://link.springer.com/journal/10800.

Jiang, H., Niu, H., Yang, X., Sun, Z., Li, F., Wang, Q., & Qu, F. (2018). Flexible Fe$_2$O$_3$ and V$_2$O$_5$ nanofibers as binder-free electrodes for high-performance all-solid-state asymmetric supercapacitors. *Chemistry − A European Journal*, *24*(42), 10683−10688. Available from https://doi.org/10.1002/chem.201800461, http://onlinelibrary.wiley.com/journal/10.1002/(ISSN)1521-3765.

Jiang, Y., & Liu, J. (2019). Definitions of pseudocapacitive materials: A brief review. *Energy & Environmental Materials*, *2*(1), 30−37. Available from https://doi.org/10.1002/eem2.12028.

Kim, C. H., Yang, C. M., Kim, Y. A., & Yang, K. S. (2019). Pore engineering of nanoporous carbon nanofibers toward enhanced supercapacitor performance. *Applied Surface Science*, *497*. Available from https://doi.org/10.1016/j.apsusc.2019.143693, http://www.journals.elsevier.com/applied-surface-science/.

Kim, Y. I., Samuel, E., Joshi, B., Kim, M. W., Kim, T. G., Swihart, M. T., & Yoon, S. S. (2018). Highly efficient electrodes for supercapacitors using silver-plated carbon nanofibers with enhanced mechanical flexibility and long-term stability. *Chemical Engineering Journal*, *353*, 189−196. Available from https://doi.org/10.1016/j.cej.2018.070.066, http://www.elsevier.com/inca/publications/store/6/0/1/2/7/3/index.htt.

Kundu, M., & Liu, L. (2015). Binder-free electrodes consisting of porous NiO nanofibers directly electrospun on nickel foam for high-rate supercapacitors. *Materials Letters*, *144*, 114−118. Available from https://doi.org/10.1016/j.matlet.2015.010.032, http://www.journals.elsevier.com/materials-letters/.

Kurtan, U., Aydın, H., Büyük, B., Şahintürk, U., Almessiere, M. A., & Baykal, A. (2020). Freestanding electrospun carbon nanofibers uniformly decorated with bimetallic alloy nanoparticles as supercapacitor electrode. *Journal of Energy Storage*, *32*. Available from https://doi.org/10.1016/j.est.2020.101671, http://www.journals.elsevier.com/journal-of-energy-storage/.

Lai, C. C., & Lo, C. T. (2015). Preparation of nanostructural carbon nanofibers and their electrochemical performance for supercapacitors. *Electrochimica Acta*, *183*, 85−93. Available from https://doi.org/10.1016/j.electacta.2015.020.143, http://www.journals.elsevier.com/electrochimica-acta/.

Lawrence, D. W., Tran, C., Mallajoysula, A. T., Doorn, S. K., Mohite, A., Gupta, G., & Kalra, V. (2015). High-energy density nanofiber-based solid-state supercapacitors. *Journal of Materials Chemistry A*, *4*(1), 160−166. Available from https://doi.org/10.1039/c5ta05552k, http://pubs.rsc.org/en/journals/journalissues/ta.

Lee, D., Jung, J. Y., Jung, M. J., & Lee, Y. S. (2015). Hierarchical porous carbon fibers prepared using a SiO_2 template for high-performance EDLCs. *Chemical Engineering Journal*, *263*, 62−70. Available from https://doi.org/10.1016/j.cej.2014.100.070, http://www.elsevier.com/inca/publications/store/6/0/1/2/7/3/index.htt.

Lee, J., Srimuk, P., Fleischmann, S., Su, X., Hatton, T. A., & Presser, V. (2019). Redox-electrolytes for non-flow electrochemical energy storage: A critical review and best practice. *Progress in Materials Science*, *101*, 46−89. Available from https://doi.org/10.1016/j.pmatsci.2018.100.005.

Li, D., Raza, F., Wu, Q., Zhu, X., & Ju, A. (2022). $NiCo_2O_4$ nanosheets on hollow carbon nanofibers for flexible solid-state supercapacitors. *ACS Applied Nano Materials*, *5*(10), 14630−14638. Available from https://doi.org/10.1021/acsanm.2c03002, https://pubs.acs.org/journal/aanmf6.

Li, J., Jia, H., Ma, S., Xie, L., Wei, X. X., Dai, L., Wang, H., Su, F., & Chen, C. M. (2023). Separator design for high-performance supercapacitors: Requirements, challenges, strategies, and prospects. *ACS Energy Letters*, *8*(1), 56−78. Available from https://doi.org/10.1021/acsenergylett.2c01853, http://pubs.acs.org/journal/aelccp.

Li, X., Zhao, Y., Bai, Y., Zhao, X., Wang, R., Huang, Y., Liang, Q., & Huang, Z. (2017). A non-woven network of porous nitrogen-doping carbon nanofibers as a binder-free electrode for supercapacitors. *Electrochimica Acta*, *230*, 445−453. Available from https://doi.org/10.1016/j.electacta.2017.020.030, http://www.journals.elsevier.com/electrochimica-acta/.

Liang, J., Zhao, H., Yue, L., Fan, G., Li, T., Lu, S., Chen, G., Gao, S., Asiri, A. M., & Sun, X. (2020). Recent advances in electrospun nanofibers for supercapacitors. *Journal of Materials Chemistry A*, *8*(33), 16747−16789. Available from https://doi.org/10.1039/d0ta05100d, http://pubs.rsc.org/en/journals/journal/ta.

Liao, Y., Fukuda, T., Wang, S. (2016). Electrospun metal oxide nanofibers and their energy applications, InTech. https://doi.org/10.5772/63414.

Liu, C., Tan, Y., Liu, Y., Shen, K., Peng, B., Niu, X., & Ran, F. (2016). Microporous carbon nanofibers prepared by combining electrospinning and phase separation methods for supercapacitor. *Journal of Energy Chemistry*, *25*(4), 587−593. Available from https://doi.org/10.1016/j.jechem.2016.030.017, http://elsevier.com/journals/journal-of-energy-chemistry/2095-4956.

Liu, K., Yu, C., Guo, W., Ni, L., Yu, J., Xie, Y., Wang, Z., Ren, Y., & Qiu, J. (2021). Recent research advances of self-discharge in supercapacitors: Mechanisms and suppressing strategies. *Journal of Energy Chemistry*, *58*, 94−109. Available from https://doi.org/10.1016/j.jechem.2020.090.041, http://elsevier.com/journals/journal-of-energy-chemistry/2095-4956.

Lu, Y., Liu, Y., Mo, J., Deng, B., Wang, J., Zhu, Y., Xiao, X., & Xu, G. (2021). Construction of hierarchical structure of Co_3O_4 electrode based on electrospinning technique for supercapacitor. *Journal of Alloys and Compounds*, 853. Available from https://doi.org/10.1016/j.jallcom.2020.157271, https://www.journals.elsevier.com/journal-of-alloys-and-compounds.

Luo, X. Y., Chen, Y., & Mo, Y. (2021). A review of charge storage in porous carbon-based supercapacitors. *Xinxing Tan Cailiao/New Carbon Materials*, 36(1), 49−68. Available from https://doi.org/10.1016/S1872-5805(21)60004-5, http://xxtcl.periodicals.net.cn/default.html.

Lyu, L., Seong, K. D., Ko, D., Choi, J., Lee, C., Hwang, T., Cho, Y., Jin, X., Zhang, W., Pang, H., & Piao, Y. (2019). Recent development of biomass-derived carbons and composites as electrode materials for supercapacitors. *Materials Chemistry Frontiers*, 3(12), 2543−2570. Available from https://doi.org/10.1039/c9qm00348g, http://rsc.li/frontiers-materials.

Ma, P. P., Zhu, B., Lei, N., Liu, Y. K., Yu, B., Lu, Q. L., Dai, J. M., Li, S. H., & Jiang, G. H. (2019). Effect of Sr substitution on structure and electrochemical properties of perovskite-type LaMn$_{0.9}$Ni$_{0.1}$O$_3$ nanofibers. *Materials Letters*, 252, 23−26. Available from https://doi.org/10.1016/j.matlet.2019.050.090, http://www.journals.elsevier.com/materials-letters/.

Mao, X., Hatton, T., & Rutledge, G. (2013). A review of electrospun carbon fibers as electrode materials for energy Storage. *Current Organic Chemistry*, 17(13), 1390−1401. Available from https://doi.org/10.2174/1385272811317130006.

Meng, Q., Cai, K., Chen, Y., & Chen, L. (2017). Research progress on conducting polymer based supercapacitor electrode materials. *Nano Energy*, 36, 268−285. Available from https://doi.org/10.1016/j.nanoen.2017.040.040, http://www.journals.elsevier.com/nano-energy/.

Miao, Y. E., Fan, W., Chen, D., & Liu, T. (2013). High-performance supercapacitors based on hollow polyaniline nanofibers by electrospinning. *ACS Applied Materials and Interfaces*, 5(10), 4423−4428. Available from https://doi.org/10.1021/am4008352.

Mishra, R. K., Choi, G. J., Choi, H. J., Singh, J., Mirsafi, F. S., Rubahn, H.-G., Mishra, Y. K., Lee, S. H., & Gwag, J. S. (2022). Voltage holding and self-discharge phenomenon in ZnO-Co$_3$O$_4$ core-shell heterostructure for binder-free symmetric supercapacitors. *Chemical Engineering Journal*, 427, 131895. Available from https://doi.org/10.1016/j.cej.2021.131895.

Mohd Abdah, M. A. A., Zubair, N. A., Azman, N. H. N., & Sulaiman, Y. (2017). Fabrication of PEDOT coated PVA-GO nanofiber for supercapacitor. *Materials Chemistry and Physics*, 192, 161−169. Available from https://doi.org/10.1016/j.matchemphys.2017.010.058, http://www.journals.elsevier.com/materials-chemistry-and-physics.

Na, W., Jun, J., Park, J. W., Lee, G., & Jang, J. (2017). Highly porous carbon nanofibers co-doped with fluorine and nitrogen for outstanding supercapacitor performance. *Journal of Materials Chemistry A*, 5(33), 17379−17387. Available from https://doi.org/10.1039/c7ta04406b, http://pubs.rsc.org/en/journals/journalissues/ta.

Pal, B., Yang, S., Ramesh, S., Thangadurai, V., & Jose, R. (2019). Electrolyte selection for supercapacitive devices: A critical review. *Nanoscale Advances*, 1(10), 3807−3835. Available from https://doi.org/10.1039/c9na00374f, http://pubs.rsc.org/en/journals/journalissues/na?_ga2.190536939.1555337663.1552312502-1364180372.1550481316#!issueidna001002&typecurrent&issnonline2516-0230.

Peng, S., Li, L., Hu, Y., Srinivasan, M., Cheng, F., Chen, J., & Ramakrishna, S. (2015). Fabrication of spinel one-dimensional architectures by single-spinneret electrospinning for energy storage applications. *ACS Nano*, 9(2), 1945−1954. Available from https://doi.org/10.1021/nn506851x, http://pubs.acs.org/journal/ancac3.

Poonam, K., Sharma, A., Arora, S. K., & Tripathi. (2019). Review of supercapacitors: Materials and devices. *Journal of Energy Storage*, *21*, 801−825. Available from https://doi.org/10.1016/j.est.2019.010.010, http://www.journals.elsevier.com/journal-of-energy-storage/.

Radhakanth, S., & Singhal, R. (2023). In−situ synthesis of MnO dispersed carbon nanofibers as binder-free electrodes for high-performance supercapacitors. *Chemical Engineering Science*, *265*.

Rahaman, M. S. A., Ismail, A. F., & Mustafa, A. (2007). A review of heat treatment on polyacrylonitrile fiber. *Polymer Degradation and Stability*, *92*(8), 1421−1432. Available from https://doi.org/10.1016/j.polymdegradstab.2007.030.023.

Saha, B., & Schatz, G. C. (2012). Carbonization in polyacrylonitrile (PAN) based carbon fibers studied by reaxff molecular dynamics simulations. *Journal of Physical Chemistry B*, *116*(15), 4684−4692. Available from https://doi.org/10.1021/jp300581b, http://pubs.acs.org/journal/jpcbfk.

Salunkhe, R. R., Kaneti, Y. V., & Yamauchi, Y. (2017). Metal-organic framework-derived nanoporous metal oxides toward supercapacitor applications: Progress and prospects. *ACS Nano*, *11*(6), 5293−5308. Available from https://doi.org/10.1021/acsnano.7b02796, http://pubs.acs.org/journal/ancac3.

Samuel, E., Joshi, B., Kim, M. W., Kim, Y. I., Swihart, M. T., & Yoon, S. S. (2019). Hierarchical zeolitic imidazolate framework-derived manganese-doped zinc oxide decorated carbon nanofiber electrodes for high performance flexible supercapacitors. *Chemical Engineering Journal*, *371*, 657−665. Available from https://doi.org/10.1016/j.cej.2019.040.065, http://www.elsevier.com/inca/publications/store/6/0/1/2/7/3/index.htt.

Shang, W., Yu, W., Xiao, X., Ma, Y., He, Y., Zhao, Z., & Tan, P. (2023). Insight into the self-discharge suppression of electrochemical capacitors: Progress and challenges. *Advanced Powder Materials*, *2*(1), 100075. Available from https://doi.org/10.1016/j.apmate.2022.100075.

Shao, Y., El-Kady, M. F., Sun, J., Li, Y., Zhang, Q., Zhu, M., Wang, H., Dunn, B., & Kaner, R. B. (2018). Design and mechanisms of asymmetric supercapacitors. *Chemical Reviews*, *118*(18), 9233−9280. Available from https://doi.org/10.1021/acs.chemrev.8b00252, http://pubs.acs.org/journal/chreay.

Shin, S., & Shin, M. W. (2021). Applied surface science nickel metal − organic framework (Ni-MOF) derived NiO/C@CNF composite for the application of high performance self-standing supercapacitor electrode. *Applied Surface Science*, *540*.

Singhal, R., & Kalra, V. (2015). Using common salt to impart pseudocapacitive functionalities to carbon nanofibers. *Journal of Materials Chemistry A*, *3*(1), 377−385. Available from https://doi.org/10.1039/c4ta05121a, http://pubs.rsc.org/en/journals/journalissues/ta.

Subramani, S., & Rajiv, S. (2022). Fabrication of poly(3-methylthiophene)/poly(ethylene oxide)/ruthenium oxide composite electrospun nanofibers for supercapacitor application. *Journal of Materials Science: Materials in Electronics*, *33*(12), 9558−9569. Available from https://doi.org/10.1007/s10854-021-07549-z, https://rd.springer.com/journal/10854.

Teli, A. M., Beknalkar, S. A., Patil, D. S., Pawar, S. A., Dubal, D. P., Burute, V. Y., Dongale, T. D., Shin, J. C., & Patil, P. S. (2020). Effect of annealing temperature on charge storage kinetics of an electrospun deposited manganese oxide supercapacitor. *Applied Surface Science*, *511*. Available from https://doi.org/10.1016/j.apsusc.2020.145466, http://www.journals.elsevier.com/applied-surface-science/.

Tian, D., Ao, Y., Li, W., Xu, J., & Wang, C. (2021). General fabrication of metal-organic frameworks on electrospun modified carbon nanofibers for high-performance asymmetric supercapacitors. *Journal of Colloid and Interface Science*, *603*, 199−209. Available

from https://doi.org/10.1016/j.jcis.2021.050.138, http://www.elsevier.com/inca/publications/store/6/2/2/8/6/1/index.htt.
Tian, D., Song, N., Zhong, M., Lu, X., & Wang, C. (2020). Bimetallic MOF nanosheets decorated on electrospun nanofibers for high-performance asymmetric supercapacitors. *ACS Applied Materials and Interfaces*, *12*(1), 1280−1291. Available from https://doi.org/10.1021/acsami.9b16420, http://pubs.acs.org/journal/aamick.
Tran, C., & Kalra, V. (2013a). Co-continuous nanoscale assembly of Nafion-polyacrylonitrile blends within nanofibers: A facile route to fabrication of porous nanofibers. *Soft Matter*, *9*(3), 846−852. Available from https://doi.org/10.1039/c2sm25976a, http://www.rsc.org/Publishing/Journals/sm/Article.asp?Type = CurrentIssue.
Tran, C., & Kalra, V. (2013b). Fabrication of porous carbon nanofibers with adjustable pore sizes as electrodes for supercapacitors. *Journal of Power Sources*, *235*, 289−296. Available from https://doi.org/10.1016/j.jpowsour.2013.010.080.
Tran, C., Singhal, R., Lawrence, D., & Kalra, V. (2015). Polyaniline-coated freestanding porous carbon nanofibers as efficient hybrid electrodes for supercapacitors. *Journal of Power Sources*, *293*, 373−379. Available from https://doi.org/10.1016/j.jpowsour.2015.050.054.
Uke, S. J., Akhare, V. P., Bambole, D. R., Bodade, A. B., & Chaudhari, G. N. (2017). Recent advancements in the cobalt oxides, manganese oxides, and their composite as an electrode material for supercapacitor: A review. *Frontiers in Materials*, *4*. Available from https://doi.org/10.3389/fmats.2017.00021, https://www.frontiersin.org/articles/10.3389/fmats.2017.00021/pdf.
Viswanathan, B. (2017). Supercapacitors. Elsevier BV, 315−328. https://doi.org/10.1016/b978-0-444-56353-8.00013-7.
Wang, G., Zhang, L., & Zhang, J. (2012). A review of electrode materials for electrochemical supercapacitors. *Chemical Society Reviews*, *41*(2), 797−828. Available from https://doi.org/10.1039/c1cs15060j.
Wang, H., Wang, W., Wang, H., Jin, X., Niu, H., Wang, H., Zhou, H., & Lin, T. (2018). High performance supercapacitor electrode materials from electrospun carbon nanofibers in situ activated by high decomposition temperature polymer. *ACS Applied Energy Materials*, *1*(2), 431−439. Available from https://doi.org/10.1021/acsaem.7b00083, http://pubs.acs.org/journal/aaemcq.
Wang, H., Wang, W., Wang, H., Li, Y., Jin, X., Niu, H., Wang, H., Zhou, H., & Lin, T. (2019). Improving supercapacitance of electrospun carbon nanofibers through increasing micropores and microporous surface area. *Advanced Materials Interfaces*, *6*(6), 1801900. Available from https://doi.org/10.1002/admi.201801900.
Wang, W., Lin, B., Zhang, H., Sun, Y., Zhang, X., & Yang, H. (2019). Synthesis, morphology and electrochemical performances of perovskite-type oxide $La_xSr_{1-x}FeO_3$ nanofibers prepared by electrospinning. *Journal of Physics and Chemistry of Solids*, *124*, 144−150. Available from https://doi.org/10.1016/j.jpcs.2018.090.011, https://www.sciencedirect.com/science/article/pii/S0022369718314367.
Wei, C., Cheng, C., Ma, L., Liu, M., Kong, D., Du, W., & Pang, H. (2016). Mesoporous hybrid NiO: X-MnOx nanoprisms for flexible solid-state asymmetric supercapacitors. *Dalton Transactions*, *45*(26), 10789−10797. Available from https://doi.org/10.1039/c6dt01025c, http://www.rsc.org/Publishing/Journals.
Xiao, Y., Xu, Y., Zhang, K., Tang, X., Huang, J., Yuan, K., & Chen, Y. (2020). Coaxial electrospun free-standing and mechanical stable hierarchical porous carbon nanofiber membrane for flexible supercapacitors. *Carbon*, *160*, 80−87. Available from https://doi.org/10.1016/j.carbon.2020.010.017, http://www.journals.elsevier.com/carbon/.

Xue, J., Wu, T., Dai, Y., & Xia, Y. (2019). Electrospinning and electrospun nanofibers: Methods, materials, and applications. *Chemical Reviews*, *119*(8), 5298–5415. Available from https://doi.org/10.1021/acs.chemrev.8b00593, http://pubs.acs.org/journal/chreay.

Xue, Y., Zheng, S., Xue, H., & Pang, H. (2019). Metal-organic framework composites and their electrochemical applications. *Journal of Materials Chemistry A*, *7*(13), 7301–7327. Available from https://doi.org/10.1039/C8TA12178H, http://pubs.rsc.org/en/journals/journal/ta.

Yanilmaz, M., Dirican, M., Asiri, A. M., & Zhang, X. (2019). Flexible polyaniline-carbon nanofiber supercapacitor electrodes. *Journal of Energy Storage*, *24*. Available from https://doi.org/10.1016/j.est.2019.100766, http://www.journals.elsevier.com/journal-of-energy-storage/.

Yao, Y., Liu, P., Li, X., Zeng, S., Lan, T., Huang, H., Zeng, X., & Zou, J. (2018). Nitrogen-doped graphitic hierarchically porous carbon nanofibers obtained: Via bimetallic-coordination organic framework modification and their application in supercapacitors. *Dalton Transactions*, *47*(21), 7316–7326. Available from https://doi.org/10.1039/c8dt00823j, http://pubs.rsc.org/en/journals/journal/dt.

Zhang, B., Kang, F., Tarascon, J. M., & Kim, J. K. (2016). Recent advances in electrospun carbon nanofibers and their application in electrochemical energy storage. *Progress in Materials Science*, *76*, 319–380. Available from https://doi.org/10.1016/j.pmatsci.2015.080.002.

Zhang, L., Jiang, Y., Wang, L., Zhang, C., & Liu, S. (2016). Hierarchical porous carbon nanofibers as binder-free electrode for high-performance supercapacitor. *Electrochimica Acta*, *196*, 189–196. Available from https://doi.org/10.1016/j.electacta.2016.020.050, http://www.journals.elsevier.com/electrochimica-acta/.

Zhang, S., & Pan, N. (2015). Supercapacitors performance evaluation. *Advanced Energy Materials*, *5*(6). Available from https://doi.org/10.1002/aenm.201401401, http://onlinelibrary.wiley.com/journal/10.1002/(ISSN)1614-6840.

Zhang, Y., Feng, H., Wu, X., Wang, L., Zhang, A., Xia, T., Dong, H., Li, X., & Zhang, L. (2009). Progress of electrochemical capacitor electrode materials: A review. *International Journal of Hydrogen Energy*, *34*(11), 4889–4899. Available from https://doi.org/10.1016/j.ijhydene.2009.040.005.

Zhi, M., Liu, S., Hong, Z., & Wu, N. (2014). Electrospun activated carbon nanofibers for supercapacitor electrodes. *RSC Advances*, *4*(82), 43619–43623. Available from https://doi.org/10.1039/C4RA05512H.

Sources of natural fibers and their physicochemical properties for textile uses

23

Abhinay Thakur[1], Ashish Kumar[2] and Valentine Chikaodili Anadebe[3,4]

[1]Department of Chemistry, School of Chemical Engineering and Physical Sciences, Lovely Professional University, Phagwara, Punjab, India, [2]Department of Science and Technology, NCE, Bihar Engineering University, Government of Bihar, Patna, Bihar, India, [3]Corrosion and Materials Protection Division, CSIR-Central Electrochemical Research Institute, Karaikudi, Tamil Nadu, India, [4]Department of Chemical Engineering, Alex Ekwueme Federal University Ndufu Alike, Abakakili, Ebonyi State, Nigeria

23.1 Introduction

The historical significance of natural fibers cannot be overstated. From the cultivation of cotton in ancient civilizations like Egypt and the Indus Valley to the weaving of silk in ancient China, these fibers have played pivotal roles in the growth and prosperity of human societies. Plant-based fibers such as flax and jute have been used for centuries to create sturdy textiles, and animal-based fibers such as wool and silk have been prized for their luxurious qualities (Bhuiyan et al., 2017; Durkin et al., 2016; Gurunathan et al., 2015). Mineral-based fibers, including asbestos and basalt, have also made their mark in history, notably for their unique properties. Asbestos, for instance, was valued for its heat-resistant characteristics and used in a variety of applications, including fireproof textiles. However, the use of asbestos has declined due to well-documented health concerns associated with its fibers. In recent times, as the world grapples with environmental challenges and an urgent need for sustainable practices, natural fibers have once again taken center stage. Cotton stands as one of the most widely cultivated natural fibers globally, with an estimated production of approximately 25.9 million metric tons in 2020 (Kabir et al., 2012; Shahinur et al., 2020). Its rich history dates back to ancient civilizations like Egypt and the Indus Valley, where it played a pivotal role in economic and trade development, with fabrics traveling along ancient trade routes to Europe and East Asia. Silk production, on the other hand, is predominantly associated with China, which is the world's leading silk producer. In 2019, China produced over 150,000 metric tons of raw silk, showcasing the enduring legacy of this luxurious and highly prized natural fiber (Jeyapragash et al., 2020; Mattiello et al., 2023). Regions like Como and Venice in Italy have also made significant contributions to the world of silk production, renowned for their craftsmanship and supply of high-quality silk fabrics to global fashion and luxury markets. In Northern Europe, including countries like Belgium and Ireland, flax and linen production have been longstanding traditions. These regions have capitalized on linen's durability and breathability, making it a sought-after material for clothing and textiles. Australia, on the other hand, boasts the production of Merino wool, celebrated for its softness and highly valued in the country's wool industry, reflecting the rich diversity of natural fibers and their regional significance (Castillo-Lara et al., 2020; Shelenkov et al., 2023; Snetkov et al., 2020). The once-extensive use of asbestos in North America for its fire-resistant properties is another facet of natural fibers' historical significance. However, the well-documented health concerns associated with asbestos exposure have led to a significant decline in its use and increased awareness of its risks. In China, hemp cultivation has an extensive history and is valued for its strength and versatility. Hemp's resurgence in recent years aligns with the global shift toward sustainable practices. Meanwhile, East Africa's sisal production and Bangladesh's jute industry have been pivotal in their respective economies, emphasizing the enduring importance of natural fibers in diverse regions. Fig. 23.1 displays images of various natural fabrics: (A) Ixtle, (B) Henequen, and (C) Jute. These global examples underscore the historical and contemporary significance of natural fibers, which continue to offer sustainable alternatives in a world increasingly focused on environmental consciousness and responsible practices Fig. 23.1.

The textile industry, often criticized for its environmental impact due to the prevalence of synthetic fibers, is witnessing a resurgence of interest in natural alternatives. The biodegradability, renewability, and lower carbon footprint of natural fibers align seamlessly with the demands of an eco-conscious consumer base. This resurgence underscores the importance of understanding the unique physicochemical properties of these fibers and their applications in textiles. In a world increasingly focused on sustainability and environmental responsibility, the textile industry is transforming. Consumers are demanding eco-friendly options, and regulations are evolving to encourage sustainable practices (Bledzki et al., 1996; Sanjay et al., 2016; Wambua et al., 2003). Natural fibers offer a compelling solution to meet

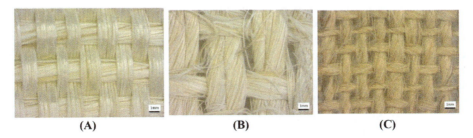

Figure 23.1 Images of various natural fabrics: (A) Ixtle, (B) Henequen, and (C) Jute. *Source*: Adapted from Torres-Arellano, M., Renteria-Rodríguez, V., & Franco-Urquiza, E. (2020). Mechanical properties of natural-fiber-reinforced biobased epoxy resins manufactured by resin infusion process. *Polymers*, *12*(12), 1−17. https://doi.org/10.3390/polym12122841 (under CCBY 4.0).

these demands. However, unlocking their full potential requires a comprehensive understanding of their properties and how these properties can be harnessed for various textile applications. The primary objectives of this chapter are multifaceted, aiming to provide a thorough examination of natural fibers and their physicochemical properties while also highlighting their crucial role in fostering sustainability and innovation in the textile sector. Natural fibers originate from a variety of sources, and each source brings its unique qualities. In this chapter, we will delve into the various categories of natural fibers, including plant-based, animal-based, and mineral-based fibers. By exploring the historical significance, contemporary relevance, and future prospects of natural fibers, we hope to inspire researchers, textile engineers, and industry professionals to embrace these eco-friendly alternatives and contribute to a more sustainable and environmentally responsible future for the textile industry.

23.2 Natural fibers: an overview

23.2.1 Definition and classification

Natural fibers are a remarkable category of materials that have deep roots in our natural world, dating back thousands of years and embodying the essence of sustainability. They are derived from the bountiful resources of our environment and offer a compelling alternative to their synthetic counterparts. These fibers can be broadly classified into three primary categories—plant-based, animal-based, and mineral-based—each with its unique characteristics and applications (Mohammed et al., 2015; Rahman & Putra, 2018).

- Plant-based fibers, which include well-known examples like cotton, flax, hemp, and jute, have been at the very foundation of the textile industry. These fibers originate from the stems, leaves, or seeds of various plants, and each possesses distinct attributes that make them invaluable in various textile applications. Cotton, for instance, is revered for its softness, breathability, and absorbency, making it a staple in clothing and other textile products. Flax, known for its durability and moisture-wicking properties, has been used for centuries to create linen textiles. Hemp, a sturdy and eco-friendly fiber, is gaining recognition for its sustainable potential, while jute's strength and affordability have made it a choice of material for various applications, from sacks to home furnishings.
- Animal-based fibers, such as silk, wool, mohair, and cashmere, are celebrated for their luxurious texture and exceptional insulating properties. These fibers are harvested from animals and have been prized for centuries for their unique qualities. Silk, often referred to as the "queen of fibers," is renowned for its softness, sheen, and smoothness, making it a symbol of opulence in fashion and textiles. Wool, derived from sheep, is cherished for its natural crimp, which traps air and provides exceptional warmth, even when damp. Specialized animal-based fibers, such as mohair and cashmere, offer additional textures and qualities that have earned them coveted positions in high-end fashion and luxury products (Jaganathan et al., 2019; Savic & Savic Gajic, 2022).
- Mineral-based fibers, while less commonly encountered, have made notable contributions. Asbestos, once celebrated for its fire-resistant qualities, was used extensively for

insulation and fireproofing purposes before its well-documented health risks became widely known. Basalt fibers have emerged as a promising material due to their exceptional strength and high-temperature resistance. Derived from molten basalt rock, these fibers have found applications in fields such as construction and aerospace, showcasing the diversity and adaptability of natural fibers.

In an era defined by increasing environmental consciousness and a growing emphasis on sustainability, natural fibers have assumed a renewed significance. Their biodegradability, renewability, and lower carbon footprint make them a compelling choice in the face of environmental challenges. As the textile industry grapples with the environmental impact of synthetic fibers, natural fibers are experiencing a resurgence, offering eco-friendly alternatives that align with the demands of consumers seeking responsible and sustainable products. This resurgence underscores the importance of understanding the unique characteristics and applications of natural fibers, as they continue to weave their enduring story into the fabric of our lives and our sustainable future (Lisitsyn et al., 2021).

23.2.2 Historical significance

The historical significance of natural fibers is a rich tapestry deeply interwoven with the development of human civilization. Throughout the annals of time, these fibers have played an integral role, in shaping cultures, economies, and societies, and their profound impact is evident in both the functional and symbolic roles they have held. In the early epochs of textile production, civilizations looked to the land for their textile needs, and two natural fibers, flax and cotton, took center stage in the cradle of civilization: ancient Egypt. Cotton, with a history dating as far back as 6000 BCE, was cultivated and spun into fabric, marking the inception of cotton textile production (Darie-Niţă et al., 2022; Prasetyaningrum et al., 2021). This ancient endeavor set in motion a trajectory that would eventually lead cotton to become one of the world's most vital and versatile natural fibers.

However, it is perhaps the legendary Silk Road that epitomizes the grandeur of natural fibers' historical importance. This intricate network of trade routes, spanning thousands of miles and connecting China to the Mediterranean, served as a vibrant conduit for the prized silk fiber. In ancient China, the secrets of silk production were closely guarded, and the acquisition of silk was a highly coveted and often clandestine endeavor. Silk, with its exquisite softness and shimmering luster, became not only a symbol of wealth and luxury but also a powerful catalyst for cultural exchange and economic development along the Silk Road. Yet, the significance of natural fibers transcended commerce. They played an indispensable role in providing essential protection against the elements. Wool, in particular, stands as a prime example. This animal-based fiber, harvested from sheep, offered warmth even when wet, making it a stalwart choice for outerwear in regions where exposure to rain and cold was commonplace (Karimah et al., 2021; Sheeba et al., 2023). Wool's natural crimp structure created air pockets that trapped warmth, rendering it an invaluable asset in the preservation of human comfort. Beyond their functional utility, natural fibers were deeply entwined with cultural symbolism. They adorned religious garments, traditional attire, and ceremonial textiles, becoming repositories of cultural identity and tradition. From intricate silk kimonos in Japan to finely woven linen garments in ancient Egypt, these fibers transcended their material nature to symbolize heritage, spirituality, and social status.

23.2.3 Modern applications

In the contemporary world, natural fibers have not only retained their historical significance but have taken on a pivotal role in various industries, driven by their eco-friendly properties and exceptional versatility. These fibers have become emblematic of sustainability, aligning seamlessly with the growing demand for responsible consumption and production. In the realm of fashion and apparel, natural fibers are cherished for their comfort and breathability. Cotton, a staple in the textile industry, remains at the forefront. Its fibers are woven into a vast array of clothing, from everyday wear to formal attire. The versatility of cotton, coupled with its natural feel, makes it a perennial favorite among consumers. Linen, characterized by its crisp texture and cooling properties, graces summer wardrobes with its elegance and freshness. Meanwhile, the lustrous and smooth silk continues to symbolize luxury in high-end fashion, adorning garments that exude opulence and refinement. In the domain of home textiles, natural fibers continue to shine. Cotton and linen are favored choices for bed linens due to their softness and excellent moisture-absorbing capabilities. Sleeping between these natural fabrics is a guarantee of comfort and a restful night's sleep. On the other hand, rugged jute lends its strength to the crafting of sturdy rugs and carpets. Its durability and natural esthetics make it an ideal choice for flooring solutions, often sought after by eco-conscious homeowners. For instance, Merais et al. (2022) delve into the extraction and evaluation of cellulose nanofibers (CNFs) from banana pseudostems through acid hydrolysis, focusing on their physicochemical and thermal attributes. Cellulose, a prominent biopolymer sourced from diverse agricultural residues, exhibits the potential for downsizing into nanocellulose, a biobased polymer characterized by nanometer-scale dimensions and unique properties. This article aimed to produce CNFs from banana pseudostems, involving multiple stages, including grinding, sieving, pretreatment, bleaching, and acid hydrolysis. The product yield was determined to be 40.5% and 21.8% for Musa acuminata and Musa balbisiana, respectively, based on the initial raw fiber weight. The weight reduction is attributed to the elimination of hemicellulose and lignin during the processing stages. Transmission electron microscopy analysis demonstrated a noteworthy reduction in the average fiber size from 180 µm to 80.3 ± 21.3 nm. Additionally, Amena et al. (2022) focused on the alkaline treatment of Ethiopian Arabica coffee husk fibers using sodium hydroxide (NaOH) to enhance their quality. A 10% (w/w) NaOH solution was employed for the alkaline treatment. Fig. 23.2 illustrates the process of preparing coffee husk fibers. Comprehensive physicochemical characterizations, encompassing proximate analysis, cellulosic composition, porosity evaluation, and structural analysis, were conducted to compare the properties of treated and untreated coffee husk fibers. The experimental findings reveal a significant reduction in lignin

Figure 23.2 Depiction of the process of preparing coffee husk fibers to attain textile fibers. *Source*: Adapted from Suárez, L., Castellano, J., Romero, F., Marrero, M.D., Benítez, A.N., & Ortega, Z. (2021). Environmental hazards of giant reed (*Arundo donax* L.) in the macaronesia region and its characterisation as a potential source for the production of natural fibre composites. *Polymers*, *13*(13). https://doi.org/10.3390/polym13132101 (under CCBY 4.0).

(72%) and hemicellulose (52%) content, signifying an overall improvement in fiber quality. Consequently, this study underscores the effectiveness of alkali treatment in enhancing the quality of Ethiopian coffee husk fibers, positioning them as promising raw materials for fiber production in the composite materials industry.

Natural fibers have also ventured into the world of technical textiles, where their unique properties are harnessed for specific purposes. Geotextiles, made from natural fibers, play an integral role in construction and civil engineering. They provide the much-needed strength and stability required for infrastructure projects while aiding in erosion control, soil stabilization, and drainage systems.

In the realm of agriculture, natural fibers take on the form of agro textiles, serving as protective shields for crops (Duarte et al., 2023; Hussain et al., 2023; Mulenga et al., 2021; Spinei & Oroian, 2022). These textiles offer protection from extreme weather conditions and minimize soil erosion, promoting optimal conditions for crop growth. As a result, agricultural yields are enhanced, and sustainable farming practices are promoted. The true essence of natural fibers lies in their contribution to sustainability. As symbols of eco-friendliness, they align perfectly with the contemporary call for environmental responsibility. Their biodegradability, renewability, and reduced carbon footprint have positioned them as sustainable alternatives to synthetic counterparts. Efforts in organic farming and responsible harvesting practices have further elevated their eco-friendly profile, making them the choice of conscious consumers and industries alike (Herrera-Franco & Valadez-González, 2004, 2005; Liu et al., 2020). In a world where environmental consciousness is burgeoning, natural fibers stand as beacons of responsible stewardship. They symbolize a commitment to preserving the planet and exemplify the principles of sustainable living. From ancient traditions to modern solutions, these fibers have not only endured but have flourished, continuing to enrich our lives and contribute to a more sustainable and eco-conscious future.

23.3 Sources of natural fibers

Natural fibers, derived from a diverse array of sources, have been integral to human civilization for millennia. These fibers, categorized into plant-based, animal-based, and mineral-based varieties, have played pivotal roles in shaping cultures, economies, and technologies throughout history (Chung et al., 2018; Nugraha et al., 2022). This comprehensive exploration delves into each category, shedding light on their origins, characteristics, and applications in contemporary industries.

23.3.1 Plant-based fibers

Plant-based fibers are a diverse and crucial category of natural materials sourced from different parts of plants, including stems, leaves, and seeds. These fibers have a rich history and offer a wide range of applications due to their versatility and sustainability. In this section, we delve into four prominent plant-based fibers, each with its unique characteristics and uses:

23.3.1.1 Cotton

Cotton, often celebrated as the "fabric of our lives," is undeniably one of the most ubiquitous and vital plant-based fibers globally. Its history is deeply intertwined with human civilization. Cotton plants, belonging to the genus Gossypium, have been cultivated for thousands of years. One of the remarkable aspects of cotton is its remarkable adaptability to various climates, making it a global staple. Cotton's journey begins with the cultivation of cotton plants, which produce delicate white or cream-colored fibers in protective bolls (Ghinea et al., 2022). These fibers are composed primarily of cellulose, a natural polymer, and are characterized by their softness, breathability, and absorbency. These qualities make cotton highly suitable for a wide range of textile applications. Cotton's applications span a broad spectrum, from everyday clothing to specialized technical textiles. In the fashion and apparel industry, it is the fabric of choice for T-shirts, denim jeans, and undergarments due to its comfort and ease of care. Cotton's use in home textiles, including bed linens and towels, is also widespread, offering softness and absorbency.

In recent years, cotton has taken on a new role as a champion of sustainability in the fashion industry. Efforts to promote organic cotton cultivation, responsible water use, and eco-friendly processing methods have positioned cotton as a leader in sustainable fashion. With the growing demand for eco-conscious products, cotton is at the forefront of the sustainable fashion movement.

23.3.1.2 Flax

Flax, the source of linen, boasts a rich history dating back thousands of years. Linen is derived from the fibers found in the stalks of the flax plant, scientifically known as Linum usitatissimum. This plant is native to the Eastern Mediterranean but has been cultivated worldwide. The production of linen involves several stages, including retting, scutching, and spinning (Oladele et al., 2020; Orlović-Leko et al., 2020). Retting is a process where the flax stalks are soaked in water to break down the pectin that binds the fibers. Once retted, the fibers are separated from the stalks through scutching, resulting in long, smooth strands that are then spun into linen threads. Linen's appeal lies in its unique properties. It has a natural luster, excellent breathability, and remarkable durability. Linen fabrics are known for their crisp texture and coolness, making them highly suitable for warm-weather clothing. Linen is often used in summer attire like shirts, dresses, and lightweight suits.

Beyond fashion, linen finds its place in home furnishings such as bed linens, tablecloths, and curtains due to its timeless elegance and practicality. Linen's eco-friendliness is another noteworthy attribute, as flax plants require minimal water and pesticides, contributing to its sustainability.

23.3.1.3 Hemp

Hemp, known for its robust and eco-friendly qualities, has witnessed a resurgence in popularity as sustainability takes center stage in various industries. Hemp is derived from the Cannabis sativa plant, and it has a long history of cultivation for its fibers, seeds, and oil. One of hemp's distinguishing features is its rapid growth. Hemp plants are hardy and can thrive in diverse climates with minimal water requirements (Ersoy et al., 2022; Nadziakiewicza et al., 2019). This resilience makes it an attractive choice for sustainable agriculture. Hemp fibers are extracted from the inner bark of the plant's stem. The fibers are strong and durable, and they have a coarse texture. Hemp textiles are highly versatile and can be used in clothing, accessories, home textiles, and industrial products. In fashion, hemp is utilized for eco-friendly clothing items such as T-shirts, jeans, and outerwear. Its durability and moisture-wicking properties make it suitable for active and outdoor wear. Additionally, hemp seeds are a source of nutritious oil, and the plant itself can be used for sustainable construction materials, paper production, and even biofuel.

Hemp's recent resurgence is driven by its potential to reduce the environmental impact of various industries, particularly in the context of sustainable fashion and responsible resource management.

Jute, celebrated for its strength and affordability, plays a significant role in various industries, particularly those requiring robust and environmentally friendly materials. Jute fibers are extracted from the Corchorus genus of plants, primarily Corchorus capsularis and Corchorus olitorius. Jute plants are annuals that grow rapidly, reaching heights of up to 12 feet (Arilla et al., 2020; Iuga et al., 2019; Prdun et al., 2021). The fibers are primarily located in the plant's stalks, which are processed to obtain the raw jute fibers. The retting process, which involves soaking the stalks in water, is crucial to separate the fibers. One of the jute's standout qualities is its strength, making it an ideal material for applications requiring durability. Jute fibers are often woven into bags, sacks, and packaging materials due to their sturdiness. Additionally, jute's natural esthetics and texture have made it popular in home textiles, including rugs, carpets, and curtains. Jute's versatility extends to the domain of technical textiles, where it is used in the creation of geotextiles. These textiles play a crucial role in construction and civil engineering, providing erosion control, soil stabilization, and reinforcement in infrastructure projects.

Furthermore, jute has gained attention as a sustainable packaging material, especially in the context of reducing plastic waste. Jute bags and packaging solutions offer an eco-friendly alternative, aligning with the global drive toward sustainability.

23.3.2 Animal-based fibers

Animal-based fibers, sourced from various animals including silkworms, sheep, goats, and rabbits, are highly valued for their luxurious feel and exceptional properties. This section delves into three prominent animal-based fibers, each with its unique characteristics and applications:

23.3.2.1 Silk

Silk, often referred to as the "queen of fibers," stands as an epitome of luxury and refinement with an illustrious history steeped in culture, trade, and fashion. The story of silk production begins with the remarkable silkworm, *Bombyx mori*. These caterpillars spin intricate cocoons made of fine silk threads as they prepare to transform into moths. The process of obtaining silk involves carefully unraveling these cocoons, a practice known as silk reeling. The appeal of silk lies in its unparalleled smoothness, delicate sheen, and exquisite drapability (Kong et al., 2022). These qualities make silk highly sought after in high-end fashion and textiles. Silk's luxurious touch and shimmering appearance elevate it to the pinnacle of elegance, making it the fabric of choice for evening gowns, lingerie, and fine garments. Beyond fashion, silk finds applications in accessories, upholstery, and home textiles.

Silk also plays a crucial role in various other industries. In the realm of technology, silk's unique properties have been harnessed for medical sutures and advanced materials. Moreover, silk's biocompatibility has led to its use in regenerative medicine and drug delivery systems.

23.3.2.2 Wool

Wool, sourced primarily from sheep but also from other animals such as goats and alpacas, boasts remarkable insulating properties and versatility. The production of wool begins with shearing, a process where the animals' fleece is carefully removed to obtain the raw wool fibers. These fibers are then cleaned, carded, and spun into yarn (Ferreira et al., 2022; Kahlaoui et al., 2022; Stevulova et al., 2018). Wool's exceptional insulating capabilities are attributed to its natural crimp, which creates air pockets that trap warmth. This inherent warmth-retaining property has made wool a staple in cold-weather clothing. Wool garments, from cozy sweaters to durable outerwear, provide unmatched comfort in chilly climates. The applications of wool extend far beyond apparel. In-home textiles, wool blankets, and rugs are prized for their warmth and durability. Moreover, wool is a key player in the realm of technical textiles, where its moisture-wicking abilities and flame resistance are valued. Wool's sustainability is another notable aspect, with efforts to promote ethical and eco-friendly practices in the wool industry, including responsible shearing and animal welfare considerations (Michalska-Ciechanowska et al., 2020; Saad Azzem & Bellel, 2022).

Wool's versatility and timeless appeal have also placed it at the forefront of sustainable and ethical fashion practices. Consumers are increasingly drawn to wool products that align with their values of responsible and eco-conscious consumption.

23.3.2.3 Mohair and Angora

Mohair, sourced from Angora goats, and Angora wool, harvested from Angora rabbits, offer unique textures and characteristics that set them apart in the world of luxury fibers. These fibers are known for their exceptional softness and fine texture. The production of mohair involves breeding and caring for Angora goats, which produce soft, lustrous fibers. Angora wool, on the other hand, is obtained from Angora rabbits. The careful and humane harvesting of these fibers is essential to ensure their quality. Both mohair and Angora are cherished materials in high-end fashion and specialty products. Mohair's long, silky fibers are used in items like luxury knitwear, suits, and plush textiles. Angora wool, known for its extreme softness, is often used in cozy sweaters, scarves, and accessories. The appeal of these fibers lies in their unmatched texture and comfort.

However, ethical and sustainability considerations are paramount in the production of mohair and Angora. Concerns have arisen regarding the treatment of animals and the environmental impact of these industries. As a result, efforts have been made to promote ethical practices and ensure the well-being of animals in these industries.

23.3.3 Mineral-based fibers

Mineral-based fibers, although less common than plant-based or animal-based fibers, offer unique properties that make them valuable in specialized applications. In this section, we explore three noteworthy mineral-based fibers, each with distinct characteristics and uses:

23.3.3.1 Asbestos

Asbestos, once celebrated for its remarkable fire-resistant properties, has a complex history marred by serious health concerns. This mineral-based fiber has its geological origins in naturally occurring asbestos deposits, which are composed of long, thin mineral fibers. These fibers possess exceptional heat resistance and fireproofing qualities (Johny et al., 2023; Karabagias et al., 2019). In the past, asbestos was widely used in various industries, including construction, shipbuilding, and manufacturing, due to its fire-resistant nature. It was incorporated into building materials like insulation, roofing, and fireproof coatings. However, the inhalation of asbestos fibers has been linked to severe health issues, including lung cancer, asbestosis, and mesothelioma. This led to the decline of asbestos use and a global effort to mitigate exposure.

Despite its health risks, the legacy of asbestos remains in many older structures, particularly in buildings constructed before the recognition of its dangers. Proper asbestos abatement and removal procedures are essential when renovating or demolishing such buildings to prevent asbestos fiber release and protect workers' health.

23.3.3.2 Basalt

Basalt fibers are derived from volcanic rock, primarily basaltic magma, which solidifies and then undergoes a process of extraction and spinning. These fibers are known for their exceptional strength and resistance to high temperatures, making them valuable in specialized applications. Basalt fibers have gained attention in industries such as construction, aerospace, and automotive manufacturing. In construction, they are used as reinforcement in concrete and as an alternative to traditional steel reinforcement due to their resistance to corrosion and high strength. In the aerospace sector, basalt composites find use in lightweight components due to their high-temperature resistance. In the automotive industry, basalt fibers are explored for their potential to reduce vehicle weight and improve fuel efficiency. One notable feature of basalt fibers is their eco-friendliness. Unlike the energy-intensive process of producing synthetic fibers, basalt fiber production relies on naturally occurring volcanic rock. This aspect makes basalt fibers an attractive option for sustainable construction and manufacturing practices.

23.3.3.3 Glass

Glass fibers, formed from molten glass, are renowned for their strength, versatility, and resistance to corrosion. These fibers are created through a process called fiberization, where molten glass is extruded through fine nozzles, producing long, thin strands that are subsequently cooled and collected. Glass fibers have a wide range of applications across various industries. In composites, they are used as reinforcement in materials like fiberglass, which is used in the construction of boats, aircraft, and automotive parts (Bachtiar et al., 2019; Kaur et al., 2021). Glass fibers also find application in the telecommunications industry, where they are used to transmit data signals as optical fibers. Additionally, in the construction sector, glass wool is utilized for thermal and acoustic insulation. One of the defining characteristics of glass fibers is their high strength-to-weight ratio, making them suitable for lightweight but robust structures. They are resistant to chemicals and moisture, making them ideal for harsh environments.

23.4 Physicochemical properties of natural fibers

23.4.1 Tensile strength

Tensile strength serves as a pivotal physicochemical property within the realm of natural fibers, offering insight into their ability to endure stretching or pulling forces without succumbing to breakage. This characteristic stands as a linchpin in assessing the overall durability and suitability of natural fibers across a broad spectrum of applications. Natural fibers, sourced from diverse origins, showcase a remarkable range of tensile strengths, an outcome of their distinct structural compositions and sources (Li et al., 2007; Suárez et al., 2021; Torres-Arellano et al., 2020). For instance, plant-based fibers such as jute and flax exhibit notable tensile strength owing to their robust structural elements derived from the stalks and stems of plants. Jute, renowned for its exceptional strength, finds favor in applications requiring rugged materials, while flax fibers, employed in linen production, contribute to the longevity of linen textiles.

Conversely, cotton, another prominent plant-based fiber, tends to display lower tensile strength when compared to fibers such as jute and flax. However, cotton's allure lies in its unmatched comfort, breathability, and softness. While cotton may not excel in scenarios necessitating high tensile strength, it remains a preferred choice in textiles where comfort and moisture management are paramount, such as in clothing and bed linens. Numerous factors influence tensile strength, encompassing fiber diameter, length, crystallinity, and the presence of defects or impurities within the fiber structure. Longer and finer fibers often exhibit superior tensile strength due to their heightened resistance to breakage (Ku et al., 2011; Malenab et al., 2017; Suriani et al., 2021). Additionally, the crystalline regions within a fiber can bolster its strength, with increased crystallinity typically correlating with enhanced tensile strength. In the textile industry, particularly in applications that subject natural fibers to mechanical stress, such as rope, fabric, and geotextile production, tensile strength assumes a position of paramount importance. Natural fibers boasting adequate tensile strength are prioritized for these applications to ensure the longevity and reliability of the final product. Consequently, comprehending the tensile strength of natural fibers empowers manufacturers and designers to make informed choices, striking a balance between strength and other desirable attributes such as comfort, breathability, and eco-friendliness. For instance, Suárez et al. (2021) presented the results of a study on the characterization of the giant reed (*Arundo donax* L.) plant and its fibers. The research was part of a project conducted in the Macaronesia region, aiming to demonstrate the feasibility of using biomass from invasive plant species in the composite materials sector to finance control campaigns and habitat conservation efforts. An experimental procedure was developed to extract fiber bundles from the giant reed plant, and the resulting material was characterized in terms of chemical composition, thermogravimetry, and infrared spectra to assess its potential application in the production of polymeric composite materials. This approach aims to valorize the residual biomass from this invasive species in Macaronesia. Thermoplastic matrix composites with fiber contents of up to 40 wt.% were manufactured, and their mechanical properties under tensile, flexural, and impact loading were determined. No prior references on the preparation of composite materials with polyolefin matrices and giant reed fibers were found. In terms of flexural loads, the addition of fibers negatively impacts the ultimate tensile strength, as illustrated in Fig. 23.3. There is a gradual reduction, with a decrease of up to 48% observed for polyethylene (PE) and 42% for polypropylene (PP) composites reinforced with treated fibers. Conversely, concerning the tensile elastic modulus, it demonstrates an upward trend with increasing content of giant reed fibers in both types of polymeric matrices. The highest modulus values are achieved when untreated fibers make up 40% of the composite by weight, surpassing the neat PE matrix by approximately 50% and the PP matrix by around 38%. Specifically, the modulus increases from 605 to 902 MPa for PE and from 857 to 1186 MPa for PP.

Torres-Arellano et al. (2020) focused on the production and mechanical characterization of biobased epoxy resins reinforced with natural fibers. Biolaminates, which are composed of low-density, biodegradable, and lightweight materials, hold great appeal for various industries. In this research, ixtle, henequen, and jute natural fibers were utilized as reinforcing fabrics for two biobased epoxy resins sourced from Sicomin. The manufacturing of these biolaminates was executed through the vacuum-assisted resin infusion process. Fig. 23.4 illustrates the arrangement used for both tensile and bending tests.

Mechanical testing revealed that Jute-reinforced biolaminates exhibit the highest stiffness and strength, while henequen-reinforced biolaminates demonstrate notable strain properties. Similar to the analysis of failures observed during tensile testing, we conducted a comprehensive examination of biolaminates following flexural tests to identify their primary failure modes. This evaluation aimed to provide insights into the modes of failure that characterize these materials. Typically, during flexural tests, we observed deflection occurring at the midpoint of the biolaminates, where they were in contact with the machine head, effectively serving as the loaded

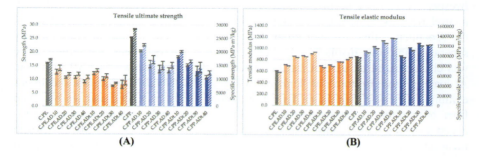

Figure 23.3 Depiction of the tensile properties (represented by solid colors) and specific tensile properties (indicated by lines) for *Arundo donax* composites. Specifically, it displays (A) ultimate strength and (B) elastic modulus.
Source: Adapted from Suárez, L., Castellano, J., Romero, F., Marrero, M.D., Benítez, A.N., & Ortega, Z. (2021). Environmental hazards of giant reed (*Arundo donax* L.) in the macaronesia region and its characterisation as a potential source for the production of natural fibre composites. *Polymers*, *13*(13). https://doi.org/10.3390/polym13132101 (under CCBY 4.0).

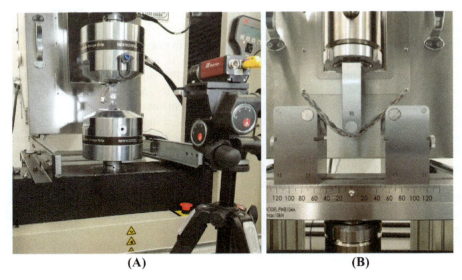

Figure 23.4 Images of the mechanical testing configuration for biolaminates, including (A) the setup for the tensile test with a video extensometer and (B) the setup for the bending test. *Source*: Adapted from Torres-Arellano, M., Renteria-Rodríguez, V., & Franco-Urquiza, E. (2020). Mechanical properties of natural-fiber-reinforced biobased epoxy resins manufactured by resin infusion process. *Polymers*, *12*(12), 1−17. https://doi.org/10.3390/polym12122841 (under CCBY 4.0).

edge. This loading condition induces maximum bending stress, leading to deflection and potential fracture of the outer fibers of the laminate under excessive loading. It is worth noting that in our experiments, the fibers did not undergo breakage. Irrespective of the type of biobased epoxy resin used, whether EVO or GP, each natural fiber examined in our study exhibited distinct failure modes. In all cases, the specimens failed through progressive deflection, and we did not observe any separation into two distinct beams. For the henequen specimens, the failure mode was characterized by sudden fracture, specifically a type known as TAB (Tension, At loading nose, Bottom), as depicted in Fig. 23.5A. Notably, henequen specimens exhibited the most significant deflection among the three fibers, and we observed fiber bifurcation at the loading point. Similarly, the ixtle specimens also experienced a sudden fracture mode, specifically TAB, as shown in Fig. 23.5B. It is worth mentioning that the resin underwent a noticeable color change, displaying localized white marks at the loading point. This particular failure mode is common in plain wave fabric with equilibrated wrap and weft tows, similar to the ixtle specimens in our study. Lastly, the jute specimens exhibited a failure mode characterized by sudden fracture (TAB), as depicted in the Fig. 23.5. In this case, the deflection was the lowest among the three fibers, which is a common characteristic of fabrics with high closure factors. These failure modes align with the typical codes outlined in ASTM D7264.

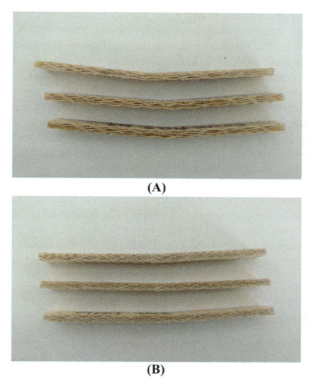

Figure 23.5 Illustration of the failure modes observed in the biolaminates following bending tests, including (A) Henequen, (B) Ixtle.
Source: From Torres-Arellano, M., Renteria-Rodríguez, V., & Franco-Urquiza, E. (2020). Mechanical properties of natural-fiber-reinforced biobased epoxy resins manufactured by resin infusion process. *Polymers*, *12*(12), 1−17. https://doi.org/10.3390/polym12122841.

Milosevic et al. (2017) assessed the performance of fused deposition modeling (FDM) using composites of natural fibers and recycled PP. Composite filaments were produced by extruding preconsumer recycled PP with varying proportions of hemp or harakeke fibers. Tensile test specimens were fabricated via FDM using these filaments. Comparative analysis was conducted with plain PP samples. The investigation revealed that the ultimate tensile strength and Young's modulus of reinforced filaments increased by over 50% and 143%, respectively, when incorporating 30 wt.% hemp or harakeke compared to pure PP filament. However, the same degree of improvement was not consistently observed in the FDM test specimens, as some compositions exhibited properties lower than unfilled PP. Scanning electron microscopy (SEM) analysis of fracture surfaces indicated uniform fiber dispersion and reasonable fiber alignment, yet it also revealed porosity and fiber pull-out. The study highlights the benefits of fiber reinforcement in terms of dimensional stability during extrusion and FDM, which is crucial for FDM implementation. Furthermore, Wang et al. (2019) focused on the modification of jute fibers through chemical treatments, including acid pretreatment, alkali pretreatment, and scouring. The research analyzes the mechanical properties, surface morphology, and Fourier transform infrared spectra of both treated and untreated jute fibers to investigate the impact of chemical modifications on the fibers. Subsequently, jute fiber/epoxy composites with unidirectional jute fiber alignment were prepared. Key composite properties such as void fraction, tensile strength, initial modulus, and elongation at break were examined. SEM images of fractured specimens demonstrated improved interfacial adhesion in treated fibers. The initial analysis, as shown in Fig. 23.6A, reveals a lack of cohesion between the fibers and the matrix, leading to a failure mechanism characterized by fiber-matrix debonding. This indicates a deficiency in the attraction between jute fibers and epoxy, highlighting the need for enhanced fiber/epoxy interface bonding. Following chemical treatment, notable improvements are observed in the adhesion interface between jute fibers and epoxy. The boundary gaps between the epoxy and fiber surface disappear, as indicated by the arrow in Fig. 23.6B. These enhancements are attributed to the development of a strong interfacial interaction subsequent to the jute fiber treatment.

23.4.2 Moisture absorption

Moisture absorption is a pivotal property of natural fibers that govern their interaction with water and humidity (Chandrasekar et al., 2017; Manaila et al., 2020; Ramadevi et al., 2012). This characteristic has far-reaching implications, influencing the comfort, dimensional stability, and resistance to microbial growth of

Figure 23.6 Depiction of the scanning electron microscopy images of fractured surfaces, showcasing (A) untreated jute and (B) jute composite after treatment.
Source: Adapted from Milosevic, M., Stoof, D., & Pickering, K. (2017). characterizing the mechanical properties of fused deposition modelling natural fiber recycled polypropylene composites. *Journal of Composites Science*, *1*(1). https://doi.org/10.3390/jcs1010007 (under CCBY 4.0).

materials made from natural fibers. Plant-based fibers, exemplified by cotton and jute, exhibit a relatively high moisture absorption capacity. This inherent quality allows these fibers to effectively wick moisture away from the body, rendering textiles made from them exceptionally comfortable to wear in hot and humid conditions. In warm weather, cotton garments, for instance, feel cool and refreshing against the skin due to their capacity to absorb perspiration and aid in its evaporation. However, it is important to note that the high moisture absorption of plant-based fibers can also be a drawback, as it may lead to fabric shrinkage when exposed to excessive moisture or improper laundering methods (Al-Maharma & Al-Huniti, 2019; Panthapulakkal & Sain, 2007).

On the other hand, animal-based fibers, such as wool, showcase a distinct ability to absorb moisture without feeling damp to the touch. This remarkable property contributes to the comfort of wool garments in a wide range of temperatures, as wool can absorb moisture vapor from the body without creating a damp or clingy sensation. This quality also aids in the regulation of body temperature, as wool acts as a moisture buffer, absorbing excess perspiration when the body is warm and releasing it when the environment cools down. Additionally, wool's capacity for moisture absorption plays a pivotal role in its natural resistance to odors, as it can effectively trap and neutralize odor molecules within its fibers. For instance, in an experiment, Manaila et al. (2020) focused on the development of natural rubber composites reinforced with hemp, flax, and wood sawdust using an electron beam irradiation process at room temperature. The aim is to create environmentally friendly materials that are safe for human health. Different composites with varying fiber content (ranging from 5 to 20 phr) and irradiation doses (ranging from 75 to 600 kGy) were produced. The research investigates the kinetics of water absorption in these materials, utilizing Fick's law for analyzing water diffusion. The findings revealed that the water absorption behavior was influenced by the type and quantity of fibers used in the composites and the irradiation dose. Interestingly, the study found that the relationship between water absorption and the cellulose and hemicellulose content varied depending on the type of fiber used, due to the effects of electron beam irradiation.

Stevulova et al. (2015) investigated the water absorption behavior of composites made from hemp hurds and an inorganic binder, which were allowed to harden for 28 days. Two types of water absorption tests were conducted on cube-shaped specimens dried beforehand, using deionized water at laboratory temperature. The tests consisted of short-term (one-hour water immersion) and long-term (up to 180 days) assessments to evaluate the composites' durability. The short-term water sorption properties of the original hemp hurd composites were influenced by the mean particle length of the hemp and the nature of the binder. A comparative analysis of long-term water sorption was performed on composites reinforced with both original and chemically modified hemp hurds in three different reagents, demonstrating that surface treatments of the filler had an impact on the sorption process. By analyzing the sorption curves and applying a model for composites containing natural fibers, it was observed that the diffusion of water molecules within these composites, whether reinforced with original or chemically modified hemp hurds, exhibited anomalous behavior rather than adhering to Fickian principles. The most significant reduction in hydrophilicity of hemp hurds was found in the case of hemp hurds modified with NaOH. This reduction was linked to changes in the chemical composition of hemp hurds, particularly a decrease in the average degree of cellulose polymerization and hemicellulose content.

Atmakuri et al. (2021) focused on investigating the mechanical and moisture-related properties of hybrid composites reinforced with Caryota and sisal fibers in an epoxy resin matrix. The primary objective is to explore the advantages of hybrid composites over those reinforced with a single type of fiber. The fabrication of these composites involved the hand lay-up technique, resulting in five different samples based on hybridization rules (40C/0S, 25C/15S, 20C/20S, 15C/25S, 0C/40S). Mechanical testing and moisture absorption studies were conducted, and the morphology was examined through SEM analysis. Fig. 23.7 illustrates the percentage weight gain by the composites over time. It was observed that all composite samples absorbed moisture when exposed to it. The weight gain increased with time, eventually reaching a constant value. Hybrid composites displayed intermediate results in terms of weight gain, while single fiber composites exhibited both the lowest and highest weight gain percentages. This difference can be attributed to the presence of voids (porosity) in the composites and the chemical composition (cellulose content) of the fibers. Cellulose in the fibers reacts with moisture, forming hydrogen-oxygen bonds. Sisal fibers, having a higher cellulose content than Caryota, absorbed more water.

SEM analysis confirmed the presence of voids in the composites, with hybrid composites showing intermediate void content compared to single-fiber composites. In Fig. 23.8, we can observe the fracture surface of the flexural test composites in relation to their hybridization. The results display a brittle fracture in both the reinforcement material and the matrix material. This brittleness can be explained by the strong interfacial bonds formed between the fibers and the epoxy resin. As depicted

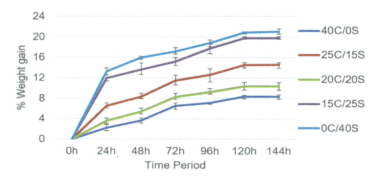

Figure 23.7 Depiction of the percentage increase in weight of the composites over time. *Source*: Adapted from Atmakuri, A., Palevicius, A., Kolli, L., Vilkauskas, A., & Janusas, G. (2021). Development and analysis of mechanical properties of caryota and sisal natural fibers reinforced epoxy hybrid composites. *Polymers*, *13*(6). https://doi.org/10.3390/polym13060864 (under CCBY 4.0).

Sources of natural fibers and their physicochemical properties for textile uses 587

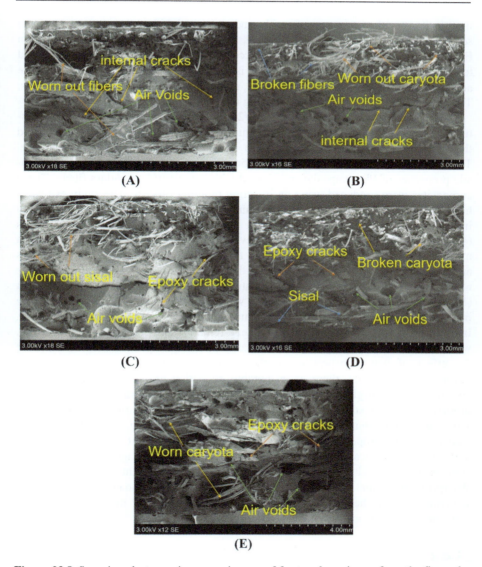

Figure 23.8 Scanning electron microscopy images of fractured specimens from the flexural tests, corresponding to different hybrid compositions: (A) 0C/40S, (B) 15C/25S, (C) 20C/20S, (D) 25C/15S, and (E) 40C/0S.
Source: Adapted Agustini, N. K. A., Triwiyono, A., Sulistyo, D., & Suyitno, S. (2021). Mechanical properties and thermal conductivity of fly ash-based geopolymer foams with polypropylene fibers. *Applied Sciences (Switzerland)*, *11*(11). https://doi.org/10.3390/app11114886 (under CCBY 4.0).

in the figures, the presence of voids and cracks significantly compromises the strength of these composites. Notably, single fiber composites (as seen in Fig. 23.8A and 23.8E) exhibit a higher occurrence of cracks and air gaps compared to hybrid composites. This discrepancy is primarily attributed to the inadequate adhesion between the fibers and the resin, resulting in poor bonding characteristics. The voids attract water molecules, leading to a higher percentage of weight gain.

Additionally, Panthapulakkal & Sain (2007) focused on investigating the moisture absorption characteristics of short hemp fiber and hybrid hemp-glass-reinforced thermoplastic composites, specifically to assess their suitability for outdoor use. The research delves into the water absorption properties and their impact on the tensile properties of PP composites reinforced with hemp and hemp/glass fibers. These composites were produced using an injection molding process. The study evaluates how hybridization affects water uptake and the kinetics of moisture absorption by immersing hybrid composite samples in distilled water at different temperatures (40°C, 60°C, and 80°C). The findings reveal that the composites follow a Fickian mode of diffusion, but deviations occur at higher temperatures. These deviations are attributed to microcracks forming at the interface and the dissolution of lower molecular weight substances from the natural fibers. Equilibrium moisture content (Mm) is highest in composites with 40 wt.% hemp fiber reinforcement, and the incorporation of glass fibers significantly reduces water uptake (by 40%). Interestingly, equilibrium moisture content remains independent of temperature, while the diffusion coefficient (D) increases with higher temperatures. Analyzing the effect of water absorption on the tensile properties of the composites, it becomes evident that there is a notable reduction in strength and stiffness. Hybridization with glass fibers does not have a significant impact on the strength properties of the aged samples. The tensile properties of the re-dried aged samples show improved retention of strength properties after drying, although complete recovery of the properties is not achieved. This suggests that water absorption is not merely a physical process but results in permanent damage to the composite after aging. In conclusion, this study provides valuable insights into the moisture absorption behavior of hemp and hemp-glass hybrid-reinforced PP composites, shedding light on their potential for outdoor applications and the factors affecting their long-term performance.

23.4.3 Thermal conductivity

Thermal conductivity is a fundamental property within the domain of natural fibers, exerting a significant influence on the comfort and thermal insulation characteristics of textiles fashioned from these fibers. This property fundamentally defines a fiber's ability to conduct heat, ultimately determining how textiles interact with the surrounding environment and the wearer's body, ensuring comfort and functionality. Natural fibers, stemming from diverse sources, encompass a wide gamut of thermal conductivity values, thereby dictating their suitability for distinct climatic conditions and purposes (Espert et al., 2004; Yang et al., 2010). This property emerges as a pivotal consideration in the selection and design of textiles, guaranteeing that they provide an optimal balance of comfort and utility. Plant-based fibers, typified by cotton and linen, typically exhibit a relatively high thermal conductivity. This quality renders them well-suited for warm-weather attire, facilitating efficient heat dissipation from the body. In hot climates, textiles crafted from these fibers excel at keeping the wearer cool and dry by dissipating excess body heat, thereby enhancing comfort. Essentially, plant-based fibers function as natural coolants for clothing, making them ideal choices for summer garments. However, their high thermal conductivity can be less advantageous in retaining warmth in colder environments (Athijayamani et al., 2009).

Conversely, animal-based fibers such as wool, mohair, and silk typically display lower thermal conductivity. This characteristic equips these fibers with exceptional insulating properties, especially in chilly conditions. Wool, in particular, shines in this regard, effectively trapping warm air close to the body, creating a natural thermal barrier against the cold (Lakshmaiya et al., 2022; Ramamoorthy et al., 2012). This unique feature positions wool as a premier choice for winter clothing, ensuring the wearer remains warm and snug even in frigid weather. In essence, thermal conductivity stands as a pivotal determinant in the thermal performance of textiles made from natural fibers. High thermal conductivity in plant-based fibers makes them an excellent choice for hot weather attire, promoting cooling and comfort, while the lower thermal conductivity of animal-based fibers like wool renders them superior insulators, providing warmth and protection in cold climates. The adaptability of natural fibers, with their varying thermal properties, empowers the creation of textiles tailored to a wide array of environmental conditions and seasonal requirements, all while ensuring comfort and safeguarding the wearer from the elements. In an experiment, Sair et al. (2017) delved into the impact of surface modifications, specifically alkali and silane treatments, on the morphological, mechanical, and thermal conductivity properties of hemp fibers. Hemp fibers underwent treatment with sodium hydroxide solutions at varying concentrations, followed by an analysis involving techniques such as infrared Fourier transform spectroscopy, X-ray diffraction (XRD), mechanical tensile testing, and SEM. Subsequently, the alkali-treated fibers underwent silane treatment and were subjected to the same analytical methods. The results of the analysis revealed significant changes induced by these treatments. Notably, the treatments led to the removal of a certain amount of lignin, wax, and oils that typically span the cell wall of the fibers. Additionally, there was a transformation of some alpha-cellulose to beta-cellulose. These alterations in the fiber's composition were responsible for considerable improvements in tensile strength and Young's modulus, with increases of approximately 39% and 23%, respectively, observed when the treatment concentration reached 8%. Furthermore, the modifications resulted in a substantial enhancement of the interfacial adhesion between the fibers and the polyurethane composite. The adhesion strength increased significantly, from 1.26 MPa for untreated fibers to 3.18 MPa for alkali-treated fibers and further to 5.16 MPa for silane-treated fibers. Moreover, thermal conductivity measurements demonstrated that the chemical treatments positively affected the fibers' thermal conductivity. Specifically, the thermal conductivity of untreated fibers was measured at 42.22 mW/m°C, while alkali-treated fibers at 8% concentration exhibited an increased thermal conductivity of 47.92 mW/m°C. Similarly, Nayab-Ul-Hossain et al. (2022) evaluated a multifunctional composite material based on graphene, natural jute fibers, and synthetic glass fibers. Traditionally, composites have relied on either natural or synthetic fibers as reinforcing agents. However, this approach often results in suboptimal mechanical and interfacial characteristics due to factors such as random fiber orientation and weak fiber-matrix interfaces. To address these limitations, the study explores the combined use of natural jute fibers and synthetic glass fibers within a graphene-based composite, with the aim of improving mechanical properties, as well as enhancing electrical and thermal conductivity. The introduction of both jute and glass fibers in the composite brings about significant enhancements, particularly in mechanical properties. The study reveals that the composite, when prepared with reduced graphene oxide (rGO), exhibited notable improvements compared to its counterpart without rGO. Specifically, tensile strength increased by 146%, flexural strength by 122%, impact energy by 144%, thermal conductivity by 108.33%, and electrical conductivity by 127.21%. These substantial enhancements in physicoelectrical properties position the composite as a promising material for applications requiring a combination of strength (maximum tensile strength of 118.20 MPa), electrical conductivity (maximum achievable current of 16.7 μA), and thermal conductivity. This study represents a departure from traditional composite manufacturing methods by embracing multifunctional properties, rather than focusing solely on specific attributes. The composite, formulated with a blend of natural jute and synthetic glass fibers along with rGO, holds the potential to usher in the next generation of intelligent, robust, and environmentally friendly multifiber composites for high-performance engineering applications.

Annie Paul et al. (2008) investigated the impact of varying banana fiber loading and different chemical treatments on the thermophysical properties of commingled composite materials consisting of PP and banana fibers. The thermophysical properties studied include thermal conductivity, thermal diffusivity, and specific heat, all assessed at room temperature. The research employs a periodic method to evaluate these properties and seeks to understand how they are influenced by changes in fiber content and the application of distinct chemical treatments to banana fibers. One notable observation from this study is that as the banana fiber loading increases in the composites, both thermal conductivity and thermal diffusivity tend to decrease. However, there is no significant alteration in specific heat with varying fiber content. Another critical finding is that the application of chemical treatments to the banana fibers results in an overall improvement in the thermophysical properties of the composites, regardless of the specific treatment employed. Among the various treatments explored, the composites containing benzoylated fibers exhibit the highest values of thermal conductivity and thermal diffusivity. Additionally, the concentration of NaOH (sodium hydroxide) is shown to influence the thermophysical properties, with composites treated with 10% NaOH displaying superior characteristics compared to those treated with 2% NaOH. To gain further insights, the study employs a theoretical series conduction model to estimate the transverse thermal conductivity of untreated banana fiber composites, arriving at a value of approximately 0.1166 $\pm$ 0.0001 Wm/K.

Agustini et al. (2021) delved into the impact of incorporating PP fibers on the mechanical properties and thermal conductivity of geopolymer foams derived from fly ash (FA). Specifically, Class C Fly ash was employed as the binding material, and the geopolymer binder was activated using a combination of sodium silicate (SS) and sodium hydroxide (SH). The foams were fabricated through mechanical means, involving the mixing of a foaming agent with distilled water under high pressure. These foams were subsequently introduced into the geopolymer mixture at volumes of 40% and 60%. Different proportions of PP fibers, ranging from 0%, 0.25%, to 0.50% by weight of FA, were incorporated into the foamed geopolymer. Figs. 23.9 and 23.10 provide a detailed insight into the influence of PP fibers on the thermal conductivity of foamed geopolymer. Interestingly, two different trends in thermal conductivity emerge between 40% and 60% foamed geopolymer, despite the similar PP fiber content. In Fig. 23.9, when examining 40% foamed geopolymer containing 0.25% PP fiber at a temperature of 50°C, a substantial increase of almost 123% in thermal conductivity is observed compared to foamed geopolymer without reinforcement. However, this value decreases notably when the temperature rises to 100°C, reaching its lowest thermal conductivity as the temperature further increases to 200°C. It is noteworthy that although there is only a moderate rise of about 45% at 50°C when compared to pure foamed geopolymer, the thermal conductivity of 40% foamed geopolymer with 0.5% PP fiber attains the highest thermal conductivity within the temperature range of 100°C–200°C. In contrast, Fig. 23.10 illustrates

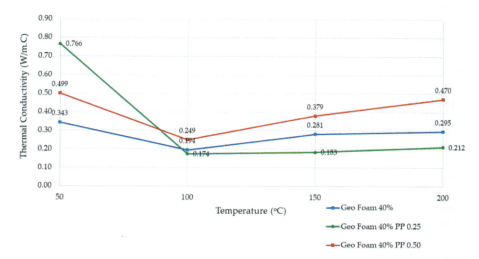

Figure 23.9 The thermal conductivity of foamed geopolymer at a 40% concentration level with different proportions of polypropylene fiber.
Source: Adapted from Agustini, N.K.A., Triwiyono, A., Sulistyo, D., & Suyitno, S. (2021). Mechanical properties and thermal conductivity of fly ash-based geopolymer foams with polypropylene fibers. *Applied Sciences (Switzerland), 11*(11). https://doi.org/10.3390/app11114886 (under CCBY 4.0).

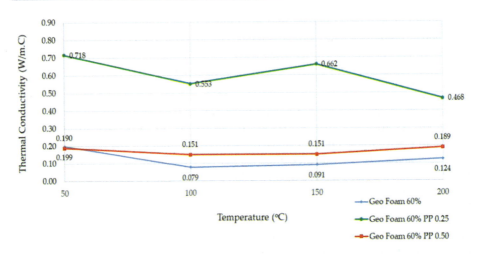

Figure 23.10 The thermal conductivity of foamed geopolymer at a 60% concentration level with different proportions of polypropylene fiber.
Source: Adapted from Agustini, N. K. A., Triwiyono, A., Sulistyo, D., & Suyitno, S. (2021). Mechanical properties and thermal conductivity of fly ash-based geopolymer foams with polypropylene fibers. *Applied Sciences (Switzerland)*, *11*(11). https://doi.org/10.3390/app11114886 (under CCBY 4.0).

that the presence of 0.25% PP fiber results in the highest thermal conductivity throughout the temperature range for 60% foamed geopolymer, exhibiting an impressive increase of up to 277%. However, the incorporation of 0.5% PP fiber in this scenario leads to only a modest increase, approximately 4.7%, compared to pure foamed geopolymer.

Agirgan et al. (2022) investigated the thermal conductivity and sound absorption properties of biocomposite materials made from rice straw fibers and polylactic acid (PLA). Agricultural waste materials such as rice straw fibers have gained popularity in recent years as reinforcements for environmentally friendly composite materials. In a previous study, we produced a nonwoven fabric using rice straw fibers. In this research, we utilized this nonwoven reinforcement material and PLA in granule form to create single- and double-layered biocomposite materials through a hot press process. The results indicate that the double-layered composite material exhibits the highest sound absorption coefficient value, nearly 0.99, while the single-layer material registers approximately 0.34. In terms of thermal conductivity, the single-layer composite material exhibits a value of 0.02564 W/m · K, whereas the nonwoven material produced from rice straw has a thermal conductivity of 0.01618 W/m · K. This suggests that the thermal conductivity of the single-layer composite material is roughly 50% higher than that of the nonwoven material. Consequently, the single-layer composite material may find application in industrial insulation settings, as its lower thermal conductivity coefficient makes it a suitable choice for thermal insulation purposes.

23.4.4 Chemical reactivity

The chemical reactivity of natural fibers is a pivotal property that significantly influences their susceptibility to chemical degradation, staining, and dyeing characteristics. Different types of natural fibers exhibit varying degrees of chemical reactivity, which, in turn, shapes their utility and adaptability in various applications (Bhuiyan et al., 2017; Durkin et al., 2016; Gurunathan et al., 2015; Li et al., 2007). Plant-based fibers, such as cotton, are known for their relatively high susceptibility to chemical degradation by acids and alkaline substances. This chemical reactivity makes them vulnerable to damage when exposed to harsh chemical environments. For instance, acids can weaken the cellulose structure of cotton fibers, leading to a loss of strength and integrity. Alkaline substances can also have a detrimental impact on cotton, potentially causing fiber degradation and reducing the fabric's lifespan. However, one notable advantage of plant-based fibers such as cotton is their excellent dyeability. They readily absorb and retain dyes, making them highly receptive to a wide spectrum of colors. This attribute is why cotton textiles come in an extensive range of vibrant and diverse hues. The combination of chemical reactivity and dyeability makes cotton a versatile choice for colorful and esthetically pleasing textiles.

Animal-based fibers, including silk and wool, possess distinct chemical properties that differentiate them from plant-based fibers. Silk, for instance, exhibits relative resistance to acidic substances, making it less susceptible to damage by acids compared to cotton. However, silk can be damaged by strong alkaline solutions, which can degrade the silk protein structure and weaken the fibers. Wool, on the other hand, displays a higher level of chemical reactivity. It is reactive with both acids and alkaline substances, making it prone to damage if not treated properly. Wool fibers can undergo a process known as felting when exposed to alkaline solutions and mechanical agitation, leading to the matting and shrinkage of the fabric (Nair & Joseph, 2014; Rong et al., 2001; Tanasă et al., 2020). However, this very reactivity also contributes to wool's unique dyeing properties. Wool readily accepts dyes, resulting in rich, vibrant, and lustrous colors that are highly sought after in high-end fashion and textiles. In summary, the chemical reactivity of natural fibers plays a crucial role in their performance and applications. Plant-based fibers such as cotton are susceptible to chemical degradation but excel in dyeability, offering a broad color palette for textiles. Animal-based fibers such as silk and wool exhibit distinct chemical properties, with silk being more resistant to acids and wool displaying a higher reactivity, but both contribute to the creation of textiles with vibrant and distinctive colors. Understanding the chemical reactivity of natural fibers is essential for their proper care, maintenance, and utilization in various industries and applications. For instance, in the experiment, Shahinur et al. (2020) delved into the impact of chemical treatments on the thermal properties of jute fibers, which are increasingly gaining attention in the realm of fiber-reinforced polymer composites. Chemical treatments are commonly applied to raw fibers to enhance the properties of resulting composites. From a composite manufacturing perspective, comprehending how these treatments influence the thermal characteristics of jute fibers is of paramount importance. The research specifically explores the effects of three distinct treatments: rot-retardant, fire-retardant, and water-retardant treatments, on the thermal properties of jute fibers. The study employed fiber samples sourced from the central region of whole jute fibers. The results unveiled intriguing insights into the thermal properties of chemically treated jute fibers. In the case of fire-retardant-treated jute fibers, there was a notable decrease in the thermal decomposition temperature compared to untreated fibers. However, these treated fibers left behind a higher residue when subjected to temperatures exceeding 400°C, a characteristic distinct from raw fibers and those subjected to other treatments. Fig. 23.11A–D presents SEM depicting both untreated and chemically modified jute fibers. A distinct contrast is evident, with the untreated jute fibers displaying a noticeably rough surface. Conversely, the figures reveal that after undergoing chemical treatment, the jute fibers exhibited a smoother surface across all three treatment scenarios. However, it is worth noting that specific chemical modifications might result in an opposite effect, potentially rendering the fiber surface rougher. This observation suggests that chemical interactions with the fiber surface may lead to the formation of a coating, contributing to these smoother surface textures.

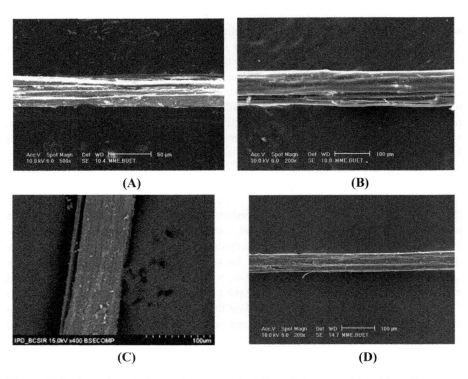

Figure 23.11 Scanning electron microscopy depicting. (A) untreated (raw) jute fibers, as well as jute fibers subjected to various chemical treatments: (B) RT (Rot-retardant Treatment), (C) FT (Fire-retardant Treatment), and (D) WT (Water-retardant Treatment). *Source*: From Shahinur, S., Hasan, M., Ahsan, Q., & Haider, J. (2020). Effect of chemical treatment on thermal properties of jute fiber used in polymer composites. *Journal of Composites Science*, *4*(3). https://doi.org/10.3390/jcs4030132.

Overall, chemically treated fibers exhibited lower heat absorption capabilities when compared to raw jute fibers. Additionally, the heat flow became negative for all treated fibers. This study holds significance as it furnishes crucial information regarding the thermal properties of chemically treated jute fibers, thus contributing to the knowledge base essential for the production of polymer-based composite materials. Understanding these properties is pivotal for tailoring jute-reinforced composites for specific applications, particularly those where thermal performance is a critical factor.

Similarly, Manimaran et al. (2019) explored the extraction and characterization of cellulosic fibers derived from the red banana peduncle (RBP) plant, aiming to assess their potential as a reinforcement material in natural fiber—reinforced polymer composites (NRPCs). NRPCs have garnered attention as a sustainable alternative to synthetic fibers due to their cost-effectiveness and environmentally friendly attributes. The chemical composition, physical properties, structural attributes, thermal behavior, and tensile characteristics of the extracted RBP fibers (RBPFs) were thoroughly investigated. RBPFs exhibited notable specific strength and binding properties, primarily attributed to their lightweight nature and composition, consisting of a substantial cellulose content (72.9 wt.%), low lignin content (10.01 wt.%), and minimal wax content (0.32 wt.%). Analysis using XRD and Fourier transform infrared spectroscopy (FTIR) revealed a significant cellulose content and a crystallinity index of 72.3%. The physical attributes of RBPFs included a density of approximately 0.896 g/cm^3 and a fiber diameter ranging from 15 to 250 μm. Furthermore, RBPFs exhibited thermal stability up to 230°C. These findings suggest that RBPFs possess promising properties for utilization as a natural reinforcement material in the development of biocomposites for various potential applications.

Rajesh and Pitchaimani (2017) investigated the mechanical properties of intraply hybrid banana/jute woven fabric composites and examined how various surface treatments (alkali, potassium permanganate, benzoyl chloride, and silane) affect their mechanical, dynamic mechanical, and free vibration characteristics. Intraply woven fabrics are created by employing banana yarn in the weft direction and jute yarn in the warp direction, forming a basket-type woven fabric. The results of the study indicate that the impact of chemical treatments on the tensile and flexural strengths of the composite is relatively minor, with the exception of benzoyl chloride treatment. However, these treatments notably enhance the impact strength when compared to untreated composites. Specifically, the tensile, flexural, and impact strengths experience a significant improvement of 10%, 30%, and 50%, respectively, with benzoyl chloride treatment in contrast to untreated composites. Fourier transform infrared spectra analysis reveals that benzoyl chloride treatment effectively removes hemicellulose and lignin content. Furthermore, both benzoyl chloride and alkali treatments lead to improved dynamic mechanical characteristics. Experimental modal analysis of intraply fabric composites shows a substantial increase in natural frequency with benzoyl chloride treatment compared to untreated composites. These findings highlight the potential for benzoyl chloride treatment as a means to enhance the mechanical and dynamic properties of intraply fabric composites.

Malenab et al. (2017) explored the potential of utilizing waste abaca (Manila hemp) fibers as reinforcement in geopolymer composites, a type of inorganic aluminosilicate material, with the aim of producing eco-friendly materials. Natural fibers are increasingly appealing for such applications due to their favorable characteristics such as cost-effectiveness, low density, and impressive mechanical properties. In this study, waste abaca fibers underwent various chemical treatments to enhance their surface properties and promote better adhesion with the FA -based geopolymer matrix. To investigate the impact of these chemical treatments on the tensile strength of the fibers, a definitive screening design of experiments was employed. The factors considered included: (Bhuiyan et al., 2017) NaOH pretreatment, (Durkin et al., 2016) immersion time in an aluminum salt solution, and (Gurunathan et al., 2015) the final pH of the slurry. The findings revealed that abaca fibers without alkali pretreatment, soaked for 12 hours in $Al_2(SO_4)_3$ solution, and adjusted to pH 6, exhibited the highest tensile strength among the treated fibers. The study results also validated that the chemical treatments effectively removed lignin, pectin, and hemicellulose from the fibers, resulting in a rougher surface with the deposition of aluminum compounds. This enhanced the interfacial bonding between the geopolymer matrix and the abaca fibers, while the geopolymer acted as a protective shield, safeguarding the treated fibers against thermal degradation. In Fig. 23.12, the energy dispersive X-ray spectroscopy (EDS) analysis on the surface of alkali-treated abaca fibers was conducted. A specific region of interest was chosen to measure the elemental composition of this surface. It is important to acknowledge that the quantification error could be relatively high because the sample volume for EDS analysis is quite small, typically on the order of a few micrometers in diameter. Nonetheless, the EDS results provide valuable indications regarding the presence of compounds with known chemical structures. For instance, in Fig. 23.12A, the measured carbon-to-oxygen (C/O) mass ratio is approximately 1.3. Comparatively, the theoretical C/O ratio for cellulose is 0.9, while for lignin, it

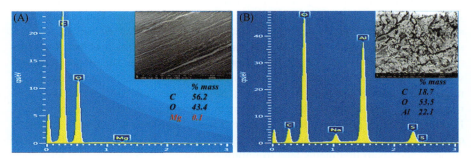

Figure 23.12 Energy dispersive X-ray spectroscopy spectra of the following samples: (A) Abaca treated with NaOH, and (B) Abaca treated with Al2(SO4)3.
Source: Adapted from Malenab, R.A.J., Ngo, J.P.S., & Promentilla, M.A.B. (2017). Chemical treatment of waste abaca for natural fiber-reinforced geopolymer composite. *Materials*, *10*(6). https://doi.org/10.3390/ma10060579 (under CCBY 4.0).

is 2.9. This observation suggests that, following alkali treatment, most of the lignin and hemicellulose components have been effectively removed from the abaca fiber. Furthermore, computed O/Al values in regions like the one depicted in Fig. 23.12B range from 2.8 to 3.2. This research underscores the potential for waste abaca fibers to serve as a sustainable reinforcing agent in geopolymer composites, furthering the development of environmentally friendly materials.

23.5 Applications of natural fibers in textiles

Natural fibers have a rich history and a wide range of applications in the textile industry, thanks to their unique properties, sustainability, and versatility. They find uses in various sectors, from apparel and fashion to technical and automotive textiles. Here is an in-depth exploration of these applications with examples:

23.5.1 Apparel and fashion

The world of fashion and apparel has, for centuries, been graced by the enduring presence of natural fibers, cherished for their unparalleled attributes of comfort, breathability, and esthetic allure (Chung et al., 2018; Herrera-Franco & Valadez-González, 2004; Oladele et al., 2020; Orlović-Leko et al., 2020). Among the array of natural fibers, four stand out prominently for their distinctive characteristics and their applications in fashion:

- Cotton: As the quintessential choice for casual and everyday wear, cotton reigns supreme, celebrated for its gentle touch against the skin and its remarkable moisture-wicking properties. It is the cornerstone of countless wardrobes, finding expression in the form of T-shirts, the timeless allure of denim jeans, and an extensive collection of summer clothing, where its ability to whisk moisture away and its airy texture provide the ideal blend of comfort and style.
- Linen: The natural luster, exceptional breathability, and crisp texture of linen make it an indispensable choice for summer clothing. It effortlessly combines sophistication with cooling comfort, gracing wardrobes with shirts that exude effortless elegance, dresses that epitomize casual chic, and lightweight suits that are the embodiment of refined summertime fashion.
- Silk: Synonymous with luxury, silk is celebrated for its unparalleled smoothness, ethereal sheen, and opulent feel. It finds its place in high-end fashion, where it adds an element of sophistication and allure. Evening gowns crafted from silk epitomize grace and glamour, silk ties are symbols of refinement and elegance in men's fashion, and silk lingerie offers a harmonious blend of comfort and sensuousness, encapsulating the very essence of luxury.
- Wool: For the realm of winter wear, wool emerges as the undisputed choice, thanks to its warmth and insulating properties that envelop the wearer in cozy comfort without the bulk. Woolen sweaters offer both protection from the chill and an array of stylish designs, wool coats provide elegant solutions to the coldest of days, and wool suits blend formality with functionality, making them ideal for winter's formal occasions.

These natural fibers, each with its distinct set of attributes, continue to weave their way into the fabric of fashion, transcending trends and seasons. Cotton, linen, silk, and wool are the pillars of comfort, functionality, and esthetic appeal in the ever-evolving world of apparel and fashion, embodying timeless elegance and modern sensibilities simultaneously.

23.5.2 Home textiles

Natural fibers play an integral role in the realm of home textiles, gracing various household items with their comfort, durability, and timeless appeal. Within this domain, several natural fibers find their unique applications:

- Cotton: Renowned for its softness, absorbency, and ease of care, cotton stands as a staple in home textiles (Balaji et al., 2022; Ramanaiah et al., 2011). It blankets bed linens in a gentle embrace, ensuring restful slumber, while its plush texture adorns towels, enveloping us in warmth and luxury after every bath. Cotton's versatility extends to curtains, where its drape and esthetic add a touch of elegance to living spaces.
- Linen: Characterized by its natural texture and exceptional breathability, linen lends its grace to a spectrum of home textiles. As tablecloths, it sets the stage for delightful gatherings, its crispness and understated elegance enhancing dining experiences. Napkins crafted from linen exude sophistication, elevating everyday meals into special occasions. Linen's breathable charm also finds its way into upholstery, imparting a sense of comfort and refinement.
- Jute: Known for its strength and rustic appearance, jute is the choice material for rugs and carpets that grace our living spaces. Its durability and earthy esthetic make it an ideal companion for high-traffic areas, where it not only withstands the test of time but also adds a touch of natural warmth to interiors.
- Hemp: With its robust nature and resilience, hemp finds its purpose in crafting durable items for the home. It serves as the canvas for wall coverings, adding texture and depth to interior décor. Upholstery made from hemp showcases its durability, ensuring that furniture remains both stylish and steadfast over time.

These natural fibers, each with its unique blend of characteristics, continue to weave themselves into the tapestry of home textiles. Cotton, linen, jute, and hemp bring comfort, durability, and a sense of timeless beauty to our living spaces, making the home a sanctuary of natural elegance and functionality.

23.5.3 Technical textiles

The versatile realm of technical textiles has welcomed natural fibers into its fold, capitalizing on their distinctive properties for specialized applications that span beyond traditional textile use (Al-Maharma & Al-Huniti, 2019; Lau et al., 2018; Liu et al., 2012). Here are instances where natural fibers find their niche:

- Geotextiles: In the world of geotextiles, where materials are employed for erosion control, soil stabilization, and landscaping, natural fibers such as jute and coir have proven invaluable. Their biodegradability aligns seamlessly with sustainable practices, ensuring that as they serve their purpose, they also leave minimal environmental footprint. Furthermore,

their inherent strength makes them reliable allies in the battle against soil erosion and in the quest to stabilize landscapes.
- Agrotextiles: Natural fibers, such as sisal, step into the domain of agrotextiles, where their properties are harnessed for crop protection and agricultural support. Sisal, known for its strength and durability, finds multifaceted use. As a trellising material, it offers vital support to climbing plants, facilitating healthy growth. Additionally, in crop protection, sisal-based materials shield plants from adverse weather conditions, promoting successful cultivation.

These natural fibers, with their inherent strength, biodegradability, and adaptability, prove that their utility transcends traditional textile applications. In the specialized arena of technical textiles, they serve as eco-friendly and reliable resources, safeguarding environments and nurturing agricultural endeavors.

23.5.4 Geotextiles

In the realm of civil engineering and construction, geotextiles fashioned from natural fibers, particularly jute and coir, have emerged as environmentally conscious solutions that address critical challenges in erosion control, soil stabilization, and embankment reinforcement. These materials, born from the earth itself, represent a harmonious synergy between human innovation and the natural world.

- Erosion Control: One of the primary applications of natural fiber geotextiles is erosion control. Erosion, driven by factors such as rainfall, wind, and the flow of water, poses a significant threat to the stability of landscapes and construction sites. Natural fiber geotextiles act as protective blankets, mitigating erosion by stabilizing soil and preventing sediment runoff. Their ability to biodegrade over time ensures that, as they fulfill their purpose, they leave behind no lasting environmental footprint (Amena et al., 2022; Simonová et al., 2022).
- Soil Stabilization: Natural fiber geotextiles also play a pivotal role in soil stabilization. Whether it is on slopes, embankments, or construction sites, these materials provide structural integrity to the soil, safeguarding against erosion and maintaining the desired landscape. Their natural strength and biodegradability are essential attributes, offering both stability and eco-friendliness.
- Embankment Reinforcement: In the construction of embankments and retaining walls, where strength and stability are paramount, natural fiber geotextiles find their niche. Jute and coir, with their robust properties, reinforce these structures, ensuring longevity and structural integrity. Importantly, their biodegradability aligns with sustainable construction practices, making them a responsible choice.
- Sustainability and Eco-Friendliness: What sets natural fiber geotextiles apart is their sustainability. In an era marked by growing environmental awareness, these materials stand as eco-friendly alternatives to synthetic geotextiles. They harness the inherent strength of natural fibers while embracing biodegradability, reducing the environmental impact of construction and civil engineering projects.
- Natural fiber geotextiles, borne from the earth and designed to protect it, exemplify the fusion of sustainability and innovation. They are more than materials; they are statements of responsible construction, where the delicate balance between human development and ecological preservation is upheld, ensuring a harmonious coexistence with the environment.

23.5.5 Automotive textiles

Natural fibers are steering their way into the automotive industry, promising a blend of lightweight design and sustainable solutions. Within the car's interiors, these fibers play a crucial role, offering eco-friendly alternatives to traditional materials and enhancing comfort (Khan et al., 2019; Peças et al., 2018).

- Flax and hemp composites: Flax and hemp composites are revolutionizing the interior components of automobiles. These natural fibers, known for their lightweight and sustainable characteristics, are replacing traditional plastics in components such as door panels, dashboards, and seat backs. By doing so, they contribute to the reduction of a car's overall weight, improving fuel efficiency and reducing carbon emissions. Furthermore, these composites are eco-friendly, aligning with the growing demand for sustainable automotive solutions. They reduce the reliance on nonrenewable materials and minimize the environmental footprint of vehicle production.
- Wool and cotton blends: Comfort and breathability are paramount in car interiors, and natural fibers such as wool and cotton excel in these aspects. They are often used in car seat upholstery, providing passengers with a comfortable and pleasant journey. Wool, with its natural insulating properties, ensures that car seats remain cozy in both hot and cold conditions. Cotton, on the other hand, offers breathability, preventing discomfort from perspiration during long drives. These natural fiber blends create a harmonious combination of comfort, sustainability, and esthetics within the automotive space (Putti et al., 2022; Valková et al., 2022).

The integration of natural fibers in the automotive sector symbolizes a commitment to sustainability, comfort, and innovation. Flax and hemp composites are redefining the industry's approach to lightweight, eco-friendly materials, while wool and cotton blends offer passengers a superior seating experience. In this evolving landscape, natural fibers are driving both change and comfort, paving the way for a more sustainable and enjoyable automotive future.

23.6 Sustainability and eco-friendliness

23.6.1 Environmental impact

In today's world, characterized by heightened environmental consciousness and growing concerns about sustainability, the choice of materials plays a crucial role in shaping our impact on the planet. Natural fibers, sourced from plants, animals, and minerals, are increasingly hailed for their eco-friendliness and reduced environmental impact. These fibers offer a range of environmental benefits that contribute to a more sustainable approach to textile production. One of the primary drivers behind the adoption of natural fibers is their relatively low carbon footprint compared to synthetic alternatives. The production of natural fibers typically requires less energy than the manufacturing of synthetic counterparts (De Matteis et al., 2021; Gänger & Schindowski, 2018). Energy-intensive processes, such as the extraction of petrochemicals for synthetic fibers, are bypassed when working with natural fibers. For instance, cotton, one of the most widely used natural fibers, requires significantly less energy to produce compared to synthetic counterparts such as polyester or nylon. Moreover, sustainable agricultural practices associated with natural fiber cultivation contribute to their reduced carbon footprint (Kaya et al., 2023; Kumar & Thakur, 2020, 2021, 2022, 2023a,; Thakur, Kaya, & Kumar, 2022; Thakur, Kaya, Abousalem, & Kumar, 2022; Thakur, Kaya, Abousalem, et al., 2022; Thakur, Kumar, Kaya, Marzouki, et al., 2022; Thakur, Kumar, Kaya, Vo, et al., 2022; Thakur, Sharma, et al., 2022). For instance, organic cotton cultivation eschews synthetic pesticides and fertilizers, reducing energy-intensive processes associated with the production and application of chemicals. This promotes soil health and biodiversity while minimizing the carbon emissions associated with chemical production.

Another compelling aspect of natural fibers is their sourcing from renewable resources. This sustainability dimension offers several benefits. Natural fibers are derived from plants (cotton, flax, and hemp), animals (wool and silk), and minerals (basalt), all of which can be replenished. This ensures a continuous and sustainable supply without depleting finite resources, a crucial consideration in a world grappling with resource scarcity. The extraction of petrochemicals for synthetic fibers, such as oil for polyester, contributes to resource depletion and environmental degradation (Milosevic et al., 2017; Wang et al., 2019). In contrast, natural fibers leverage the regenerative capacity of ecosystems, placing less strain on natural resources. Furthermore, sustainable cultivation practices for natural fibers, such as organic farming, typically involve fewer chemicals compared to conventional agriculture. For instance, organic cotton farming avoids the use of synthetic pesticides and fertilizers, promoting soil health and minimizing the release of harmful pollutants into the environment. This results in reduced chemical usage and mitigates the environmental impact associated with synthetic fiber production.

Natural fibers are also celebrated for their biodegradability. When natural fiber products reach the end of their life cycle, they decompose naturally over time, returning to the environment without causing long-term harm. This aligns with the principles of sustainability and responsible resource management. The biodegradability of natural fibers significantly reduces the accumulation of nonbiodegradable waste in landfills, unlike synthetic fibers that can persist for centuries. Additionally, natural fibers do not shed microplastics when laundered, contributing to cleaner water ecosystems and reduced environmental harm. While natural fibers offer several environmental benefits, it is important to acknowledge that challenges and considerations exist. The cultivation of natural fibers, particularly in large quantities, can require significant land use, potentially leading to deforestation or habitat loss if not managed sustainably. Responsible land management practices are essential to mitigate such impacts. Additionally, water usage in agriculture, even for natural fibers, remains a concern. Sustainable water management practices, such as efficient irrigation methods and rainwater harvesting, can help address this issue.

23.6.2 Biodegradability

Biodegradability is a hallmark environmental attribute of natural fibers, and it holds significant implications for waste management and environmental conservation.

Natural fibers, unlike their synthetic counterparts, possess the remarkable ability to return to the environment in a natural and sustainable way at the end of their useful life. This inherent property aligns closely with the principles of sustainability and responsible resource management (Atmakuri et al., 2021; Panthapulakkal & Sain, 2007; Sair et al., 2017). When products made from natural fibers, such as clothing or textiles, reach the end of their lifecycle, they undergo a decomposition process that allows them to break down naturally over time. This decomposition is facilitated by the actions of microorganisms, bacteria, and environmental factors, eventually reintegrating the fibers into the ecosystem without causing long-term harm. One of the key advantages of biodegradability is its potential to reduce the burden on landfills. Unlike synthetic fibers, which can persist in landfills for centuries, natural fibers significantly reduce the accumulation of nonbiodegradable waste. This translates into reduced pressure on landfill capacity and contributes to more sustainable waste management practices. Additionally, the environmental concern of microplastic pollution, commonly associated with synthetic fibers, is mitigated by the biodegradability of natural fibers. Synthetic garments shed tiny plastic particles called microplastics into water bodies during laundering, posing risks to aquatic ecosystems and even entering the food chain. In contrast, natural fibers, being biodegradable, do not release microplastics during their decomposition, contributing to cleaner water ecosystems and reducing environmental harm. Moreover, the biodegradability of natural fibers is integral to the broader concept of circular fashion (Gurunathan et al., 2015; Shahinur et al., 2020). In a circular fashion model, garments and textiles are designed with their end-of-life considerations in mind. Natural fibers, through their biodegradable nature, can seamlessly fit into this model by providing a sustainable way to cycle products back into the environment. This aligns with the principles of reducing waste, conserving resources, and promoting responsible consumption.

While the biodegradability of natural fibers is a significant advantage from an environmental perspective, it is important to note that the speed of decomposition can vary depending on several factors, including environmental conditions, fiber type, and the presence of other materials in the product. However, the overarching principle remains clear: Natural fibers offer an environmentally responsible way to reduce the burden of waste on landfills and mitigate the environmental harm associated with persistent synthetic fibers (Chatterjee et al., 2018; Chirică et al., 2021; Prasad et al., 2021). This property underscores the importance of considering biodegradability as a critical factor when evaluating the environmental impact of textiles and materials, contributing to a more sustainable and eco-friendly approach to fashion and product design.

23.6.3 Role in sustainable fashion

Natural fibers play a pivotal role in driving sustainability within the fashion industry, reshaping the way we produce, consume, and perceive clothing. As the world grapples with the environmental and ethical challenges posed by fast fashion and synthetic materials, natural fibers emerge as sustainable champions, fostering a more responsible approach to fashion. Their role in sustainable fashion can be examined through several key dimensions. First and foremost, natural fibers are integral to reducing the environmental impact of fashion. The fashion industry has been criticized for its heavy carbon footprint, with synthetic fibers such as polyester contributing to greenhouse gas emissions (Grabska-Zielińska et al., 2021; Ku et al., 2011). Natural fibers, however, often have a lower carbon footprint due to more energy-efficient production processes and sustainable agricultural practices. For instance, organic cotton cultivation avoids synthetic pesticides and fertilizers, reducing energy-intensive processes and minimizing carbon emissions (Dhonchak et al., 2023; Sharma, Thakur, Sharma, Jakhar, et al., 2023; Sharma, Thakur, Sharma, Sharma, et al., 2023; Thakur & Kumar, 2023b; 2023c; Thakur et al., 2023; Verma et al., 2023). The use of renewable resources in natural fiber production further lowers the industry's overall carbon footprint. Biodegradability is another cornerstone of natural fibers' contribution to sustainable fashion. In a circular fashion model, where products are designed with their end-of-life considerations in mind, biodegradable materials such as natural fibers play a crucial role. Clothing made from natural fibers can naturally decompose, returning to the environment without causing long-term harm. This aligns with the principles of reducing waste and conserving resources, promoting responsible consumption and disposal. Moreover, natural fibers promote ethical practices within the fashion supply chain. Sustainability in fashion extends beyond environmental concerns to encompass fair labor practices and ethical treatment of workers. Natural fiber production often involves more transparent and responsible labor practices, from cotton pickers to artisans producing handmade textiles. This shift toward ethical fashion aligns with consumer demands for transparency and accountability in the industry.

Natural fibers also offer versatile applications sustainably. Cotton, linen, and hemp are staples in clothing, known for their comfort and breathability. These fibers are used in a wide range of clothing items, from casual wear to high-fashion pieces. Their natural esthetics and tactile appeal contribute to the popularity of sustainable fashion (Nair & Joseph, 2014; Rajesh & Pitchaimani, 2017). Additionally, wool and silk, renowned for their luxurious feel and insulation properties, find applications in sustainable and ethical practices, offering warmth and comfort while adhering to eco-friendly principles. The role of natural fibers in sustainable fashion extends beyond clothing to accessories and home textiles. Linen, for instance, is used in home furnishings like tablecloths and upholstery, reflecting a holistic approach to sustainability that encompasses lifestyle choices beyond apparel. Jute, prized for its strength and rustic appearance, is employed in sustainable accessories such as bags and footwear, aligning with the ethos of responsible consumption and production. Furthermore, natural fibers support biodiversity and responsible land management practices. Sustainable cultivation of these fibers often encourages practices that promote biodiversity and reduce monoculture. For example, organic cotton farming may involve the growth of diverse crops alongside cotton plants, creating habitats for beneficial insects and mitigating the risks associated with monoculture. This biodiversity-focused approach contributes to the conservation of ecosystems and the responsible stewardship of land resources.

23.7 Challenges and future directions

23.7.1 Challenges in natural fiber production

Despite their numerous advantages, the production of natural fibers is not without challenges. One of the significant challenges lies in the agricultural aspects of fiber cultivation. Natural fibers such as cotton and flax are susceptible to pests and diseases, which can necessitate the use of pesticides and impact yields. Conventional cotton farming, in particular, has faced criticism for its heavy reliance on synthetic pesticides and fertilizers, leading to environmental concerns. Addressing these challenges often involves transitioning to more sustainable and organic farming practices, which can be a costly and resource-intensive process. Water usage is another critical concern, especially in regions with water scarcity (Felföldi et al., 2021; Gänger & Schindowski, 2018; Khan et al., 2019; Simonová et al., 2022; Valková et al., 2022). Cotton cultivation, for instance, is water-intensive, and the overextraction of water resources can lead to ecological imbalances and water scarcity issues. Sustainable water management practices, such as efficient irrigation methods and rainwater harvesting, are essential to mitigate these challenges. In some cases, land use and deforestation can be associated with natural fiber production. Expanding cultivation areas can encroach on natural habitats and lead to biodiversity loss. Ensuring responsible land management practices and promoting agroforestry can help address these concerns. Furthermore, labor practices in some natural fiber-producing regions may not align with ethical standards, raising questions about fair labor practices and worker conditions. Efforts to improve labor conditions and transparency within the supply chain are essential to address these challenges.

23.7.2 Innovations and future trends

The future of natural fibers is poised for a transformative journey, underpinned by innovation and sustainability. Researchers and industry experts are actively exploring various avenues to not only improve the environmental footprint of natural fiber production but also enhance their performance and utility in a rapidly changing world. One promising frontier in this evolution is the development of genetically modified (GM) natural fibers (Amena et al., 2022; Vera-Cespedes et al., 2023). GM cotton, for instance, has been engineered to exhibit resistance to specific pests and diseases, significantly reducing the need for chemical pesticides. This innovation not only benefits the environment by reducing chemical usage but also brings economic advantages to farmers through decreased input costs and increased yields. It exemplifies how science and technology can be harnessed to make natural fiber cultivation more sustainable and economically viable. Another avenue of progress lies in harnessing waste and byproducts generated during the processing of natural fibers. For instance, cotton stalks and jute waste, traditionally considered agricultural residues, can be repurposed to create sustainable building materials or serve as feedstock for biofuel production. This not only minimizes waste but also promotes resource efficiency, aligning with the principles of a circular economy.

The integration of digital technology and data-driven farming practices, often referred to as precision agriculture, is gaining significant traction in the realm of natural fiber production. Precision agriculture leverages a network of sensors, data analytics, and automation to optimize farming practices. By precisely managing factors like irrigation, fertilization, and pest control, precision agriculture not only reduces resource consumption but also enhances crop yields. This approach minimizes the environmental impact of farming while ensuring the sustainability and profitability of natural fiber cultivation.

23.7.3 Prospects for natural fiber blends

Blending natural fibers with other materials, notably synthetic fibers, represents a promising avenue for advancing the properties of textiles while upholding sustainability principles. This approach enables the creation of hybrid materials that leverage the inherent strengths of both natural and synthetic fibers, resulting in textiles that offer a well-rounded package of desirable characteristics. Natural-synthetic fiber blends provide a prime example of this synergy. They can merge the comfort, breathability, and biodegradability of natural fibers with the robustness, moisture-wicking capabilities, and resilience of synthetic fibers (Khalili et al., 2019; Paturel & Dhakal, 2020; Suriani et al., 2021). For instance, a blend of cotton and polyester combines cotton's softness and moisture absorption with polyester's durability and quick-drying properties, resulting in a fabric that is comfortable to wear and less prone to wrinkles. Furthermore, the incorporation of recycled and regenerated fibers into textiles offers an eco-friendly alternative to virgin natural fibers. Materials such as recycled cotton and regenerated cellulose fibers such as lyocell and modal are often produced using closed-loop manufacturing processes. These processes minimize waste, reduce water and chemical usage, and have a significantly lower environmental impact compared to traditional fiber production methods. Thus, incorporating these sustainable fibers into textile blends contributes to resource conservation and promotes circular economy principles.

Innovative textile technologies also play a pivotal role in enhancing the performance of natural fibers. Fabric coatings and treatments, for example, can impart stain resistance, moisture repellency, and increased wear resistance to textiles made from natural fibers. These advancements expand the practical applications of natural fiber textiles by making them more versatile and adaptable to various environments and usage scenarios. Moreover, they contribute to the durability and longevity of these textiles, ultimately reducing the need for frequent replacements and thus lessening the environmental impact associated with textile waste.

23.8 Conclusion

The exploration of natural fibers and their physicochemical properties for textile uses reveals a captivating journey through the diverse and sustainable world of textiles. In the contemporary landscape, natural fibers continue to play a pivotal role across various industries, driven by their eco-friendly properties and versatility. In the realm of fashion and apparel, cotton remains a cornerstone, celebrated for its comfort and breathability. Linen graces summer wardrobes with its crisp texture and cooling properties. Silk, synonymous with luxury, adorns high-end fashion. Meanwhile, at home, textiles, cotton, and linen find favor in bed linens due to their softness and absorbency, while rugged jute lends strength to rugs and carpets. Natural fibers have even ventured into the world of technical textiles, including geotextiles for construction and civil engineering and agro textiles for agriculture, where they play roles in erosion control and crop protection. In this rapidly evolving landscape, understanding the sources and physicochemical properties of natural fibers is paramount. The exploration of tensile strength, moisture absorption, thermal conductivity, chemical reactivity, and surface morphology sheds light on the intricate characteristics that influence the performance and functionality of textiles. These properties serve as critical considerations for textile engineers, designers, and industry professionals, guiding the selection and application of natural fibers in various contexts. Moreover, the world of natural fibers is a testament to the enduring synergy between nature and human creativity. As we navigate the challenges and opportunities of the textile industry, natural fibers stand as enduring symbols of sustainability, innovation, and the rich tapestry of our shared heritage. This chapter serves as a valuable resource for researchers, textile engineers, and industry professionals, offering a comprehensive understanding of the sources and physicochemical properties of natural fibers, fostering innovation and sustainability in the textile sector.

References

Agirgan, M., Agirgan, A. O., & Taskin, V. (2022). Investigation of thermal conductivity and sound absorption properties of rice straw fiber/polylactic acid biocomposite material. *Journal of Natural Fibers*, *19*(16), 15071−15084. Available from https://doi.org/10.1080/15440478.2022.2070323, http://www.tandfonline.com/toc/wjnf20/current.

Agustini, N. K. A., Triwiyono, A., Sulistyo, D., & Suyitno, S. (2021). Mechanical properties and thermal conductivity of fly ash-based geopolymer foams with polypropylene fibers. *Applied Sciences (Switzerland)*, *11*(11). Available from https://doi.org/10.3390/app11114886, https://www.mdpi.com/2076-3417/11/11/4886/pdf.

Al-Maharma, A. Y., & Al-Huniti, N. (2019). Critical review of the parameters affecting the effectiveness of moisture absorption treatments used for natural composites. *Journal of Composites Science*, *3*(1). Available from https://doi.org/10.3390/jcs3010027, https://www.mdpi.com/2504-477X/3/1/27/pdf.

Amena, B. T., Altenbach, H., Tibba, G. S., & Hossain, N. (2022). Physico-chemical characterization of alkali-treated ethiopian arabica coffee husk fiber for composite materials production. *Journal of Composites Science*, *6*(8). Available from https://doi.org/10.3390/jcs6080233, http://www.mdpi.com/journal/jcs.

Annie Paul, S., Boudenne, A., Ibos, L., Candau, Y., Joseph, K., & Thomas, S. (2008). Effect of fiber loading and chemical treatments on thermophysical properties of banana fiber/polypropylene commingled composite materials. *Composites Part A: Applied Science and Manufacturing*, *39*(9), 1582−1588. Available from https://doi.org/10.1016/j.compositesa.2008.06.004.

Arilla, E., Igual, M., Martínez-Monzó, J., Codoñer-Franch, P., & García-Segovia, P. (2020). Impact of resistant maltodextrin addition on the physico-chemical properties in pasteurised orange juice. *Foods, 9*(12), 1832. Available from https://doi.org/10.3390/foods9121832.

Athijayamani, A., Thiruchitrambalam, M., Natarajan, U., & Pazhanivel, B. (2009). Effect of moisture absorption on the mechanical properties of randomly oriented natural fibers/polyester hybrid composite. *Materials Science and Engineering: A, 517*(1−2), 344−353. Available from https://doi.org/10.1016/j.msea.2009.04.027.

Atmakuri, A., Palevicius, A., Kolli, L., Vilkauskas, A., & Janusas, G. (2021). Development and analysis of mechanical properties of caryota and sisal natural fibers reinforced epoxy hybrid composites. *Polymers, 13*(6). Available from https://doi.org/10.3390/polym13060864, https://www.mdpi.com/2073-4360/13/6/864/pdf.

Bachtiar, E. V., Kurkowiak, K., Yan, L., Kasal, B., & Kolb, T. (2019). Thermal stability, fire performance, and mechanical properties of natural fibre fabric-reinforced polymer composites with different fire retardants. *Polymers, 11*(4). Available from https://doi.org/10.3390/polym11040699, https://res.mdpi.com/polymers/polymers-11-00699/article_deploy/polymers-11-00699.pdf?filename = &attachment = 1.

Balaji, N., Natrayan, L., Kaliappan, S., Patil, P. P., & Sivakumar, N. S. (2022). Annealed peanut shell biochar as potential reinforcement for aloe vera fiber-epoxy biocomposite: Mechanical, thermal conductivity, and dielectric properties. *Biomass Conversion and Biorefinery*. Available from https://doi.org/10.1007/s13399-022-02650-7, http://www.springer.com/engineering/energy + technology/journal/13399.

Bhuiyan, M. A. R., Islam, A., Ali, A., & Islam, M. N. (2017). Color and chemical constitution of natural dye henna (*Lawsonia inermis* L.) and its application in the coloration of textiles. *Journal of Cleaner Production, 167*, 14−22. Available from https://doi.org/10.1016/j.jclepro.2017.08.142, https://www.journals.elsevier.com/journal-of-cleaner-production.

Bledzki, A. K., Reihmane, S., & Gassan, J. (1996). Properties and modification methods for vegetable fibers for natural fiber composites. *Journal of Applied Polymer Science, 59*(8), 1329−1336, https://doi.org/10.1002/(SICI)1097-4628(19960222)59:8 < 1329::AID-APP17 > 3.0.CO;2-0.

Castillo-Lara, J. F., Flores-Johnson, E. A., Valadez-Gonzalez, A., Herrera-Franco, P. J., Carrillo, J. G., Gonzalez-Chi, P. I., & Li, Q. M. (2020). Mechanical properties of natural fiber reinforced foamed concrete. *Materials, 13*(14). Available from https://doi.org/10.3390/ma13143060, https://res.mdpi.com/d_attachment/materials/materials-13-03060/article_deploy/materials-13-03060.pdf.

Chandrasekar, M., Ishak, M. R., Sapuan, S. M., Leman, Z., & Jawaid, M. (2017). A review on the characterisation of natural fibres and their composites after alkali treatment and water absorption. *Plastics, Rubber and Composites, 46*(3), 119−136. Available from https://doi.org/10.1080/14658011.2017.1298550.

Chatterjee, S., Hui, P. C. L., & Kan, Cw (2018). Thermoresponsive hydrogels and their biomedical applications: Special insight into their applications in textile based transdermal therapy. *Polymers, 10*(5). Available from https://doi.org/10.3390/polym10050480, http://www.mdpi.com/2073-4360/10/5/480/pdf.

Chirică, I. M., Enciu, A. M., Tite, T., Dudău, M., Albulescu, L., Iconaru, S. L., Predoi, D., Pasuk, I., Enculescu, M., Radu, C., Mihalcea, C. G., Popa, A. C., Rusu, N., Niță, S., Tănase, C., & Stan, G. E. (2021). The physico-chemical properties and exploratory real-time cell analysis of hydroxyapatite nanopowders substituted with Ce, Mg, Sr, and Zn (0.5−5 at.%). *Materials, 14*(14). Available from https://doi.org/10.3390/ma14143808, https://www.mdpi.com/1996-1944/14/14/3808/pdf.

Chung, T. J., Park, J. W., Lee, H. J., Kwon, H. J., Kim, H. J., Lee, Y. K., & Yin Tze, W. T. (2018). The improvement of mechanical properties, thermal stability, and water absorption resistance of an eco-friendly PLA/kenaf biocomposite using acetylation. *Applied Sciences (Switzerland)*, *8*(3). Available from https://doi.org/10.3390/app8030376, http://www.mdpi.com/2076-3417/8/3/376/pdf.

Darie-Niță, R. N., Irimia, A., Grigoraş, V. C., Mustaţă, F., Tudorachi, N., Râpă, M., Ludwiczak, J., & Iwanczuk, A. (2022). Evaluation of natural and modified castor oil incorporation on the melt processing and physico-chemical properties of polylactic acid. *Polymers*, *14*(17). Available from https://doi.org/10.3390/polym14173608, http://www.mdpi.com/journal/polymers.

Dhonchak, C., Agnihotri, N., Kumar, A., Kamal, R., Thakur, A., & Kumar, A. (2023). Spectrophotometric investigation and computational studies of zirconium(IV)-3-hydroxy-2-[1'-phenyl-3'-(p-methoxyphenyl)-4'-pyrazolyl]-4H-chromen-4-one complex. *Journal of Analytical Chemistry*, *78*(7), 856−865. Available from https://doi.org/10.1134/S1061934823070055, https://www.springer.com/journal/10809.

Duarte, S., Betoret, E., Barrera, C., Seguí, L., & Betoret, N. (2023). Integral recovery of almond bagasse through dehydration: Physico-chemical and technological properties and hot air-drying modelling. *Sustainability (Switzerland)*, *15*(13). Available from https://doi.org/10.3390/su151310704, http://www.mdpi.com/journal/sustainability/.

Durkin, D. P., Ye, T., Larson, E. G., Haverhals, L. M., Livi, K. J. T., De Long, H. C., Trulove, P. C., Fairbrother, D. H., & Shuai, D. (2016). Lignocellulose fiber- and welded fiber- supports for palladium-based catalytic hydrogenation: A natural fiber welding application for water treatment. *ACS Sustainable Chemistry and Engineering*, *4*(10), 5511−5522. Available from https://doi.org/10.1021/acssuschemeng.6b01250, http://pubs.acs.org/journal/ascecg.

Ersoy, O., Rençberoğlu, M., Karapınar Güler, D., & Özkaya, Ö. F. (2022). A novel flux that determines the physico-chemical properties of calcined diatomite in its industrial use as a filler and filter aid: Thenardite (Na_2SO_4). *Crystals*, *12*(4). Available from https://doi.org/10.3390/cryst12040503, https://www.mdpi.com/2073-4352/12/4/503/pdf.

Espert, A., Vilaplana, F., & Karlsson, S. (2004). Comparison of water absorption in natural cellulosic fibres from wood and one-year crops in polypropylene composites and its influence on their mechanical properties. *Composites Part A: Applied Science and Manufacturing*, *35*(11), 1267−1276. Available from https://doi.org/10.1016/j.compositesa.2004.04.004.

Felföldi, Z., Ranga, F., Socaci, S. A., Farcas, A., Plazas, M., Sestras, A. F., Vodnar, D. C., Prohens, J., & Sestras, R. E. (2021). Physico-chemical, nutritional, and sensory evaluation of two new commercial tomato hybrids and their parental lines. *Plants*, *10*(11). Available from https://doi.org/10.3390/plants10112480, https://www.mdpi.com/2223-7747/10/11/2480/pdf.

Ferreira, R. M., Amaral, R. A., Silva, A. M. S., Cardoso, S. M., & Saraiva, J. A. (2022). Effect of high-pressure and thermal pasteurization on microbial and physico-chemical properties of *Opuntia ficus-indica* juices. *Beverages*, *8*(4). Available from https://doi.org/10.3390/beverages8040084, http://www.mdpi.com/journal/beverages.

Ghinea, C., Prisacaru, A. E., & Leahu, A. (2022). Physico-chemical and sensory quality of oven-dried and dehydrator-dried apples of the starkrimson, golden delicious and florina cultivars. *Applied Sciences (Switzerland)*, *12*(5). Available from https://doi.org/10.3390/app12052350, https://www.mdpi.com/2076-3417/12/5/2350/pdf.

Grabska-Zielińska, S., Gierszewska, M., Olewnik-Kruszkowska, E., & Bouaziz, M. (2021). Polylactide films with the addition of olive leaf extract—Physico-chemical characterization. *Materials*, *14*(24). Available from https://doi.org/10.3390/ma14247623, https://www.mdpi.com/1996-1944/14/24/7623/pdf.

Gurunathan, T., Mohanty, S., & Nayak, S. K. (2015). A review of the recent developments in biocomposites based on natural fibres and their application perspectives. *Composites Part A: Applied Science and Manufacturing*, 77, 1−25. Available from https://doi.org/10.1016/j.compositesa.2015.06.007.

Gänger, S., & Schindowski, K. (2018). Tailoring formulations for intranasal nose-to-brain delivery: A review on architecture, physico-chemical characteristics and mucociliary clearance of the nasal olfactory mucosa. *Pharmaceutics*, 10(3). Available from https://doi.org/10.3390/pharmaceutics10030116, http://www.mdpi.com/1999-4923/10/3/116/pdf.

Herrera-Franco, P. J., & Valadez-González, A. (2004). Mechanical properties of continuous natural fibre-reinforced polymer composites. *Composites Part A: Applied Science and Manufacturing*, 35(3), 339−345. Available from https://doi.org/10.1016/j.compositesa.2003.09.012.

Herrera-Franco, P. J., & Valadez-González, A. (2005). A study of the mechanical properties of short natural-fiber reinforced composites. *Composites Part B: Engineering*, 36(8), 597−608. Available from https://doi.org/10.1016/j.compositesb.2005.04.001.

Hussain, M., Levacher, D., Leblanc, N., Zmamou, H., Djeran-Maigre, I., Razakamanantsoa, A., & Saouti, L. (2023). Analysis of physical and mechanical characteristics of tropical natural fibers for their use in civil engineering applications. *Journal of Natural Fibers*, 20(1). Available from https://doi.org/10.1080/15440478.2022.2164104, http://www.tandfonline.com/toc/wjnf20/current.

Iuga, M., Ávila Akerberg, V. D., González Martínez, T. M., & Mironeasa, S. (2019). Consumer preferences and sensory profile related to the physico-chemical properties and texture of different maize tortillas types. *Foods*, 8(11). Available from https://doi.org/10.3390/foods8110533, https://www.mdpi.com/2304-8158/8/11/533/pdf.

Jaganathan, S. K., Mani, M. P., & Khudzari, A. Z. M. (2019). Electrospun combination of peppermint oil and copper sulphate with conducive physico-chemical properties for wound dressing applications. *Polymers*, 11(4). Available from https://doi.org/10.3390/polym11040586, https://res.mdpi.com/polymers/polymers-11-00586/article_deploy/polymers-11-00586.pdf?filename = &attachment = 1.

Jeyapragash, R., Srinivasan, V., & Sathiyamurthy, S. (2020). Mechanical properties of natural fiber/particulate reinforced epoxy composites − A review of the literature. *Materials Today: Proceedings*, 22, 1223−1227. Available from https://doi.org/10.1016/j.matpr.2019.12.146.

Johny, V., Kuriakose Mani, A., Palanisamy, S., Rajan, V. K., Palaniappan, M., & Santulli, C. (2023). Extraction and physico-chemical characterization of pineapple crown leaf fibers (PCLF). *Fibers*, 11(1). Available from https://doi.org/10.3390/fib11010005, http://www.mdpi.com/journal/fibers.

Kabir, M. M., Wang, H., Lau, K. T., & Cardona, F. (2012). Chemical treatments on plant-based natural fibre reinforced polymer composites: An overview. *Composites Part B: Engineering*, 43(7), 2883−2892. Available from https://doi.org/10.1016/j.compositesb.2012.04.053.

Kahlaoui, M., Bertolino, M., Barbosa-Pereira, L., Ben Haj Kbaier, H., Bouzouita, N., & Zeppa, G. (2022). Almond Hull as a functional ingredient of bread: Effects on physico-chemical, nutritional, and consumer acceptability properties. *Foods*, 11(6). Available from https://doi.org/10.3390/foods11060777, https://www.mdpi.com/2304-8158/11/6/777/pdf.

Karabagias, I. K., Karabagias, V. K., & Riganakos, K. A. (2019). Physico-chemical parameters, phenolic profile, in vitro antioxidant activity and volatile compounds of ladastacho (*Lavandula stoechas*) from the region of saidona. *Antioxidants*, 8(4). Available from https://doi.org/10.3390/antiox8040080, https://www.mdpi.com/2076-3921/8/4/80/pdf.

Karimah, A., Ridho, M. R., Munawar, S. S., Adi, D. S., Ismadi., Damayanti, R., Subiyanto, B., Fatriasari, W., & Fudholi, A. (2021). A review on natural fibers for development of eco-friendly bio-composite: Characteristics, and utilizations. *Journal of Materials Research and Technology*, *13*, 2442−2458. Available from https://doi.org/10.1016/j.jmrt.2021.06.014, http://www.elsevier.com/journals/journal-of-materials-research-and-technology/2238-7854.

Kaur, L., Lamsar, H., López, I. F., Filippi, M., Min, D. O. S., Ah-Sing, K., & Singh, J. (2021). Physico-chemical characteristics and in vitro gastro-small intestinal digestion of new zealand ryegrass proteins. *Foods*, *10*(2). Available from https://doi.org/10.3390/foods10020331, https://www.mdpi.com/2304-8158/10/2/331/pdf.

Kaya, S., Thakur, A., & Kumar, A. (2023). The role of in silico/DFT investigations in analyzing dye molecules for enhanced solar cell efficiency and reduced toxicity. *Journal of Molecular Graphics and Modelling*, *124*. Available from https://doi.org/10.1016/j.jmgm.2023.108536.

Khalili, P., Liu, X., Zhao, Z., & Blinzler, B. (2019). Fully biodegradable composites: Thermal, flammability, moisture absorption and mechanical properties of natural fibre-reinforced composites with nano-hydroxyapatite. *Materials*, *12*(7). Available from https://doi.org/10.3390/ma12071145, https://res.mdpi.com/materials/materials-12-01145/article_deploy/materials-12-01145.pdf?filename = &attachment = 1.

Khan, A. R., Nadeem, M., Aqeel Bhutto, M., Yu, F., Xie, X., El-Hamshary, H., El-Faham, A., Ibrahim, U. A., & Mo, X. (2019). Physico-chemical and biological evaluation of PLCL/SF nanofibers loaded with oregano essential oil. *Pharmaceutics*, *11*(8). Available from https://doi.org/10.3390/pharmaceutics11080386, https://www.mdpi.com/1999-4923/11/8/386/pdf.

Kong, I., Degraeve, P., & Pui, L. P. (2022). Polysaccharide-based edible films incorporated with essential oil nanoemulsions: Physico-chemical, mechanical properties and its application in food preservation—A review. *Foods*, *11*(4). Available from https://doi.org/10.3390/foods11040555, https://www.mdpi.com/2304-8158/11/4/555/pdf.

Ku, H., Wang, H., Pattarachaiyakoop, N., & Trada, M. (2011). A review on the tensile properties of natural fiber reinforced polymer composites. *Composites Part B: Engineering*, *42*(4), 856−873. Available from https://doi.org/10.1016/j.compositesb.2011.01.010.

Kumar, A., & Thakur, A. (2020). Encapsulated nanoparticles in organic polymers for corrosion inhibition. *Corrosion protection at the nanoscale* (pp. 345−362). India: Elsevier. Available from https://www.sciencedirect.com/book/9780128193594, 10.1016/B978-0-12-819359-4.00018-0.

Lakshmaiya, N., Ganesan, V., Paramasivam, P., & Dhanasekaran, S. (2022). Influence of biosynthesized nanoparticles addition and fibre content on the mechanical and moisture absorption behaviour of natural fibre composite. *Applied Sciences (Switzerland)*, *12*(24). Available from https://doi.org/10.3390/app122413030, http://www.mdpi.com/journal/applsci/.

Lau, Kt, Hung, Py, Zhu, M. H., & Hui, D. (2018). Properties of natural fibre composites for structural engineering applications. *Composites Part B: Engineering*, *136*, 222−233. Available from https://doi.org/10.1016/j.compositesb.2017.10.038.

Li, X., Tabil, L. G., & Panigrahi, S. (2007). Chemical treatments of natural fiber for use in natural fiber-reinforced composites: A review. *Journal of Polymers and the Environment*, *15*(1), 25−33. Available from https://doi.org/10.1007/s10924-006-0042-3.

Lisitsyn, A., Semenova, A., Nasonova, V., Polishchuk, E., Revutskaya, N., Kozyrev, I., & Kotenkova, E. (2021). Approaches in animal proteins and natural polysaccharides application for food packaging: Edible film production and quality estimation. *Federation Polymers*, *13*(10). Available from https://doi.org/10.3390/polym13101592, https://www.mdpi.com/2073-4360/13/10/1592/pdf.

Liu, K., Takagi, H., Osugi, R., & Yang, Z. (2012). Effect of lumen size on the effective transverse thermal conductivity of unidirectional natural fiber composites. *Composites Science and Technology*, *72*(5), 633−639. Available from https://doi.org/10.1016/j.compscitech.2012.01.009.

Liu, R., Pan, Y., Bao, H., Liang, S., Jiang, Y., Tu, H., Nong, J., & Huang, W. (2020). Variations in soil physico-chemical properties along slope position gradient in secondary vegetation of the hilly region, Guilin, southwest China. *Sustainability (Switzerland)*, *12*(4). Available from https://doi.org/10.3390/su12041303, https://res.mdpi.com/d_attachment/sustainability/sustainability-12-01303/article_deploy/sustainability-12-01303-v2.pdf.

Malenab, R. A. J., Ngo, J. P. S., & Promentilla, M. A. B. (2017). Chemical treatment of waste abaca for natural fiber-reinforced geopolymer composite. *Materials*, *10*(6). Available from https://doi.org/10.3390/ma10060579, http://www.mdpi.com/journal/materials.

Manaila, E., Craciun, G., & Ighigeanu, D. (2020). Water absorption kinetics in natural rubber composites reinforced with natural fibers processed by electron beam irradiation. *Polymers*, *12*(11), 1−20. Available from https://doi.org/10.3390/polym12112437, https://www.mdpi.com/2073-4360/12/11/2437/pdf.

Manimaran, P., Sanjay, M. R., Senthamaraikannan, P., Jawaid, M., Saravanakumar, S. S., & George, R. (2019). Synthesis and characterization of cellulosic fiber from red banana peduncle as reinforcement for potential applications. *Journal of Natural Fibers*, *16*(5), 768−780. Available from https://doi.org/10.1080/15440478.2018.1434851.

De Matteis, V., Rojas, M., Cascione, M., Mazzotta, S., Di Sansebastiano, G. P., & Rinaldi, R. (2021). Physico-chemical properties of inorganic nps influence the absorption rate of aquatic mosses reducing cytotoxicity on intestinal epithelial barrier model. *Molecules (Basel, Switzerland)*, *26*(10). Available from https://doi.org/10.3390/molecules26102885, https://www.mdpi.com/1420-3049/26/10/2885/pdf.

Mattiello, S., Guzzini, A., Del Giudice, A., Santulli, C., Antonini, M., Lupidi, G., & Gunnella, R. (2023). Physico-chemical characterization of keratin from wool and chicken feathers extracted using refined chemical methods. *Polymers*, *15*(1). Available from https://doi.org/10.3390/polym15010181, http://www.mdpi.com/journal/polymers.

Merais, M. S., Khairuddin, N., Salehudin, M. H., Mobin Siddique, M. B., Lepun, P., & Chuong, W. S. (2022). Preparation and characterization of cellulose nanofibers from banana pseudostem by acid hydrolysis: Physico-chemical and thermal properties. *Membranes*, *12*(5). Available from https://doi.org/10.3390/membranes12050451, https://www.mdpi.com/2077-0375/12/5/451/pdf.

Michalska-Ciechanowska, A., Majerska, J., Brzezowska, J., Wojdyło, A., & Figiel, A. (2020). The influence of maltodextrin and inulin on the physico-chemical properties of cranberry juice powders. *ChemEngineering*, *4*(1), 1−12. Available from https://doi.org/10.3390/chemengineering4010012, https://www.mdpi.com/2305-7084/4/1/12/pdf.

Milosevic, M., Stoof, D., & Pickering, K. (2017). Characterizing the mechanical properties of fused deposition modelling natural fiber recycled polypropylene composites. *Journal of Composites Science*, *1*(1). Available from https://doi.org/10.3390/jcs1010007.

Mohammed, L., Ansari, M. N. M., Pua, G., Jawaid, M., & Islam, M. S. (2015). A review on natural fiber reinforced polymer composite and its applications. *International Journal of Polymer Science*, *2015*. Available from https://doi.org/10.1155/2015/243947, http://www.hindawi.com/journals/ijps/.

Mulenga, T. K., Ude, A. U., & Vivekanandhan, C. (2021). Techniques for modelling and optimizing the mechanical properties of natural fiber composites: A review. *Fibers*, *9*(1), 1−17. Available from https://doi.org/10.3390/fib9010006, https://www.mdpi.com/2079-6439/9/1/6.

Nadziakiewicza, M., Kehoe, S., & Micek, P. (2019). Physico-chemical properties of clay minerals and their use as a health promoting feed additive. *Animals*, *9*(10). Available from https://doi.org/10.3390/ani9100714, https://www.mdpi.com/2076-2615/9/10/714/pdf.

Nair, A. B., & Joseph, R. (2014). *Eco-friendly bio-composites using natural rubber (NR) matrices and natural fiber reinforcements* (pp. 249−283). Elsevier BV. Available from 10.1533/9780857096913.2.249.

Nayab-Ul-Hossain, A. K. M., Sela, S. K., Hasib, M. A., Alam, M. M., & Shetu, H. R. (2022). Preparation of graphene based natural fiber (jute)-synthetic fiber (glass) composite and evaluation of its multifunctional properties. *Composites Part C: Open Access*, *9*. Available from https://doi.org/10.1016/j.jcomc.2022.100308, http://www.journals.elsevier.com/composites-part-c-open-access/.

Nugraha, A. D., Nuryanta, M. I., Sean, L., Budiman, K., Kusni, M., & Muflikhun, M. A. (2022). Recent progress on natural fibers mixed with CFRP and GFRP: Properties, characteristics, and failure behaviour. *Polymers.*, *14*(23). Available from https://doi.org/10.3390/polym14235138, http://www.mdpi.com/journal/polymers.

Oladele, I. O., Michael, O. S., Adediran, A. A., Balogun, O. P., & Ajagbe, F. O. (2020). Acetylation treatment for the batch processing of natural fibers: Effects on constituents, tensile properties and surface morphology of selected plant stem fibers. *Fibers*, *8*(12), 1−19. Available from https://doi.org/10.3390/fib8120073, https://www.mdpi.com/2079-6439/8/12/73.

Orlović-Leko, P., Vidović, K., Ciglenečki, I., Omanović, D., Sikirić, M. D., & Šimunić, I. (2020). Physico-chemical characterization of an urban rainwater (Zagreb, Croatia). *Atmosphere*, *11*(2). Available from https://doi.org/10.3390/atmos11020144, https://res.mdpi.com/d_attachment/atmosphere/atmosphere-11-00144/article_deploy/atmosphere-11-00144-v2.pdf.

Panthapulakkal, S., & Sain, M. (2007). Studies on the water absorption properties of short hemp—Glass fiber hybrid polypropylene composites. *Journal of Composite Materials*, *41*(15), 1871−1883. Available from https://doi.org/10.1177/0021998307069900.

Paturel, A., & Dhakal, H. N. (2020). Influence of water absorption on the low velocity falling weight impact damage behaviour of flax/glass reinforced vinyl ester hybrid composites. *Molecules (Basel, Switzerland)*, *25*(2). Available from https://doi.org/10.3390/molecules25020278, https://www.mdpi.com/1420-3049/25/2/278/pdf.

Peças, P., Carvalho, H., Salman, H., & Leite, M. (2018). Natural fibre composites and their applications: A review. *Journal of Composites Science*, *2*(4). Available from https://doi.org/10.3390/jcs2040066, https://www.mdpi.com/2504-477X/2/4/66/pdf.

Prasad, K., Nikzad, M., Nisha, S. S., & Sbarski, I. (2021). On the use of molecular dynamics simulations for elucidating fine structural, physico-chemical and thermomechanical properties of lignocellulosic systems: Historical and future perspectives. *Journal of Composites Science*, *5*(2). Available from https://doi.org/10.3390/jcs5020055, https://www.mdpi.com/2504-477X/5/2/55/pdf.

Prasetyaningrum, A., Utomo, D. P., Raemas, A. F. A., Kusworo, T. D., Jos, B., & Djaeni, M. (2021). Alginate/κ-carrageenan-based edible films incorporated with clove essential oil: Physico-chemical characterization and antioxidant-antimicrobial activity. *Polymers*, *13*(3), 1−16. Available from https://doi.org/10.3390/polym13030354, https://www.mdpi.com/2073-4360/13/3/354/pdf.

Prdun, S., Svečnjak, L., Valentić, M., Marijanović, Z., & Jerković, I. (2021). Characterization of bee pollen: Physico-chemical properties, headspace composition and ftir spectral profiles. *Foods*, *10*(9). Available from https://doi.org/10.3390/foods10092103, https://www.mdpi.com/2304-8158/10/9/2103/pdf.

Putti, F. F., Nogueira, B. B., Vacaro de Souza, A., Festozo Vicente, E., Zanetti, W. A. L., de Lucca Sartori, D., & Pigatto de Queiroz Barcelos, J. (2022). Productive and physicochemical parameters of tomato fruits submitted to fertigation doses with water treated with very low-frequency electromagnetic resonance fields. *Plants*, *11*(12). Available from https://doi.org/10.3390/plants11121587, https://www.mdpi.com/2223-7747/11/12/1587/pdf?version = 1655362549.

Rahman, R., & Putra, S. Z. F. S. (2018). Tensile properties of natural and synthetic fiber-reinforced polymer composites. *Mechanical and physical testing of biocomposites, fibre-reinforced composites and hybrid composites* (pp. 81−102). Malaysia: Elsevier. Available from http://www.sciencedirect.com/science/book/9780081022924, 10.1016/B978-0-08-102292-4.00005-9.

Rajesh, M., & Pitchaimani, J. (2017). Mechanical characterization of natural fiber intra-ply fabric polymer composites: Influence of chemical modifications. *Journal of Reinforced Plastics and Composites*, *36*(22), 1651−1664. Available from https://doi.org/10.1177/0731684417723084, http://jrp.sagepub.com/content/by/year.

Ramadevi, P., Sampathkumar, D., Srinivasa, C. V., & Bennehalli, B. (2012). Effect of alkali treatment on water absorption of single cellulosic abaca fiber. *BioResources*, *7*(3), 3515−3524. Available from https://doi.org/10.15376/biores.7.3.3515-3524.

Ramamoorthy, S. K., Di, Q., Adekunle, K., & Skrifvars, M. (2012). Effect of water absorption on mechanical properties of soybean oil thermosets reinforced with natural fibers. *Journal of Reinforced Plastics and Composites*, *31*(18), 1191−1200. Available from https://doi.org/10.1177/0731684412455257.

Ramanaiah, K., Ratna Prasad, A. V., & Hema Chandra Reddy, K. (2011). Mechanical properties and thermal conductivity of typha angustifolia natural fiber−reinforced polyester composites. *International Journal of Polymer Analysis and Characterization*, *16*(7), 496−503. Available from https://doi.org/10.1080/1023666X.2011.598528.

Rong, M. Z., Zhang, M. Q., Liu, Y., Yang, G. C., & Zeng, H. M. (2001). The effect of fiber treatment on the mechanical properties of unidirectional sisal-reinforced epoxy composites. *Composites Science and Technology*, *61*(10), 1437−1447. Available from https://doi.org/10.1016/S0266-3538(01)00046-X.

Saad Azzem, L., & Bellel, N. (2022). Thermal and physico-chemical characteristics of plaster reinforced with wheat straw for use as insulating materials in building. *Buildings*, *12*(8). Available from https://doi.org/10.3390/buildings12081119, http://www.mdpi.com/journal/buildings.

Sair, S., Oushabi, A., Kammouni, A., Tanane, O., Abboud, Y., Oudrhiri Hassani, F., Laachachi, A., & El Bouari, A. (2017). Effect of surface modification on morphological, mechanical and thermal conductivity of hemp fiber: Characterization of the interface of hemp − Polyurethane composite. *Case Studies in Thermal Engineering*, *10*, 550−559. Available from https://doi.org/10.1016/j.csite.2017.10.012.

Sanjay, M. R., Arpitha, G. R., Laxmana Naik, L., Gopalakrishna, K., & Yogesha, B. (2016). Applications of natural fibers and its composites: An overview. *Natural Resources*, *07*(03), 108−114. Available from https://doi.org/10.4236/nr.2016.73011.

Savic, I. M., & Savic Gajic, I. M. (2022). Determination of physico-chemical and functional properties of plum seed cakes for estimation of their further industrial applications. *Sustainability (Switzerland)*, *14*(19). Available from https://doi.org/10.3390/su141912601, http://www.mdpi.com/journal/sustainability/.

Shahinur, S., Hasan, M., Ahsan, Q., & Haider, J. (2020). Effect of chemical treatment on thermal properties of jute fiber used in polymer composites. *Journal of Composites Science*, *4*(3). Available from https://doi.org/10.3390/jcs4030132.

Sharma, D., Thakur, A., Sharma, M. K., Jakhar, K., Kumar, A., Sharma, A. K., & Hari, O. M. (2023). Synthesis, electrochemical, morphological, computational and corrosion inhibition studies of 3-(5-naphthalen-2-yl-[1,3,4]oxadiazol-2-yl)-pyridine against mild steel in 1M HCl. *Asian Journal of Chemistry*, *35*(5), 1079−1088. Available from https://doi.org/10.14233/ajchem.2023.27711, https://doi.org/10.14233/ajchem.2023.27711.

Sharma, D., Thakur, A., Sharma, M. K., Sharma, R., Kumar, S., Sihmar, A., Dahiya, H., Jhaa, G., Kumar, A., Sharma, A. K., & Om, H. (2023). Effective corrosion inhibition of mild steel using novel 1,3,4-oxadiazole-pyridine hybrids: Synthesis, electrochemical, morphological, and computational insights. *Environmental Research*, *234*. Available from https://doi.org/10.1016/j.envres.2023.116555.

Sheeba, K. R. J., Priya, R. K., Arunachalam, K. P., Avudaiappan, S., Maureira-Carsalade, N., & Roco-Videla, Á. (2023). Characterisation of sodium acetate treatment on acacia pennata natural fibres. *Polymers*, *15*(9). Available from https://doi.org/10.3390/polym15091996, http://www.mdpi.com/journal/polymers.

Shelenkov, P. G., Pantyukhov, P. V., Poletto, M., & Popov, A. A. (2023). Influence of vinyl acetate content and melt flow index of ethylene-vinyl acetate copolymer on physico-mechanical and physico-chemical properties of highly filled biocomposites. *Polymers*, *15*(12). Available from https://doi.org/10.3390/polym15122639, http://www.mdpi.com/journal/polymers.

Simonová, M. P., Chrastinová, L., & Lauková, A. (2022). Enterocin 7420 and sage in rabbit diet and their effect on meat mineral content and physico-chemical properties. *Microorganisms*, *10*(6). Available from https://doi.org/10.3390/microorganisms10061094, https://www.mdpi.com/2076-2607/10/6/1094/pdf?version = 1653556573.

Snetkov, P., Zakharova, K., Morozkina, S., & Olekhnovich, R. (2020). Hyaluronic acid: The influence of molecular weight and degradable properties of biopolymer. **12**.

Spinei, M., & Oroian, M. (2022). The influence of extraction conditions on the yield and physico-chemical parameters of pectin from grape pomace. *Polymers*, *14*(7). Available from https://doi.org/10.3390/polym14071378, https://www.mdpi.com/2073-4360/14/7/1378/pdf.

Stevulova, N., Cigasova, J., Purcz, P., Schwarzova, I., Kacik, F., & Geffert, A. (2015). Water absorption behavior of hemp hurds composites. *Materials*, *8*(5), 2243−2257. Available from https://doi.org/10.3390/ma8052243, http://www.mdpi.com/1996-1944/8/5/2243/pdf.

Stevulova, N., Cigasova, J., Schwarzova, I., Sicakova, A., & Junak, J. (2018). Sustainable bio-aggregate-based composites containing hemp hurds and alternative binder. *Buildings*, *8*(2). Available from https://doi.org/10.3390/buildings8020025, http://www.mdpi.com/2075-5309/8/2/25/pdf.

Suriani, M. J., Rapi, H. Z., Ilyas, R. A., Petrů, M., & Sapuan, S. M. (2021). Delamination and manufacturing defects in natural fiber-reinforced hybrid composite: A review. *Polymers*, *13*(8). Available from https://doi.org/10.3390/polym13081323, https://www.mdpi.com/2073-4360/13/8/1323/pdf.

Suárez, L., Castellano, J., Romero, F., Marrero, M. D., Benítez, A. N., & Ortega, Z. (2021). Environmental hazards of giant reed (*Arundo donax* L.) in the macaronesia region and its characterisation as a potential source for the production of natural fibre composites. *Polymers*, *13*(13). Available from https://doi.org/10.3390/polym13132101, https://www.mdpi.com/2073-4360/13/13/2101/pdf.

Tanasă, F., Zănoagă, M., Teacă, C. A., Nechifor, M., & Shahzad, A. (2020). Modified hemp fibers intended for fiber-reinforced polymer composites used in structural applications—A review. I. Methods of modification. *Polymer Composites*, *41*(1), 5−31. Available from https://doi.org/10.1002/pc.25354, http://onlinelibrary.wiley.com/journal/10.1002/(ISSN)1548-0569.

Thakur, A., Kaya, S., Abousalem, A. S., & Kumar, A. (2022). Experimental, DFT and MC simulation analysis of *Vicia sativa* weed aerial extract as sustainable and eco-benign corrosion inhibitor for mild steel in acidic environment. *Sustainable Chemistry and Pharmacy*, 29. Available from https://doi.org/10.1016/j.scp.2022.100785.

Thakur, A., Kaya, S., Abousalem, A. S., Sharma, S., Ganjoo, R., Assad, H., & Kumar, A. (2022). Computational and experimental studies on the corrosion inhibition performance of an aerial extract of *Cnicus benedictus* weed on the acidic corrosion of mild steel. *Process Safety and Environmental Protection*, *161*, 801−818. Available from https://doi.org/10.1016/j.psep.2022.03.082, http://www.elsevier.com/wps/find/journaldescription.cws_home/713889/description#description.

Thakur, A., Kaya, S., & Kumar, A. (2022). Recent innovations in nano container-based self-healing coatings in the construction industry. *Current Nanoscience*, *18*(2), 203−216. Available from https://doi.org/10.2174/1573413717666210216120741, http://www.eurekaselect.com/article/114345.

Thakur, A., Kaya, S., & Kumar, A. (2023). Recent trends in the characterization and application progress of nano-modified coatings in corrosion mitigation of metals and alloys. *Applied Sciences (Switzerland)*, *13*(2). Available from https://doi.org/10.3390/app13020730, http://www.mdpi.com/journal/applsci/.

Thakur, A., Kumar, A., Kaya, S., Marzouki, R., Zhang, F., & Guo, L. (2022). Recent advancements in surface modification, characterization and functionalization for enhancing the biocompatibility and corrosion resistance of biomedical implants. *Coatings*, *12*(10). Available from https://doi.org/10.3390/coatings12101459.

Thakur, A., Kumar, A., Kaya, S., Vo, D. V. N., & Sharma, A. (2022). Suppressing inhibitory compounds by nanomaterials for highly efficient biofuel production: A review. *Fuel*, *312*. Available from https://doi.org/10.1016/j.fuel.2021.122934, http://www.journals.elsevier.com/fuel/.

Thakur, A., Sharma, S., Ganjoo, R., Assad, H., & Kumar, A. (2022). Anti-corrosive potential of the sustainable corrosion inhibitors based on biomass waste: A review on preceding and perspective research. *Journal of Physics: Conference Series*, *2267*(1). Available from https://doi.org/10.1088/1742-6596/2267/1/012079.

Thakur, A., & Kumar, A. (2021). Sustainable inhibitors for corrosion mitigation in aggressive corrosive media: A comprehensive study. *Journal of Bio- and Tribo-Corrosion*, *7*(2). Available from https://doi.org/10.1007/s40735-021-00501-y, http://www.springer.com/materials/surfaces + interfaces/journal/40735.

Thakur, A., & Kumar, A. (2022). Recent advances on rapid detection and remediation of environmental pollutants utilizing nanomaterials-based (bio)sensors. *Science of The Total Environment*, *834*. Available from https://doi.org/10.1016/j.scitotenv.2022.155219.

Thakur, A., & Kumar, A. (2023a). Computational insights into the corrosion inhibition potential of some pyridine derivatives: A DFT approach. *European Journal of Chemistry*, *14*(2), 246−253. Available from https://doi.org/10.5155/eurjchem.14.2.246-253.2408.

Thakur, A., & Kumar, A. (2023b). Recent trends in nanostructured carbon-based electrochemical sensors for the detection and remediation of persistent toxic substances in real-time analysis. *Materials Research Express*, *10*(3). Available from https://doi.org/10.1088/2053-1591/acbd1a.

Thakur, A., & Kumar, A. (2023c). Ecotoxicity analysis and risk assessment of nanomaterials for the environmental remediation. *Macromolecular Symposia*, *410*(1). Available from https://doi.org/10.1002/masy.202100438, http://onlinelibrary.wiley.com/journal/10.1002/(ISSN)1521-3900.

Torres-Arellano, M., Renteria-Rodríguez, V., & Franco-Urquiza, E. (2020). Mechanical properties of natural-fiber-reinforced biobased epoxy resins manufactured by resin infusion process. *Polymers*, *12*(12), 1−17. Available from https://doi.org/10.3390/polym12122841, https://www.mdpi.com/2073-4360/12/12/2841/pdf.

Valková, V., Ďúranová, H., Havrlentová, M., Ivanišová, E., Mezey, J., Tóthová, Z., Gabríny, L., & Kačániová, M. (2022). Selected physico-chemical, nutritional, antioxidant and sensory properties of wheat bread supplemented with apple pomace powder as a by-product from juice production. *Plants*, *11*(9). Available from https://doi.org/10.3390/plants11091256, https://www.mdpi.com/2223-7747/11/9/1256/pdf?version = 1651827132.

Vera-Cespedes, N., Muñoz, L. A., Rincón, M. Á., & Haros, C. M. (2023). Physico-chemical and nutritional properties of chia seeds from latin american countries. *Foods*, *12*(16). Available from https://doi.org/10.3390/foods12163013, http://www.mdpi.com/journal/foods.

Verma, C., Thakur, A., Ganjoo, R., Sharma, S., Assad, H., Kumar, A., Quraishi, M. A., & Alfantazi, A. (2023). Coordination bonding and corrosion inhibition potential of nitrogen-rich heterocycles: Azoles and triazines as specific examples. *Coordination Chemistry Reviews*, *488*. Available from https://doi.org/10.1016/j.ccr.2023.215177.

Wambua, P., Ivens, J., & Verpoest, I. (2003). Natural fibres: Can they replace glass in fibre reinforced plastics? *Composites Science and Technology*, *63*(9), 1259−1264. Available from https://doi.org/10.1016/S0266-3538(03)00096-4, http://www.journals.elsevier.com/composites-science-and-technology/.

Wang, H., Memon, H., Hassan, E. A. M., Miah, M. S., & Ali, M. A. (2019). Effect of jute fiber modification on mechanical properties of jute fiber composite. *Materials*, *12*(8). Available from https://doi.org/10.3390/ma12081226, https://res.mdpi.com/materials/materials-12-01226/article_deploy/materials-12-01226.pdf?filename = &attachment = 1.

Yang, Z., Peng, H., Wang, W., & Liu, T. (2010). Crystallization behavior of poly(ε-caprolactone)/layered double hydroxide nanocomposites. *Journal of Applied Polymer Science*, *116*(5), 2658−2667, China. Available from https://doi.org/10.1002/app, http://www3.interscience.wiley.com/cgi-bin/fulltext/123249591/PDFSTART.

Index

Note: Page numbers followed by "*f*" and "*t*" refer to figures and tables, respectively.

A

AACA. *See* Aluminum acetylacetonate (AACA)
AB. *See* Ammonia borane (AB)
Absorbents, 475
AC. *See* Alternating current (AC)
Acetic acid-treated exfoliated and sintered sheets of graphitic carbon nitride (AAs-gC$_3$N$_4$), 452–454
Acetylation, 59
ACF. *See* Activated carbon microfibers (ACF)
ACNFs. *See* Activated carbon nanofibers (ACNFs); Amorphous carbon nanofibers (ACNFs)
Activated carbon (AC), 379–380
Activated carbon microfibers (ACF), 486
Activated carbon nanofibers (ACNFs), 380–383
Active alumina, 237
Adsorbent materials, 377, 379
Adsorption, 237–240, 379, 386
  chemical surface absorption, 238–239
  exchange surface adsorption, 239–240
  physical surface absorption, 237–238
Adult stem cells, 167–168
Aerogels, 84
  formation, 84
  nanofibers, 388
AFM. *See* Atomic force microscopy (AFM)
Agricultural/agriculture, 574–575
  pesticides, 244–245
  residues, 239
  wastes, 239, 592
Agrotextiles, 599
Air compressor lubricants, 340–341, 340*f*
Air filters, 476–477
  nanofibers materials for, 479–481, 481*t*
  systems, 476

Air filtration, 477
  classification of nanofibers materials in air filter application, 479*f*
  nanofibers in, 476–478
Air pollutants, 472
Air pollution mitigation, 476
Air purification systems, 386, 472
Airbrushing method, 2
ALD. *See* Atomic layer deposition (ALD)
Aldehyde groups, 101
Algae, 240
Alginate, 36–37, 98–100
  structural, physicochemical, and mechanical properties of, 36–37
Aligned nanofiber array, 507
Aliphatic polyester, 44
Alkaline substances, 593
All-solid state lithium-ion batteries (ASS-LIBs), 308
Alpha-Fe$_2$O$_3$ nanoparticles, 289
Alternating current (AC), 282–283
Alumina (Al$_2$O$_3$), 329
Aluminum (Al), 329
Aluminum acetylacetonate (AACA), 306
Ambient pressure chemical vapor deposition (APCVD), 3
American National Science Foundation, 135
Amines, 350, 385
Amino acids, 97
Ammonia, 428–430
Ammonia borane (AB), 392
Ammonium fluoride (NH$_4$F), 17–18
Ammonium persulfate (APS), 10, 78–79
Amniotic membrane, 169
Amorphous carbon nanofibers (ACNFs), 392
Amphiphile copolymers, 161
Amphoteric cellulose, 205
Angora, 578

Animal fibers, 33, 571, 577−578, 585, 589−590, 593−594
  mohair and angora, 578
  silk, 577
  wool, 577−578
Animal sources, 96
Anion-curing electrospinning (ACE), 121
Anode materials, 276−278, 281−282, 512
Anodic aluminum oxide (AAO), 13−14
Anodic oxidation. *See* Electrochemical oxidation (ECO)
Anodization. *See* Electrochemical oxidation (ECO)
Anti-wear additives, 345−346, 345*f*
Anticancer drugs, 170
Antifoam additives, 346
Antioxidants, 350
APCVD. *See* Ambient pressure chemical vapor deposition (APCVD)
Aqueous electrolytes, 541
Aqueous solutions by composites, 231−244
Aromatic compounds, 387
Arsenic contamination, 227
Artificial human tissues, 142
Artificial photosynthesis, 417
Artificial silk, 34
Arundo donax L. *See* Giant reed (*Arundo donax* L.)
Asbestos, 569−570, 578−579
ASS-LIBs. *See* All-solid state lithium-ion batteries (ASS-LIBs)
Asymmetric SCs, 543
Atomic force microscopy (AFM), 21−23, 450−452
Atomic layer deposition (ALD), 457
Automotive industry, 438
Automotive textiles, 600

B

Bacteria, 232, 240
Bacterial cells, 232
Bacterial cellulose, 95
Banana fiber, 590
Basalt fibers, 579
Basaltic magma, 579
Basement membrane, 139
Batteries, 298, 539
  advances in nanofibers for, 308−309
  nanofibers for, 280−309

Benzene, 411−412
BET analysis. *See* Brunauer−Emmet−Teller analysis (BET analysis)
Betaine-modified cationic cellulose, 205
β-Chitin, 93
Bimetal oxide nanofibers, 304
Bimetallic oxides, 452
Bio-absorption, 240−244
Bio-based lubricants, 338−339
Bio-ink, 172
Bio-thermal inkjet printers, 171−172
Bioabsorbent particles, 241−242
Bioaccumulation, 240
Biocides, 342
Biocompatibility of NFs, 332−333
Biocompatible polymers, 116−117, 168
Biodegradability, 601−602
Biodegradable lubricants, 338−339, 339*f*, 350
  comparison between conventional and synthetic lubricants, 339*f*
Biodegradable materials, 487−488
Biodegradable plastics, 230
Biodegradable polyester, 42
Biodegradable polymers, 246−247
Biodegradable scaffold, 169
Biodegradable sources, nanofibers extracted from, 305−306
Biodegradation, 248
Biolaminates, 581
Biological wastewater treatment, 243−244
Biomasses, 241
Biomaterials, 143
Biomedical applications for nanofibers, 61
Biopolymers, 91, 98, 479
  for green synthesis of nanofibers, 114−117
    eco-friendly alternatives to replace traditional organic solvents, 118*f*
    ranking of solvents, 120*t*
  removal of pollutants using, 246−248
Bioprinting, 172
Bioremediation, 69
Biosensors, 423−424
Biosorption by plant composites, influential functional groups in, 244−246
Biotechnology, 44
Bisphenol A (BPA), 191
Bisphenol F (BPF), 191

Index   619

Bisphenol S (BPS), 191
Blending natural fibers, 605
Blood–brain barrier, 160
BN. *See* Boron nitride (BN)
Bombyx mori, 39–40, 577
Bone tissue engineering, nanotechnology methods in, 144–146
Bones, 144–145, 176
Borazine ($B_3N_3H_6$), 392
Boron carbide (B_4C), 329
Boron carbon nitride nanofiber, 388
Boron nitride (BN), 326–327, 392
Bottom-up techniques, 256
BPA. *See* Bisphenol A (BPA)
BPF. *See* Bisphenol F (BPF)
BPS. *See* Bisphenol S (BPS)
Brunauer–Emmet–Teller analysis (BET analysis), 450–452
1-butyl-3-methylimidazolium chloride (BMIMCl), 122

C

CA. *See* Cellulose acetate (CA)
Cadmium oxide (CdO), 513
Calcination, 203, 428–430
Calcium alginate ions, 99–100
Calcium compounds, 347–348
Cancer tissues, 147, 170
Carbon, 528–529
Carbon based materials, 481
Carbon black (CB), 72–73, 292
  particle-containing polymer nanofibers, 72–73
Carbon dioxide (CO_2), 549
  adsorption, 378, 380, 385
  capture, 378–386
    activated carbon nanofibers, 380–383
    carbon nanotubes, 383
    graphene and graphene oxide, 383–384
    ionic liquid, 385–386
    metal-organic frameworks, 384–385
    polymer nanofibers, 385
Carbon fibers (CF), 278, 357
Carbon hybrids (CHs), 5
Carbon monoxide, 3–4
Carbon nanocones (CNCs), 4
Carbon nanofiber web (CNFW), 11, 276
Carbon nanofiber-graphene nanosheet materials (CNF–GN materials), 5

Carbon nanofibers (CNFs), 4, 76–78, 258, 273, 381, 393–394, 414–415, 431, 449–450, 473, 528–529, 541, 548–553
  activation of, 549
  composites, 330, 330f, 410, 413–414, 553–556
    metal oxide-carbon nanofiber composites, 553–554
    metal-organic framework-carbon nanofiber composites, 554–556
  heteroatom doping of, 552–553
  pore engineering, 548–552
Carbon nanotubes (CNTs), 3–4, 71–72, 366, 371, 379–380, 383, 411, 423–424, 446, 448–449, 473, 497, 529, 541
Carbon to-oxygen (C/O), 596–597
Carbon-based adsorbents, 378
Carbon-based composite nanofibers, 283–286
Carbon-based materials, 394–395, 448–449
Carbon-free metal oxide nanofibers, 556
Carbonaceous materials, 380, 542
Carbonized nanofibers, 276–278
Carbonous materials, 328–329
Carboxyl groups, 385
Carboxylate, 263
Carboxylated chains, 61
Carboxylic acid, 263
Carboxymethyl cellulose (CMC), 114
Cardiac tissue engineering, 146
Carmellose, 101
Carriers, 159–160
Cassie–Baxter model, 86
Catalysis, 405, 438
  application of nanofiber composites, 411–416
Catalyst, 4
Catalytic activity enhancement, 447–448
Catalytic electrode, 452–454
Catalytic engineering, 457–458
Cathode materials, 512
Cation exchange capacity, 239–240
CB. *See* Carbon black (CB)
CD. *See* Cyclodextrin (CD)
CDA. *See* Cellulose diacetate (CDA)
CE. *See* Counter electrode (CE)
Cell culture, 140

Cell wall polymers, 241
Cellulose (CL), 33, 94–96, 114, 573–574
  bacteria, 95–96
  based air filter, 479–480
  nanocomposites, 56–58
  nanofibrils, 96
  structure of glucose production, 95f
Cellulose acetate (CA), 72–73, 114
Cellulose diacetate (CDA), 74–75
Cellulose nanocrystals (CNCs), 9, 56–58, 387
Cellulose nanofibers (CNFs), 38–39, 122, 205–208, 256, 361, 473–476, 573–574
  aerogel, 84
  cellulose nanofiber-based adsorbents, 207f, 474–475
  composites, 413–414
  structural, physicochemical, and mechanical properties of, 38–39
Cellulose nitrate, 72–73
Centrifugal spinning, 261
Ceramics, 326
  ceramic-polymer composites, 327
  fibers, 197–204
    application of TiO_2 fibers for water remediation, 200–204, 206t
    ceramic nanofibers by electrospinning technique, 200f
    ceramic nanofibers for water treatment applications, 201t
  fields, 177
  nanofiber, 135–136
Cerium oxide (CeO_2), 394
Cetrimonium bromide (CTAB), 11
CF. *See* Carbon fibers (CF)
Chelation, 53–56
Chemical adsorption, 388–389
Chemical functionalization, 56–63
Chemical oxidation, 78–79, 244
Chemical precipitation, 233–234
Chemical surface absorption, 238–239
Chemical treatment of nanocellulose, 56–60
Chemical vapor deposition (CVD), 1, 3–6, 197, 259, 405, 441–442, 457
Chemotherapeutic agents, 169
Chickpea (*Cicer arietinum*), 486
Chitin (CTN), 114

Chitosan (CHN), 36, 93–94, 114, 149–150
  structural, physicochemical, and mechanical properties of, 36
Chlorodimethyl isopropyl silane, 58
Chlorpyrifos, 244–245
CHN. *See* Chitosan (CHN)
Chondroitin sulfate (C_4S), 115
CHs. *See* Carbon hybrids (CHs)
Cicer arietinum. See Chickpea (*Cicer arietinum*)
CL. *See* Cellulose (CL)
Climate change, 378–379
CMC. *See* Carboxymethyl cellulose (CMC)
CNCs. *See* Carbon nanocones (CNCs); Cellulose nanocrystals (CNCs)
CNF–GN materials. *See* Carbon nanofiber-graphene nanosheet materials (CNF–GN materials)
CNFs. *See* Carbon nanofibers (CNFs); Cellulose nanofibers (CNFs)
CNFW. *See* Carbon nanofiber web (CNFW)
CNTs. *See* Carbon nanotubes (CNTs)
CO_2-assisted electrospinning (CO_2-AE), 121
Coagulation process, 233–234
Coal gas, 388
Coatings, 63, 368–370
Coaxial electrospinning technique, 2, 390
Cobalt, 315
Cobalt oxide, 316
Codelivery, 161–164
  block-copolymers, 162f
  pharmaceutical carriers, 161f
Collagen, 37–38, 102–103
  structural, physicochemical, and mechanical properties of, 37–38
Colloidal carriers, 161
Compliant batteries, nanofibers in, 302–303
Composite nanofibers, 69–70, 148–151, 313. *See also* Medical nanofibers
  comments and ideas for development of polymer matrix-based nanocomposites, 150–151
  pollutants, 150f
  polymer nanotechnology, 148–150
Composites, 1–2, 148
  additive application of nanofibers and composites for enhanced performance, 522–525
  oil–water separation, 522–524

Index

volatile organic compounds, 524–525
application of, 504–509
  applications of electrospun nanofibers in soft electronics, 507–508
  electronics in textiles, 504–505
  energy harvesting and storage devices, 509
  energy storage, 505–506
  sensing in textiles, 505
  strategies for electrospun nanofibers in soft electronics, 506–507
  yarn, 507
formation and characteristics, 443–444
fundamentals of, 439–444
materials, 327, 485
  synergistic effects on, 446
membranes, 385
nanofibers, 175–179
  comments and ideas for development of polymer matrix-based nanocomposites, 177–179
  polymer nanotechnology, 175–177
nanomaterials, 60–61
processing and fabrication technologies, 2–21
property characterization, 23–25
  optoelectrical properties, 24–25
  thermoelectric behaviour, 23–24
  UV-visible light sensitivity, 24
sequential methods of removing contaminants from aqueous solutions by, 231–244
structure assessment of, 21–23
  atomic force microscopy, 22–23
  scanning electron microscopy, 21
  transmission electron microscopy, 21–22
  X-ray diffraction, 22
in water splitting, 444–449
Computational modeling, 457
Concentrated hydrochloric acid, 9
Conducting polymers (CPs), 542–543, 558–559
  nanofibers, 558–559
Conductive fluid, 499
Conductors, 508
Congo red (CR), 200
Contamination of water sources, 185
Conventional lubricants, 343

Copper ions, 207
Core-charged electrospun fibers, 364
Core-shell electrospinning, 501
Core–shell structured nanofiber composites, 2
Corrosion inhibitors, 342, 349–350
Corrosion resistance, 150
Cotton, 33, 569–570, 573–575, 580–581, 597–598
Counter electrode (CE), 510
COVID-19 pandemic, 437–438
CPs. *See* Conducting polymers (CPs)
CR. *See* Congo red (CR)
Cross-linking, 53–56
  hydrophobic covalently bonded zein nanofibers, 55*f*
Crystalline cellulose, 38–39
CTN. *See* Chitin (CTN)
CV. *See* Cyclic voltammetry (CV)
CVD. *See* Chemical vapor deposition (CVD)
Cyclic voltammetry (CV), 543–544
Cyclodextrin (CD), 387

D

d-CA. *See* Deacetylated cellulose acetate (d-CA)
d-MXene. *See* Delaminated-MXene (d-MXene)
DC voltage. *See* Direct current voltage (DC voltage)
Deacetylated cellulose acetate (d-CA), 80–82
Deacetylation, 80–84
  process, 94
  reaction of cellulose acetate, 83*f*
Deep eutectic solvents (DES), 119, 122–123
Deep wounds, 141
Degradable synthetic nanofiber, 51–52
Degree of substitution (DS), 59
Deionized water (DI water), 10, 73–74
Delaminated-MXene (d-MXene), 305–306
Delignification, 79–80
  process, 80
  three-dimensional cellulose nanofiber structure made from, 79*f*
Delivery of new drugs, 158–160
Dendrimer, 146–147, 169–171
DES. *See* Deep eutectic solvents (DES)
Detergents, 347–348
Diallyl dimethyl ammonium chloride, 53–54

Diffusion coefficient, 588
Dimethyl carbonate (DMC), 290
Dimethyl sulfoxide (DMSO), 119—121, 309—310
Dimethylacetamide (DMAC), 74—75
Dimethylformamide (DMF), 71—72, 118—119
Dipolar—dipolar interactions, 238
Direct carbonization from polymer nanofiber template, 276—278, 277f
Direct current voltage (DC voltage), 6—7, 273—274
Direct methanol fuel cells (DMFCs), 511
Direct oxidation-reduction reactions, 246
Disaccharide unit, 100—101
Dispersants, 347—348
DMC. See Dimethyl carbonate (DMC)
DMFCs. See Direct methanol fuel cells (DMFCs)
Domestic electrospun nanofibers, 524—525
Doped-zirconia-based balls, 257
Drain electrodes, 507
Drawing, 258
Drugs, 31—32
  delivery, 143, 146
    agents, 93
    systems, 159, 165
  delivery of new, 158—160
  drug-carrying polymers, 160
  molecules, 147
  release process, 163
  screening, 166
Dry cyclic corrosion tests, 365
DSSCs. See Dye-sensitized solar cells (DSSCs)
Dye-sensitized solar cells (DSSCs), 278—279, 509—510, 527

E

EC. See Ethylene carbonate (EC)
ECM. See Extracellular matrix (ECM)
ECO. See Electrochemical oxidation (ECO)
EDLCs. See Electric double-layer capacitors (EDLCs)
EDS analysis. See Energy dispersive X-ray spectroscopy analysis (EDS analysis)
Efficiency (E), 477
EIA. See Environmental Impact Assessment (EIA)

EIS. See Electrochemical impedance spectroscopy (EIS)
EK remediation. See Electrokinetic remediation (EK remediation)
EKG. See Electrokinetic geosynthetics (EKG)
Elastane, 42
Electric double-layer capacitors (EDLCs), 505, 542, 545—546
Electric field, 407
Electric vehicles, 539
Electrical conductivity, 70—71, 357
Electricity, 499
Electrocatalysis, 405, 413—414
Electrochemical capacitors, 509
Electrochemical deposition, 457
Electrochemical hydrogen storage, 393
Electrochemical impedance spectroscopy (EIS), 358, 545
Electrochemical oxidation (ECO), 1, 17—18
Electrochemical performance improvement, 448—449
Electrochemical reactions, 525—527
Electrodes, 302—303, 512, 541, 545—546, 548—549
  materials, 263—265, 539—540
  nanofibers as, 548—559
Electrohydrodynamic casting, 6—7
Electrohydrodynamic process, 260
Electrokinetic geosynthetics (EKG), 485
Electrokinetic remediation (EK remediation), 472—473
Electrokinetic technique, 482—485
Electrolysis, 437—438, 444—445
Electrolytes, 541
Electron donor polymers, 310—311
Electronic-based nanofibers
  application of, 504—509
  applications of electrospun nanofibers in soft electronics, 507—508
  electronics in textiles, 504—505
  energy harvesting and storage devices, 509
  energy storage, 505—506
  sensing in textiles, 505
  strategies for electrospun nanofibers in soft electronics, 506—507
  yarn, 507
Electronized nanofibers, 245

Index

Electrospinning, 6–8, 70–76, 157, 245, 273–275, 283, 295–296, 417, 428–431, 441–442, 455, 472, 476, 501, 506, 513, 546–547, 559–560
  fiber systems produced by, 188–197
    electrospun natural fibers, 192–197
    electrospun synthetic fibers, 190–191
  jet, 7, 274
  methods, 31–32, 51, 91, 406–407, 449–450, 507
  nanofiber membranes, 70
  polyvinylidene fluoride nanofibers, 366
Electrospraying, 7, 274
Electrospun (ES), 426
  CNFs, 548–549
  electrolyte membrane, 511
  fibers, 6–7, 273–274, 500
    sensors, 505
    systems, 190
  mesoporous inorganic nanofibers, 406–407
  nanofibers, 51–52, 260–261, 281, 505, 508, 548
    electrospun nanofiber-based membrane, 281
    electrospun nanofiber-based separators, 559–560
    materials, 275
    methods, 525–527
  nanofibrous scaffolds, 146
  nanofibrous substances, 522–524
  natural fibers, 192–197
    adsorption and desorption of methylene blue, 198f
    applications in energy devices of, 509–513
    electrospun natural nanofibers for wastewater treatment, 199t
    polyacrylonitrile, 54–56
    polymer nanofiber mats, 507
    protein-based fibers, 192
  synthetic fibers, 190–191
    synthetic polymer–based electrospun nanofibers for water treatment, 193t
  TiO_2 nanofibers, 311–312
  ZnO nanofibers, 310
Electrostatic forces, 70–71, 200
Electrostatic repulsion, 407
Electrostatic spinning, 364

Electsrocatalysts, 449–450
Embryonic stem cells (ES), 167
Emulsifiers, 342
Energy applications of nanofibers and composites, 262–267
  fabrication methods, 256–261
  field effect transistors, 267
  fuel cells, 262–263
  functional nanofibers, 262
  history, 256
  hydrogen storage, 266
  lithium-ion batteries, 263–265, 264f, 265f
  solar cells, 266–267
  supercapacitors, 266
Energy conversion, 525–532
  applications, 529–532, 529f
  process, 512–513
  sodium batteries, 528–529
  solar cells, 527
  solid oxide fuel cell, 525–527
Energy devices of electrospun NFs
  applications, 509–513
  energy harvesting and conversion devices, 509–511
    fuel cells, 511
    mechanical energy harvesters, 511
    solar cells, 509–510
    solar-driven hydrogen generation, 511
  energy storage devices, 512–513
    hydrogen storage, 512–513
    rechargeable lithium-ion batteries, 512
    supercapacitors, 512
Energy dispersive X-ray spectroscopy analysis (EDS analysis), 596–597
Energy efficiency and reduced temperature accumulation, 334
Energy harvesting and storage devices, 509
  nanogenerators, 509
  supercapacitors, 509
  transistors, 509
Energy production, 513
Energy storage, 439, 505–506, 525–532
  applications, 529–532, 529f
  hydrogen storage, 506
  lithium-ion batteries, 505
  sodium batteries, 528–529
  solar cells, 527
  solid oxide fuel cell, 525–527
  supercapacitors, 505

Energy storage systems (ESSs), 528−529
Engine oil, 330
　additives, 345−346, 348−349
Enhanced permeability and retention (EPR), 147, 170
Entangled porous nanofiber sheet, 506
Environmental Impact Assessment (EIA), 124−126
　system boundaries to obtain cellulose nanofiber materials, 125f
Environmental pollution, 244−245
Environmental threat of pollutants, 229−231
Enzymatic catalysis, 411, 413−414
EPDM. See Ethylene−propylene−diene monomer (EPDM)
EPR. See Enhanced permeability and retention (EPR)
Equivalent series resistance (ESR), 545
ES. See Embryonic stem cells (ES)
Escherichia coli, 115, 231−232, 480
ESR. See Equivalent series resistance (ESR)
ESSs. See Energy storage systems (ESSs)
Ethylammonium nitrate, 122
Ethylene carbonate (EC), 290
3,4-ethylene dioxyethiophene (EDOT), 301
Ethylene−propylene−diene monomer (EPDM), 347
Ethyne (C_2H_2), 5
Eucalyptus pulp-derived CNFs, 207
Exchange surface adsorption, 239−240
Extracellular matrix (ECM), 114−115, 136, 157
Extraction methods, 246−247
Extreme pressure additives, 346−347
Extrusion, 6−7, 172

F

FA. See Fly ash (FA)
Fabrication, 256, 586
　methods, 91, 256−261, 443
　　drawing, 258
　　grinding, 257
　　milling, 257
　　nanofibers, 257f
　　phase separation, 258
　　refining, 257
　　self-assembly, 258
　　spinning, 259
　　template-based fabrication, 258
　　vapor deposition methods, 259
　processing and fabrication technologies, 2−21
　　chemical vapor deposition, 3−6
　　electrochemical oxidation, 17−18
　　electrospinning, 6−8
　　high-temperature annealing, 18−19
　　hydrothermal synthesis, 8−10
　　liquid phase deposition, 11−17
　　spraying pyrolysis, 20−21
　　template assembling, 10−11
Faradaic reactions, 545−546
Fashion industry, 602−603
Fashion model, 601−602
Fatty acid, 346
FDM. See Fused deposition modeling (FDM)
Felting, 593−594
Fermentation, 94−95
Fiber systems produced by electrospinning, 188−197
Fiberglass, 579
Fiberization, 579
Fibers, 273−274, 547, 573−574, 576, 580−581
　fiber-making methods, 487−488
Fibrin, 101−102
　fibrinogen chains, 103f
　tissue, 101
Fibrinogen, 101
Fibroblast cells, 116
Fibrous materials, 503−504
Fillers, 178−179
Filter membranes, 234
Filtration systems, 204, 233−234, 438
Flax, 576
Flexible batteries, nanofibers in, 302−303
Flexible packaging materials, 177−178
Flexible transparent electrodes, 508
Flocculants, 474
Flocculation process, 233−234
Fluidized bed column, 242
Fluorescence emission, 426
Fluoride coatings, 299
Fluoropolymers, 298
Fly ash (FA), 591−592
Folate, 147
Folic acid, 147
Formic acid, 415

Index

Fossil fuels, 388, 509
 fossil fuel-derived polymers, 192
Fourier transform infrared spectroscopy (FTIR), 82–84, 595
Friction modifiers, 348, 349*f*
Fuel cells, 262–263, 414, 511, 525–527, 539
 nanofibers for, 313–316
Fuel purification, gas adsorption for, 387–390
Functional nanofibers, 262
Fungi, 240
Fused deposition modeling (FDM), 583–584

G
Galvanostatic charge/discharge (GCD), 543–544
Gas adsorption, 377
 and storage of nanofibers
  carbon dioxide capture, 378–386
  gas adsorption for fuel purification, 387–390
  hydrogen storage, 391–395
  volatile organic compound adsorption for indoor air pollution control, 386–387
Gas sensitization, 426
Gaseous by-products, 3
GCD. *See* Galvanostatic charge/discharge (GCD)
GCP. *See* Green chemistry principles (GCP)
GE. *See* Gelatin (GE)
Gear lubricants, 341
Gear oil, 330
Gel electrospinning, 75
Gel polymer electrolyte (GPE), 301
Gelatin (GE), 96–97, 114
 gelatin-calcium alginate composite membranes, 197
Gelation process, 258
Genetically modified natural fibers (GM), 604
Geotextiles, 574, 598–599
GHG emissions. *See* Greenhouse gas emissions (GHG emissions)
Giant reed (*Arundo donax* L.), 580–581
Glass fibers, 579
Global warming potential (GWP), 123

Glucometer, 424
Gluconacetobacter xylinus, 95–96
Glycosaminoglycan, 100
GM. *See* Genetically modified natural fibers (GM)
Gold nanosystems, 146
GPE. *See* Gel polymer electrolyte (GPE)
Grafting, 53–54
 surface grafting, 53–56
Graphene, 331, 357, 379–380, 383–384, 446, 448–449, 507
Graphene oxide (GO), 5, 40–41, 295, 383–384, 555
 graphene oxide-coated nanofiber membranes, 473
Graphite anodes, 281–282
Graphitic layers, 381–382
Greases, 340
Green chemistry principles (GCP), 117–118
Green energy conversion technology, 262
Green methods, 113
Green solvents for nanofiber synthesis, 117–124
 deep eutectic solvents, 122–123
 ionic liquids, 122
 supercritical fluids, 123–124
Green synthesis of nanofibers
 biopolymers for green synthesis of nanofibers, 114–117
 environmental impact assessment, 124–126
 green solvents for nanofiber synthesis, 117–124
Greenhouse gas emissions (GHG emissions), 378
Grinding, 257
Groundwater system, 228
GWP. *See* Global warming potential (GWP)

H
HA. *See* Humic acid (HA)
Hansen solubility parameters theory, 119
Hard template method, 551–552
HAS. *See* High-amylose starch (HAS)
HBA. *See* Hydrogen-bond acceptor (HBA)
HBD. *See* Hydrogen-bond donor (HBD)
HCNFs. *See* Hollow carbon nanofibers (HCNFs)
Heart tissue engineering, 146
Heat deterioration, 332

Heat-treatment process, 550–551, 556
Heavy metals, 190, 233–234
  ions, 207, 423–424
Heavy transition metal (HTM), 423–424
Helical nanotubes, 145
Hemp, 576–577, 598
  cultivation, 569–570
  fibers, 589–590
  seeds, 576
  textiles, 576
Hepatocytes, 166
HER. See Hydrogen evolution reaction (HER)
Heteroatom doping of carbon nanofibers, 552–553
Heteroatomic doping, 381
Heterogeneous catalysis, 411–412
Hexadecyl trimethyl ammonium bromide (CTAB), 388
1,1,1,3,3,3-hexafluoro-2-propanol (HFP), 37–38
1,1,1,3,3,3-hexafluoroisopropanol (HFIP), 123
Heyrovsky reaction, 413
HF. See Hydrofluoric acid (HF)
High aspect ratio, 331
High molecular weight (HMW), 35–36
High surface area, 331
High thermal conductivity, 333–335
High vapor pressures, 8–9
High voltage electric supply, 261
High-amylose starch (HAS), 550–551
High-resolution transmission electron microscopic image (HRTEM image), 283–286
High-temperature annealing technique, 18–19
High-voltage solid lithium-metal batteries (HVSLMBs), 304
HMW. See High molecular weight (HMW)
Hollow carbon nanofibers (HCNFs), 554
  from quartz fiber template, 278, 279f
Hollow fiber membranes, 235
Home textiles, 598
HRTEM image. See High-resolution transmission electron microscopic image (HRTEM image)
HTM. See Heavy transition metal (HTM)
Humic acid (HA), 61–62

HVSLMBs. See High-voltage solid lithium-metal batteries (HVSLMBs)
Hyaluronan, 35–36
Hyaluronic acid (HA), 35–36, 100–101
  physiochemical and biological properties, 100f
  structural, physiochemical, and mechanical properties of, 35–36
Hybrid carbon nanofiber-graphene nanosheet materials, 5
Hybrid nanocomposites, 456
Hybrid nanofiber composites, 286–287
Hybrid photocatalysts, 452–454
Hydrazine-modified polyacrylonitrile fiber, 186
Hydrocarbons, 337, 410
Hydrochloric acid (HCl), 74–75
Hydrodynamic lubrication, 329
Hydrofluoric acid (HF), 278
Hydrogels, 163
Hydrogen (H_2), 266, 392, 437–438, 506
  bonding, 36
  composites reinforced with nanocellulose, 61
  energy, 511
  generators, 511
  production, 445
  storage, 266, 391–395, 506, 512–513
    carbon nanofibers, 393–394
    metal oxide nanofibers, 394
    polymer nanofibers, 394–395
Hydrogen evolution reaction (HER), 412–413, 439
Hydrogen sulfide (H_2S), 388
Hydrogen-bond acceptor (HBA), 122–123
Hydrogen-bond donor (HBD), 122–123
Hydrolysis process, 94–96
Hydrophilic surfaces, 85
Hydrophobic fiber coatings, 364
Hydrothermal synthesis, 8–10, 410
Hydroxyapatite nanoparticles, 157–158
Hydroxyapatite nanophase, 145
Hydroxypropylated cellulose, 63
Hyperaccumulator plants, 485
Hypofibrinogenemia, 101

I

ILs. See Ionic liquids (ILs)
In situ casting method, 328

In situ forming implants, 163−164
In situ spectroscopy, 457
Indium nitrate, 278−279
Indium tin oxide (ITO), 278−279
Indoor air pollution control, volatile organic compound adsorption for, 386−387
Induced pluripotent stem cells (iPS), 167
Infiltration systems, 442−443
Inkjet printers, 174
Inorganic nanofibers, 258, 292
INPs. *See* Iron nanoparticles (INPs)
Intermediate temperature (IT), 525−527
International Energy Agency, 437−438
International Renewable Energy Agency, 437−438
Intramolecular hydrogen bonds, 93
Ion exchange resins, 235−237
Ion-balancing resins, 236
Ionic liquids (ILs), 119, 122, 385−386, 541
iPS. *See* Induced pluripotent stem cells (iPS)
Iron, 289−290
 iron-based carbon nanofibrous composites, 390
Iron alkoxide, 191
Iron hydroxides, 232
Iron nanoparticles (INPs), 227
Iron oxide (Fe$_2$O$_3$), 394
Isocyanic acid, 58−59
IT. *See* Intermediate temperature (IT)
ITO. *See* Indium tin oxide (ITO)

J

Joint Research Center of European Commission (JCR), 227−228
Jute fibers, 33−34, 576−577, 593−594, 598

K

KAIST. *See* Korea Advanced Institute of Science and Technology (KAIST)
Kamlet−Taft parameters (K−T parameters), 117−118
KIST. *See* Korea Institute of Science and Technology (KIST)
Klebsiella pneumoniae, 231−232
Korea Advanced Institute of Science and Technology (KAIST), 445
Korea Institute of Science and Technology (KIST), 445

L

Laminaria hyperborea, 36−37, 98−99
Large-scale membrane processes, 235
Layer-by-layer process, 172
Layered double hydroxide nanoparticles (LDH), 158
LCA. *See* Life cycle assessment (LCA)
LDH. *See* Layered double hydroxide nanoparticles (LDH)
LE. *See* Leucoemeraldine (LE)
Leucoemeraldine (LE), 361
LIBs. *See* Lithium-ion batteries (LIBs)
Life cycle assessment (LCA), 114, 124
LiFePO$_4$/C cathodes. *See* Lithium iron phosphate/carbon nanofiber cathodes (LiFePO$_4$/C cathodes)
Lignocellulosic materials, 236−237
Linear sweep voltammetry (LSV), 450−452
Linen, 573−574, 576, 597−598
Linum usitatissimum, 576
Liposomal drug delivery systems, 160
Liposomes, 162
Liquid absorption, 298−299
Liquid phase deposition (LPD), 1, 11−17
Lithium batteries, oxide nanofibers for, 304
Lithium cobalt phosphate (LiCoPO$_4$), 292
Lithium hexafluorophosphate (LiPF$_6$), 295
Lithium iron phosphate/carbon nanofiber cathodes (LiFePO$_4$/C cathodes), 280−281
Lithium lanthanum titanate oxide (LLTO), 280−281
Lithium polyacrylate (PAALi), 298
Lithium-ion batteries (LIBs), 15, 263−265, 278−279, 298, 300, 505
Lithium-ion full-cells, 303
Lithium-ion half-cells, 303
Lithium-sulfur batteries (Li−S batteries), 276
 nanfibers for, 300−302
 rechargeable batteries, 300
LLTO. *See* Lithium lanthanum titanate oxide (LLTO)
LMW. *See* Low molecular weight (LMW)
Long-chain HA polymer, 35−36
Low molecular weight (LMW), 35−36
Low-pressure CVD (LPCVD), 3
LPD. *See* Liquid phase deposition (LPD)
LSV. *See* Linear sweep voltammetry (LSV)

Lubricant additives, 325, 346–347
  nanofibers and composites as, 344–351
  types of, 345–351
    anti-wear additives, 345–346
    antifoam additives, 346
    antioxidants, 350
    biodegradable lubricants, 350
    detergents and dispersants, 347–348
    extreme pressure additives, 346–347
    friction modifiers, 348
    nanoadditives, 350–351
    rust and corrosion inhibitors, 349–350
    self-healing lubricants, 348–349
    viscosity index improver additives, 347
Lubricants, 330, 335–344, 349
  types of, 337–344
    air compressor lubricants, 340–341
    biodegradable lubricants, 338–339
    gear lubricants, 341
    greases, 340
    lubricant in sliding and movable parts, 337f
    metalworking fluids, 342–343
    mineral oil-based lubricants, 337
    solid lubricants, 343–344
    synthetic lubricants, 338
    tribological applications of nanocomposites, 336t

M

Macromolecules, 162
Magnetic NPs, 232
Magnetic stirring, 71–72
Manganese oxides, 553–554
Massachusetts Institute of Technology (MIT), 445
MB. See Methylene blue (MB)
MC. See Methylcellulose (MC)
MEAs. See Membrane-electrode assemblies (MEAs)
Mechanical energy harvesters, 511
  nanogenerators, 511
Mechanical testing, 581–582
Mechano-electrospinning (MES), 428–430
Medical nanofibers, 139–148. See also Composite nanofibers
  dendrimer, 146–147
  heart tissue engineering, 146
  nanofibers as three-dimensional scaffolds for tissue regeneration, 148
  nanomedicine, 139–141
  nanotechnology methods in bone tissue engineering, 144–146
  relationship between nanomedicine and tissue engineering, 143
  tissue engineering, 141–143
MEK. See Methyl ethyl ketone (MEK)
Melt spinning process, 136–137
Membrane bioreactor, 243
Membrane-electrode assemblies (MEAs), 457
MES. See Mechano-electrospinning (MES)
Mesenchymal stem cells (MSCs), 157
Mesopores, 306
Mesoporous CNFs, 393
Mesoporous electrospun ZnO NFs, 200
Metal atoms, 266
Metal fiber composites, 328
Metal hydrides, 393–394
Metal lubricant composites, 329
Metal matrix composites, 327–328
Metal nanoparticles, 326, 394–395, 416–417, 445–448
Metal oxide nanofibers, 394, 556–558
  composites, 410–412
  metal oxide–metal oxide composite, 557–558
  perovskite metal oxides, 557
  spinel metal oxides, 557
  unitary metal oxides, 556
Metal oxide nanoparticles, 414
Metal oxide-carbon nanofiber composites, 553–554
Metal oxides, 389–390, 410, 445–446, 553–554, 556
Metal powder, 175
Metal substrates, 448–449
Metal-lubricant composites, 328–329
Metal-metal composites, 327–330
  metal fiber composites, 328
  metal matrix composites, 327–328
  metal-lubricant composites, 328–329
  metal-metal intermetallic alloys, 329–330
Metal-metal intermetallic alloys, 329–330
Metal-organic frameworks (MOFs), 378, 384–385, 456, 553
  metal-organic framework-carbon nanofiber composites, 554–556

Index 629

Metalworking fluids (MWFs), 341*f*, 342–343
   gear lubricant in gear boxes of vehicles, 342*f*
   mineral oil-based metalworking fluids, 342
   semisynthetic metalworking fluids, 342
   synthetic metalworking fluids, 342
   water-based metalworking fluids, 342–343
Methane, 415
Methanol, 415
Methyl ethyl ketone (MEK), 387
Methyl orange (MO), 190
Methyl-β-cyclodextrin, 387
Methylcellulose (MC), 59–60
Methylene blue (MB), 190
Micelles, 161
Micro-cellular polymer foam-supported polyaniline, 78–79
Micro-pipette, 136–137
Microbial cellulose, 95
Microbial fuel cells, 263
Microorganisms, 94–95, 241
Microphase separation, 42
Microplastics, 601–602
Micropore confinement, 383
Microporous nanofibers, 258
Microwave plasma chemical vapor deposition (MPCVD), 4
Milling, 257
Mineral oil-based compounds, 346
Mineral oil-based lubricants, 337, 341
Mineral oil-based metalworking fluids, 342
Mineral-based fibers, 569–572, 578–579
   asbestos, 578–579
   basalt, 579
   glass, 579
MIT. *See* Massachusetts Institute of Technology (MIT)
Modified cellulose nanofiber photoinitiators (MCNFI), 361
MOFs. *See* Metal-organic frameworks (MOFs)
Mohair, 578
Moisture absorption, 584–588, 587*f*
   depiction of percentage increase in weight of composites over time, 586*f*
Molybdenum disulfide (MoS$_2$), 328–329

Molybdenum disulfide nanofiber (MSNF), 301
Monoglyceride, 371–373
Monometallic alloys, 329–330
Montmorillonite, 178
MPCVD. *See* Microwave plasma chemical vapor deposition (MPCVD)
MSCs. *See* Mesenchymal stem cells (MSCs)
MSNF. *See* Molybdenum disulfide nanofiber (MSNF)
Multifunctional nanofiber composites, 417
MWFs. *See* Metalworking fluids (MWFs)
Mxenes, 326–327
Mycobacterium tuberculosis, 426–428

N
N,N-dimethylacetamide (DMAc), 286–287
N,N-dimethylformamide (DMF), 74–75, 283
Nacetyl-D-glucosamine (NAG), 93
Nano wood (NW), 80
Nano-reinforcement component, 176
Nano-sized carriers, 162
Nanoadditives, 350–351
Nanocages, 255–256
Nanocellulose, 60–61, 236–237
   applications, 60–63, 60*f*
     biomedical applications, 61
     coatings, 63
     water treatment, 61–62
   chemical functionalization, properties, and applications of, 56–63
   chemical treatment, 56–60
   modification techniques, 58*f*
Nanocellulosic coatings, 63
Nanocomposites, 56, 136, 148–150, 175–178, 229, 371–373
Nanocrystals, 255–256, 292
Nanodendrimers, 146, 170
Nanofabrication, 425–426
Nanofibers (NFs), 1–2, 31–32, 51, 135, 185, 245, 255–256, 257*f*, 302, 325, 378–380, 391–392, 438–439, 474–475, 485, 513, 521–522
   additive application of nanofibers and composites for enhanced performance, 522–525
   anode materials, 281–282
   anodes, 281–290, 284*f*

Nanofibers (NFs) (*Continued*)
  application of electronic-based nanofibers and composites, 504−509
  applications in energy devices of electrospun NFs, 509−513
  for batteries, 280−309
  for battery applications, 304−309
    advances in nanofibers for batteries, 308−309
    nanofibers extracted from biodegradable sources, 305−306
    nanofibers for sodium batteries, 305
    nanofibers with tuned structures, 306−308
    oxide nanofibers for lithium batteries, 304
  cathodes, 290−292
  classification of nanofibers based on chemical composition, 503, 504*f*
  coatings, 299
  in compliant and flexible batteries, 302−303
  composites, 1, 262, 326−330, 358−373, 378, 405−407, 410−412, 417, 443−444, 446, 521−522
    air purification application of, 476−481
    biocompatibility, 332−333
    carbon nanofiber-based composites, 330
    catalysis application of, 411−416, 415*f*
    ceramic-polymer composites, 327
    cost-effectiveness, 333
    energy efficiency and reduced temperature accumulation, 334
    enhanced heat transfer, 333−334
    enhanced performance, 335
    excellent mechanical properties, 331
    high aspect ratio, 331
    high surface area, 331
    high thermal conductivity, 333−335
    hydrogen evolution reaction mechanism on Ni/Gd$_2$O$_3$/NiO NF surface, 413*f*
    low density, 332
    as lubricant additives, 344−351
    metal-metal composites, 327−330
    nanofibers in air filtration, 476−478
    nanofibers materials for air filters, 479−481
    polymer−metal composites, 326−327
    properties of nanofibers and composites in tribology, 330−335
    reduced maintenance costs, 334−335
    schematic illustration showing structure of FeNi/N-CPCF membrane, 412*f*
    thermal stability, 332
    types of lubricant additives, 345−351
    typical polymeric composite as lubricant additive, 344*f*
  as electrodes, 548−559
    carbon nanofiber-based composites, 553−556
    carbon nanofibers, 548−553
    conducting polymer nanofibers, 558−559
    metal oxide nanofibers, 556−558
  energy conversion and storage, 525−532
  evaluating structure measuring or chemical composition methods, 522*t*
  evolution of nanofiber technologies over time, 499−501
  fabrication, 117
  for fuel cells, 313−316
  fundamentals of, 439−444
  future perspectives, 533
  green solvents for, 117−124
  for Li−S batteries, 300−302
  mat, 555−556
  materials, 380, 385−386, 502−503, 522−524
  membranes, 52, 292−295, 442−443, 472−473
  methods of surface functionalization of, 52−56
  modification of nanofibers for water remediation applications, 186−188
  molecules, 524−525
  morphology, 275
  nanofiber-based air filters, 476, 479
  nanofiber-based catalyst composites, 443−444, 447−448, 456
  nanofiber-based electrode materials, 448−449
  nanofiber-based materials, 335, 454−456
  nanofiber-based membranes, 439
  nanofiber-based sensors, 438
  nanofiber-based tissue engineering, 439

Index

nanofiber-based water-splitting technologies, 455–458
nanofiber-reinforced polymer composites, 443
and nanofibrous material types, 502–503, 502f, 503f
NF-based thermal interface materials, 334
NF-reinforced composite-based antiwear additives, 346
NF-reinforced composites, 343–345, 349
processing and fabrication technologies, 2–21
processing technologies, 273–279
   electrospinning, 273–275
   spray pyrolysis, 278–279
   template-assisted chemical deposition, 275–278
production processes, 499
properties of, 442–443
property characterization, 23–25
   optoelectrical properties, 24–25
   thermoelectric behaviour, 23–24
   UV-visible light sensitivity, 24
segments, 2
in sensors, 425–430
   manufacturing of DNA biosensors, 428f
   synthesis of polymeric unit, 427f
   ways for synthesizing carbon nanofibers, 425f
separators, 281, 292–300, 559–560
soil purification applications of, 482–487, 483t
   role of nanofiber in remediation by using electrokinetic technique, 482–485
   role of nanofibers in phytoremediation, 485–487
for solar cells, 309–313
   photothermal solar cells, 312–313
   photovoltaic solar cells, 309–312
structure assessment of, 21–23
   atomic force microscopy, 22–23
   scanning electron microscopy, 21
   transmission electron microscopy, 21–22
   X-ray diffraction, 22
in supercapacitors, 546–547
   synthesis of nanofibers, 546–547, 547f
synthesis methods, 136–139, 407–411, 408f, 441–442
nanofibers by stretching and fiber diameter method, 138f
nanofibers in form of beads strung together, enclosing drug particles, 409f
of nanofibers, 504, 504f
production of nanofibers, 139f
standard method for production of arranged porous inorganic threads by electrospinning, 406f
type materials and diameters of individual nanofibers, 138t
terminology for nanotechnology, 498–499
as three-dimensional scaffolds for tissue regeneration, 148, 171–175
water purification application of, 473–476, 474f
in water splitting, 444–449
   characterization techniques applied to $Co_3O_4\#TiO_2$ composite, 451f
   recent developments and advances of, 445
   recent developments and advances of, 449–455, 454f
   role of, 444–445
   X-ray diffraction patterns of CMO-650 following stability test, 453f
Nanofibrillated cellulose (NFC), 58
Nanofillers, 357
Nanofilms, 255–256
Nanogenerators, 509, 511
"NanoHydroChem" project, 445
Nanomaterials (NMs), 94, 135, 228, 255–256, 423, 482, 502
Nanomedicine, 139–141, 143, 165
   relationship between tissue engineering and, 143
Nanoobject, 498
Nanoparticles (NPs), 1, 163, 170, 228
   composites, 262
   production, 497
Nanopore arrays, 13
Nanoporous drug delivery systems, 146
Nanorod arrays, 13
Nanorods, 255–256

Nanoscaffolds, 165–168
  classification of nanomedicine, 167f
Nanoscale carbon fibers, 500
Nanoscale fiber, 477
Nanoscale materials, 143, 444–445
Nanosheet, 557–558
Nanosized lubricant additives, 350–351
Nanospheres, 255–256
Nanostructured materials, 145
Nanosystems, 147, 170
Nanotechnology, 149, 185, 227, 423–424, 437–438, 497, 499
  methods in bone tissue engineering, 144–146
  terminology for, 498–499
    nano, 498
    nanocomposite, 498
    nanofiber, 498
    nanomaterials, 498
    nanoparticle, 498
    nanoscale, 498
    nanoscience, 498–499
    nanotechnology, 499
Nanotubes, 255–256, 507
Nanowires (NW), 497
National Aeronautics Administration, 378
Natural crimp, 577–578
Natural fiber–reinforced polymer composites (NRPCs), 595
Natural fibers, 33–34, 230–231, 569–575, 588–589, 597–598, 600–603
  animal fibers, 33
  applications of natural fibers in textiles, 597–600
    apparel and fashion, 597–598
    automotive textiles, 600
    geotextiles, 599
    home textiles, 598
    technical textiles, 598–599
  challenges and future directions, 604–605
    challenges in natural fiber production, 604
    innovations and future trends, 604–605
    prospects for natural fiber blends, 605
  cotton, 33
  definition and classification, 571–572
  geotextiles, 599
  historical significance, 572–573
  jute, 33–34
  modern applications, 573–575, 574f
  physicochemical properties of, 580–597
    chemical reactivity, 593–597
    moisture absorption, 584–588
    tensile strength, 580–584, 582f
    thermal conductivity, 588–592
  plant fibers, 33
  silk, 33
  sources of, 575–579
    animal-based fibers, 577–578
    mineral-based fibers, 578–579
    plant-based fibers, 575–577
Natural materials, 91, 575
Natural nanofibers, 34–40, 69. *See also* Synthetic nanofibers
  alginate, 98–100, 98f, 99f
  cellulose, 94–96
  chitin/chitosan, 93–94, 94f
  collagen, 102–103
  fibrin, 101–102
  gelatin, 96–97, 97f
  hyaluronic acid, 100–101
  natural and synthetic polymers, 92f
  natural polymer-based nanofibers, 35–40
  natural polymers and sources, 93t
  scope of biopolymer, 92f
  structural, physicochemical, and mechanical properties of
    alginate, 36–37
    cellulose nanofibers, 38–39
    chitosan, 36
    collagen, 37–38
    hyaluronic acid, 35–36
    silk nanofibers, 39–40
Natural polymers, 91–92, 473–474, 521–522
Neoteric solvents, 114, 119
NFC. *See* Nanofibrillated cellulose (NFC)
NFs. *See* Nanofibers (NFs)
Nickel (Ni), 79, 391–392, 525–527
Nickel metal coated with graphite and aluminum powder (Ni-Gr/Al), 329–330
Nickel titanate nanofibers (NiTiO$_3$ NFs), 452–454
Nitrides, 410
Nitrogen (N), 276, 380–381, 549
  doping, 552–553
  groups, 382–383

NMs. *See* Nanomaterials (NMs)
Nontransparent conductors, 508
NPs. *See* Nanoparticles (NPs)
NRPCs. *See* Natural fiber—reinforced polymer composites (NRPCs)
Nutrient microenvironment, 157
NW. *See* Nano wood (NW); Nanowires (NW)
Nylon, 34
Nyquist plots, 365, 370

O

Octadecyltetrachlorosilane-self-assembled monolayers (OTS-SAMs), 14
Octafluorocyclobutane (C_4F_8), 552—553
OER. *See* Oxygen evolution reaction (OER)
Oil absorption nanofibers, 69—70
Oil emulsions, 522—524
Oil industry, 151
Oil pipelines, 357
Oil-spill cleaning
　classical theories on contact angles of water and oil drops to solid surfaces, 85—86
　processing technologies, 70—84
　　aerogel formation, 84, 85*f*
　　chemical oxidation, 78—79
　　deacetylation, 80—84, 82*f*
　　delignification, 79—80
　　electrospinning, 70—76, 72*f*, 73*f*, 74*f*
　　vacuum filtration, 76—78, 78*f*
Oil—water mixture, 522—524
Oil—water separation, 84—85, 522—524
One-dimension (1D), 539—540
　architecture, 539—540
　CNFs, 283—286, 414—415
　nanofibers, 497
　nanomaterials, 423
　nanostructures, 506
Operando microscopy, 457
Optical chemical sensors, 428—430
Optimization of synthesis methods, 417
Organic coatings, 357
Organic compounds, 63, 411—412
Organic materials, 117
Organic semiconducting nanofibers, 290—292
Organic solar cells, 309—310
Organic solvents, 101

ORR. *See* Oxygen reduction reaction (ORR)
Osmosis, 235
OTS-SAMs. *See* Octadecyltetrachlorosilane-self-assembled monolayers (OTS-SAMs)
Oxidation, 332
Oxidative stabilization, 548
Oxide nanofibers, 311—312
　for lithium batteries, 304
4,4-oxydianilline (ODA), 286—287
Oxygen, 552
Oxygen evolution reaction (OER), 411, 439
Oxygen reduction reaction (ORR), 411

P

PA. *See* Polyamide (PA)
PA6. *See* Polyamide 6 (PA6)
PAA. *See* Polyamic acid (PAA)
PAMD. *See* Polymer-assisted metal deposition (PAMD)
Particulate matter (PM), 472
Particulate matter 2.5 (PM2.5), 116
Passive targeting, 147, 170
Pathogenic microorganisms, 231
PCs. *See* Pseudocapacitors (PCs)
PDMS. *See* Polydimethylsiloxane (PDMS)
PE. *See* Polyethylene (PE)
PECVD. *See* Plasma-enhanced CVD (PECVD)
PEEK. *See* Polyetherketone (PEEK)
Pegylated dendrimers, 147, 170
PEM. *See* Proton exchange membrane (PEM)
PEMFCs. *See* Proton exchange membrane fuel cells (PEMFCs)
Perkin Elmer Elemental Analyzer, 283
Perovskite metal oxides, 557
Perovskite nanofibers (PNFs), 121—122
Pesticides, 244—245
Petroleum-based polymers, 114
Petryanov filters, 500
PF. *See* Phenol formaldehyde (PF)
Pharmaceutical industry, 165—166
Phase separation, 258, 442
Phenol formaldehyde (PF), 278
Phenols, 350
(6,6)-phenyl C61 butyric acid methyl ester (PCBM), 310—311
Phosphoric acid (H_3PO_4), 549

Photoanode, 510
Photocatalysis, 198–200, 411, 413, 444–445
Photocatalysts, 413, 511
Photocatalytic oxidation, 203
Photodegradation, 203–204
Photothermal solar cells, 275–276, 312–313
Photovoltaic solar cells, 309–312
Physical adsorption, 388–389
Physical surface absorption, 237–238
Phytoremediation
  nanofiber for remediation of contaminated soil with phytoremediation, 487*f*
  role of nanofibers in, 485–487
Piezoelectric nanomaterials, 511
Plant composites
  influential functional groups in biosorption by plant composites, 244–246
  pollutant absorption by, 244–248
  removal of pollutants using biopolymers, 246–248
Plant fibers, 33, 569–571, 575–577, 580, 584–585
  cotton, 575
  flax, 576
  hemp, 576–577
Plant-derived natural fibers, 394
Plasma treatment, 52–53
Plasma-enhanced CVD (PECVD), 3
Plastic materials, 230
Plastic pollution, 246
Plasticization, 40–41
PM. *See* Particulate matter (PM)
PMDA. *See* Pyromellitic dianhydride (PMDA)
PMIA. *See* Poly-m-phenylene isophthalamide (PMIA)
PNFs. *See* Perovskite nanofibers (PNFs)
Pollutants
  absorption by plant composites, 244–248
  environmental threat of, 229–231
  removal of pollutants using biopolymers, 246–248
Poly (L-lactic acid) (PLLA), 387
Poly lacto-co-glycolic acid (PLGA), 42–43
Poly-m-phenylene isophthalamide (PMIA), 292–294, 301

poly(3-hexylthiophene) (P3HT), 310–311
Poly(3-hydroxybutyrate-co-3-hydroxy valerate) (PHBV), 44, 123
Poly(3-methyl thiophene) (PMT), 558–559
Poly(3,4-ethylene dioxythiophene) (PEDOT), 558–559
Poly(3,4-ethylenedioxythiophene):Poly (styrenesulfonate) (PEDOT:PSS), 309–310
Poly(acrylaminophosphonic-carboxyl-hydrazide), 187
Poly(acrylic acid)-sodium alginate nanofibrous hydrogels, 197
Poly(amic acid) (PAA), 558–559
Poly(amidoamine) (PAMAM), 53–54
Poly(dimethylsiloxane) (PDMS), 303, 364
Poly(ethylene oxide) (PEO), 558–559
Poly(ethylene-co-vinylacetate) (PEVA), 44–45
Poly(hydroxyalkanoates) (PHAs), 246–247
Poly(lactic acid) (PLA), 40–41, 59, 114, 480, 592
Poly(methyl methacrylate) (PMMA), 550–551
Poly(methyl vinyl ether-alt-maleic acid) (PMVEA), 426–428
Poly(N-isopropylacrylamide)-co-(N-methylolacrylamide)-co-(acrylic acid) (poly (NIPAAm-co-NMA-co-AA)), 426
Poly(p-phenylene terephthalamide) (PPTA), 300–301
Poly(vinyl alcohol-coethylene) (PVA-coPE), 188
Poly(vinyl alcohol)-silica (PVA-SiO$_2$), 295
Poly(vinylidene fluoride membrane) (PVDF membrane), 73–74
Poly(vinylidene fluoride-co-hexafluoropropylene) (PVDF-HFP), 298–299, 559–560
Poly(vinylidene fluoride) (PVDF), 52–53, 260, 294–295, 390, 428–430, 559–560
Polyacrylic acid (PAA), 56
Polyacrylonitrile (PAN), 54–56, 119–121, 190–191, 280–282, 379–380, 382–383, 386–387, 390, 477, 548
Polyalphaolefin, 338
Polyamic acid (PAA), 286–287

Index

Polyamide (PA), 379−380
Polyamide 6 (PA6), 306−308
Polyamide-56 (PA56), 480
Polyaniline (PANI), 10, 78−79, 258, 558−559
  gas sensors, 428−430
  nanofibers, 358−359
Polyaniline/reduced cationic graphene oxide (P-RGO$^+$), 358−359
Polycaprolactone (PCL), 42, 52−53, 114, 137, 157−158, 480
  fibers, 42
  nanocomposites, 158
Polydimethylsiloxane (PDMS), 76−78, 366, 390, 506
Polyesters, 157−158
Polyetherimide (PEI), 190
Polyetherketone (PEEK), 299−300
Polyethylene (PE), 44−45, 294−295, 479, 580−581
Polyethylene glycol (PEG), 72−73, 162−163
Polyethylene oxide (PEO), 115, 305−306
Polyethylene terephthalate (PET), 246, 506
Polyethyleneimine grafting (PEI grafting), 80, 290, 385
Polyglycolic acid (PGA), 42−43
Polyglycols, 338, 346
Polyimide (PI), 286−287, 327, 479, 506, 548
Polyindole
  nanofiber, 290−291
  polymer, 290
Polyisobutenes, 347−348
Polymer-assisted metal deposition (PAMD), 79
Polymeric materials, 559−560
Polymeric nanofibers, 135−136
Polymers, 144−145, 228−229, 286−287, 326, 479−480, 550−551
  comments and ideas for development of polymer matrix-based nanocomposites, 150−151, 177−179
  composite fiber, 314
  compounds, 63
  gel electrolyte, 303
  nanocomposites, 151, 176−177
  nanofibers, 385, 394−395
    direct carbonization from, 276−278
  modification of, 56
  surface modification of electroextruded polymer nanofibers, 57f
  nanotechnology, 148−150, 175−177
  pellet, 178−179
  polymer-based composites, 150−151, 334−335
  polymer-based textile sensors, 428−430
  polymer/ceramic nanofiber materials, 525−527
  polymer−metal composites, 326−327, 326f
Polymethacrylates, 347−348
Polyolefin films, 292−294
Polyphenylene oxide-based nanofiber separator, 299
Polypropylene (PP), 295, 479, 522−524, 580−581
Polypyrrole (PPy), 11, 276, 382, 558−559
Polysaccharide, 93
Polystyrene (PS), 71−72, 326−327, 387, 550−551
Polysulfides, 299
Polysulfone (PSF), 550−551
Polytetrafluoroethylene (PTFE), 327
Polythiophene (PTh), 558−559
Polyurethane (PU), 42, 78−79, 387
Polyvinyl alcohol (PVA), 56, 115, 446, 475, 548
Polyvinyl alcohol/poly(lithium acrylate) (C-PVA/PAA-Li), 300
Polyvinyl pyrrolidone (PVP), 56, 548
Polyvinylacetate (PVA), 260
Polyvinylidene fluoride-co-chlorotrifluoroethylene (PVDF-co-CTFE), 299
Polyvinylpyrrolidone (PVP), 115, 260, 283−286, 479−480, 550−553
Pore engineering, 548−552
  activation of carbon nanofibers, 549
  addition of fillers and pore-inducing materials, 551−552
  addition of sacrificial polymer, 550−551
Porous carbon materials, 380
Porous graphite nanofibers, 383−384
Porous materials, 377
Porous nitrogen-doped carbon nanofibers (NCPF), 415−416
Porous silicon (PSi nanocomposite), 16

Porous wood (PW), 79
Powder adsorbents, 381
Powder metallurgy, 328
Powder X-ray diffraction (PXRD), 450−452
Power density, 545
PPNFs. *See* Preoxidized PAN nanofibers (PPNFs)
Preoxidized PAN nanofibers (PPNFs), 555
Pressure sensors, 508
Pressure swing adsorption (PSA), 380
Pristine organic nanotubes, 145
Protective clothing, 439
Proteins, 34
Proton exchange membrane (PEM), 511
Proton exchange membrane fuel cells (PEMFCs), 262, 511
PSA. *See* Pressure swing adsorption (PSA)
Pseudo capacitive materials, 553
Pseudo-capacitive electrode materials, 505
Pseudocapacitors (PCs), 505, 542−543
Pseudomonas aeruginosa, 231−232
PSF. *See* Polysulfone (PSF)
PTFE. *See* Polytetrafluoroethylene (PTFE)
Purification methods, 229
PVDF-co-CTFE. *See* Polyvinylidene fluoride-co-chlorotrifluoroethylene (PVDF-co-CTFE)
PW. *See* Porous wood (PW)
PXRD. *See* Powder X-ray diffraction (PXRD)
Pyrolysis, 313
Pyromellitic dianhydride (PMDA), 286−287

Q
QF. *See* Quality factors (QF)
Quality factors (QF), 476−477
Quartz fiber template, hollow carbon nanofiber from, 278, 280*f*

R
Radio frequency (RF), 4
Ragone plot, 539
Rate-determining step (RDS), 412−413
Rayon fiber, 34
RBP. *See* Red banana peduncle (RBP)
RBPFs. *See* Red banana peduncle fibers (RBPFs)
RCCVD. *See* Reductive-catalytic chemical vapor deposition (RCCVD)
RDS. *See* Rate-determining step (RDS)

Real-world testing, 417
Rechargeable lithium-ion batteries, 512
  anode materials, 512
  cathode materials, 512
  separator materials, 512
Recyclable resources, 527
Red banana peduncle (RBP), 595
Red banana peduncle fibers (RBPFs), 595
Redox reactions, 558−559
Reduced graphene oxide (rGO), 589−590
Reduction cationic graphene oxide (RGO$^+$), 358−359
Reductive-catalytic chemical vapor deposition (RCCVD), 5
Refining, 257
Regenerated cellulose fibers, 605
Reinforcement, 507
Relative humidity (RH), 274−275
Remediation by using electrokinetic technique, role of nanofiber in, 482−485
Removal of pollutants using biopolymers, 246−248
Renewable energy sources, 350, 407, 417
Reservoirs, 357
Retting process, 576−577
Reverse osmosis, 234−235
RF. *See* Radio frequency (RF)
rGO. *See* Reduced graphene oxide (rGO)
RGO$^+$. *See* Reduction cationic graphene oxide (RGO$^+$)
RH. *See* Relative humidity (RH)
RhB. *See* Rhodamine B (RhB)
Rhodamine B (RhB), 40−41
Rust inhibitors, 349−350

S
SA. *See* Succinic anhydride (SA)
Sacrificial polymers, 551−552
  addition of, 550−551
SAED. *See* Selected area electron diffraction (SAED)
SARS-CoV-2. *See* Severe acute respiratory syndrome coronavirus 2 (SARS-CoV-2)
Scaffold, 136
Scanning electron microscopy (SEM), 16, 19, 21, 80, 276−278, 283, 290−291, 450−452, 583−584

Index

Scanning probe microscopy (SPM), 22−23
Scanning TEM (STEM), 21−22
SCFs. *See* Supercritical fluids (SCFs)
SCs. *See* Supercapacitors (SCs)
SDS. *See* Sodium dodecyl sulfonate (SDS)
SEI. *See* Solid-electrolyte-interphase (SEI)
Selected area electron diffraction (SAED), 286−287
Self-assembled template synthesis, 409−410
Self-assembly techniques, 258, 442
Self-healing
   coatings, 364
   lubricants, 348−349
SEM. *See* Scanning electron microscopy (SEM)
Semiconductor metal oxide gas sensors, 423−424
Semisynthetic lubricants, 341
Semisynthetic metalworking fluids, 342
Sensing, 438
Sensors, 423−425, 508
   nanofibers in, 425−430
Separation processes, 381, 522−524
Separator materials, 512
Sequential methods of removing contaminants from aqueous solutions by composites, 231−244
   adsorption, 237−240
   bio-absorption and factors affecting, 240−244, 242f
   chemical precipitation and filtration, 233−234
   ion exchange, 235−237
   reverse osmosis, 234−235
Sericins, 39−40
Severe acute respiratory syndrome coronavirus 2 (SARS-CoV-2), 479
Sewage flotation, 234
SF. *See* Silk fibroin (SF)
SFE. *See* Solvent-free electrospinning (SFE)
Short-term water sorption, 585−586
Shuttle effect, 300−301
SIBs. *See* Sodium-ion batteries (SIBs)
Silane coupling agents, 58
Silica (SiO_2), 11−12, 551−552
   carbon composite fiber, 278
   silica-carbon-silica tri-layered nanofiber, 312−313

Silicon carbide (SiC), 329
   silicon carbide-reinforced aluminum matrix composites, 327
Silicone-based compounds, 346
Silk, 33, 572−573, 577, 593−594, 597
   production, 569−570, 572−573
   reeling, 577
Silk fibroin (SF), 39−40
Silk nanofibers, 39−40
Silkworms, 33
Siloxane, 390
Silver nanocrystals, 231
Silver nanoparticles (AgNPs), 115, 507
Single semiconducting nanofiber, 506
Skin, 168
   tissue engineering, 168−169
Sodium batteries, 291−292, 528−529
   nanofibers for, 305
Sodium compounds, 347−348
Sodium dodecyl sulfonate (SDS), 5−6
Sodium hyaluronate, 101
Sodium hydroxide (NaOH), 573−574, 590−592
Sodium ions, 99−100
Sodium silicate (SS), 591−592
Sodium-ion batteries (SIBs), 528−529
SOFCs. *See* Solid oxide fuel cell (SOFCs)
Soft electronics
   applications of electrospun nanofibers in, 507−508
   conductors, 508
   nontransparent conductors, 508
   pressure sensors, 508
   sensors, 508
   transparent electrodes, 508
   strategies for electrospun nanofibers in, 506−507
   aligned nanofiber array, 507
   entangled porous nanofiber sheet, 506
   reinforcement, 507
   single semiconducting nanofiber, 506
   temporary template, 507
Soft energy storage devices, 509
Soft template method, 550−551
Soft tissues, 141
Softwood pulp sheets, 61
Soil particles, 485
Soil remediation, 472
Soil stabilization, 599

Sol-gel method, 260
Sol-gel reaction, 124
Solar cells, 266−267, 509−510, 527
  dye-sensitized solar cells, 510
  nanofibers for, 309−313
Solar energy technology, 509−510
Solar-driven hydrogen generation, 511
  hydrogen generators, 511
Sol−gel spinning, 274
Sol−gel synthesis, 410, 441−442
Solid electrolyte system, 525−527
Solid lubricants, 343−344, 343f
Solid lubrication, 329
Solid oxide fuel cell (SOFCs), 525−527
Solid polymer electrolyte (SPE), 302
Solid surfaces, classical theories on contact angles of water and oil drops to, 85−86
Solid-electrolyte-interphase (SEI), 287−289
Solution-based synthesis, 410
Solvent-free electrospinning (SFE), 121
Spandex, 42
SPANI. See Sulfonated polyaniline (SPANI)
SPE. See Solid polymer electrolyte (SPE)
Spinel metal oxides, 557
Spinning process, 70, 259
SPM. See Scanning probe microscopy (SPM)
Spraying pyrolysis, 20−21, 278−279
Standard drawing processes, 136−137
Staphylococcus sp., 231−232
  S. aureus, 115
Starch, 114
STEM. See Scanning TEM (STEM)
Stem cells, 141, 143, 167
Stimuli-sensitive drug delivery systems, 159−160
Stir casting, 328
Stress-strain curve, 39−40
Stretching technique, 137
Succinic anhydride (SA), 282
Sulfonated polyaniline (SPANI), 359
Sulfur compounds, 390
Supercapacitors (SCs), 266, 505, 509, 512, 539, 543−544
  and components, 540−542, 541f
  nanofibers in, 546−547
  performance metrics of, 543−546
  types of, 542−543, 542f

Supercritical carbon dioxide (scCO$_2$), 123
Supercritical fluids (SCFs), 119, 123−124
Surface absorption process, 235−237
Surface functionalization of nanofibers
  methods of, 52−56, 53f
  updates in surface modifications, 52−53
    plasma treatment, 52−53
    surface grafting, cross-linking, and chelation, 53−56
Surface modification techniques, 457
Surface science, 457−458
Surfactants, 228−229
Sustainability and eco-friendliness, 600−603
  biodegradability, 601−602
  environmental impact, 600−601
  role in sustainable fashion, 602−603
Sustainable biomaterials, 61−62
Sustainable energy systems, 113
Sustainable materials, 113
Synergy, 169, 417
Synthetic ester-based lubricants, 338−339
Synthetic fibers, 34, 601−602
  nylon, 34
  rayon, 34
Synthetic lubricants, 337−338, 338f, 341, 350
Synthetic materials, 205
Synthetic metalworking fluids, 342
Synthetic nanofibers, 40−45. See also Natural nanofibers
  synthetic polymer-based nanofibers, 40−45
    poly lacto-co-glycolic acid, 42−43, 43f
    poly(3-hydroxybutyrate-co-3-hydroxy valerate) fibers, 44, 44f
    poly(ethylene-co-vinylacetate), 44−45, 45f
    polycaprolactone fibers, 42
    polylactic acid fibers, 40−41
    polyurethane fibers, 42
Synthetic polymeric nanofibers, 69
Synthetic polymers, 91−92, 190, 521−522

T
TAB. See Tension, at loading nose, bottom (TAB)
Tafel reaction, 413
TANF. See Tetraaniline-based conductive nanofibers (TANF)

Index

Targeted delivery, 160
Targeted drug delivery, 147
Taylor cone, 114–115, 274–275
  formation, 7–8, 70–71
TCEs. *See* Transparent conducting electrodes (TCEs)
Technical textiles, 598–599
TEM. *See* Transmission electron microscopy (TEM)
Temperature swing adsorption (TSA), 380
Template assembling, 10–11
Template synthesis, 409–410, 441–442
Template-assisted chemical deposition, 275–278
  direct carbonization from polymer nanofiber template, 276–278
  hollow carbon nanofiber from quartz fiber template, 278
Template-based fabrication, 258, 259*f*
Tensile strength, 580–584, 581*f*, 582*f*, 584*f*
Tension, at loading nose, bottom (TAB), 581–582
Tetraaniline, 358
Tetraaniline-based conductive nanofibers (TANF), 358
Tetraethylorthosilicate (TEOS), 74–75, 278, 551–552
Tetrahydrofuran (THF), 72–73
2,2,6,6-tetramethylpiperidinyloxyl (TEMPO), 205
Textiles
  applications of natural fibers in, 597–600
  electronics in, 504–505
  industry, 570–572
  production, 572
  sensing in, 505
Thermal conductivity, 588–592
  of foamed geopolymer, 591*f*
  of foamed geopolymer concentration level with different proportions of polypropylene fiber, 592*f*
Thermal plastic elastomer ester, 191
Thermal stability, 332
Thermo-curing electrospinning (TCE), 121
Thermoelectric behaviour, 23–24
Thermoplastic polyurethane (TPU), 366
THF. *See* Tetrahydrofuran (THF)
Thiol-functionalized cellulose nanofiber membrane, 474–475
Thiosemicarbazide-modified polyacrylonitrile fiber, 187
Three-dimension (3D)
  bioprinting, 171
  frameworks, 31–32
  materials, 423
  nanofibers as three-dimensional scaffolds for tissue regeneration, 148, 171–175
    bioprinter nozzle, 173*f*
    nozzle of bioprinters based on extrusion, 174*f*
  printing, 7
  scaffolds, 143
  structures, 256
Tissue engineering, 139, 141–143, 165, 439, 442–443
  controlled release, 168–175
    dendrimer, 169–171
    nanofibers as three-dimensional scaffolds for tissue regeneration, 171–175
    skin tissue engineering, 168–169
  relationship between nanomedicine and, 143
  3D carriers in tissue engineering, 144*f*
  patterns of tissue engineering, 144*f*
  types of drug delivery systems in, 166*f*
  volume transition of hydrogel, 165*f*
Tissue regeneration, 148, 171–175
Titanium dioxide, 446
Titanium oxide (TiO$_2$), 394
  application of TiO$_2$ fibers for water remediation, 200–204
  nanofibers, 413, 557–558
Titanium oxide-loaded hollow carbon nanofibers (TiO$_2$-HCFs), 304
Titanium tetra-isopropoxide (TTIP), 75
TMDs. *See* Transition metal dichalcogenides (TMDs)
Toluene, 411–412
Top-down techniques, 256
TPU. *See* Thermoplastic polyurethane (TPU)
Traditional drug delivery systems, 158–160
Traditional electrical sensors, 428–430
Transistors, 509
Transition metal dichalcogenides (TMDs), 326–327
Transition metals, 3–4

Transmission electron microscopy (TEM), 18−19, 21−22, 276−278, 450−452
Transparent conducting electrodes (TCEs), 309−310
Transparent electrodes, 508
Tribology, 325, 332
   components, 331
   properties of nanofibers and composites in, 330−335
2,2,2-trifluoroethanol (TFE), 37−38
TSA. *See* Temperature swing adsorption (TSA)

U

UEA. *See* Urushiol epoxy acrylate (UEA)
Ultracapacitors, 539
Ultraviolet (UV), 14
   UV-visible absorption spectra, 24
   UV-visible light sensitivity, 24
   visible light absorption, 315−316
Underground water, 228−229
Union of Soviet Socialist Republics (USSR), 500
Unitary metal oxides, 556
Unsulfonated polysaccharide, 100
Urbanization, 522−524
Urushiol epoxy acrylate (UEA), 361
US Food and Drug Administration, 157−158
USSR. *See* Union of Soviet Socialist Republics (USSR)
UV-curing electrospinning (UV-CE), 121

V

VACNFs. *See* Vertically aligned carbon nanofibers (VACNFs)
Vacuum filtration, 76−78, 257
Vanadium oxide (V oxide), 12
Vapor deposition methods, 259
Vapor-phase synthesis, 410
Vegetable fiber, 33−34
Vegetable oil-based lubricants, 338−339
Vertically aligned carbon nanofibers (VACNFs), 5
VI. *See* Viscosity index (VI)
VIIs. *See* Viscosity index improvers (VIIs)
Viscoelastic materials, 258
Viscosity index (VI), 347
Viscosity index improvers (VIIs), 347

VOCs. *See* Volatile organic compounds (VOCs)
Volatile molecules, 63
Volatile organic compounds (VOCs), 524−525
   adsorption for indoor air pollution control, 386−387
Volmer reaction, 413

W

Wastewater, 233, 473
   treatments, 200−203, 243
Water, 233, 512−513, 522−524
   absorption, 588
   classical theories on contact angles of water and oil drops to solid surfaces, 85−86
   emulsions, 522−524
   pollution, 227
   purification
      processes, 472−473
      systems, 245
   remediation, 185
      application of TiO_2 fibers for, 200−204
      modification of nanofibers for, 186−188, 189*t*
   treatment, 61−62
      nanocellulose membranes to filter out waterborne pollutants, 62*f*
   water-based metalworking fluids, 342−343
   water-mediated electrospinning, 74−75
Water contact angle (WCA), 74
Water splitting, 407, 413−414, 437−439
   challenges and future prospects, 455−458
      opportunities for further research, 457−458
      strategies for optimization, 456−457
      technical challenges and limitations, 455−456
   complete process of water splitting occurring on Ni2P/rGO/NF electrode, 439*f*
   device, 452
   fundamentals of nanofibers and composite, 439−444
      composite formation and characteristics, 443−444

Index

nanofiber synthesis techniques, 441–442
potential photocatalytic degradation mechanism by nanofibers, 441f
properties of nanofibers, 442–443
time course of H_2 and O_2 evolution over different catalysts, 440f
nanofibers and composites in water splitting, 444–449
catalytic activity enhancement, 447–448
electrochemical performance improvement, 448–449
recent developments and advances of nanofibers in water splitting, 445
role of nanofibers in water splitting, 444–445
synergistic effects on composite materials, 446
reactions, 444–445
recent developments and advances of nanofibers in water splitting, 449–455
Water-based polyurethane (WPU), 358–359
WCA. See Water contact angle (WCA)
Wentzel model, 86
Wet cyclic corrosion tests, 365
Woods, 229–230
Wool, 33, 577–578, 589–590, 597
Wound healing, 149
core–shell nanofiber composite, 2
Woven fabrics, 595
Woven nanofibers, 148
WPU. See Water-based polyurethane (WPU)

X

X-ray diffraction (XRD), 18–19, 22, 286–287, 428–430, 589–590
XRD. See X-ray diffraction (XRD)

Y

Yarn, 507
Yeasts, 240
Young's equation, 85
Young's modulus, 260–261, 583–584

Z

ZB. See Zinc borate (ZB)
Zeolite-imidazolate framework-8 (ZIF-8), 384–385
Zeolites, 378
ZIF-8. See Zeolite-imidazolate framework-8 (ZIF-8)
Zinc borate (ZB), 345
Zinc oxide (ZnO), 310, 394, 428–430
Zirconium oxide, 446

Printed and bound by CPI Group (UK) Ltd, Croydon, CR0 4YY
02/12/2024
01798480-0010